AF775674

Wastewater Conveyance and Nonpoint Source Pollution Control

Wastewater Conveyance and Nonpoint Source Pollution Control: Theory and Design Examples presents the fundamentals of fluid mechanics related to wastewater collection, conveyance, and pumping; the principles of surface water hydrology and watershed management; and the control of nonpoint source pollution. Furthermore, it examines the important design parameters in conveyance, stormwater runoff, nonpoint source pollution control, and combined sewer overflows in the overall context of watershed management. It provides numerous design examples and step-by-step solutions to enhance the readers' comprehension and deeper understanding of the basic concepts of the topics discussed. It serves as a useful guide for researchers, professionals, students, and those who work for engineering firms, for equipment manufacturers, and as technical personnel involved with wastewater conveyance.

This book

- Provides 351 theory and design examples and step-by-step solutions to design wastewater collection and conveyance systems.
- Aids students and professionals in learning and mastering the problem-solving skills needed for professional engineering (P.E.) exams.
- Includes useful appendices that include various standards, constants, and properties.

Wastewater Conveyance and Nonpoint Source Pollution Control

Theory and Design Examples

Syed R. Qasim and Habib Ahmari

CRC Press
Taylor & Francis Group
Boca Raton London New York

CRC Press is an imprint of the
Taylor & Francis Group, an **informa** business

Designed cover image: Shutterstock

First edition published 2026
by CRC Press
2385 NW Executive Center Drive, Suite 320, Boca Raton FL 33431

and by CRC Press
4 Park Square, Milton Park, Abingdon, Oxon, OX14 4RN

CRC Press is an imprint of Taylor & Francis Group, LLC

© 2026 Syed R. Qasim and Habib Ahmari

Reasonable efforts have been made to publish reliable data and information, but the author and publisher cannot assume responsibility for the validity of all materials or the consequences of their use. The authors and publishers have attempted to trace the copyright holders of all material reproduced in this publication and apologize to copyright holders if permission to publish in this form has not been obtained. If any copyright material has not been acknowledged please write and let us know so we may rectify in any future reprint.

Except as permitted under U.S. Copyright Law, no part of this book may be reprinted, reproduced, transmitted, or utilized in any form by any electronic, mechanical, or other means, now known or hereafter invented, including photocopying, microfilming, and recording, or in any information storage or retrieval system, without written permission from the publishers.

For permission to photocopy or use material electronically from this work, access www.copyright.com or contact the Copyright Clearance Center, Inc. (CCC), 222 Rosewood Drive, Danvers, MA 01923, 978-750-8400. For works that are not available on CCC please contact mpkbookspermissions@tandf.co.uk

For Product Safety Concerns and Information please contact our EU representative GPSR@taylorandfrancis.com. Taylor & Francis Verlag GmbH, Kaufingerstraße 24, 80331 München, Germany.

Trademark notice: Product or corporate names may be trademarks or registered trademarks and are used only for identification and explanation without intent to infringe.

ISBN: 978-1-041-04672-1 (hbk)
ISBN: 978-1-041-04671-4 (pbk)
ISBN: 978-1-003-62939-9 (ebk)

DOI: 10.1201/9781003629399

Typeset in Times
by codeMantra

Contents

List of Examples and Solutions

Preface

In the past two decades, there have been rapid developments and changes in the field of sanitary wastewater and stormwater collection, treatment, and reuse. Municipal wastewater is collected from the point of origin and conveyed to a resource recovery facility. Stormwater is collected from urban streets and conveyed to a disposal point. These underground pipe networks, pumps, and appurtenances are invaluable assets, and are an integral part of wastewater management and pollution control. However, several issues within these systems can adversely affect the operation of wastewater treatment facilities. These include infiltration and inflow, illegal system hook-ups from stormwater runoffs and sump pumps, and combined sewer overflows (CSOs). These sources lead to increased flows and pollutant loadings by placing additional stress on treatment infrastructure. Additionally, uncontrolled nonpoint sources (NPSs) of stormwater runoffs from urban and rural watersheds are well documented as significant contributors to surface water quality impairment of the nation's waters and frequent permit violations. As a result, watershed management has become an integral part of modern water pollution control strategies.

There are several published books that cover well the fundamentals and scientific principles of wastewater collection and conveyance, watershed management, and control of NPS pollution. The actual planning and design calculation steps in the form of numerical examples and solutions with intense focus on practical application of theory and practice into facility design are not fully covered in these publications. The intent of the authors in developing this book is threefold: *first*, to present briefly the knowledge of applied fluid mechanics and surface water hydrology to the problems of collection, conveyance, watershed management, flow measurement, and pumping of wastewater generated from point and nonpoint sources; *second*, to develop and apply the important design parameters involved in generation, characterization, collection, conveyance, and pumping of NPS pollution in the overall context of watershed management, and NPS pollution; and *third*, to apply a design approach by presenting step-by-step design examples and solutions. This text is an outgrowth of the authors' widely accepted book *Wastewater Treatment and Reuse: Theory and Design Examples* (CRC Press, Taylor & Francis Group, 2018).

In this book, 351 illustrative example problems and solutions have been worked out to cover the complete spectrum of wastewater collection and conveyance, flow measurement, pumping, urban and rural watershed management, and NPS pollution. These examples and solutions enhance the readers' comprehension and deeper understanding of the basic concepts. They also serve as a good source of information for more experienced engineers, and aid in the formal design training and instruction of engineering students. These practice examples with step-by-step solutions provide skills to engineering students and professionals of diverse backgrounds in learning and mastering the problem-solving techniques needed for professional engineering (P.E.) exams. Also, these solved examples can be applied by the designers to design various components of collection and conveyance systems, watershed management, NPS pollution, and select piping and equipment for conveyance and pumping facilities. Thus, this book is a consolidated resource of valuable, quick-and-easy access to a myriad of theory and practice information and solved examples on wastewater collection and conveyance, watershed management, and nonpoint pollution control.

This work is divided into 12 chapters and 4 appendices. Chapter 1 is an overview of historical development in wastewater collection and conveyance, pumping, urban and rural watershed management, and NPS pollution: past, present, and future directions. In engineering and science, significant figures deal with how accurately we can measure, understand, and communicate with others. For this reason, Chapter 2 is included on significant figures, units of expression, and principles for applying SI units. Sampling and data analysis is a technique used to select, manipulate, and analyze a representative subset of data points to identify patterns and trends in the larger data set being examined. Also, precision, reliability, and data quality indicators for process and equipment selection are important considerations. These topics are covered in detail in Chapter 3. Sources, flow rates, and characteristics of municipal wastewater, and flow equalization are provided in Chapter 4. The fundamentals of fluid mechanics and applied hydraulics, fluid flow in pressure pipe, and open channel hydraulics are fully covered in Chapter 5. Chapter 6 provides in-depth coverage of wastewater collection and conveyance objectives, source-control best management practices (BMPs) of watershed management, NPS pollution control, and their design considerations. Measurement of water and wastewater flows; purpose, location, and elements of flow measurement; and principles of flow measurement devices are presented in Chapter 7. Chapters 8 and 9 are devoted to hydraulics of full and partially full conduits, sanitary and storm sewers design, pump selection, and pumping stations. Chapter 10 provides an in-depth coverage of various methods of stormwater runoff estimation, hydrograph, source-control BMPs and their performance evaluation in rural and urban watershed management, and NPS pollution. Chapter 11 is devoted to combined sewers and CSOs, quantity and quality characterization, flow regulators, and storage and treatment of CSOs. Finally, the conveyance and design of outfall sewers is covered in Chapter 12. To further increase the utility of this book, four appendices have been included. These appendices contain abbreviations and symbols, basic information about elements, useful constants, and statistical tables; physical constants and properties of water, solubility of dissolved gases in water, hydrological information, and equipment design data; minor head loss coefficients for pressure conduits and open channels, and normal commercial pipe sizes; and units of conversion.

This book will serve the needs of students, teachers, consulting engineers, equipment manufacturers, and technical personnel in city, state, and federal organizations who are involved in the review of designs, permitting, and enforcement. To maximize the usefulness of this book, the technical information is summarized in many tables that have been developed from a variety of sources. The list of numerical examples is included after the table of contents, followed by a list of examples and solutions. This gives additional benefit to the users of this book to identify and locate the solved examples that deal with the step-by-step calculations on the specific subject matter.

Enough material is presented in this text to develop and teach three one-semester courses entitled (1) Wastewater Collection and Conveyance, (2) Watershed Management and Control of Nonpoint Source Pollution, and (3) Water Resource Engineering. The specific topics to be covered will depend on the time available, depth of coverage, and the course objectives. It is suggested that these courses can be offered as an undergraduate technical elective or one-semester introductory graduate courses.

Course A – wastewater collection and conveyance: The suggested course contents of a one-semester introductory course on wastewater collection and conveyance:

Topic	Chapter	Section
Overview of wastewater collection and conveyance	1	All
Significant figures and units of expression	2	All
Sampling and data analysis	3	3.2.1–3.3.6
Sources, flow rates, and characteristics of municipal wastewater, and regulatory status	4	All
Fundamentals of applied hydraulics	5	All
Wastewater conveyance and NPS pollution design considerations	6	All
Flow measurement	7	All
Sanitary sewer design	8	All
Pumps and pumping station design	9	All
Management of combined sewers and CSOs	11	All
Effluent discharge into natural systems	12	All

Course B – watershed management and control of NPS pollution: The suggested course contents of a one-semester introductory course on NPS pollution control:

Topic	Chapter	Section
Overview of nonpoint source pollution	1	1.2.2 and 1.3
Significant figures and units of expression	2	All
Sampling and data analysis	3	3.1–3.3 and 3.4.1–3.4.3
Best watershed management practices for control of NPS pollution control	10	10.1 and 10.2
Stormwater runoff causes and effects	10	10.3
Hydrograph, unit hydrograph, and derivation of unit hydrograph	10	10.4.1, 10.4.2, and 10.4.3
Stormwater runoff estimation	10	10.5
Watershed management and NPS discharge	10	10.6.1–10.6.3
Source-control BMP, treatment of stormwater runoff, and NPS pollution	10	10.7.1–10.7.16
Application of watershed modeling for control of NPS pollution	10	10.8
Performance evaluation of source-control BMP	10	10.9
Quality of stormwater runoff from urban areas	10	10.10.1 and 10.10.2
Manufactured equipment and devices for the treatment of stormwater	10	10.10.3
Description of manufactured equipment and devices	10	10.11

Course C – water resource engineering: The suggested course contents of a one-semester introductory undergraduate course on water resource engineering:

Topic	Chapter	Section
Overview of water resources	1	All
Significant figures and units of expression	2	All
Sources, flow rates, and characteristics of municipal wastewater	4	All
Fundamentals of applied hydraulics	5	All
Elements of flow measurement systems	7	All
Frequency and intensity of rainfall, time of concentration, and stormwater runoff	8	8.3.2–8.3.4
Design equations and design of sanitary and storm sewers	8	All
Pumps and pumping station design	9	All
Hydrograph and unit hydrograph	10	10.4
Stormwater runoff estimation	10	10.5.1–10.5.4
Watershed management and nonpoint pollution sources	10	10.6 and 10.7
Detention/retention basins and storage ponds	10	10.7.5

To make this textbook useful as a teaching and reference text, detailed example problems have been prepared with units carried out through all computational steps to aid the reader's understanding of principles and steps involved. In the solution of an example, full expressions are usually provided to demonstrate step-by-step calculations.

The International System (SI) units are used in this book. This is consistent with the teaching practices in most universities in the U.S. and around the world. Most tables in this book have dual units and include conversion from SI to U.S. customary units in footnotes. Significant figures and units of expression are presented in Chapter 2, and useful conversion units are covered in Appendix D.

CHAPTERS SUMMARY

Chapter 1: Introduction

This chapter provides an overview of the historical development, current practices, and future directions in wastewater collection, conveyance, and NPS pollution control. The development of sewerage systems from ancient civilizations to the 19th-century innovations is discussed. Today's urban collection infrastructures typically include separate sanitary and storm sewers, while legacy combined sewer systems remain a concern due to overflow risks during heavy rainfalls. The NPS pollution generated from widespread human activity and discharge of urban, rural, and agricultural runoffs continue to impair the water quality of nations receiving waters. The importance of watershed management, implementation of BMPs, and growing use of watershed-scale models to identify pollution sources, support planning of BMPs, and develop water quality policies are introduced.

Chapter 2: Significant Figures and Units of Expression

This chapter provides an overview of significant figures; U.S. customary and SI units; and the fundamentals of engineering dimensions, quantities, and unit conversions. The importance of significant figures, measurement accuracy, proper use of units and consistency in conversions, how to apply them correctly when rounding and reporting values, and their importance in engineering practice are presented in practical examples. These examples demonstrate their application in real-world engineering problems.

Chapter 3: Sampling and Data Analysis

This chapter presents key concepts and practical guidance for wastewater sampling and analysis. The appropriate sampling locations, frequency, methods, and equipment used to collect samples and determine both flow and characteristics are discussed. Data collection; common statistical methods; importance of data quality indicators such as accuracy, precision, representativeness, and completeness; and their role in ensuring the validity of sampling results are emphasized. Additionally, precision and reliability considerations are discussed to support informed selection of processes and equipment for wastewater monitoring and treatment. Numerical examples are provided to illustrate the application of these principles to real-world settings.

Chapter 4: Sources, Flow Rates, and Characteristics of Municipal Wastewater and Flow Equalization

This chapter focuses on the flow characteristics and concentrations of major components of municipal wastewater. The sources of municipal wastewater are residential, commercial, industrial, and institutional, as well as infiltration and inflow during wet weather conditions. The relationship between municipal water demand and wastewater flow, and key components of water use are identified. Topics included are flow variations, infiltration and inflow, common flow-related terminology, initial and design flow projections, sustained flows, and flow equalization. The physical, chemical, and biological characteristics of municipal wastewater are also *discussed*. Several practical examples are provided to illustrate the application of theoretical concepts in real-world scenarios.

Chapter 5: Fundamentals of Applied Hydraulics

This chapter provides an overview of applied hydraulic concepts that are fundamental to the design and analysis of wastewater conveyance systems. The essential hydraulic principles and their relevance to engineering applications are demonstrated. Key topics include the distinction between pressurized and open channel flows, classification of flow types, and governing principles that influence fluid movement in these systems. The hydraulics of pipes and open channels in the context of sewer infrastructure are presented. To enhance the understanding, several practical examples are presented to illustrate the application of these concepts in the context of real-world system design and operation.

Chapter 6: Wastewater Conveyance, Watershed Management, and Nonpoint Source Pollution: Objectives and Design Considerations

This chapter covers wastewater conveyance, watershed management, and NPS pollution. It highlights regulatory requirements, wastewater characteristics, and design considerations such as population growth, design period, and service area capacity. Types and components of collection and conveyance systems are outlined, along with site selection, hydraulics, and equipment. Broader aspects include energy and resource needs, economics, environmental impacts, and project planning steps such as cost estimation, bidding, and construction management.

Chapter 7: Flow Measurements

In this chapter, an outline of the purpose, ideal locations, and key components of flow measurement systems is provided. The types and operating principles of flow measurement devices are explained; those most commonly used in wastewater applications are highlighted; and their design considerations, along with practical applications in the field, are discussed.

Chapter 8: Sanitary and Storm Sewer Design

The theory, design principles, and construction practices for both sanitary and storm sewer systems are presented in this chapter. The fundamentals of sanitary sewerage systems, the hydraulics of conduits operating under full and partial flow conditions, and the theoretical and practical aspects of designing sanitary and storm sewers are covered. Construction methods and considerations involved in developing a community sewerage system are also addressed.

Chapter 9: Pumps and Pumping Stations

In this chapter, the types of pumps and pumping stations are described, and their applications at wastewater treatment plants are presented. Hydraulic terms and definitions are introduced, while the principles, characteristics, design, and selection of centrifugal pumps are explained. Pumping station design calculations, along with their operation and maintenance, are also addressed.

Chapter 10: Best Watershed Management Practices for Control of Nonpoint Source Pollution

In this chapter, the regulatory framework governing stormwater runoff is outlined. The characteristics of stormwater runoff, its impacts, and methods for its management are discussed. Hydrographs, unit hydrographs, and methods of stormwater runoff estimation are presented. Source-control BMPs for stormwater runoff and watershed management are described, and the performance and effectiveness of these practices are evaluated. In addition, urban runoff quality and flow are discussed, and treatment by source-control BMPs and proprietary systems is addressed.

Chapter 11: Management of Combined Sewer Overflow

In this chapter, the definition and significance of CSOs on surface water quality are explained, and their regulatory status is outlined. The sources, impacts, and characterization of CSOs are discussed, along with approaches for computer simulation and modeling. Overflow and diversion structures, including side weirs, baffled side weirs, transverse weirs, leaping weirs, and relief siphons, are described, as well as flow regulators such as mechanical devices, tipping plates, and hydro-brakes. Management strategies for CSOs and storage facilities are presented, including source control, in-system storage, off-line storage within the collection system, and on-site storage. Finally, the treatment of CSOs is addressed.

Chapter 12: Conveyance and Disposal Structures for Outfall Sewer

In this chapter, issues related to effluent reuse and disposal are discussed, and the principal uses of treated wastewater effluent are presented. Storage facilities for reclaimed water and effluent disposal considerations are described, along with methods of effluent disposal in surface waters. Design considerations for outfalls and methods for designing marine disposal infrastructure are also addressed.

Acknowledgments

A project of this magnitude requires the cooperation and collaboration of many people and organizations. We are indebted to many professionals, faculty members, students, and friends who have helped and provided constructive suggestions. We must acknowledge the support, encouragement, and stimulating discussion by Dr. Guang Zhu, a professional and consultant in the Dallas-Fort Worth area, who provided the technical support. He reviewed several chapters and made valuable comments and suggestions that have been included. Many students also assisted with typing, artwork, literature search, and proofreading. The support of the Civil Engineering Department at the University of Texas at Arlington is greatly appreciated for making many departmental resources available.

Finally, we must acknowledge with deep appreciation the support, encouragement, and patience of our families.

Although the portions of this book have been reviewed by professionals and students, the real test will not come until this book is used in classes, and by professionals in the design of conveyance systems, watershed management, and nonpoint source pollution control. We shall appreciate it very much if all who use this book will let us know of any errors and changes that they believe would improve the usefulness of this work.

Syed R. Qasim and Habib Ahmari
Arlington, Texas

Authors

Syed R. Qasim is a Professor Emeritus in the Department of Civil Engineering. He received his PhD and MSCE degrees from West Virginia University and his BSCE degree from India. His principal research and teaching interests include water and wastewater quality; treatment processes and plant design; industrial pollution control; solid and hazardous waste management; and environmental impact assessment. Dr. Qasim has served nationally and internationally as a consultant with government agencies and private concerns, and has over 50 years of experience as an educator, researcher, and practitioner in the related fields of environmental engineering. He is the recipient of many distinguished research awards, including a Fulbright Fellowship; an invited participant by the United Nations Development Program (UNDP) under the Transfer of Knowledge through Expatriate Nationals (TOKTEN-India) program; and a reviewer of concept papers and research proposals for the National Science Foundation and many national and international organizations. Syed Qasim Endowed Professorship was established in the Environmental Engineering area of the Civil Engineering Department to honor his service to the department, as well as furthering the Environmental Engineering profession. Dr. Qasim has authored/co-authored five books, over 200 technical papers, encyclopedia and book chapters, and research reports, and has conducted over 50 short courses, seminars, and training programs nationally and internationally. Dr. Qasim is a life member and Fellow of the American Society of Civil Engineers, a life member of the Water Environment Federation, a member of the American Water Works Association, and a registered professional engineer in Texas.

Habib Ahmari is an Associate Professor in the Department of Civil Engineering at the University of Texas at Arlington (UTA). He received his Ph.D. in Hydrotechnical Engineering from Queen's University, Canada, and his M.Sc. and B.Sc. degrees in Civil Engineering from Sharif University of Technology, Iran. He is a licensed professional engineer in Texas, USA, and Ontario, Canada. Dr. Ahmari's principal research and teaching interests include hydraulics and hydrology, sediment transport and stream geomorphology, and stormwater management and nonpoint source pollution. He has more than 20 years of experience as an educator, researcher, and practitioner in environmental water resources engineering. He has served nationally and internationally as a consultant with government agencies and private organizations, and has led research projects funded by federal and state agencies. He regularly reviews proposals and manuscripts for national and international funding agencies and journals. Dr. Ahmari has authored or co-authored one book, two book chapters, more than 40 peer-reviewed technical papers, and numerous research reports. He has also contributed to professional training, seminars, and outreach programs in water resources engineering. He is an active member of the American Society of Civil Engineers (ASCE) and the International Association for Hydro-Environment Engineering and Research (IAHR).

1 Introduction

The modern sewerage system is an essential infrastructure in any developed area. It keeps cities clean and prevents health issues associated with poor hygiene and exposure to waste. The sewerage systems that we encounter on a day-to-day basis have been developed and refined over the course of thousands of years. In fact, many of the same technologies that were used in ancient societies are still present in some form today in developing countries. Following is a brief discussion on wastewater collection and conveyance systems, watershed management, and nonpoint source pollution that have evolved through time.

1.1 HISTORICAL DEVELOPMENT IN WASTEWATER COLLECTION, CONVEYANCE, AND DISPOSAL SYSTEMS

As societies moved from nomadic cultures to building more permanent settlements, the disposal of human waste became an important concern. Over thousands of years, this issue has been dealt with in many ways, and knowledge has been lost and regained. When humans were living as hunter-and-gatherer groups, their waste was deposited over the land. Decomposition took place naturally over the earth and the soil. As communities developed, it was necessary to address this issue in the context of sanitation and minimizing the health risks, primarily from infectious diseases.

1.1.1 Early Developments

In the ancient world, as cultures or societies grew larger, newer waste treatment technologies were adapted as the skills of various cultures developed. The early developments cover a period from 4000 to 200 B.C.[1,2] Mesopotamia (modern-day Iraq) was the location of three ancient civilizations that overlapped each other. Evidence of pipes and cesspits has been found in several larger ancient cities; the earliest date back to 4000 B.C. The Mesopotamians were well known for their pipe-making skills. They made pipes from terracotta, stone, copper, and lead. In the same period, the Neolithic village of Skara Brae in Scotland's Orkney Islands had evidence of plumbing, such as pipes that carried excrement from the village to nearby waterways.[1,3] Beginning around 3000 B.C., the Indus Valley civilization (present-day Pakistan) created large open sewage channels that flowed with water drawn off from nearby rivers or streams.[1,2] This was probably the most extensive use of a gravity-fed waste drainage system. There is evidence that some houses had bathrooms with water flushing toilets. In the elite and religious quarters, there was a deliberate effort made to remove all waste from living and commercial areas. The waste was transported to the rivers and streams.[1–3]

The Minoan civilization between 2000 and 1500 B.C. was an Aegean Bronze Age civilization on the Mediterranean island of Crete. It had a highly developed waste management system. They had highly advanced plumbing systems using terracotta pipes. This piping system is similar to the techniques used today. They had flushing toilets with wooden seats and an overhead reservoir. Excavations revealed four large drainage systems that emptied into large sewers built of stone. These sewers discharged wastewater into a nearby stream.[1–3]

1.1.2 Medieval Period

The Medieval period (Middle Ages) includes the period from 200 B.C. to 1500 A.D. Greek cities in 200 B.C. incorporated sewer systems to prevent contamination of incoming drinking water. The Minoan royals were the last group to use toilets until the redevelopment of that technology in late

DOI: 10.1201/9781003629399-1

1500 A.D. The Romans' waste management practices were by far the most developed of any civilization prior to the 19th century. In fact, the Romans were better than those in the Middle Ages.[4,5] The Romans' concern for water was best illustrated by their aqueducts to provide water to their cities. The water was used for baths, fountains, public conveniences, and flushing toilets and sewers. The water-carrying devices were open, and sewers were covered to transport most of the waste to the nearby River Tiber. The construction projects continued, and by the 4th century, Rome had many public and private baths with water-flush toilets.[6,7]

The fall of the Roman Empire turned the urban society into a rural society. By 500 A.D., the water taps were being turned off all over Europe, and they would not be turned on again for nearly a thousand years. Sanitation technology entered its dark ages. Epidemics raged in the cities, but the relationship between excrement and disease was not recognized. Dysentery, typhus, and typhoid fever became common. By the end of the 12th and beginning of the 13th centuries, changes began to take place. Following the major Bubonic Plague of the 12th century, which killed nearly 50 million people or 60% of Europe's population, waste management became a priority.[8,9] Laws were passed to impose severe penalties for throwing waste in streets, pollution of streams, and discharging animal and industrial wastes because of the danger of diseases. The cesspool was one of the technical developments of the Renaissance. Restrictions were placed on cesspools location and disposal of cesspool waste. By the end of the 15th century, all these restrictions brought new awareness to human, animal, and industrial wastes that enhanced associated human health.[5,10]

1.1.3 Early Modern Period

Research and development continued in the field of human waste treatment and disposal. From 1500 A.D. up until the middle of the 19th century, people were still doing their business in pits, outhouses, and chamber pots. In 1860, John-Louis Mouros of France invented the septic tank. The septic tanks removed gross solids before discharging the effluent into the nearest stream or river. Public health and diseases were driving rapid changes.[4,5] The scourge of the 19th century was cholera. In the mid-19th century, a worldwide epidemic of cholera occurred. The relationship of cholera to drinking water was discovered by English physician John Snow. He traced the contamination of public wells by privy vaults during the epidemic of 1854 in London.[10] This initiated the development of sewerage system. Work on the first sanitary sewer in the United States began in Chicago, 12 years after the world's first sanitary sewerage system was completed in Hamburg, Germany. The sewer diverted the wastewater from the communities to the natural waters. The pollution of streams and rivers became very serious, and the need to build a sewage treatment plant was recognized. The need to solve the pollution of natural waters and the health care concerns in large cities grew rapidly. Thus, the 19th century became the beginning of wastewater treatment technologies and pollution control in the United States.[11]

1.2 CURRENT STATUS OF WASTEWATER COLLECTION AND CONVEYANCE, AND NONPOINT SOURCE POLLUTION

Today, wastewater and stormwater collection and conveyance are essential components of modern urban infrastructure. They come in a variety of shapes and sizes. In an urban watershed, the collection and conveyance systems consist of a series of interconnected pipes that are designed to collect and transport municipal wastewater to a treatment plant. A separate system of pipes and channels collects and directs stormwater runoff away from roads, sidewalks, and other paved surfaces to a disposal site. The nonpoint source (NPS) pollution is widely a dispersed runoff that carries contaminants and deposits them directly into natural waters.[11]

1.2.1 Wastewater and Stormwater Collection and Conveyance

The current sewerage system in urban areas includes a pipe network of sewers, manholes, chambers and appurtenances, pumping stations and force mains, and many types of pipe connections. There are three types of sewers: sanitary, storm, and combined sewers. Sanitary sewers are designed to collect and convey municipal wastewater. Storm sewers are designed to carry storm runoff. Combined sewers are designed to carry both municipal and storm flows. When combined sewers were introduced in 1855, they were hailed as a vast improvement over open drains. These open drains ran along the city streets and carried drainage from cesspools, newly invented flush toilets, pet wastes, and spilled-over runoffs when it rained. However, the combined sewers are now considered undesirable because under heavy precipitation, they carry very large flows exceeding the designed capacity of the wastewater treatment facilities, causing combined sewer overflow. For these reasons, combined sewers are not constructed these days in the United States.[10,11]

Because of the importance and significance of pollution control from stormwater, Chapter 11 is exclusively devoted to the management of combined sewer overflows. Additionally, Chapter 12, *Conveyance and Disposal Structures for Outfall Sewer*, is included to present the design and site selection criteria to minimize the water quality impact from effluent disposal in freshwater and marine environments.

1.2.2 Watershed Management and Control of Nonpoint Source Pollution

The Clean Water Act has been very successful at reducing pollution discharges from point sources (industries and municipalities). NPS pollution is widely dispersed in the environment and comes from a variety of human open-land activities. These activities include feedlots, on-site wastewater treatment and disposal, landfills, agriculture, and land development projects. Urban watersheds contribute oil, pet waste, pesticides, herbicides, nutrients, road salt, bacteria, sediment, and any other contaminant that ends up on the ground naturally or from human activity. Rainwater and snowmelt pick up these contaminants as they wash over different land uses in the watershed, such as fields, agricultural land, over yards, sidewalks, driveways, and parking lots, and deposit them into streams as NPS pollution.[12,13]

Effective watershed management is a critical aspect of controlling NPS pollution. Best management practices (BMPs) are essential for mitigating the risks associated with human activities. Sixteen source control BMPs and their effectiveness, many of which are priority systems modified for urban settings to reduce their footprint, installed underground and in combination with other systems, are presented in Chapter 10. It is a systematic approach that routes runoff to specific conveyances where energy-reducing BMPs and specialized groundcovers are employed. These covers capture and treat contaminant-laden overland flow, create potential for infiltration, and reduce stormwater runoff. This is the predominant approach to be adopted by developers and contractors. Among the watershed rules and regulations controlling NPS pollution program, three basic mechanisms are followed. These mechanisms are (1) strict performance standards applied to activities that produce NPS pollution, (2) a review and approval process for activities that produce NPS pollution, and (3) prohibition of certain activities in a "setback" region between the activity and nearby waterbodies.[14,15]

1.3 FUTURE DIRECTIONS

It is well recognized that NPS pollution is a persistent and pervasive source of water pollution in the United States. In practice, a watershed is a fundamental unit of monitoring and management of NPS pollution. To better understand the complex mechanisms and processes of NPS pollution, various watershed-scale NPS models and tools have been developed to understand NPS pollution and evaluate sources of pollution and their impact on surface water quality.[12] These models also assist in the planning and implementation of BMPs, support the development of water quality criteria or standards, and provide informed decisions to support policy-makers.[13–16] With the development of model theory, computer technology, environmental legislation, and advancements in water quality, hydrology, rainfall-runoff, and NPS pollution, different models will be integrated into the watershed models.[16,17] These NPS models will become essential tools in watershed management and will be widely used in the future to identify critical sources of pollution, evaluate the effect, and combat NPS pollution.[12,17,18]

1.4 ROLE OF ENGINEERS AND SCIENTISTS IN POLLUTION CONTROL

Wastewater engineering and pollution control is a branch of environmental engineering in which various science and engineering disciplines are needed to achieve the goals of water quality improvement, pollution control, and environmental protection. Wastewater collection and conveyance systems require expertise in environmental engineering, hydraulics, and fluid mechanics. Environmental engineers and scientists are actively involved in wastewater treatment, disposal, and reuse.[12,19]

In the United States, NPS pollution is still considered and reported as a leading cause of water quality impairment of the nation's waters. It is a challenge from both regulatory and assessment perspectives. In practice, a watershed is a fundamental source of monitoring and management of NPS pollution. To better understand the complex mechanisms and processes of NPS pollution, environmental engineers, hydrologists, water resources engineers, and scientists will develop watershed-scale NPS models to combat the leading causes of NPS pollution.[1]

REFERENCES

1. Angelakis, A. N., and J. B. Rose (Editors), *Evolution of Sanitation and Wastewater Technologies Through the Centuries*, IWA Publishing, London, August 2014.
2. Zeldovich, L., The Early History of Human Excreta, *JSTOR Newsletter,* November 10, 2021.
3. Yannopoulos, S., C. Yapijakis, A. Kaiafa-Saropoulou, G. Antoniou, and A. N. Angelakis, History of Sanitation and Hygiene Technologies in the Hellenic World, *Journal of Water, Sanitation and Hygiene for Development*. vol. 7, no. 2, February 2017, pp. 163–180. 10.2166/washdev.2017.178.
4. Taylor, C., The Disposal of Human Waste: A Comparison Between Ancient Rome and Medieval London, *Past Imperfect*, vol. 11, 2005, pp. 53–72.
5. Halliday, S., *An Underground Guide to Sewers: Or Down, Through and Out in Paris*, MIT Press, London, New York, 2019.
6. Havlicek, F., and M. Morcinek, Waste and Pollution in the Ancient Roman Empire, *Journal of Land Ecology,* vol. 9, no. 3, December 2016, pp. 33–49.
7. Koloski-Ostrow, A. O., *The Archaeology of Sanitation in Roman Italy: Toilets, Sewers, and Water Systems (Studies in the History of Greece and Rome),* The University of North Carolina Press, Chapel Hill, NC, 2015.
8. Bell, D. P. (Editor), *Plague in the Early Modern World A Documentary History*, Routledge, New York, 2019.
9. Kennedy, J., *Pathogenesis: A History of the World in Eight Plagues*, Crown, New York, 2024.
10. Herbert, L., Century History of Waste and Waste Managers in London and South East England, *The Chartered Institution of Wastes Management,* June 2002, pp. 1–52.
11. Qasim, S. R., *Wastewater Treatment Plants: Planning, Design, and Operation,* 2nd ed., CRC Press LLC, Boca Raton, FL, 1999.
12. Yuan, L., T. Sinshaw, and K. J. Forshay, Review of Watershed-Scale Water Quality and Nonpoint Source Pollution Models, *Geosciences,* vol. 10, no. 25, January 2020, pp. 1–36, 10.3390/geosciencesl0010025.
13. Liu R, F. Xu, P. Zhang, W. Yu, and C. Men, Identifying Non-point Source Critical Source Areas Based on Multi-factors at a Basin Scale With SWAT, *Journal of Hydrology*, vol. 533, no. 5, February 2016, pp. 379–388. 10.1016/j.jhydrol.2015.12.024.

14. Mittelstet, A. R., D. E. Storm, and M. J. White, Using SWAT to Enhance Watershed-based Plans to Meet Numeric Water Quality Standards, *Sustainability of Water Quality and Ecology*, vol. 7, January 2016, pp. 5–21. 10.1016/j.swaqe.2016.01.002.
15. Kim S., D. J. Seo, H. Riazi, and C. Shin, Improving Water Quality Forecasting via Data Assimilation – Application of Maximum Likelihood Ensemble Filter to HSPF, *Journal of Hydrology,* vol. 519, 2014, pp. 2797–2809.
16. Liu Y, L. M. Ahiablame, V. F. Bralts, and B. A. Engel, Enhancing a Rainfall-runoff Model to Assess the Impacts of BMPs and LID Practices on Storm Runoff, *Journal of Environmental Management,* vol. 147, January 2015, pp. 12–23. 10.1016/j.jenvman.2014.09.005.
17. Gene Y., Using GWLF for Development of "Reference Watershed Approach" TMDLs. In *Proceedings of the ASAE/CSAE Annual International Meeting,* Ottawa, ON, Canada, August 2004, pp. 1–4.
18. Corrales J., G. M. Naja, M. G. Bhat, and F. Miralles-Wilhelm, Water Quality Trading Opportunities in Two Sub-watersheds in the Northern Lake Okeechobee Watershed, *Journal of Environmental Management,* vol. 196, July 2017, pp. 544–559. 10.1016/j.jenvman.2017.03.061.
19. Metcalf & Eddy, Inc., *Wastewater Engineering: Collection, and Pumping of Wastewater,* McGraw-Hill Book Company, New York, NY, 1981.

2 Significant Figures and Units of Expression

2.1 CHAPTER OBJECTIVES

The accuracy of a number is specified by the significant figure it contains. The significant figures (also called significant digits) are those digits that give the indication of accuracy and are used for rounding off the numbers. All systems of units and measures are linked through a network of the *International System of Units (SI)*.[1]

The chapter objectives are to review:

- Use of significant figures
- Principles for applying U.S. customary and SI units
- Engineering dimensions, quantities, and unit conversions

2.2 USE OF SIGNIFICANT FIGURES

There are four basic rules of significant figures:

1. Zero at the beginning of a number is not counted
2. Zero at the end is counted only if there is a written decimal point
3. The digits 1–9 are counted, and
4. Zeros between the digits 1 and 9 are counted.

The numbers 1,825 and 18.25 each have four significant figures. Leading zeros that hold place are not significant since they can be eliminated without loss of information. For example, the number 0.065 has one zero that holds a place. It can be eliminated by writing 6.5×10^{-2}. Thus, there are two significant figures. Zeros at the end of a number may require careful consideration. As an example, the estimated population of a city is 12,500. Since it is an estimate, the last two digits (zeros) may be any number. Therefore, there are only three significant figures (1.25×10^4). On the other hand, suppose 1,800 tickets were sold for a symphony, and someone will be holding that last ticket. Therefore, in this case, 1,800 has four significant figures (1.800×10^3).[2,3]

2.3 ROUNDING OFF NUMBERS

These days, handheld calculators or computers provide more figures in the answer than the number of significant figures needed. For this reason, a calculated result should be "rounded off" to an appropriate number of significant figures. *One must remember that the accuracy obtained from the solution of a problem generally can never be better than the accuracy of the problem data.* Thus, the resultant calculation will be significant to the least significant data. The calculated value should be rounded accordingly. Suppose the total average daily water demand of a city is 10,600 m^3/d and the estimated population is 25,000. The average water demand is (10,600 m^3/d × 1,000 L/m^3)/25,000 person = 424 or 420 liters per capita per day (Lpcd). Notice that the calculated value is 424 Lpcd. In this case, both the water demand and population are estimated numbers (three and two significant figures). Therefore, 420 Lpcd (two significant figures) would be a more realistic representation than even 424 Lpcd.[2–4]

DOI: 10.1201/9781003629399-2

Example 2.1: Significant Figures

Give the significant figures of the following numbers.

a. \$27.36: Price of a calculator
b. 18.2030 g: Weight of a crucible weighed over an analytical balance that has an accuracy of ±0.1 mg
c. 250 lb: Weight of a person taken over a weighing scale that has an accuracy of ±1 lb
d. 46,200: Fans in a football stadium
e. 0.0027 lb: Weight of a crucible

Solution

Give the significant figures

a. Significant figures are 4
b. Significant figures are 6
c. Significant figures are 3
d. Significant figures are 3
e. Significant figures are 2 because $0.0027 = 2.7 \times 10^{-3}$

Example 2.2: Rounding Off Project Cost

The unit cost of laying a pipe is \$12.18 per foot. What will be the project cost if the total length installed is ¼ mile. Add 15% contingency to get the total cost.

Solution

1. Calculate the total cost of pipe installation
   Installation = \$12.18/ft × 0.25 mile × 5,280 ft/mile = \$16,077.60
   Contingency = 0.15 × \$16,077.60 = \$2,411.64
   Total = \$16,077.60 + \$2,411.64 = \$18,489.24
2. Round off number
   Consider how silly it would be to report the estimated cost of \$18,489.24. Can you estimate the cost of this project in cents? A reasonable estimate in this case would be **\$18,500** (three significant figures).

Example 2.3: Rounding Off Flow Estimate

A city has a population of 65,800. The average water demand is 630 Lpcd. It is estimated that 15% water is lost due to cracks in pipes. Estimate the average daily water lost in cubic meters.

Solution

1. Calculate average water demand
   Average water demand = 65,800 people × 630 L/c.d × 10^{-3} m^3/L = 41,454 m^3/d
2. Estimate the lower and upper bounds of water loss
   Lower bound = 41,454 m^3/d × 0.1 = 4,145 m^3/d
   Upper bound = 41,454 m^3/d × 0.2 = 8,290 m^3/d
3. Round off number
   Since water lost from the distribution system is a gross estimate, a reasonable round off value for 15% loss will lie near the average of two bounds. In this case, it would be 6,200 m^3/d.

2.4 PRINCIPLES OF APPLYING U.S. CUSTOMARY AND SI UNITS

Proper conversion relationships between U.S. customary units and the SI units (International System) have been well established.

2.4.1 U.S. Customary Units

The fundamental units in U.S. customary units are foot (ft) for length, pound (lb_f) for force (or pound weight), and second (sec) for time. All other U.S. customary units are derived from these units.

2.4.2 International System of Units

The SI system of units comes from the French name, *Systéme International d' Unités* (SI units). It is adopted by many international bodies. In SI system, the fundamental units are meter (m), kilogram (kg), and second (s).[1] The reference SI units and conversion to U.S. customary units are summarized in Table 2.1.

2.4.3 Multipliers of Standard Units

To express quantities larger or smaller than the standard units, systems of multipliers are used. These multipliers, along with prefixes and abbreviations, are listed in Table 2.2.

TABLE 2.1
SI Units and Conversion to U.S. Units

Physical Quantity	Unit Name	Unit Symbol	U.S. Customary Unit and Conversion
Length	meter	m	ft = 0.3048 m
Mass	kilogram	kg	slug ($lb_f \cdot s^2$/ft) = 14.5938 kg
Time	second	s	sec
Area	square meter	m^2	$ft^2 = 0.0929\ m^2$
Volume	cubic meter	m^3	$ft^3 = 0.0283\ m^3$, gal $= 3.875 \times 10^{-3}\ m^3$

Note: The abbreviation for *second* in SI units is "s," and in U.S. customary unit is "sec." In this book "s" for second is used in both systems.

TABLE 2.2
Multipliers of Standard SI Units

Prefix	Abbreviation	Multiplier	Prefix	Abbreviation	Multiplier
deka	da	10	deci	d	10^{-1}
hecto	h	10^2	centi	c	10^{-2}
kilo	k	10^3	milli	m	10^{-3}
mega	M	10^6	micro	μ	10^{-6}
giga	G	10^9	nano	n	10^{-9}
tera	T	10^{12}	pico	p	10^{-12}
peta	P	10^{15}	femto	f	10^{-15}
Exa	E	10^{18}	atto	a	10^{-18}

Example 2.4: Multipliers of Standard SI Units

Express the following SI units as indicated: (a) 7,835 mg in g, (b) 6,598 N in kN, (c) 0.968 kg in g, (d) 586×10^5 mm in km, (e) 0.00564 mg in µg, and (f) 3.21 M^3g in g.

Solution

Express in SI units

a. 7,835 mg $=7.835\times10^3$ mg$\times(10^{-3})$ g/mg $=7.835$ g
b. 6,598 N $=6,598$ N$\times10^{-3}$ kN/N $=6.598$ kN
c. 0.968 kg $=0.968$ kg $\times10^3$ g/kg $=968$ g
d. 586×10^5 mm $=586\times10^5$ mm$\times10^{-3}$ m/mm$\times10^{-3}$ km/m $=58.6$ km
e. 0.00564 mg $=0.00564$ mg$\times10^{-3}$ g/mg $\times10^6$ µg/g $=5.64$ µg
f. $3.21\ \mathrm{M}^3\ \mathrm{g} = 3.21\ \mathrm{M}^3\ \mathrm{g}\times\left(10^6\right)^3\ \mathrm{g/M}^3\ \mathrm{g} = 3.21\times10^{18}$ g or 3.21 Eg

2.4.4 Temperature

Temperature determines the flow of heat. It is independent of dimension. The SI unit of temperature is Kelvin (K). One Kelvin is –273.15 °C; it is also the absolute zero. The SI or Kelvin scale is often called the absolute temperature scale. The Rankine (°R) scale is the absolute temperature scale related to the Fahrenheit scale. The relationship between Kelvin, Celsius, and Fahrenheit scales is given in Table 2.3.

Temperature conversions are commonly expressed by Eqs. (2.1)–(2.4).

$$^{\circ}\mathrm{F}(\mathrm{Fahrenheit}) = {}^{\circ}\mathrm{C} + 32 \tag{2.1}$$

$$^{\circ}\mathrm{C}(\mathrm{Celsius}) = ({}^{\circ}\mathrm{F} - 32) \times \frac{5}{9} \tag{2.2}$$

$$^{\circ}\mathrm{K}(\mathrm{Kelvin}) = 273 + {}^{\circ}\mathrm{C} \tag{2.3}$$

$$^{\circ}\mathrm{R}(\mathrm{Rankin}) = 460 + {}^{\circ}\mathrm{F} \tag{2.4}$$

TABLE 2.3
Relationship between Kelvin, Celsius, Rankine, and Fahrenheit Scales

Reference Point	Kelvin, °K	Celsius, °C	Rankine, °R	Fahrenheit, °F
Boiling point of water	373.15 (373)	100	671.67 (672)	212
Freezing point of water	273.15 (273)	0	491.67 (492)	32
Absolute zero temperature	0	–273.15 (–273)	0	–459.67 (–460)

Note: Values in parentheses are rounded values.

Example 2.5: Temperature Conversion

Convert 85°F in °C, °K, and °R.

Solution

Convert 85° in different scales

Apply Eqs. (2.2)–(2.4)

°C=(85−32)×5/9=29

°K=273+29=302

°R=460+85=545

2.5 ENGINEERING DIMENSIONS, QUANTITIES, AND UNIT CONVERSIONS

2.5.1 Length, Area, Volume, Velocity, and Discharge

The standard unit of length (L) is m or ft. The area and volume are expressed by L^2 and L^3, respectively. The velocity and discharge units are length and volume per unit time.

Example 2.6: Length Conversion

Distance between two reference points is 1.2 km. Express the distance in meters, feet, and miles. List all conversion factors used.

Solution

1. Convert the distance in meters

$$\text{Distance} = 1.2\text{ km} \times 1{,}000\text{ m/km} = 1{,}200\text{ m}$$

2. Express the distance in feet

$$\text{Distance} = 1{,}200\text{ m} \times \frac{1}{0.3048\text{ m/ft}} = 3{,}937\text{ ft}$$

3. Express the distance in miles

$$\text{Distance} = 3{,}937\text{ ft} \times \frac{1}{5{,}280\text{ ft/mile}} = 0.75\text{ miles}$$

4. List the conversion factors used

   The conversions utilized are

   1 ft=0.3048 m; 1 mile=5,280 ft; 1 mile=1.609 km; 1 km=0.622 mile.

Example 2.7: Area Conversion (Channel Cross-Section)

The cross-sectional area of a channel is 0.50 ft². Express the area in square inch (in²), cm², and m². List all conversion factors used.

Solution

1. Convert area in in²

   Area=0.5 ft²×(12 in/ft)²=72 in²

2. Convert area in cm^2
   Area = 72 $in^2 \times (2.54\ cm/in)^2 = 464.5\ cm^2$
3. Express the area in m^2

$$\text{Area} = 464.5\ \text{cm}^2 \times \frac{1}{(100\ \text{cm/m})^2} = 0.0465\ \text{m}^2$$

4. List the conversion factors used
   1 $ft^2 = 144\ in^2$; 1 $in^2 = 6.452\ cm^2$; 1 $ft^2 = 0.0929\ m^2$

Example 2.8: Area Conversion (Land Area)

A land area is 1.45 ha. Express the area in m^2 and acres. List all conversion factors used.

Solution

1. Express the area in m^2
   Area = 1.45 ha × 10,000 m^2/ha = 14,500 m^2
2. Express the area in acre

$$\text{Area} = 14{,}500\ \text{m}^2 \times \frac{1}{(0.3048\ \text{m/ft})^2} \times \frac{1}{43{,}560\ \text{ft}^2/\text{acre}} = 3.58\ \text{acres}$$

3. List the conversion factors used
   1 ha = 10,000 m^2; 1 ft = 0.3048 m; 1 acre = 43,560 ft^2; 1 ha = 2.471 acre

Example 2.9: Volume Conversion (Tank Volume)

A mixing chamber has a volume of 0.75 m^3. Express the volume in liters, gallons, and ft^3. List all conversion factors used.

Solution

1. Express volume in liters
   Volume = 0.75 m^3 × 1,000 L/m^3 = 750 L
2. Express volume in gallons

$$\text{Volume} = 750\ \text{L} \times \frac{1}{3.785\ \text{L/gal}} = 198.2\ \text{gal}$$

3. Express volume in ft^3

$$\text{Volume} = 198.2\ \text{gal} \times \frac{1}{7.48\ \text{gal/ft}^3} = 26.5\ \text{ft}^3$$

4. List the conversion factors used
   1 m^3 = 1,000 L; 1 gal = 3.785 L; 1 ft^3 = 7.48 gal

Example 2.10: Velocity Conversion

The water velocity in a pipe is 2 m/s. Express the velocity as km/h, ft/s, and miles/h. List all conversion factors used.

Solution

1. Express velocity as km/h
   Velocity=2 m/s×(60×60) s/h×10^{-3} km/m=7.2 km/h
2. Express velocity as ft/s

$$\text{Velocity} = 2 \text{ m/s} \times \frac{1}{0.3048 \text{ m/ft}} = 6.56 \text{ ft/s}$$

3. Express velocity as miles/h

$$\text{Velocity} = 6.56 \text{ ft/s} \times (60 \times 60) \text{ s/ h} \times \frac{1}{5{,}280 \text{ ft/mile}} = 4.5 \text{ miles/h}$$

4. List all conversion factors used
   m/s=3.6 km/h; m/s=3.28 ft/s; ft/s=0.68 miles/h; miles/h=1.609 km/h
   Note: The abbreviation for *second* in SI unit is "s," and in U.S. customary system is "sec." In this book, "s" for second is used in both systems.

Example 2.11: Flow, Demand, and Discharge Conversions

A water distribution main is carrying a flow of 0.5 m^3/s. Express the demand in m^3/d, million L/d (mLd), ft^3/s (cfs), million gallons/d (mgd), and gallons per min (gpm).

Solution

1. Express flow (discharge or demand) in m^3/d
   Flow=0.5 m^3/s×(60×60×24) s/d=43,200 m^3/d
2. Express flow in mLd (million liters per day)
   Flow=43,200 m^3/d×1,000 L/m^3×10^{-6} million liters/L=43.2 mLd
3. Express flow in ft^3/s
   Flow=0.5 m^3/s×(1 ft/0.3048 m)^3=17.66 ft^3/s (cfs)
4. Express flow in mgd
   Flow=17.66 ft^3/s×7.48 gal/ft^3×(60×60×24) s/d×10^{-6} million gallon/gal=11.41 mgd
5. Express flow as gallons per minute (gpm)

$$\text{Flow} = 11.41 \text{ mgd} \times 10^6 \text{ gal/million gallon} \times \frac{1}{(24 \times 60) \text{ min/d}} = 7{,}924 \text{ gpm}$$

6. List all conversion factors used
   1 m^3/s=86,400 m^3/d; 1 m^3/s=35.32 cfs; 1 m^3/d=10^3 Lpd; 1 m^3/d=10^{-3} mLd; 1 mgd=1.547 ft^3/s; and 694.4 gpm

2.5.2 Force, Weight, and Mass

In the SI system, the unit for force is a Newton (N), and mass is in kg. In the U.S. customary system, the unit for force is lb_f, and mass is in slug. Force and weight are derived from Eqs. (2.5a) and (2.5b), respectively.[4,5]

$$F = ma \tag{2.5a}$$

$$W = mg \tag{2.5b}$$

where

F = force, N (lb_f). The units for N = kg m/s², and units for lb_f = slug ft/s²
m = mass, kg (slug)
a = acceleration, m/s² (ft/s²)
W = weight or force on a mass when acceleration is due to gravity, g = 9.81 m/s² (32.2 ft/s²)
From Eq. (2.5a), 1 N = force when 1 kg mass accelerates at 1 m/s² (1 N = 1 kg × (a = 1 m/s²)).
Weight is force (Eq. 2.5b), when 1 kg mass accelerates at 9.81 m/s² (1 W = 1 kg × (g = 9.81 m/s²)).
From Eq. (2.5a), 1 lb_f = force when 1 slug accelerates at 1 ft/s² (1 lb_f = 1 slug × (a = 1 ft/s²))
Weight from Eq. (2.5b) is force when 1 slug accelerates at 32.2 ft/s² (1 W = 1 slug × (g = 32.2 ft/s²))

From the force, the units of pressure and work are derived. The units of force, weight, mass, pressure, and work in both units are summarized in Table 2.4.

The direct conversions between SI and U.S. customary units are provided in Table 2.5.

TABLE 2.4
Units of Force, Mass, Weight, Pressure, and Work

System of Units	Force	Mass	Weight	Pressure or Stress	Work
SI Unit	Newton (N) = kg m/s²	m (kg)	W = mg (kg × 9.81 m/s²)	Pascal (Pa) = 1 N/m²	Joule (J) = N m
U.S. Customary	lb_f = slug ft/s²	slug = lb_f s²/ft	W = lb_f (slug g) = lb_f (slug × 32.2 ft/s²)	lb_f/ft², lb_f/in²	lb_f ft

Note: It is not uncommon to use lb symbol instead of lb_f for force in U.S. customary unit.

TABLE 2.5
Conversion Factors

Quantity	US Customary Unit	SI Unit
Force (weight)	lb_f	4.4482 N (kg.m/s²)
Mass	slug (lb_f s²/ft)	14.5938 kg
Pressure or Stress	lb_f/ft²	47.88 Pa
	lb_f/in²	6.895 kPa
Work	lb_f ft	1.356 J

Example 2.12: Mass to Weight Conversion in SI Units

An object has a mass of 1.0 kg. What is its weight in SI units?

Solution

Express the weight in SI units. Apply Eq. (2.5b).

$W = mg$
$W = 1.0\ \text{kg} \times 9.81\ \text{m/s}^2 = 9.81\ \text{kg m/s}^2 = 9.81\ \text{N}$

Example 2.13: Force to Mass Conversion in U.S. Customary Units

Convert 5 lb_f weight or force to slug.

Solution

1. Use Eq. (2.5b)
   $W = mg$
   m (mass) $= W/g$

$$5 \text{ lb mass} = \frac{5 \text{ lb}_f}{32.2 \text{ ft/s}^2}$$

$$= \frac{5 \text{ lb}_f \text{ s}^2}{32.2 \text{ ft}} = 0.155 \text{ slug}$$

2. State the conversion

$$1 \text{ slug} = \frac{\text{lb}_f}{32.2 \text{ ft/s}^2} = \frac{0.031 \text{ lb}_f \text{ s}^2}{\text{ft}}$$

   1 lb_f weight or force $= 32.2$ slug ft/s^2

Example 2.14: Force (Weight) Conversion from U.S. Customary Unit to SI Unit

Convert 500 lb_f in appropriate SI unit.

Solution

Convert lb_f in N
   Apply conversions given in Table 2.5
   500 $lb_f = 500$ $lb_f \times (4.4482$ N/$lb_f) = 2224.1$ N or 2.224 kN

Example 2.15: Slug to SI Unit Conversion

Convert the density of 80 slug/ft^3 in kg/m^3.

Solution

1. Convert density from U.S. customary units to SI units.
   Apply the conversions given in Tables 2.1 and 2.5
   80 slug/ft$^3 = 80$ slug/ft$^3 \times (14.5938$ kg/slug$) \times ($ft/0.3048 m$)^3 = 41{,}230$ kg/m$^3 = 41.23 \times 10^3$ kg/m^3
2. State the conversion
   1 slug/ft$^3 = 515.4$ kg/m^3

Example 2.16: Weight to Mass Conversion

An object weighs 175 lb_f. Determine the mass in (a) slug, and (b) kg.

Solution

1. Convert weight to mass. Use Eq. (2.5b)
   $W = mg$

$$175\ lb_f\ mass = \frac{175\ lb_f}{32.2\ ft/s^2} = 5.44\ lb_f\ s^2/ft = 5.44\ slug$$

$$= 5.44\ slug \times 14.5938\ kg/slug\ (Table\ 2.5)$$

$$= 79.4\ kg$$

2. List the conversions
   1 lb_f=0.031 slug=0.4536 kg

Example 2.17: Conversions of Mass to Newton (N), and kg to Slug

An object has a mass of 50 kg. Express this mass as (a) weight in Newton, and (b) in slug.

Solution

1. Express mass as weight. Use the conversions given in Tables 2.4 and 2.5
   W=50 kg×9.81 m/s^2=490.5 N
2. Express mass as slug

$$50\ kg \times \frac{slug}{14.5938\ kg} = 3.426\ slug$$

3. List the conversions
   1 kg=9.81 N; 1 kg=0.0685 slug

Example 2.18: Conversion of Force from U.S. Customary Units to SI Units, and Mass to Weight

Make the following conversions: (a) 250 lb_f ft to N m, and (b) 10 kg mass to weight in Newtons.

Solution

1. Convert force from U.S. customary units to SI units. Use the conversions given in Tables 2.1 and 2.5
   250 lb_f ft=250 lb_f ft×(4.4482 N/lb_f)×(0.3048 m/ft)=339 N m
2. Convert mass to weight in SI units
   10 kg mass=10 kg×9.81 m/s^2=98.1 N (weight)
3. List the conversions
   1 lb_f ft=1.356 N m; 1 kg mass=9.81 N

2.5.3 Pressure, Work, and Power

The unit of pressure is defined as the unit of force acting upon a unit area. The common units of pressure are kg/m^2, lb_f/ft^2, or lb_f/in^2. Work is done by a force in moving a body through a unit distance in the direction of the force. The unit of work is joule (J) or ft.lb_f (lb_f ft). One joule of work is done when one Newton is displaced for a distance of 1 m. Power is defined as the amount of work performed per unit of time. The basic units of power used in the SI and U.S. customary systems are watt (W) and horsepower (hp), respectively. These units are defined as

1 W=1 J/s=1 N.m/s

1 hp=550 ft lb_f/s

1 hp=746 W

Example 2.19: Conversion of Pressure to Several Units

One atmosphere is 14.7 lb/in^2. Convert this pressure in: (a) lb_f/ft^2, (b) Pa, (c) ft of water, (d) m of water, and (e) cm of mercury.

Solution

1. Make conversions as required. Use the relationships given in Tables 2.1 and 2.5
   a. $1\ atm = 14.7\ lb_f/in^2 = 14.7\ lb_f/in^2 \times 144\ in^2/ft^2 = 2{,}116.8\ lb_f/ft^2$
   b. $1\ atm = 2{,}116.8\ lb_f/ft^2 \times 47.88\ Pa/(lb_f/ft^2) = 101{,}352\ Pa = 101.35\ kPa$
   c. $1\ \text{atm} = 2{,}116.8\ \text{lb}_\text{f}/\text{ft}^2 \times \dfrac{1}{62.4\ \text{lb}_\text{f}/\text{ft}^3(\text{water})} = 33.9\ \text{ft of water}$
   d. $33.9\ ft \times 0.3048\ m/ft = 10.34\ m$ of water
   e. $1\ \text{atm} = 10.34\ \text{m water} \times \dfrac{\text{sp.gr. of water}}{\text{sp.gr. of mercury}}$

$$= 10.34\ \text{m water} \times \frac{1}{13.55\ \text{m water/m Hg}}$$

$$= 0.760\ \text{m of mercury} = 760\ \text{cm of mercury}$$

2. List the conversions
   $1\ atm = 14.7\ lb_f/in^2 = 2{,}116.8\ lb_f/ft^2 = 101.35\ kPa$; 33.9 ft of water = 10.34 m of water = 760 cm of mercury

Example 2.20: Work and Power Conversion

A loaded truck weighs $20 \times 10^3\ lb_f$. It accelerates uniformly on a level road from 15 to 30 ft/s in 5 s. If the frictional resistance to motion is 400 lb_f, determine: (a) the maximum power that must be delivered to the wheels in ft lb_f/s, and (b) express the power in hp, kW, and J/s. List all conversions used.

Solution

1. Calculate force, F

$$\text{Acceleration} = \frac{(30\ \text{ft/s} - 15\ \text{ft/s})}{5\ \text{s}} = 3\ \text{ft/s}^2$$

$$W = mg\ (\text{from Eq. (2.5b)})$$

$$m = W/g$$

$$F = m \times a\ \ (\text{from Eq. (2.5a)})$$

$$F = W/g \times a$$

$$F = \frac{20 \times 10^3\ \text{lb}_\text{f}}{32.2\ \text{ft/s}^2} \times 3\ \text{ft/s}^2$$

$$F = 1{,}863.4 \text{ lb}_f$$

$$(F - 400 \text{ lb}_f) = 1{,}863.4 \text{ lb}_f$$

$F = 1{,}863.4 \text{ lb}_f + 400.0 \text{ lb}_f = 2{,}263.4 \text{ lb}_f$

2. Calculate power

   $P = F \times V = 2{,}263.4 \text{ lb}_f \times 30 \text{ ft/s} = 67{,}902 \text{ ft lb}_f\text{/s}$

3. Express power in other units

$$\text{hp} = \frac{67{,}902 \text{ ft lb}_f/\text{s}}{550 \text{ ft lb}_f/\text{hp}} = 123.5 \text{ hp}$$

   kW = 123.5 hp × 0.746 kW/hp = 94.4 kW

   94.4 kW × 1,000 W/kW = 94,400 W or 94,400 J/s

4. List all conversion factors used

   1 W = 1 J/s = 1 N m/s; 1 hp = 0.746 W

2.5.4 Specific (or Unit) Weight, Mass Density, and Specific Gravity

Specific Weight or Unit Weight. The *specific (or unit) weight* γ of a substance is the weight per unit volume. In SI units, it is expressed in kilonewton per cubic meter (kN/m^3). The specific weight of water at 10°C is 62.4 lb$_f$/ft^3 or 9.81 kN/m^3.

Mass Density or Density. The mass density (density), ρ (rho) is mass per unit volume (Eq. 2.6).

$$\rho = \gamma / g \tag{2.6}$$

where

ρ = mass density or density, kg/m^3, (slug/ft^3)

γ = specific (or unit) weight, kN/m^3, (lb$_f$/ft^3)

g = acceleration due to gravity, m/s^2, (ft/s^2)

In SI units, the mass density ρ (or density) of water is 1,000 kg/m^2 at 4°C.

In U.S. customary units, the mass density ρ (or density) is expressed in slug/ft^3. The density of water at 60°F is (62.4 lb$_f$/ft^3)/(32.2 ft/s^2) or 1.94 lb$_f$ s^2/ft^{3}ft or 1.94 slug/ft^3.

Specific Gravity. The specific gravity is the relative density of a substance with respect to that of water at a given temperature. It is a dimensionless ratio. It can be expressed by Eqs. (2.7) in different ways.

$$\text{Sp. gr.} = \frac{\text{Weight of a substance}}{\text{Weight of equal volume of water}} = \frac{W}{W_w} \tag{2.7a}$$

$$\text{Sp. gr.} = \frac{\text{Specific weight of a substance}}{\text{Specific weight of water}} = \frac{\gamma}{\gamma_w} \tag{2.7b}$$

$$\text{Sp. gr.} = \frac{\text{Density of a substance}}{\text{Density of water}} = \frac{\rho}{\rho_w} \tag{2.7c}$$

The commonly used unit for the density of water is g/cm^3 or kg/m^3 (lb$_f$/ft^3).

Dynamic Viscosity. The dynamic viscosity μ is a measure of fluid resistance to tangential or shear stress. Dynamic viscosity in SI units is expressed in Newton second per square meter (N s/m^2). In U.S. customary units, it is expressed as lb_f s/ft^2. The typical value of μ at 10°C is 1.307×10^{-3} N s/m^2 (2.735×10^{-5} lb_f s/ft^2).[5,6]

Kinematic Viscosity. Kinematic viscosity υ (nu) is the ratio of dynamic viscosity to density. $\upsilon=\mu/\rho$. The unit of expression for kinematic viscosity is m^2/s in SI units, and ft^2/s in U.S. customary units. The typical values of kinematic viscosity in SI and U.S. customary units are 1.307×10^{-6} m^2/s and 1.410×10^{-5} ft^2/s.

Surface Tension. Surface tension σ is a physical property. It is created by the attraction of the molecules to each other in a liquid. Many common observations of surface tension are: (1) a glass filled with water has surface slightly above the brim before it spills; (2) a metal needle floats on the surface of water; and (3) a drop of water is held in suspension at the tip of a pipette. The unit of expression of surface tension is N/m (lb_f/ft). The surface tension of water at 15°C is 0.0735 N/m (0.00504 lb_f/ft). The surface tension decreases with increase in temperature of a liquid.[5,6]

The definition and values of specific weight (γ), mass density (ρ), dynamic viscosity (μ), kinematic viscosity (υ), and surface tension (σ) of water at 10°C (50°F) are summarized in Table 2.6. These values for water at different temperatures are provided in Appendix B.

The surface tension σ of a liquid is expressed by Eqs. (2.8a) and (2.8b). The derivation and application of Eqs. (2.8a) and (2.8b) are given in Examples (2.31) – (2.33).

$$\sigma = \frac{\Delta F}{\Delta L} \quad \text{or} \quad F = \sigma L \tag{2.8a}$$

$$\sigma = 1/4pd \tag{2.8b}$$

where

σ = surface tension, N/m (lb_f/ft^2)
ΔF = elastic force transverse to a length element ΔL in the surface
F = force over an object, N (lb_f)
L = contact length, m (ft)
p = differential pressure, Pa (lb_f/ft^2)
d = diameter, m (ft)

TABLE 2.6
Specific Weight γ and Mass Density ρ, Dynamic Viscosity μ, Kinematic Viscosity υ, and Surface Tension σ of Water at Temperature 10°C (50°F)

Definition	SI Unit	U.S. Customary Unit
Specific (or unit) weight γ is the weight of unit volume	9.804 kN/m^3	62.41 lb_f/ft^3
Mass density ρ (density) is mass of unit volume	999.7 kg/m^3	1.940 slug/ft^3
Dynamic viscosity μ is the fluid resistance to tangential or shear stress	1.307×10^{-3} N s/m^2	2.735×10^{-5} lb_f s/ft^2
Kinematic viscosity υ is the ratio of dynamic viscosity and density (μ/ρ)	1.307×10^{-6} m^2/s	1.410×10^{-5} ft^2/s
Surface tension σ is due to attraction of molecules to each other	74.2×10^{-3} N/m	5.09×10^{-3} lb_f/ft

Example 2.21: Weight of a Given Volume of Solution

The specific gravity of a salt solution is 1.15. How many lb_f are in 50 gal of this solution?

Solution

Calculate the weight of 50 gal of the salt solution

Weight of 1 gal of water = 8.34 lb_f

$$\text{Weight of 50 gal of salt solution} = 50\ \text{gal} \times 8.34\ \text{lb}_f\ \text{water/gal} \times \frac{1.15\ \text{g/cc salt sol.}}{1\ \text{g/cc water}}$$

$$= 479.55\ \text{lb}_f$$

Example 2.22: Density and Specific Gravity Calculated from Mass and Volume

A body weighs 500 g and has a volume of 60 cm^3. Calculate the density in: (a) g/cm^3 and kg/m^3, and (b) specific gravity.

Solution

1. Calculate density

$$\text{Density} = \frac{\text{Mass}}{\text{Volume}} = \frac{500\ \text{g}}{60\ \text{cm}^3} = 8.33\ \text{g/cm}^3$$

$$= 8.33\ \text{g/cm}^3 \times \frac{10^{-3}\ \text{kg/g}}{10^{-6}\ \text{m}^3/\text{cm}^3} = 8{,}333\ \text{kg/m}^3$$

2. Calculate specific gravity of the substance

$$\text{Specific gravity} = \frac{\text{Density of the substance}}{\text{Density of water at 4°C}} = \frac{8{,}333\ \text{kg/m}^3}{1{,}000\ \text{kg/m}^3} = 8.33$$

Example 2.23: Specific Gravity Measurement

An empty 100 mL graduated cylinder weighs 110.675 g. The cylinder filled with water at 20°C weighs 210.495 g. The same graduated cylinder filled with an oil at 20°C weighs 209.475 g. Calculate the density of the oil and its specific gravity.

Solution

1. Calculate the density of water at 20°C

   Weight of water = 210.495 − 110.675 = 99.820 g

$$\text{Density of water at 20°C} = \frac{\text{Mass}}{\text{Volume}} = \frac{99.820\ \text{g}}{100\ \text{mL}} = 0.9982\ \text{g/cc}$$

2. Calculate the density of oil at 20°C

   Weight of oil = 209.475 − 110.675 = 98.800 g

$$\text{Density of oil at 20°C} = \frac{98.800\ \text{g}}{100\ \text{mL}} = 0.9880\ \text{g/cc}$$

3. Calculate the specific gravity of oil

$$\text{Specific gravity of oil at } 20°\text{C} = \frac{0.9880 \text{ g/cc}}{0.9982 \text{ g/cc}} = 0.990$$

Note: The specific gravity of oil can also be calculated directly from the weights

$$\text{Specific gravity of oil} = \frac{\text{Wt. of 100 mL of oil}}{\text{Wt. of 100 mL of water}} = \frac{98.800 \text{ g}}{99.820 \text{ g}} = 0.990$$

Example 2.24: Relative Weight of a Substance

The specific gravity of seawater is 1.25. By what percentage is it heavier than fresh water?

Solution

Calculate the relative weight of seawater with respect to fresh water

$$\text{The relative weight} = \frac{\text{Sp.gr. of seawater} - \text{Sp.gr. of fresh water}}{\text{Sp.gr. of fresh water}}$$

$$= \frac{1.25 - 1.00}{1.00} = 0.25$$

Seawater is 25% heavier than fresh water

Example 2.25: Specific Gravity and Density in lb_f/ft^3

Calculate the following: (a) the volume of 300 g of mercury; the specific gravity of mercury is 13.55, (b) the specific weight of cast iron in kg/m^3 if its specific gravity is 7.2, and (c) the mass in grams of 9.00 cm^3 of zinc if the specific gravity of zinc is 8.01.

Solution

1. Calculate the volume of mercury

   $\rho_{Hg} = \rho_{H2O} \times \text{sp. gr. of Hg}$

   $\rho_{Hg} = 1.0 \text{ g/cm}^3 \times 13.55 = 13.55 \text{ g/cm}^3$

$$\rho_{Hg} = \frac{\text{Mass}}{\text{Vol.}}$$

$$\text{Volume of Hg} = \frac{\text{Mass}}{\rho_{Hg}} = \frac{300 \text{ g}}{13.55 \text{ g/cm}^3} = 22.14 \text{ cm}^3$$

2. Calculate density of cast iron (CI)

   $\rho_{CI} = \rho_{H2O} \times \text{Sp. gr. of CI}$

   $\rho_{CI} = 1.0 \text{ g/cm}^3 \times 7.2 = 7.2 \text{ g/cm}^3$

$$\rho_{CI} = 7.2 \text{ g/cm}^3 \times 10^{-3} \text{ kg/g} \times \frac{1}{10^{-6} \text{ cm}^3/\text{m}^3}$$

$$= 7{,}200 \text{ kg/m}^3$$

3. Mass of zinc (Zn)

   $\rho_{Zn} = \rho_{H2O} \times$ Sp. gr. of Zn

   $\rho_{Zn} = 1.0\ g/cm^3 \times 8.01 = 8.01\ g/cm^3$

$$\rho_{Zn} = \frac{\text{Mass}}{\text{Vol.}}$$

Mass of Zn $= \rho_{Zn} \times$ vol of Zn $= 8.01\ g/cm^3 \times 9.00\ cm^3 = 72.09\ g$

Example 2.26: Relationship Between γ, ρ and Specific Gravity

If 5 m^3 of an oil weighs 40 kN, calculate its: (a) specific weight γ, (b) density ρ, and (c) specific gravity.

Solution

Calculate the specific weight, density, and specific gravity of oil

$$\text{Specific weight, } \gamma = \frac{\text{Weight}}{\text{Volume}} = \frac{40\text{ kN}}{5\text{ m}^3} = 8.00\text{ kN/m}^3$$

$$\text{Density, } \rho = \frac{\gamma}{g} = \frac{8{,}000\text{ N/m}^3}{9.81\text{ m/s}^2} \cdot \frac{\text{kg m/s}^2}{\text{N}} = 816\text{ kg/m}^3$$

$$\text{Specific gravity} = \frac{\gamma_{oil}}{\gamma_{water}} = \frac{8.00\text{ kN/m}^3}{9.81\text{ kN/m}^3} = 0.816$$

Example 2.27: Density of Seawater

The weight of 35.06 mL of seawater is 35.93 grams. Calculate the density of seawater in: (a) g/cc (g/cm^3), and (b) kg/m^3.

Solution

a. Calculate the density of seawater in g/cc

$$\text{Density} = \frac{\text{Mass}}{\text{Volume}} = \frac{35.93\text{ g}}{35.06\text{ m}} = 1.025\text{ g/cc } (1\text{ mL} = 1\text{ cm}^3)$$

b. Calculate density in kg/m^3

$$1.025\text{ g/cc} \times \frac{\text{kg}}{1{,}000\text{ g}} \times 10^6\text{ cc/m}^3 = 1{,}025\text{ kg/m}^3$$

Example 2.28: Sludge Volume

Sludge from a wastewater treatment plant has 3% solids. The dry solids processed are 3,000 kg/d. Calculate the volume of liquid sludge in m^3. Specific gravity of liquid sludge is 1.020.

Solution

1. Determine the density of liquid sludge
   Density of liquid sludge = 1.020 × 1,000 kg/m³ (water) = 1,020 kg/m³
2. Determine the weight of liquid sludge

$$\text{Weight of liquid sludge} = \frac{100 \text{ kg liquid sludge}}{3 \text{ kg dry solids}} \times 3{,}000 \text{ kg/d dry solids}$$

$$= 100{,}000 \text{ kg/d liquid sludge}$$

3. Determine the volume of liquid sludge

$$\text{Volume of liquid sludge} = \frac{100{,}000 \text{ kg/d liquid sludge}}{1{,}020 \text{ kg/m}^3 \text{liquid sludge}} = 98 \text{ m}^3/\text{d}$$

Example 2.29: Expression of γ, μ, and σ in SI Units

The common unit for specific weight γ is N/m³ (kN/m³), dynamic viscosity μ is N s/m², and surface tension σ is kN/m. Express these units in g, s, and m units.

Solution

1. Convert the units of specific weight γ
   Unit for γ = N/m³
   N is a force and is equal to mass × acceleration

$$\gamma = \text{kg m/s}^2 \times \frac{1}{\text{m}^3} = \text{kg/m}^2 \text{ s}^2$$

2. Convert the units of dynamic viscosity μ
   μ = N s/m²
   μ = kg m/s² × s/m² = kg/m s
3. Convert the units of surface tension σ
   σ = N/m
   σ = kg m/s² × (1/m) = kg/s²

Example 2.30: Conversion Units for of γ, μ, and σ from SI Units to U.S. Customary Units

At 20°C the specific weight γ, dynamic viscosity μ, kinematic viscosity υ, and surface tension σ for water are, respectively, 9.789 kN/m³, 1.002×10^{-3} N s/m², 1.004×10^{-6} m²/s, and 72.8×10^{-3} N/m. Convert these values into U.S. customary units, and list the conversions.

Solution

The conversion factors for N and m in U.S. customary units are 1 N = 0.22481 lb_f (Table 2.5) and 1 m = 3.28 ft (Table 2.1).

1. Make the desired conversions
   a. Specific weight γ

$$9.789 \times \frac{\text{kN}}{\text{m}^3} \times \frac{10^3 \text{ N/kN} \times 0.2241 \text{ lb/N}}{(3.28 \text{ ft/m})^3} = 62.17 \text{ lb}_f/\text{ft}^3$$

b. Dynamic viscosity μ

$$1.002 \times 10^{-3} \frac{\text{N s}}{\text{m}^2} \times \frac{0.2241 \text{ lb}_f \text{ s/N}}{(3.28 \text{ ft/m})^2} = 2.09 \times 10^{-5} \text{ lb}_f \text{ s/ft}^2$$

c. Kinematic viscosity υ

$1.004 \times 10^{-6} \text{m}^2/(\text{s} \times (3.28 \text{ ft/m})^2) = 1.08 \times 10^{-5} \text{ft}^2/\text{s}$

d. Surface tension σ

$$74.2 \times 10^{-3} \text{ N/m} \times \frac{0.2241 \text{ lb}_f/\text{N}}{3.28 \text{ ft/m}} = 5.070 \times 10^{-3} \text{ lb}_f/\text{ft} = 0.0051 \text{ lb}_f/\text{ft}$$

2. List the conversions

$1 \text{ kN/m}^3 = 6.351 \text{ lb}_f/\text{ft}^3$; $1 \text{ N s/m}^2 = 2.086 \times 10^{-2} \text{ lb}_f \text{ s/ft}^2$; $1 \text{ m}^2/\text{s} = 1.076 \text{ ft}^2/\text{s}$; $1 \text{ N/m} = 0.07 \text{ lb}_f/\text{ft}$.

Example 2.31: Relationship Between Inside Pressure and Surface Tension over a Droplet

Develop the relationship between the surface tension σ and the gauge pressure p inside a liquid droplet of diameter d (Eq. 2.8b).

Solution

Equate the sum of forces due to surface tension around the perimeter and the sum of forces inside the drop

Surface tension around perimeter = pressure × projected area

$$\sigma \int dL = \int p \, dA$$

$$\sigma (\pi d) = p(\pi/4 \cdot d^2)$$

$$\sigma = \frac{1}{4} pd$$

where

σ = surface tension, N/m (lb_f/ft)
dL = elemental length, m
d = diameter, m (ft)
p = differential pressure, N/m^2 (lb_f/ft^2)
dA = projected area, m^2 (ft^2)

Example 2.32: Force Due to Surface Tension over a Needle

A 4-cm steel needle floats over the surface of the water. Calculate the force required over the needle to float on the surface. The surface tension due to attraction of molecule of water at 10°C is 74.2×10^{-3} N/m.

Solution

Determine the force on the needle to float on the surface

From Eq. (2.8a), $F = \sigma \times L = 74.2 \times 10^{-3} \text{ N/m} \times 0.04 \text{ m} = 0.00297 \text{ N}$

Example 2.33: Calculation for Surface Tension

Calculate the surface tension on a droplet of water in contact with air. The diameter of the droplet is 0.015 in and temperature is 70°F. The pressure inside the droplet is 0.079 psi greater than the atmosphere.

Solution

Calculate the surface tension on a droplet of water

Apply Eq. (2.8b), and substitute the values of p and d

$$\sigma = \frac{1}{4} \times 0.079 \text{ lb}_f/\text{in}^2 \times 144 \text{ in}^2/\text{ft}^2 \times 0.015 \text{ in} \times \frac{1}{12} \text{ ft/in}$$

$$= 0.00356 \text{ lb}_f/\text{ft}$$

2.5.5 Concentrations

The concentration of a substance may be expressed in aqueous solution or dry mass. The commonly used units for aqueous solutions are:[5–7]

mg/L, μg/L, g/m^3, kg/m^3, grammole/L (gmole/L), millimole/L (mmole/L), gram equivalent/L (geq/L), and milliequivalent/L (meq/L).

The commonly used units on dry mass basis are:

μg/kg, mg/kg, g/kg, parts per million (ppm), parts per billion (ppb), and percent (%)

It may be noted that for an aqueous solution if $\rho = 1.0$ g/cm^3, the following conversions are valid.

mg/L = g/m^3 = mg/kg = mg/g = ppm = 0.0001% (1% = 10,000 mg/L)

lead /m^3 (soil)

Example 2.34: Concentration in Aqueous Solution, $\rho = 1.0$ g/cm^3

5 lb/d of chlorine is added to one million gallons per day (mgd) of treated effluent. Calculate chlorine concentration in: (a) mg/L, (g/m^3), (b) μg/L, and (c) kg/m^3.

Solution

a. Calculate the concentration of chlorine in mg/L (lb implies lb_f)

$$\frac{5 \text{ lb/d}}{\text{mgd}} = \frac{5 \text{ lb/d} \times 453.6 \text{ g/lb} \times 1{,}000 \text{ mg/g}}{10^6 \text{ gal/d} \times 3.79 \text{ L/gal}} = 0.60 \text{ mg/L } (0.60 \text{ g/m}^3)$$

b. Calculate the concentration of chlorine in μg/m^3

   Concentration in μg/m^3 = 0.60 mg/L × 1,000 μg/mg = 600 μg/L

c. Calculate the concentration of chlorine in kg/m^3

   Concentration in kg/m^3 = 0.60 mg/L × (10^{-6} kg/mg) × 1,000 L/m^3 = 6.0×10^{-4} kg/m^3

Example 2.35: Chemical Solution in Water

10 gal of disinfectant solution containing 1,000 mg/L of chlorine is added to a tank that has 1,500 gal of water. Calculate the concentration of chlorine in mg/L.

Solution

1. Calculate the quantity of chlorine added. Assume $\rho = 1.0$ g/cm^3
   Chlorine added = 10 gal × 3.79 L/gal × 1,000 mg/L = 37,900 mg
2. Calculate the concentration of chlorine in the tank

$$\text{Concentration of chlorine in the tank} = \frac{37{,}900\ \text{mg}}{(1{,}500+10)\ \text{gal total volume} \times 3.79\ \text{L/gal}}$$

$$= 6.62\ \text{mg/L}$$

Example 2.36: Conversion of mg/L in ppm

A water sample contains 5 mg/L of sugar solution. Calculate the concentration of sugar in ppm. Assume ρ for sugar solution = 1.0 g/cm^3.

Solution

$$\text{Concentration in ppm} = 5\ \text{mg/L} \times \frac{1}{1{,}000\ \text{g/L} \times 1{,}000\ \text{mg/g}}$$

$$= \frac{5}{10^6} = 5\ \text{ppm}$$

Example 2.37: Conversion of Percent Concentration in mg/L

Express 1% concentration in mg/L (sp. gr. = 1.0)

Solution

Convert 1% concentration in mg/L

$$1\%\ \text{concentration in mg/L} = \frac{1\ \text{part by wt}}{100\ \text{parts by wt}} = \frac{1\ \text{g}}{100\ \text{g}}$$

$$= \frac{1\ \text{g} \times 1{,}000\ \text{mg/g}}{100\ \text{g} \times \dfrac{1\ \text{L}}{1{,}000\ \text{g}}}$$

$$= 10{,}000\ \text{mg/L}\ (10{,}000\ \text{ppm})$$

Example 2.38: Concentration in Aqueous Solution, Sp.gr. Not Equal to 1.0

Seawater has salinity (TDS) of 3.5%. If sp. gr.= 1.025, express the salinity in mg/kg, ppm, mg/L, and g/m^3.

Solution

1. Calculate the salinity in ppm

$$\text{Concentration of } 3.5\% = \frac{3.5\ \text{g}}{100\ \text{g}} \times \frac{1{,}000\ \text{mg/g}}{10^{-3}\text{kg/g}} = 35{,}000\ \text{mg/kg}$$

$$= \frac{35{,}000\ \text{mg}}{\text{kg} \times 10^6\ \text{mg/kg}} = \frac{35{,}000\ \text{mg}}{10^6\ \text{mg}} = \frac{35{,}000\ \text{parts}}{\text{million parts}} = 35{,}000\ \text{ppm}$$

2. Calculate the concentration in mg/L and g/m³

$$\text{Volume of 100 g sea water} = \frac{100\ \text{g sea water}}{1.025\ \text{g/mL sea water}} = 97.561\ \text{mL}$$

$$\text{Concentration in mg/L} = \frac{3.5\ \text{g} \times 1{,}000\ \text{mg/g} \times 1{,}000\ \text{mL/L}}{97.561\ \text{mL}} = 35{,}875\ \text{mg/L}$$

$$\text{Concentration in g/m}^3 = 35{,}875\ \text{mg/L} \times \frac{10^{-3}\ \text{g/mg}}{10^{-3}\ \text{m}^3\text{/L}} = 35{,}875\ \text{g/m}^3$$

Example 2.39: Sludge Solids and Volume

Sludge from a wastewater treatment plant has 2% dry solids, and sp. gr. = 1.02. Calculate the volume of sludge in m^3/d if dry solids are 2,500 kg/d.

Solution

1. Calculate the weight of wet sludge
   Weight of dry solids = 2,500 kg/d

$$\text{Weight of wet sludge at 2\% dry solids} = 2{,}500\ \text{kg/d dry solids} \times \frac{100\ \text{kg wet sludge (dry solids + water)}}{2\ \text{kg dry solids}}$$

$$= 125{,}000\ \text{kg wet sludge per day (dry solids + water)}$$

2. Calculate density of wet sludge

$$\text{Sp. gr. of wet sludge} = \frac{\text{Density of wet sludge}}{\text{Density of water}}$$

Density of wet sludge = 1.02 × 1,000 kg/m³ (density of water) = 1,020 kg/m³ (density of wet sludge)

3. Calculate the volume of wet sludge

$$\text{Volume of wet sludge produced} = \frac{125{,}000\ \text{kg/d wet sludge}}{1{,}020\ \text{kg/m}^3\ (\text{density of wet sludge})} = 122.6\ \text{m}^3\text{/d}$$

Example 2.40: Concentration in mole/L

Concentration of NaCl in a solution is 35 mg/L. Calculate the concentration in: (a) mole/L, and (b) millimole/L (mmole/L).

Solution

1. Calculate the molecular weight of NaCl
   Gram molecular weight of NaCl = 58.5 g/mole
2. Calculate the concentration of NaCl

a. Molar concentration of NaCl $= 35\ \text{mg/L} \times 10^{-3}\ \text{g/mg} \times \dfrac{1}{58.5\ \text{g/g mole}}$

$= 0.0005983$ g mole/L (or mole/L)

b. Concentration of NaCl in mmole/L $=0.0005983$ g mole/L $\times 10^3$ mg/g $=0.60$ mg mole/L (mmole/L)

Example 2.41: Concentration in Equivalent/L (eq/L)

30 mg/L Ca^{2+} is dissolved in water. Express the concentration in: (a) eq/L and (b) meq/L.

Solution

1. Calculate the equivalent weight of Ca^{2+}

   Gram molecular weight of a substance is the molecular weight of the species divided by the number of equivalents = (g/mole)/(eq/mole) = g/eq

$$= \frac{40\ \text{g}}{\text{mole}} \times \frac{1}{2\ \text{eq/mole}} = 20\ \text{g/eq}$$

2. Calculate the concentration of Ca^{2+}

   a. Equivalent concentration of $Ca^{2+} = 30\ \text{mg/L} \times 10^{-3}\ \text{g/mg} \times \dfrac{1}{20\ \text{g/g eq}}$

   $= 0.0015$ g eq/L (or eq/L)

   b. Milliequivalent concentration of $Ca^{2+} = \dfrac{0.0015\ \text{g eq}}{\text{L}} \times \dfrac{10^3\ \text{mg}}{\text{g}}$

   $= 1.5$ mg eq/L (or m eq/L)

Example 2.42: meq/L Expressed as mg/L $CaCO_3$

A water sample contains 45 mg/L Ca^{2+}. Express the result as mg/L $CaCO_3$.

Solution

1. Convert mg/L Ca^{2+} as meq/L Ca^{2+}

$$Ca^{2+}\text{in meq/L} = 45\ \text{mg/L}\ Ca^{2+} \times \frac{1}{20\ \text{mg}\ Ca^{2+}/\text{meq}} = 2.25\ \text{meq/L}$$

2. Calculate the concentration as mg/L $CaCO_3$

   Concentration as mg/L $CaCO_3$ = 2.25 meq/L × 50 mg $CaCO_3$/ meq = 112.5 mg/L as $CaCO_3$

Example 2.43: Concentration Expressed in Solid Media

Soil contains 10 lb of lead per ton. Express the results in: (a) g/kg, (b) ppm, and (c) g/m^3 (sp. gr. of soil = 1.38).

Solution

1. Calculate the quantity of lead in soil in g and kg
   Quantity of lead = 10 lb × 453.6 g/lb = 4,536 g
   Quantity of soil = 1 ton × 2,000 lb/ton × 453.6 g/lb × 10^{-3} kg/g = 907.2 kg
2. Calculate the concentration of lead on a weight basis

   a. $$\text{Concentration of lead in g/kg} = \frac{4{,}536 \text{ g lead}}{907.2 \text{ kg soil}} = 5 \text{ g (lead)/kg (soil)}$$

   b. $$\text{Concentration of lead in ppm} = 5 \text{ g/kg} \times 1{,}000 \text{ mg/g} \times \frac{1}{10^6 \text{ mg/kg}}$$

   $$= \frac{5{,}000 \text{ mg}}{10^6 \text{ mg}} = 5{,}000 \text{ ppm}$$

   c. Concentration of lead in g/m^3 (soil) = 5 g lead/kg (soil) × 1,380 kg (soil)/m^3 (soil) = 6,900 g

2.6 RESIDENCE TIME

Residence time (also called theoretical detention time or retention time) is the average time a fluid particle stays in a continuous flow sedimentation basin. It is also equal to the time it takes to fill a tank at a given constant flow. The residence time is expressed by Eq. (2.9)

$$D_t = V/Q \tag{2.9}$$

where

D_t = residence time, theoretical detention time, or hydraulic residence time; min, h, or day
V = volume, L^3 (m^3 or ft^3)
Q = flow, L^3/t (m^3/s or ft^3/s)

Example 2.44: Residence Time in a Primary Sedimentation Basin

The volume of a primary sedimentation basin is 1,650 gal. The average flow is 20,000 gal/d. Calculate the residence time in hours.

Solution

$$\text{The residence time, } D_t = \frac{1{,}650 \text{ gal}}{20{,}000 \text{ gal/d}} \times 24 \text{ h/d} = 2.0 \text{ h}$$

Example 2.45: Volume of an Aeration Basin

An aeration basin has an aeration period (T) of 8 h. The average flow is 15,000 m^3/d. Calculate the volume of the aeration basin in m^3.

Solution

$$\text{Volume of the aeration basin} = 15{,}000 \text{ m}^3\text{/d} \times \frac{8 \text{ h}}{24 \text{ h/d}} = 5{,}000 \text{ m}^3$$

2.7 REVIEW PROBLEMS

2.1 Round off the following numbers to three significant figures: (a) 4.6348 m, (b) 8,352 N, (c) 25.724 s, and (d) 4,838 kg.

2.2 Give the significant figures of the following numbers.
a. 18.7 tons of a loaded vehicle weighed on a scale that has an accuracy of ± 1 ton
b. $64.35 price of a book
c. 38.7 million gallons per day average capacity of a water treatment plant
d. 6,834 spectators in a parade

2.3 The average unit cost of laying a drainage pipe is $9.25/ft. Calculate the project cost if the total length of the pipe is 1.30 miles.

2.4 The cost of constructing a primary sedimentation basin is expressed by the equation:
Construction cost, $ = $-0.0002\ Q^2 + 19.29\ Q + 220{,}389$
where Q = average design flow in m^3/d
Calculate the construction cost of the basin if the average design flow is 35,670 m^3/d.

2.5 A stabilization pond has a surface area of 25 acres and storage capacity of 3.27×10^6 ft^3. Calculate: (a) area in ha, and (b) capacity in m^3.

2.6 A train is traveling at a speed of 65 mi/h. Determine the speed in km/h and m/s.

2.7 Represent the following combination of units in SI units with appropriate prefix:
(a) µ MN, (b) kN/m s, (c) µ m Mg, (d) mN/kg µs, (e) (208 mm) (0.053 kg)/(31.9 N) to three significant figures.

2.8 Calculate each of the following expressions in three significant figures. Express each answer in SI unit using appropriate prefix.
a. (38 mm)^2 (42 kg)^3
b. (21 ms) (0.0451 Mm)/361 mg
c. (2.81 mm) (421 Mg)

2.9 A wastewater treatment plant has a design capacity of 35 mgd. Calculate the flow in: (a) gpm, (b) cfs, (c) m^3/d, and (d) L/s.

2.10 Make the following conversions of mass, weight, and force
a. Mass of 200 kg in kN, and in slug
b. Mass of 600 lbs in slug
c. Force of 800 lb in kN
d. Mass of 864 kg/m^3 in slug/ft^3

2.11 A car tire has a pressure of 29 lb/in^2. Convert this pressure into: (a) atmosphere, (b) Pa, (c) N/m^2, and (d) m of water.

2.12 A raw wastewater pump has a discharge of 1,850 gpm, and total dynamic head developed is 43 ft. Calculate waterpower in: (a) lb ft/s, (b) hp, (c) kW (N.m/s), and (d) kJ/s. Use conversions given in Appendix D and Chapter 2. Assume a specific weight of water $\gamma = 62.4$ lb/ft^3.

2.13 A piece of brass has a mass of 151.73 g. The volume of water displaced in a graduated cylinder is 18.22 mL. Calculate: (a) density in g/cc, and kg/m^3, (b) sp. gr., and (c) specific weight γ, lb/ft^3.

2.14 At a temperature of 4°C, water is at its greatest density. Calculate this temperature in: (a) °F, (b) °K, and (c) °R.

2.15 400 lb/d chlorine is added to 10 mgd effluent for disinfection. Calculate the concentration of chlorine in: (a) mg/L and g/m^3, (b) µg/L, and (c) ppm.

2.16 The feed rate of liquid alum solution is 15 L/min. The concentration of alum in the feed solution is 2% (sp. gr. = 1.01). The raw water is flowing to the rapid mix basin at a rate of 3,000 gpm. Calculate the concentration of alum and feed rate in: (a) mg/L, (b) kg/d, and (c) lb/million gallon.

2.17 A 250 mg/L stock solution of ferric chloride is used to prepare 5 mg/L standard iron solution. Calculate the volume of stock solution to be mixed to prepare one liter of standard solution. Also, calculate the molarity of standard iron solution. Atomic weight of iron is 55.85.

2.18 Water analysis of municipal water supply showed 40 and 10 mg/L concentrations of Ca^{2+} and Mg^{2+}. Determine total hardness as (a) milliequivalent/L, and (b) mg/L as $CaCO_3$.

2.19 An aeration basin has an aeration period of 6 h. The flow to the basin is 38,000 m^3/d. Calculate the volume in: (a) m^3, (b) ft^3, and (c) million gallons.

2.20 A stabilization pond has a surface area of 5 ha, and the average depth is one meter. If a flow of 1.5 mgd is treated, calculate: (a) detention time in days, (b) volume in acre ft, and (c) hydraulic loading in million gallons/acre d.

REFERENCES

1. Mechtly, E. A., *The International System of Units, Physical Constants and Conversion Factors*, 2nd Revision, Scientific and Technical Information Office, National Aeronautics and Space Administration, Washington, DC, 1973.
2. Hibbeler, R. C., *Engineering Mechanics: Statics & Dynamics*, 9th ed., Prentice-Hall, Upper Saddle River, NJ, 2001.
3. Vesiland, P. A., *Introduction to Environmental Engineering*, PWS Publishing Company, Boston, MA, 1997.
4. Doresi, A. P., and R. J. Schmidt, *Engineering Mechanics: Statics*, Brooks/Cole, A Division of Thomson Learning, Pacific Grove, CA, 2001.
5. Franzini, J. B., and E. J. Finnermore, *Fluid Mechanics with Engineering Applications*, 9th ed., The McGraw-Hill Companies, Inc., New York, 1997.
6. Lindeburg, M. R., *Civil Engineering Reference Manual for the P.E. Exam*, 8th ed., Professional Publications, Inc., Belmont, CA, 2001.
7. Lee, C. C., and S. D. Lin, *Handbook of Environmental Engineering Calculations*, McGraw-Hill, New York, 2000.

3 Sampling and Data Analysis

3.1 CHAPTER OBJECTIVES

Collection of representative samples, determination of flow and constituents, and statistical data analyses are the integral parts of planning, design, and operation of wastewater conveyance and treatment facilities. The objectives of this chapter are to present:

- Sampling, location, frequency, methods, and equipment for flow and quality determination
- Data presentation and common statistical methods used to determine the flow and characteristics of wastewater
- Precision and reliability considerations for process and equipment selection
- Data quality indicators, precision, and reliability considerations.

3.2 SAMPLING

Careful analytical work on sophisticated equipment has little value or meaning if the sample is not representative of true flow condition or quality. Therefore, selection of sample location, sampling device, and frequency, flow data logging, labeling, preservation, and prompt analytical work are important considerations. A comprehensive discussion on sampling may be found in the Standard Methods.[1]

3.2.1 Sampling Location and Frequency

Sampling Location. Selection of sampling points is important in obtaining a flow measurement or a representative sample. Accordingly, the sampling points must be properly located. The following criteria should be considered when selecting a sampling point: (1) The location should have good mixing and homogeneous conditions to accurately represent the true condition and quality of the flow, (2) excessive mixing should be avoided, as it may strip gases and volatile compounds, (3) there should be continuous flow at the sampling location, and it should not be influenced by sludge or debris deposits, (4) the flow and sample should not be affected by other side of the stream, (5) in sewers or channels, the sample should be collected from a point one-third of the water depth from the bottom, and (6) flow measurements in sewers or channels should be taken in straight sections or at manholes.

Sampling and Flow Measurement Frequency. The frequency of sampling and flow measurement depends on the flow rate and flow characteristics. If there is low variability in flow and composition, manual grab samples at suitable time intervals may serve the purpose. For highly variable streams, the automatic samplers may be necessary to collect samples and measure flow at the desired frequency.

3.2.2 Methods of Sampling

Sampling and flow measurement may be *grab*, *continuous*, *composite*, or *continuous measurements in situ*. Grab samples or instantaneous flow measurements are single-batch samples or flow measurements taken at a given time. Composite or integrated samples are taken at constant time intervals, usually over a 24-h period, and then pooled together to obtain one representative sample of the day. Continuous flow measurement is done for a desired instance, such as wet weather conditions,

DOI: 10.1201/9781003629399-3

or to develop a diurnal (24 h) flow pattern. Continuous flow measurement and sampling are done by installing a measuring device in situ. These sampling techniques are described below.[2–5]

Grab Sample or Instantaneous Flow Measurement. Grab samples or instantaneous flow measurements provide valuable information at a low cost in many cases. These grab samples give the best results under the following situations:

1. When flow condition or quality remains relatively uniform over long periods. Examples include groundwater, lakes, large rivers, water distribution systems, sludge digester, and others. In all these cases, the quality and flow variations may be seasonal or due to some known conditions such as precipitation, flood, drought, equipment failure, and others.
2. Where the effect of *slug loads* or discharge is desired, the grab sample or instantaneous flow measurement taken at the proper time will give the best results.
3. Where the concentration of certain constituents is needed for the adjustment of the chemical feed.
4. When the sample requires immediate analysis due to the instability of the constituents. Pooling over 24-h period may change the results.
5. When the flow is due to intermittent discharges.

Composite Sample and Continuous Flow Measurement. Continuous flow measurement in a conveyance system is needed for system design, plant operation, and process control. Flow measurement locations, needs, and devices are discussed in detail in Chapter 7. Composite or integrated sampling is done to obtain the average quality data for the day. Continuous sampler that removes the volume of sample in proportion to the flow is desirable. Another type of composite sampler removes a sample every hour and deposits it into different bottles. Volumes of samples from each bottle are mixed in proportion to flow at that time to obtain the composite average sample for the day. Individual hourly samples are the grab samples representing the condition at that instant. Sources, flow rates, and characteristics of municipal wastewater are covered in detail in Chapter 4.

Continuous Measurement in Situ. Many measurements are taken continuously using in-line measurement devices. Examples of such measurements are velocity, flow, temperature, pH, conductivity, dissolved oxygen, chlorine residual, ammonia nitrogen, and others. Such measurements are generally recorded continuously and require sophisticated sensing, transmitting, and recording devices.

Sampling Procedure. Wastewater samples are normally collected by dipping the sampling container into the source. For microbiological sample, the container must be sterilized. In all cases, the container must be completely filled (without leaving any air) and then capped tightly. For specialized applications, various types of apparatus are used to collect water and sediment samples.

Grab water sampler:

- Nansen bottle or Alpha style vertical bottle
- Kemmerer bottle
- Van Dorn bottle
- Niskin bottle

Composite water sampler:

- Constant flow rate pumping
- Variable flow rate pumping proportional to flow in the source
- Discrete or hourly samples in separate bottles over 24-h period

Bottom sediment grab sampler for surface sample:

- Wildco–Eckman dredge
- Ponar grab dredge
- Peterson grab dredge

Bottom sediment grab sampler for deeper samples:

- Phleger cover
- KB cover
- Wilding or stovepipe

Example 3.1: Composite Sample

One liter of hourly wastewater samples was collected every hour at an industrial plant. The hourly flow record is given below. Calculate the volume of samples to be pooled into a container to give 2 L of 24-h composite sample.

	PM							AM				
Time	6	7	8	9	10	11	12 MN	1	2	3	4	5
Flow, m^3/h	24	18	15	12	6	3	3	3	3	3	3	3
	AM							PM				
Time	6	7	8	9	10	11	12 N	1	2	3	4	5
Flow, m^3/h	3	6	12	15	18	24	24	30	24	24	36	30

MN = midnight, N = noon.

Solution

1. Calculate the proportional flow ratio at different times.
   The proportional flow for the hour is calculated by dividing the hourly flow by the daily flow. These values are calculated in Col. (4) of Table 3.1. Total ratio of 24 h is 1.00.
2. Calculate the volume of sample mixed to give 2 L of 24-h composite sample.
   These values are also provided in Col. (5) of Table 3.1. Total volume of 24 h is 2,000 mL.

Example 3.2: Average Concentration

An industrial waste stream is equipped with an in-situ continuous measurement of flow and the concentration of a dye used in the coloring process. Determine the average daily flow and average daily concentration of dye. The in-situ continuous measurement data are plotted in Figure 3.1.

Solution

The average daily flow and concentration are determined by integrating the areas under the respective curves divided by 24 h. Apply graphical integration.

1. Tabulate the data given in Figure 3.1 as follows:
   a. Divide hourly flow and concentration data into segments of Δt
   b. Determine average flow and average concentration in each segment

TABLE 3.1
Hourly Flow Data and Proportional Volume Mixed for a Composite Sample (Example 3.1)

Time	Flow, m³/h	Total Flow in Each Hour, m³	Proportional Hourly Flow Ratio	Proportional Volume Pooled, mL
(1)	(2)	(3)	(4)	(5)
6 PM	24	24[a]	0.070[b]	140[c]
7 PM	18	18	0.053	106
8 PM	15	15	0.044	88
9 PM	12	12	0.035	70
10 PM	6	6	0.018	36
11 PM	3	3	0.009	18
12MN	3	3	0.009	18
1 AM	3	3	0.009	18
2 AM	3	3	0.009	18
3 AM	3	3	0.009	18
4 AM	3	3	0.009	18
5 AM	3	3	0.009	18
6 AM	3	3	0.009	18
7 AM	6	6	0.018	36
8 AM	12	12	0.035	70
9 AM	15	15	0.044	88
10 AM	18	18	0.053	106
11 AM	24	24	0.070	140
12N	24	24	0.070	140
1 PM	30	30	0.088	176
2 PM	24	24	0.070	140
3 PM	24	24	0.070	140
4 PM	36	36	0.105	210
5 PM	30	30	0.088	176
Total flow in 24h (one day)		= 342 m³/24h	1.003	2,006 mL = 2 L

[a] $24\,m^3/h \times 1\,h = 24\,m^3$.

[b] $24\,m^3/342\,m^3 = 0.070$.

[c] $0.070 \times 2{,}000\,mL = 140\,mL$.

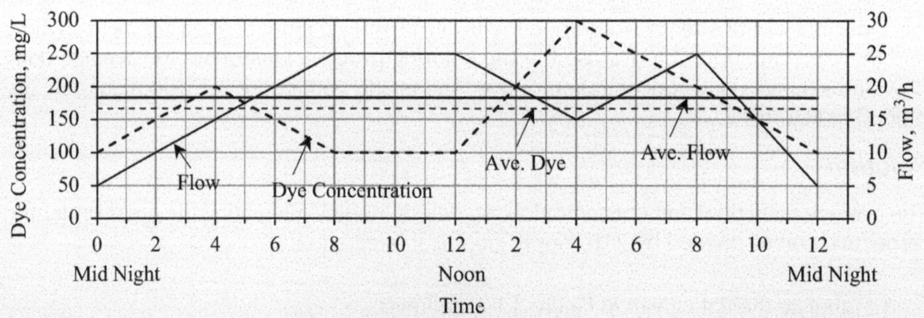

FIGURE 3.1 Continuous in-situ measurement of flow and dye concentration data (Example 3.2).

TABLE 3.2
Calculation Steps (Example 3.2)

Time	Flow, m^3/h	Avg. Flow in Segment $(Q_1+Q_2)/2$, m^3/h	Conc., mg/L	Avg. Conc. in Segment $(C_1+C_2)/2$, mg/L	Δt, h	Avg. Flow × Δt, m^3	Avg. Conc. × Δt, (mg/L) h
0 MN	5	—	100	—	—	—	—
2 AM	10	7.5[a]	150	125[b]	2	15[c]	250[d]
4 AM	15	12.5	200	175	2	25	350
6 AM	20	17.5	150	175	2	35	350
8 AM	25	22.5	100	125	2	45	250
10 AM	25	25.0	100	100	2	50	200
12 N	25	25.0	100	100	2	50	200
2 PM	20	22.5	200	150	2	45	300
4 PM	15	17.5	300	250	2	35	500
6 PM	20	17.5	250	275	2	35	550
8 PM	25	22.5	200	225	2	45	450
10 PM	15	20.0	150	175	2	40	350
12 MN	5	10.0	100	125	2	20	250
Total					24	440	4,000

[a] $((5+10)\ m^3/h)/2 = 7.5\ m^3/h$.

[b] $((100+150)\ mg/L)/2 = 125\ mg/L$.

[c] $7.5\ m^3/h \times 2\ h = 15\ m^3$.

[d] $125\ mg/L \times 2\ h = 250\ mg\ h/L$.

c. Determine the area of each segment, (average flow)×Δt, and (average concentration)×Δt

These values are provided in Table 3.2

2. Determine average daily flow

$$\text{Average daily flow} = \frac{\sum (\text{ave. flow}) \times \Delta t}{\sum \Delta t} = \frac{440\ m^3}{24\ h} = 18.3\ m^3/h$$

3. Determine average daily dye concentration

$$\text{Average daily dye concentration} = \frac{\sum (\text{ave. conc.} \times \Delta t)}{\sum \Delta t} = \frac{4{,}000\ mg\ h/L}{24\ h} = 166.7\ mg/L$$

3.3 DATA PRESENTATION AND STATISTICAL ANALYSIS

The statistical analyses are necessary to develop useful results from numerous individual data points. The statistical procedures are helpful in comparing the sampling methods and procedures, and changes resulting from flow, concentration, process modifications, waste loadings, and chemicals.

These methods are also beneficial in identifying errors and mistakes, accuracy and precision of estimates, and variability and correlation. Statistics and data analysis are very broad topics. Only basic information commonly applied to wastewater flow and treatment is covered in this section. Readers are referred to several textbooks for additional information.[6–9]

3.3.1 Graphical Presentation of Data

A well-planned and executed monitoring program will provide valuable information for plant operation and future expansion. Therefore, the parameters monitored and the significant results obtained must be graphically presented to identify the trends prior to making the decisions. Time series and histograms or frequency plots are commonly used for data presentation.

Time Series Distribution. The observations, when arranged in the order of their occurrence in time, are called time series. The graph may show a valuable trend, such as linear, cyclic, or fluctuating. The cyclic trend may show peaks and valleys. The fluctuating time series may go up or down with no apparent pattern. Moving averages are often used to suppress the random irregularities in time series data to give long-term trends.

Linear Time Series Distribution. The linear or arithmetic relationship is expressed by a straight line. This is the most common and useful relationship in process analysis and model calibration. A linear equation in terms of X and Y coordinates is given by Eq. (3.1).

$$Y = a + mX \tag{3.1}$$

where

a, m = intercept, and slope

Y = dependent variable, that is, the variable to be estimated

X = independent or measured variable

Cyclic Time Series Distribution. The cyclic time distribution may show peaks and valleys, or some well-defined pattern. The diurnal hourly wastewater flow may have cyclic distribution that may repeat daily under dry weather conditions.

Fluctuating Time Series Distribution and Moving Average. The fluctuating time series distributions may go up and down with no apparent pattern. The flow and waste loadings in a wastewater treatment plant have fluctuating patterns daily and seasonally. Moving averages are often used to suppress the unwanted fluctuations. The irregular patterns are thus smoothed out. The moving averages may be simple or weighted, and may be the average of K order (where $K = 2, 3, 4$, or any number). Moving averages with proper order (to include daily or seasonal effect) may provide a trend that can be used for management purposes. One disadvantage of moving average method is that the beginning and the end portions of the series are lost.

Example 3.3: Linear Data Trend

A pipeline discharges a constant flow of 0.2 m^3/s into a storage tank. The storage tank is rectangular with vertical walls and surface area of 2,880 m^2. Plot the water depth in the tank with respect to time 1, 2, 3, 4, 5, and 6 h. Initial depth of water in the tank is 0.5 m. Determine the value of the constants in Eq. (3.1).

Solution

1. Identify the variables in Eq. (3.1). Dependent variable Y is depth, and the independent or measured variable X is time.
2. Calculate constant flow in the pipe, m^3/h

$$\text{Constant flow} = 0.2\ \frac{\text{m}^3}{\text{s}} \times \frac{(60 \times 60)\ \text{s}}{\text{h}} = 720\ \text{m}^3/\text{h}$$

3. Calculate the increase in water depth *Y* per hour in the tank

$$\text{Increase in depth} = \frac{\text{Flow}}{\text{Area of tank}}$$

$$= \frac{720\ \text{m}^3/\text{h}}{2{,}880\ \text{m}^2} = 0.25\ \text{m/h}$$

4. Summarize the time series depth data in Table 3.3.
5. Plot the time series of depth data
   The time series of depth data is plotted in Figure 3.2
6. Determine the constants *a* and *m* in the linear equation (Eq. 3.1)
   a = intercept = 0.5 m
   m = slope = 0.25 m/h
   The linear equation is Depth = 0.5 + 0.25 × time or $d = 0.5 + 0.25t$

TABLE 3.3
The Water Depth in Storage Basin at Different Time Intervals (Example 3.3)

Time Since Start, h	Time Interval, h	Increase in Water Depth, m	Total Water Depth Since Start, m
0	0	0.5	0.50
1	1	0.25	0.75
2	1	0.25	1.00
3	1	0.25	1.25
4	1	0.25	1.50
5	1	0.25	1.75
6	1	0.25	2.00

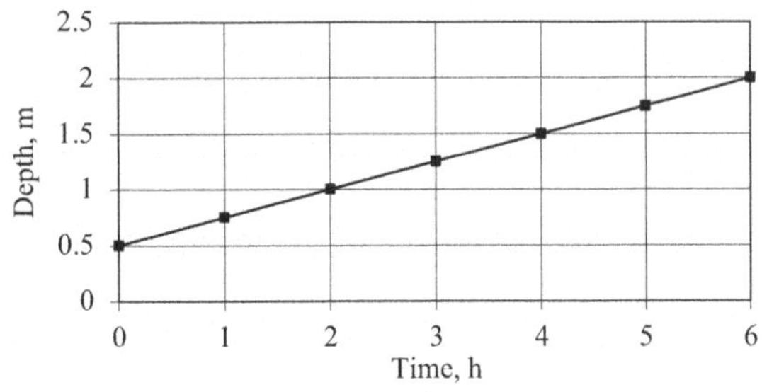

FIGURE 3.2 Linear time series plot of the depth data (Example 3.3).

Example 3.4: Simple Moving Average

The flow values of 11 samples from an industrial wastewater stream are given below. Determine the simple moving averages of the order 3 by sequence. Also, plot the original data and the moving average values.

Sample No.	1	2	3	4	5	6	7	8	9	10	11
Flow, m^3/h	180	305	210	60	180	250	110	320	68	310	201

Solution

1. Calculate the moving averages of the order 3 by sequence

$$\frac{(180+305+210)\ m^3/h}{3} = 232\ m^3/h, \quad \frac{(305+210+60)\ m^3/h}{3} = 192\ m^3/h$$

$$\frac{(210+60+180)\ m^3/h}{3} = 150\ m^3/h, \quad \frac{(60+180+250)\ m^3/h}{3} = 163\ m^3/h$$

$$\frac{(180+250+110)\ m^3/h}{3} = 180\ m^3/h, \quad \frac{(250+110+320)\ m^3/h}{3} = 227\ m^3/h$$

$$\frac{(110+320+68)\ m^3/h}{3} = 166\ m^3/h, \quad \frac{(320+68+310)\ m^3/h}{3} = 233\ m^3/h$$

$$\frac{(68+310+201)\ m^3/h}{3} = 193\ m^3/h$$

2. Plot the individual values and moving averages.
   The individual data points and moving averages are shown in Figure 3.3.

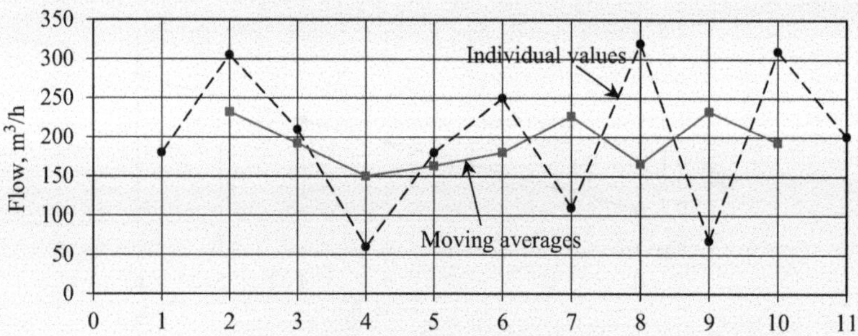

FIGURE 3.3 Graphical presentation of individual data points and moving averages (Example 3.4).

Example 3.5: Weighted Moving Averages

A sampling program was conducted at an industrial plant. Samples were collected at 1 AM, 9 AM, and 5 PM over a period of five consecutive days. The phenol concentration in every sample was determined. Manufacturing activity for three shifts is low at night, medium in the morning, and high in the afternoon. Based on flow, the weights for night, morning, and afternoon samples were assigned as 1, 2, and 3. Calculate the weighted moving averages of order 3 by sequence. Plot the original data and weighted moving average values.

	Day 1			Day 2			Day 3			Day 4			Day 5		
Sampling Time	1 AM	9 AM	5 PM	1 AM	9 AM	5 PM	1 AM	9 AM	5 PM	1 AM	9 AM	5 PM	1 AM	9 AM	5 PM
Phenol Concentration, mg/L	10	12	9	12	18	15	12	10	8	16	15	19	19	9	6

Solution

1. Calculate the moving averages of the third order with the assigned weights

$$\frac{(1\times10+2\times12+3\times9)\text{ mg/L}}{1+2+3}=10\text{ mg/L}$$

$$\frac{(2\times12+3\times9+1\times12)\text{ mg/L}}{2+3+1}=10.5\text{ mg/L}$$

$$\frac{(3\times9+1\times12+2\times18)\text{ mg/L}}{3+1+2}=12.5\text{ mg/L}$$

$$\frac{(1\times12+2\times18+3\times15)\text{ mg/L}}{1+2+3}=15.5\text{ mg/L}$$

$$\frac{(2\times18+3\times15+1\times12)\text{ mg/L}}{2+3+1}=15.5\text{ mg/L}$$

$$\frac{(3\times15+1\times12+2\times10)\text{ mg/L}}{3+1+2}=12.8\text{ mg/L}$$

$$\frac{(1\times12+2\times10+3\times8)\text{ mg/L}}{1+2+3}=9.3\text{ mg/L}$$

$$\frac{(2\times10+3\times8+1\times16)\text{ mg/L}}{2+3+1}=10\text{ mg/L}$$

$$\frac{(3\times 8+1\times 16+2\times 15)\text{ mg/L}}{3+1+2}=11.7\text{ mg/L}$$

$$\frac{(1\times 16+2\times 15+3\times 19)\text{ mg/L}}{1+2+3}=17.2\text{ mg/L}$$

$$\frac{(2\times 15+3\times 19+1\times 19)\text{ mg/L}}{2+3+1}=17.7\text{ mg/L}$$

$$\frac{(3\times 19+1\times 19+2\times 9)\text{ mg/L}}{3+1+2}=15.7\text{ mg/L}$$

$$\frac{(1\times 19+2\times 9+3\times 6)\text{ mg/L}}{1+2+3}=9.2\text{ mg/L}$$

2. Plot the individual values and the weighted moving averages.
   The individual data points and the weighted moving averages are shown in Figure 3.4.
3. Determine the data trend from the weighted moving average plot.
   The weighted moving average plot shows a cyclic trend reaching peaks and valleys on alternate days.

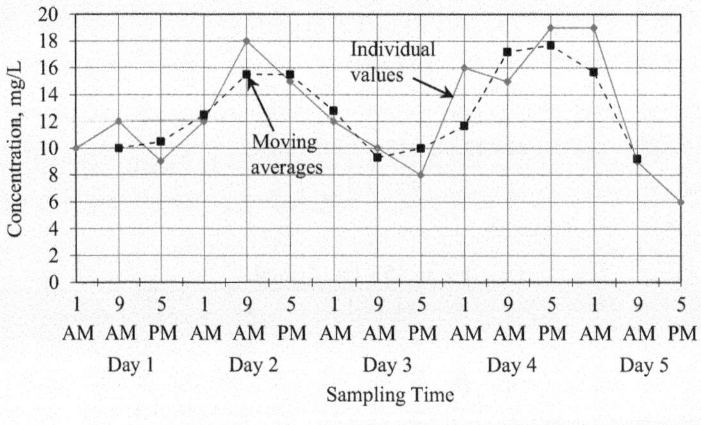

FIGURE 3.4 Graphical presentation of individual data points and weighted moving averages (Example 3.5).

Example 3.6: Cyclic Distribution

The diurnal hourly wastewater flow rates from an industrial plant are given below. Calculate the simple moving averages of order 2 by sequence. Also, plot the original data and the moving averages.

Time of Day	12 MN	1:00 AM	2:00 AM	3:00 AM	4:00 AM	5:00 AM	6:00 AM	7:00 AM	8:00 AM	9:00 AM	10:00 AM	11:00 AM
WW flow rate, m^3/h	36.3	27.5	28.8	21.9	28.7	26.2	38.1	33.7	48.7	40.0	48.6	35.6
Time of day	12 N	1:00 PM	2:00 PM	3:00 PM	4:00 PM	5:00 PM	6:00 PM	7:00 PM	8:00 PM	9:00 PM	10:00 PM	11:00 PM
WW flow rate, m^3/h	42.4	22.7	35.0	17.6	38.7	21.2	44.9	28.2	53.1	34.9	53.0	31.8

MN = midnight, N = noon.

Solution

1. Calculate the moving averages of order 2 by sequence (Table 3.4)
2. Plot the individual values and moving averages (Figure 3.5)

TABLE 3.4
Calculation Table to Determine the Simple Moving Averages of a Cyclic Diurnal Distribution (Example 3.6)

Time of Day	WW Flow, m^3/h	Moving Average, m^3/h	Moving Average Value, m^3/h
12 MN	36.3		
1:00 AM	27.5	(36.3+27.5)/2	31.9
2:00 AM	28.8	(27.5+28.8)/2	28.2
3:00 AM	21.9	(28.8+21.9)/2	25.4
4:00 AM	28.7	(21.9+28.7)/2	25.3
5:00 AM	26.2	(28.7+26.2)/2	27.5
6:00 AM	38.1	(26.2+38.1)/2	32.1
7:00 AM	33.7	(38.1+33.7)/2	35.9
8:00 AM	48.7	(33.7+48.7)/2	41.2
9:00 AM	40.0	(48.7+40.0)/2	44.4
10:00 AM	48.6	(40.0+48.6)/2	44.3
11:00 AM	35.6	(48.6+35.6)/2	42.1
12:00 N	42.4	(35.6+42.4)/2	39.0
1:00 PM	22.7	(42.4+22.7)/2	32.6
2:00 PM	35.0	(22.7+35.0)/2	28.9
3:00 PM	17.6	(35.0+17.6)/2	26.3
4:00 PM	38.7	(17.6+38.7)/2	28.1
5:00 PM	21.2	(38.7+21.2)/2	30.0
6:00 PM	44.9	(21.2+44.9)/2	33.1
7:00 PM	28.2	(44.9+28.2)/2	36.5
8:00 PM	53.1	(28.2+53.1)/2	40.7
9:00 PM	34.9	(53.1+34.9)/2	44.0
10:00 PM	53.0	(34.9+53.0)/2	43.9
11:00 PM	31.8	(53.0+31.8)/2	42.4

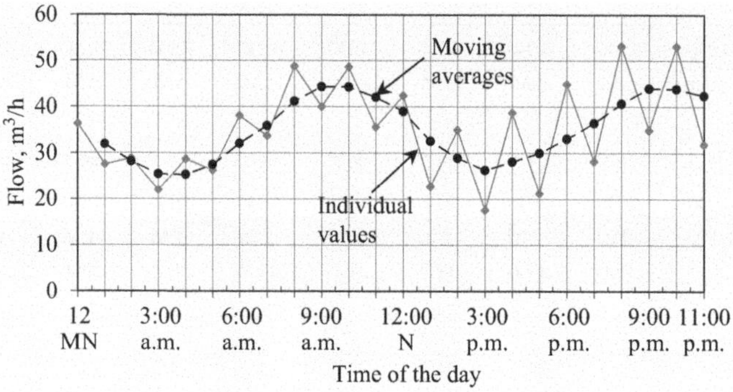

FIGURE 3.5 Graphical presentation of moving averages showing cyclic distribution of diurnal dry weather wastewater flow (Example 3.6).

3.3.2 Frequency Distribution

Large masses of raw data are normally organized into classes or categories. The number of observations or data points that fall into each class is called *class frequency*, *frequency distribution*, or *frequency table*. The interval between the class frequencies is called *frequency interval* that has *lower* and *upper class limits* or *boundaries*. The difference between upper and lower class boundaries is the *class width*, *class size*, or *class length*. The class mark is the *midpoint* of the class interval. It is obtained by adding upper and lower class limits and dividing by 2. The midpoint of a class interval is also called the *class midpoint*.

Histogram and Frequency Polygon. The histogram and frequency polygon are two graphic representations of frequency distribution. A histogram or frequency histogram consists of a set of rectangles having bases on the horizontal axis (*X*-axis) and height (*Y*-axis) equal to the frequency. A frequency polygon is a line graph of the class frequency plotted against the class mark. It is obtained by connecting the midpoint of the top of the rectangles in the histogram (Figure 3.6).

Relative-Frequency Distribution. The relative frequency is obtained by dividing each class frequency by the total frequency of all classes. It is expressed as a percentage or fraction. The sum of the relative frequencies of all classes is 100 percent or fraction 1. The data may be presented as relative-frequency table or relative-frequency histogram (Table 3.7).

Cumulative and Relative-Frequency Distributions and Ogives. The total frequency of all values below the upper class boundary of a given class interval is the cumulative frequency. The data may be presented as cumulative-frequency table, cumulative-frequency distribution, or cumulative distribution. The graph showing the cumulative-frequency data is called a *cumulative-frequency polygon* or *ogives*. These polygons (ogives) show cumulative-frequency distribution or percentage ogives (less than ogives). Quite often, the cumulative-frequency distribution of all values greater than or equal (or more) to the lower class boundary is tabulated or plotted (Figures 3.8 and 3.9).

Frequency Curves and Smoothed Ogives. The frequency curves arising from large data may have a characteristic shape. These shapes are of many types.

a. **The symmetrical, bell-shaped, or normal curve**: The observations are equidistant from the central maximum.
b. **The moderately asymmetrical or skewed**: The tail of the frequency distribution curves to one side of the central maximum is longer than the other. The curve may be skewed to the right or have positive skewness, or be skewed to the left or have negative skewness.
c. **J-shaped or reverse J-shaped curve**: The frequency curve has a maximum at one end.
d. **U-shaped curve**: The frequency curve has a maximum on both sides.
e. **Bimodal curve**: The frequency curve has two maxima.
f. **Multimodal curve**: The frequency curve has more than two maxima.

Example 3.7: Frequency Distribution

A sampling program was conducted in a pharmaceutical plant. Grab samples were collected four times a day and 5 days a week over 5 weeks (total 100 samples). The chemical oxygen demand (COD) values in mg/L of each sample are given below.

Week	Day 1				Day 2				Day 3				Day 4				Day 5			
1	68	73	70	66	96	79	65	86	84	79	65	78	78	74	80	67	75	88	75	81
2	89	67	73	73	82	73	87	75	61	97	57	81	68	60	74	92	75	78	88	72
3	90	91	74	77	95	85	78	63	74	71	95	69	60	76	62	76	88	59	78	74
4	79	65	76	75	76	85	103	68	83	71	53	85	91	75	72	60	71	75	74	77
5	80	80	81	81	83	83	83	84	84	84	84	86	87	89	89	92	93	94	94	101

Using class frequency of 4, prepare a table to arrange the COD data of each class into an array and the frequency. With reference to this table, give the numerical values of the following terms

a. The number of COD values higher than 75 but less than 79 mg/L
b. The percentage of COD values higher than 65 but less than 85 mg/L
c. The lower class limit of the sixth class
d. The frequency interval
e. The class width
f. The class mark of the seventh class

Solution

1. Arrange the array and frequency of COD data.
   An array is the arrangement of raw numerical data in ascending or descending order of magnitude. The array and frequency of COD data in a class size of 4 are arranged in ascending order in Table 3.5.
2. Determine the numerical values of various terms
   a. COD values higher than 75 but less than 79 mg/L = 11
   b. Percent of COD values higher than 65 but less than 85 mg/L is = (61/100) × 100 = 61%
   c. Lower class limit of the sixth class = 75
   d. The frequency interval = 4

TABLE 3.5
The Array and Frequency of COD Data (Example 3.7)

Class	Array	Frequency
50–54	53	1
55–59	57, 59	2
60–64	60, 60, 60, 61, 62, 63	6
65–69	65, 65, 65, 66, 67, 67, 68, 68, 68, 69	10
70–74	70, 71, 71, 71, 72, 72, 73, 73, 73, 73, 74, 74, 74, 74, 74, 74	16
75–79	75, 75, 75, 75, 75, 75, 75, 76,76, 76, 76, 77, 77, 78, 78,78, 78, 78, 79, 79, 79	21
80–84	80, 80, 80, 81, 81, 81, 81, 82, 83, 83, 83, 83, 84, 84, 84, 84, 84	17
85–89	85, 85, 85, 86, 86, 87, 87, 88, 88, 88, 89, 89, 89	13
90–94	90, 91, 91, 92, 92, 93, 94, 94	8
95–99	95, 95, 96, 97	4
100–104	101, 103	2

e. The class width = 4
f. The class mark of the seventh class = (80 + 84)/2 = 82

Example 3.8: Histogram and Frequency Polygon

Prepare a histogram and frequency polygon using the COD data of the pharmaceutical plant in Example 3.7.

Solution

1. **Prepare the Histogram**: Draw the set of rectangles having (1) bases on the horizontal axis (*x*-axis), with centers at the class marks, and length equal to class interval sizes, and (2) heights on the vertical axis (*y*-axis) equal to the class frequencies. The histogram is shown in Figure 3.6.
2. **Prepare Frequency Polygon**: The frequency polygon is a line graph of the class frequency plotted against the class mark. It can be obtained by connecting the midpoints of the tops of the rectangles in the histogram. The frequency polygon is also shown in Figure 3.6.

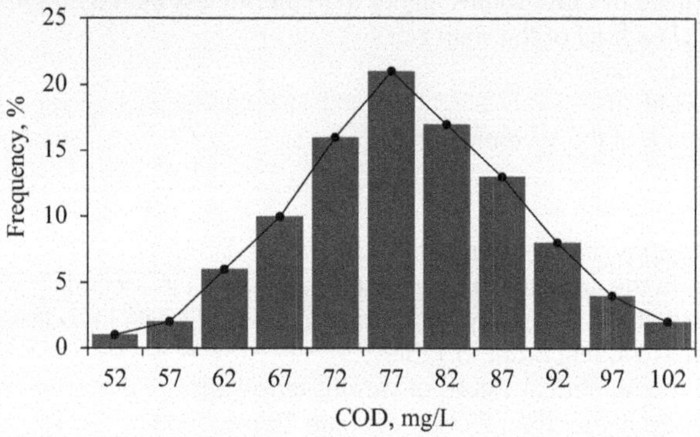

FIGURE 3.6 Histogram and frequency polygon (Example 3.8).

Example 3.9: Relative-Frequency Distribution Table and Histogram

The frequency distribution of wastewater flow to a treatment plant is given below. Prepare a percent relative-frequency distribution table, a histogram, and a frequency polygon.

Class	900–999	1,000–1,099	1,100–1,199	1,200–1,299	1,300–1,399	1,400–1,499	1,500–1,599	1,600–1,699	1,700–1,799	1,800–1,899	1,900–1,999
Frequency	0	4	9	14	11	7	4	2	0	1	0

Solution

1. Prepare the percent frequency distribution table
   The class, frequency, and relative and percent frequency distribution of wastewater flow to the treatment plant are summarized in Table 3.6
2. Prepare the percent frequency histogram and polygon.
   The histogram and polygon are shown in Figure 3.7

TABLE 3.6
Summary Data of Example 3.9

Class	Frequency	Relative Frequency	Percent Frequency
900–999	0	0	0
1,000–1,099	4	0.0769	7.7
1,100–1,199	9	0.1731	17.3
1,200–1,299	14	0.2692	26.9
1,300–1,399	11	0.2115	21.1
1,400–1,499	7	0.1346	13.5
1,500–1,599	4	0.0769	7.7
1,600–1,699	2	0.0386	3.9
1,700–1,799	0	0	0
1,800–1,899	1	0.0192	1.9
1,900–1,999	0	0	0
	52	1.000	100.0

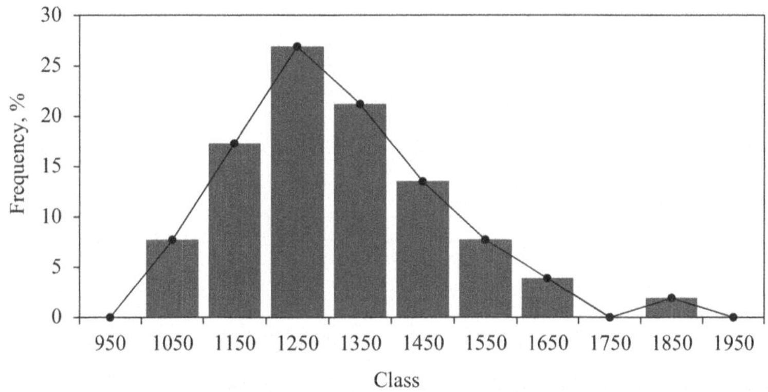

FIGURE 3.7 The percent frequency histogram and percent frequency polygon (Example 3.9).

Example 3.10: Cumulative-Frequency Distributions and Ogives

Use the wastewater flow data given in Example 3.9 to develop the cumulative and relative cumulative-frequency distribution table and ogives. Also, prepare cumulative-frequency distribution of all values greater than or equal to the lower class boundary.

Solution

1. Tabulate the cumulative and percent frequency distributions
   The cumulative and percent cumulative distributions of Example 3.10 are summarized in Table 3.7.
3. Plot the cumulative and percent cumulative ogives
   The cumulative-frequency distribution less than or equal to the upper class limit is shown in Figure 3.8.
4. Plot the cumulative and percent cumulative-frequency distribution of values greater than or equal to the lower class boundary.
   The greater than or equal to the lower class limit ogives are plotted in Figure 3.9.

TABLE 3.7
Cumulative and Percent Cumulative-Frequency Distributions (Example 3.10)

Class	Frequency	Cumulative Frequency Less Than or Equal to Upper Class Limit	Percent Cumulative Frequency	Cumulative Frequency Greater Than or Equal to Lower Class Limit	Percent Cumulative Frequency
900–999	0	0	0	52	100
1,000–1,099	4	4[a]	8[b]	48[c]	92[d]
1,100–1,199	9	13	25	39	75
1,200–1,299	14	27	52	25	48
1,300–1,399	11	38	73	14	27
1,400–1,499	7	45	87	7	14
1,500–1,599	4	49	94	3	6
1,600–1,699	2	51	98	1	2
1,700–1,799	0	51	98	1	2
1,800–1,899	1	52	100	0	0
1,900–1,999	0	52	100	0	0

[a] $0+4=4$.

[b] $(4/52)\times 100=8\%$.

[c] $52-4=48$.

[d] $(48/52)\times 100=92\%$.

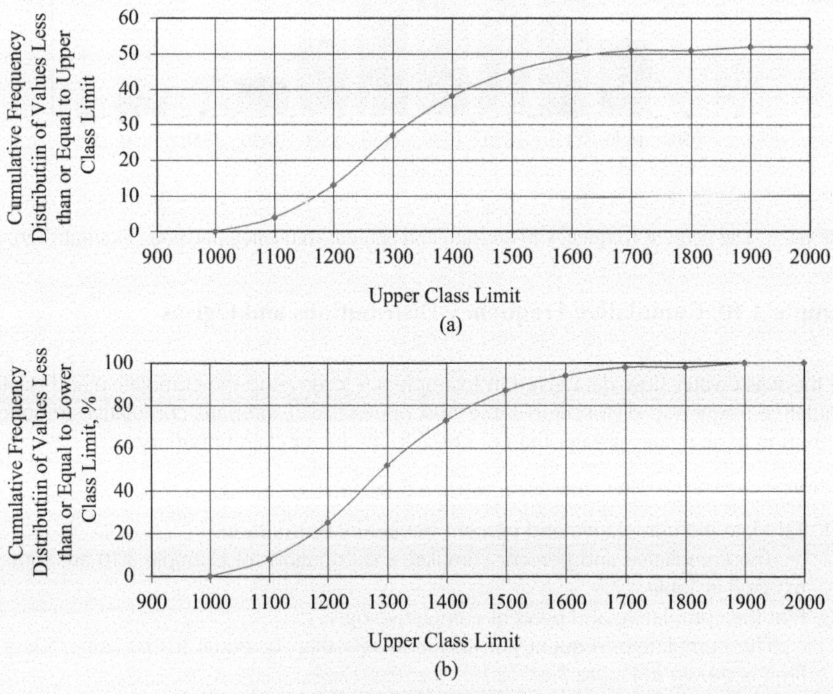

FIGURE 3.8 The less than ogives. (a) Cumulative. (b) Percent (Example 3.10).

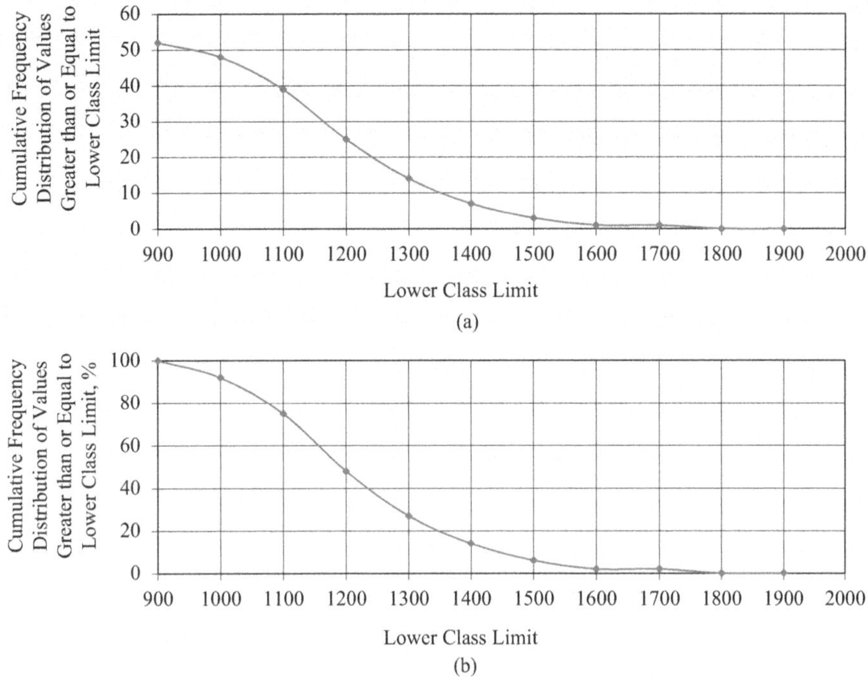

FIGURE 3.9 The greater than ogives. (a) Cumulative. (b) Percent (Example 3.10).

Example 3.11: Shapes of Frequency Curves

Draw the shapes of normal, skewed, J-shaped, U-shaped, bimodal, and multimodal frequency distribution curves.

Solution

The various frequency distribution curves are shown in Figure 3.10.

Example 3.12: Shapes and Description of Frequency Distribution

Four smoothed shapes of frequency distribution of flow profiles from an industry are given in Figure 3.11. Explain each flow profile.

Solution

The frequency distribution curves of flow pattern are shown in Figure 3.11. Each curve is explained below.

a. The distribution is normal which is indicative of random flow pattern.
b. The flow pattern has bimodal distribution. It is indicative of two distinctly different flow activities. An example of such a pattern is two-shift operation with second shift having lower flow generation rate.
c. The distribution is skewed to the right, or positive skewness. This represents the flow generation pattern operating at varying levels, but with some heavy flow generation activity.
d. The flow generation pattern represents a uniform distribution. This is characteristic of uniformly spaced flow generation at a constant rate.

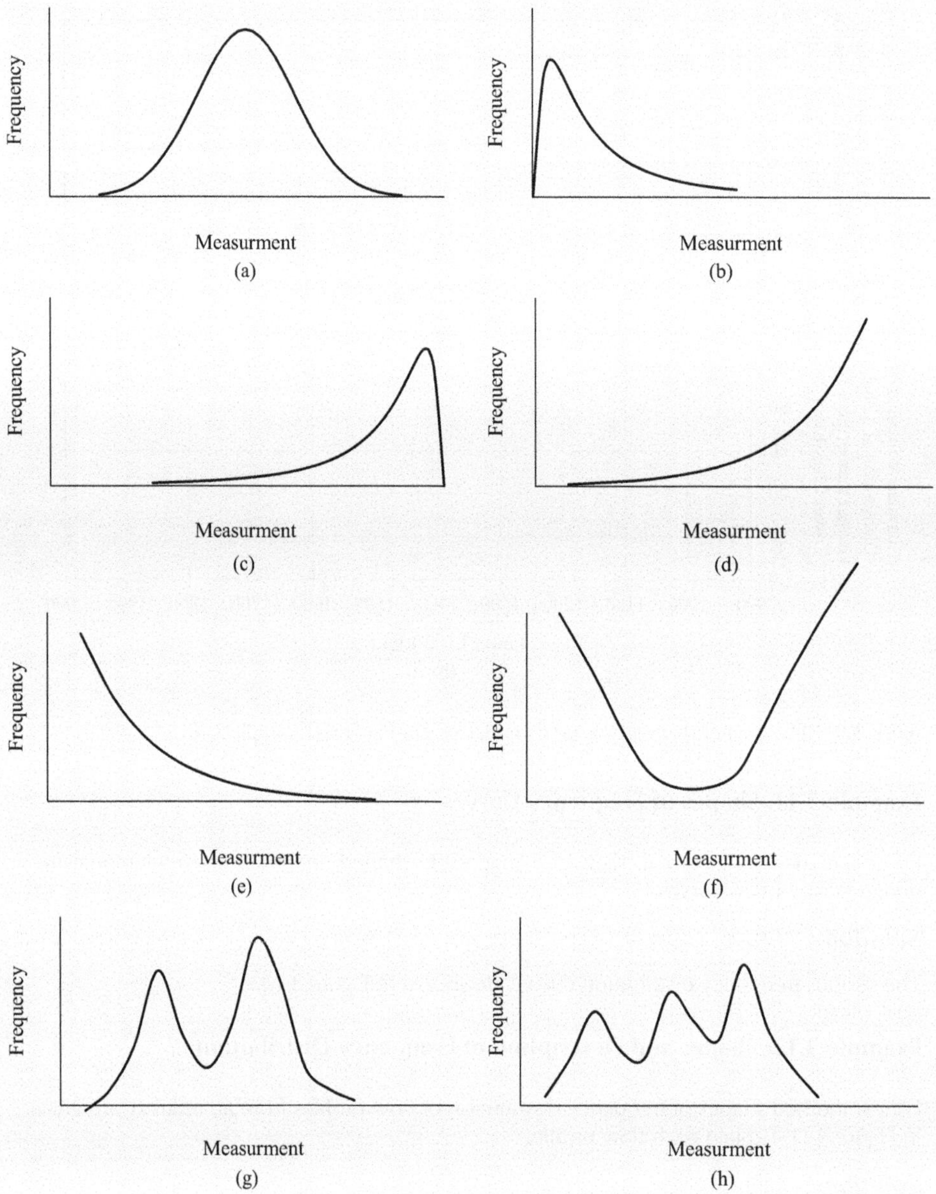

FIGURE 3.10 Different shapes of frequency distribution curves. (a) Normal, symmetrical or bell-shaped. (b) Skewed to the right (positive skewness). (c) Skewed to the left (negative skewness). (d) J-shaped. (e) Reverse J-shaped. (f) U-shaped. (g) Bimodal. (h) Multimodal frequency distribution (Example 3.11).

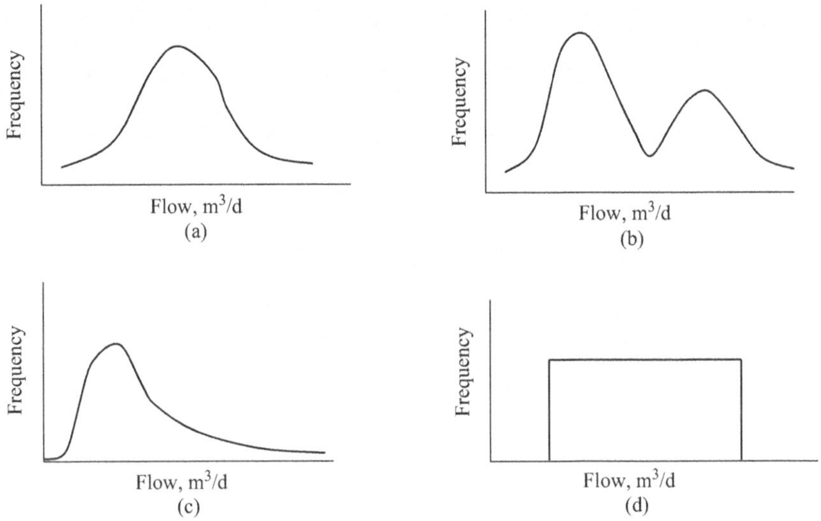

FIGURE 3.11 Frequency distribution curves of flow profiles (Example 3.12).

3.3.3 Measure of Central Tendency

Mean, Median, and Mode. The measures of central tendency or central distribution are expressed by the central value. These are the mean, median, and mode. The arithmetic mean (or average) of a number of observations is the sum of values divided by the number of observations. The geometric mean of a set of *N* observations is the *N*th root of the product of all observations. The median of a set of observations is the middle value, or the mean of the two middle values when the observations are arranged in order of magnitude. The mode of a set of observations is the value that occurs most frequently. It corresponds to the peak of the frequency distribution curve. There may be no mode, or a distribution may have more than one mode.

Empirical Relationship between Mean, Median, and Mode. For unimodal frequency curves that are moderately skewed (asymmetrical), an empirical relationship often applies (Eq. 3.2).

$$(\text{Mean} - \text{Mode}) = 3(\text{Mean} - \text{Median}) \tag{3.2}$$

Example 3.13: Arithmetic Mean

24-h composite samples from an industrial waste stream were collected for seven days. The biological oxygen demand (BOD) results are given below. Calculate the 7-day arithmetic mean value from Eq. (3.3).

Day	1	2	3	4	5	6	7
BOD_5, mg/L	180	230	165	210	150	190	220

Solution

The number of observations $N = 7$

$$\text{Arithmetic mean } X = \frac{1}{N}\left(\sum X_1 + X_2 + \cdots + X_7\right) \quad (3.3)$$

$$= \frac{1}{7}(180 + 230 + 165 + 210 + 150 + 190 + 220)$$

$$= \frac{1}{7}(1{,}345) = 192 \text{ mg/L}$$

Example 3.14: Weighted Arithmetic Mean

The weighted arithmetic mean is associated with weighing factors depending upon the importance attached to the observations. In a BOD_5 test, five measurements were made. Measurements 1, 2, and 5 were single test values. Measurements 3 and 4 were averages of triplicates and duplicate values. Determine the weighted arithmetic mean.

Measurement	1	2	3	4	5
Values	120	150	93	91	89

Solution

The weighted arithmetic mean is calculated below:

$$\text{Weighted arithmetic mean of BOD}_5 = \frac{(120 + 150 + 3 \times 93 + 2 \times 91 + 89) \text{ mg/L}}{8} = 103 \text{ mg/L}$$

Example 3.15: Geometric Mean

Four random samples were analyzed for coliform count. The following MPN values were obtained. Calculate the geometric mean.

Sample	1	2	3	4
MPN/100 mL	120	200	280	150

Solution

The geometric mean is *N*th root of the observations; it is expressed by Eq. (3.4)

$$X_g = (X_1 \times X_2 \times X_3 \times X_4)^{1/N} \quad (3.4)$$

$$= (120 \times 200 \times 280 \times 150)^{1/4}$$

$$= \left(1.008 \times 10^9\right)^{1/4} = 178 \text{ MPN/100 mL}$$

ALTERNATE METHOD

If the number of observations is large, the geometric mean can be calculated from Eq. (3.5)

$$\ln X_g = \frac{1}{N}[\ln X_1 + \ln X_2 + \ln X_3 + \ln X_4] \quad (3.5)$$

$$= \frac{1}{4}[\ln 120 + \ln 200 + \ln 280 + \ln 150]$$

$$= \frac{1}{4}[4.79 + 5.30 + 5.64 + 5.01]$$

$$= 5.183$$

X_g = inverse ln 5.183 = 178 MPN/100 mL

Example 3.16: Median of a Set of Numbers

Determine the median of the following numbers:

a. 12, 10, 18, 25, 30, 28, 16, 33, 21
b. 18, 17, 15, 25, 21, 28, 19, 30

Solution

a. Arrange the numbers in order of magnitude
   10, 12, 16, 18, 21, 25, 28, 30, 33
   The median value is the middle value
   Median = 21
b. Arrange the numbers in order of magnitude
   15, 17, 18, 19, 21, 25, 28, 30
   The median value is the average of the two middle numbers

$$\text{Median} = \frac{19+21}{2} = 20$$

Example 3.17: Mode of a Set of Numbers

Determine the mode of the following numbers:

a. 6, 8, 11, 13, 13, 15, 17, 21
b. 8, 10, 12, 15, 16, 19, 23
c. 11, 12, 13, 14, 14, 14, 16, 18, 18, 18, 23, 28, 30

Solution

a. The mode is the most frequently occurring number
   Mode = 13
b. There is no frequently occurring number
   Therefore, there is no mode
c. There are two frequently occurring numbers
   14 and 18, therefore, the modes are 14 and 18, or the distribution is bimodal.

Example 3.18: Mean, Median, and Mode of a Frequency Distribution

The frequency distribution of BOD_5 data of a wastewater treatment plant is given below. Determine mean, median, and mode. Calculate the approximate value of the mode from Eq. (3.2) and compare with the value obtained above.

BOD_5, mg/L	255	265	275	285	295	305	315
Frequency	8	10	16	14	10	5	2

Solution

1. Prepare Table 3.8 for data analysis
2. Calculate the mean from Eq. (3.6)

TABLE 3.8
Data Analysis (Example 3.18)

BOD_5 (*X*), mg/L	Frequency, *f*	*fX*
255	8	2,040
265	10	2,650
275	16	4,400
285	14	3,990
295	10	2,950
305	5	1,525
315	2	630
Total	65	18,185

$$\bar{X} = \frac{\sum f_i X_i}{N} \tag{3.6}$$

$$\bar{X} = \frac{18,185}{65} = 279.8 \text{ mg/L}$$

3. Calculate the median
   The median value is the middle value
   Median = 285 mg/L
4. Calculate the mode
   Mode is the frequently occurring value
   Mode = 275 mg/L
5. Calculate the mode from Eq. (3.2)
   Mean − Mode = 3 (Mean − Median)
   279.8 − Mode = 3 × 279.8 − 3 × 275
   Mode = 279.8 − 839.4 + 825 = 265.4 mg/L or 265 mg/L

3.3.4 Variance, Standard Deviation, Coefficients of Variability, Skewness, and Kurtosis

Variance. The error of an individual measurement in a set is defined as the difference between the arithmetic mean and the value of the measurement. The variance is the sum of squares of the deviation of the results from their arithmetic mean divided by one less than the total number of observations. It is expressed by Eq. (3.7).

$$s^2 = \frac{\sum f_i (X_i - \bar{X})^2}{N-1} \tag{3.7}$$

where

s^2 = variance
f_i = frequency (for ungrouped data $f_i = 1$)
X_i = the midpoint of the *i*th data range (for ungrouped data X_i = the *i*th observation)
$\bar{X}$ = arithmetic mean
N = number of observations

Standard Deviation. The standard deviation s of a set of numbers is the square root of the variance. It is expressed by Eq. (3.8)

$$s=\sqrt{\frac{\sum f_i(X_i-\bar{X})^2}{N-1}} \tag{3.8}$$

where

s = standard deviation

For normal distribution, 68.27% of the observations are in the range of $(\bar{X}\pm s)$. Likewise, 95.45% and 99.73% of the observations are in the ranges of $(\bar{X}\pm 2s)$ and $(\bar{X}\pm 3s)$, respectively.

Coefficient of Variation (CV). Standard deviation gives an indication of the absolute dispersion of measured values. It does not tell how large or small the dispersion is. CV (Eq. 3.9) is a relative measure of the dispersion

$$\text{CV}=\frac{s}{\bar{X}},\quad \frac{s}{\bar{X}}\times 100 \tag{3.9}$$

where

CV = coefficient of variation, fraction or percent

s and $\bar{X}$ = standard deviation and arithmetic mean

Typically, the CV in a biological field is 10%–30%. In chemical analyses, the CV is 2%–10%.

Coefficient of Skewness. The coefficient of skewness is a measure of a skewed frequency distribution. It is expressed by Eqs. (3.10) and (3.11).

$$\alpha_3=\frac{\sum f_i(X_i-\bar{X})^3/(N-1)}{s^3} \tag{3.10}$$

$$\alpha_3=\frac{\bar{X}-\text{Mode}}{s} \tag{3.11}$$

where

α_3 = coefficient of skewness

Other terms have been defined previously.

The positive and negative values of the coefficient of skewness indicate skewness to the right or left, respectively. Also, the larger the value, the greater is the skewness.

Coefficient of Kurtosis. The coefficient of kurtosis is a measure of peaked or flat-topped distribution in comparison to the normal distribution. It is expressed by Eq. (3.12).

$$\alpha_4=\frac{\sum f_i(X_i-\bar{X})^4/(N-1)}{s^4} \tag{3.12}$$

where

α_4 = coefficient of kurtosis

Other terms have been defined previously.

The value of the coefficient of kurtosis for the normal distribution is 3. A value higher than 3 indicates a peaked or mound-shaped curve, while a lower value indicates a flatter curve. For U-shaped curves, the value is less than 1.75.

Graphical Solution. The statistical analysis of the data can be performed using a graphical procedure. The probability paper is used to develop the plot of the measured parameter on the Y axis versus the percent probability on the X axis. The probability plot is a variation of the "Gaussian" or normal curve. Instead of obtaining a bell-shaped curve, the probability plot yields a linear relationship. If the resulting plotting points of a data set lie on a straight line on the arithmetic probability paper, the data will be normally distributed, and normal statistics can be applied. On the contrary, if the resulting plotting points have a linear relationship on the logarithmic probability paper, the distribution will be skewed. The graphical procedure is summarized below.

1. Arrange the data in increasing order.
2. Assign each data point a number m ($m=1, 2, 3, \ldots$, N).
3. The data is then plotted on arithmetic and logarithmic probability papers. The probability scale is labeled as "percent of values equal or less than the indicated value."
4. The plotting position is completed using Eq. (3.13).

$$\text{Plotting postion, \%} = \left(\frac{m}{N+1}\right)100 \tag{3.13}$$

where

m=rank serial number

N=number of observations

5. From this plot on the probability paper, the 50% value is the arithmetic mean.

   The standard deviation s is half the difference in the values which occur at the 15.87 (50.00−34.13) and 84.13 (50.00+34.13) percentile levels.

   Eq. (3.14) is used to determine s from the plot.

$$s = \frac{\text{Value at 84.13\%} - \text{Value at 15.87\%}}{2} \tag{3.14}$$

Example 3.19: Statistical Analysis

Determine the statistical characteristics of COD data of the pharmaceutical industry in Example 3.7. Describe the distribution and its statistical properties.

Solution

1. Tabulate the data and develop the relationships.
   The computations are summarized in Table 3.9.
2. Determine the statistical characteristics
   i. Arithmetic mean (Eq. 3.6)

$$\bar{X} = \frac{7{,}830}{100} = 78.3 \text{ mg/L}$$

   ii. Median (the midmost value)=77 mg/L
   iii. Mode (most frequent value)=77 mg/L
   iv. Variance (Eq. 3.7)

$$s^2 = \frac{10{,}381}{99} = 104.86$$

   v. Standard deviation (Eq. 3.8)
   s=10.24 mg/L

TABLE 3.9
The Statistical Characteristics of Pharmaceutical COD Data (Example 3.19)

Class	X_i, mg/L	f_i	f_iX_i	$X_i - \bar{X}$	$f_i(X_i - \bar{X})^2$	$f_i(X_i - \bar{X})^3$	$f_i(X_i - \bar{X})^4$
50–54	52[a]	1	52	−26.3	692	−18,192	478,450
55–59	57	2	114	−21.3	907	−19,327	411,666
60–64	62	6	372	−16.3	1,594	−25,985	423,556
65–69	67	10	670	−11.3	1,277	−14,429	163,048
70–74	72	16	1,152	−6.3	635	−4,001	25,206
75–79	77	21	1,617	−1.3	36	−46	60
80–84	82	17	1,394	3.7	233	861	3,186
85–89	87	13	1,131	8.7	984	8,561	74,481
90–94	92	8	736	13.7	1,502	20,571	281,823
95–99	97	4	388	18.7	1,399	26,157	489,136
100–104	102	2	204	23.7	1,123	26,624	630,989
Total	847	100	7,830		10,381	794	2,981,601

[a] Class midpoint.

vi. Coefficient of variance (Eq. 3.9)

$$\text{CV} = \frac{10.24}{78.3} \times 100 = 13.08\%$$

vii. Coefficient of skewness (Eq. 3.10)

$$\alpha_3 = \frac{794/99}{(10.24)^3} = 0.0075$$

viii. Coefficient of kurtosis (Eq. 3.12)

$$\alpha_4 = \frac{2,981,601/99}{(10.24)^4} = 2.7$$

Comments. The statistical analysis conducted in this example shows that the mean value of COD is 78 mg/L with a standard deviation of 10. It is expected that the COD of the subsequent samples collected and analyzed under the identical conditions will have values between

a. 68 and 88 mg/L, 68% of the time ($\bar{X} \pm s = 78$ mg/L ± 10 mg/L $= 68 - 88$ mg/L)
b. 58 and 98 mg/L, 95% of the time ($\bar{X} \pm 2s = 78$ mg/L $\pm 2 \times 10$ mg/L $= 58 - 98$ mg/L)

The coefficient of skewness is positive and is very small in magnitude (0.0075). This means that the distribution is distorted very slightly to the right. For all practical purposes, with such a small coefficient of skewness, the distribution is considered normal bell shaped. The coefficient of kurtosis of 2.7 indicates that the distribution is less peaked than the normal curve. This means that the data are less bunched together than for a normal curve. The CV of 13.08 is well within the range of the biological field.

The statistical information developed in this example is useful for the design of wastewater treatment facilities and may serve as a check against future analyses. The COD values falling within 68% confidence limit (68–88 mg/L) should be used for the facility design. COD values falling outside this range would be a suspect of possibly being caused by other than normal conditions, such as in-plant spills, change in normal industrial process, difference in handling of samples, and others. The process should be designed to handle short surges without significant loss in the effluent quality as long as the extreme values remain within the 95% confidence limit (58–98 mg/L). Beyond this range and under sustained conditions, the process may suffer adverse effluent quality or operational problems.

Example 3.20: Graphical Solution of Normally Distributed Data

The wastewater flow from an industrial plant was measured at the end of each shift for several days. The sample number and flow data are given below. Determine graphically the arithmetic mean, standard deviation, and CV. Also, prepare the histogram and calculate the mean, standard deviation, CV, and other statistical relationships using various statistical equations.
Solution

Sample No.	1	2	3	4	5	6	7	8	9	10	11	12	13
Flow, ft^3/min	0.5	2.5	4.1	6.7	7.5	10.5	11.3	12.0	12.5	15.6	17.1	20	22.2

Solution

1. Tabulate the data in increasing magnitude.
   Assign a rank serial, set up the class mark, frequency, plotting position, and other statistical parameters. This information is summarized in Table 3.10.
2. Calculate $\bar{X}$

$$\bar{X} = \frac{142.5}{13}\ \text{ft}^3/\text{min} = 11.0\ \text{ft}^3/\text{min}$$

3. Prepare arithmetic and logarithmic probability plots.
   The arithmetic and logarithmic probability plots are shown in Figure 3.12a and b.
   The plots show a linear relationship on arithmetic probability paper. Therefore, the data are normally distributed. The geometric probability plot clearly does not fall on a straight line.
4. Determine the arithmetic mean, standard deviation, and CV from the graphical results.
   $\bar{X}$ (at 50% values equal to or less than the stated value) = 11.0 ft^3/min
   The values at 15.87 and 84.13 percentile levels from this plot are 3.8 and 17.0 ft^3/min, respectively.

$$s\ (\text{Eq. } 3.14) = \frac{17.0 - 3.8}{2} = 6.6\ \text{ft}^3/\text{min}$$

$$\text{CV (Eq. 3.9)} = \frac{6.6 \times 100}{11.0} = 60\%$$

5. Prepare the histogram
   The histogram (Figure 3.13) shows nearly a normal distribution

TABLE 3.10
Rank Serial Number and Plotting Position as %, and Other Statistical Parameters (Example 3.20)

Rank Serial No. (m)	Flow (X_i), ft³/min	Plotting Position, %	$(X_i - \bar{X})$	$(X_i - \bar{X})^2$	$(X_i - \bar{X})^3$	$(X_i - \bar{X})^4$
1	0.5	7.1[a]	−10.5	110.3	−1,158	12,155
2	2.5	14.3	−8.5	72.3	−614	5,220
3	4.1	21.4	−6.9	47.6	−329	2,267
4	6.7	28.6	−4.3	18.5	−80	342
5	7.5	35.7	−3.5	12.3	−43	150
6	10.5	42.9	−0.5	0.3	0	0
7	11.3	50.0	0.3	0.1	0	0
8	12.0	57.1	1.0	1.0	1	1
9	12.5	64.3	1.5	2.3	3	5
10	15.6	71.4	4.6	21.2	97	448
11	17.1	78.6	6.1	37.2	227	1,385
12	20.0	85.7	9.0	81.0	729	6,561
13	22.2	92.9	11.2	125.4	1,405	15,735
Total	142.5			529.5	238	44,269

a Plotting position (Eq. 3.13) $= \left(\frac{1}{13+1}\right) \times 100 = 7.1.$

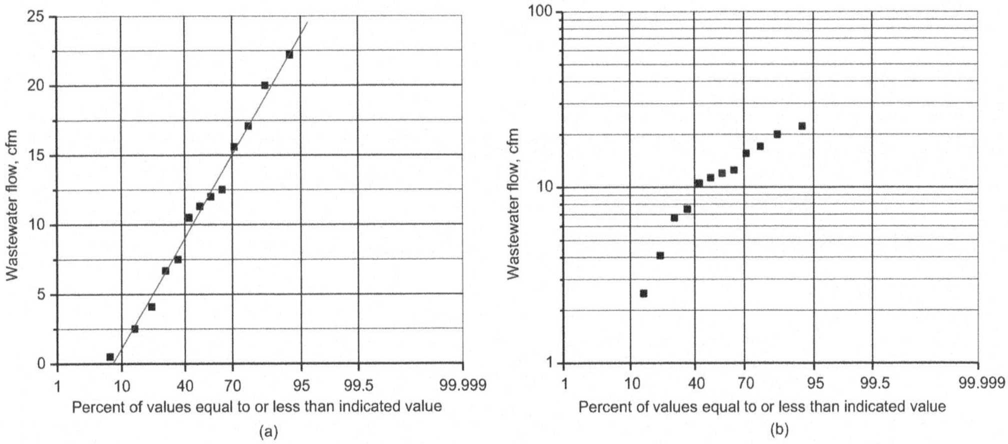

FIGURE 3.12 Probability plots of flow data. (a) Arithmetic. (b) Geometric (Example 3.20).

6. Calculate the arithmetic mean, standard deviation, coefficient of variance, coefficient of skewness, and coefficient of kurtosis from statistical equations.

$$\text{Arithmetic mean } \bar{X} \text{ (Eq. 3.6)} = \frac{142.5}{13} = 1.0 \text{ ft}^3/\text{min}$$

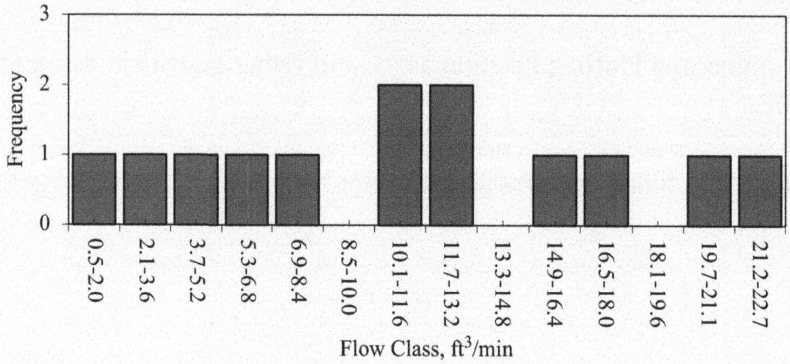

FIGURE 3.13 Histogram of flow data (Example 3.20).

$$\text{Standard deviation (Eq. 3.8)} = \sqrt{\frac{529.5}{12}} = 6.6\ \text{ft}^3/\text{min}$$

$$\text{CV (Eq. 3.9)} = \frac{100 \times 6.6}{11.0} = 60\%$$

$$\text{Coefficient of skewness (Eq. 3.10)} = \frac{238/12}{(11.0)^3} = 0.015$$

$$\text{Coefficient of kurtosis (Eq. 3.11)} = \frac{44{,}269/12}{(11.0)^4} = 0.25$$

The coefficient of skewness is small, showing a nearly normal distribution. The small value of the coefficient of kurtosis shows that data is spread out with little peaking. These results can be verified from the histogram given in Figure 3.13.

Example 3.21: Graphical Solution of Skewed Distribution

Given below is the wastewater flow data for a period of 10 days. Determine graphically the type of distribution, mean, standard deviation, and CV. Compare these results with the values obtained from the statistical equations.

Day	1	2	3	4	5	6	7	8	9	10
Flow data, m³/h	34	48	61	75	90	110	120	170	205	290

Solution

1. Tabulate the data

   Table 3.11 contains the data to be used for graphical and numerical solutions and to prepare the histogram.

2. Calculate $\bar{X}$

$$\bar{X} = 1{,}203/10 = 120.3\ \text{m}^3/\text{h}$$

TABLE 3.11
Rank Serial Number and Plotting Position as %, and Frequency Distribution (Example 3.21)

Rank Serial No. (m)	Flow (X_i), m³/h	Plotting Position, %	$(X_i - \bar{X})$	$(X_i - \bar{X})^2$
1	34	9.1[a]	−86.3	7,448
2	48	18.2	−72.3	5,227
3	61	27.3	−59.3	3,517
4	75	36.4	−45.3	2,052
5	90	45.5	−30.3	918
6	110	54.5	−10.3	106
7	120	63.6	0.3	0
8	170	72.7	49.7	2,470
9	205	81.8	84.7	7,174
10	290	90.9	169.7	28,798
Total	1,203			57,710

[a] From Eq. (3.13), $(1/(10+1)) \times 100 = 9.1$.

3. Develop the arithmetic and logarithmic probability plots.
   The arithmetic and logarithmic probability plots are shown in Figure 3.14a and b, respectively.
   The resulting plotting points on the arithmetic probability plot do not lie on a straight line. Therefore, the distribution is not normal. Contrary to this, the resulting points on the logarithmic probability plot lie on a straight line (Figure 3.14b). Therefore, the distribution is log-normal or skewed.
4. Plot the histogram
   The histogram is shown in Figure 3.15. The distribution is clearly skewed. This is in agreement with the log-normal presentation.
5. Determine graphically the arithmetic mean, standard deviation s, and coefficient of variance.

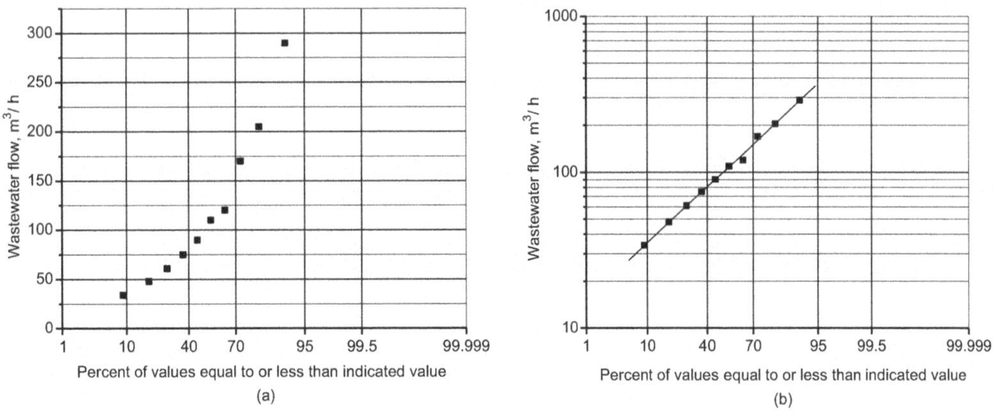

FIGURE 3.14 Arithmetic and logarithmic probability plots of flow data. (a) Arithmetic. (b) Geometric (Example 3.21).

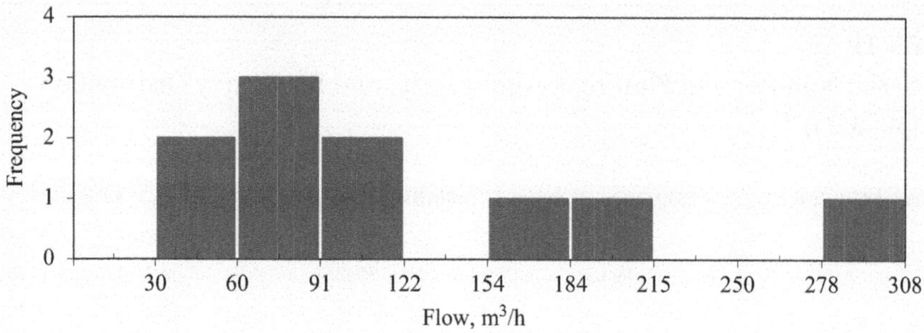

FIGURE 3.15 Histogram of daily flow data (Example 3.21).

Arithmetic mean $\bar{X}$ (at 50% value equal to or less than stated value) = 120.3 m³/h.
The value 15.87% and 84.13% levels from Figure 3.14 are 50 and 210 m³/h, respectively.

$$s \text{ (Eq. 3.14)} = \frac{210-45}{2} = 82.5 \text{ m}^3/\text{h}$$

$$\text{CV (Eq. 3.9)} = \frac{100 \times 80}{120.3} = 68.6\%$$

6. Determine the arithmetic mean, standard deviation, and CV from equations

$$\bar{X} \text{ (Eq.3.6)} = \frac{1{,}203}{10} = 120.3 \text{ m}^3/\text{h}$$

$$s \text{ (Eq. 3.8)} = \sqrt{\frac{57{,}710}{9}} = 80.1 \text{ m}^3/\text{h}$$

$$\text{CV (Eq. 3.9)} = \frac{100 \times 80.1}{120.3} = 66.6 \text{ m}^3/\text{h}$$

3.3.5 The Student's *t* Test

The student's t test is normally used for statistical comparison of two means ($\bar{X}$) to determine whether or not they are the same. The test is commonly used for comparative evaluation of the performance of two processes, equipment, chemicals, or analytical laboratory results. To test the difference of two means, the t value is calculated from Eqs. (3.15) and (3.16).

$$t = \frac{\bar{X}_1 - \bar{X}_2}{\sigma \sqrt{1/N_1 + 1/N_2}} \quad (3.15)$$

where

$$\sigma = \sqrt{\frac{(N_1-1)s_1^2 + (N_2-1)s_2^2}{\text{DF}}} \quad (3.16)$$

N_1, N_2=the number of sample sizes in the first and second sets
$\bar{X}_1$, $\bar{X}_2$=the arithmetic means of two sets
σ=universal standard deviation, %. It is calculated from Eq. (3.16)
where
s_1, s_2=the standard deviations of two sets
DF=degree of freedom=(N_1+N_2-2)

The critical value of t is found from the t distribution table for the desired confidence level and degrees of freedom. This table may be found in many statistical books and is included in Appendix A.[6–9] The results of two processes are different if calculated t is greater than that obtained from the table.

Example 3.22: Student's *t* Test

Phosphorus precipitation tests with two coagulants *A* and *B* were conducted for 12 days. The results of percent phosphorus removal for both coagulants are given below. Compare the performance of both coagulants.

	Phosphorus Removal, %											
Coagulant *A*	95	96	97	96	92	97	87	89	93	92	96	91
Coagulant *B*	87	95	83	88	90	91	87	84	90	86	87	90

Solution

1. Tabulate the data and determine the standard deviation for both sets of data
   The calculated values are summarized in Table 3.12
2. Calculate the arithmetic means $\bar{X}_A$ and $\bar{X}_B$ from Eq. (3.6)

TABLE 3.12
The Results of Phosphorus Removal by Two Coagulants (Example 3.22)

	Coagulant *A*			Coagulant *B*		
Day	X_{Ai}	$(X_{Ai}-\bar{X})$	$(X_{Ai}-\bar{X})^2$	X_{Bi}	$(X_{Bi}-\bar{X})$	$(X_{Bi}-\bar{X})^2$
1	95	1.6	2.6	87	−1.2	1.4
2	96	2.6	6.8	95	6.8	46.2
3	97	3.6	13.0	83	−5.2	27.0
4	96	2.6	6.8	88	−0.2	0
5	92	−1.4	2.0	90	1.8	3.2
6	97	3.6	13.0	91	2.8	7.8
7	87	−6.4	41.0	87	−1.2	1.4
8	89	−4.4	19.4	84	−4.2	17.6
9	93	−0.4	0.2	90	1.8	3.2
10	92	−1.4	2.0	86	−2.2	4.8
11	96	2.6	6.8	87	−1.2	1.4
12	91	−2.4	5.8	90	1.8	3.2
Total	1,121	0.2	119.4	1,058	−0.4	117.2

$$\bar{X}_A = \frac{1{,}121}{12} = 93.4\%$$

$$\bar{X}_B = \frac{1{,}058}{12} = 88.2\%$$

3. Calculate standard deviations s_A and s_B from Eq. (3.8)

$$s_A = \sqrt{\frac{119.4}{11}} = 3.3\%$$

$$s_B = \sqrt{\frac{117.2}{11}} = 3.3\%$$

4. Calculate the degree of freedom, DF

$$DF = N_1 + N_2 - 2 = 12 + 12 - 2 = 22$$

5. Calculate the universal standard deviation σ (Eq. 3.16)

$$\sigma = \sqrt{\left[\frac{(12-1)\times 3.3^2 + (12-1)\times 3.3^2}{22}\right]} = 3.3\%$$

6. Calculate the student's t value from Eq. (3.15)

$$t = \frac{93.4\% - 88.2\%}{3.3\% \sqrt{1/12 + 1/12}} = \frac{5.2\%}{3.3\% \left(\sqrt{0.167}\right)} = 3.86$$

7. Determine the critical value of t

   The critical value of t is found from the t distribution table given in Appendix A (Table A.4) for 95% confidence level for DF=22.

   $t_{0.05(22)} = 1.717$

8. Compare the actual and critical values of t and state the conclusion.

   The calculated value of $t=3.86$ is greater than the critical value of $t=1.717$ from Table A.4. Therefore, it can be concluded that the results from two coagulants used for phosphorus removal are significantly different at the 95% confidence level. The coagulant A is better than coagulant B.

3.3.6 Regression Analysis

Linear Relationship. One of the most commonly used statistical functions is the correlation between two parameters, such as BOD_5 and COD, removal versus dose of chemicals or reaction time, and meter reading versus flow. This can be achieved by relating the paired parameters to each other in order to obtain a functional relationship given by Eq. (3.1).

By the method of least square curve fitting, the coefficients a and m can be calculated. Many computer programs are available that can be used to determine the constants and the coefficient of correlation r, which is a measure of the "goodness of fit." Eqs. (3.17)–(3.19) are used to determine the constants a, m, and r.

$$a = \frac{\left(\sum X^2\right)\left(\sum Y\right) - \left(\sum X\right)\left(\sum XY\right)}{N\left(\sum X^2\right) - \left(\sum X\right)^2} \tag{3.17}$$

$$m = \frac{N\sum XY - \sum X \sum Y}{N\sum X^2 - \left(\sum X\right)^2} \tag{3.18}$$

$$r = \frac{N\sum XY - \sum X \sum Y}{\sqrt{\left[N\sum X^2 - \left(\sum X\right)^2\right]\left[N\sum Y^2 - \left(\sum Y\right)^2\right]}} \tag{3.19}$$

The correlation coefficient must be in the range $0 < |r| \leq 1$. The value of $r = 1$ corresponds to a perfect fit, and $r = 0$ indicates that no relationship exists between X and Y functions. To interpret the meaning of r expected at levels of significance of 1% and 5%, tables are given in many textbooks.[6–9] The levels of significance, that is, 1% and 5% represent a 99% and 95% level of probability of true correlation. If the absolute value of the calculated r exceeds the value r chosen from the table, it can be concluded that correlation does exist.

Nonlinear Equations. Common nonlinear relationships or mathematical models used to develop regression are quadratic or parabolic, logarithmic, geometric, and multiple linear regression.[10] These relationships are given below (Eqs. 3.20–3.23).

$$\text{Quadratic or parabolic} \quad Y = a + bX + cX^2 \tag{3.20}$$

$$\text{Logarithmic} \quad Y = aX^b, \quad Y = ae^{bX} \tag{3.21}$$

$$\text{Geometric} \quad Y = a\log X + b \tag{3.22}$$

$$\text{Multiple linear} \quad Z = C_0 + C_1U_1 + C_2U_2 + C_3U_3 + \cdots + C_nU_n \tag{3.23}$$

where

Y = dependent variable

X = independent variable

a, b, c = constants determined by regression analysis

Z = a parameter that may be affected by numerous variables. Examples of Z are flow in pipe or channel, phosphorus or BOD_5 removal, diffuser efficiency, and gas generation in a digester.

$C_0, C_1, \ldots, C_n$ = regression constants

$U_1, U_2, \ldots, U_n$ = independent variables. Examples are roughness coefficient, conduit slope, pH, chemical dose, temperature, reaction time, mixing, loading rates, and others.

The regression analysis by any mathematical model can be tedious and time-consuming. The analysis can be easily performed with many readily available computer programs.

Example 3.23: Linear Regression Analysis

A 15-min COD surrogate test has been developed for quick estimation of 5-day BOD values. The test is conducted by using vials containing pre-measured reagents. Exact volume of sample is transferred into the vials. A blank is also prepared using distilled water. The vials are digested in a preheated block digester for 15 min, cooled, and then absorbance is measured. The concentration is obtained from a calibration curve. The results of 15-min COD test and 5-d BOD values of raw municipal wastewater are given below. Determine the regression line of *Y* (BOD_5) and *X* (15-min COD), and the coefficient of regression for the data. Estimate BOD_5 of a sample if 15-min COD value is 380 mg/L.

15-min COD, mg/L	485	381	364	402	377	412	343	408	184	151	220	306	331	345
BOD_5, mg/L	222	204	210	195	177	186	138	216	108	105	117	120	185	150

Solution

1. Tabulate various values (Table 3.13) to determine constants *a* and *m* in the Eq. $Y = a + mX$
2. Determine the constants *a* (intercept) and *m* (slope)

$$a \text{ from Eq. (3.17)} = \frac{(1{,}697{,}751)(2{,}333) - (4{,}709)(830{,}219)}{14 \times (1{,}697{,}751) - (4{,}709)^2}$$

$$a = 32.2$$

$$m \text{ from Eq. (3.18)} = \frac{14 \times (830{,}219) - (4{,}709)(2{,}333)}{14 \times (1{,}697{,}751) - (4{,}709)^2}$$

$$m = 0.4$$

TABLE 3.13
BOD_5 and 15-min COD Data, and Statistical Values (Example 3.23)

X, COD, mg/L	*Y*, BOD_5, mg/L	*XY* $(mg/L)^2$	X^2 $(mg/L)^2$	Y^2 $(mg/L)^2$
485	222	107,670	235,225	49,284
381	204	77,724	145,161	41,616
364	210	76,440	132,496	44,100
402	195	78,390	161,604	38,025
377	177	66,729	142,129	31,329
412	186	76,632	169,744	34,596
343	138	47,334	117,649	19,044
408	216	88,128	166,464	46,656
184	108	19,872	33,856	11,664
151	105	15,855	22,801	11,025
220	117	25,740	48,400	13,689
306	120	36,720	93,636	14,400
331	185	61,235	109,561	34,225
345	150	51,750	119,025	22,500
$\sum X = 4{,}709$	$\sum Y = 2{,}333$	$\sum XY = 830{,}219$	$\sum X^2 = 1{,}697{,}751$	$\sum Y^2 = 412{,}153$

3. Determine the regression constant *r*, and compare with a 99% level of probability for true correlation.

$$r \text{ from Eq. (3.19)} = \frac{14\times(830{,}219)\times(4{,}709)(2{,}333)}{\sqrt{\left[14\times(1{,}697{,}751)-(4{,}709)^2\right]\left[14\times(412{,}153)-(2{,}333)^2\right]}}$$

   $r = 0.88$

   The values of r^2 required to establish statistical significance of a simple regression equation for various sample sizes are summarized in Table A.5 of Appendix A. For a sample size of 14, the r^2 value for 99% level of probability is 0.44, and the r value is 0.66. Since the calculated value of r exceeds the value in the table, it can be concluded that at 99% level of probability, correlation does exist.
4. Express the linear relationship

   $Y\ (BOD_5) = 32.2 + 0.4X$ (measured 15-min COD)
5. Plot the line

   The linear relationship is shown in Figure 3.16
6. Determine BOD_5 from 15-min COD value of 380

   The estimated BOD_5 value is 184 mg/L.

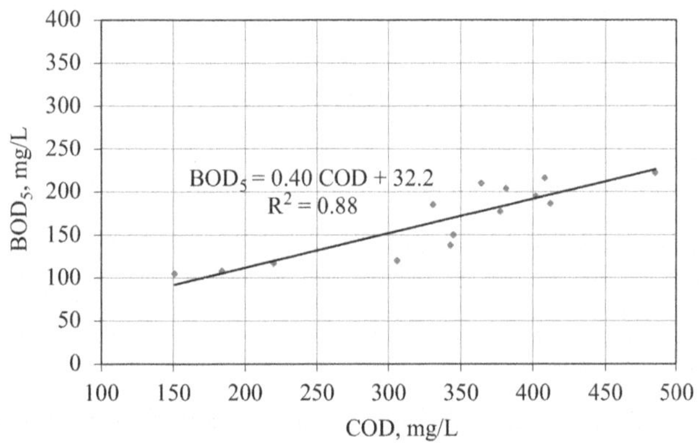

FIGURE 3.16 Linear plot of BOD_5 and 15-min COD test data (Example 3.23).

3.3.7 Data Quality Indicators

Quality Assurance (QA). It is essential that all data generated have a high level of confidence. QA is a set of operating principles that, if strictly followed during flow measurement and sample collection and analysis, will produce data of known and defensible quality. Included in QA are *quality control* and *quality assessment*. A QA plan should include sampling and analysis procedures, equipment, training of analysts, performance audits, data assessment procedures for bias and precision, data reduction, validation, and reporting.[1]

Quality Control (QC). QC may be either internal or external. A good QC program consists of at least seven elements: (1) Certification of operator compliance, (2) recovery of known additions, (3) analysis of externally supplied standards, (4) analysis of reagent blanks, (5) calibration with standards, (6) analysis of duplicates, and (7) maintenance of control charts.[1]

Quality Assessment. Quality assessment measures the quality of data produced by the laboratory. The data are applied to test the recovery, bias, precision, detection limit, and adherence to the standard operating procedures.[1]

Precision. Precision is a measure of the closeness with which multiple analyses of a given sample agree with each other. Precision is specified by the standard deviation of the results.

All data collection efforts performed for U.S. Environmental Protection Agency are required to have well defined QA project plan to produce data with high level of confidence. The recommended guidelines for QA program may be found in several sources.[1,10–13]

Example 3.24: Relative Percent Difference (RPD) of Replicate Tests

A wastewater effluent stream was measured for average flow. The results of two replicates of flow measurements of the stream are provided below. Determine the relative percent difference (RPD)

Replicate	Measured flow, m^3/h
1	9.8
2	10.3

Solution

1. State the equation for RPD

   The RPD is calculated from Eq. (3.24)

$$\text{RPD} = \frac{(C_1 - C_2) \times 100\%}{(C_1 + C_2)/2} \tag{3.24}$$

   where
   RPD = relative percent difference
   C_1 = larger of the measured value
   C_2 = smaller of the measured value

2. Calculate the RPD of two measured flows

$$\text{RPD} = \frac{(10.3\ \text{m}^3/\text{h} - 9.81\ \text{m}^3/\text{h}) \times 100\%}{(10.3\ \text{m}^3/\text{h} + 9.81\ \text{m}^3/\text{h})/2} = 5\%$$

   The percent difference between two measured flow replicates is 5%.

Example 3.25: Relative Standard Deviation (RSD) of Replicates

A wastewater effluent sample was tested for total nitrogen. Four replicate tests were conducted. The results are given below. Calculate the relative standard deviation (RSD).

Replicates	1	2	3	4
Total nitrogen, mg/L	38	40	37	39

Solution

1. State the equation for RSD

   The RSD is calculated from Eq. (3.25)

$$\text{RSD} = \frac{s \times 100\%}{\overline{X}} \tag{3.25}$$

where
RSD = relative standard deviation of replicates
s = standard deviation
$\bar{X}$ = mean of replicate analysis

2. Calculate the mean and standard deviation from Eqs. (3.3) and (3.8)

$$\bar{X} = (38 + 40 + 37 + 39)/4 = 38.5 \text{ mg/L}$$

$$s = \sqrt{[(38.0 - 38.5)^2 + (40.0 - 38.5)^2 + (37.0 - 38.5)^2 + (39.0 - 38.5)^2]/3}$$

$$= \sqrt{\left[0.5^2 + 1.5^{\,2} + 1.5^2 + 0.5^2\right]/3}$$

$$= 1.30 \text{ mg/L}$$

3. Calculate the RSD of four replicates (Eq. 3.25)

$$\text{RSD} = \frac{1.3 \times 100\%}{38.5} = 3.4\%$$

Example 3.26: Percent Deviation from Spike Recovery

The concentration of nitrate–nitrogen in a water sample is 5.40 mg/L. The sample was spiked with 1.20 mg/L of nitrate–nitrogen. The measured concentration in the spiked sample was 6.75 mg/L. Calculate the percent deviation of recovery from spike, and the percent recovery of spike.

Solution

1. State the equation for the percent deviation of recovery from spike, and the percent recovery of spike.

   The percent deviation of recovery from spike, and the percent recovery of spike are expressed by Eqs. (3.26a) and (3.26b).

$$\%D = \frac{\left[(S - U) - C\right] \times 100\%}{C} \tag{3.26a}$$

$$\%R = \frac{[S - U] \times 100\%}{C} \tag{3.26b}$$

where
$\%D$ = percent deviation of recovery from spike, %
$\%R$ = percent recovery of spike, %
S = measured concentration in spiked aliquot, mg/L
U = measured concentration in unspiked aliquot, mg/L
C = measured concentration of known spike, mg/L
$(S - U)$ = spike recovery, mg/L
$(S - U) - C$ = deviation from spike recovery

2. Calculate percent deviation of recovery from spike

   Apply Eq. (3.26a)

$$\%D = \frac{[(6.75 - 5.40) - 1.20] \text{ mg/L} \times 100}{1.20 \text{ mg/L}} = 12.5\%$$

3. Calculate percent recovery of spike from Eq. (3.26b)

$$\%\text{Recovery of spike} = \frac{(6.75 - 5.40)\ \text{mg/L} \times 100}{1.20\ \text{mg/L}} = 112.5\%$$

Example 3.27: Precision of an Analytical Test with Spiked Samples

An analytical lab was given 10 samples for total phosphorus analysis. Each sample had two portions, one with no spike and the other had a spike of 1.3 mg/L total phosphorus. Thus, the analytical lab received a total of 20 sample bottles. Each bottle was labeled one through 20 with no indication of spiked or unspiked bottles. The lab results after decoding and arranging the results are summarized below. Determine the precision of the analytical procedure based on spiking data.

Sample No.	1	2	3	4	5	6	7	8	9	10
Total *P*, mg/L Unspiked Portion	6.15	7.32	8.12	7.60	6.03	6.76	7.34	7.82	6.80	7.24
Total *P*, mg/L Spiked Portion	7.30	8.78	9.50	8.86	7.43	7.97	8.53	9.14	8.08	8.55

Solution

1. Calculate the spike recovery
   The spike recovery calculations are summarized in Table 3.14
2. Calculate the standard deviation using Eq. (3.8)

TABLE 3.14
Calculation Table to Determine Precision of Spiked Sample (Example 3.27)

Sample No.	Concentration in Spiked Portion, mg/L	Concentration in Unspiked Portion, mg/L	Spike Recovery, mg/L	Spike Concentration, mg/L	Deviation from Spike Recovery, mg/L	(Deviation)^2, (mg/L)^2
1	7.30	6.15	1.15	1.30	−0.15	0.0225
2	8.78	7.32	1.46	1.30	0.16	0.0256
3	9.50	8.12	1.38	1.30	0.08	0.0064
4	8.86	7.60	1.26	1.30	−0.04	0.0016
5	7.43	6.03	1.40	1.30	0.10	0.0100
6	7.97	6.76	1.21	1.30	−0.09	0.0081
7	8.53	7.34	1.19	1.30	−0.11	0.0121
8	9.14	7.82	1.32	1.30	0.02	0.0004
9	8.08	6.80	1.28	1.30	−0.02	0.0004
10	8.55	7.24	1.31	1.30	0.01	0.0001
						Σ(deviation)^2=0.0872

$$s = \sqrt{\frac{\left(\sum \text{deviation}^2\right)}{N-1}}$$

$$= \sqrt{\frac{(0.0872)}{10-1}} = 0.10\ \text{mg/L}$$

3. Calculate $\bar{X}$ and CV from Eqs. (3.3) and (3.9)

$$\bar{X} = \frac{(1.15+1.46+1.38+1.26+1.40+1.21+1.19+1.32+1.28+1.31)\text{ mg/L}}{10}$$

$$= \frac{12.96\text{ mg/L}}{10} = 1.3\text{ mg/L}$$

$$\text{CV} = \frac{100 \times 0.10\text{ mg/L}}{1.3\text{ mg/L}} = 7.7\%$$

4. Comment on the result

   The standard deviation of 0.10 mg/L gives the absolute dispersion of the set of readings. A CV of 7.7% is on the higher side for precision of analytical tests, but it is acceptable for chemical analyses since the acceptable CV range is 2%–10%.

Example 3.28: Detection Limit

The detection limit of an analytical procedure or instrument is needed to determine the accuracy of the lower limit of a measurement. A standard solution with 0.15 mg/L of total organic carbon (TOC) concentration was analyzed five times in a TOC measuring instrument. The measured concentrations for 5 replicates are given below. Find the method detection limit (MDL).

Replicates	1	2	3	4	5
Measured concentration, mg/L	0.143	0.174	0.169	0.139	0.144

Solution

1. State the MDL equation

   The MDL is calculated from Eq. (3.27)

$$\text{MDL} = t_{(n-1,1-\alpha=0.99)} \cdot s \tag{3.27}$$

where

MDL = method detection limit

s = standard deviation of replicate analyses

$t_{(n-1,1-\alpha=0.99)}$ = student's t values for a one-sided 99% confidence level. Tables containing these values may be found in Appendix A (Table A.4). For different degrees of freedom, these values are provided in Table 3.15.[6]

n = number of observations

TABLE 3.15
Student's *t* Values Corresponding to Various Degrees of Freedom (Example 3.28)

Degree of freedom $(n-1)$	2	3	4	5	6	7	8	10	20
$t_{(n-1,1-\alpha=0.99)}$	6.965	4.541	3.747	3.365	3.143	2.998	2.896	2.764	2.528

TABLE 3.16
Calculation Table to Determine Detection Limits (Example 3.28)

Sampling No.	Measured Concentration, mg/L	Known Concentration, mg/L	Deviation, mg/L	(Deviation)^2, (mg/L)^2
1	0.143	0.150	−0.007	4.9×10^{-5}
2	0.174	0.150	0.024	57.6×10^{-5}
3	0.169	0.150	0.019	36.1×10^{-5}
4	0.139	0.150	−0.011	12.1×10^{-5}
5	0.144	0.150	−0.006	3.6×10^{-5}
				Total = 114.3×10^{-5}

2. Determine $t_{(4,1-\alpha=0.99)}$

$$\text{DF} = (n-1) = 4$$

$$t_{(4,1-\alpha=0.99)} = 3.747$$

3. Calculate the standard deviation from Eq. (3.8)
   The standard deviation calculations are summarized in Table 3.16.

$$s = \sqrt{\frac{\Sigma(\text{deviation})^2}{n-1}}$$

$$s = \sqrt{\frac{114.3\times10^{-5}}{5-1}} = 0.017 \text{ mg/L}$$

4. Calculate MDL from Eq. (3.27)
   MDL = (3.747)(0.017) mg/L = 0.064 mg/L
   The lower limit or detection limit of the TOC measuring instrument is 0.06 mg/L

3.3.8 Reliability Analysis

Design of a wastewater treatment facility should employ the reliability concept along with environmental and cost considerations. Reliability is defined as the probability of achieving the permit requirements (or a level of performance) for at least a specified period of time under specified conditions. For example, a reliability of 99% means (a) facility is expected to meet the level of performance 99% of the time, or (b) 1% – or three to four times in a year – the performance or permit may not be met. The reliability analysis should also include (1) capital cost of achieving the specified level of reliability, (2) operational and maintenance costs, and (3) adverse environmental effects of not meeting the performance.

The reliability is calculated from Eqs. (3.28) and (3.29)

$$M_x = S_F(\text{COR}) \tag{3.28}$$

$$\text{COR} = (\text{CV}_x^2+1)^{1/2} \times e^{\left\{-Z_{(1-\alpha)}[\ln(\text{CV}_x^2+1)]^{1/2}\right\}} \tag{3.29}$$

where

M_x = mean constituent value such as BOD_5, total suspended solids (TSS), phosphorus, nitrogen (ammonia, nitrate), chlorine residual (all in mg/L), or total coliform count (number/100mL)

S_F = a fixed standard such as $BOD_5 \leq 10$ mg/L (this may be given in the effluent standards)

COR = coefficient of reliability

CV_x = coefficient of variation (Eq. 3.9). In the reliability concept, it is $(s_x/\overline{X}_x)$.

Selection of CV_x should be based on actual data from pilot testing, or the operational data of a similar facility

$(1-\alpha)$ = cumulative probability of occurrence (reliability level such as 99% of the time).

$Z_{(1-\alpha)}$ = number of standard deviations away from the mean of a normal distribution. The values of $Z_{(1-\alpha)}$ for various cumulative probability levels $(1-\alpha)$, may be found in several sources. Common values of $Z_{(1-\alpha)}$ for various cumulative probability levels are summarized in Table 3.17.[14–16]

The coefficient of reliability (COR) is calculated from Eq. (3.29) for different cumulative probability levels $(1-\alpha)$. Tabular values of this relationship are available in several sources. Table 3.18 provides the typical values commonly used for wastewater treatment facilities.[16,17]

TABLE 3.17
Values of $Z_{(1-\alpha)}$ for Different Cumulative Probability $(1-\alpha)$

Cumulative Probability $(1-\alpha)$	Percentile $Z_{(1-\alpha)}$
99.9	3.090
99	2.326
95	1.645
90	1.282
80	0.842
70	0.525
60	0.253
50	0

Source: Adapted in part from Refs 14–16.

TABLE 3.18
Coefficient of Reliability (COR), for Different Values of Coefficient of Variation CV_X

	Coefficient of Reliability, %					
CV_X	50	80	90	95	99	99.9
0.3	1.04	0.81	0.71	0.64	0.53	0.42
0.4	1.08	0.78	0.66	0.57	0.44	0.33
0.5	1.12	0.75	0.61	0.51	0.37	0.26
0.6	1.17	0.73	0.57	0.47	0.32	0.21
0.7	1.22	0.72	0.54	0.43	0.28	0.17
0.8	1.28	0.71	0.52	0.40	0.25	0.15
0.9	1.35	0.70	0.50	0.38	0.22	0.12
1.0	1.41	0.70	0.49	0.36	0.20	0.11
1.2	1.56	0.70	0.46	0.33	0.17	0.08
1.5	1.80	0.70	0.45	0.30	0.14	0.06

Source: Adapted in part from Refs. 16,17

Example 3.29: Reliability Analysis

A nitrification facility must meet the new effluent NH_3–N requirement of 1 mg/L or less. Calculate the maximum allowable NH_3–N concentration in the effluent to achieve a reliability level of 95%. One-year pilot plant test data is given below.

Month	Jan	Feb	Mar	Apr	May	Jun	Jul	Aug	Sep	Oct	Nov	Dec
Monthly Average NH_3–N Concentration, mg/L	1.9	2.1	1.7	1.4	1.1	0.9	0.7	0.8	1.2	1.4	1.6	1.8

Solution

1. Tabulate the pilot plant test data (Table 3.19)
2. Calculate mean and standard deviation using Eqs. (3.3) and (3.8)

$$\bar{X} = 16.6/12 = 1.4 \text{ mg/L}$$

$$s = \sqrt{\frac{2.26}{12 - 1}} = 0.4 \text{ mg/L}$$

3. Calculate the CV using the results of statistical analysis

$$CV_x = 0.4/1.4 = 0.3$$

4. Calculate percentile $Z_{(1-\alpha)}$

   From Table 3.17 corresponding to cumulative probability of 95%, value of $Z_{(1-\alpha)} = 1.645$

TABLE 3.19
Calculation Table to Determine Standard Deviations (Example 3.29)

Month	Monthly Average NH_3–N Concentration X_i, mg/L	$(X_i - \bar{X})$, mg/L	$(X_i - \bar{X})^2$, $(mg/L)^2$
Jan	1.9	0.5	0.25
Feb	2.1	0.7	0.49
Mar	1.7	0.3	0.09
Apr	1.4	0.0	0.0
May	1.1	−0.3	0.09
Jun	0.9	−0.5	0.25
Jul	0.7	−0.7	0.49
Aug	0.8	−0.6	0.36
Sep	1.2	−0.2	0.04
Oct	1.4	0.0	0.0
Nov	1.6	0.2	0.04
Dec	1.8	0.4	0.16
Total	16.6	−0.2	2.26

5. Calculate the COR using Eq. (3.29)

$$\text{COR} = (0.3^2 + 1)^{1/2} \times e^{\left\{-1.65\,[\ln(0.3^2+1)]^{1/2}\right\}} = 0.64$$

Note: COR=0.64 may also be obtained from Table 3.18.

6. Calculate the maximum allowable NH_3–N concentration (M_x) using Eq. (3.28)
   M_x=(1 mg/L)(0.64)=0.64 mg/L
   The nitrification facility should be designed to achieve the maximum effluent NH_3–N concentration of 0.64 mg/L or less to meet the effluent NH_3–N concentration of 1 mg/L or less.

3.4 REVIEW PROBLEMS AND DISCUSSION TOPICS

3.1 One liter of wastewater samples was collected every hour at an industrial plant. The hourly flow record is given below. Calculate the volume of samples to be pooled into a container to give 3 L of 24-h composite sample.

Time	12:00 MN	1:00 AM	2:00 AM	3:00 AM	4:00 AM	5:00 AM	6:00 AM	7:00 AM	8:00 AM	9:00 AM	10:00 AM	11:00 AM
Flow rate, ft^3/h	120	100	100	90	80	80	75	125	270	380	500	725
Time	12:00 N	1:00 PM	2:00 PM.	3:00 PM	4:00 PM	5:00 PM	6:00 PM	7:00 PM	8:00 PM	9:00 PM	10:00 PM	11:00 PM
Flow rate, ft^3/h	775	750	800	900	950	875	850	685	550	400	250	150

MN = midnight, N = noon.

3.2 An industrial wastewater stream is equipped with an in-situ continuous monitoring of flow and ammonia nitrogen concentration. The 2-h interval flow and ammonia nitrogen concentration obtained from the in-situ monitoring are summarized below:

Time	2:00 AM	4:00 AM	6:00 AM	8:00 AM	10:00 AM	12:00 AM	2:00 PM	4:00 PM	6:00 PM	8:00 PM	10:00 PM	12:00 MN
Flow, m^3/h	20	30	40	50	60	60	40	35	45	55	35	10
NH_3–N conc., mg/L	45	50	60	45	70	80	90	65	55	75	85	35

Determine (1) the average daily flow and ammonia nitrogen concentration, (2) develop the 24-h in-situ flow and ammonia nitrogen concentration plots, and (3) indicate on this plot the average daily flow, and average ammonia nitrogen concentration.

3.3 A pipeline discharges 0.5 m^3/s of constant flow into a circular collection tank with a surface area of 4,500 m^2. The initial depth of water in the tank is 0.2 m. Determine the value of constants a and b in equation $D = a + bt$ by plotting the water depth with respect to 1, 2, 3, 4, and 5 h.

3.4 10 flow measurements of a wastewater stream were recorded in an interceptor. The flow data are given below. Determine the simple moving average of order 3 by sequence, and plot individual data points and moving averages.

Measurement no.	1	2	3	4	5	6	7	8	9	10
Recorded flow, L/s	220	160	190	150	200	185	215	165	230	210

3.5 Draw the frequency distribution curve of the following 40 flow measurement data. Select a class interval of 5 L/s.

138	164	150	132	144	125	149	157
146	158	140	147	136	148	152	144
168	126	138	176	163	119	154	165
146	173	142	147	135	153	140	135
161	145	135	142	150	150	145	128

3.6 Calculate mean and standard deviation of 100 data samples of phenol in a wastewater stream. Frequency table is given below. Also, calculate the range of phenol concentration in which 68% of the time the data will fall. Draw the histogram.

Phenol Conc., mg/L	Class-Mark	Frequency
60–62	61	5
63–65	64	18
66–68	67	42
69–71	70	27
72–74	73	8

3.7 The frequency distribution of TSS concentration in a wastewater stream is given below. Prepare the percent relative-frequency distribution table, histogram, and frequency polygon.

TSS conc., mg/L	170–174	175–179	180–184	185–189	190–194	195–199	200–204	205–209	210–214	215–219	220–224
Frequency	0	3	6	9	11	14	12	8	5	2	0

3.8 Develop the cumulative and percent cumulative-frequency distribution table and ogives for the TSS data given below. Also, prepare cumulative-frequency distribution of all values greater than or equal to the lower class boundary.

TSS concentration, mg/L	181–185	186–190	191–195	196–200	201–205	206–210	211–215
Frequency	0	3	8	10	7	5	0

3.9 Calculate the geometric mean of the following data set, which represents effluent NH_3–N concentration from a wastewater treatment plant.

Sample	1	2	3	4	5	6	7
Effluent NH_3–N, mg/L	1.6	2.2	2.4	1.9	1.4	2.1	1.8

3.10 Calculate the median, mode, and arithmetic mean of the following numbers:
80, 76, 54, 42, 95, 42, 87, 86, 85, 63

3.11 The following flow data was obtained from a wastewater treatment plant. Determine graphically the arithmetic mean, standard deviation, and CV. Also, prepare the histogram and calculate the mean, standard deviation, CV, and other statistical relationships using various statistical equations.

Day	1	2	3	4	5	6	7	8
Flow data, gpm	360	410	372	423	394	408	430	443

3.12 BOD removal was carried out by two processes: (a) Trickling filter, and (b) activated sludge. Results are given below. Compare the performance of the two processes.

	BOD Removal, %							
Trickling filter	88	87	86	92	89	87	79	84
Activated sludge	84	80	87	91	82	90	85	88

3.13 Determine the regression line of X (15 min COD), Y (BOD_5), and coefficient of regression. The results of 15-min COD and BOD_5 tests are given below:

15 min COD, mg/L	240	350	420	300	400	180	220	430	250	320	260	190
BOD_5, mg/L	170	250	370	250	290	140	160	375	190	275	180	140

3.14 The results of two duplicate tests for effluent TSS in a wastewater sample are given below. Calculate the RPD.
Result 1: 8.4 mg/L Result 2: 7.8 mg/L

3.15 A wastewater effluent sample was tested for total phosphorus. The results of four replicates are given below. Calculate RSD.

Result	1	2	3	4
Total phosphorus, mg/L	6.5	7.2	7.6	6.9

3.16 Sixteen wastewater samples were analyzed in the lab to determine the analytical precision. Eight samples were spiked with 0.5 mg/L NH_3–N, while the other eight were not spiked. The total NH_3–N concentrations were measured in all 16 samples. Calculate the precision of the analytical procedure based on spiking data.

Sample No.	**1**	**2**	**3**	**4**	**5**	**6**	**7**	**8**
Concentration of NH_3–N in unspiked portion of sample, mg/L	2.1	1.8	2.4	3.1	3.4	2.7	1.9	2.2
Concentration of NH_3–N in spiked portion of sample, mg/L	2.6	2.3	2.7	3.5	4.0	2.9	2.2	2.4

3.17 The concentration of chloroform in a wastewater sample is 0.50 mg/L. The sample was spiked with 0.10 mg/L chloroform. The measured concentration of chloroform in the spiked sample was 0.62 mg/L. Calculate the percent deviation recovery from spike, and percent recovery of spike.

3.18 A standard solution with a 0.20 mg/L concentration of total organic carbon was analyzed six times in a laboratory. The results are as given below. Determine the MDL

Replicates	1	2	3	4	5	6
Measured concentration, mg/L	0.193	0.220	0.196	0.188	0.211	0.205

3.19 A wastewater treatment plant must meet the new effluent NO_3–N requirement of 10 mg/L or less. Calculate the maximum allowable NO_3–N concentration in the effluent to achieve the reliability level of 99%. A one-year pilot plant test data is given below

Month	Jan	Feb	Mar	Apr	May	Jun	Jul	Aug	Sept	Oct	Nov	Dec
Monthly avg. NO_3–N conc., mg/L	8.5	9.2	11.3	13.4	11.6	12.4	14.1	9.8	10.7	11.5	12.1	10.5

REFERENCES

1. APHA, AWWA, and WEF, *Standard Methods for the Examination of Water and Wastewater*, 19th ed., American Public Health Association, Washington, DC, 1991.
2. U.S. Environmental Protection Agency, *Handbook for Monitoring Industrial Wastewater*, Technology Transfer by Associated Water and Air Resources Engineers, Inc., Nashville, TN, 1973.
3. Tomar, M., *Quality Assessment of Water and Wastewater*, Lewis Publishers, Boca Raton, FL, 1999.
4. Water Quality Division, TNRCC, *Surface Water Quality Monitoring Procedures Manual*, Texas Natural Resource Conservation Commission, June 1999.
5. Fresenius W., K. E. Quentin, and W. Schneider (Editors), *Water Analysis, Deutsche Gesellschaft für*, Technische Zusammenarbeit, Germany, 1988.
6. Berthouex, P. M., and L. C. Brown, *Statistics for Environmental Engineers*, Lewis Publishers, Boca Raton, FL, 1994.
7. Winerm, B. J., D. R. Brown, and K. M. Michels, *Statistical Principles in Experimental Design*, 3rd ed., McGraw-Hill, Inc., New York, 1991.
8. Spiegel, M. R., and L. J. Stephens, *Statistics*, 3rd ed., Schaum's Outline Series, McGraw-Hill, a Division of McGraw-Hill Companies, New York, 1999.
9. Walpole, R. E., and R. H. Myers, *Probability and Statistics for Engineers and Scientists*, Macmillan Publishing Co., New York, 1993.
10. Lee, C. C., and S. D. Lin, *Handbook of Environmental Engineering Calculations,* McGraw-Hill, A Division of McGraw-Hill Companies, New York, 1999.
11. Simes, G. F., *Preparation Aids for the Development of Category I, Quality Assurance Project Plans*, EPA/600/8–91/003, Risk Reduction Engineering Laboratory, Office of Research and Development, US Environmental Protection Agency, February 1991.
12. U. S. Environmental Protection Agency, *Preparation Aids for the Development of Category III Quality Assurance Project Plans*, EPA/600/8–91/005, Office of Research and Development, Washington, DC, February 1991.
13. U. S. Environmental Protection Agency, *EPA Guidance For Quality Assurance Project Plans, External Working Draft*, EPA QA/G-5, Quality Assurance Division, U.S. EPA, Washington, DC, November 1996.
14. Metcalf and Eddy, Inc., *Wastewater Engineering: Treatment, Disposal, and Reuse*, 4th ed., Mc-Graw-Hill, Inc. New York, 1991.
15. Crites, R., and G. Tchobanoglous, *Small and Decentralized Wastewater Management Systems*, WCB McGraw-Hill, Boston, MA, 1998.
16. Niku, S., E. D. Shroeder, and F. J. Samaniego, Performance of Activated Sludge Processes and Reliability-Based Design, *Journal Water Pollution Control Federation*, vol. 51, no. 12, 1979, pp. 2841–2857.
17. Qasim, S. R., and G. Zhu, *Wastewater Treatment and Reuse: Theory and Design Examples*, vol. 1, *Principles and Basic Treatment*, CRC Press, Taylor & Francis Group, Baca Raton, FL, 2018.

4 Sources, Flow Rates and Characteristics of Municipal Wastewater, and Flow Equalization

4.1 CHAPTER OBJECTIVES

Municipal wastewater is the general term applied to the liquid wastes collected from residential, commercial, and industrial areas of a city. In addition to these wastes, groundwater and surface runoff may also enter the collection system. All these flows are conveyed by means of a sewerage system to a central location for treatment. There is a wide variation in seasonal, daily, and hourly flow rates of the municipal wastewater, and several flow terms are used to identify these flow variations and characterize the municipal wastewaters.

The purpose of this chapter is to present the flow and major contaminants of municipal wastewater. The following topics are covered:

- Relationship between municipal water demand and wastewater flow
- Sources and components of water usage and wastewater generation
- Wastewater flow variations, infiltration and inflow, and definition of common terms
- Design flow projections and sustained flows
- Physical, chemical, and biological quality of municipal wastewater
- Flow equalization

4.2 RELATIONSHIP BETWEEN MUNICIPAL WATER DEMAND AND WASTEWATER FLOW

Under dry weather conditions, municipal wastewater is derived largely from municipal water supply. A close relationship exists between water usage and wastewater flow in a community. The following observations outline the interdependence between water usage and wastewater flow[1–5]:

1. Under dry weather conditions, the daily water demand and wastewater flow show a diurnal flow pattern.
2. The diurnal wastewater flow curve closely parallels that of water demand, with a lag of several hours. This relationship is shown in Figure 4.1.
3. The fluctuations in wastewater flow are less than those of the water supply.
4. A considerable portion of the water supply may not reach the sewers. This includes (1) lawn sprinkling, (2) street washing and firefighting, (3) leakage from the water mains and service pipes, (4) water consumed in products and manufacturing processes, and (5) homes and establishments that may use city water but utilize on-site wastewater treatment and disposal systems.
5. Establishments using private water supply may increase average wastewater flow.
6. Infiltration/inflow may enter the sewers during wet weather conditions and may increase the flow substantially over a short period.

DOI: 10.1201/9781003629399-4

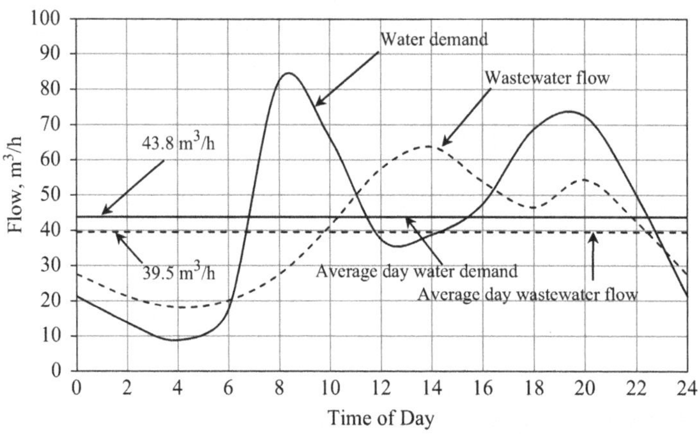

FIGURE 4.1 Typical diurnal municipal water demand and wastewater flow. (Printed from Ref. 1 with permission of CRC Press, Taylor & Francis Group)

7. Exfiltration may occur during dry weather, causing a reduction in average wastewater flow.
8. Annual average wastewater flow may vary from 60% to 130% of the annual average water demand. Designers frequently assume that annual average wastewater flow equals 80%–100% percent of annual average water usage in a community.

Example 4.1: Relationship between Diurnal Water Demand and Wastewater Flow

The diurnal wastewater flow pattern measured at a wastewater treatment plant is similar to that of water demand but exhibits less fluctuation and shows a lag of few hours. Explain why this happens?

Solution

The diurnal wastewater flow has a lag of few hours due to the following:

a. The water lines are under pressure. Any usage in the distribution system transmits a pressure wave to the supply point. The pressure wave travels at the speed of sound. As a result, the water withdrawal is almost instantly recorded.
b. The sewer lines flow under gravity (open channel flow).
c. The flow from different parts of the city takes different time to reach a central location (treatment plant or lift station).
d. As the flow increases, the depth of flow in the sewer lines also increases. As a result, the increase in velocity in sewer is relatively small. This results in a built-in storage capacity within the sewer lines and has a damping effect upon the peaking factor.
e. As the length of collection system increases, these effects further reduce the peaking factor.

Example 4.2: Ratios of Water Demand and Wastewater Flow from Diurnal Flow Data

The diurnal flow pattern of municipal water demand and wastewater flow for a typical dry day is given in Figure 4.1. Calculate the following:

a. Average-day water demand,
b. Average-day wastewater flow,
c. Ratio of average wastewater flow and water demand,
d. Ratio of peak and average wastewater flow, and
e. Ratio of peak and average water demand.

Solution

1. The trapezoidal rule is generally used to calculate the areas under the demand curves. The trapezoidal rule is given by Eq. (4.1).

$$A = \frac{1}{2} \times \Delta T \left[q_1 + 2\left(q_2 + q_3 + q_4 + \cdots + q_{n-1}\right) + q_n \right] \tag{4.1}$$

where
A = area under the curve, m^3
ΔT = constant time increment, h
$q_1, q_2, \ldots, q_n$ = flows at various time intervals, m^3/h

2. Calculate the daily average water demand from the area under water demand curve.

$$A_1 = \frac{1}{2} \times 2\text{ h}\left[21\text{ m}^3/\text{h} + 2\times(12+8+18+85+65+36+39+48+69+74+50)\text{ m}^3/\text{h} + 21\text{ m}^3/\text{h}\right]$$

$$= 1{,}050\text{ m}^3 \text{ total in 24 h or } 43.8\text{ m}^3/\text{h in daily average}$$

3. Calculate the daily average water flow from the area under wastewater flow curve.

$$A_2 = \frac{1}{2} \times 2\text{ h}\left[28\text{ m}^3/\text{h} + 2\times(21+18+20+27+41+58+64+53+46+55+43)\text{ m}^3/\text{h} + 28\text{ m}^3/\text{h}\right]$$

$$= 948\text{ m}^3 \text{ total in 24 h or } 39.5\text{ m}^3/\text{h in daily average}$$

4. Calculate daily average flows (DAFs) and various ratios.
   a. Daily average water demand = 43.8 m^3/h
   b. Daily average wastewater flow = 39.5 m^3/h
   c. Ratio of average wastewater flow and water use $= \dfrac{39.5\text{ m}^3/\text{h}}{43.8\text{ m}^3/\text{h}} = 0.90$
   d. Ratio of peak and average wastewater flow $= \dfrac{64\text{ m}^3/\text{h}}{39.5\text{ m}^3/\text{h}} = 1.62$
   e. Ratio of peak and average water demand $= \dfrac{85\text{ m}^3/\text{h}}{43.8\text{ m}^3/\text{h}} = 1.94$

4.3 COMPONENTS OF MUNICIPAL WATER DEMAND

Municipal water demand in the United States cities varies from 300 to 1,800 liters per capita per day (Lpcd) or 79–476 gallons per capita per day (gpcd).[1,2,6,7] The various components of municipal water use are (1) residential or domestic, (2) commercial, (3) industrial, (4) institutional and public, and (5) water lost or unaccounted for.

The relative proportions of these uses with respect to overall municipal water demand are summarized in Table 4.1. General discussion is presented below.

TABLE 4.1
Components of Municipal Water Demand and Percent Distribution

Components of Municipal Water Demand	Percent of Total
Residential or domestic	35–50
Commercial	10–25
Industrial	15–30
Institutional and public	5–20
Water lost or unaccounted for	5–10

Source: Adapted in part from Refs.1,2

TABLE 4.2
Typical Breakdown and Flow Rates of Residential Water Uses

Types of Water Use	Non-conserving Home Usage, %	Flow Rate
Toilet flush, including toilet leakage	33	Tank type 19–27 L/use Valve type 90–110 L/min
Shower and bathing		Shower head 90–110 L/use or 19–40 L/min
Shower	20	Tub bath 60–90 L/use
Bathing	8	
Wash basin	11	4–8 L/use
Kitchen	9	
Drinking, cooking	2–6	Kitchen sink
Dishwashing	3–5	15–30 L/dishwasher load
Garbage disposal	0–6	6,000–7,500 L/week, 4–8 L/person d
Laundry and washing machine	16	110–200 L/load
Lawn	3	Sprinkler system

Source: Adapted in part from Refs. 2,4,8–10

4.3.1 Residential or Domestic Water Use

The residential or domestic water uses include toilet flush, bathing and washing, cooking and drinking, lawn watering, and others. Typical breakdown of residential water uses and flow rates is summarized in Table 4.2. The average residential water demand varies from 300 to 450 Lpcd. With flow reduction devices, the residential water demand for many types of residential establishments is decreasing.[1,2]

4.3.2 Commercial Water Use

Commercial establishments include motels, hotels, office buildings, shopping centers, service stations, movie houses, airports, and the like. The water use in commercial establishments varies greatly depending upon the size of operation. The commercial water demand may be estimated from unit loading or floor area. Typical unit loadings of residential and commercial establishments are provided in Table 4.3.

TABLE 4.3
Average Water Demand in Residential, Institutional, Commercial, and Industrial Establishments

Source	Unit	Unit Flow, m^3/unit·d[a]
Residential		
Single-family detached, low income to high income	Person	0.25–0.38
Apartment	Person	0.18–0.23
Trailer park	Person	0.15
Commercial		
Country club		
Resident	Member	0.38
Nonresident	Member	0.10
Hotel/Motel	Unit guest	0.38
Resort	Unit guest	0.19
Restaurant	Customer	0.03
Store	Toilet room	1.52
	Employee	0.04
Department store and shopping center	Employee	0.04
	Per m^2 floor area	0.001–0.002
Office building and complex	Employee	0.065
	Per m^2 floor area	0.015
Movie theater	Seat	0.008
Laundromat	Machine	2.5
Barber shop	Chair	0.065
Beauty salons	Station	1.026
Service station	First bay	3.8
	Additional bays	1.9
Industrial		
Industrial building	Employee	0.055
Factories		
With shower	Employee-shift	0.133
Without shower	Employee-shift	0.095
Light industrial zone	ha	9–14
Medium industrial zone	ha	14–30
Heavy industrial zone	ha	30–100
Industrial products		
Cattle	Head	0.04–0.05
Dairy	Head	0.07–0.08
Chicken	Head	0.03–0.04
Canning	Metric ton/d	30–60 m^3/unit
Dairy, milk	Metric ton/d	2–3 m^3/unit
Meat packing	Metric ton/d	15–25 m^3/unit
Pulp and paper	Metric ton/d	200–800 m^3/unit
Steel	Metric ton/d	260–300 m^3/unit
Tannery	Metric ton of raw hide processed/d	60–70 m^3/unit
Institutional and Public Uses		
Nursing homes	Bed	0.38
Hospital	Bed	0.95

(Continued)

TABLE 4.3 (*Continued*)
Average Water Demand in Residential, Institutional, Commercial, and Industrial Establishments

Source	Unit	Unit Flow, m^3/unit·d[a]
Prison	Inmate	0.45
School		
Boarding	Student	0.3
Day	Student	0.076

Source: Adapted in part from Refs. 1–5,8–10

[a] All unit flows are in m^3/unit·d unless indicated.

$m^3 \times 264.5$ = gallon.

4.3.3 Institutional and Public Water Use

Water uses in public buildings and institutional establishments (city halls, prisons, hospitals, schools, etc.) – as well as water used for public services (fire protection, street washing, park irrigation, and the like) – are considered public water uses. Some residential, commercial, and institutional water uses are provided in Table 4.3.

4.3.4 Industrial Water Use

Industrial water demands in the U.S. are very large. Generally, large industries develop their own water supply systems. Only small industries purchase water and therefore impose demand on local municipal system. The industrial water demand may be estimated on the basis of proposed industrial zoning and unit loadings for specific industries. Unit loadings for many industries are summarized in Table 4.3.

4.3.5 Water Unaccounted for or Lost

Major sources of water unaccounted for are leaks from mains, faulty meters, and unauthorized connections. This loss mostly depends on the condition of the water distribution system. In many developing nations where individual water connections are unmetered, as much as 30%–50% water supplied may be lost or unaccounted for due to old and leaky distribution systems and wastage from unmetered connections. To reduce these losses in many countries, water is supplied only for a few hours in the mornings and in the evenings. This practice does save water, but may cause serious contamination of the water supply due to the entry of polluted surface and groundwater into the distribution system during the periods when the supply is interrupted.

4.3.6 Factors Affecting Water Use

The water usage in a community depends upon many factors. Many of these factors are as follows:

1. Geographical location and climate. More water is used in a hot and dry climate or region than in a wet, humid, and cold climate or region.
2. Size, population density, and economic conditions of the community. Wealthier and sparsely populated communities have higher water demand.

3. Industrialization increases water demand. Industries use large amounts of water, especially seasonal industries such as vegetable canneries and others.
4. Metered water supply and cost have a significant effect on water demand. Less water is used where water supply is individually metered. Also, water usage is closely related to the cost of water.
5. Water pressure has a significant effect on water usage. Higher supply pressure may cause unnecessary waste due to spurts from faucets, leaks, and drips.
6. Water conservation efforts of general public and regulatory agencies in water conservation have resulted in 20%–30% savings. These savings are achieved by pressure-reducing valves, faucet aerators, flow-limiting shower heads, shallow-trap water closet, level-controlled washing machines, and education of consumers.[1,2,7–10]

Example 4.3: Municipal Water Demand and Components

A city has a population of 30,000 residents. The residential water demand is 210 Lpcd. The commercial, industrial, institutional, and public water uses, and water lost or unaccounted for, are respectively 18%, 23%, 12%, and 6% of the total municipal water demand. Calculate total municipal water demand in m^3/d and mgd, and the demand for each component.

Solution

1. Calculate residential water demand.

$$\text{Residential water demand} = 30{,}000\ \text{residents} \times \frac{210\ \text{L}}{\text{person} \cdot \text{d}} \times \frac{\text{m}^3}{1{,}000\ \text{L}} = 6{,}300\ \text{m}^3/\text{d or } 1.67\ \text{mgd}$$

2. Calculate total municipal water demand.

   Total demand for commercial, industrial, institutional, and public uses, and water lost or unaccounted for = (18% + 23% + 12% + 6%) = 59% of total municipal water demand

   Residential water demand = (100 − 59)% = 41% of municipal water demand

$$\text{Total municipal water demand} = \frac{6{,}300\ \text{m}^3/\text{d residential}}{0.41\ \text{residential/total municipal}} = 15{,}366\ \text{m}^3/\text{d}$$

3. Calculate water demand for each component.

   Water demand for each component is tabulated below.

Water Use	Percent of Total	Water Demand	
		m^3/d	Mgd
Residential	41	6,300	1.67
Commercial	18	2,766	0.73
Industrial	23	3,534	0.93
Institutional and public use	12	1,844	0.49
Water lost or unaccounted for	6	922	0.24
Total	100	15,366	4.06

1 mgd = $3.78 \times 10^3\ m^3/d$.

Example 4.4: Water Demand and Wastewater Flow

A subdivision of a suburban city is being developed. The ultimate zoning plan shows the following residential, commercial, industrial, and institutional establishments. Using the average values given in Table 4.3, estimate the following:

a. Annual average water demand for the entire subdivision in m^3/d;
b. Residential, commercial, industrial, and institutional water demands and their percentages with respect to total municipal water demand; and
c. Annual average wastewater flow if 85% average water supply is returned into the municipal sewers.

Water Using Establishment	Sources of Water Demand
Residential	
Single-family detached residential population	1,000 residents
Apartments	1,650 residents
Trailer park	400 residents
Commercial	
One hotel/motel	250 units
Five restaurants, total seating	250 customers
Shopping centers	288 employees
Office buildings	500 employees
Office complexes	600 m^2
One movie theater	200 seats
One commercial laundry	40 machines
Barber shops	10 chairs
Beauty salons	20 stations
One service station	6 bays
Industrial	
Factories with shower	100 employee-shifts
Light industrial zone	2 ha
One feedlot	1,000 heads
One canning plant	4 metric tons
One meat packaging plant	3 metric tons
Institutional	
One nursing home	50 beds
One hospital	200 beds
Two daytime schools	1,200 students

Solution

1. Estimate the residential population of the subdivision.
   Single-family detached housing residents = 1,000
   Apartment residents = 1,650
   Trailer park residents = 400
   Total = 3,050 residents
2. Estimate residential demand.
   Single-family detached housing = 1,000 residents × 0.315 m^3/person d = 315 m^3/d

Apartments = 1,650 residents × 0.205 m^3/person d = 338 m^3/d
Trailer park = 400 residents × 0.15 m^3/person d = 60 m^3/d
Total residential demand = 713 m^3/d

3. Estimate commercial water demand.
Hotel/motel = 250 units × 0.38 m^3/unit d = 95 m^3/d
Restaurant = 250 customers × 0.03 m^3/customer d = 7.5 m^3/d
Shopping centers = 288 employees × 0.04 m^3/employee d = 11.5 m^3/d
Office building = 500 employees × 0.065 m^3/employee d = 32.5 m^3/d
Office complexes = 600 m^2 × 0.015 m^3/employee d = 9 m^3/d
Movie theater = 200 seats × 0.008 m^3/seat d = 1.6 m^3/d
Laundromat = 40 machines × 2.5 m^3/machine d = 100 m^3/d
Barber shops = 10 chairs × 0.065 m^3/chair d = 0.65 m^3/d
Beauty salons = 20 stations × 1.026 m^3/station d = 20.5 m^3/d
Service station
Fist bay = 1 bay × 3.8 m^3/bay d = 3.8 m^3/d
Additional bays = 5 bays × 1.9 m^3/bay d = 9.5 m^3/d
Total commercial demand = 292 m^3/d

4. Estimate industrial water demand.
Factories with shower = 100 employee-shifts × 0.133 m^3/employee-shift = 13 m^3/d
Light industrial zone = 2 ha × 11.5 m^3/ha d = 23 m^3/d
Feedlot = 1,000 heads × 0.045 m^3/head d = 45 m^3/d
Canning plant = 4 metric tons/d × 45 m^3/metric ton = 180 m^3/d
Meat packaging plant = 3 metric tons/d × 20 m^3/metric ton = 60 m^3/d
Total industrial water demand = 321 m^3/d

5. Estimate institutional water demand.
Nursing home = 50 beds × 0.38 m^3/bed d = 19 m^3/d
Hospital = 100 beds × 0.95 m^3/bed d = 95 m^3/d
School = 1,200 students × 0.076 m^3/student d = 91.2 m^3/d
Total institutional water demand = 205 m^3/d

6. Estimate annual municipal water demand and its percentages.

Components	Demand	Percentage
Residential	= 713 m^3/d	47
Commercial	= 292 m^3/d	19
Industrial	= 321 m^3/d	21
Institutional	= 205 m^3/d	13
Total	**= 1,531 m^3/d**	**100**

$$\text{Average municipal water demand} = 1{,}531\ \text{m}^3/\text{d} \times \frac{1{,}000\ \text{L}}{\text{m}^3} \times \frac{1}{3{,}050\ \text{residents}} = 502\ \text{Lpcd (133 gpcd)}$$

Note: 1 gal = 3.78 L

7. Estimate annual wastewater flow

$$\text{Annual wastewater flow} = 0.85 \times 1{,}531\ \text{m}^3/\text{d} = 1{,}301\ \text{m}^3/\text{d or 427 Lpcd (112 gpcd)}$$

Example 4.5: Water Saving from Supply Pressure Reduction

In a community of 50,800 residents, the average water demand is 620 Lpcd, and supply pressure is 414 kN/m^2. The average water lost through cracks at this pressure is 15%. How much savings would be achieved if the supply pressure is reduced to 276 kN/m²? Assume cracks behave like orifices.

Solution

1. Estimate average water demand and average water lost through the cracks at a pressure of 414 kN/m².

$$\text{Average water demand} = 50{,}800\text{ people} \times \frac{620\text{ L}}{\text{person} \cdot \text{d}} \times \frac{\text{m}^3}{1{,}000\text{ L}} = 31{,}500\text{ m}^3/\text{d}$$

$$\text{Water lost through cracks} = 0.15 \times 31{,}500\text{ m}^3/\text{d} = 4{,}725\text{ m}^3/\text{d}$$

2. Select orifice equation
   Velocity and discharge through an orifice are expressed by Eqs. (4.2) and (4.3).

$$V = \sqrt{2gh} \tag{4.2}$$

$$q = C_d AV = C_d A\sqrt{2gh} \tag{4.3}$$

where

V = velocity through an orifice, m/s
q = discharge through an orifice, m³/s
C_d = coefficient of discharge (typical value for orifice is 0.61)
A = area of orifice, m²
g = acceleration due to gravity, m/s²
h = water head over the orifice, m

3. Calculated water heads at pressures of 414 and 276 kN/m².
   There is a direct conversion between the pressure head in kN/m² and water head in m based on specific weight of water. At 4°C (39°F), the specific weight for water is 9.81 kN/m³ (62.4 lb/ft³) and the static head of water in m = $\frac{\text{Pressure in kN/m}^2}{9.81\text{ kN / m}^3}$

   Water head at pressure of 414 kN/m², $h_{414} = \frac{414\text{ kN/m}^2}{9.81\text{ kN/m}^3} = 42.2\text{ m}$

   Water head at pressure of 276 kN/m², $h_{276} = \frac{276\text{ kN/m}^2}{9.81\text{ kN/m}^3} = 28.1\text{ m}$

4. Calculate the loss of water through the cracks at pressures 414 and 276 kN/m².

$$q_{414} = C_d A\sqrt{2gh_{414}}$$

$$q_{276} = C_d A\sqrt{2gh_{276}}$$

$$\frac{q_{414}}{q_{276}} = \frac{C_d A\sqrt{2gh_{414}}}{C_d A\sqrt{2gh_{276}}} = \sqrt{\frac{h_{414}}{h_{276}}} = \sqrt{\frac{42.2\text{ m}}{28.1\text{ m}}} = 1.23$$

   Water lost through the cracks at 414 kN/m², $q_{414} = 4{,}725\text{ m}^3/\text{d}$

Water lost through the cracks at 276 kN/m²,

$$q_{276} = \frac{q_{414}}{1.23} = \frac{4{,}725 \text{ m}^3/\text{d}}{1.23} = 3{,}841 \text{ m}^3/\text{d}$$

5. Calculate savings

   Savings achieved by the reduction of supply pressure from 414 to 276 kN/m² = 4,725 − 3841 m³/d = 884 m³/d or 17.4 Lpcd or 2.8%

Example 4.6: Water Saving from Pressure-Reducing Valve (PRV)

A home has installed a pressure-reducing value (PRV) at the point of entry into the home. The PRV reduces supply pressure from 500 to 300 kN/m². The water usage before PRV is 400 Lpcd. Calculate the typical water saving. Make use of the information provided in Table 4.2.

Solution

1. Select from Table 4.2 the residential water uses that are affected by PRV.

   The toilet flush tank has a fixed volume, and the washing machine and dishwasher have a level controller. Therefore, installation of the PRV will not reduce water use in these categories. Only those uses that are directly from the faucet will be affected. These uses are (1) shower 20% (bathing 8% not affected), (2) wash basin 11%, (3) kitchen assume 4% (out of 9%), and (4) lawn sprinkling 3%.
2. Calculate water uses in the home directly affected by PRV.

   Total percentage affected = (20 + 11 + 4 + 3)% = 38%

   Total usage affected = 0.38 × 400 Lpcd = 152 Lpcd
3. Calculate ratio of usage before and after PRV.

$$\frac{q_{500}}{q_{300}} = \frac{C_d A\sqrt{2gh_{500}}}{C_d A\sqrt{2gh_{300}}} = \sqrt{\frac{\left(\dfrac{P_{500}}{9.81 \text{ kN/m}^3}\right)}{\left(\dfrac{P_{300}}{9.81 \text{ kN/m}^3}\right)}} = \sqrt{\frac{P_{500}}{P_{300}}} = \sqrt{\frac{500 \text{ kN/m}^2}{300 \text{ kN/m}^2}} = 1.29$$

4. Calculate water saving due to PRV.

$$q_{300} = \frac{q_{500}}{1.29} = \frac{152 \text{ Lpcd}}{1.29} = 118 \text{ Lpcd}$$

$$\text{Water saving, } \Delta q = q_{500} - q_{300} = 152 - 118 \text{ Lpcd} = 34 \text{ Lpcd}$$

$$\text{Overall water saving} = \frac{\Delta q}{q_{500}} = \frac{34 \text{ Lpcd}}{400 \text{ Lpcd}} \times 100\% = 8.5\%$$

Example 4.7: Water Conservation in Homes

A home has changed the conventional devices to water conservation devices. These devices and expected percent savings over conventional devices are given in Table 4.4. Calculate total water saving in Lpcd and overall percent saving. The average water use in the home before installation of these devices was 400 Lpcd.

TABLE 4.4
Water Saving Devices Expected Savings over Conventional Devices (Example 4.7)

Device	Expected Saving, %
Shallow-trap water closet (toilet)	35
Sink faucet aerator	2
Limiting shower head	10
Water-efficient dishwasher	5
Level-controlled washing machine	7

Solution

1. Select from Table 4.2 the residential water uses that are affected by new devices.
   The residential water uses that are affected by new devices are (1) toilet 33%, (2) wash basin 11%, (3) shower use is 20% (bathing 8% not affected), (4) dishwasher 4%, and (5) washing machine 16%.
2. Calculate savings achieved by new devices.
   The savings achieved from different new devices are summarized below.

Water Uses	Usage		Saving	
	%	Lpcd	%	Lpcd
Water closet (toilet)	33	132[a]	35	46.2[b]
Wash basins and sinks	11	44	2	0.9
Shower	20	80	10	8.0
Dishwasher	4	16	5	0.8
Washing machine	16	64	7	4.5
Total	84	332	—	60

[a] 400 Lpcd × 0.33 = 132 Lpcd.
[b] 132 Lpcd × 0.35 = 46.2 Lpcd.

Overall savings from water conservation devices = 60 Lpcd

$$\text{Percent overall saving} = \frac{60\ \text{Lpcd}}{400\ \text{Lpcd}} = 15\%$$

4.4 WASTEWATER FLOW

Municipal wastewater is derived largely from the water supply. A considerable portion of the water supply, however, does not reach the sewers. This includes water used for street washing, lawn sprinkling, firefighting, and leakages from water mains and service pipes. A small portion of water may also be consumed in products and manufacturing processes. Also, many homes and other establishments that are not served by a sewerage system may use the city water supply but utilize on-site wastewater treatment and disposal. On the other hand, infiltration/inflow, and water used by industries and residences that is obtained from privately owned sources, may increase the quantity of wastewater larger than the public water supply. In general, the average wastewater flow may vary from 60% to 130% of the water used in the community. Many designers frequently assume that the average rate of wastewater flow, including a moderate allowance for infiltration/inflow, equals the average rate of water consumption. Average wastewater flows from residential, commercial, industrial, institutional, and other sources may be obtained from careful consideration of local water consumption data or by using wastewater generation rates for other similar cities in that region.

4.5 WASTEWATER FLOW VARIATION

4.5.1 Dry Weather Flow

Similar to water demand, wastewater flow also varies with respect to time of day, day of the week, weather conditions, and season of the year. Under dry weather conditions, the daily wastewater flow shows a diurnal pattern, exhibiting a peak and a minimum in a 24-h period. The diurnal wastewater flow parallels that of water demand with a lag of a few hours (Figure 4.1).

Hourly peak and minimum dry weather flows are generally estimated from several equations and graphical relationships developed from case studies. The ratio of hourly peak to daily average, and hourly minimum to DAFs, depends upon the contributing population. Commonly used relationships are given by Eqs. (4.4)–(4.8).[1,2,4,5] Ratios of extreme flows to average daily flow (peaking factors) developed from various sources are given in Figure 4.2. This relationship clearly shows that the peaking factor decreases as population increases and approaches 1.0 when the population or size of the collection system is very large.

$$M_{max} = 1 + \frac{14}{4 + \sqrt{P}} \tag{4.4}$$

$$M_{max} = \frac{5}{P^{0.167}} \tag{4.5}$$

$$Q'_{max} = 3.2\left(Q'_{ave}\right)^{5/6} \tag{4.6}$$

$$\log M_{max} = -0.19 \log P' + 0.74 \tag{4.7}$$

$$\log M_{min} = 0.142 \log P' - 0.682 \tag{4.8}$$

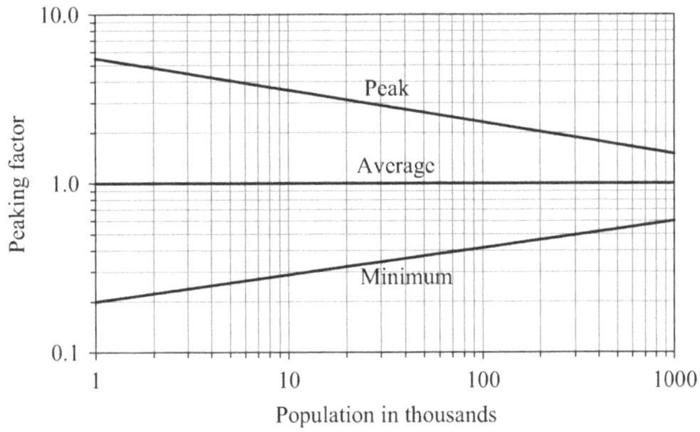

FIGURE 4.2 Ratios of extreme flows to average daily flow (peaking factor) for municipal wastewater under dry weather conditions. (Printed from Ref. 11 with permission of Water Environment Federation, and American Society of Civil Engineers.)

where

M_{max} = peaking factor, or ratio of hourly maximum to DAFs
P = population in thousands
Q'_{max} = hourly maximum dry weather flow, mgd
Q'_{ave} = daily average dry weather flow, mgd
M_{min} = ratio of hourly minimum to DAFs
P' = population in thousands, range $1 < P' < 10{,}000$

Example 4.8: Peak Dry Weather Wastewater Flow Calculated from Different Equations

A city has a population of 20,000 residents. The average wastewater flow is 380 Lpcd. Calculate (1) daily average and hourly peak dry weather flows in m³/d and mgd, (2) hourly minimum flow in m³/d, and (3) compare the results.

Solution

1. Calculate daily average wastewater flow.

$$\text{Daily average wastewater flow} = \frac{380\text{ L}}{\text{person d}} \times 20{,}000\text{ persons} \times \frac{\text{m}^3}{1{,}000\text{ L}} = 7{,}600\text{ m}^3/\text{d}$$

$$= 7{,}600\text{ m}^3/\text{d} \times \frac{1{,}000\text{ L}}{\text{m}^3} \times \frac{\text{gal}}{3.8\text{ L}} = 2\times 10^6\text{ gpd}$$

$$= 2\times 10^6\text{ gpd} \times \frac{\text{millon gallon (MG)}}{10^6\text{ gal}} = 2\text{ mgd}$$

2. Calculate hourly peak dry weather flows from Eqs. (4.4) through (4.7).

Peaking factor from Eq. (4.4), $M_{max} = 1 + \dfrac{14}{4+\sqrt{P}} = 1 + \dfrac{14}{4+\sqrt{20}} = 2.65$

Hourly peak dry weather flow = 2.65 × 7,600 m³/d = 20,410 m³/d or 5.30 mgd

Peaking factor from Eq. (4.5), $M_{max} = \dfrac{5}{P^{0.167}} = \dfrac{5}{(20)^{0.167}} = 3.03$

Hourly peak dry weather flow $= 3.03 \times 7{,}600\text{ m}^3/\text{d} = 23{,}028\text{ m}^3/\text{d}$ or 6.09 mgd

Hourly peak dry weather flow Eq. (4.6),

$$Q'_{max} = 3.2(Q'_{ave})^{5/6} = 3.2(2.0)^{5/6} = 5.70\text{ mgd}$$

$$= 5.70\text{ mgd} \times \frac{10^6\text{ gal}}{\text{MG}} \times \frac{3.8\text{ L}}{\text{gal}} \times \frac{\text{m}^3}{1{,}000\text{ L}} = 21{,}660\text{ m}^3/\text{d}$$

Peaking factor from Eq. (4.7), $\log M_{max} = -0.19\log P' + 0.74 = -0.19\log 20 + 0.74$

$$M_{max} = 3.11$$

Hourly peak dry weather flow $= 3.11 \times 7{,}600\text{ m}^3/\text{d} = 23{,}636\text{ m}^3/\text{d}$ or 6.22 mgd

3. Calculate hourly minimum flow.
   Ratio of hourly minimum to DAFs from Eq. (4.8)

$$\log M_{\min} = 0.142\log P' - 0.682 = 0.142\log 20 - 0.682 = -0.497$$

$$M_{\min} = 0.32$$

Hourly minimum dry weather flow $= 0.32 \times 7{,}600\ \text{m}^3/\text{d} = 2{,}432\ \text{m}^3/\text{d}$ or 0.64 mgd

4. Compare the results.

Flow Condition	m³/d	Mgd
Hourly peak dry weather flow from Eq. (4.4)	20,140	5.30
Hourly peak dry weather flow from Eq. (4.5)	23,028	6.09
Hourly peak dry weather flow from Eq. (4.6)	21,660	5.70
Hourly peak dry weather flow from Eq. (4.7)	23,636	6.22
Hourly minimum dry weather flow from Eq. (4.8)	2,432	0.64

The results of hourly peak dry weather flow from four equations are comparable. The average hourly peak dry weather flow is 5.8 mgd. The results also indicate that the ratio of hourly peak and minimum dry weather flows is approximately 9 to 1.

4.5.2 Infiltration and Inflow (I/I)

Infiltration is the groundwater that enters the sewers through service connections, cracked pipes, defective joints, and cracked manhole walls. Inflow is the direct flow due to surface runoff that may enter through a manhole cover, roof area drains, and cross connections from storm sewers and combined sewers. Flow from cellar and foundation drains, cooling water discharges, and drainage from springs and other steady flows are included with the infiltration. The conditions that apply to infiltration/inflow (I/I) are listed below.[1,2,5]

1. The amount of I/I reaching a sewer system depends upon
   a. the length and age of the sewer;
   b. the construction material, methods, and workmanship;
   c. number of roof or drain connections;
   d. the groundwater table relative to sewer position; and
   e. type of soil, ground cover, and topographic conditions.
2. Permissible I/I allowance is 780 Lpcd (205 gpcd) as long as there are no operational problems such as surcharges, bypasses, or poor treatment performance resulting from hydraulic overloading of publicly owned treatment works (POTWs) during storm events.

3. In older sewers, infiltration is high because of deteriorated joints and masonry mortar. Newer sewers use joints sealed with rubber gaskets or synthetic material and use precast manholes. The I/I rate is significantly smaller.
4. When designing sewers, allowance must be made for old and new sewers. Average values of I/I allowance are 100–10,000 L/d/cm (diameter) per km (linear length) (110–11,000 gpd/in·mile) or 200–28,000 L/d/ha (20–3,000 gpd/acre).[1,2,5,12]
5. In the absence of flow records, many states specify 1,500 Lpcd (400 gpcd) for design of laterals and submains, and 1,300 Lpcd (350 gpcd) for mains and trunk sewers. These design flows include normal I/I allowance.
6. For new sewers or recently constructed sewer systems having precast manholes and pipe joints made with gaskets of rubber or rubber-like material, the average infiltration allowance may be found from Figure 4.3.[1,2,5] The permissible infiltration is 90 L/cm·km·d (100 gpd/in·mile). The permissible exfiltration rate should not exceed the permissible infiltration rate. The allowable limit of manhole leakage is 12.4 Lpd/m vertical depth (10.5 gpd/ft vertical depth) in the manhole.

4.5.3 Common Terms Used to Express Flow Variations

Several terms are commonly used to express the flow variations in the overall context of planning, design, and operation of wastewater treatment and conveyance facilities. These terms and their significance are briefly presented below.[1,2]

Daily Average Flow. The DAF (also called the annual average flow) is based on annual flow rate data. It is the average flow occurring over 24-h period under dry weather conditions. This flow is very important and is used to (1) evaluate treatment plant capacity, (2) develop flow rate ratios, (3) size many treatment units, (4) calculate organic loadings, (5) estimate sludge solids, (6) estimate chemical needs, and (7) calculate pumping and treatment unit costs.

Maximum Dry Weather Flow. Maximum dry weather flow is the hourly peak of the diurnal flow curve. This flow is important as it brings about daily flushing action in sewers, conduits, and channels. It is used to design and check the retention time of several critical components of a treatment facility.

Peak Wet Weather Flow. Peak wet weather flow occurs after or during precipitation, and includes a substantial amount of I/I. Some regulatory agencies define peak wet weather flow as the highest 2-h flow encountered under any operational condition, including times of highest rainfall

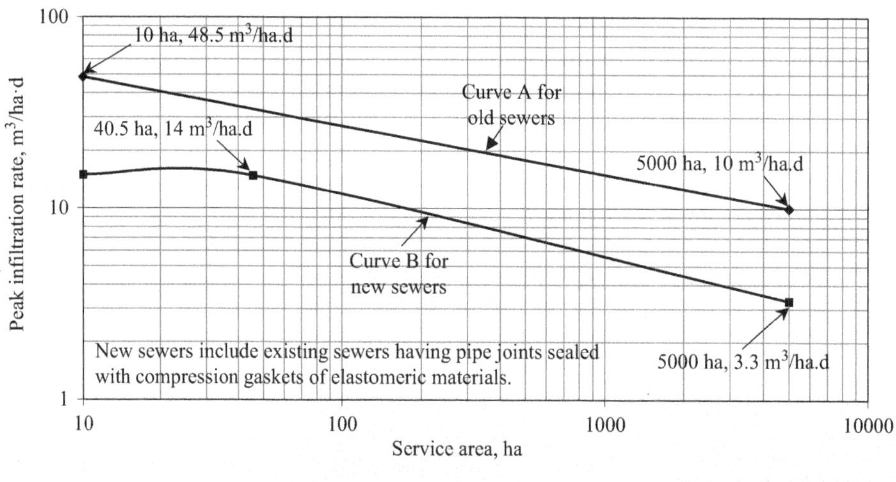

FIGURE 4.3 Average allowance for infiltration rate in new sewers. *Note*: ha × 2.4711 = acre; m³/ha·d × 106.9 = gal/acre·d. (Printed from Ref. 1 with permission of CRC Press, Taylor & Francis Group.)

(generally the 2-year, 24-h storm is assumed), and prolonged period of wet weather. The highest 2-h flow is very important in the design of wastewater treatment facilities. The ratio of wet weather flow with respect to average daily flow may reach three to five. Peak hourly flow (or highest 2-h flow) is needed to design (1) the interceptor sewers and collection system, (2) conduits and connecting pipes and channels in a treatment plant, (3) pumping stations, (4) flow meters, (5) bar screen and grit channels, (6) the hydraulics of the influent and effluent structures of most treatment units, and (7) to check the retention period of many critical treatment units.

Minimum Hourly Flow. The minimum hourly flow is the lowest flow on a typical dry weather diurnal flow curve (Figure 4.1). Low flows may cause settling of solids in pipes and channels. These flow rates are needed for sizing of (1) flow meters, (2) lower range of chemical feeder, and (3) pumping equipment. Often re-circulation is needed during low flows to maintain hydraulic loadings in some treatment units.

Sustained Flows. Sustained flows are the flows that persist over several days. Sustained flows could be on the low or high sides. As an example, extraordinarily dry or hot weather may cause sustained low extremes. Likewise, prolonged wet weather and special events – such as fairs, conventions, games, and exhibitions – in a community may be a source of population surges and could all cause high sustained flows. Data on sustained flow rates may be needed in sizing equalization basins and other plant hydraulic components, and to check the design of many units under extreme conditions. The sustained-flow envelope should be developed from the longest available period of record. The typical sustained-flow envelope at a treatment plant is shown in Figure 4.4.[1–5]

Emergency Flow Diversion. Many POTWs are equipped with a flow diversion facility to hold the excess or entire flow under emergencies. Such conditions may develop if (1) power failure occurs, (2) the combined sewer overflows are directed, and (3) excessive I/I is diverted from the sanitary sewers. The diverted flow is temporarily stored in a basin and rerouted through the plant after restoration of power or when the plant is able to handle the stored flows. Various types of flow-regulating devices are used to achieve proper flow diversion. Common diversion devices include side weir, transverse weir, leaping weir, orifice, relief siphon, tripping plate, and hydro-brake regulators. Chapter 11 and Refs. 1,5,13 should be referred to for design information on these diversion devices.

Design Flows. The design flows are the ultimate flows when the facility is expected to reach its full design capacity. Design flows occur at the end of the design period, which may be 10–20 years after initial construction is completed. The design flows are peak wet weather, average day, and minimum flows.

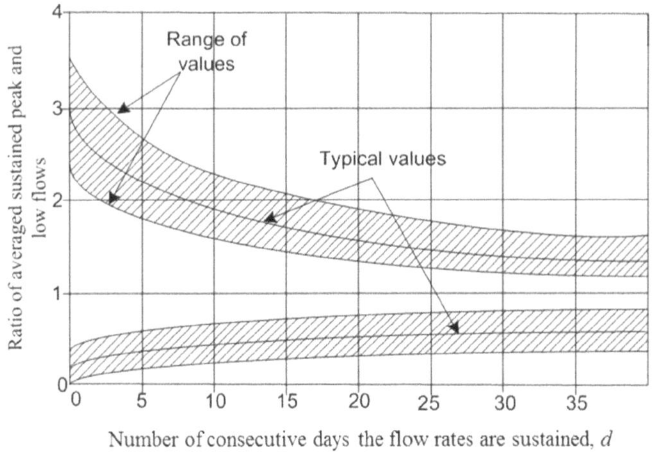

FIGURE 4.4 Typical sustained-flow envelope at a wastewater treatment plant. (Printed from Ref. 1 with permission of CRC Press, Taylor & Francis Group.)

Example 4.9: Peak I/I

A city subdivision has a population density of 4,000 residents/km^2. During wet weather conditions, the expected average I/I from this subdivision is 4,500 L/ha.d and the peaking factor is 1.5. Estimate average and peak I/I in Lpcd.

Solution

1. Estimate population per ha.

$$\text{Population density} = \frac{4{,}000 \text{ persons}}{\text{km}^2} \times \frac{\text{km}^2}{10^6 \text{ m}^2} \times \frac{10{,}000 \text{ m}^2}{\text{ha}} = 40 \text{ persons/ha}$$

2. Estimate average I/I.

$$\text{Average I/I} = \frac{4{,}500 \text{ L}}{\text{ha d}}$$

$$\text{Average I/I per person} = \frac{4{,}500 \text{ L}}{\text{ha d}} \times \frac{1}{40 \text{ person/ha}} = 113 \text{ Lpcd}$$

3. Estimate peak I/I.

$$\text{Peak I/I} = 113 \text{ Lpcd average I/I} \times \frac{1.5 \text{ peak I/I}}{\text{average I/I}} = 170 \text{ Lpcd}$$

Example 4.10: Permissible I/I for a New Development

A new housing subdivision is being developed over a land with a total area of 4,900 acres. Using Figure 4.3, estimate the average I/I allowance for design of an intercepting sewer.

Solution

1. Determine the area served in ha.

$$\text{Land area} = 4{,}900 \text{ acres} \times \frac{1 \text{ ha}}{2.4711 \text{ acre}} = 2{,}000 \text{ ha}$$

2. Estimate from Figure 4.3 the I/I allowance.

$$\text{I/I allowance} = 4.4 \text{ m}^3/\text{ha d}$$

3. Determine average I/I.

$$\text{Average I/I} = 4.4 \text{ m}^3/\text{ha d} \times 2{,}000 \text{ ha} = 8{,}800 \text{ m}^3/\text{d}$$

Example 4.11: Permissible I/I Allowance in a Sewer and Peak Wet Weather Flow

A city's population is 40,000. Average wastewater flow is 380 Lpcd, and permissible average I/I allowance is 890 L/d/cm/km (961 gpd/in/mile). The average length of sanitary sewer is 8 m per capita. The sewer-size distribution is as follows: 15 cm dia. = 4%, 20 cm dia. = 79%, 31 cm dia. = 10%,

46 cm dia. = 5%, and 61 cm dia. = 2%. Estimate I/I in Lpcd and peak wet weather flow in m^3/d. Assume the peaking factor for I/I is 1.6.

Solution

1. Calculate daily average and hourly peak dry weather flows.

$$\text{Daily average flow} = \frac{380\text{ L}}{\text{person d}} \times 40{,}000\text{ persons} \times \frac{\text{m}^3}{1{,}000\text{ L}} = 15{,}200\text{ m}^3/\text{d}$$

Peaking factor from Eq. (4.4), $M_{\max} = 1 + \frac{14}{4+\sqrt{P}} = 1 + \frac{14}{4+\sqrt{40}} = 2.36$

$$\text{Hourly peak dry weather flow} = 2.36 \times 15{,}200\text{ m}^3/\text{d} = 35{,}872\text{ m}^3/\text{d}$$

$$\text{Hourly peak dry weather flow rate in Lpcd} = 2.36 \times 380\text{ Lpcd} = 897\text{ Lpcd}$$

2. Calculate length and equivalent sewer diameter in the collection system.

$$\text{Total length of sanitary sewer} = \frac{8\text{ m}}{\text{person}} \times 40{,}000\text{ persons} \times \frac{\text{km}}{1{,}000\text{ m}} = 320\text{ km}$$

Equivalent diameter (weighted average)

$$= \frac{0.04 \times 15\text{ cm} + 0.79 \times 20\text{ cm} + 0.1 \times 31\text{ cm} + 0.05 \times 46\text{ cm} + 0.02 \times 61\text{ cm}}{0.04 + 0.79 + 0.1 + 0.05 + 0.02}$$

$$= \frac{23.02\text{ cm}}{1} = 23\text{ cm}$$

3. Calculate peak I/I.

$$\text{Daily average I/I allowance} = \frac{890\text{ L}}{\text{d cm km}} \times (23\text{ cm} \times 320\text{ km}) \times \frac{\text{m}^3}{1{,}000\text{ L}} = 6{,}550\text{ m}^3/\text{d}$$

$$\text{Daily average I/I allowance in Lpcd} = 6{,}550\text{ m}^3/\text{d} \times \frac{1{,}000\text{ L}}{\text{m}^3} \times \frac{1}{40{,}000\text{ persons}} = 164\text{ Lpcd}$$

$$\text{Hourly peak I/I allowance} = 164\text{ Lpcd average} \times \frac{1.6\text{ peak I/I}}{\text{average I/I}} = 262\text{ Lpcd}$$

$$\text{Hourly peak I/I allowance} = 40{,}000\text{ persons} \times \frac{262\text{ L}}{\text{person d}} \times \frac{\text{m}^3}{1{,}000\text{ L}} = 10{,}480\text{ m}^3/\text{d}$$

4. Calculate hourly peak wet weather flow.
   Total hourly peak wet weather flow to the plant = hourly peak dry weather flow + hourly peak I/I allowance = 35,872 m^3/d + 10,480 m^3/d = 46,352 m^3/d
   Total hourly peak wet weather flow in Lpcd = (897 + 262) Lpcd = 1159 Lpcd

Example 4.12: I/I, Diurnal Flow Curve, and 2-h Peak Wet Weather Flow

A sewer evaluation study of a service area was conducted to determine the infiltration inflow reaching the collection system, and to estimate the 2-h peak wet weather flow. A flow measurement device was installed in the interceptor that carried the entire flow from the service area. The hourly flow measurements on a typical dry day gave the diurnal weather flow pattern. The diurnal flow pattern of a typical dry day is shown in Figure 4.5. After a long dry spell, the first high-intensity storm caused heavy downpour in the service area. Since the groundwater table was low, the infiltration was negligible, and the excess flow over the dry weather flow was the inflow. The measured hourly inflow under this condition is illustrated in Figure 4.5. After this downpour, high-intensity storms and intermittent showers continued to occur over a 10-day period. Under these wet weather conditions, the ground was completely saturated and the recorded flow data provided a measure of infiltration. The hourly variation of recorded infiltration plus dry weather flow curve is also shown in Figure 4.5. Determine (1) daily average dry weather flow, (2) hourly peak dry weather flow, (3) daily average infiltration, (4) 2-h peak inflow, (5) 2-h peak wet weather flow, and (6) 2-h peaking factor, which is the ratio of 2-h peak wet weather flow and daily average dry weather flow.

Solution

1. Calculate daily average and hourly peak dry weather flows.

   The daily average dry weather flow is calculated from the diurnal dry weather flow curve given in Figure 4.5. Apply the trapezoidal rule (Eq. 4.1).

Daily average dry weather flow

$$= \frac{1}{2} \times 2\text{ h} \left[40\text{ m}^3/\text{h} + 2 \times (24+20+26+40+75+110+105+51+54+77+60)\text{ m}^3/\text{h} + 40\text{ m}^3/\text{h}\right]$$

$$= 1{,}364\text{ m}^3 \text{ total in 24 h or } 56.8\text{ m}^3/\text{h in daily average}$$

Hourly peak dry weather flow, as read between 12:00 and 1:00 PM in Figure 4.5 = 111 m^3/h

2. Calculate daily average infiltration when soil is completely saturated.

   The combined infiltration and dry weather flow curve is also shown in Figure 4.5.

   The average infiltration is also calculated from the curve using the trapezoidal rule (Eq. 4.1).

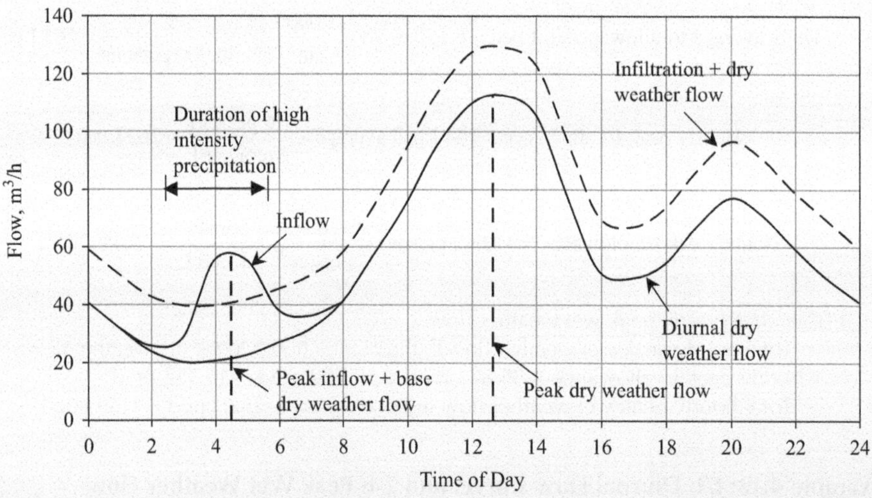

FIGURE 4.5 Diurnal dry weather flow, infiltration and inflow recorded in an interceptor (Example 4.12). (Printed from Ref. 1 with permission of CRC Press, Taylor & Francis Group.)

Area under infiltration flow curve (includes dry weather flow)

$$= \frac{1}{2} \times 2\ \text{h}\left[59\ \text{m}^3/\text{h} + 2 \times (42+39+45+58+94+128+123+70+73+97+79)\ \text{m}^3/\text{h} + 59\ \text{m}^3/\text{h}\right]$$

$$= 1{,}814\ \text{m}^3\ \text{total in 24 h}$$

$$\text{Daily average infiltration plus dry weather flow} = \frac{1{,}814\ \text{m}^3}{24\ \text{h}} = 75.6\ \text{m}^3/\text{h}$$

Daily average infiltration flow = 75.6 m^3/h − 56.8 m^3/h = 18.8 m^3/h

3. Estimate 2-h peak inflow.

   The inflow is estimated from the inflow curve between 3:30 and 5:30 AM in Figure 4.5.

   2-h peak inflow plus base dry weather flow = 54 m^3/h

   2-h base dry weather flow = 21 m^3/h

   2-h peak inflow = 54 m^3/h − 21 m^3/h = 33 m^3/h

4. Estimate 2-h peak wet weather flow.

   In the worst-case scenario, let's assume peak inflow occurs between 12:00 and 2:00 PM when hourly peak dry weather flow plus infiltration is reached.

   2-h peak wet weather flow = hourly peak dry weather flow + daily average infiltration + 2-h peak inflow = 111 + 18.8 + 33 m^3/h = 162.8 m^3/h

5. Calculate 2-h peaking factor.

$$\text{2-hour peaking factor} = \frac{162.8\ \text{m}^3/\text{h}}{56.8\ \text{m}^3/\text{h}} = 2.9$$

Example 4.13: Peak Design Wastewater Flow and Peaking Factor for a City Development Project

The future land-use plan of a city has been developed. The development project has residential, commercial, and industrial areas, and a day school. Data on the expected saturation population densities, daily average wastewater flows and peaking factors for various types of land-use plan of the development project are given in Table 4.5. The day school

TABLE 4.5
Land Area Developed, Expected Saturation Population Densities, Wastewater Flows, and Peaking Factors for Various Land-Use Plan of the City Development Project (Example 4.13)

Land Use	Area, ha	Expected Population Density per ha	Unit	Unit Loading	Peaking Factor
Single-family dwellings	170	35	Lpcd	350	Eq. (4.4)
Mixed residential dwellings	120	45	Lpcd	260	Eq. (4.4)
Apartments	150	130	Lpcd	220	Eq. (4.4)
School	50	—	Lpcd	76	4.8
Commercial	140	—	m^3/ha·d	30	1.8
Industrial	110	—	m^3/ha·d	60	2.2
Infiltration/inflow	Total area	—	m^3/ha·d	5	6.0

has an enrollment of 3,000 students. Estimate the daily average and peak wastewater flow rates and the overall peaking factor.

Solution

1. Calculate population and flows.
   Set up a computational table for estimating the population and flows. The summary of computations is provided in Table 4.6.
2. Estimate DAFs.
   Daily average dry weather flow = 18,757 m^3/d
   Daily average I/I flow = 3,700 m^3/d
   DAF including average I/I = 22,457 m^3/d
3. Estimate peak flows.
   Peak dry weather flow = 45,686 m^3/d
   Peak I/I flow = 22,200 m^3/d
   Peak wet weather flow = 67,886 m^3/d
4. Estimate the overall peaking factor.

$$\text{Overall peaking factor} = \frac{67{,}886\ \text{m}^3/\text{d}}{22{,}457\ \text{m}^3/\text{d}} = 3.0$$

TABLE 4.6
Computation Table for Population and Flow Estimation (Example 4.13)

Land Use	Area, ha	Expected Population Density per ha	Population	Unit	Unit Loading	Daily Average Flow, m^3/d	Peaking Factor	Peak Flow, m^3/d
Single-family dwellings	170	35	5,950	Lpcd	350	2,083	3.2[a]	6,666
Mixed residential dwellings	120	45	5,400	Lpcd	260	1,404	3.2[a]	4,493
Apartments	150	130	19,500	Lpcd	220	4,290	2.7[a]	11,583
School	50	—	3,000	Lpcd	60	180	4.8	864
Commercial	140	—	—	m^3/ha·d	30	4,200	1.8	7,560
Industrial	110	—	—	m^3/ha·d	60	6,600	2.2	14,520
Subtotal	740	—	33,850	—	—	**18,757**	2.4[b]	**45,686**
I/I	740	—	—	m^3/ha·d	5	**3700**	6.0	**22,200**
Total/overall	**740**	**—**	**33,850**	—	**—**	**22,457**	**3.0**[c]	**67,886**

[a] Values calculated from Eq. (4.4).

[b] $\dfrac{45{,}686\ \text{m}^3/\text{d}}{18{,}757\ \text{m}^3/\text{d}} = 2.4.$

[c] See the calculation in Step 4.

Bold values are subtotal and total values of each column. The subtotal of each column + I/I gives the total value for each column. They are made bold for emphasis.

Example 4.14: Sustained Average Flow Ratio (SAFR) Envelope

Describe the procedure for developing sustained average flow ratio (SAFR) envelope. DAF data during 30 consecutive days at a wastewater treatment plant are summarized in Table 4.7. The annual average-day flow (AADF) at the plant is 0.58 m^3/s. Develop the maximum and minimum SAFR envelope up to a 10-day period.

Solution

1. Describe the procedure for developing the SAFR envelope.
   a. Select the longest available period of consecutive DAF data.
   b. Determine the AADF.
   c. Calculate the ratio of DAF to AADF for each day during the period.
   d. Determine the $SAFR_i$ for the i consecutive days, $i = 1, 2, 3, \ldots, n$. To be reliable, the value of n is usually much less than the maximum number of consecutive days during the period. $SAFR_1$ equals the ratio of DAF to AADF, while $SAFR_i$ ($i > 1$) is obtained from the running average of the ratios for the i consecutive days, $i = 2, 3, \ldots, n$.
   e. Identify the maximum value of $SAFR_i$ ($i = 1, 2, 3, \ldots, n$). It is the maximum ratio with respect to i consecutive days the flow rate is sustained.
   f. Identify the minimum value of $SAFR_i$ ($i = 1, 2, 3, \ldots, n$). It is the minimum ratio with respect to i consecutive days the flow rate is sustained.
   g. Plot the maximum and minimum $SAFR_i$ ($i = 1, 2, 3, \ldots, n$) as well as the AADF ratio with respect to the i consecutive days the flow rate is sustained. The AADF line will be horizontal and pass through the ratio of 1.0. The maximum and the minimum SAFR curves will be above and below the AADF line.
2. Develop the maximum and minimum SAFR envelope for a 10-day period.

   Calculation results are tabulated in Table 4.8. A brief description is provided below for each column of Table 4.8.
   a. *Columns (1) and (2).* 30 consecutive days and DAF data obtained from Table 4.7 are arranged.
   b. *Column (3).* The ratio of DAF and AADF are calculated and listed. AADF = 0.58 m^3/s. $\frac{\text{DAF}}{\text{AADF}}$ ratio for day $1 = \frac{0.25\ \text{m}^3/\text{s}}{0.58\ \text{m}^3/\text{s}} = 0.43$. $\frac{\text{DAF}}{\text{AADF}}$ ratio for day $2 = \frac{0.16\ \text{m}^3/\text{s}}{0.58\ \text{m}^3/\text{s}} = 0.28$. Similarly, these ratios for 30-d period are calculated and listed.

TABLE 4.7
Daily Average Flow (DAF) Data during 30 Consecutive Days (Example 4.14)

Day	DAF, m^3/s	Day	DAF, m^3/s	Day	DAF, m^3/s
1	0.25	11	0.65	21	0.83
2	0.16	12	0.66	22	0.80
3	0.15	13	0.67	23	0.72
4	0.13	14	0.68	24	0.66
5	0.14	15	0.69	25	0.66
6	0.25	16	0.70	26	0.65
7	0.35	17	0.75	27	0.50
8	0.48	18	1.12	28	0.32
9	0.52	19	1.20	29	0.20
10	0.60	20	1.15	30	0.18

TABLE 4.8
Summary of Calculations of Maximum and Minimum of SAFRs (Example 4.14)

		Ratio of DAF	SAFR$_i$								
Day	DAF, m³/s	to AADF	i = 2	i = 3	i = 4	i = 5	i = 6	i = 7	i = 8	i = 9	i = 10
(1)	(2)	(3)	(4)	(5)	(6)	(7)	(8)	(9)	(10)	(11)	(12)
1	0.25	0.43	0.36	0.32	0.30	0.29	**0.31**	**0.35**	**0.41**	**0.47**	**0.52**
2	0.16	0.28	0.27	0.25	**0.25**	**0.29**	0.34	0.41	0.47	0.53	0.59
3	0.15	0.26	0.24	**0.24**	0.29	0.35	0.43	0.50	0.56	0.63	0.68
4	**0.13**	**0.22**	**0.23**	0.30	0.38	0.47	0.54	0.61	0.67	0.72	0.77
5	0.14	0.24	0.34	0.43	0.53	0.60	0.67	0.74	0.79	0.83	0.86
6	0.25	0.43	0.52	0.62	0.69	0.76	0.82	0.86	0.90	0.93	0.96
7	0.35	0.60	0.72	0.78	0.84	0.90	0.94	0.97	0.99	1.02	1.03
8	0.48	0.83	0.86	0.92	0.97	1.00	1.03	1.05	1.07	1.08	1.10
9	0.52	0.90	0.97	1.02	1.05	1.07	1.09	1.10	1.11	1.13	1.21
10	0.60	1.03	1.08	1.10	1.11	1.12	1.14	1.15	1.16	1.25	1.33
11	0.65	1.12	1.13	1.14	1.15	1.16	1.16	1.18	1.28	1.36	1.43
12	0.66	1.14	1.15	1.16	1.16	1.17	1.19	1.30	1.39	1.46	1.46
13	0.67	1.16	1.16	1.17	1.18	1.20	1.32	1.43	1.50	1.49	1.48
14	0.68	1.17	1.18	1.19	1.22	1.36	1.48	1.55	1.53	1.52	1.49
15	0.69	1.19	1.20	1.23	1.41	1.54	1.61	1.59	1.56	1.52	**1.49**
16	0.70	1.21	1.25	1.48	1.63	1.70	1.65	1.61	**1.57**	**1.52**	1.48
17	0.75	1.29	1.61	1.76	1.82	1.74	**1.68**	**1.62**	1.56	1.51	1.47
18	1.12	1.93	2.00	**1.99**	**1.85**	**1.76**	1.67	1.60	1.54	1.49	1.43
19	**1.20**	**2.07**	**2.03**	1.83	1.72	1.62	1.54	1.48	1.44	1.37	1.29
20	1.15	1.98	1.71	1.60	1.51	1.43	1.39	1.35	1.29	1.20	1.12
21	0.83	1.43	1.41	1.35	1.30	1.27	1.24	1.19	1.11	1.02	0.95
22	0.80	1.38	1.31	1.25	1.22	1.20	1.15	1.06	0.97	0.90	—
23	0.72	1.24	1.19	1.17	1.16	1.10	1.01	0.91	0.84	—	—
24	0.66	1.14	1.14	1.13	1.06	0.96	0.86	0.78	—	—	—
25	0.66	1.14	1.13	1.04	0.92	0.80	0.72	—	—	—	—
26	0.65	1.12	0.99	0.84	0.72	0.64	—	—	—	—	—
27	0.50	0.86	0.71	0.59	0.52	—	—	—	—	—	—
28	0.32	0.55	0.45	0.40	—	—	—	—	—	—	—
29	0.20	0.34	0.33	—	—	—	—	—	—	—	—
30	0.18	0.31	—	—	—	—	—	—	—	—	—
Maximum	1.20	2.07	2.03	1.99	1.85	1.76	1.68	1.62	1.57	1.52	1.49
Minimum	0.13	0.22	0.23	0.24	0.25	0.29	0.31	0.35	0.41	0.47	0.52

Bold values are the min and max values in each column that are plotted in Fig. 4.6.

c. *Column (4)*. SAFR are calculated from the average ratios of $\frac{\text{DAF}}{\text{AADF}}$ for two consecutive sustained days (i = 2). SAFR_2 for day $1 = \frac{0.43+0.28}{2} = 0.36$. SAFR_2 for day $2 = \frac{0.28+0.26}{2} = 0.27$. Similarly, the SAFR_2 for 30-day is calculated from average SARF for two consecutive sustained days. The results are listed.

d. *Column (5).* SAFR are calculated from the average ratios of $\frac{\text{DAF}}{\text{AADF}}$ for three consecutive sustained days ($i = 3$). SAFR_3 for day $1 = \frac{0.43+0.28+0.26}{3} = 0.32$. For day 2, $\text{SAFR}_3 = \frac{0.28+0.26+0.22}{3} = 0.25$. Similarly, the $SAFR_3$ for 30 days are calculated from average *SARF* for three consecutive days of sustained flow.

e. *Column (6)–Column (12).* The procedure is repeated using the number of days the flow is sustained. As an example, day 1 and day 2 of SAFR_4 and SAFR_{10} are calculated.

$$\text{Day 1 of SAFR}_4 = \frac{0.43+0.28+0.26+0.22}{4} = 0.3$$

$$\text{Day 2 of SAFR}_4 = \frac{0.28+0.26+0.22+0.24}{4} = 0.25$$

$$\text{Day 1 of SAFR}_{10} = \frac{0.43+0.28+0.26+0.22+0.24+0.43+0.60+0.83+0.90+1.03}{10} = 0.52$$

$$\text{Day 2 of SAFR}_{10} = \frac{0.28+0.26+0.22+0.24+0.43+0.60+0.83+0.90+1.03+1.12}{10} = 0.54$$

The results of SAFR_4 – SAFR_{10} are summarized in Table 4.8.

3. Identify and list the maximum and minimum SAFR_i (i = 1, 2, 3, ….,10)

   Identify maximum and minimum DAF, ratio of DAF/AADF, and SAFR_i (i = 1, 2, 3, …,10) in Columns (2) through (12) of Table 4.8. List these values in the bottom two rows of the table.

4. Plot the SAFR envelope

   Plot the maximum and minimum of SAFRs with respect to the number of consecutive days during the 10-day period. These plotted lines form an envelope that tends to converge to the AADF ratio passing through 1.0 with increasing number of consecutive days. The SAFR envelope is shown in Figure 4.6.

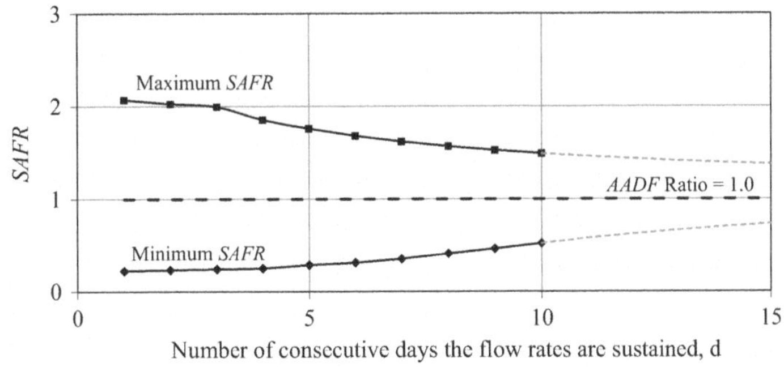

FIGURE 4.6 The maximum and minimum SAFR plotted with respect to number of consecutive days (Example 4.14). (Printed from Ref. 1 with permission of CRC Press, Taylor & Francis Group.)

Example 4.15: Forecasting Design Flow Rates

A city has a current total population of 40,000 residents. The estimated design population at the end of a design period of 20 years is 55,000. Determine the design average, peak, and minimum flow rates. Use the following data.

Current daily average flow to the wastewater treatment plant = 18,400 m^3/d

Daily average I/I flow = 80 L/capita d

Hourly peak I/I flow = 180 L/capita d

A new junior college will attract 1,875 nonresident students per day, and the estimated daily average wastewater flow will be 40 L/capita d. A new industry will add a daily average flow of 900 m^3/d for a 24 h/d operation. An hourly peak flow of 1,500 m^3/d will occur during the day shift. The plant will be shut down one day a week.

It is expected that 10% per capita reduction in wastewater flow will occur with the future population growth. The minimum to average flow ratio is 0.4 for the current dry weather wastewater flow.

Solution

1. Calculate the current wastewater flow in Lpcd.

$$\text{Daily average I/I flow} = \frac{80\text{ L}}{\text{person d}} \times 40{,}000\text{ people} \times \frac{\text{m}^3}{1{,}000\text{ L}} = 3{,}200\text{ m}^3\text{/d}$$

Daily average dry weather wastewater flow = (18,400 − 3,200) m^3/d = 15,200 m^3/d

$$\text{Daily average dry weather flow rate in Lpcd} = 15{,}200\text{ m}^3\text{/d} \times \frac{1{,}000\text{ L}}{\text{m}^3} \times \frac{1}{40{,}000\text{ persons}} = 380\text{ Lpcd}$$

2. Calculate the expected future wastewater flow rate in Lpcd.

Future daily average flow rate after 10% reduction = (1 − 0.1) × 380 Lpcd = 342 Lpcd

3. Calculate design daily average wastewater flow.

Current daily average dry weather flow = 15,200 m^3/d

$$\text{Future addition daily average residential flow} = \frac{342\text{ L}}{\text{person d}} \times 15{,}000\text{ people} \times \frac{\text{m}^3}{1{,}000\text{ L}} = 5{,}310\text{ m}^3\text{/d}$$

$$\text{Daily average students' contribution} = \frac{40\text{ L}}{\text{student d}} \times 1{,}875\text{ people} \times \frac{\text{m}^3}{1{,}000\text{ L}} = 75\text{ m}^3\text{/d}$$

Subtotal daily average residential flow = (15,200 + 5,130 + 75) m^3/d = 20,405 m^3/d

Daily average industrial flow = 900 m^3/d

$$\text{Daily average I/I flow} = \frac{80\text{ L}}{\text{person d}} \times 55{,}000\text{ people} \times \frac{\text{m}^3}{1{,}000\text{ L}} = 4{,}400\text{ m}^3\text{/d}$$

Design daily average flow = (20,405 + 900 + 4,400) m^3/d = 27,705 m^3/d

4. Calculate design hourly peak flow.

$$\text{(Peaking factor from Eq. [4.4],} = 1 + \frac{14}{4+\sqrt{P}} = 1 + \frac{14}{4+\sqrt{55}} = 2.23$$

Hourly peak dry weather residential flow = 2.23 × 20,405 m^3/d = 45,500 m^3/d
Hourly peak industrial flow = 1,500 m^3/d

$$\text{Hourly peak I/I flow} = \frac{180\ \text{L}}{\text{person d}} \times 55{,}000\ \text{people} \times \frac{\text{m}^3}{1{,}000\ \text{L}} = 9{,}900\ \text{m}^3/\text{d}$$

Design hourly peak flow = (45,500 + 1,500 + 9,900 m^3/d) = 56,900 m^3/d

5. Calculate design minimum flow.
   Minimum dry weather flow = 0.4 × 15,200 m^3/d = 6,080 m^3/d
   Minimum industrial flow = 0 (during the plant shut down)
   Total design minimum flow = 6,080 m^3/d

4.6 CHARACTERISTICS OF MUNICIPAL WASTEWATER

The physical, chemical, and biological quality of municipal wastewater is briefly presented in this section to describe various contaminants that are commonly used to characterize pollution of natural waters. Many of these contaminants are also present in nonpoint source (NPS) discharges that cause serious water pollution problems.

The municipal wastewater contains approximately 99.9% water. The remaining constituents include suspended (settleable and non-settleable) and dissolved solids that have organic, inorganic, and nutrient components, including microorganisms. These constituents give physical, chemical, and biological qualities that characterize residential, commercial, industrial, and municipal wastewaters. The strength of wastewater is based on the concentrations of these constituents.

4.6.1 Physical Quality

The physical quality of municipal wastewater is generally reported in terms of temperature, color, turbidity, odor, as well as suspended, settleable, non-settleable, and dissolved solids. The significance of these parameters is briefly summarized below.[1–4]

Temperature. The temperature of municipal wastewater is slightly higher than that of the water supply, and stays in the range of 10°C–21°C. The average temperature varies slightly with the season and is higher than the average air temperature most of the year except during the hot summer months. The wastewater temperature has a significant effect on the microbiological activity, treatability, solubility of gases, density, and viscosity.

Color, Turbidity, and Odor. The color of fresh wastewater is slightly gray. Stale or septic wastewater is dark gray or black. Turbidity is due to suspended and colloidal particles. In general, stronger wastewaters have higher turbidity. Fresh wastewater has a soapy or oily odor that is somewhat disagreeable. Stale wastewater has putrid odor due to hydrogen sulfide, indol, and skatol, and other decomposition products.

Settleable and Suspended (Non-filterable), Dissolved (Filterable), Volatile, and Fixed Solids. The settleable solids contain organic and inorganic matter. They settle under low velocity and may block the channels and pipes. Organic content of settleable solids will undergo decomposition and cause odors. The other components of solids are suspended and dissolved solids. Solids in municipal wastewater contain 50%–80% volatile, and 20%–50% fixed solids. The typical concentrations of these components and their significance are summarized in Table 4.9.

TABLE 4.9
The Concentration of Many Physical and Chemical Quality Constituents of Municipal Wastewater

Chemical Quality Parameters	Description	Concentration Range	Concentration Typical
Total solids (TS), mg/L	Organic and inorganic, suspended, and dissolved matter.	400–1,200	750
Total suspended solids (TSS), mg/L	The total suspended solids in raw wastewater include settleable and non-settleable (suspended) organic and inorganic solids.	150–400	250
Fixed, mg/L	Non-combustible or mineral components of TSS	40–100	65
Volatile,[a] mg/L	Combustible or organic components of TSS	115–310	185
Settleable suspended solids (SS), mL/L (mg/L)	Portion of organic and inorganic solids that settles in 1 h in an *Imhoff* cone. These solids are an approximate measure of sludge that is removed in a sedimentation basin.	5–20 (90–260)	10 (150)
Fixed, mg/L	Non-combustible or mineral components of settleable SS	20–60	40
Volatile,[a] mg/L	Combustible or organic components of settleable SS	70–200	110
Non-settleable SS, mg/L	Colloidal solids that do not settle in a sedimentation basin. They are removed by filtration through a glass-fiber or membrane filter.	65–150	100
Fixed, mg/L	Non-combustible or mineral portion of non-settleable SS	20–40	25
Volatile, mg/L	Combustible or organic fraction of non-settleable SS	45–110	75
Total dissolved or filterable solids (TDS), mg/L	Portion of organic and inorganic solids that are not filterable. Solids smaller than one millimicron (mμ)[b] fall in this category.	250–700	500
Fixed, mg/L	Non-combustible or mineral components of TDS	100–300	200
Volatile, mg/L	Combustible or organic fraction of TDS	150–400	300
BOD_5, mg/L	Biochemical oxygen demand (5-d at 20°C). It represents the biodegradable portion of organic component. It is a measure of dissolved oxygen required by microorganisms to stabilize the organic matter in 5 days at 20°C.	110–400	210
COD, mg/L	Chemical oxygen demand. It is a measure of organic matter and represents the amount of oxygen required to oxidize the organic matter by strong oxidizing chemicals (potassium dichromate) under acidic condition.	200–780	400
TOC, mg/L	Total organic carbon is a measure of organic matter. TOC is determined by converting organic carbon to carbon dioxide. It is done in a high-temperature furnace in the presence of a catalyst. Carbon dioxide is quantitatively measured.	80–290	150
Total nitrogen (TN)[c], mg/L as N	Total nitrogen includes organic, ammonia, nitrite, and nitrate nitrogen. Nitrogen and phosphorus along with carbon and other trace elements serve as nutrients. They accelerate the aquatic plant growth in natural waters.	20–85	40

(Continued)

TABLE 4.9 (*Continued*)
The Concentration of Many Physical and Chemical Quality Constituents of Municipal Wastewater

	Description	Concentration	
Chemical Quality Parameters		**Range**	**Typical**
Organic nitrogen (ON), mg/L as N	It binds nitrogen into protein, amino acids, and urea.	8–35	20
Ammonia nitrogen (AN), mg/L as N	Ammonia nitrogen is produced as the first stage of decomposition of organic nitrogen	12–40	20
Nitrite and nitrate nitrogen, mg/L as N	Nitrite and nitrate nitrogen are the higher oxidized states of ammonia nitrogen. Both forms of nitrogen are absent in raw domestic wastewater.	0–small	0
Total phosphorus (TP)[d], mg/L as P	Total phosphorus exists in organic and inorganic form. Phosphorus in natural water is a source of eutrophication.	4–8	6
Organic, mg/L as P	Organic phosphorus is bound in proteins and amino acids.	1–3	2
Inorganic,[e] mg/L as P	Inorganic form of phosphorous exists as orthophosphate and polyphosphate.	3–6	4
pH	pH is an indication of acidic or basic nature of wastewater. A solution is neutral at pH 7.	6.7–7.5	7.0
Alkalinity, mg/L as $CaCO_3$	Alkalinity in wastewater is due to presence of bicarbonate, carbonate, and hydroxide ion.	80–350	220
Hardness, mg/L as $CaCO_3$	Hardness in wastewater is primarily due to calcium and magnesium ions. Hardness in wastewater depends on the hardness of water supply.	120–350	200
Chloride, mg/L	Chloride in wastewater comes from water supply, human wastes, and domestic water softeners.	30–100	50
Fats, oils, and grease (FOG), mg/L	These are the soluble portions of organic matter in hexane. Their primary sources are fats and oils used in foods.	50–150	100

Source: Refs.[1–4,14–16].

[a] Volatile fraction is obtained after ignition in a muffle furnace at 550°C ± 50°C.

[b] Micrometer (μm) = 10^{-6} m [also referred to as micron (μ)]. Other measures are nanometer (nm) = 10^{-9}, also referred to as millimicron (mμ); angstrom (Å) = 10^{-10} m.

[c] TN = ON + AN + NO_2–N + NO_3–N. Total Kjeldahl nitrogen (TKN) = ON + AN.

[d] The concentration of phosphorous in municipal wastewaters in the United States has, in general, been falling over the past decade. In the late 1960s, typical total P concentrations in raw wastewater were 10–12 mg/L. Currently, concentrations are usually in the range of 4–8 mg/L where phosphorous-based detergents are regulated. It is expected that in future the concentration of TP in municipal wastewater will decrease.

[e] The predominate form of inorganic phosphorus in raw wastewater is orthophosphate (Ortho-P).

Example 4.16: Settleable Solids in an Imhoff Cone

A 1-L sample of raw municipal wastewater is settled for one hour in an *Imhoff* cone (Figure 4.7). The volume of settled solids in the bottom of the cone is 15 mL. Express the result in mL/L and mg/L. Assume the bulk density of settled solids in the cone is 1,020 kg/m^3.

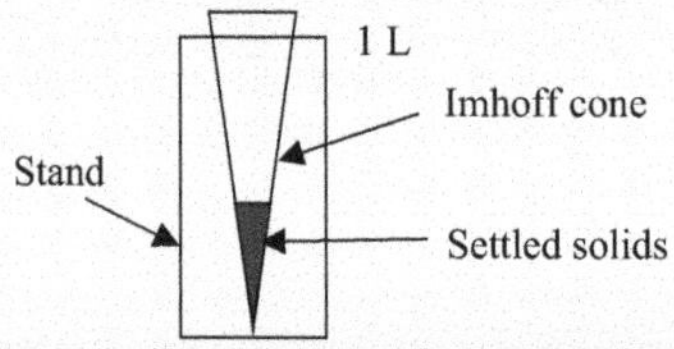

FIGURE 4.7 Determination of settleable solids in an Imhoff cone (Example 4.16).

Solution

1. Express the result in mL/L.
   Since the volume of sample in the *Imhoff* cone is 1 L, the volume of settleable solids = 15 mL/L
2. Express the result in mg/L
   Based on bulk density, the concentration of settled solids in the *Imhoff* cone =

$$15\ \text{mL/L} \times \frac{\text{m}^3}{10^6\ \text{mL}} \times 1{,}020\ \text{kg/m}^3 \times \frac{10^6\ \text{mg}}{\text{kg}} = 15{,}300\ \text{mg/L}$$

Example 4.17: Settleable, Non-settleable, and Dissolved (Filterable) Solids

50 mL of raw wastewater sample was filtered through a dried and pre-weighed glass-fiber filter to determine the settleable suspended solids (SS) and total suspended solids (TSS). One liter of raw wastewater sample was settled in an *Imhoff* cone for one hour to determine the volume of settleable SS. 50 mL of the settled sample was filtered through another pre-dried and weighed glass-fiber filter to determine the non-settleable SS. 50 mL of filtrate was evaporated in a pre-dried and weighed petri dish over a steam bath to determine the total dissolved or filterable solids. Calculate (1) TSS, (2) non-settleable (or non-filterable) SS, (3) settleable SS, and (4) total dissolved (filterable) solids (TDS).

The results of the tests are given below:

TSS, including settleable and non-settleable SS

Weight of dried filter paper in an aluminum dish = 12.3478 g

Weight of dried filter paper in the aluminum dish and TSS from raw wastewater = 12.3590 g

Non-settleable SS

Weight of dried filter paper in an aluminum dish = 12.3480 g

Weight of dried filter paper in the aluminum dish, and non-settleable SS = 12.3536 g

Settleable SS

Volume of settled SS in the *Imhoff* cone = 10 mL/L

TDS

Weight of dried petri dish = 65.4711 g

Weight of dried petri dish and filterable residue = 65.5080 g

Solution

1. Calculate the concentration of TSS in raw wastewater sample, including settleable and non-settleable SS.

$$\text{Concentration of TSS} = \frac{(12.3590 - 12.3478)\text{ g}}{50\text{ mL}} \times \frac{10^3\text{ mg}}{\text{g}} \times \frac{10^3\text{ mL}}{\text{L}} = 244\text{ mg/L}$$

2. Calculate the concentration of non-settleable SS.

$$\text{Concentration of non-settleable SS} = \frac{(12.3536\ - 12.3480)\text{ g}}{50\text{ mL}} \times \frac{10^3\text{ mg}}{\text{g}} \times \frac{10^3\text{ mL}}{\text{L}} = 112\text{ mg/L}$$

3. Determine concentration of settleable SS.
   Concentration of settleable SS = TSS − non-settleable SS = (244 − 112) mg/L = 132 mg/L
4. Calculate concentration of TDS.

$$\text{Concentration of TDS} = \frac{(65.5080 - 65.4711)\text{ g}}{50\text{ mL}} \times \frac{10^3\text{ mg}}{\text{g}} \times \frac{10^3\text{ mL}}{\text{L}} = 738\text{ mg/L}$$

5. Calculate concentration of settled solids in the *Imhoff* cone.
   Volume of settled SS in 1-L *Imhoff* cone = 10 mL/L

$$\text{Concentration of solids in the sludge} = \frac{132\text{ mg/L}}{10\text{ mL/L}} \times \frac{10^3\text{ mL}}{\text{L}} = 13{,}200\text{ mg/L}$$

Example 4.18: Volatile and Fixed Solids

50 mL of municipal wastewater sample was filtered through a dried pre-weighed glass-fiber filter. The filter after filtration of solids was dried and weighed. 50 mL of the sample was evaporated in a pre-weighed petri dish. The dish was dried and weighed. Both the filter paper and the dish were ignited in a muffle furnace for 15 min, cooled, and weighed. Determine (a) non-settleable SS, and fixed and volatile solids components, (b) filterable and non-filterable solids, and fixed and volatile solids components, and (c) dissolved or filterable solids, fixed and volatile solids components.

The experimental values are given below:

i. Weight of dried filter paper in the aluminum dish = 12.3469 g
ii. Weight of aluminum dish, filter paper, and SS after drying = 12.3611 g
iii. Weight of aluminum dish and ash after ignition = 12.3507 g
iv. Weight of ignited petri dish = 65.3821 g
v. Weight of petri dish, and total (filterable and non-filterable) solids after drying = 65.4221 g
vi. Weight of petri dish and ash after ignition = 65.3920 g

Solution

1. Determine TSS, fixed suspended solids (FSS), and volatile suspended solids (VSS).

$$\text{TSS} = \frac{(12.3611 - 12.3469)\text{ g}}{50\text{ mL}} \times \frac{10^3\text{ mg}}{\text{g}} \times \frac{10^3\text{ mL}}{\text{L}} = 284\text{ mg/L}$$

$$\text{FSS} = \frac{(12.3507 - 12.3469)\text{ g}}{50\text{ mL}} \times \frac{10^3\text{ mg}}{\text{g}} \times \frac{10^3\text{ mL}}{\text{L}} = 76\text{ mg/L}$$

$$\text{VSS} = \frac{(12.3611 - 12.3507)\text{ g}}{50\text{ mL}} \times \frac{10^3\text{ mg}}{\text{g}} \times \frac{10^3\text{ mL}}{\text{L}} = 208\text{ mg/L}$$

2. Determine total solids (TS), total fixed solids (TFS), and total volatile solids (TVS).

$$\text{TS} = \frac{(65.4221 - 65.3821)\text{ g}}{50\text{ mL}} \times \frac{10^3\text{ mg}}{\text{g}} \times \frac{10^3\text{ mL}}{\text{L}} = 800\text{ mg/L}$$

$$\text{TFS} = \frac{(65.3920 - 65.3821)\text{ g}}{50\text{ mL}} \times \frac{10^3\text{ mg}}{\text{g}} \times \frac{10^3\text{ mL}}{\text{L}} = 198\text{ mg/L}$$

$$\text{TVS} = \frac{(65.4221 - 65.3920)\text{ g}}{50\text{ mL}} \times \frac{10^3\text{ mg}}{\text{g}} \times \frac{10^3\text{ mL}}{\text{L}} = 602\text{ mg/L}$$

3. Determine TDS, fixed dissolved solids, and volatile dissolved solids.
   TDS = TS − TSS = 800 − 284 mg/L = 516 mg/L
   Fixed dissolved solids = TFS − FSS = 198 − 76 mg/L = 122 mg/L
   Volatile dissolved solids = TVS − VSS = 602 − 208 mg/L = 394 mg/L

4.6.2 Chemical Quality

The chemical quality of wastewater is expressed in terms of organic and inorganic constituents. The organic components may be biodegradable, non-biodegradable, or toxic. The inorganic constituents may be essential, non-essential, or toxic. The concentration of many common chemical quality constituents of municipal wastewater and their significance are summarized in Table 4.9. Many constituents in this table are discussed in several sections.

4.7 ORGANIC MATTER AND ORGANIC STRENGTH

Major components of organic matter in municipal wastewater are proteins, carbohydrates, and fats. These components are approximately 50%, 40%, and 10%, respectively. Proteins and carbohydrates are readily biodegradable. Fats are not readily decomposed by bacteria. Many surfactants, phenolic, and refractory organic compounds resist breakdown by microorganisms. The volatile organic compounds are emitted into the atmosphere, cause health risks, and lead to the formation of photochemical oxidants. Non-biodegradable and toxic organic compounds are presented in Section 4.7.5.

4.7.1 Biochemical Oxygen Demand (BOD)

The biochemical oxygen demand (BOD) is the amount of oxygen utilized in 5 days and at 20°C by a mixed population of microorganisms under aerobic conditions to stabilize the organic matter in the sample. It is mostly expressed as BOD_5. The BOD_5 of municipal wastewater is normally 110–400 mg/L. Higher BOD_5 indicates higher organic content and greater organic-pollutional strength. The ultimate BOD is also called ultimate or total carbonaceous oxygen demand, or first-stage BOD. The ultimate BOD is reached in approximately 20 days. At this point, most of the biodegradable carbonaceous organic matter is stabilized. For municipal wastewater, the 5-day BOD is 60%–70% of ultimate BOD.[1–4]

4.7.2 Nitrogenous Oxygen Demand (NOD)

Nitrogen and phosphorus are macronutrients and are required for the growth of living systems. The sources of these nutrients in wastewater are proteins, amino acids, and urea. Nitrogen exists in various states: organic, ammonia, nitrite, and nitrate nitrogen. Microorganisms, after consuming carbonaceous organic compounds, oxidize nitrogenous compounds. The oxygen required for oxidation of ammonia to nitrite and then to nitrate is called *nitrogenous BOD*, simply *nitrogenous oxygen demand (NOD)*, or *second-stage BOD*. In a BOD test, nitrification normally starts after 6–10 days.[1–4]

4.7.3 Chemical Oxygen Demand (COD)

Chemical oxygen demand (COD) represents the amount of oxygen required to oxidize the organic matter by a strong oxidizing chemical in acid solution. There is generally a linear relationship between COD and BOD_5 results. This relationship depends entirely on the composition of the wastewater. The COD test is performed in 2 h, while BOD_5 test takes 5 days. The ratio of COD to BOD_5 in raw municipal wastewater is 1.5–2.5, and the ratio for secondary effluent is 5.0–7.0. A higher ratio means more non-biodegradable organics. Secondary effluent contains a high proportion of non-biodegradable organics; therefore, COD to BOD_5 ratio is 2.5–3.0 times higher than that of raw municipal wastewater.[1–4]

4.7.4 Total Organic Carbon (TOC)

The total organic carbon (TOC) is a measure of total carbon bound in organic matter in a sample. This test is performed very rapidly and is becoming a popular test for small concentrations of organic matter. The ratio of BOD_5 and TOC in municipal wastewater is 1.2.[1,2]

4.7.5 Non-biodegradable and Toxic Compounds

Most inorganic and organic compounds in municipal wastewater come from water supply, domestic water use, and toilet flush. Nitrogen and phosphorus are macronutrients. Many organic compounds are non-biodegradable and toxic. Examples of these compounds are heavy metals, solvents, pesticides, and herbicides. These compounds are *priority pollutants*. These compounds are selected on the basis of their known or suspected *carcinogenicity*, *mutagenicity*, *teratogenicity*, or high *acute toxicity*.[2,3] These pollutants may interfere with the treatment processes, pass through unchanged, and be transformed, generated, or accumulated in the sludge. The U.S. EPA has identified approximately 129 pollutants in 65 classes to be regulated. It is anticipated that this list will continue to change in the future. In the original inorganic list of priority pollutants, there are 13 toxic metals and non-metals, fibrous asbestos, and total cyanide. The non-metals and metals are antimony, arsenic, beryllium, cadmium, chromium, copper, lead, mercury, nickel, selenium, silver, thallium, and zinc.

The 114 representative priority organic compounds in the original list can be subdivided into aliphatics (36), aromatics (59), and pesticides (19). Approximately 30 of these organic compounds can be considered volatile, and 69 contain chlorine.[18] Examples of common organic priority pollutants are: benzene, carbon tetrachloride, dichloroethane, vinyl chloride, bromoform, aldrin, dieldrin, polychlorinated biphenyl (PCB), and many more. A complete list of priority pollutants may be found in many sources.[2,17,18]

Under the National Pollutant Discharge Elimination System (NPDES), effluent as well as industrial pretreatment standards are developed. Under the pretreatment regulations, two types of federal pretreatment standards are established: (1) prohibited discharges, and (2) categorical standards. The prohibited discharges are those that cause fire or explosion hazard, corrosion, obstruction, slug discharges, and heated discharge to sewers or POTWs. The categorical standards are developed for priority pollutants to limit their discharge into natural waters.[2,17–20]

Example 4.19: BOD Definition

If a wastewater sample has BOD_5 of 120 mg/L, what does it mean?

Solution

BOD_5 of 120 mg/L means that microorganisms will consume under aerobic conditions 120 mg of oxygen in 5 days and at 20°C to stabilize the organic matter contained in one liter of wastewater. The oxygen consumed is therefore an indirect measurement of organic content in the wastewater.

Example 4.20: Ratio of 5-Day BOD and Ultimate BOD

The ultimate BOD (L_0) of a municipal wastewater is 310 mg/L. The ratio of 5-d and ultimate BOD is 0.68. Calculate the 5-d BOD.

Solution

1. Use the ultimate BOD (L_0) of the municipal wastewater

   $L_0 = 310$ mg/L

2. Calculate BOD_5

   $$\frac{BOD_5}{L_0} = 0.68$$

   BOD_5 is 68% of the ultimate BOD

   $BOD_5 = 0.68 \times 310$ mg/L $= 210.8$ mg/L

Example 4.21: Carbonaceous and Nitrogenous Oxygen Demand

An organic waste sample has molecular formula $C_5H_7O_2N$. Its concentration in a waste stream is 300 mg/L. Calculate carbonaceous, nitrogenous, and ultimate oxygen demands in mg/L.

Solution

1. Write the balanced reaction for carbonaceous oxygen demand.

   $$C_5H_7O_2N + 5O_2 \rightarrow NH_3 + 5CO_2 + 2H_2O$$

2. Calculate the molecular weight of the waste material.

$$C_5H_7O_2N = (5\times 12 + 7\times 1 + 2\times 16 + 1\times 14)\ \text{g/mole} = 113\ \text{g/mole waste}$$

3. Calculate the ultimate carbonaceous oxygen demand (L_0).

$$L_0 = \frac{5\ \text{mole}\ O_2}{\text{mole waste}} \times \frac{32\ \text{g}\ O_2}{\text{mole}\ O_2} \times \frac{\text{mole waste}}{113\ \text{g waste}} \times 300\ \text{mg/L waste} = 425\ \text{mg/L}$$

4. Write balanced reactions for NOD.

$$NH_3 + \frac{3}{2}O_2 \rightarrow HNO_2 + H_2O$$

$$HNO_2 + \frac{1}{2}O_2 \rightarrow HNO_3$$

$$\overline{NH_3 + 2O_2 \rightarrow HNO_3 + H_2O}$$

5. Calculate the ultimate NOD (L_N).

$$L_N = \frac{2\ \text{mole}\ O_2}{\text{mole waste}} \times \frac{32\ \text{g}\ O_2}{\text{mole}\ O_2} \times \frac{\text{mole waste}}{113\ \text{g waste}} \times 300\ \text{mg/L waste} = 170\ \text{mg/L}$$

6. Calculate ultimate oxygen demand *UOD*.

$$UOD = 425\ \text{mg/L} + 170\ \text{mg/L} = 595\ \text{mg/L}$$

Example 4.22: Relationship between COD and BOD_5 of Raw Municipal Wastewater

BOD_5 and COD tests were conducted on 10 samples of same raw wastewater obtained on different days. The results are summarized below. Determine the COD/BOD_5 ratio and comment on the result.

BOD_5, mg/L	191	135	140	120	229	193	150	102	238	212
COD, mg/L	418	298	309	266	499	422	330	228	518	462

Solution

1. Plot the data
   Plot the COD on *Y* axis and BOD_5 on *X* axis. The graph is shown in Figure 4.8.
2. Develop the relationship.
   The slope and intercept of the linear relationship are obtained.

$$\text{Slope or}\left(\frac{\text{COD}}{\text{BOD}_5}\ \text{ratio}\right) = 2.13\ \text{and intercept} = 10.1\ \text{mg/L}$$

The linear relationship is expressed by the following equation:
COD = 2.13 BOD_5 + 10.1
The correlation ratio, r^2 = 1.0.

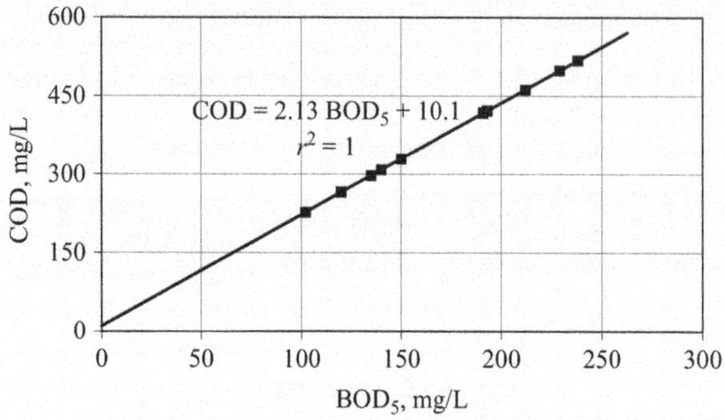

FIGURE 4.8 BOD_5 and COD relationship in raw municipal wastewater (Example 4.22). (Printed from Ref. 1 with permission of CRC Press, Taylor & Francis Group.)

3. Comments on the result.

   In raw wastewater, a significant portion of organic matter is biodegradable. The ratio of COD to BOD_5 is slightly more than 2. Also, at $BOD_5 = 0$ mg/L, the concentration of COD is about 10 mg/L. It indicates that the residual non-biodegradable organic matter exerts COD.

Example 4.23: Relationship Between Various Parameters and BOD_5 Test for Raw Municipal Wastewater

A number of studies have been conducted to establish the relationship between BOD_5 and many other tests commonly used for municipal wastewater. Tabulate these results.

Solution

The relationship between COD, TOC, and BOD_5 is obtained from many sources. These relationships are summarized in Table 4.10.

TABLE 4.10
Concentrations of COD, TOC and BOD_5 in Raw Municipal Wastewater and Their Ratios with BOD_5 Ratios

Test	Concentration, mg/L		BOD_5
	Range	Typical	Ratio
COD (standard dichromate)	330–350	340	1.62
TOC	190–280	250	1.19
BOD (ultimate)	286–330	300	1.43
BOD_5	190–220	210	1.0

Source: Adapted in part from Ref. 1

4.8 MICROBIOLOGICAL QUALITY

The municipal wastewater contains microorganisms that play an important role in biological waste treatment and public health and safety. The principal groups of microorganisms of significance in wastewater treatment include bacteria, fungi, protozoa, and algae.

4.8.1 Basic Concepts

The principal groups of organisms found in water and wastewater are *eukaryotes* and *prokaryotes*. *Eukaryotes* include algae, fungi, protozoa, and multicellular plants and animals. Each cell has a clearly defined nucleus containing nucleic acid (deoxyribonucleic acid (DNA)). The DNA has genetic information that is vital for cell reproduction. *Prokaryotes* (mainly bacteria and blue-green algae) have poorly defined nuclei. Viruses fall between living and non-living. They are not complete organisms, being made up of protein-protective coating surrounding a strand of nucleic acid.[1,2] Different organisms have different nutritional and environmental requirements. These requirements include carbon source, energy source, oxygen, temperature, and others. Basic classification of organisms based on carbon source, energy source, and oxygen and temperature requirements is presented below.[1,2,21]

Carbon Source. The **heterotrophic** organisms utilize organic carbon for cell synthesis. The **autotrophic** organisms derive their cell carbon from carbon dioxide.

Energy Source. The **phototrophic** organisms derive energy from sunlight for cell synthesis. The **chemotrophic** organisms derive energy from chemical reactions.

Oxygen Requirements. The **aerobic** organisms require molecular oxygen (O_2) for their metabolism. The **anaerobic** microorganisms grow in the absence of molecular oxygen. For strict anaerobes, molecular oxygen may be toxic. The **facultative** microorganisms grow in presence or absence of molecular oxygen.

Temperature Requirement. The optimum temperature range for **psychrophilic** or **cryophilic** organisms is 12°C–18°C. The optimum temperature range for **mesophilic** organisms is 25°C–40°C. The optimum temperature range for **thermophilic** organism is in the range of 55°C–65°C.

Brief discussion of virus, bacteria, fungi, algae, protozoa, and plants and animals in the context of water and wastewater treatment and reuse is provided below.[1,2,21]

Virus. Viruses are the smallest of infectious agents, some being as small as a single protein molecule. They are obligate parasites, and, as such, require a host. Common virus-caused diseases are infectious hepatitis, gastroenteritis, and respiratory diseases. Viruses are more resistant to disinfection than bacteria. In natural waters, there are fewer virus pathogens than bacteria.

Bacteria. Bacteria play a very important role in wastewater treatment. They are single-celled prokaryotic eubacteria. Based on morphology bacteria are (1) spherical (*cocci*) 1–3 μm, (2) rod shaped (*bacilli*) 0.3–1.5 μm (in diameter) and 1–10 μm long, (3) curved rod shaped (*vibros*) 0.6–1 μm (in diameter) and 1–6 μm long, (4) spiral (*spirilla*) 50 μm long, and (5) filamentous 100 μm long. The pathogenic bacteria that are excreted by humans may cause many diseases. Bacteria that caused common waterborne diseases are gastroenteritis, typhoid fever, dysentery, diarrhea, cholera, and others.

Fungi. Fungi are aerobic, single to multicellular, nonphotosynthetic, heterotrophic *eucaryotic protists*. Most fungi and molds are *saprophytes* (obtain food from dead organic matter). Fungi, along with bacteria, play principal role in waste treatment. Fungi can compete and perform better than bacteria at lower pH, nutrients, and moisture in wastes.

Algae. Algae are autotrophic and photosynthetic and contain chlorophyll. Algae play an important role in waste stabilization ponds.

Protozoa. Protozoa are a group of unicellular, nonphotosynthetic, and aerobic organisms. Many protozoa cause disease. *Entamoeba histolytica* (amoebic dysentery), *Giardia lamblia* (giardiasis), and *Cryptosporidium parvum* (cryptosporidiasis), are transmitted by drinking water, while *Naegleria fowleri* may enter by nasal inhalation from swimming in polluted water and causes amoebic meningoencephalitis. Protozoa feed upon bacteria, and their presence in wastewater treatment plants indicates healthy operation.

Plants and Animals. Plants and animals range in size from microscopic to larger sizes. Many plants and animals play an important role in waste treatment and effluent quality control. Larger aquatic plants are used in constructed wetlands. Many smaller animals also play an important role in natural purification and food chain. Rotifers and crustaceans are lower-order animals that prey on bacteria, protozoa, and algae. They help to maintain a balance in population of primary producers and become a part of the food chain. Sludge worms and other insect larvae feed on sludge and organic debris. Parasitic worms (*helminths*) cause many waterborne diseases such as intestinal roundworm, Guinea worm, lung fluke, and schistosomiasis.[1,2,21]

4.8.2 Indicator Organisms

Water has long served as a mode of transmission of diseases. The most important waterborne diseases are those of the intestinal tract. Pathogens (disease-causing bacteria, viruses, protozoa, and parasitic helminths) are excreted in the feces of patients and carriers. In order to determine the presence of pathogenic organisms in water, the microbiologist must have a reliable measurement technique. Unfortunately, the analytical procedures for the detection of pathogenic organisms are not clear-cut. Hence, rather than look for the specific pathogens, the microbiologists have found a group of organisms called **coliform organisms** that can measure the potential of a water to transmit disease. The coliform organisms are nonpathogenic bacteria whose origin is in fecal matter. The presence of these bacteria in water is an indication of fecal contamination and probably unsafe water. The coliform bacteria are gram-negative, non-spore-forming *bacilli* capable of fermenting lactose with the production of acids and gases. The coliform bacteria are members of the *Enterobacteriaceae* family and include genera *Escherichia*, *Klebsiella*, *Citrobacter*, and *Enterobacter*. The *Escherichia coli (E. coli)* species appears to be most representative of fecal contamination. Fecal *Streptococci* and *Enterococci* are also used as indicator organisms under specific conditions.[1–4,14,21]

The coliform organisms were originally believed to be entirely of fecal origin, but it has been shown that certain genera can grow in soil. Therefore, the presence of coliform organisms in surface water may be due to fecal wastes from human and animal sources and from soil erosion. To separate *E. coli* species from possible soil type, special tests are generally run. Frequently, for this purpose, fecal *Streptococci* are also used as indicator organisms, because they also originate from the intestinal tract of warm-blooded animals.[1–4,21]

Each person discharges from 100 to 400 billion coliform organisms per day, in addition to many other types of bacteria. The numbers of specific indicator organisms found in untreated municipal wastewater are: total coliform (TC) 10^5–10^8, fecal coliform (FC) 10^4–10^5, fecal *Streptococci* (FS) 10^3–10^4, and *Enterococci* 10^2–10^3 per mL. The standard techniques used to enumerate the coliform organisms are multiple-tube fermentation and the membrane filter.[21]

Example 4.24: Common Waterborne Diseases and Causative Agents

Two lists are given below. Match the terms.

List A		List B	
i	Giardiasis	a	Virus
ii	Schistosomiasis	b	Helminth
iii	Dysentery	c	G. *lamblia*
iv	Hepatitis	d	Bacteria
v	Typhoid fever	e	Entamoeba histolytica
vi	Cholera	f	Continued fever
vii	Amoebic meningoencephalitis	g	Naegleria fowleri

Solution

The matching terms are (i–c), (ii–b), (iii–e), (iv–a), (v–f), (vi–d), and (vii–g).

Example 4.25: Classification Based on Nutritional and Environmental Requirements of Microorganisms

Match the terms in lists A and B.

List A		List B	
i	DNA	a	Need oxygen
ii	Phototrophs	b	12°C–18°C
iii	Prokaryotic cell	c	Do not need oxygen
iv	Anaerobic	d	Need organic carbon
v	Heterotrophic	e	Genetic information
vi	Aerobic	f	Nucleus clearly defined
vii	Psychrotrophic	g	Sunlight
viii	Mesophilic	h	Nucleus poorly defined
ix	Actinomycetes	i	25°C–40°C
x	Protozoa	j	Filamentous
xi	Eukaryotic cell	k	*Guardia lamblia*

Solution

The matching terms are (i–e), (ii–g), (iii–h), (iv–c), (v–d), (vi–a), (vii–b), (viii–i), (ix–j), (x–k), and (xi–f).

Example 4.26: Coliform Number in Raw Wastewater

It is estimated that each person discharges 150 billion total coliform organisms per day. If the average wastewater flow is 120 gpcd, what is the average coliform density per mL in raw fresh wastewater?

Solution

$$\text{Average total coliform density} = \frac{150 \times 10^9 \text{ coliform organism/person d}}{120 \text{ gal/person d} \times 3.8 \text{ L/gal}}$$

$$= 3.3 \times 10^8 \text{ coliform organisms/L or } 3.3 \times 10^5 \text{coliform organisms/mL}$$

4.9 TOXICITY AND BIOMONITORING

On March 9, 1984, the U.S. EPA published a new national policy based on the 1972 Clean Water Act, prohibiting the discharge of toxic pollutants in toxic amounts into natural waters.[19,20] The effluents from the municipal wastewater treatment facility may contain numerous toxic chemicals that may cause synergistic effects. For this reason, instead of specific chemical tests, biomonitoring was introduced. Biomonitoring requirements involve the following steps: (1) assess the toxicity of wastewater effluent to natural aquatic life, (2) assess the effectiveness and degree of wastewater treatment to meet water pollution control requirements, and (3) determine compliance with federal and state water quality standards and water quality criteria associated with NPDES permits.[21,22] The biomonitoring protocols have been published and commented on extensively in the literature. The results of acute toxicity tests are reported in 48 or 96 h. The lethal concentration LC_{50} is the concentration of effluent in dilution water that causes 50% mortality of the test organisms. The chronic toxicity is measured over a long period or generations. The results are based on mortality, reduced growth, or reproductive behavior.[1,2,22–30]

Based on toxicity test evaluation, if the toxicity level is found and verified, *toxicity reduction evaluation* is required. These studies involve *toxicity investigation examination* and *effluent toxicity treatability*. The objective is to determine the best method to reduce or eliminate the toxicity of the effluent. The plan may include process modification, chemical change, construction, or industrial pretreatment. After reduction or elimination of toxicity, the plant is considered in compliance by the U.S. Environmental Protection Agency.[1–4,31]

Example 4.27: Biomonitoring and LC_{50} of Wastewater Effluent

What is the purpose of effluent biomonitoring? Define acute and chronic toxicity tests and LC_{50}. A wastewater treatment plant effluent was tested for toxicity using a biomonitoring protocol. A series of dilutions from volume of effluent and volume of receiving water was made. Freshwater organism *Ceriodaphina dubia* was used in the test. The results of acute toxicity were reported in 48 h. At 30% and 80% dilutions, the recorded mortality of the organisms was 15% and 80%, respectively. Calculate LC_{50} using an arithmetic relationship.

Solution

1. Describe the purpose of biomonitoring

   The effluent from a wastewater treatment plant may contain numerous toxic chemicals that may have synergistic effects. For this reason, instead of evaluating the effects of each chemical on aquatic life, biomonitoring was introduced.
2. Define acute and chronic toxicity tests, and LC_{50}

   The acute toxicity is measured by death of organism in 48 or 96 h.

   The chronic toxicity is measured over a long period or generations. The results are based on mortality, reduced growth, mobility, or reproductive behavior.

The lethal concentration LC_{50} is the concentration of effluent in dilution water or water from specific site of receiving water that has 50% mortality of test organisms.

3. Calculate LC_{50}

At 30% dilution mortality is 15%, and at 80% dilution, mortality is 80%.

$$\text{Using arithmetic relationship, slope } m \text{ of the line} = \frac{(80-30)\%}{(80-15)\%} = \frac{50}{65} = 0.769$$

At $y = 80\%$, $x = 80 \times 0.769 = 61.52\%$. Intercept $C = (80 - 61.52)\ \% = 18.48\%$

The linear equation is $y = 0.769x + 18.48$

x at 50% mortality $= 0.769 \times 50 + 18.48 = 56.93\%$

$LC_{50} = 56.93\%$. At 56.93% dilution, mortality is 50%.

4.10 POLLUTED RUNOFF: NONPOINT SOURCE POLLUTION

NPS pollution generally results from land runoff, precipitation, atmospheric deposition, drainage, seepage, or hydrologic modification. NPS pollution, unlike pollution from point sources such as municipal and industrial wastewater treatment plants, comes from many diffuse sources. NPS pollution is caused by rainfall or snowmelt moving over and through the ground. As the runoff moves, it picks up and carries away natural and human-made pollutants, finally depositing them into lakes, rivers, wetlands, coastal waters, and groundwater. NPS pollution can include (1) excess fertilizers, herbicides, and insecticides from agricultural lands and residential areas; (2) oil, grease, and toxic chemicals from urban runoff; (3) sediment from improperly managed construction sites, crop and forest lands, and eroding streambanks; (4) salt from roads and acid drainage from abandoned mines; (5) bacteria and nutrients from livestock, pet wastes, and faulty septic systems; and (6) atmospheric deposition and hydromodification, and habitat alteration. Many states report that NPS pollution is the leading leftover cause of water quality problems. The effects of NPS pollutants on specific waters vary and may not always be fully assessed. However, we know that these pollutants have harmful effects on drinking water supplies, recreation, fisheries, and wildlife, including human health.[32]

Discharges from NPS may often have toxic contaminants and chemicals, bacteria, and viruses that can adversely affect the quality of receiving waters. These toxic contaminants and chemical compounds include heavy metals such as lead, mercury, zinc, and cadmium; organics like polychlorinated biphenyls (PCBs) and polycyclic aromatic hydrocarbons (PAHs); and other substances resistant to breakdown. The adverse health effects associated with many of these toxic substances are blood and brain disorders, kidney and liver damage, tremors, impaired cognitive skills, damage to nerve receptors, lung damage, skin disorders, lowered immune responses, fatigue, headaches, and coughs. Some compounds are probable human carcinogens. Pathogens found in contaminated runoff may include: *Cryptosporidium parvum*, *Giardia lamblia*, *Salmonella*, *E. coli*, and parasitic worms (helminths). Primary symptoms of *C. parvum* infection are diarrhea. Other symptoms can include anorexia, nausea/vomiting, and abdominal pain.[32]

4.11 FLOW EQUALIZATION

4.11.1 Need and Types

All wastewater treatment facilities experience hourly, daily, and seasonal flow and strength variations. Flow equalization is damping of the flow rate variations so that a relatively constant flow rate may be maintained through the downstream facilities. Depending upon the situation, flow equalization facilities therefore are provided to (1) overcome the operational problems caused by the flow variations, (2) reduce the surge through the units, (3) equalize the strength, and dilute toxic inhibitory constituents, and maintain uniform concentrations and pH, (4) improve the performance of the process, and (5) reduce the size and cost of treatment facilities.[32–34]

The flow equalization may be of two types: (1) in-line and (2) off-line. In the in-line system, the entire flow is passed through the equalization basin. A constant flow is pumped from the basin. In an off-line system, only the excess flow rate is diverted to the equalization basin and is rerouted through the plant under low-flow situations. In-line equalization basins offer better damping of the constituent mass loading, while only slight damping of constituents is achieved with off-line equalization basins.[1–4]

4.11.2 Design Considerations

The design of equalization basin should utilize mixing and aeration to prevent solids deposition and odor problems. Location below grit removal or primary sedimentation basin is desirable. Rectangular, square, and circular shapes with 4–6 m liquid depth have been adapted by designers. Controlled flow pumping is necessary from the equalization basin.

Example 4.28: Process Train with Equalization Basins

Equalization basin is provided between grit and primary sedimentation basins. Draw the process trains to show in-line and off-line arrangements of the equalization basins.

Solution

The process trains for in-line and off-line flow equalization basins, and flow patterns before and after equalization are shown in Figure 4.9.

4.11.3 Design Volume of Equalization Basin

The volume of an equalization basin is established for flow equalization. Basin sizing for flow equalization is given below.

The volume or storage capacity of an equalization basin is determined by either graphical or analytical technique. The graphical technique requires preparation of a mass diagram in which cumulative inflow volume and cumulative average-day volume are plotted on the same plot against the time of the day.[1,2,33–35] The analytical technique is based on cumulative change in volume deficiency between inflow and outflow as expressed by Eqs. (4.9) and (4.10).

$$\Delta V = V_{in} - V_{out} \tag{4.9}$$

$$\Delta V = \left(Q_{in} - Q_{out}\right)\Delta t \tag{4.10}$$

where

ΔV = change in storage volume during a specified time interval, (volume)
V_{in} = total inflow volume during the specified time interval, (volume)
V_{out} = total outflow volume during the specified time interval, (volume)
Q_{in} = total inflow, (volume/time)
Q_{out} = total outflow, (volume/time)
Δt = time interval, (time)

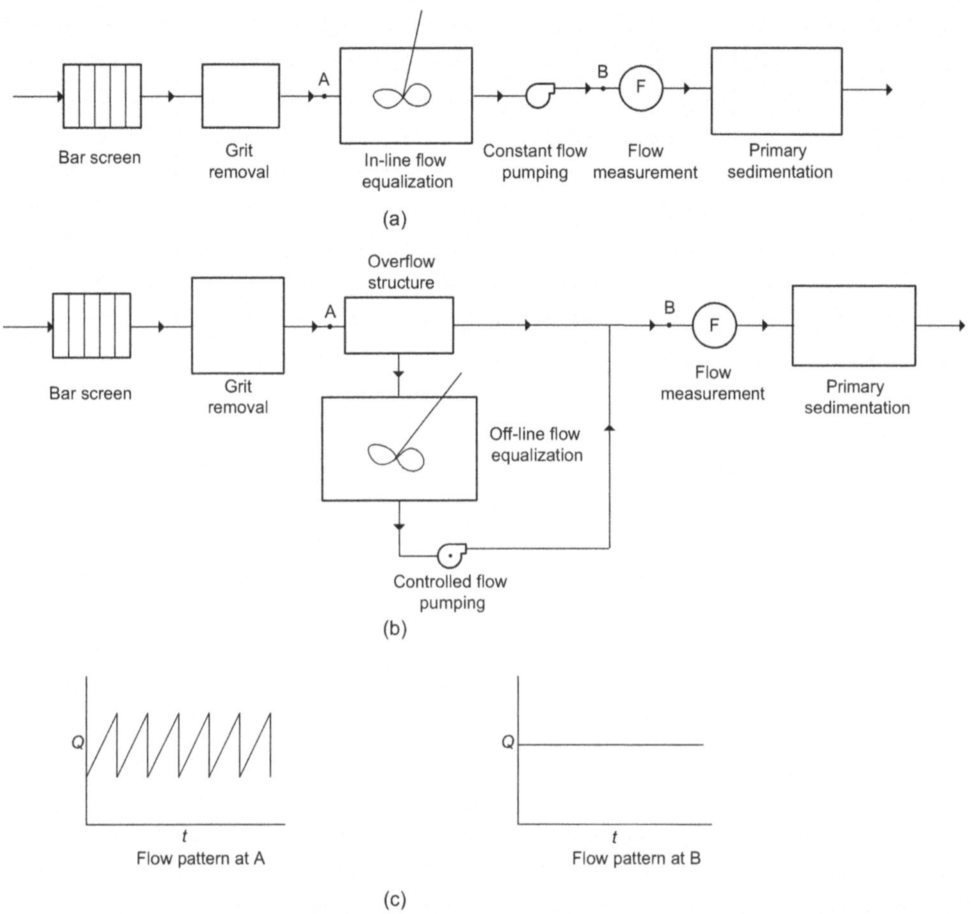

FIGURE 4.9 Process trains and flow pattern with equalization basins, (a) in-line and (b) off-line, and (c) flow patterns (Example 4.28). (Printed from Ref. 1 with permission of CRC Press, Taylor & Francis Group.)

Example 4.29: Storage Volume Based on Mass Diagram and Change in Volume

The hourly flow pattern of an industrial plant is given below. The treatment facility is designed for a constant flow rate of 1,500 m^3/h. Determine the volume of an in-line equalization basin using mass diagram and change in volume calculations. Also, draw the line representing the volume remaining in the basin.

Time period	Midnight–1 AM	1–2	2–3	3–4	4–5	5–6	6–7	7–8	8–9	9–10	10–11	11 AM–Noon
Flow, m^3/h	1,050	960	930	930	975	1,080	1,170	1,350	1,570	1,800	2,100	2,235

Time period	Noon–1 PM	1–2	2–3	3–4	4–5	5–6	6–7	7–8	8–9	9–10	10–11	11 PM–Midnight
Flow, m^3/h	2,250	2,235	2,100	1,920	1,680	1,380	1,230	1,350	1,574	1,574	1,383	1,174

Solution

1. Develop the calculation table for the mass diagram and change in volume.
   The calculations are provided in Table 4.11.
2. Develop the mass diagram.
   a. Plot cumulative inflow and outflow with respect to time of the day. This plot is shown in Figure 4.10.
   b. Draw two tangents at points *A* and *B* on cumulative inflow line. Points *A* and *B* are at the valley and hump of the curvatures. The tangents are parallel to cumulative outflow or pumping line.
3. Determine the required volume of equalization basin from mass diagram.
   The vertical distance between the two tangents is the required theoretical volume of the equalization basin. This volume is 4,390 m³.

TABLE 4.11

Calculation Table for Determination of Theoretical Volume of Equalization Basin from Mass Diagram and Change in Volume (Example 4.29)

Time	Δt, h	Q_{in}, m³/h	Q_{out}, m³/h	$\sum \Delta V_{in}$, m³	$\sum \Delta V_{out}$, m³	ΔV, m³	$\sum \Delta V$, m³
(1)	(2)	(3)	(4)	(5)	(6)	(7)	(8)
Midnight	—	—	—	—	—	—	—
1 AM	1	1,050	1,500	1,050	1,500	–450	3,105
2	1	960	1,500	2,010	3,000	–540	2,565
3	1	930	1,500	2,940	4,500	–570	1,995
4	1	930	1,500	3,870	6,000	–570	1,425
5	1	975	1,500	4,845	7,500	–525	900
6	1	1,080	1,500	5,925	9,000	–420	480
7	1	1,170	1,500	7,095	10,500	–330	150
8	1	1,350	1,500	8,445	12,000	–150	0[a]
9	1	1,570	1,500	10,015	13,500	70[b]	70
10	1	1,800	1,500	11,815	15,000	300	370
11	1	2,100	1,500	13,915	16,500	600	970
Noon	1	2,235	1,500	16,150	18,000	735	1,705
1 PM	1	2,250	1,500	18,400	19,500	750	2,455
2	1	2,235	1,500	20,635	21,000	735	3,190
3	1	2,100	1,500	22,735	22,500	600	3,790
4	1	1,920	1,500	24,655	24,000	420	4,210
5	1	1,680	1,500	26,335	25,500	180	4,390[c]
6	1	1,380	1,500	27,715	27,000	–120	4,270[d]
7	1	1,230	1,500	28,945	28,500	–270	4,000
8	1	1,350	1,500	30,295	30,000	–150	3,850
9	1	1,574	1,500	31,869	31,500	74	3,924
10	1	1,574	1,500	33,443	33,000	74	3,998
11	1	1,383	1,500	34,826	34,500	–117	3,881
Midnight	1	1,174	1,500	36,000	36,000	–326	3,555

[a] Basin is empty.
[b] Basin starts filling.
[c] Basin full. Required theoretical capacity of the basin is reached.
[d] Basin starts emptying.

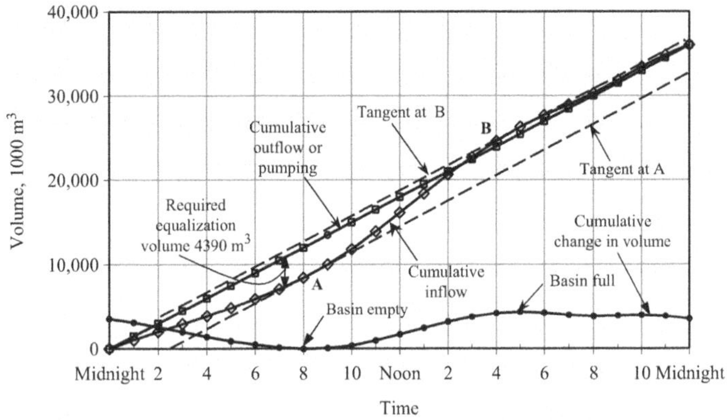

FIGURE 4.10 Mass diagram for determination of the capacity of the equalization basin, and volume remaining (Example 4.29). (Printed from Ref. 1 with permission of CRC Press, Taylor & Francis Group.)

4. Determine the required volume of the equalization basin using change in volume calculations.

   The change in volume calculations is summarized in Table 4.11. The following information may be drawn:

   a. The basin is empty at 8:00 AM.
   b. The basin starts filling after 8:00 AM with the incoming flow to the plant. The cumulative inflow in the basin starts with 70 m³ at 9:00 AM and continues to increase.
   c. It is completely full to its maximum volume of 4,390 m³ at 5:00 PM.
   d. The basin starts emptying after 5:00 PM from cumulative volume of 4,270 m³ at 6:00 PM.
   e. The basin is totally empty at 8:00 AM.
   f. This cycle repeats.

5. Draw the line representing the basin volume remaining based on cumulative volume change (Figure 4.10).

   a. From 8:00 AM to 5:00 PM is the filling cycle. The basin is filled in 9 h. From 5:00 PM to 8:00 AM is the emptying cycle. The basin is emptied in 15 h.
   b. During the filling and emptying cycles, a constant flow of 1,500 m³/h is withdrawn from the basin.
   c. The maximum volume of 4,390 m³ on the curve represents the required theoretical volume of the equalization basin.

Example 4.30: Volume of an In-line Equalization Basin

An in-line equalization basin receives variable flow. Determine the volume of an equalization basin to withdraw a constant flow of 652 m³/min into a biological wastewater treatment plant. Use mass diagram and volume change calculations to determine the volume. Indicate what time of the day the basin will be full and empty.

Time period	Midnight–1 AM	1–2	2–3	3–4	4–5	5–6	6–7	7–8	8–9	9–10	10–11	11 AM–Noon
Average flow rate during time period, m^3/min	582	468	348	276	222	210	252	432	750	870	900	912
Time period	Noon–1 PM	1–2	2–3	3–4	4–5	5–6	6–7	7–8	8–9	9–10	10–11	11 PM–Midnight
Average flow rate during time period, m^3/min	900	858	817	744	690	690	696	774	864	846	804	745

Solution

1. Determine the constant withdrawal rate from the equalization basin during each hour of pumping.
   Constant flow rate = 652 $m^3/min \times 60$ min/h = 39.1×10^3 m^3/h
2. Determine the average variable flow during each hour time period.
   Flow during midnight and 1 AM = 582 $m^3/min \times 60$ min/h $\times$ 1 h (duration) = 34.9×10^3 m^3
   Repeat calculations for each hour time period.
3. Prepare a calculation table to determine the volume of the basin.
   The cumulative inflow and cumulative outflow, and change in volume are summarized in Table 4.12.
4. The basin is empty at 8:00 AM and starts filling. At 9:00 AM inflow volume is 5.88 $\times 10^3$ m^3. At 12:00 midnight, a maximum volume of 146 $\times 10^3$ m^3 is released. From 1:00 AM, the volume starts to decrease. This cycle repeats.
4. Determine the basin volume from the mass diagram.
   The basin mass diagram and volume remaining at any time of the day are plotted in Figure 4.11. The basin volume when full is 146 $\times 10^3$ m^3.
5. Determine basin volume based on volume change.
   The theoretical volume of the equalization basin is 146 $\times 10^3$ m^3. The basin will be empty at 8 am and it will be full at midnight.
6. Draw the volume change curve.
   The equalization basin starts filling at 8:00 AM and is full at midnight in a period of 16 h. The basin starts emptying at midnight and becomes empty at 8:00 AM in a period of 8 h. The volume change curve is also shown in Figure 4.11.

TABLE 4.12
Calculation Table to Determine Equalization Basin Volume (Example 4.30)

Time of Day	Q_{in}, m^3/min	V_{in}, 10^3 m^3	$\sum \Delta V_{in}$, 10^3 m^3	V_{out}, 10^3 m^3	$\sum \Delta V_{out}$, 10^3 m^3	ΔV, 10^3 m^3	$\sum \Delta V$, 10^3 m^3
Midnight	—	—	—	—	—	—	—
1 AM	582	34.9	34.9	39.1	39.1	–4.20[a]	141
2	468	28.1	63.0	39.1	78.2	–11.0	130
3	348	20.9	83.9	39.1	117	–18.2	112

(Continued)

TABLE 4.12 (*Continued*)
Calculation Table to Determine Equalization Basin Volume (Example 4.30)

Time of Day	Q_{in}, m³/min	V_{in}, 10³ m³	$\sum \Delta V_{in}$, 10³ m³	V_{out}, 10³ m³	$\sum \Delta V_{out}$, 10³ m³	ΔV, 10³ m³	$\sum \Delta V$, 10³ m³
4	276	16.6	100	39.1	156	−22.6	89.5
5	222	13.3	114	39.1	196	−25.8	63.7
6	210	12.6	126	39.1	235	−26.5	37.2
7	252	15.1	141	39.1	274	−24.0	13.2
8	432	25.9	167	39.1	313	−13.2	0.00[b]
9	750	45.0	212	39.1	352	5.88[c]	5.88
10	870	52.2	265	39.1	391	13.1	19.0
11	900	54.0	319	39.1	430	14.9	33.8
Noon	912	54.7	373	39.1	469	15.6	49.4
1 PM	900	54.0	427	39.1	509	14.9	64.3
2	858	51.5	479	39.1	548	12.4	76.7
3	817	49.0	528	39.1	587	9.90	86.5
4	744	44.6	572	39.1	626	5.52	92.1
5	690	41.4	614	39.1	665	2.28	94.3
6	690	41.4	655	39.1	704	2.28	96.6
7	696	41.8	697	39.1	743	2.64	99.2
8	774	46.4	743	39.1	782	7.32	107
9	864	51.8	795	39.1	822	12.7	119
10	846	50.8	846	39.1	861	11.6	131
11	804	48.2	894	39.1	900	9.12	140
Midnight	745	44.7	939	39.1	939	5.58	146[d]

[a] Basin starts emptying cycle.
[b] Basin is empty.
[c] Basin starts filling.
[d] Basin is full. Required theoretical capacity of the basin is reached.

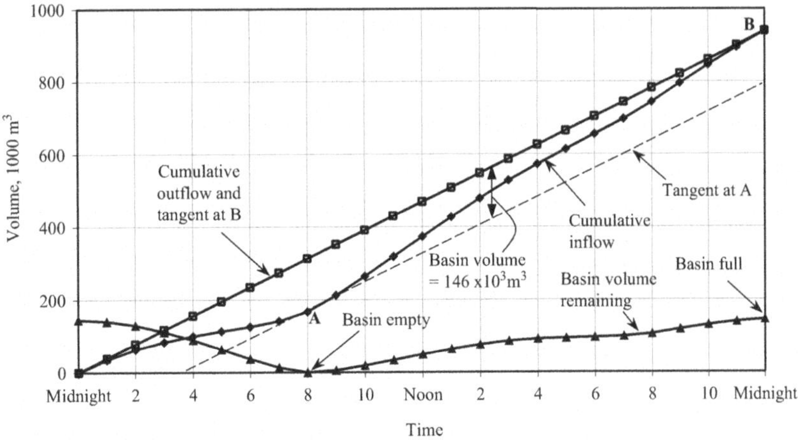

FIGURE 4.11 Mass diagram for determination of the capacity of the equalization basin, and volume remaining (Example 4.30). (Printed from Ref. 1 with permission of CRC Press, Taylor & Francis Group.)

4.12 REVIEW PROBLEMS AND DISCUSSION TOPICS

4.1 The hourly diurnal water demand and wastewater flow data of a city are given below. Calculate the average water demand and average wastewater flow. What percentage of the water is returned to the wastewater treatment facility?

	AM						PM						
Time	**12**	**2**	**4**	**6**	**8**	**10**	**12**	**2**	**4**	**6**	**8**	**10**	**12**
Water demand, m^3/h	21.3	13.8	8.8	17.5	82.5	66.3	37.5	38.8	47.5	68.8	72.5	50.0	21.3
Wastewater flow, m^3/h	27.5	21.3	18.1	20.1	27.5	41.3	57.5	63.8	53.8	46.5	54.4	42.5	27.5

4.2 A 200-home residential subdivision has been developed. Calculate the annual water saving that can be achieved if, instead of conventional plumbing, all homes are installed with faucet aerators, water-efficient washing machines and dishwashers, and shallow-trap water closets. Assume that average non-conserving water consumption is 380 Lpcd, and there are an average of 3.5 residents per home. Make use of the information provided in Table 4.2. Average savings from conservation devices are given in Table 4.4.

4.3 A home has installed a PRV at the point of entry into the home. If water usage is 350 Lpcd and there are four residents, calculate the percent water saving. Typical breakdown of residential water uses is given in Table 4.2. Average savings from PRV are assumed to be 25%.

4.4 Total average water loss in a home caused by toilet leaks, faucet drips, and leakage, and from the lawn sprinkler is approximately 4% of total water demand. The water supply pressure is 345 kPa. A PRV is installed, which drops the supply pressure in the home to 110 kPa. Assuming that water demand is 304 Lpcd and all leaks behave like orifices, calculate the water saving that will be achieved from these sources after PRV is installed.

4.5 A wastewater treatment plant receives an average 87% of the water supply during dry weather conditions. Average expected municipal water demand is 400 Lpcd. Calculate average and maximum dry weather flows. The population of the town is 45,000.

4.6 A small subdivision of a city is being developed. The ultimate zoning plan shows the following: residential, institutional, and commercial establishments.

Zoning Component	Units	Loading Lpd/unit
Single-family high income	1,500 residents	380
Single-family medium income	800 residents	310
Single-family low income	500 residents	270
One apartment complex	500 residents	230
One hotel/motel	500 units	380
One hospital	300 beds	950
One nursing home	150 beds	380
One boarding school	1,500 students	300
Five restaurants	1,900 customers	30

(Continued)

(Continued)

One bar	150 customers	8
Shopping centers	250 employees	40
Office building	1,500 employees	65
Three barber shops	30 chairs	210
Two beauty salons	20 stations	1,026
One commercial laundry	20 machines	2,500
One service station	10 bays	3,800
One movie theater	500 seats	8

Calculate the residential water demand, and institutional and commercial water demand. What is the average water demand for the entire community?

4.7 The residential population of a city is 60,000. The average water demand is 350 Lpcd. The institutional and commercial, and industrial average areas in the city are 200 and 300 ha, and the water demand expected is 20 and 23 m^3/ha/d. The public water use and water unaccounted for are 10% and 6% of total municipal water demand, respectively. Calculate the total municipal water demand, and that of each component as a percent of total municipal water demand.

4.8 The service area of a community is 1,650 ha. Average estimated infiltration/inflow allowance is 3,500 L/ha/d. Calculate the I/I quantity per capita per day. Assume that the population density of the service area is 25 persons/ha.

4.9 The total length of the sewerage system of a city is 300 km. The weighted equivalent diameter of the sewers is 21 cm. Calculate the I/I flow per capita per day that may reach the wastewater treatment plant if the I/I allowance is 1,500 L/d/cm/km. The average sewerage length per capita is 8 m.

4.10 Total population of a city is 68,000, and average wastewater flow rate is 380 Lpcd. The equivalent length of wastewater collection system is 10 m/capita. The distribution of collection system is as follows.

Sewer diameter, cm	15	20	31	46	61
Distribution, %	4	75	12	6	3

The permissible I/I allowance is 1,200 L/d/cm/km (1,300 gpd/in/mile). The peaking factor for I/I is 1.5. Determine I/I flow in Lpcd, and peak wet weather flow in m^3/d.

4.11 The total population of a city is 25,000. The design average flow to a wastewater treatment plant is 100 gpcd. Determine the average annual sustained one-day and five-consecutive-days peak and low flows in mgd. Use the information given in Figure 4.4.

4.12 The design period and estimated population of a city are 15 years and 35,000 residents. The expected average per capita wastewater flow and peak I/I allowances are 400 and 150 L/capita/d. The estimated peak industrial flow is 3,500 m^3/d. Calculate the design peak flow.

4.13 SSs test was performed on a wastewater sample. Determine TSS, VSS, and FSS. Use the following data:
Volume of sample filtered = 60 mL
Weight of dried filter paper in aluminum dish = 12.7324 g
Weight of dried filter paper + dish + TSS = 12.7459 g
Weight of dish + ash after ignition= 12.7423 g

4.14 A wastewater sample is analyzed for solids. The total solids (TS) concentration is 960 mg/L. TSS is 35% and total volatile suspended solids (TVSS) is 70% of TSS. TDS has a 45% volatile fraction. Calculate the concentrations of TSS, TVSS, TFSS, TDS, TVDSS, and TFDS.

4.15 Calculate the carbonaceous BOD of a sample. The BOD_5 is 120 mg/L, and the ratio of BOD_5 and BOD_L is 0.68.

4.16 The BOD_5 and TOC test results of a wastewater stream over 7-day of data collection are given below. Develop the linear relationship and average BOD_5/TOC ratio.

Sampling Day	1	2	3	4	5	6	7
BOD_5, mg/L	122	97	215	250	100	139	144
TOC, mg/L	102	80	180	210	85	115	120

4.17 Two lists are given below. Match the terms.

List A	List B
i. Phototroph	a. Oxygen must be absent
ii. Autotroph	b. G. *lamblia*
iii. Anaerobes	c. 12°C–18°C
iv. Giardiasis	d. Poorly defined nucleus
v. Cryophilic	e. Derive energy from sunlight
vi. Prokaryotes	f. *Eukaryotic protists*
vii. Fungi	g. Derive cell carbon from CO_2

4.18 Identify the characteristics that apply to coliform organisms.
- a. Nonpathogenic
- b. Gram negative
- c. Members of *Enterobacteriaceae* family
- d. Present in wastewater
- e. Non-spore-forming cells
- f. Include genera *Klebsiella*
- g. Incapable of fermenting lactose
- h. Presence means safe water

4.19 An equalization basin is designed for a wastewater treatment plant. The hourly average flow data over a 24-h period are summarized below. The cumulative inflow in a day is equal to cumulative outflow from the basin.

Time of Day	1 AM	2	3	4	5	6	7	8	9	10	11	Noon
Flow, m^3/h	10	10	11	12	14	16	20	30	35	40	45	50
Time of Day	1 PM	2	3	4	5	6	7	8	9	10	11	Midnight
Flow, m^3/h	80	80	100	80	60	40	35	30	20	16	12	10

Determine (a) the theoretical volume of the equalization basin using mass diagram and change in volume, (b) maximum constant outflow to the plant, and (c) the times when the basin will be empty and full. Also, draw a flow line representing the volume remaining in the basin.

REFERENCES

1. Qasim, S. R., and G. Zhu, Wastewater Treatment and Reuse: Theory and Design Examples, vol. 1, Principles and Basic Treatment, CRC Press, Taylor & Francis Group, Boca Raton, FL, 2018.
2. Qasim, S. R., *Wastewater Treatment Plants: Planning Design and Operation*, 2nd ed., CRC Press, Boca Raton, FL, 1999.
3. Metcalf and Eddy|AECOM, *Wastewater Engineering: Treatment and Resource Recovery*, 5th ed., McGraw-Hill, New York, 2014.
4. Tchobanoglous, G., and E. D. Shroeder, *Water Quality: Characteristics, Modeling, and Modification*, Addison-Wesley Publishing Co., Reading, MA, 1985.
5. Metcalf and Eddy, Inc. *Wastewater Engineering: Collection and Pumping of Wastewater*, McGraw – Hill Book Company, New York, 1981.
6. Murrary, C. R., and E. B. Reeves, *Estimated Use of Water in the United States in 1970*, U.S. Geological Survey, Department of Interior, Geological Survey Circular 676, Washington, DC, 1972.
7. Dufor, C. N, and E. Becker, *Public Water Supplies of the 100 Largest Cities in the United States*, U.S. Geological Survey, Water Supply Paper 1812, 1964, p. 35.
8. Bailey, J. R., R. J. Benoit, J. L. Dodson, J. M. Robb, and H. Wallman, *A study of Flow Reduction and Treatment of Wastewater from Households*, EPA Water Pollution Control Research Series Report 11050 FKE 12/69, Advanced Water Treatment Research Laboratory, Cincinnati, OH, December 1969.
9. U. S. Department of Housing and Urban Development, Residential Water Conservation, Summary Report, June 1984.
10. U. S. Department of Housing and Urban Development, Water Saved by Low-Flush Toilets and Low-Flow Shower Heads, March 1984.
11. Joint Committee of the Water Pollution Control Federation and American Society of Civil Engineers, Design and Construction of Sanitary and Storm Sewer, WPCF Manual of Practice No. 9, Water Pollution Control Federation, Washington, DC, 1970.
12. U. S. Environmental Protection Agency, *Handbook for Sewer System Evaluation and Rehabilitation*, Municipal Construction Division, Office of Water Program Operations, Washington, DC, 1975.
13. Water Environment Federation, *Prevention and Control of Sewer System Overflow, 3e-MOP FD-17*, 3rd ed., Water Environment Federation, August 9, 2011.
14. APHA, AWWA, and WEF, *Standard Methods for Examination of Water and Wastewater*, 22nd ed., American Public Health Association, Washington, DC, 2012.
15. Faust, D. D., and O. M. Aly, *Chemistry of Water Treatment*, 2nd ed., Lewis Publishers, Boca Raton, FL, 1997.
16. Perry, H. S., D. R. Rowe, and G. Tchobanoglous, *Environmental Engineering*, McGraw-Hill Book Co., New York, 1985.
17. Miller, L. A., R. S. Taylor, and L. A. Monk, *NPDES Permit Handbook*, Government Institutes, Inc., Swidler & Berlin Chartered, Washington, DC, May 1991.
18. Davis, M. L., and D. A. Cornell, *Introduction to Environmental Engineering*, 2nd ed., McGraw-Hill, Inc., New York, 1991.
19. Federal Water Pollution Control Act Amendments of 1972 (PL 92–500), 92nd Congress, October 18, 1972.
20. Clean Water Act (PL 92–217), 95th Congress, December 27, 1977.
21. Wheelis, M., *Principles of Modern Microbiology*, Jones & Bartlett Publishers, Sudburr, MA, 2011.
22. U. S. Environmental Protection Agency, Biomonitoring to Achieve Control of Toxic Effluents, Technology Transfer, EPA/625/8–87/013, Duluth, MN, September 1987.
23. McIntyre, D., M. DeGraere, and J. Fava, Monitoring Effluent for Toxicity, Specialty Conference on Toxicity Based Permits for NPDES Compliance and Laboratory Techniques, Water Pollution Control Federation, April 16–19, 1989.
24. U. S. Environmental Protection Agency, Technical Support Document for Water Quality-Based Toxics Control, U.S. EPA Office of Water, EPA-440/4–85/032, Washington, DC, 1985.
25. Peltier, W. and C. I. Weber, Methods for Measuring the Acute Toxicity of Effluents to Aquatic Organisms, 3rd ed., EPA-600/4–85-013, Office of Research and Development, Cincinnati, OH, 1992.
26. U. S. Environmental Protection Agency, Short Term Methods for Estimating Chronic Toxicity of Effluents and Receiving Waters to Freshwater Organisms, 3rd ed., EPA-600/4-89/001, Cincinnati, OH, 1992.

27. U. S. Environmental Protection Agency, Methods for Measuring the Acute Toxicity of Effluents to Freshwater and Marine Organisms, U.S. EPA Environmental Monitoring and Support Laboratory, EPA-600/4–85/013, Cincinnati, OH, 1985.
28. U. S. Environmental Protection Agency, User's Guide to Conduct and Interpretation of Complex Effluent Toxicity Tests at Estuarine/Marine Sites, EPA-600/X-86/224, Washington, DC, 1985.
29. U. S. Environmental Protection Agency, Short Term Methods for Estimating the Chronic Toxicity of Effluents and Receiving Waters to Marine and Estuarine Organisms, EPA-600/4-88/028, Cincinnati, OH, 1988.
30. Stephen, C. E., Methods for Calculating LC50, In *Aquatic Toxicology and Hazard Evaluation*, F. L. Mayer, and J. L. Hamelink (Editors), ASTM STP 634, American Society for Testing and Materials, Philadelphia, PA, 1984, pp. 65–84.
31. Louvar, J. F. and B. D. Louvar, *Health and Environmental Risk Analysis: Fundamental Management with Applications*, Environmental Management and Engineering Series, vol. 2, Prentice Hall PTR, Upper Saddle River, NJ, 1998.
32. U. S. Environmental Protection Agency, Polluted Runoff: Nonpoint Source (NPS) Pollution, https://www.epa.gov/nps/basic-information-about-nonpoint-source-nps-pollution, 2019.
33. U. S. Environmental Protection Agency, Flow Equalization, Office of Technology Transfer, Washington, DC, 1974.
34. MacInnes, C. D., K. Adamowaski, and A. C. Middleton, Stochastic Design of Flow Equalization Basin, *Journal of the Environmental Engineering Division, American Society of Civil Engineers*, vol. 104, no. EE6, 1978, pp. 1275–1289.
35. Rudolfs, W., and J. N. Millar, A Method for Accelerated Equalization of Industrial Wastes, *Sewage Works Journal*, vol. 18, no. 4, 1946, pp. 686–689.

5 Fundamentals of Applied Hydraulics

5.1 CHAPTER OBJECTIVES

This chapter provides an overview of topics in applied hydraulics that are extensively used to design wastewater conveyance systems. The objectives of this chapter are to review the fundamentals of applied hydraulics. The following topics are briefly covered:

- Pressurized and open channel flows
- Flow types
- Hydraulic principles
- Hydraulics of pipe flow
- Hydraulics of open channel flow

5.2 PRESSURIZED AND OPEN CHANNEL FLOWS

Wastewater is carried from one location to another through closed conduits and open channel systems. The flow in a closed conduit is termed as pipe flow if the fluid fills the entire cross section, is pressurized, and there is no free fluid surface. The flow in a pipe, when it is flowing full, is shown in Figure 5.1a. In open channels and closed conduits with free water surface, the fluid surface is subject to atmospheric pressure, the flow is due to gravity, and flow conditions are greatly influenced by the bottom slope of the channel. Examples include the flow in streams, rivers, and sanitary and storm sewers.[1] The open channel and closed conduit flows with free water surface are shown in Figure 5.1bi and bii.

5.3 FLOW TYPES

The flow can be classified by how it changes with time and space, or based on predominant forces affecting the hydraulic behavior of the flow. These flow types are discussed below.

5.3.1 Steady and Unsteady Flow

The steady flow occurs when the mean flow velocity and discharge at a given cross section of a conduit or channel do not change with time. An example of steady flow through a pipe is the flow under a constant pressure head (e.g., reservoir or tank). Under unsteady flow, the mean flow velocity at a given cross section of a conduit or channel changes with time. An example of unsteady flow is the flow through a pipe of constant or variable diameter under variable pressure due to an increasing/decreasing water level in the reservoir, opening/closing of a valve, or starting/stopping of a hydraulic machine such as a pump connected to the pipe.

5.3.2 Uniform and Non-uniform Flows

The uniform flow occurs when the velocity and discharge do not change with space along a given length of a pipe or channel. An example of a uniform flow is liquid flowing under pressure through a long pipeline that has a constant diameter. The non-uniform flow occurs when the velocity of flow

DOI: 10.1201/9781003629399-5

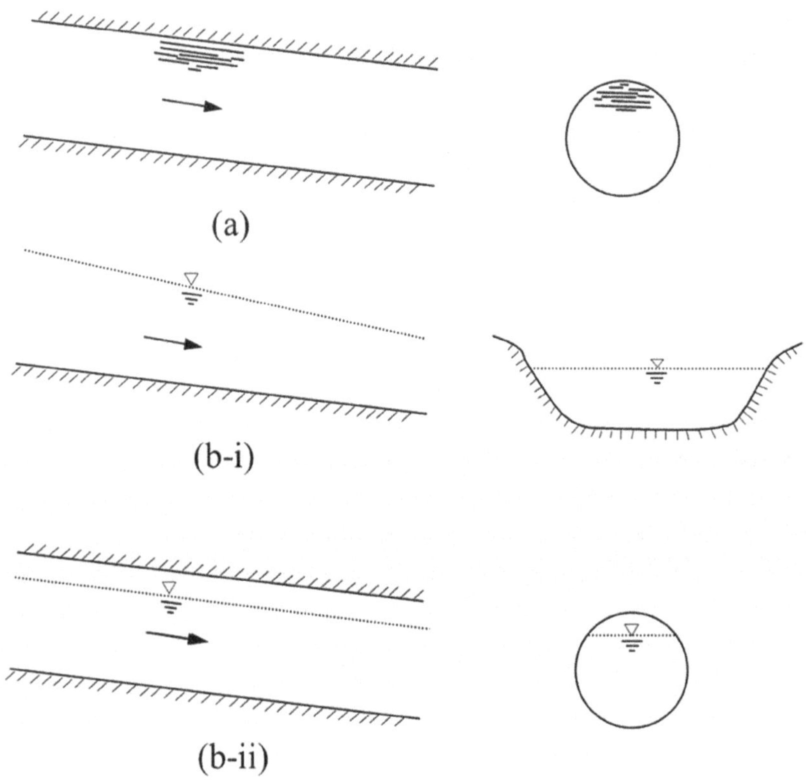

FIGURE 5.1 Open channel and pressurized pipe flows: (a) pipe flowing full or pressurized flow and (b) open channel flow: (i) streams and (ii) closed circular conduit flow with free water surface.

varies at different points in a flowing fluid, for a given time. An example of non-uniform flow is the flow of liquid under pressure through a long pipeline of varying diameter.

A combination of different flow conditions may exist in a pipe or open channel. These combinations are steady-uniform, steady–non-uniform, unsteady-uniform, and unsteady–non-uniform flows. The flow through a pipe of constant diameter under a constant pressure head (e.g., reservoir) is an example of steady-uniform flow. If the pipe diameter varies, the flow is steady–non-uniform. An example of unsteady-uniform flow is the flow through a pipe of constant diameter under variable pressure due to an increasing/decreasing water level in the reservoir. If the pipe diameter varies, the flow is unsteady–non-uniform.

In an open channel, non-uniform flow may be referred to as varied flow. The varied flow occurs if the flow depth changes along the channel length. If this change occurs gradually, it is called gradually varied flow (GVF); a flow with an abrupt change in depth within a short distance along the channel length is called rapidly varied flow (RVF). These flow classifications are illustrated in Figure 5.2. The example of GVF is the backwater caused by storm sewer inlet in a storm drainage channel through a sluice gate, followed by RVF and a hydraulic jump. A hydraulic jump is an example of RVF that may occur in a channel downstream of sluice gates, weirs, and spillways. These flow conditions are shown in Figure 5.3ai and aii.

Flow is referred to as spatially varied flow (SVF) if the flow rate varies along the channel length. Flow passing through a channel with a side weir is an example of SVF, because the flow rate decreases in the flow direction due to the fluid spilling over the side weir. The flow over a side weir is shown in Figure 5.3bi. Another example of SVF is the overland drainage into a collection channel (Figure 5.3bii).

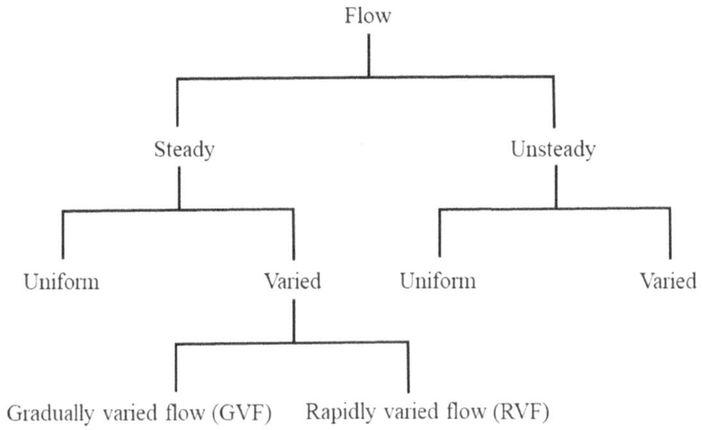

FIGURE 5.2 Flow classification in an open channel.

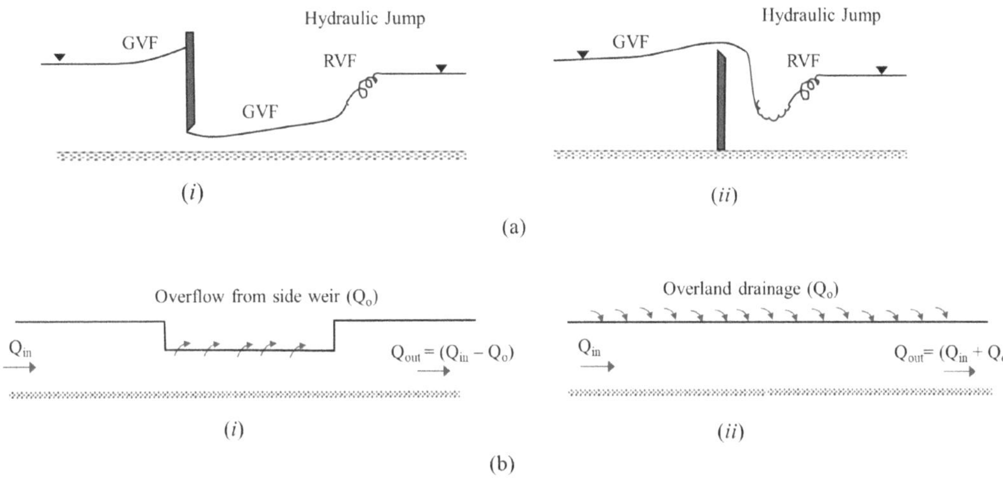

FIGURE 5.3 Type of open channel varied flows: (a) gradually varied flow (GVF) and rapidly varied flow (RVF), (i) sluice gate and (ii) weir, and (b) spatially varied flow (SVF), (i) overflow from a side weir and (ii) overland drainage from a watershed.

5.3.3 Laminar and Turbulent Flows

In laminar flow, fluid particles move in layers, sliding over each other, causing a small energy exchange to occur between layers. The turbulent flow, on the other hand, is characterized by random movements and intermixing of fluid particles, with a great exchange of energy throughout the fluid. The dimensionless Reynolds number is used to determine whether the flow is laminar or turbulent. The general form of the Reynolds number is expressed by Eq. (5.1)[1,2]

$$Re = \frac{\rho VD}{\mu} = \frac{VD}{\upsilon} \tag{5.1}$$

where

Re = Reynolds number, dimensionless
V = mean flow velocity, m/s (ft/s)
D = diameter of pipe, m (ft)

ρ=fluid mass density, kg/m^3 (slug/ft^3)
μ=fluid dynamic viscosity, N.s/m^2 (lb$_f$.s/ft^2)
υ=fluid kinematic viscosity, m^2/s (ft^2/s)

In an open channel flow, the hydraulic radius (R) or hydraulic depth (D_h) is used instead of the diameter of pipe to calculate Reynolds number. Hydraulic radius is defined as the flow area (A) divided by the channel-wetted perimeter (P), i.e. $R = A/P$. The hydraulic depth is defined as the flow area divided by the top water surface width of the channel (T), i.e. $D_h = A/T$.

In pipe flows, if the Reynolds number is less than 2,000 ($\text{Re} < 2,000$), the flow is called laminar; if it is greater than 4,000 ($\text{Re} > 4,000$), the flow is considered turbulent. If the Reynolds number is greater than 2,000 and less than 4,000 ($2,000 < \text{Re} < 4,000$), the flow is referred to as transitional.[3] In open channel flows, if the calculated Re, based on the *hydraulic radius*, is less than 500, the flow is considered laminar. The transitional range of Re from laminar to turbulent, in open channel flow, is from 500 to 2,000. The flow is considered turbulent when the Re is greater than 2,000.[2] Some references consider hydraulic depth (D_h) of an open channel 4 times its hydraulic radius (R). The lower limit of Reynolds number calculated using hydraulic depth is 2,000.[3] The flow of fluids with high viscosity and low velocity in a pipe is an example of laminar flow. Flow in natural streams is an example of turbulent flow.

Example 5.1: Flow Types in an Open Channel

A rectangular channel is carrying water at a constant rate of 7.0 m^3/s. The channel is 3.0-m wide along its length, and the flow depth at both upper and lower sections of the channel is 1.5 m. Determine if the flow is (1) steady or unsteady, (2) uniform or non-uniform, and (3) laminar or turbulent. Water temperature is T=20°C.

Solution

1. Determine the flow type
   Flow is steady since the discharge and flow depth at any section of the channel do not change with time.
2. Determine if the flow is uniform or non-uniform
   Flow is uniform since the flow depth does not change along the channel.
3. Determine if the flow is laminar or turbulent
   Calculate the Reynolds number to determine whether the flow is laminar or turbulent.
   From Table B.2, at T=20°C, the density of water $\rho = 998.2\ \text{kg/m}^3$ and its dynamic viscosity $\mu = 1.002 \times 10^{-3}\ \text{N.s/m}^2$.
   a. Calculate hydraulic radius

$$A = 3.0\ \text{m} \times 1.5\ \text{m} = 4.5\ \text{m}^2, P = (1.5\ \text{m} \times 2) + 3.0\ \text{m} = 6.0\ \text{m}$$

$$R = \frac{A}{P} = \frac{4.5\ \text{m}^2}{6.0\ \text{m}} = 0.75\ \text{m}$$

   b. Calculate flow velocity

$$V = \frac{Q}{A} = \frac{7.0\ \text{m}^3/\text{s}}{4.5\ \text{m}^2} = 1.56\ \text{m/s}$$

   c. Calculate the Reynolds number using Eq. (5.1) by replacing D with the flow hydraulic radius R

$$\text{Re} = \frac{\rho VR}{\mu} = \frac{998.2\ \text{kg/m}^3 \times 1.56\ \text{m/s} \times 0.75\ \text{m}}{1.002 \times 10^{-3}\ \text{N.s/m}^2}$$

$$= \frac{998.2\ \text{kg/m}^3 \times 1.56\ \text{m/s} \times 0.75\ \text{m}}{1.002 \times 10^{-3}\ (\text{kg. m/s}^2) \times (\text{s/m}^2)} = 1.2 \times 10^6$$

The Reynolds number is greater than 4,000; therefore, the flow is turbulent.
Note: $N\,(\text{force}) = \text{kg}\,(\text{mass}) \times \text{m/s}^2\,(\text{acceleration}) = \text{kg.m/s}^2$.

5.4 HYDRAULIC PRINCIPLES

The hydraulics of pipe and open channel flows is governed by three fundamental equations: *continuity equation*, *energy equation*, and *momentum equation*. In solving pipe or open channel problems, we may use the combined application of these equations. However, in problems with fewer unknowns, one or two of these equations might be sufficient.

5.4.1 Continuity Equation

The continuity equation can be derived based on the principle of conservation of mass, i.e., mass can neither be created nor destroyed. Consider an element of the fluid volume (or control volume (CV)) in Figure 5.4.

The continuity equation is expressed by Eq. (5.2), and flow relationship is developed by Eqs. (5.3) and (5.4)

$$\rho_1 V_1 A_1 = \rho_2 V_2 A_2 \tag{5.2}$$

For incompressible fluids $\rho_1 = \rho_2$; therefore

$$V_1 A_1 = V_2 A_2, \quad \text{or} \tag{5.3}$$

$$Q_1 = Q_2 = \text{constant}\, Q \tag{5.4}$$

where

V_1, V_2 = mean flow velocity at sections (1) and (2), m/s (ft/s)
ρ_1, ρ_2 = fluid mass density at sections (1) and (2), kg/m^3, (slug/ft^3)
A_1, A_2 = cross-sectional area at sections (1) and (2), m^2 (ft^2)
Q_1, Q_2 = flow rate at sections (1) and (2), m^3/s (ft^3/s)
y_1, y_2 = depth of flow in an open channel at sections (1) and (2), m/s (ft/s) (see Figure 5.4b)

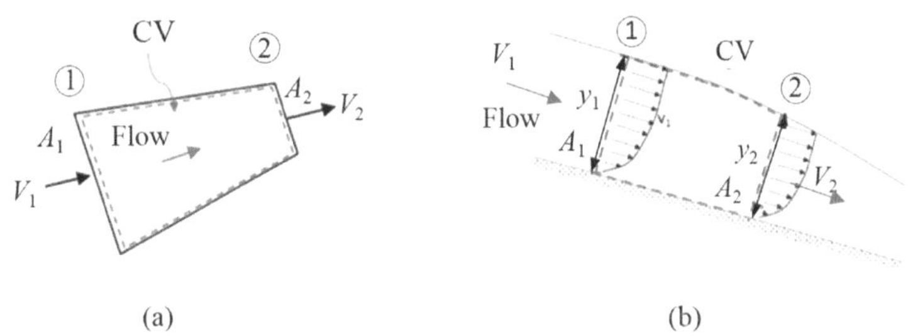

FIGURE 5.4 Definition sketch showing control volume (CV) between sections (1) and (2), and flow variables (velocity V and cross-sectional area A): (a) flow in a pipe under pressure and (b) open channel flow.

5.4.2 Energy Equation

When liquid flows in a pipe or open channel, the energy is present in the form of pressure head, velocity head, and fluid elevation or potential head. This is true if no energy exchange occurs due to viscous dissipation, heat transfer, or shaft work (pump or some other devices). The relationship among these three forms of energy is known as *Bernoulli's equation* or *energy equation*. The concept is based on the conservation of energy principle. Three assumptions apply: steady flow, incompressible fluid, and no losses from fluid friction. Bernoulli's equation is expressed by Eq. (5.5) at any two sections of the pipe or open channel. Figure 5.5 is the definition sketch that shows the components of the energy equation at sections (1) and (2) in a pipe flowing under pressure.

$$\frac{P_1}{\gamma}+\frac{V_1^2}{2g}+z_1=\frac{P_2}{\gamma}+\frac{V_2^2}{2g}+z_2=H \tag{5.5}$$

where

P_1, P_2 = static pressure at sections (1) and (2), N/m² (lb$_f$/ft²)
γ = specific weight (or unit weight), N/m³ (lb$_f$/ft³)
g = acceleration due to gravity, m/s² (ft/s²)
V_1, V_2 = mean flow velocity at sections (1) and (2), m/s (ft/s)
z_1, z_2 = vertical elevation of the fluid above datum at sections (1) and (2), m (ft)
H = total energy, m (ft)

In Eq. (5.5), the terms $\frac{P_1}{\gamma}$, $\frac{V^2}{2g}$, and z are pressure head, velocity head, and elevation head, respectively.

If the energy loss or head loss (h_L) between sections (1) and (2) in Figure 5.5 is taken into account, the energy equation between these two sections can be written by Eq. (5.6).

$$\frac{P_1}{\gamma}+\frac{V_1^2}{2g}+z_1=\frac{P_2}{\gamma}+\frac{V_2^2}{2g}+z_2+h_L \tag{5.6}$$

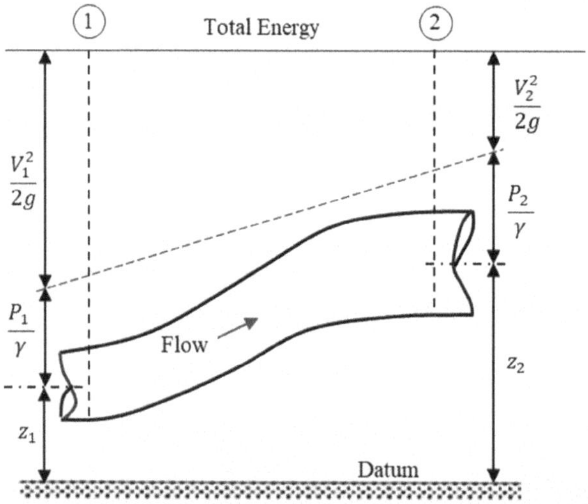

FIGURE 5.5 Definition sketch showing the components of the energy equation in a pipe flowing under pressure.

In open channel flow, $\frac{P_1}{\gamma}$ and $\frac{P_2}{\gamma}$ are replaced by y_1 and y_2 that are the depths of flow at sections (1) and (2) (see Figure 5.4b). The energy equation for open channel flow is expressed by Eq. (5.7).

$$y_1 + \frac{V_1^2}{2g} + z_1 = y_2 + \frac{V_2^2}{2g} + z_2 + h_L \tag{5.7}$$

5.4.3 Momentum Equation

Moving fluid in open channels or pipe systems exerts forces on objects in contact with it. These forces are determined from Newton's second law or *momentum equation*. These forces are calculated from the net mass rate that flows into and out of a finite region of the CV of the fluid (Figure 5.6). For steady and incompressible flow, the momentum equation is expressed by Eq. (5.8).

$$\Sigma \boldsymbol{F} = \rho Q(\boldsymbol{V}_{\text{out}} - \boldsymbol{V}_{\text{in}}) \tag{5.8}$$

where

$\boldsymbol{V}_{\text{in}}$ and $\boldsymbol{V}_{\text{out}}$ = flow velocity vectors, m/s (ft/s)

$(\boldsymbol{V}_{\text{out}} - \boldsymbol{V}_{\text{in}})$ = change in velocity vector, m/s (ft/s)

$\Sigma \boldsymbol{F}$ = all external forces exerted on a CV, N (lb_f). The external forces may include weight, pressure, and shear forces as shown in Figure 5.6.

The $\boldsymbol{F}$ and $\boldsymbol{V}$ vectors may have three components in the longitudinal (x), vertical (y), and transverse (z) directions. Therefore, Eq. (5.8) may be written as Eqs. (5.9a)–(5.9c).

$$\Sigma F_x = \rho Q(V_{x\text{out}} - V_{x\text{in}}) \tag{5.9a}$$

$$\Sigma F_y = \rho Q(V_{y\text{out}} - V_{y\text{in}}) \tag{5.9b}$$

$$\Sigma F_z = \rho Q(V_{z\text{out}} - (V_{z\text{in}}) \tag{5.9c}$$

where ΣF_x, ΣF_y, and ΣF_z represent the components of all external forces exerted on a CV in x, y, and z directions, respectively. Similarly, V_x, V_y, and V_z are the x, y, and z components of the flow velocity vector. The subscripts "*in*" and "*out*" denote the input section and output section of the CV. Figure 5.6 shows a pipe bend flowing under pressure and all forces exerted on the CV. In this figure,

$\boldsymbol{F}_{P1}$ = pressure force at section (1), N (lb_f)

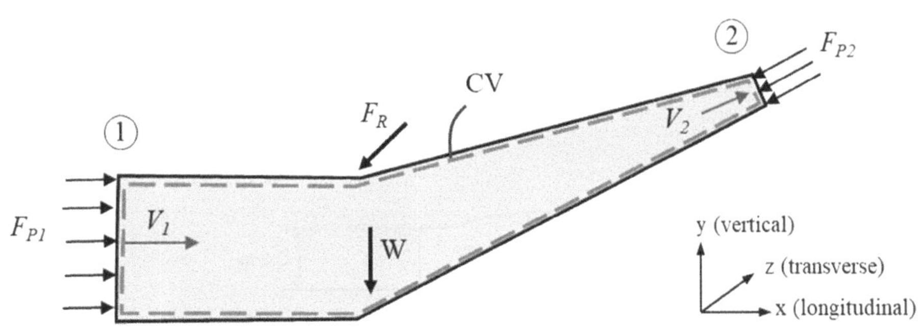

FIGURE 5.6 Example of the components of the momentum equation in a pipe flowing under pressure. F_P, W, and F_R are the pressure force, weight of control volume (CV), and reaction force exerted on the CV; V is the flow velocity.

F_{P2} = pressure force at section (2), N (lb_f)
F_R = reaction force exerted by the pipe on the CV, N (lb_f)
W = weight of the water within the CV, N (lb_f)
V_1 = flow velocity vector at the input section (1), m/s (ft/s)
V_2 = flow velocity vector at the out section (2), m/s (ft/s)

Example 5.2: Application of the Continuity Equation in Pipe Flows

A 2.0-m diameter sewer pipe is connected to a 2.5-m diameter pipe through a gradual transition section, as shown in Figure 5.7. If the mean flow velocity in the smaller pipe is 1.5 m/s, determine the mean flow velocity in the larger pipe.

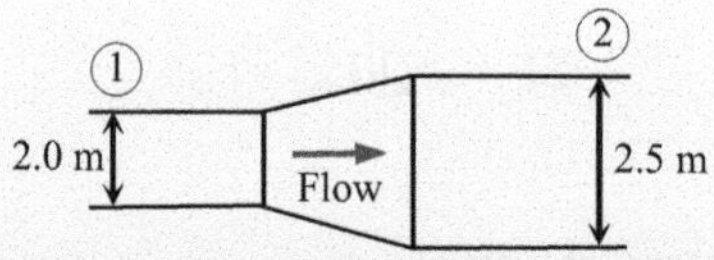

FIGURE 5.7 Plan view of flow through a pipe with a gradual expansion (Example 5.2).

Solution

Determine the mean velocity in the larger pipe
Apply the continuity equations (Eqs. 5.3 and 5.4)

$$Q_1 = Q_1 \quad \text{and} \quad V_1A_1 = V_2A_2$$

$$A_1 = \frac{\pi}{4}D_1^2 = \frac{\pi}{4}(2.0\text{ m})^2 = 3.14\text{ m}^2$$

$$A_2 = \frac{\pi}{4}(2.5\text{ m})^2 = 4.91\text{ m}^2$$

$$1.5\text{ m/s} \times 3.14\text{ m}^2 = V_2 \times 4.91\text{ m}^2$$

$$V_2 = 0.96\text{ m/s}$$

Example 5.3: Application of Energy Equation in an Open Channel Flow

A 3.0-m wide rectangular channel with a depth of flow of 1.0 m has a mean velocity of 2.0 m/s. The channel undergoes a small and smooth contraction to a width of 2.8 m. Determine the depth of the flow and mean velocity at section (2). The energy loss between sections (1) and (2) is negligible, and the channel bottom is horizontal between these two sections (Figure 5.8).

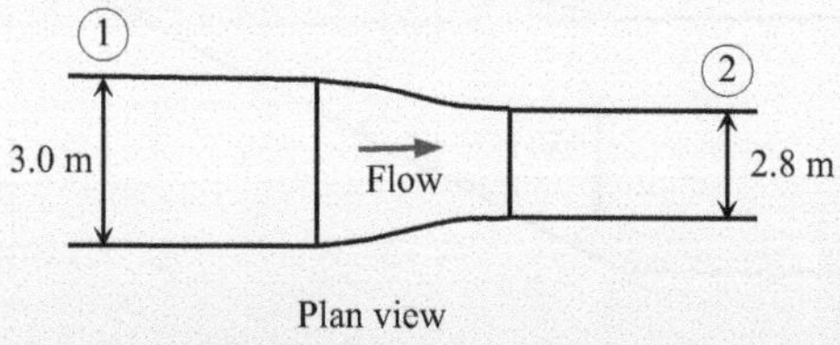

FIGURE 5.8 Plan view of a smooth contraction in an open channel (Example 5.3).

Solution

1. Set up equations to determine the mean flow velocity V_2 at section (2)
   Apply the continuity equations (Eqs. 5.3 and 5.4) between sections (1) and (2)
   $Q_1 = Q_1$ and $V_1A_1 = V_2A_2$
   Assume water depth at section (2) = y_2

$$A_1 = 3.0 \text{ m} \times 1.0 \text{ m} = 3.0 \text{ m}^2$$

$$A_2 = 2.8 \text{ m} \times y_2 = 2.8y_2 \text{ m}^2$$

$$Q_1 = 2.0 \text{ m/s} \times 3.0 \text{ m}^2 = 6.0 \text{ m}^3/\text{s}$$

$$6.0 \text{ m}^3/\text{s} = V_2\left(2.8y_2 \left(\text{m}^2\right)\right)$$

$$V_2 = \frac{6.0 \text{ m}^3/\text{s}}{2.8y_2 \left(\text{m}^2\right)} = \frac{2.143 \text{ m}^2/\text{s}}{y_2 \text{ (m)}}$$

2. Set up equation to determine y_2
   Apply energy equation (5.7) between sections (1) and (2)

$$y_1 + \frac{V_1^2}{2g} + z_1 = y_2 + \frac{V_2^2}{2g} + z_2 + h_L$$

The channel is horizontal ($z_1 = z_2$), and the energy loss is negligible ($h_L = 0$). Therefore,

$$1.0 \text{ m} + \frac{(2.0 \text{ m/s})^2}{2 \times 9.81 \text{ m/s}^2} = y_2 + \frac{V_2^2}{2 \times 9.81 \text{ m/s}^2}$$

Replace $V_2 = \dfrac{2.143 \text{ m}^2/\text{s}}{y_2}$ from step 1

$$1.0 \text{ m} + \frac{(2.0 \text{ m/s})^2}{2 \times 9.81 \text{ m/s}^2} = y_2 + \frac{\left(\dfrac{2.143 \text{ m}^2/\text{s}}{y_2}\right)^2}{2 \times 9.81 \text{ m/s}^2}$$

$$1.0 \text{ m} + 0.204 \text{ m} = y_2 + \frac{0.234 \text{ m}^3}{y_2^2}$$

$$1.204 \text{ m} \times y_2^2 = y_2^3 + 0.234 \text{ m}$$

$$y_2^3 - 1.204 \text{ m} \times y_2^2 + 0.234 \text{ m} = 0$$

3. Solve the cubic equation to determine the depth of flow and mean velocity at sections (1) and (2)

   This equation can be solved by trial and error or by using an equation solver. There are two positive and one negative solutions. The three values of y_2 are −0.38, 0.65, and 0.94 m. The negative flow depth does not have a physical meaning. Since the flow is subcritical before the contraction section, the depth corresponding to the subcritical flow is the solution.[4] The depth after the contraction section is

$$y_2 = 0.94 \text{ m}$$

The mean flow velocity is

$$V_2 = \frac{6.0 \text{ m}^3/\text{s}}{2.8 \text{ m} \times 0.94 \text{ m}} = 2.28 \text{ m/s}$$

Example 5.4: Application of Momentum Equation in Pipe Flow through a Bend

Water flows through a pipe bend, as shown in Figure 5.9. Determine horizontal and vertical components of force that are necessary to hold the pipe in place. The pipe diameter is 0.5 m, the flow rate is 0.3 m^3/s, and the pressure at both ends of the pipe is 30.0 kPa (kN/m^2). Water temperature is T=20°C.

Solution

1. Identify forces acting on the CV, and the direction of velocity vectors.
   a. Select the CV.
      The CV is selected between sections (1) and (2), as shown in Figure 5.10.
   b. Identify forces.
      The pressure forces acting on the CV and the direction of the mean flow velocity vectors shown in Figure 5.10 are F_{P1} and F_{P2}, and V_1 and V_2, respectively. F_{Rx} and F_{Ry} are the components of the total force exerted on the CV by the pipe in x and y directions. The directions of F_{Rx} and F_{Ry} are arbitrary.

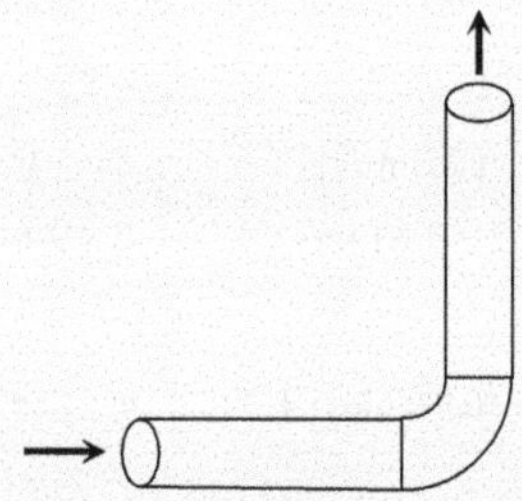

FIGURE 5.9 Flow through a 90° bend (Example 5.4).

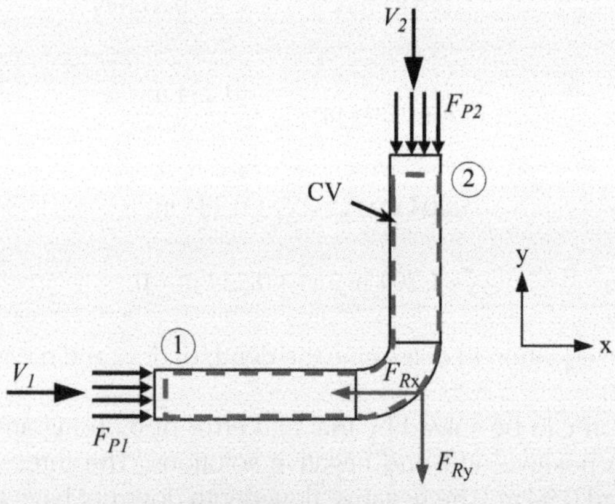

FIGURE 5.10 Forces exerted on the control volume (CV) of the bend and velocity vectors (Example 5.4).

2. Calculate the force exerted by the pipe on the CV in x direction
   Apply the momentum equation in x direction (Eq. 5.9a).

$$\Sigma F_x = \rho Q(V_{x2} - V_{x1})$$

$$F_{P1} - F_{Rx} = \rho Q(V_{x2} - V_{x1})$$

Replace velocity V_{x1} with $\frac{Q}{A_1}$, $V_{x2} = 0$, and $F_{P1} = P_1 A_1$

$$P_1 A_1 - F_{Rx} = \rho Q\left(0 - \frac{Q}{A_1}\right)$$

where
ΣF_x = all external forces exerted on the CV in x direction, N
F_{P1} = pressure force at section (1), N
P_1 = pressure at section (1), N/m^2
A_1 = cross-sectional area of the pipe at section (1), m
F_{Rx} = reaction force exerted by the pipe on the CV in x direction, N
V_{x1} = component of flow velocity vector in x direction at section (1), m/s
V_{x2} = component of flow velocity vector in x direction at section (2), m/s

The density of water ρ at T=20°C is 998.2 kg/m^3 (Table B.2).

$$A_1 = \frac{\pi}{4} D^2 = \frac{\pi}{4}(0.5 \text{ m})^2 = 0.196 \text{ m}^2$$

$$30{,}000 \text{ N/m}^2 \times 0.196 \text{ m}^2 - F_{Rx} = 998.2 \text{ kg/m}^3 \times 0.3 \text{ m}^3/\text{s} \times \left(0 - \frac{0.3 \text{ m}^3/\text{s}}{0.196 \text{ m}^2}\right)$$

$$5{,}880 \text{ N} - F_{Rx} = -458.36 \text{ N}, \quad \left(1 \text{ N} = 1 \text{ kg.m/s}^2\right)$$

$F_{Rx} = 6{,}338$ N. Direction is from right to left
   Since F_{Rx} is positive, the direction of the force is the same as that shown in Figure 5.10.

3. Calculate the force exerted by the pipe on the CV
   Apply the momentum equation in y direction (Eq. 5.9b).

$$\Sigma F_y = \rho Q(V_{y2} - V_{y1}) - F_{P2} - F_{Ry} = \rho Q\left(V_{y2} - V_{y1}\right)$$

Replace velocity V_{y2} with $\frac{Q}{A_2}$, $V_{y1} = 0$, and $F_{P2} = P_2 A_2$

$$-P_2 A_2 - F_y = \rho Q\left(\frac{Q}{A_2} - 0\right)$$

where
ΣF_y = all external forces exerted on the CV in y direction, N
F_{P2} = pressure force at section (2), N
P_2 = pressure at section (2), N/m^2
A_2 = cross-sectional area of the pipe at section (2), m
F_{Ry} = reaction force exerted by the pipe on the CV in y direction, N
V_{y1} = component of flow velocity vector in y direction at section (1), m/s

V_{y2} = component of flow velocity vector in x direction at section (2), m/s
$A_2 = A_1 = 0.196\ \text{m}^2$

$$-30{,}000\ \text{N/m}^2 \times 0.196\ \text{m}^2 - F_{Ry} = 998.2\ \text{kg/m}^3 \times 0.3\ \text{m}^3/\text{s} \times \frac{0.3\ \text{m}^3/\text{s}}{0.196\ \text{m}^2}$$

$$-5{,}880\ \text{N} - F_{Ry} = 458.36\ \text{N}, \quad \left(1\ \text{N} = 1\ \text{kg. m/s}^2\right)$$

$$F_{Ry} = -6{,}338\ \text{N}$$

Since F_{Ry} is negative, the direction of the force is opposite to the direction shown in Figure 5.10.
$F_{Ry} = 6{,}338$ N. Direction is upward

4. Calculate the magnitude of the force exerted by the pipe

$$F = \left(F_{Rx}^2 + F_{Ry}^2\right)^{1/2} = \left((6{,}338\ \text{N})^2 + (6{,}338\ \text{N})^2\right)^{1/2} = 8{,}963\ \text{N}$$

The magnitude of the force exerted by water on the pipe is 8963 N.

5.5 PIPE FLOW

Pipes are frequently used in water distribution and wastewater conveyance systems. The flow of fluids in pressurized pipes is maintained by the pressure exerted by a reservoir or a pump. The pressure decreases in the flow direction in the pipe because of energy loss. Two types of energy losses predominate in fluid flow through a pipe network. These are major losses and minor losses. Major losses are associated with frictional energy loss that is caused by the viscous effects of the fluid and the roughness of the pipe wall. Minor losses are due to entrance and exit, valves and fittings, and changes in the flow direction and flow area.

Pipes are produced with different diameters and from a variety of materials such as concrete, cast iron, PVC, and steel, with considerably different internal surface roughness. The standard diameters of pipe sizes that are commercially available in the U.S. market, and equivalent normal pipe sizes in SI units, are summarized in Table C.1.

The type of fluid carried by the pipe and the flow regime also contribute to energy loss in a pipe system. Flow in pipes carrying water or wastewater is generally turbulent.[5] While the flow of sludge may be laminar at practical design velocities, sludge is non-Newtonian fluid and creates turbulent conditions. Head loss associated with the flow of sludge in pipes should be calculated considering the thixotropic behavior and rheological parameters of sludge.

5.5.1 Pipe Flow Equation

The total energy in a pipe can be determined by applying the energy equation, also known as Bernoulli's equation (Eq. 5.5). This equation is presented in Section 5.4.2. For ideal fluids, in which no head losses occur, the total energy throughout the pipe remains constant. In systems carrying real fluid, the total energy reduces along the direction of flow. Figure 5.11 shows a segment of a pipeline and components of the energy equation. Applying the energy equation between sections (1) and (2) gives Eq. (5.10)

$$\frac{P_1}{\gamma} + \frac{V_1^2}{2g} + z_1 = \frac{P_2}{\gamma} + \frac{V_2^2}{2g} + z_2 + (\text{total losses}) \tag{5.10}$$

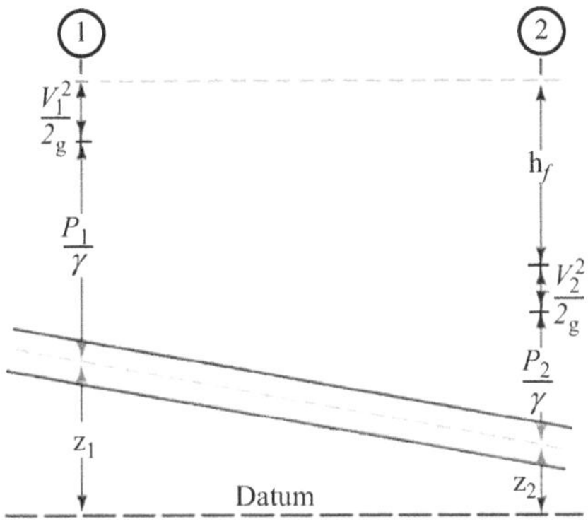

FIGURE 5.11 Definition sketch showing the components of energy equation and friction loss (h_f) in a pipe flowing under pressure.

In a pipe flow, total losses (h_L) are the sum of the major losses (h_f) and minor losses (h_m). The major losses occur over the length of the pipe L due to friction. The head losses resulting from fittings, valves, and pipe transitions are minor or local losses. The energy equation can be modified to include both major losses (h_f) and minor losses (h_m), and is expressed by Eq. (5.11).

$$\frac{P_1}{\gamma}+\frac{V_1^2}{2g}+z_1=\frac{P_2}{\gamma}+\frac{V_2^2}{2g}+z_2+h_f+h_m \tag{5.11}$$

In water and wastewater conveyance systems, the major head loss in pipes (h_f) may be significantly larger than the sum of minor losses (h_m), and minor losses may be ignored.

The minor head losses are discussed in Section 5.5.3.

Example 5.5: Friction Head Loss and Velocity in a Pipe

A 12.0-in. diameter horizontal water pipe discharges 10.0 ft^3/s into the atmosphere. Determine the friction head loss in this pipe if the pressure at the pipe inlet is 10 psi. Also, calculate the velocity. The minor losses due to entrance and exit are small and are ignored.

Solution

1. Write the energy equation (5.11) between the pipe inlet and outlet.

$$\frac{P_{in}}{\gamma}+\frac{V_{in}^2}{2g}+z_{in}=\frac{P_{out}}{\gamma}+\frac{V_{out}^2}{2g}+z_{out}+h_f+h_m$$

Since the pipe discharges into the atmosphere, the gage pressure $P_{out}=0$. The minor losses are ignored; therefore, $h_m=0$. Also, $z_{in}=z_{out}$ because the pipe is horizontal, and $V_{in}^2/2g=V_{out}^2/2g$ since the pipe diameter and discharge are constant.

$$\frac{(10\ \text{lb/in}^2)\times(12\ \text{in/ft})^2}{62.4\ \text{lb/ft}^3}+\frac{V_{in}^2}{2g}+z_{in}=\frac{0}{\gamma}+\frac{V_{out}^2}{2g}+z_{out}+h_f+0$$

$$h_f=23.1\ \text{ft}$$

2. Determine velocity in the pipe

$$V = \frac{Q}{A}$$

$$A_1 = \frac{\pi}{4}D^2 = \frac{\pi}{4}\left(\frac{12\text{ in.}}{12\text{ in./ft}}\right)^2 = 0.785\text{ ft}^2$$

$$V = \frac{10.0\text{ ft}^3/\text{s}}{0.785\text{ ft}^2} = 12.74\text{ ft/s}$$

The friction loss and velocity in the pipe are 23.1 ft and 12.74 ft/s.

Example 5.6: Friction Head Loss in a Pipe Draining a Reservoir

From a large reservoir, water discharges into the atmosphere through a pipe at a rate of 127.0 ft³/s. The diameter of the pipe is 48.0 in. The reservoir and pipe arrangement are shown in Figure 5.12. Calculate the friction head loss in the pipe. Neglect minor losses.

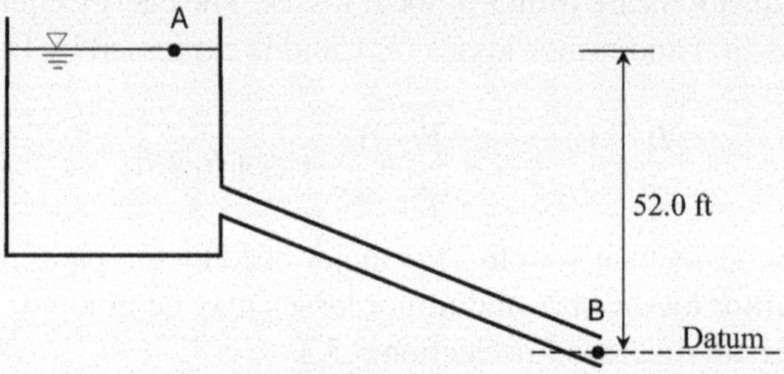

FIGURE 5.12 Head loss through a pipe (Example 5.6).

Solution

1. Calculate the area of the pipe

$$A = \frac{\pi}{4}D^2 = \frac{\pi}{4}\left(\frac{48.0\text{ in.}}{12\text{ in./ft}}\right)^2 = 12.56\text{ ft}^2$$

2. Determine the flow velocity through the pipe

$$V = \frac{Q}{A} = \frac{127.0\text{ ft}^3/\text{s}}{12.56\text{ ft}^2} = 10.11\text{ ft/s}$$

3. Apply energy equation (Eq. 5.11) between points *A* and *B*

$$\frac{P_A}{\gamma} + \frac{V_A^2}{2g} + z_A = \frac{P_B}{\gamma} + \frac{V_B^2}{2g} + z_B + h_f + h_m$$

Since pressure at points *A* and *B* is atmospheric, the gauge pressure $P_A = P_B = 0$. The water velocity at the surface of a large reservoir (point *A*) is negligible, $V_A = 0$. The minor losses are negligible, therefore $h_m = 0$.

$$0 + 0 + 52.0\ \text{ft} = 0 + \frac{(10.11\ \text{ft/s})^2}{2 \times (32.2\ \text{ft/s}^2)} + 0 + h_f + 0$$

$$h_f = 50.4\ \text{ft}$$

The friction loss in the pipeline is 50.4 ft.

5.5.2 Major Head Loss

Major losses are associated with frictional energy loss that is caused by the viscous effects of the fluid and the roughness of the pipe wall. Major losses create a pressure drop along the pipe since the pressure must work to overcome the frictional resistance. The energy loss in a pipe due to friction can be determined by applying Eq. (5.11) to a segment of a straight pipe in Figure 5.11 with a uniform cross section. In a pipe with uniform cross section $V_1 = V_2$, and if the pipe is horizontal ($z_1 = z_2$), and minor losses are negligible ($h_m = 0$); Eq. (5.11) can be solved for h_f (Eq. 5.12).

$$h_f = \frac{(P_2 - P_1)}{\gamma} \tag{5.12}$$

The pressure difference ($P_2 - P_1$) between two sections in the pipe is due to the frictional resistance. The frictional head loss (h_f) is directly proportional to the pressure difference.

The Darcy–Weisbach, Hazen–Williams, or other empirical equations may be used to determine friction head loss (h_f) in a pressurized pipe.[5–7]

Darcy–Weisbach equation: The Darcy–Weisbach equation was developed for pipe friction from the experimental results conducted on various pipes. The formula now known as the Darcy–Weisbach equation for circular pipes in terms of velocity and flow rate is given by Eqs. (5.13a) and (5.13b).

$$h_f = f\frac{L}{D}\frac{V^2}{2g} \tag{5.13a}$$

$$h_f = \frac{8fLQ^2}{\pi^2 g D^5} \tag{5.13b}$$

where

h_f = head loss due to friction, m (ft)
f = Darcy–Weisbach coefficient of friction, dimensionless
L = pipe length, m (ft)
D = pipe diameter, m (ft)
g = acceleration due to gravity, m/s^2 (ft/s^2)
V = mean flow velocity, m/s (ft/s)
Q = flow rate, m^3/s (ft^3/s)

In these equations, the friction factor (f) is used to describe the friction loss in a pipe. The value of f has been found to vary with the Reynolds number (Eq. 5.1), pipe wall roughness, pipe size, and other factors. In laminar flows, f is only a function of the Reynolds number and is independent of the surface roughness of the pipe. This relationship is given by Eq. (5.14)

$$f = \frac{64}{\text{Re}} \tag{5.14}$$

In turbulent flows, f depends on both the Reynolds number and the relative roughness of the pipe wall. Other factors such as wall roughness and spacing and shape of roughness may also affect the value of f. However, these effects are not well understood and may be negligible in many cases. Therefore, f must be determined experimentally. The Moody diagram relates f to the Reynolds number and the pipe wall relative roughness (ε/D); ε is the pipe wall roughness equivalent height in mm (ft), and D is the pipe diameter in mm (ft). The common values of pipe wall roughness equivalent height (ε) for new commercial pipes of different material types are given in Table 5.1.

The Moody diagram is shown in Figure 5.13. If the Reynolds number and ε/D are known, f can be obtained from this diagram. The Moody diagram is divided into four different regions: laminar flow, critical zone, transition zone, and complete turbulence.[3,5,7]

Laminar flow: In this region, the Reynolds number is less than 2,000 ($\text{Re} < 2{,}000$). The friction factor (f) is independent of the pipe roughness and varies inversely with the Reynolds number, as expressed by Eq. (5.14). Flow is fully stable in this region.

Critical zone: The Reynolds number is greater than 2,000 and less than 4,000 ($2{,}000 < \text{Re} < 4{,}000$). Flow is unstable in this zone.

Transitional zone: In this region, flow is turbulent ($\text{Re} > 4{,}000$). The viscosity and relative roughness are both significant; therefore, f is a function of the Reynolds number and ε/D ratio.

Complete turbulence: Reynolds number is greater than 4,000 ($\text{Re} > 4{,}000$) in this region. Complete turbulent flow occurs in pipes with high relative wall roughness. In this type of flow, the effect of viscosity is insignificant, and f is not a function of the Reynolds number.[6]

Several empirical equations are also used to calculate friction factor (f) for turbulent flow when $\text{Re} > 4{,}000$.[7] Some of these empirical relationships are expressed by Eqs. (5.15)–(5.17).

$$\frac{1}{\sqrt{f}} = 2\log\left(\text{Re}\sqrt{f}\right) - 0.8 \quad \text{(smooth pipe)} \tag{5.15}$$

$$\frac{1}{\sqrt{f}} = 1.14 - 2\log\left(\frac{\varepsilon}{D} + \frac{9.35}{\text{Re}\sqrt{f}}\right) \quad \text{(transitional flow)} \tag{5.16}$$

$$\frac{1}{\sqrt{f}} = 1.14 - 2\log\left(\frac{\varepsilon}{D}\right) \quad \text{(complete turbulent or rough pipe)} \tag{5.17}$$

TABLE 5.1

Values of Pipe Wall Roughness (ε) for New Commercial Pipes

Material	ft	mm
Glass, plastic (smooth)	0.0	0.0
Brass, lead, cement, bituminous lining	0.000005	0.0015
Welded steel pipe, wrought iron	0.00015	0.046
Asphalt-dipped cast iron	0.0004	0.12
Galvanized iron	0.0005	0.15
Cast iron (average)	0.00085	0.25
Wood stave	0.0006–0.003	0.18–0.9
Concrete	0.001–0.01	0.3–3
Riveted steel	0.003–0.03	0.9–9

Source: Adopted in part from Refs. 5–9

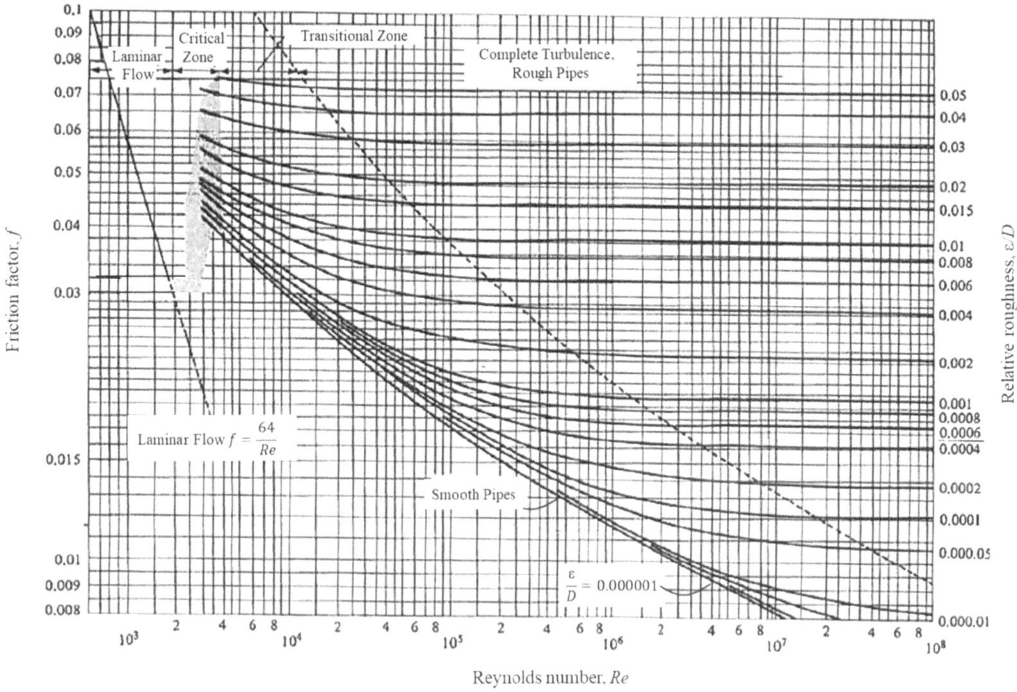

FIGURE 5.13 Moody diagram to determine the Darcy–Weisbach friction factor (f) in pressurized pipes.

Finding f from the Moody diagram or the use of Eqs. (5.15)–(5.17) requires knowledge about the flow velocity, pipe material, and pipe diameter to calculate the Reynolds number and ε/D. If either the flow velocity or pipe diameter is unknown, the application of the Moody diagram and these equations involves a trial and error solution. Because the Darcy–Weisbach equation is applicable only to pressurized circular pipes, additional formulas and charts are required for partially full pipe flow and pipes with a non-circular section. For these reasons, several equations have been developed for calculating friction loss in pressurized pipes and open channels. The Hazen–Williams equation is most commonly used for pressure pipes for the conveyance of water and wastewater.

Hazen–Williams equation: The Hazen–Williams equation is most commonly used for pressurized pipes. This equation, in various forms for circular pipes, is given below

$$h_f = 6.82\left(\frac{V}{C}\right)^{1.85} \times \frac{L}{D^{1.167}} \quad \text{or} \quad V = 0.355CD^{0.63}\left(\frac{h_f}{L}\right)^{0.54} \quad \text{(SI units)} \tag{5.18}$$

$$h_f = 3.02\left(\frac{V}{C}\right)^{1.85} \times \frac{L}{D^{1.167}} \quad \text{or} \quad V = 0.550CD^{0.63}\left(\frac{h_f}{L}\right)^{0.54} \quad \text{(U.S. customary units)} \tag{5.19}$$

$$Q = 0.278CD^{2.63}\left(\frac{h_f}{L}\right)^{0.54} \quad \text{(SI units)} \tag{5.20}$$

$$Q = 0.432CD^{2.63}\left(\frac{h_f}{L}\right)^{0.54} \quad \text{(U.S. customary units)} \tag{5.21}$$

where

C = coefficient of roughness in the Hazen–Williams equation, dimensionless
D = equivalent diameter of the pipe, m (ft)
g = acceleration due to gravity, 9.81 m/s^2 (32.2 ft/s^2)

h_f = total friction head loss, m (ft)
L = length of pipe, m (ft)
Q = flow rate, m^3/s (ft^3/s)
V = velocity in the pipe, m/s (ft/s)

The value of C depends on the pipe material, age of the pipe, and characteristics of the liquid being conveyed. Common values of C, for the Hazen–Williams friction coefficient, are given in Table 5.2.

The friction head loss (h_f) in pipes can be calculated from one of Eqs. (5.18)–(5.21). These equations for h_f are derived from Eqs. (5.20) and (5.21) in terms of Q, D, and C.

$$h_f = \frac{10.7Q^{1.85}\ L}{C^{1.85}D^{4.87}} \quad \text{(SI units)} \tag{5.22}$$

$$h_f = \frac{4.52Q^{1.85}\ L}{C^{1.85}D^{4.87}} \quad \text{(U.S. customary units)} \tag{5.23}$$

Several nomographs have been developed for the solution of the Hazen–Williams equation. One such nomograph is given in Figure 5.14. It provides the solution of the Hazen–Williams equation, based on the roughness coefficient $C = 100$. For pipes with different roughness coefficients, Q, D, and friction slope ($S = h_f/L$) are adjusted from Eqs. (5.24)–(5.26).

$$Q = Q_{100}\left(\frac{C}{100}\right) \quad \text{(adjustment for discharge)} \tag{5.24}$$

$$D = D_{100}\left(\frac{100}{C}\right)^{0.38} \quad \text{(adjustment for pipe diameter)} \tag{5.25}$$

$$S = S_{100}\left(\frac{100}{C}\right)^{1.85} \quad \text{(adjustment for friction slope)} \tag{5.26}$$

where the subscript 100 refers to the values obtained from Figure 5.14 for a pipe with $C = 100$.

TABLE 5.2
Hazen–Williams Friction Coefficient

Description of Pipe	C
Smooth and Straight Cast Iron	
New	140
5 years old	130
10 years old	120
15 years old	110
20 years old	90–100
30 years old	75–90
Concrete or cement-lined	120
Asbestos cement	120–140
Plastic	150

Source: Adopted in part from Refs. 6–9

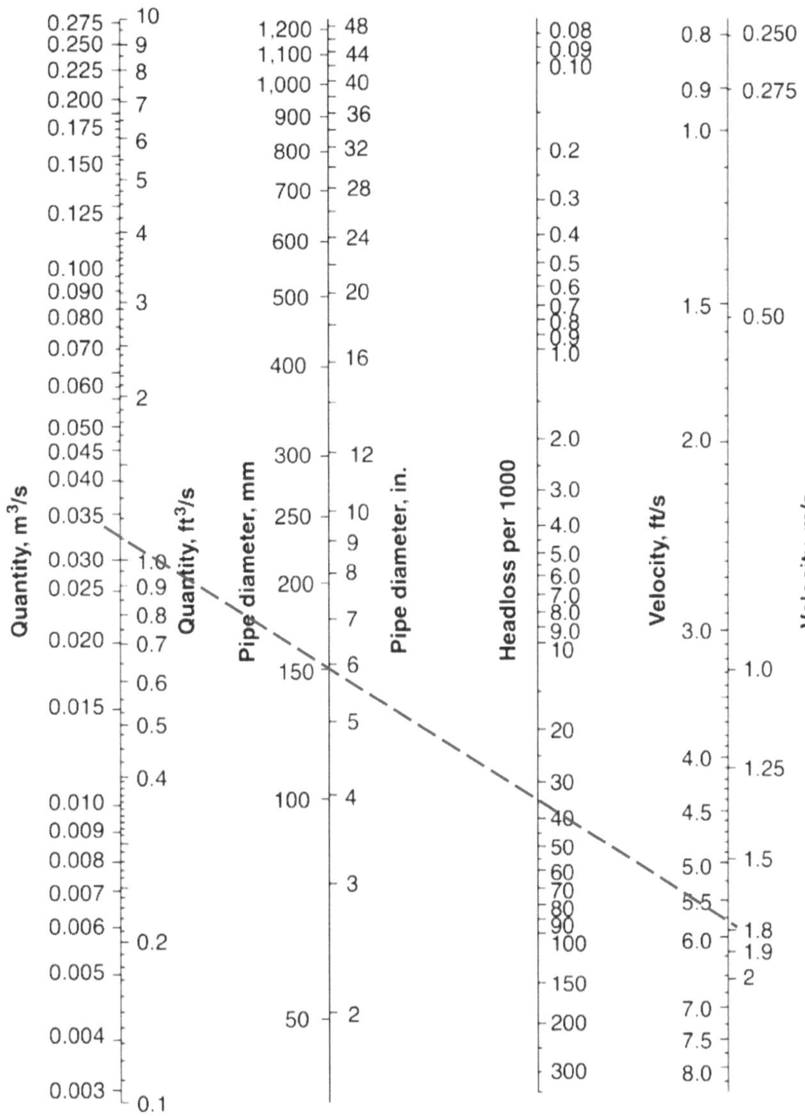

FIGURE 5.14 Nomograph for the Hazen–Williams equation for $C = 100$. (Printed from Ref. 10 with permission of John Wiley and Sons, Inc.)

Example 5.7: Application of the Darcy–Weisbach Equation and the Moody Diagram

Calculate the flow rate through a 130.0-ft section of a 24.0-in. diameter cast iron pipe. The energy loss is 27.0 ft over this section of the pipe. Water temperature is 70°F.

Solution

1. Determine ε/D

From Table 5.1 for cast iron pipe, $\varepsilon = 0.0085$ ft.

$$\frac{\varepsilon}{D} = \frac{(0.00085 \text{ ft}) \times (12 \text{ in./ft})}{24 \text{ in}} = 0.000425$$

2. Select a value of friction factor f.
   For the first trial, choose a value of $f = 0.0162$ from Figure 5.13. Assume rough turbulent flow for $\varepsilon/D = 0.000425$.
3. Calculate the flow velocity from the Darcy–Weisbach equation (Eq. 5.13a)

$$h_f = f \frac{L}{D} \frac{V^2}{2g}$$

$$27.0 \text{ ft} = 0.0162 \times \frac{(130.0 \text{ ft}) V^2}{\left(\dfrac{24.0 \text{ in.}}{12 \text{ in./ft}}\right) \times 2 \times 32.2 \text{ ft/s}^2}$$

$$V = 41.40 \text{ ft/s}$$

4. Check whether the flow is turbulent
   Since the calculated flow velocity is based on an assumed value for f, calculate the Reynolds number from Eq. (5.1). The dynamic viscosity and density of water at 70°F are 2.050×10^{-5} $lb_f.s/ft^2$ and 1.936 slug/ft^3, or 1.936 lb_f s^2/ft, respectively (Table B.2).
   Note: 1 slug = 1 lb_f s^2/ft

$$\text{Re} = \frac{\rho VD}{\mu} = \frac{(1.936 \text{ lb}_f \text{s}^2/\text{ft}) \times (41.40 \text{ ft/s}) \times \left(\dfrac{24.0 \text{ in.}}{12 \text{ in./ft}}\right)}{2.050 \times 10^{-5} \text{ lb}_f.\text{s/ft}^2} = 7.8 \times 10^6$$

5. Obtain a new value for f
   From the Moody diagram (Figure 5.13), obtain a new value for f corresponding to $\text{Re} = 7.8 \times 10^6$ and $\varepsilon/D = 0.000425$.
   The new value of $f = 0.0166$
   There is a small change in the value of f from that in step 2. This change is acceptable, and the final value of friction factor in the pipe is 0.0166.
6. Calculate the flow rate through the pipe.

$$Q = VA$$

$$A = \frac{\pi}{4} D^2 = \frac{\pi}{4} \left(\frac{24.0 \text{ in.}}{12 \text{ in./ft}}\right)^2 = 3.14 \text{ ft}^2$$

$$Q = 41.40 \text{ ft/s} \times 3.14 \text{ ft}^3$$

$$= 130 \text{ ft}^3/\text{s}$$

The flow rate through the pipe is 130 ft^3/s.

Example 5.8: Darcy–Weisbach Friction Factor (*f*) from Empirical Equations

Determine the friction factor for water flowing at a rate of 30.0 L/s in a galvanized iron pipe of 5.0 cm in diameter. The water temperature is 20°C.

Solution

1. Calculate the area of pipe cross section

$$A = \frac{\pi}{4}D^2 = \frac{\pi}{4}\left(\frac{5.0\text{ cm}}{100\text{ cm/m}}\right)^2 = 0.00196\text{ m}^2$$

2. Calculate flow velocity through the pipe

$$V = \frac{Q}{A} = \frac{\left(\dfrac{30.0\text{ L}}{1{,}000\text{ L/m}^3}\right)}{0.00196\text{ m}^2} = 15.31\text{ m/s}$$

3. Calculate the Reynolds number from Eq. (5.1)
   The dynamic viscosity and density of water at 20°C are 1.002×10^{-3} N.s/m^2 and 998.2 kg/m^3, respectively (Table B.2)

$$\text{Re} = \frac{\rho\, VD}{\mu} = \frac{(998.2\text{ kg/m}^3)\times(15.31\text{ m/s})\times\left(\dfrac{5.0\text{ cm}}{100\text{ cm/m}}\right)}{1.002\times(10^{-3}\text{ kg.m/s}^2.\text{ s/m}^2)} = 7.6\times10^5$$

   Since Re > 4,000, the flow is turbulent.
   Note: 1 N = 1 kg.m/s^2

4. Determine ε/D
   From Table 5.1 for galvanized iron pipe, $\varepsilon = 0.15$ mm

$$\frac{\varepsilon}{D} = \frac{0.15\text{ mm}}{5.0\text{ cm}\times(10\text{ mm/cm})} = 0.003$$

5. Determine f from Eq. (5.16)

$$\frac{1}{\sqrt{f}} = 1.14 - 2\log\left(\frac{\varepsilon}{D} + \frac{9.35}{\text{Re}\sqrt{f}}\right)$$

$$\frac{1}{\sqrt{f}} = 1.14 - 2\log\left(0.003 + \frac{9.35}{7.6\times10^5\sqrt{f}}\right)$$

$$f = 0.026$$

6. Check f value from Eq. (5.17)

$$\frac{1}{\sqrt{f}} = 1.14 - 2\log\left(\frac{\varepsilon}{D}\right)$$

$$\frac{1}{\sqrt{f}} = 1.14 - 2\log(0.003)$$

$$f = 0.026$$

Since the flow in the pipe is rough turbulent, the value of f calculated from Eqs. (5.16) and (5.17) is the same.

Example 5.9: Application of Darcy–Weisbach Equation and the Moody Diagram

A welded steel force main is conveying wastewater from a pumping station to a grit chamber. The equivalent pipe length is 500 m (640 ft) and the diameter is 1.0 m (3.28 ft). There is a drop of 2.0 m (6.6 ft) in the energy grade line over the length of the force main. Determine the pump discharge in m^3/s. Liquid temperature is 15°C.

Solution

Since the discharge Q and friction factor f are unknown, it will be a trial and error solution.

1. Select a value of friction factor f
   From Table 5.1 for welded steel pipe $\varepsilon = 0.046$ mm

$$\frac{\varepsilon}{D} = \frac{0.046\ \text{mm}}{1\ \text{m} \times (1{,}000\ \text{mm/m})} = 0.000046$$

   For the first approximation, assume $f = 0.0105$ and the flow is rough turbulent.
2. Calculate the flow rate from Eq. (5.13b)

$$h_f = \frac{8fLQ^2}{\pi^2 g D^5}$$

$$Q = \left[\frac{h_f \pi^2 g D^5}{8fL}\right]^{1/2}$$

$$Q = \left[\frac{(2.0\ \text{m}) \times \pi^2 \times (9.81\ \text{m/s}^2) \times (1.0\ \text{m})^5}{8 \times 0.0105 \times (500\ \text{m})}\right]^{1/2} = 2.15\ \text{m}^3/\text{s}$$

3. Calculate the fluid velocity

$$V = \frac{Q}{A}$$

$$A = \frac{\pi}{4} D^2 = \frac{\pi}{4}(1.0\ \text{m})^2 = 0.785\ \text{m}^2$$

$$V = \frac{2.15\ \text{m}^3/\text{s}}{0.785} = 2.74\ \text{m/s}$$

4. Calculate the Reynolds number using Eq. (5.1)
   At 15°C, the kinematic viscosity υ is $1.14 \times 10^{-6} m^2/s$ (Table B.2)

$$\text{Re} = \frac{VD}{\upsilon} = \frac{(2.74\ \text{m/s}) \times (1.0\ \text{m})}{1.14 \times 10^{-6}\ \text{m}^2/\text{s}} = 2.4 \times 10^6$$

5. Obtain the revised value of f from Figure 5.13 using the calculated Re and ε/D

$$f = 0.0115$$

6. Compare the calculated f in step 5 with that in step 1
   The values of f from steps 1 and 5 are not the same. The final results are obtained when the difference between the adjusted values of f is within an acceptable limit.
7. Carry out another iteration
   Repeat steps 2 to 6 with new values of f and Re.

$$Q = \left[\frac{(2.0\text{ m})\times\pi^2\times(9.81\text{ m/s}^2)\times(1.0\text{ m})^5}{8\times 0.0115\times(500\text{ m})}\right]^{1/2} = 2.05\text{ m}^3/\text{s}$$

$$V = \frac{2.05\text{ m}^3/\text{s}}{0.785} = 2.61\text{ m/s}$$

$$\text{Re} = \frac{VD}{\upsilon} = \frac{(2.61\text{ m/s})\times(1\text{ m})}{1.14\times 10^{-6}\text{ m}^2/\text{s}} = 2.3\times 10^6$$

$f = 0.0115$ from Figure 5.13
   The values of f from steps 5 and 7 are the same
8. Summarize the final results
   The final values of Q, V, Re, and f are given below

$$Q = 2.05\text{ m}^3/\text{s}$$

$$V = 2.61\text{ m/s}$$

$$\text{Re} = 2.3\times 10^6$$

$$f = 0.0115$$

Example 5.10: Application of the Hazen–Williams Equation

Calculate the head loss in a 1,300-m section of a new cast iron pipe. The diameter of the pipe is 500 mm and is conveying water at a rate of 0.25 m³/s.

Solution

1. Determine the Hazen–Williams coefficient
   From Table 5.2, the Hazen–Williams coefficient $C = 140$ for new cast iron pipe
2. Calculate the head loss using Eq. (5.22)

$$h_f = \frac{10.7Q^{1.85}L}{C^{1.85}D^{4.87}}$$

$$= \frac{10.7\times(0.25\text{ m}^3/\text{s})^{1.85}\times(1{,}300\text{ m})}{(140)^{1.85}\times\left(\dfrac{500\text{ mm}}{1{,}000\text{ mm/m}}\right)^{4.87}}$$

$$= 3.35\text{ m}$$

The friction head loss is 3.35 m.

Example 5.11: Application of the Hazen–Williams Equation and Nomograph

Calculate the discharge capacity of a 150-mm diameter existing cast iron force main to carry sewage. The pressure loss due to friction may not exceed 35 m/1,000 m of the pipe length.

Solution

1. Determine the Hazen–Williams coefficient
   From Table 5.2, the Hazen–Williams coefficient $C = 120$ for a 10-year-old cast iron force main
2. Calculate the head loss using Eq. (5.20)

$$Q = 0.278CD^{2.63}\left(\frac{h_f}{L}\right)^{0.54}$$

$$= 0.278 \times 120 \times \left(\frac{150 \text{ mm}}{1{,}000 \text{ mm/m}}\right)^{2.63}\left(\frac{35 \text{ m}}{1{,}000 \text{ m}}\right)^{0.54}$$

$$= 0.037 \text{ m}^3/\text{s}$$

3. Calculate the head loss from the nomograph in Figure 5.14
   a. Mark a point on the pipe diameter scale for $D = 150$ mm
   b. Mark another point on the loss of head for 1,000 m scale at 35 m
   c. Draw a straight line connecting these points and extend the line to the discharge scale. This line is drawn on Figure 5.14. Read the discharge for $C = 100$

$$Q_{100} = 0.0323 \text{ m}^3/\text{s}$$

4. Adjust the discharge for $C = 120$ using Eq. (5.24)

$$Q = Q_{100}\left(\frac{C}{100}\right)$$

$$= \left(0.0325 \text{ m}^3/\text{s}\right)\left(\frac{120}{100}\right)$$

$$= 0.039 \text{ m}^3/\text{s}$$

The flow rates calculated using the Hazen–Williams equation and the nomograph are 0.037 and 0.039 m³/s, respectively.

5.5.3 Minor Head Loss

Minor head losses are produced because of fittings, valves, bends, expansion, contraction, entrances, and exits. They are usually calculated as a function of the velocity head ($V^2/2g$) and expressed by Eq. (5.27)

$$h_m = K\frac{V^2}{2g} \tag{5.27}$$

where

h_m = minor loss, m (ft)

K = head loss coefficient

Because of the complexity of the flow in many fittings, K is usually derived empirically to calculate minor head losses. In short pipe sections with many fittings and valves, such as the suction piping of a pump, the minor losses may be greater than the major (friction) losses. Thus, the accurate K values for all fittings in a pipe system are necessary to predict the total head loss across the pipe system. In practice, K values assist engineers in totaling all minor losses by multiplying the sum of the K values by the velocity head to determine the total minor head loss due to all fittings. Knowing the K value for each fitting enables engineers to use the proper fitting when designing an efficient piping system that can minimize the head loss and maximize the flow rate. The typical values of minor head loss coefficient for valves, elbows, tees, entrance, exit, increaser, and reducer used in plant hydraulic calculations are presented in Appendix C.

Example 5.12: Flow Rate in a Pipeline with Minor Head Losses

Determine the flow rate from reservoir A to reservoir B shown in Figure 5.15. The connecting pipe is 2.0-in. diameter and 300 ft long. A gate valve is located 100 ft from the reservoir A and is fully open. Assume friction factor $f = 0.02$.

Solution

1. Calculate the total head loss, including major and minor losses ($h_{\text{loss}} = h_f + h_m$)
   Apply the energy equation (Eq. 5.11) between points 1 and 2

$$\frac{P_1}{\gamma} + \frac{V_1^2}{2g} + z_1 = \frac{P_2}{\gamma} + \frac{V_2^2}{2g} + z_2 + h_f + h_m$$

Pressure at points (1) and (2) is atmospheric, and water velocity at the surface of reservoirs (points (1) and (2)) is negligible; therefore,

$$0 + 0 + 1{,}000 \text{ ft} = 0 + 0 + 930 \text{ ft} + h_f + h_m$$

$$h_{\text{loss}} = h_f + h_m = 70 \text{ ft}$$

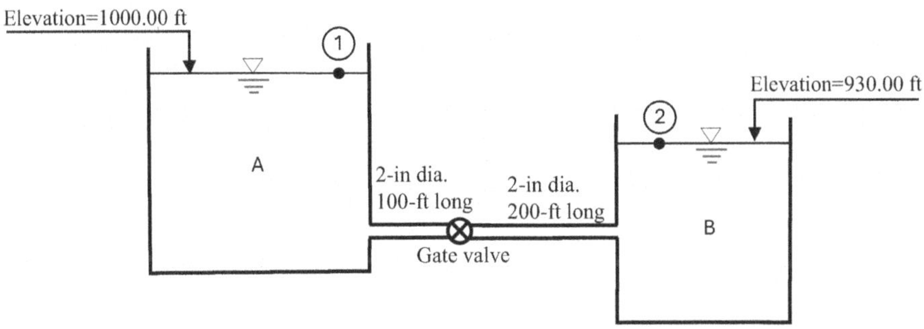

FIGURE 5.15 Reservoirs and the pipe system (Example 5.12).

2. Calculated friction loss (h_f) in the pipes

$$h_f = h_{100} + h_{200}$$

where h_{f100} and h_{f200} are the friction losses in the 100-ft-long and 200-ft-long pipe sections

Apply the Darcy–Weisbach equation (Eq. 5.13a)

$$h_f = \sum f \frac{L}{D}\frac{V^2}{2g}$$

$$h_{f100} = 0.02 \frac{100 \text{ ft}}{\left(\dfrac{2.0 \text{ in.}}{12 \text{ in./ft.}}\right)} \frac{V^2}{2g} = 12\frac{V^2}{2g}$$

$$h_{f200} = 0.02 \frac{200 \text{ ft}}{\left(\dfrac{2.0 \text{ in.}}{12 \text{ in./ft}}\right)} \frac{V^2}{2g} = 24\frac{V^2}{2g}$$

3. Calculated minor losses (h_m)

From Appendix C, the loss coefficients of $K_{\text{entrance}} = 0.5$, $K_{\text{gate valve}} = 0.19$ (fully opened), and $K_{\text{exit}} = 1$

$$h_m = \sum K \frac{V^2}{2g} = h_{\text{entrance}} + h_{\text{gate valve}} + h_{\text{exit}}$$

$$= (0.5)\frac{V^2}{2g} + (0.19)\frac{V^2}{2g} + (1)\frac{V^2}{2g} = \frac{V^2}{2g}(0.5 + 0.19 + 1) = 1.69\frac{V^2}{2g}$$

4. Calculate flow velocity V and velocity head $\dfrac{V^2}{2g}$

$$h_{\text{loss}} = \sum h_f + \sum h_m$$

$$70 \text{ ft} = \left(12\frac{V^2}{2g} + 24\frac{V^2}{2g}\right) + 1.69\frac{V^2}{2g}$$

$$70 \text{ ft} = 37.69 \times \frac{V^2}{2g}$$

$$\frac{V^2}{2g} = \frac{70 \text{ ft}}{37.69} = 1.86 \text{ ft}$$

$$V = (1.86 \text{ ft} \times 2g)^{1/2} = (1.86 \text{ ft} \times 2 \times 32.2 \text{ ft/s}^2)^{1/2} = 10.94 \text{ ft/s}$$

5. Calculate the flow rate

$$Q = VA = 10.94 \text{ ft/s} \times \frac{\pi}{4}\left(\frac{2.0 \text{ in.}}{12 \text{ in./ft}}\right)^2 = 0.24 \text{ ft}^3/\text{s}$$

The flow rate through the pipes is $0.24 \text{ ft}^3/\text{s}$.

5.5.4 Energy Grade Line and Hydraulic Grade Line

The *energy grade line* (EGL) is a plot of total energy or head at different sections of a pipe. This is the sum of the three terms in the energy equation $\left(P/\gamma+V^2/2g+z\right)$. The *hydraulic grade line* (HGL) is a plot of only the pressure and elevation heads $(P/\gamma+z)$.

The HGL is also called the *piezometric head.* If a series of piezometer tubes is inserted into a pressurized pipe, a line joining the top of the liquid surface in the piezometers is the HGL. Figure 5.16 shows the EGL and HGL for a segment of a pipeline. If the velocity head $\left(V^2/2g\right)$ is added to the HGL, and the resulting points are connected by a line, this line is the EGL.

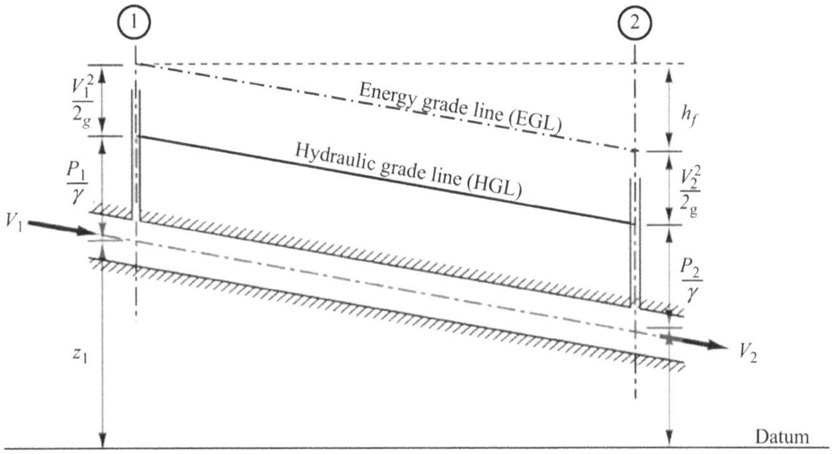

FIGURE 5.16 Energy grade line and hydraulic grade line in a pressurized pipe flow. Two piezometer tubes are tapped into the pipe at sections (1) and (2).

Example 5.13: Energy and Hydraulic Grade Lines in the Pipe System Presented in Example 5.12

Draw the EGL and HGL for the pipe system in Figure 5.15. The velocity and discharge through the connecting pipe were calculated in Example 5.12. The calculated velocity and velocity head through the connecting pipe are 10.94 ft/s and 1.86 ft.

Solution

1. Determine the total heads at the surface of reservoirs A and B

   Establish the datum at the surface of reservoir B (point (2)). At the surface of both reservoirs, the velocity and pressure heads are zero (points (1) and (2)). The total heads at these two points are calculated below

$$H_1=\frac{P}{\gamma}+\frac{V^2}{2g}+z=0+0+1{,}000\text{ ft}=1{,}000\text{ ft}$$

$$H_2=\frac{P}{\gamma}+\frac{V^2}{2g}+z=0+0+930\text{ ft}=930\text{ ft}$$

2. Determine the head loss in the pipe connecting reservoirs A and B

   From point (1) to point (2), the total head drops because of friction losses in the pipe and minor losses due to the entrance, gate valve, and exit.

   a. Major head losses in 100 and 200 ft pipe sections

      From Example 5.12 $\frac{V^2}{2g} = 1.86 \text{ ft}$

      Major head loss in the 100-ft-long pipe $h_{f100} = 12\frac{V^2}{2g} = 12 \times 1.86 \text{ ft} = 22.30 \text{ ft}$

      Major head loss in the 200-ft-long pipe $h_{f200} = 24\frac{V^2}{2g} = 24 \times 1.86 \text{ ft} = 44.60 \text{ ft}$

   b. Minor head losses due to the entrance, fully opened gate valve, and exit loss

      From Example 5.12

$$h_{\text{entrance}} = (0.50)\frac{V^2}{2g} = 0.5 \times 1.86 \text{ ft} = 0.93 \text{ ft}$$

$$h_{\text{gate valve}} = (0.19)\frac{V^2}{2g} = 0.19 \times 1.86 \text{ ft} = 0.35 \text{ ft}$$

$$h_{\text{exit}} = (1)\frac{V^2}{2g} = 1.86 \text{ ft}$$

3. Calculate the total head at different points of connecting piping

   The total head at points (*a*) to (*d*) along the pipe is calculated as follows (Figure 5.17)

$$H_a = H_1 - h_{\text{entrance}} = 1{,}000 \text{ ft} - 0.93 \text{ ft} = 999.07 \text{ ft}$$

$$H_b = H_a - h_{f100} = 999.07 \text{ ft} - 22.30 \text{ ft} = 976.77 \text{ ft}$$

$$H_c = H_b - h_{\text{gate valve}} = 976.77 \text{ ft} - 0.35 \text{ ft} = 976.42 \text{ ft}$$

$$H_d = H_c - h_{f200} = 976.42 \text{ ft} - 44.60 \text{ ft} = 931.82 \text{ ft}$$

$$H_2 = H_d - h_{\text{exit}} = 931.82 \text{ ft} - 1.86 \text{ ft} = 930 \text{ ft}$$

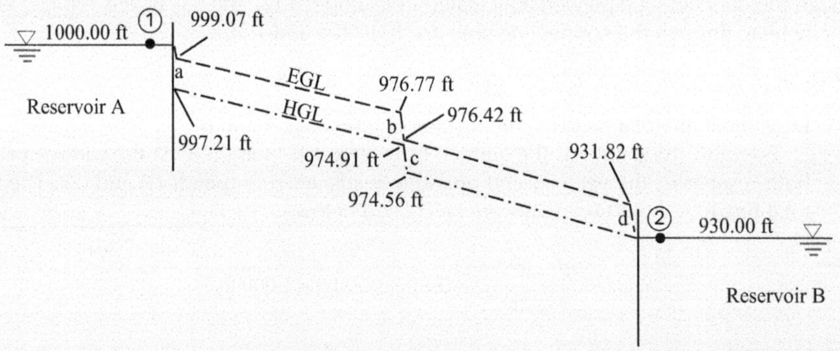

FIGURE 5.17 The EGL and HGL for the pipe system (Example 5.13).

4. Draw the EGL and HGL

   Friction head losses in the connecting pipes and minor head losses are calculated in step 2. The total head at points (1) and (2), and at points (*a*) to (*d*), is shown in Figure 5.17. The line connecting these points represents the EGL. The HGL is located below the EGL at a distance equal to the velocity head. The velocity head in the pipe is 1.86 ft. Use these values to draw EGL and HGL. These grade lines are shown in Figure 5.17.

Example 5.14: Wastewater Conveyance at a Treatment Plant

Wastewater is pumped from a wet well to the primary sedimentation basin. The static discharge head from the pump to the discharge point is 21 m. The length of the cast iron force main is 70 m, and the diameter is 0.45 m (18 in.). The force main has a check valve, and three 90° and two 45° elbows. The pump discharge is 0.35 m³/s. Calculate the total head developed by the pump at the discharge pressure gauge. The Hazen–Williams friction coefficient for new cast iron pipe is 140.

Solution

1. Calculate the velocity in the force main

$$\text{Cross sectional area of force main } A = \frac{\pi}{4}D^2 = \frac{\pi}{4}(0.45\text{ m})^2 = 0.16\text{ m}^2$$

$$\text{Velocity } V = \frac{Q}{A} = \frac{0.35\text{ m}^3/\text{s}}{0.16\text{ m}^2} = 2.20\text{ m/s}$$

2. Calculate friction head loss h_f

   Apply the Hazen–Williams equation (Eq. 5.18)

$$V = 0.355CD^{0.63}\left(\frac{h_f}{L}\right)^{0.54}$$

$$2.20\text{ m/s} = 0.355\times140\times(0.45\text{ m})^{0.63}\left(\frac{h_f}{L}\right)^{0.54}$$

$$\left(\frac{h_f}{L}\right)^{0.54} = 0.0732$$

$$\frac{h_f}{70\text{ m}} = 0.0079$$

$$h_f = 0.55\text{ m}$$

3. Calculate minor head losses

$$\text{Velocity head } \frac{V^2}{2g} = \frac{(2.20\text{ m})^2}{2\times(9.81\text{ m/s}^2)} = 0.247\text{ m}$$

   K for 90° elbow = 0.3, K for 45° elbow = 0.2, K for check valve = 2.0, K for exit loss = 1.0 (Appendix C)

$$h_{m\,(\text{elbow }90^\circ)} = 3\times0.3\times0.247\text{ m} = 0.22\text{ m}$$

$$h_{m\ (\text{elbow } 45°)} = 2 \times 0.2 \times 0.247\ \text{m} = 0.10\ \text{m}$$

$$h_{m\ (\text{check valve})} = 1 \times 2 \times 0.247\ \text{m} = 0.49\ \text{m}$$

$$h_{m\ (\text{exit})} = 1 \times 1.0 \times 0.247\ \text{m} = 0.25\ \text{m}$$

$$\Sigma h_m = 0.22\ \text{m} + 0.1\ \text{m} + 0.49\ \text{m} + 0.25\ \text{m} = 1.1\ \text{m}$$

4. Determine total head developed by the pump at the discharge pressure gauge

The head developed at the discharge pressure gauge = Static head + $h_f + \Sigma h_m$

$$= 21 + 0.55\ \text{m} + 1.1\ \text{m} = 22.65\ \text{m}$$

5.5.5 Pipes in Series

A pipe system consisting of several pipes of different sizes and roughness coefficients is connected in series. This system is shown in Figure 5.18. The following relationships can be derived from the continuity and the energy equations for pipes in series:

$$Q = Q_1 = Q_2 = Q_3 = \cdots \quad (5.28)$$

$$h_f = h_{f_1} + h_{f_2} + h_{f_3} + \cdots \quad (5.29)$$

Equations (5.28) and (5.29), along with the head loss equations such as the Darcy–Weisbach or Hazen–Williams equations, can be used to analyze the flow through such a pipe system.

5.5.6 Pipes in Parallel

In a pipe system where two or more pipes are in parallel, the flow is divided among the pipes and is rejoined. This system is shown in Figure 5.19. The continuity and energy equations apply. The total incoming flow to a node is equal to the total outgoing flow. Also, the sum of total head loss with proper sign in a closed loop is zero.

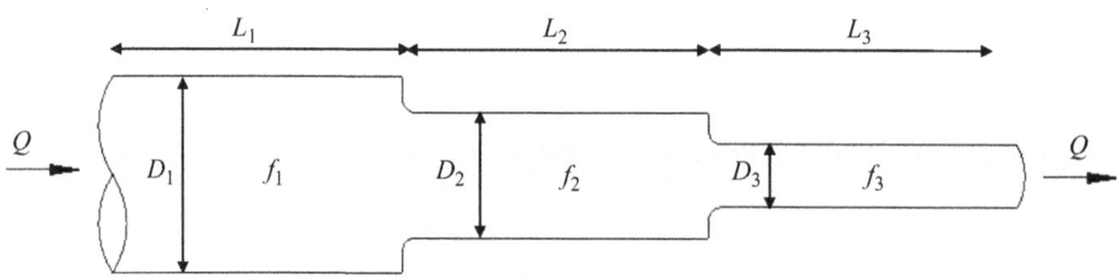

FIGURE 5.18 Pipes in series with different lengths (L), diameters (D), and roughness (f).

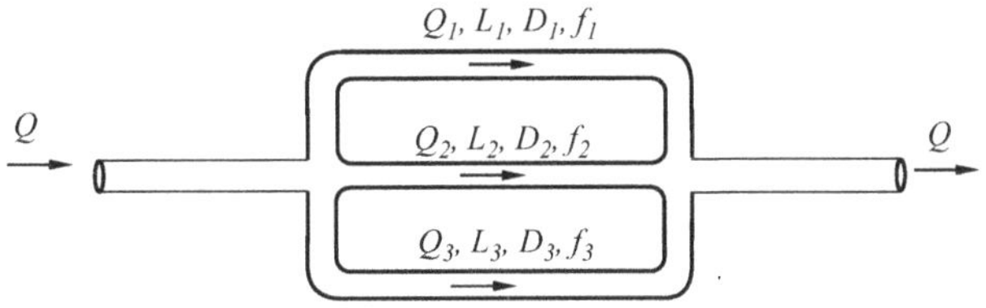

FIGURE 5.19 Pipes in parallel with different lengths (L), diameters (D), and roughness coefficient (f).

$$Q = Q_1 + Q_2 + Q_3 + \cdots \tag{5.30}$$

$$h_f = h_{f1} = h_{f2} = h_{f3} = \cdots \tag{5.31}$$

Pumps and pumping stations utilize complex long and short sections of piping, fittings, elbows, tees, reducers, enlargers, and specials. Readers are referred to Chapter 9 to study many complex examples and solutions. In these examples, the Darcy–Weisbach and Hazen–Williams equations, minor head loss equations, and EGL and HGL are demonstrated to develop the design of pumping stations and associated piping.

Example 5.15: Calculation of Flow Rate, Head Loss, Velocity, and EGL and HGL in Pipes Connected in Series

Determine the flow rate from reservoir A to reservoir B through two pipes connected in series. The reservoirs and the pipe arrangement in series, and pipe diameters, lengths, and friction factors for pipes 1 and 2 are given in Figure 5.20. Also, draw the EGL and HGL. Assume friction factor $f = 0.02$ for pipe 1 and $f = 0.03$ for pipe 2.

Solution

1. Determine the differential water surface elevations in reservoirs A and B

   At the surface of both reservoirs (points (1) and (2)), the velocity and pressure heads are zero, and $z_1 = 1{,}290$ m, and $z_2 = 1{,}205$ m.

$$H_1 = \frac{P_1}{\gamma} + \frac{V_1^2}{2g} + z_1 \quad \text{and} \quad H_2 = \frac{P_2}{\gamma} + \frac{V_2^2}{2g} + z_2$$

$$H_1 - H_2 = \text{differential elevation } \Delta H = 1{,}290 \text{ m} - 1{,}205 \text{ m}$$

$$\Delta H = 85 \text{ m}$$

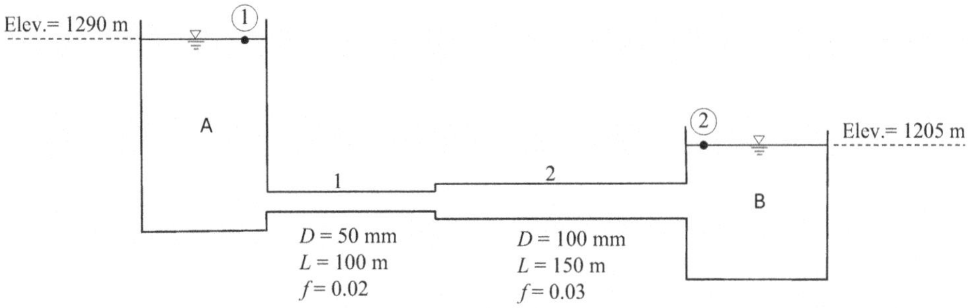

FIGURE 5.20 Reservoirs and the pipe system (Example 5.15).

This is the differential water surface elevations in reservoirs A and B.

2. Develop friction loss relationship and velocity in 50-mm diameter pipe

   Apply Eq. (5.29) and the Darcy–Weisbach equation

$$h_f = h_{f50} + h_{f100} = 0.02 \times \frac{(100\text{ m})}{\left(\dfrac{50\text{ mm}}{1{,}000\text{ mm/m}}\right)} \times \frac{V_{50}^2}{2g} + 0.03 \times \frac{(150\text{ m})}{\left(\dfrac{100\text{ mm}}{1{,}000\text{ mm/m}}\right)} \times \frac{V_{100}^2}{2g}$$

$$= 2.039V_{50}^2 + 2.294V_{100}^2$$

where V_{50} and V_{100} are the velocities in 50-mm and 100-mm diameter pipes, respectively.

3. Determine velocity in pipes

   Use Eq. (5.28) to obtain velocity in the 100-mm and 50-mm diameter pipes

$$Q = Q_{50} = Q_{100} = V_{50}A_{50} = V_{100}A_{100}$$

$$V_{100} = V_{50}\frac{A_{50}}{A_{100}}$$

$$A_{50} = \frac{\pi}{4}D_{50}^2 = \frac{\pi}{4}\left(\frac{50\text{ mm}}{1{,}000\text{ mm/m}}\right)^2 = 0.00196\text{ m}^2$$

$$A_{100} = \frac{\pi}{4}D_{100}^2 = \frac{\pi}{4}\left(\frac{100\text{ mm}}{1{,}000\text{ mm/m}}\right)^2 = 0.00785\text{ m}^2$$

$$V_{100} = V_{50}\frac{0.00196\text{ m}^2}{0.00785\text{ m}^2}$$

$$V_{100} = 0.25V_{50}$$

4. Determine total friction head loss in pipes

$$h_f = 2.039V_{50}^2 + 2.294(0.25V_{50})^2 = 2.182V_{50}^2$$

5. Develop minor head loss relationship with respect to velocity in 50-mm diameter pipe

   Head loss coefficients of $K_{\text{entrance}} = 0.5$, $K_{\text{expansion}} = 0.33$ (for $d/D = 50\text{ mm}/100\text{ mm} = 0.5$), and $K_{\text{exit}} = 1$ (Appendix C)

$$h_m = \sum K\frac{V^2}{2g} = h_{\text{entrance}} + h_{\text{expansion}} + h_{\text{exit}}$$

$$= (0.5)\frac{V_{50}^2}{2g} + (0.33)\frac{V_{50}^2}{2g} + (1)\frac{V_{100}^2}{2g}$$

   Replace V_{100} with $0.25V_{50}$ from step 3

$$h_m = (0.5)\frac{V_{50}^2}{2g} + (0.33)\frac{V_{50}^2}{2g} + \frac{(0.25V_{50})^2}{2g}$$

$$= 0.0455V_{50}^2$$

6. Calculate the velocity in a 50-mm diameter pipe

$$h_{\text{loss}} = h_f + h_m$$

$$85 = 2.182V_{50}^2 + 0.0455V_{50}^2 = 2.227V_{50}^2$$

$$V_{50} = 6.177 \text{ m/s}$$

7. Calculate velocity in a 100-mm pipe

$$V_{100} = 0.25V_{50}$$

$$V_{100} = 0.25 \times 6.177 \text{ m/s} = 1.544 \text{ m/s}$$

8. Calculate friction head losses in 50- and 100-mm diameter pipes and all components of minor head losses

$$h_{f50} = 2.039V_{50}^2 = 2.039 \times (6.177 \text{ m/s})^2 = 77.80 \text{ m}$$

$$h_{f100} = 2.294V_{100}^2 = 2.294 \times (1.544 \text{ m/s})^2 = 5.47 \text{ m}$$

$$\text{Velocity head in 50-mm pipe } \frac{V_{50}^2}{2g} = \frac{(6.177 \text{ m/s})^2}{2 \times 9.81 \text{ m/s}^2} = 1.94 \text{ m}$$

$$\text{Velocity head in 100-mm pipe } \frac{V_{100}^2}{2g} = \frac{(1.544 \text{ m/s})^2}{2 \times 9.81 \text{ m/s}^2} = 0.12 \text{ m}$$

$$h_{\text{entrance}} = 0.5 \times \frac{V_{50}^2}{2g} = 0.5 \times 1.94 \text{ m} = 0.97 \text{ m}$$

$$h_{\text{expansion}} = 0.33 \times \frac{V_{50}^2}{2g} = 0.33 \times 1.94 \text{ m} = 0.64 \text{ m}$$

$$h_{\text{exit}} = 1 \times \frac{V_{100}^2}{2g} = 1 \times 0.12 \text{ m} = 0.12 \text{ m}$$

9. Calculate the flow rate

$$Q = VA$$

$$A = 0.00196 \text{ m}^2 \text{ from step 3}$$

$$Q = 6.18 \text{ m/s} \times 0.00196 \text{ m}^2$$

$$= 0.0121 \text{ m}^3/\text{s}$$

The flow rate through the pipes is 0.0121 m^3/s or 12.1 L/s.

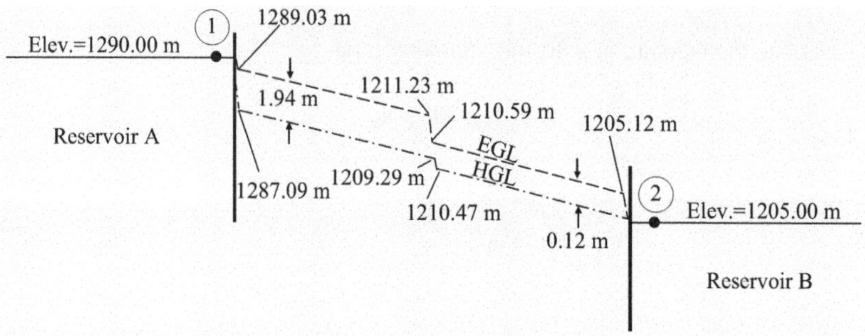

FIGURE 5.21 The EGL and HGL for the pipe system (Example 5.15).

10. Draw EGL and HGL
    Friction head losses in both connecting pipes and minor head losses are calculated in step (8). Use these values to draw EGL and HGL. These grade lines are shown in Figure 5.21.

Example 5.16: Flow Rate and Head Loss in Two Pipes Connected in Parallel

A 2.0-in. diameter, 300-ft long pipe has friction factor $f = 0.02$. It is connected in parallel with a 4.0-in. diameter, 250-ft long pipe with friction factor $f = 0.025$. The total discharge entering the pipe system at point A is 1.0 cfs. Determine the flow through each pipe. Ignore the minor head losses. Also, calculate the head loss in each pipe. The pipe arrangement in parallel is shown in Figure 5.22.

Solution

1. Develop a relationship between V_1 and V_2
   Write the continuity equation for pipes connected in parallel (Eq. 5.30)

$$Q = Q_1 + Q_2$$

$$1.0\ \text{ft}^3/\text{s} = V_1 A_1 + V_2 A_2$$

$$A_1 = \frac{\pi}{4} D_1^2 = \frac{\pi}{4}\left(\frac{2.0\ \text{in.}}{12\ \text{in./ft}}\right)^2 = 0.0218\ \text{ft}^2$$

$$A_2 = \frac{\pi}{4} D_2^2 = \frac{\pi}{4}\left(\frac{4.0\ \text{in.}}{12\ \text{in./ft}}\right)^2 = 0.0872\ \text{ft}^2$$

$$1.0\ \text{ft}^3/\text{s} = \left(0.0218\ \text{ft}^2\right) V_1 + \left(0.0872\ \text{ft}^2\right) V_2$$

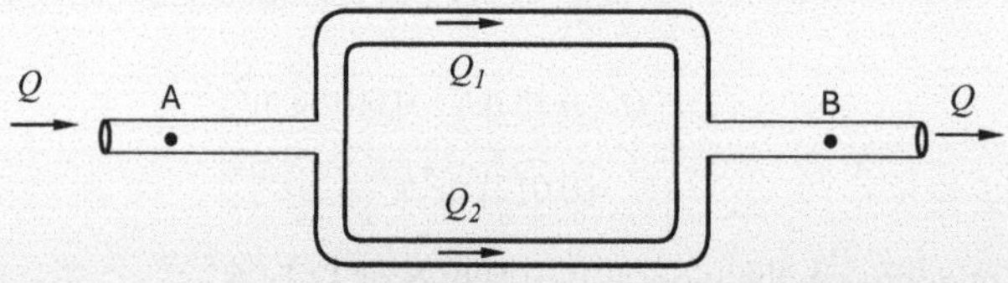

FIGURE 5.22 Parallel pipe system (Example 5.16).

$$V_1 = \frac{1.0 \text{ ft}^3/\text{s}}{0.0218 \text{ ft}^2} - \frac{0.0872 \text{ ft}^2}{0.0218 \text{ ft}^2} V_2$$

$$V_1 = 45.87 \text{ ft/s} - 4V_2$$

Apply Eq. (5.31) and the Darcy–Weisbach equation (Eq. 5.13)

$$h_f = h_{f1} = h_{f2}$$

$$h_{f1} = (0.02) \frac{300 \text{ ft}}{\left(\frac{2.0 \text{ in.}}{12 \text{ in./ft}}\right)} \frac{V_1^2}{2g} = 0.56V_1^2$$

$$h_{f2} = (0.025) \frac{250 \text{ ft}}{\left(\frac{4.0 \text{ in.}}{12 \text{ in./ft}}\right) 2g} = 0.29V_2^2$$

$$0.56V_1^2 = 0.29V_2^2$$

2. Calculate velocity in both pipes

$$0.56\left(45.87 \text{ ft/s} - 4V_2\right)^2 = 0.29V_2^2$$

$$V_2 = 9.72 \text{ ft/s}$$

$$V_1 = 45.87 \text{ ft/s} - 4 \times 9.72 \text{ ft/s} = 7.0 \text{ ft/s}$$

3. Calculate discharge through each pipe

$$Q_1 = V_1 A_1 = \left(7.0 \text{ ft/s}\right)(0.0218 \text{ ft}^2) = 0.15 \text{ ft}^3/\text{s}$$

$$Q_2 = V_2 A_2 = \left(9.72 \text{ ft/s}\right)(0.0872 \text{ ft}^2) = 0.85 \text{ ft}^3/\text{s}$$

4. Calculate head loss in each pipe

$$h_{f1} = 0.56V_1^2 = 0.56(7.0 \text{ ft/s})^2 = 27.4 \text{ ft}$$

$$h_{f2} = 0.29V_2^2 = 0.29(9.72 \text{ ft/s})^2 = 27.4 \text{ ft}$$

5. Check the requirements of Eqs. (5.30) and (5.31)
   The requirement of Eq. (5.30) is

$$Q = Q_1 + Q_2 = 0.15 \text{ ft}^3/\text{s} + 0.85 \text{ ft}^3/\text{s} = 1.00 \text{ ft}^3/\text{s}$$

The requirements of Eq. (5.31) is

$$+h_{f1} - h_{f2} = 0$$

$$27.4 \text{ ft} - 27.4 \text{ ft} = 0$$

The requirements of Eqs. (5.30) and (5.31) for flow at point B (in Figure 5.22) and head loss in each pipe connected in parallel are satisfied. $h_{f1} = h_{f2}$, and $Q = Q_1 + Q_2$.

5.6 OPEN CHANNEL FLOW

Water is transported by pressurized pipes or open channels. The basic hydraulic principles apply in both systems, but the presence of a free water surface in open channels makes flow calculations more complicated. An open channel is a conveying system that has a free surface. These systems could be natural or artificial channels. Examples of natural channels are rivers, creeks, and streams. The artificial channels are irrigation canals, sewers, and drainage ditches. The driving force in open channels is the gravitational force caused by the channel bottom slope. The flow in an open channel may be steady or unsteady, uniform or non-uniform (varied), or laminar or turbulent. The steady and uniform flow in channels carrying water and wastewater is very rare. The flow is almost always turbulent in these channels. The state of flow in an open channel can also be classified as subcritical, critical, or supercritical. The dimensionless *Froude number* is used to classify the flow. The Froude number is expressed as the ratio of the inertia force to the gravitational force

$$\mathrm{Fr} = \frac{V}{\sqrt{gD_h}} \tag{5.32}$$

where

Fr = Froude number, dimensionless

V = velocity in the channel, m/s (ft/s)

D_h = hydraulic depth, m (ft). The hydraulic depth is the ratio of cross-sectional area and top width of the channel (A/B), shown in Figure 5.23a

A = cross-sectional area, m^2 (ft^2)

B = top width, m (ft)

g = acceleration due to gravity, 9.81 m/s^2 (32.2 ft/s^2)

If the Froude number is less than 1 ($\mathrm{Fr} < 1$), the flow is considered *subcritical*. If the Froude number is greater than 1 ($\mathrm{Fr} > 1$), the flow is considered *supercritical*. The flow is referred to as *critical* if the Froude number is equal to 1 ($\mathrm{Fr} = 1$).

The shape and slope of an open channel may vary along its length. A channel with a constant cross-section and bottom slope throughout its length is known as a *prismatic channel*. The channel cross-section and/or bottom slope vary throughout the length of a *non-prismatic channel*. The basic hydraulic principles of open channel flow that are required for the design of storm and sanitary sewer systems are covered in the following sections.

5.6.1 Geometric Elements of Open Channel Flow

Several parameters are used to characterize flow in an open channel. The definition of these parameters or geometric elements is given below.[4] Figure 5.23 shows the geometric elements of an open channel.

- **Depth of flow (d):** Depth of flow is normal to the direction of flow, measured perpendicular to the channel bottom (Figure 5.23b).
- **Vertical depth of flow (y):** Vertical distance is between the channel bottom and the free surface ($d = y\cos\alpha$ from Figure 5.23b).
- **Top width (B):** Top width of the channel is the cross-sectional width at the free water surface (Figure 5.2a).
- **Flow area (A):** Flow area is the cross-section normal to the direction of flow (Figure 5.23a).
- **Wetted perimeter (P):** It is the length of the channel surface in contact with flow (Figure 5.35a).
- **Hydraulic radius (R):** It is the ratio of the flow area to the wetted perimeter ($R = A/P$).
- **Hydraulic depth (D_h):** It is the ratio of the flow area to the top width ($D_h = A/B$).

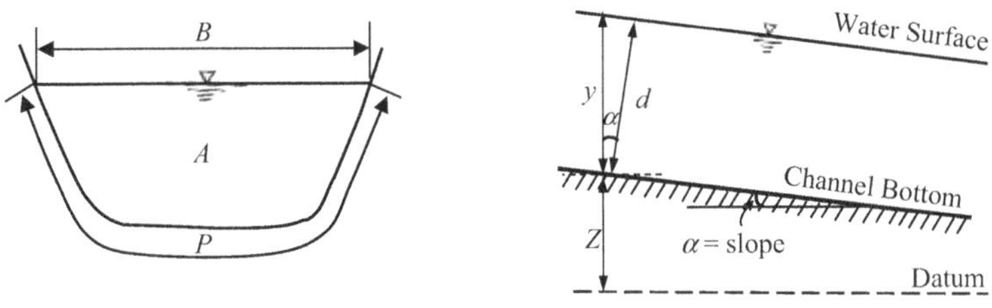

(a) (b)

FIGURE 5.23 Open channel geometry definition: (a) channel section and (b) longitudinal profile.

The geometric elements of rectangular, trapezoidal, triangular, and circular open channel cross sections are listed in Table 5.3. In a circular pipe under partial flow, the central angle θ is used to express all hydraulic elements of the pipe. The values of θ depend upon the partial flow depth in the sewer, and are expressed by Eq. (5.33).

$$\cos\frac{1}{2}\theta = \left(1 - \frac{2d}{D}\right) \quad \text{or} \quad \theta = 2\cos^{-1}\left(1 - \frac{2d}{D}\right) \tag{5.33}$$

where

θ = central angle, degree (π rad = 180°)
d = partial flow depth, m (ft)
D = diameter of sewer, m (ft)

It is a common practice to use d, v, q, a, and p (lower case) notations for depth of flow, velocity, discharge, area, and wetted perimeter for pipes under partial flow conditions, while D, V, Q, A, and P (upper case) notations are used for pipes flowing full.

TABLE 5.3
Geometric Elements of Typical Open Channel Cross Sections

Channel Cross Section	Area	Wetted Perimeter	Hydraulic Radius	Top Width	Hydraulic Depth
Rectangular 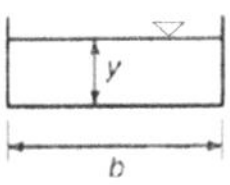	by	$b + 2y$	$\frac{by}{b + 2y}$	b	y
Trapezoidal 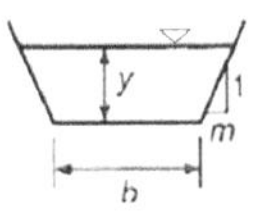	$(b + my)y$	$b + 2y\sqrt{1 + m^2}$	$\frac{(b + my)y}{b + 2y\sqrt{1 + m^2}}$	$b + 2my$	$\frac{(b + my)y}{b + 2my}$
Triangular 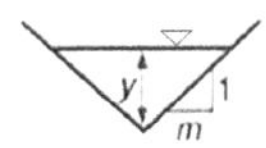	my^2	$2y\sqrt{1 + m^2}$	$\frac{my}{2y\sqrt{1 + m^2}}$	$2my$	$\frac{y}{2}$
Circular 	$\frac{1}{8}(\frac{\pi\theta}{180} - \sin\theta)D^2$	$(\frac{\pi\theta}{360})D$	$\frac{1}{4}(1 - \frac{180\sin\theta}{\pi\theta})D$	$\sin(\frac{\theta}{2})D$	$\frac{1}{8}\frac{\frac{2\pi\theta}{180} - \sin 2\theta}{\sin\theta}D$

Source: Adopted in part from Ref. 4

5.6.2 Uniform Flow

In uniform channel flow, the depth and velocity remain constant along the channel length. Therefore, the EGL and HGL are parallel to the channel bottom. Figure 5.24 shows the EGL and HGL in a uniform flow. In this figure, S_f, S_w, and S_o are the slopes of the EGL, HGL, and channel bottom, respectively. Under uniform channel flow, these slopes are equal, i.e., $S_f = S_w = S_o$. The corresponding depth to a uniform flow is called the *normal depth*. The uniform flow can be observed in laboratory flumes and artificial channels with prismatic cross sections. The uniform flow rarely occurs in natural channels due to hydraulic structures such as weirs, sluice gates, and others, or changes due to the geometry and bottom slope that cause the flow depth to deviate from the normal depth. The uniform flow calculation is a suitable approximation for irregular channels without hydraulic structures. The concept of uniform flow is also central to the analysis of GVFs. The flow resistance relations developed for uniform flows predict satisfactory channel friction in GVFs.[11]

5.6.3 Uniform Flow Calculation

Uniform flow equations proposed by Manning, Chezy, Gangrullet, Kutter, and Scobey have been used for designing sewers and drains.[2,8,12,13] Manning's equation, however, has received the most widespread application. Manning's equation in various forms is expressed by Eqs. (5.34)–(5.37)[8,9]

$$V = \frac{1}{n} R^{2/3} S^{1/2} \quad \text{(SI unit)} \tag{5.34}$$

$$Q = \frac{1}{n} A R^{2/3} S^{1/2} \quad \text{(SI unit)} \tag{5.35}$$

$$V = \frac{1.486}{n} R^{2/3} S^{1/2} \quad \text{(U.S. customary unit)} \tag{5.36}$$

$$Q = \frac{1.486}{n} A R^{2/3} S^{1/2} \quad \text{(U.S. customary unit)} \tag{5.37}$$

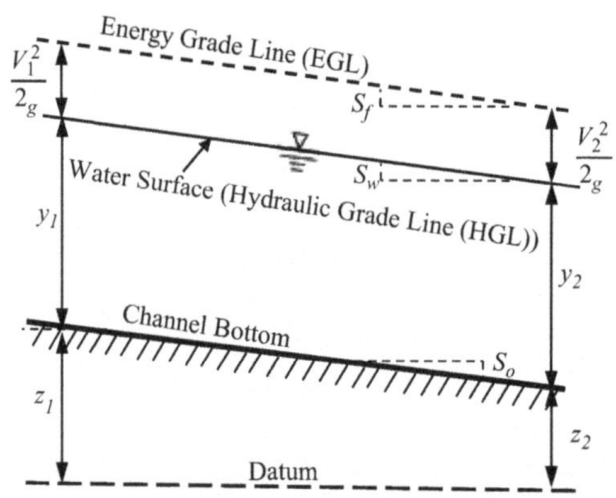

FIGURE 5.24 Energy grade line and hydraulic grade line in an open channel with uniform flow.

where

V = flow velocity, m/s (ft/s)
R = hydraulic radius, m (ft)
A = flow area, m^2 (ft^2)
Q = discharge, m^3/s (ft^3/s)
S = slope of EGL or channel bottom, m/m (ft/ft)
n = coefficient of roughness used in Manning's equation, dimensionless

The coefficient of roughness depends on the material of the channel. Commonly used values of n for different materials are given in Table 5.4.

Circular pipes flowing full: Manning's equation can be derived for pipes flowing full by substituting the flow area = $\pi D^2/4$, wetted perimeter $P = \pi D$, and hydraulic radius $R = A/P = D/4$ in Eqs. (5.35) and (5.37).[8] The discharge in a pipe flowing full is expressed by Eqs. (5.38) and (5.39)

$$Q = \frac{0.312}{n} D^{8/3} S^{1/2} \quad \text{(Circular pipe flowing full, SI unit)} \tag{5.38}$$

$$Q = \frac{0.464}{n} D^{8/3} S^{1/2} \quad \text{(Circular pipe flowing full, U.S. customary unit)} \tag{5.39}$$

where D is the pipe diameter in m or ft.

A nomograph based on Manning's equation for circular pipes flowing full is provided in Figure 5.25.[14]

Circular pipes flowing partially full: Sanitary and storm sewers are designed to flow partially full. The geometric element equations for circular pipes flowing under partially full flow conditions are given in Table 5.3. Various types of nomographs have been developed for the solution of problems involving pipes flowing partially full.[8,12,13] Nomographs based on Manning's equation for circular pipes and variable n values are provided in Figure 5.25 Similar nomographs may be found in Refs. 8,9,13–15 To use such nomographs, the flow parameters should first be calculated for the pipe flowing full. It may be noted that the value of n decreases with the depth of flow (Figure 5.26b); however, in most designs, n is assumed constant for all flow depths. As shown in Figure 5.26a, d is the depth of partial full flow and D is the diameter of the pipe. A sewer line that is flowing 80% full has a d/D ratio of 0.8.

TABLE 5.4
Common Values of Roughness Coefficient Used in Manning's Equation

Material	Commonly Used Values for n
Concrete	0.013–0.015
Vitrified clay	0.013–0.015
Cast iron	0.013–0.015
Brick	0.015–0.017
Corrugated metal pipe	0.022–0.025
Asbestos cement	0.013–0.015
Earthen channels	0.025–0.030

Source: Adopted in part from Refs. 2,8,12,13

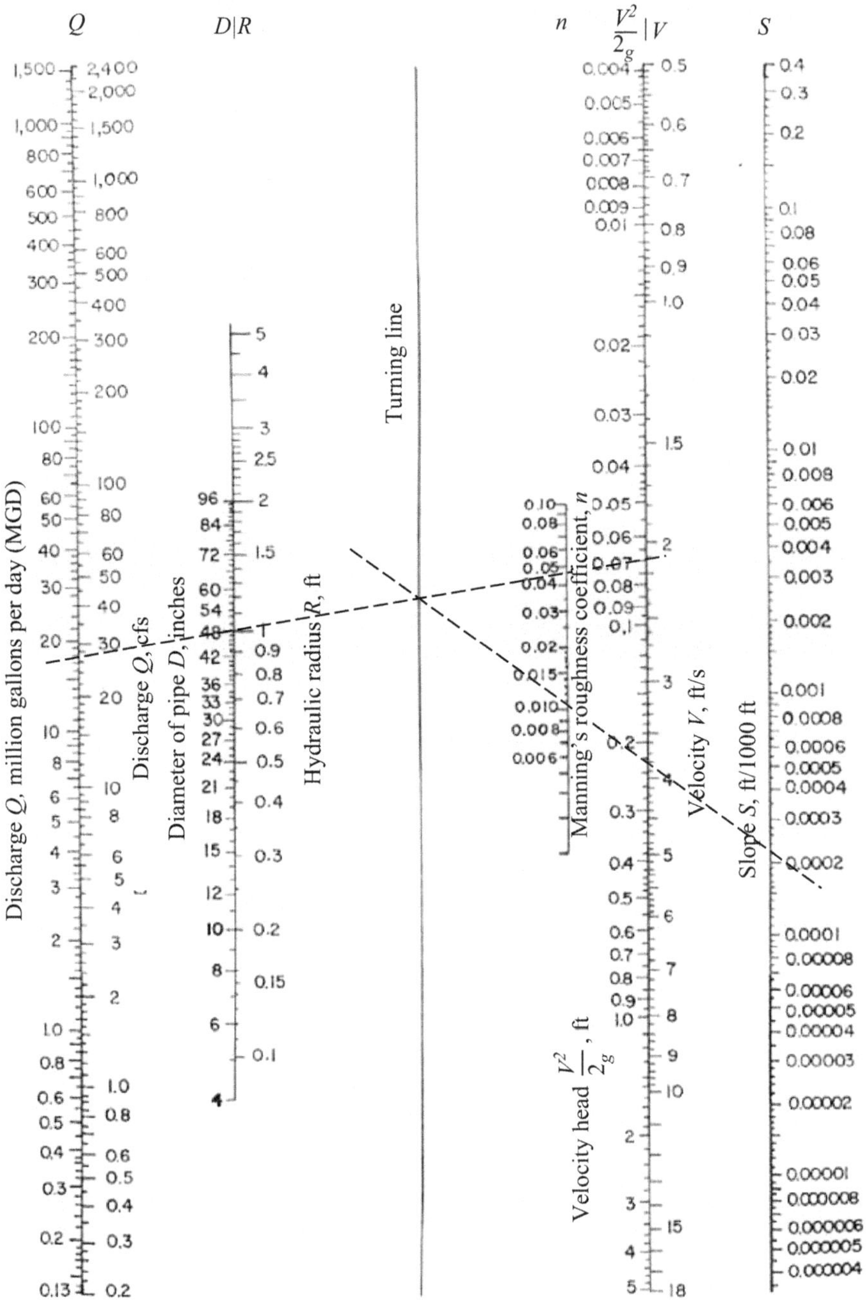

FIGURE 5.25 Nomogram for solving Manning's equation for pipes flowing full.

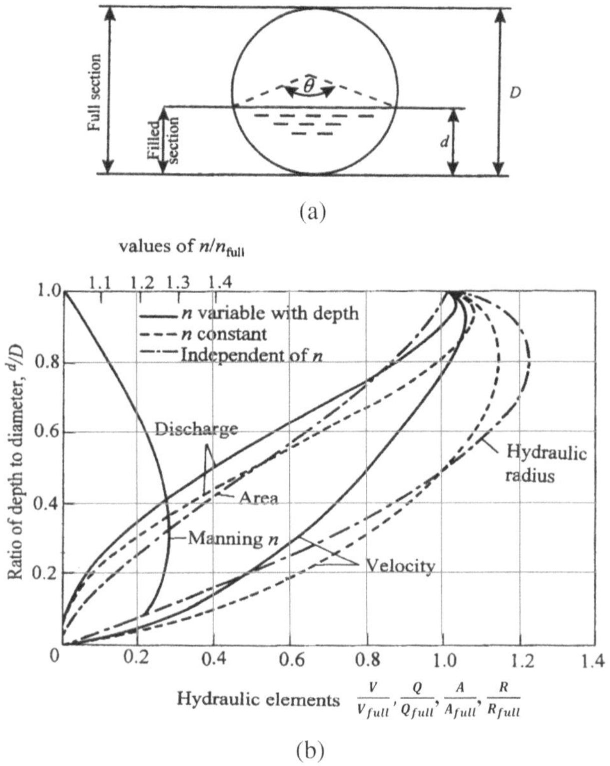

FIGURE 5.26 Hydraulic properties of circular sewer: (a) hydraulic elements and (b) partial flow condition nomograph.

Example 5.17: Uniform Flow in a Rectangular Channel

Calculate the average velocity and the discharge through a 3.0-m-wide rectangular concrete aqueduct. The bottom slope of the aqueduct is 0.003, and the flow depth in the aqueduct is 1.1 m.

Solution

1. Calculate flow velocity
   Apply Manning's equation (5.34)
   $n = 0.015$ from Table 5.4

$$V = \frac{1}{n} R^{2/3} S^{1/2} = \frac{1}{n}\left(\frac{A}{P}\right)^{2/3} S^{1/2}$$

$$= \frac{1}{0.015}\left(\frac{3.0\text{ m} \times 1.1\text{ m}}{2.0\text{ m} \times 1.1\text{ m} + 3.0\text{ m}}\right)^{2/3} (0.003)^{1/2}$$

$$= 2.7\text{ m/s}$$

2. Calculate the discharge

$$Q = VA$$
$$= (2.7\ \text{m/s}) \times (1.1\ \text{m} \times 3.0\ \text{m})$$
$$= 8.91\ \text{m}^3/\text{s}$$

The average velocity and discharge are 2.7 m/s and 8.91 m³/s.

Example 5.18: Minimum Slope of a Circular Sewer Flowing Full with Uniform Flow

Calculate the minimum slope of a 54 in. diameter concrete sewer to maintain a minimum average velocity of 2.0 ft/s. Also, calculate the discharge when the pipe is flowing full. Manning's $n = 0.013$.

Solution

1. Rearrange Manning's equation (5.34) for slope

$$V = \frac{1.486}{n} R^{2/3} S^{1/2}$$

$$S = \frac{\left(\dfrac{V \times n}{1.486}\right)^2}{R^{4/3}} = \frac{V^2 n^2}{2.208 R^{4/3}}$$

2. Substitute $n = 0.013$, $V = 2.0$ ft/s, and $R = D/4$

$$S = \frac{1}{2.208} \times (2.0\ \text{ft/s})^2 \times (0.013)^2 \times \left(\frac{D}{4}\right)^{-4/3}$$

$$S = 0.001943\left(D^{-4/3}\right)$$

3. Substitute the diameter $D = \left(\dfrac{54.0\ \text{in.}}{12\ \text{in./ft}}\right)$ and solve for S

$$S = 0.001943\left(\frac{54.0\ \text{in.}}{12\ \text{in./ft}}\right)^{-4/3}$$
$$= 0.00026\ \text{ft/ft}$$

The minimum slope of the sewer to maintain a minimum average velocity of 2.0 ft/s is 0.00026 ft/ft.

4. Calculate the discharge when the sewer is flowing full

$$Q = VA$$

$$A = \frac{\pi}{4}\left(\frac{54.0\ \text{in.}}{12\ \text{in./ft}}\right)^2 = 15.9\ \text{ft}^2$$

$$Q = 2.0 \text{ ft/s} \times 15.9 \text{ ft}^2$$

$$= 31.80 \text{ ft}^3/\text{s}$$

The discharge is 31.80 ft³/s when the sewer is flowing full.

Example 5.19: Application of Geometric Element Equations for Circular Pipe Flowing Partially Full

The concrete sewer in Example 5.18 is flowing partially full. The depth of flow is 44.0 in., and sewer diameter is 54.0 in. Manning's $n = 0.013$ and slope = 0.00026 ft/ft. Compute the average velocity and the discharge through the sewer under partial flow.

Solution

1. Calculate the geometric elements for circular pipes flowing partially full
   Apply partial flow equations from Table 5.3
   a. Calculate θ using Eq. (5.33)

$$\theta = 2\cos^{-1}\left(1 - \frac{2d}{D}\right)$$

$$= 2\cos^{-1}\left(1 - \frac{2 \times 44.0 \text{ in.}}{54.0 \text{ in.}}\right)$$

$$= 4.5037 \text{ rad or } 258°$$

   b. Calculate the partial flow area α

$$a = \frac{1}{8}\left(\frac{\pi\theta}{180} - \sin\theta\right)D^2$$

$$= \frac{1}{8}\left(\frac{\pi(258°)}{180} - \sin 258°\right)\left(\frac{54.0 \text{ in.}}{12 \text{ in./ft}}\right)^2$$

$$= 13.88 \text{ ft}^2$$

   c. Calculate the partial flow hydraulic radius r

$$r = \frac{1}{4}\left(1 - \frac{180\sin\theta}{\pi\theta}\right)D$$

$$= \frac{1}{4}\left(1 - \frac{180\sin 258°}{\pi(258°)}\right)\left(\frac{54.0 \text{ in.}}{12 \text{ in./ft}}\right)$$

$$= 1.37 \text{ ft}$$

2. Calculated the velocity from Eq. (5.36)

$$v = \frac{1.486}{n} r^{2/3} S^{1/2}$$

$$= \frac{1.486}{0.013}(1.37 \text{ ft})^{2/3}(0.00026 \text{ ft/ft})^{1/2}$$

$$= 2.27 \text{ ft/s}$$

3. Calculate the discharge from velocity and area

$$q = va$$

$$= 2.27 \text{ ft/s} \times 13.88 \text{ ft}^2$$

$$= 31.51 \text{ ft}^3\text{/s}$$

The discharge and average velocity are 31.51 ft^3/s and 2.27 ft/s when the pipe is flowing partially full.

Example 5.20: Application of Nomographs for Circular Pipes

Compute the average velocity and the discharge through a pipe flowing full and partially full. Use Manning's nomograph and partial flow nomograph. The pipe diameter is 48 in., and partial depth of flow is 39.0 in. Manning's n=0.011 and slope S=0.00022 ft/ft.

Solution

1. Estimate the discharge and velocity when the sewer is flowing full
   Apply Manning's nomograph in Figure 5.25
   a. Mark a point on the slope scale for $S = 0.00022$ ft/ft
   b. Mark another point on Manning's roughness coefficient scale for n=0.011
   c. Draw a straight line connecting these points. Extend this line to the turning line and mark this point. This line is shown in Figure 5.25
   d. Mark a point on the pipe diameter scale for $D = 48$ in.
   e. Draw a straight line connecting the points on the pipe diameter scale and the point on the turning line, and extend this line to the discharge scale. These lines and the velocity scale are drawn on Figure 5.25. Read Q_{full} from the discharge scale, and velocity from the velocity scale.

$$Q_{\text{full}} = 26.8 \text{ ft}^3\text{/s}$$

$$V_{\text{full}} = 2.1 \text{ ft/s}$$

2. Determine the flow (q) and velocity (v) under partial conditions
   a. Calculate d/D ratio

$$\frac{d}{D} = \frac{39 \text{ in.}}{48 \text{ in.}} = 0.81$$

   b. From Figure 5.26b for $d/D = 0.81$, $v/V_{\text{full}} = 1.13$ and $q/Q_{\text{full}} = 0.99$
   c. Estimate the average velocity and discharge for partial flow in the pipe

$$Q_{\text{full}} = 26.8 \text{ ft}^3\text{/s} \quad \text{and} \quad V_{\text{full}} = 2.1 \text{ ft/s from step 1}$$

$$\frac{v}{V_{\text{full}}} = 1.13$$

$$v = 1.13 \times 2.1 \text{ ft/s}$$

$$= 2.4 \text{ ft/s}$$

$$\frac{q}{Q_{\text{full}}} = 0.99$$

$$q = 0.99 \times 26.8 \text{ ft}^3/\text{s}$$
$$= 26.5 \text{ ft}^3/\text{s}$$

The discharge and average velocity are 26.5 ft³/s and 2.4 ft/s when the pipe is flowing partially full.

In Chapter 8, the design of complex sanitary and storm sewers for city subdivisions is developed. Readers are referred to this chapter to find design examples and solutions of partial flow sewers, appurtenances, and hydraulic profile.

5.6.4 Varied Flow

The flow is called varied or non-uniform if the water surface in an open channel deviates from its normal depth. This may happen due to hydraulic structures such as weirs and sluice gates, or changes to the channel geometry. In such cases, the downslope component of the fluid weight may not locally balance the channel bed friction. As a result, the slope of the EGL and the channel bottom slope are different, and the flow depth and velocity change in the flow direction. As discussed in Section 5.3.2 and shown in Figure 5.3, varied flows in open channels may be grouped as GVF and RVF.

Gradually varied flow (GVF): In GVF, the depth changes gradually in the flow direction. Therefore, the flow is treated as one-dimensional, with no transverse pressure distribution except for that caused by gravity. The uniform flow formulas cannot be applied to determine the flow depth and velocity. Both flow depth and velocity are essential for planning, design, and operation of storm sewers and drainage channels. The GVF in an open channel is shown in Figure 5.27. The water surface elevation of GVF can be calculated from the energy equation and uniform flow relations such as Manning's equation, and is solved from the local flow depth along the channel length. The total head (H) at any particular cross section may be written by Eq. (5.40) or (5.41)

$$H = y + \frac{V^2}{2g} + z \tag{5.40}$$

$$H = y + \frac{Q^2}{2gA^2} + z \tag{5.41}$$

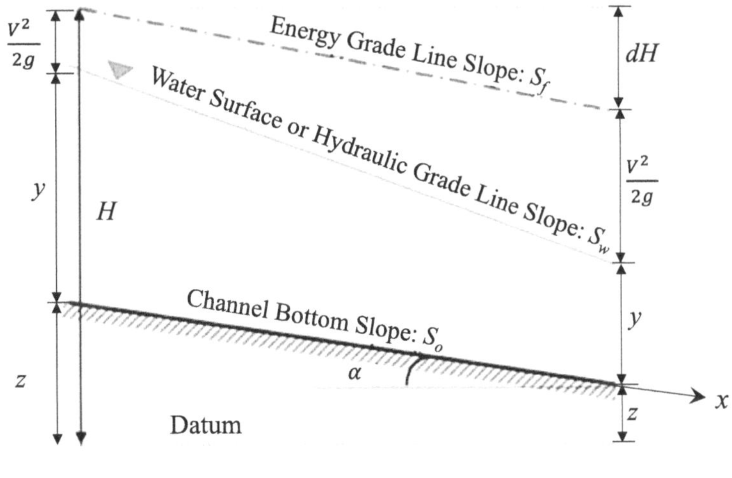

FIGURE 5.27 Gradually varied flow (GVF) in an open channel.

where

y = flow depth, m (ft)
V = flow mean velocity, m/s (ft/s)
Q = discharge, m^3/s (ft^3/s)
A = flow area, m^2 (ft^2)
g = acceleration due to gravity, 9.81 m/s^2 (32.2 ft/s^2)
z = channel bed elevation from a datum, m (ft)

Differentiate Eq. (5.41) with respect to flow direction x in Figure 5.27. The results are expressed by Eq. (5.42)

$$\frac{dH}{dx} = \frac{dy}{dx} + \frac{Q^2}{2g}\frac{d}{dx}\left(\frac{1}{A^2}\right) + \frac{dz}{dx} \tag{5.42}$$

By rearranging and simplifying Eq. (5.42), Eq. (5.43) is obtained.[1,3] Derivation of Eq. (5.43) is presented in Example 5.21.

$$\frac{dy}{dx} = \frac{S_o - S_f}{1 - \mathrm{Fr}^2} \tag{5.43}$$

where

$\frac{dy}{dx}$ = change in water depth in the channel (S_w), dimensionless
$\frac{dH}{dx}$ = change in energy grad line (S_f), dimensionless
$\frac{dz}{dx}$ = change in channel bottom elevation (S_o), dimensionless
Fr = Froude number, dimensionless

The definition of the Froude number (Fr) is given by Eq. (5.32). For a given discharge, channel geometry, slope, and roughness, the water depth in the channel (y) can be determined as a function of distance along the channel length (x) by solving Eq. (5.43). The solution of Eq. (5.43) is obtained graphically or by numerical techniques such as the direct step and standard step methods.[1,4]

Rapidly varied flow (RVF) and hydraulic jump: This type of flow is characterized by sudden changes in depth and substantial energy loss in a short distance along the length of the channel. The *hydraulic jump* is a classic example of RVF in an open channel. It is formed when a shallow flow with high velocity (supercritical flow) impinges on a body of deep slow-moving fluid (subcritical flow). Figure 5.28 shows a hydraulic jump downstream of a spillway. There is a sharp discontinuity in the water surface profile within the jump, and a substantial loss of energy takes place due to turbulence at the jump location. It is impractical to accurately estimate the energy losses involved

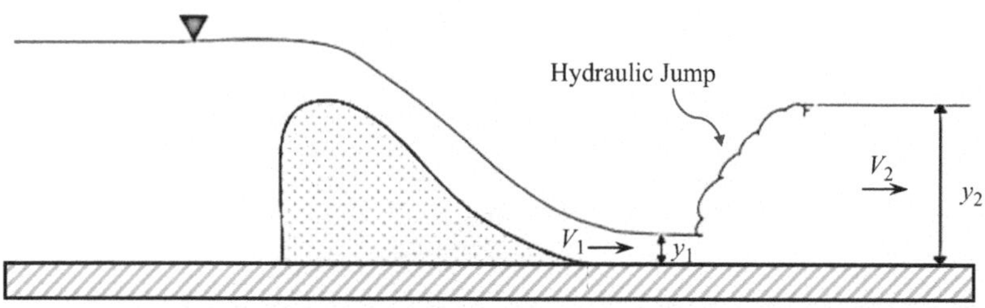

FIGURE 5.28 Hydraulic jump downstream of a spillway.

in a jump. For this reason, the momentum equation in open channel flows is applied to estimate the flow parameters, including depth and velocity at sections upstream and downstream of the jump. The relationships derived for a rectangular channel from the momentum equation are expressed by Eqs. (5.44) and (5.45).

$$y_2 = \frac{y_1}{2}\left(\sqrt{1+8\,\mathrm{Fr}_1^2} - 1\right) \tag{5.44}$$

or

$$y_1 = \frac{y_2}{2}\left(\sqrt{1+8\mathrm{Fr}_2^2} - 1\right) \tag{5.45}$$

where

y_1, y_2 = flow depth upstream and downstream of the jump, m (ft)

Fr_1, Fr_2 = flow Froude number upstream and downstream of the jump, dimensionless

The upstream and downstream depths of a hydraulic jump are called *sequent depths*; y_2 is the sequent depth of y_1.

Equation (5.44) can be solved for the flow depth downstream of the jump (y_2) if the upstream depth (y_1) and the Froude number (Fr_1) are known. If y_2 and Fr_2 are given, Eq. (5.45) can be solved for y_1. Dimensionless charts for channels with trapezoidal, circular, triangular, and parabolic cross sections have been developed to calculate the values of y_1 and y_2. Such charts may be found in Refs. 1,4

Example 5.21: Derivation of Gradually Varied Flow Equation

Derive the governing equation (Eq. 5.42) for GVF.

Solution

1. Write the total head equation at a channel section in Figure 5.27

$$H = y + \frac{V^2}{2g} + z$$

2. Express V in terms of discharge Q

$$V = Q/A$$

$$H = y + \frac{Q^2}{2gA^2} + z$$

3. Differentiate both sides of the total head equation with respect to x

$$\frac{dH}{dx} = \frac{dy}{dx} + \frac{Q^2}{2g}\frac{d}{dx}\left(\frac{1}{A^2}\right) + \frac{dz}{dx}$$

4. Replace $\frac{dH}{dx}$ and $\frac{dz}{dx}$ with the slope of energy grad line and channel bottom (see Figure 5.27)

$$\frac{dH}{dx} = -S_f \quad \text{and} \quad \frac{dz}{dx} = -S_o$$

Note: Since both H and z decrease as x increases, there is a negative sign with S_f and S_o

$$-S_f = \frac{dy}{dx} + \frac{Q^2}{2g}\frac{d}{dx}\left(\frac{1}{A^2}\right) - S_o$$

5. Calculate $\frac{d}{dx}\left(\frac{1}{A^2}\right)$

$$\frac{d}{dx}\left(\frac{1}{A^2}\right) = \frac{d}{dA}\left(\frac{1}{A^2}\right)\frac{dA}{dx} = \frac{d}{dA}\left(\frac{1}{A^2}\right)\frac{dA}{dy}\frac{dy}{dx}$$

If the channel is prismatic, $\frac{dA}{dy} = B$, as shown in Figure 5.29; therefore,

$$\frac{d}{dx}\left(\frac{1}{A^2}\right) = \frac{d}{dA}\left(\frac{1}{A^2}\right)B\frac{dy}{dx} = \frac{-2B}{A^3}\frac{dy}{dx}$$

6. Substitute $\frac{d}{dx}\left(\frac{1}{A^2}\right)$ in the last equation in step 4 and solve for $\frac{dy}{dx}$

$$-S_f = \frac{dy}{dx} + \frac{Q^2}{2g}\left(\frac{-2B}{A^3}\right)\frac{dy}{dx} - S_o$$

$$\frac{dy}{dx} = \frac{S_o - S_f}{1 - \left(BQ^2/gA^3\right)}$$

7. Express Froude number from Eq. 5.32 in terms of B, A, and Q

$$\text{Fr} = \frac{V}{\sqrt{gD_h}} = \frac{Q/A}{\sqrt{gA/B}}$$

8. Calculate Fr^2

$$Fr^2 = \left(\frac{Q/A}{\sqrt{gA/B}}\right)^2 = \frac{BQ^2}{gA^3}$$

9. Substitute Fr^2 from step 8 into the last equation in step 6

$$\frac{\text{dy}}{\text{dx}} = \frac{S_o - S_f}{1 - Fr^2}$$

This is Eq. (5.43) that can be used to determine the water surface profile of GVF.

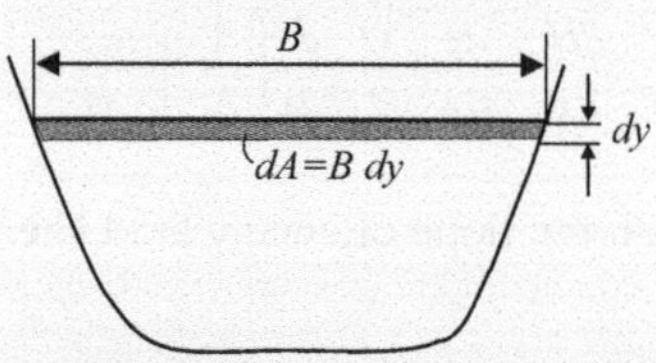

FIGURE 5.29 Definition of $\frac{dA}{dy}$ in a prismatic channel (Example 5.21).

Example 5.22: Gradually Varied Flow in a Rectangular Channel

A rectangular channel is carrying water at a constant rate of 7.0 m^3/s. The channel is 3.0-m wide along its length. The bottom slope is 0.001, and Manning's is 0.015. The channel terminates as a free fall. Calculate the water surface profile.

Solution

1. Identify control sections

   Note: A control point is a location in the channel where the relationship between the flow depth and discharge is known. Normal depth and critical depth are common control points in channels with backwater or drawdown water surface profiles. A depth set by a hydraulic control structure and a hydraulic jump can also be considered as control points. Control points are usually the boundary conditions for GVF water surface profile calculations. Different boundary conditions result in different water surface profiles.[4]

   The control sections for this channel are the *normal depth* upstream of the channel, far from the free fall, and the *critical depth* at the free fall. The control sections are shown in Figure 5.30.
2. Determine the normal depth of flow (y_n)

   Apply Manning's equation (Eq. 5.34)

$$Q = \frac{1}{n} A R^{2/3} S^{1/2}$$

Replace in this equation the flow area (A) and hydraulic radius (R) of the rectangular channel from Table 5.3

$$A = b y_n = (3.0 \text{ m}) y_n$$

wetted perimeter $P = b + 2y$

$$R = \frac{b y_n}{b + 2y_n} = \frac{(3.0 \text{ m}) y_n}{(3.0 \text{ m}) + 2y_n}$$

$$7 \text{ m}^3/\text{s} = \frac{1}{0.015}(3.0 \text{ m}) y_n \left(\frac{(3.0 \text{ m})\ y_n}{3.0 \text{ m} + 2y_n} \right)^{\frac{2}{3}} (0.001)^{1/2}$$

$$0.532 = \frac{y_n^{5/3}}{(3.0 \text{ m} + 2y_n)^{2/3}}$$

Determine the normal depth (y_n) by trial and error or by using an equation solver.

$$y_n = 1.38 \text{ m}$$

This is the depth of the flow upstream of the free-fall section, where the depth is no longer affected by the drawdown. The normal depth (y_n) is shown in Figure 5.30.

3. Determine the critical depth at the free fall.

   At the free-fall section, flow depth is critical, and the Froude number (Eq. 5.32) is equal to 1

$$\mathrm{Fr} = \frac{V}{\sqrt{gD_h}} = 1$$

Replace in this equation $V = \dfrac{Q}{A}$ and $D_h = y$ (for a rectangular channel)

$$\mathrm{Fr} = \frac{Q}{A\sqrt{gy}} = \frac{Q}{by\sqrt{gy}} = 1$$

At the control section with critical flow $y = y_c$, the critical velocity y_c is expressed by Eq. (5.46)

$$\mathrm{Fr} = \frac{Q}{by_c\sqrt{gy_c}} = 1$$

Rearrange Froude number equation for critical depth y_c

$$y_c = \left(\frac{(Q/b)^2}{g}\right)^{1/3} \tag{5.46}$$

$$y_c = \left(\frac{\left(7.0\ \mathrm{m^3/s}\,/\,3.0\ \mathrm{m}\right)^2}{9.81\ \mathrm{m/s^2}}\right)^{1/3}$$

$$y_c = 0.82\ \mathrm{m}$$

The critical depth (y_c) is shown in Figure 5.30.

4. Determine water surface profile between control points, i.e., between the normal depth and critical depth.

   Apply Eq. (5.43) to calculate water surface profile

$$\frac{dy}{dx} = \frac{S_o - S_f}{1 - \mathrm{Fr}^2}$$

Write Eqs. (5.47) and (5.48) in the form of discrete differences

$$\Delta x = \frac{1 - \mathrm{Fr}^2}{S_0 - S_f}\Delta y \tag{5.47}$$

where

Δx = incremental channel length, m (ft)

Δy = incremental water depth in the channel, m (ft)

$$(x_{i+1} - x_i) = \frac{1 - \overline{\mathrm{Fr}}^2}{S_o - \overline{S}_f}(y_{i+1} - y_i) \tag{5.48}$$

where $\overline{\mathrm{Fr}}$ and $\overline{S}_f$ are the arithmetic mean of Froude number and slope of the EGL at stations x_i and x_{i+1} along the channel with flow depth of y_i and y_{i+1}, respectively.

5. Start the computation with the known depth of $y = y_c = 0.82$ m at the free fall and proceed in the upstream direction. At the free fall $x = 0$, and the distance in the upstream flow direction is considered positive

The water surface calculation is an iterative process that can easily be done in a *spreadsheet*. The calculations are shown in Table 5.5.

a. **Column 1:** The value of y is selected in Column 1. The size of the increment in the depth ($\Delta y = y_{i+1} - y_i$) determines the length ($\Delta x = x_{i+1} - x_i$) in each computational step and the accuracy of the calculated water surface profile. In each iteration, y_{i+1} is the flow depth for the current step, and y_i is the depth for the previous step.
Note: The water surface profile approaches normal depth asymptotically. The length of backwater or drawdown curves from a control point to the normal depth is theoretically infinite. Stop calculations at 99% of the normal depth to avoid numerical problems in calculating the water surface profile of a GVF that approaches normal depth asymptotically.

b. **Columns 2–6**: Calculate flow area A, wetted perimeter P, hydraulic radius R, and flow velocity V, and Fr^2 for the depth in Column 1.

c. **Column 7:** Determine the slope of EGL (S_f) from the Manning's equation and compute R and V of Columns 4 and 5.

d. **Columns 8–9:** Calculate $\bar{S}_f$ and $\overline{\mathrm{Fr}}^2$. These are the average of S_f and Fr^2 for the current depth and for the depth in the previous step.

e. **Column 10:** Calculate $S_o - \bar{S}_f$ by subtracting $\bar{S}_f$ of Column 9 from the S_o (channel bottom slope).

f. **Column 11:** Calculate $1 - \overline{\mathrm{Fr}}^2$.

g. **Column 12:** Calculate the depth increment $\Delta y = (y_{i+1} - y_i)$. This Column is the difference between the flow depth in the current step and the depth in the previous step.

h. **Column 13:** Calculate $\dfrac{1 - \overline{\mathrm{Fr}}^2}{S_o - \bar{S}_f}$ by dividing Column 11 by Column 10.

TABLE 5.5
Computation of Water Surface Profile in a Rectangular Channel (Example 5.22)

y, m	A, m²	P, m	R, m	V, m/s	Fr^2	S_f	$\overline{\mathrm{Fr}}^2$	$\bar{S}_f$	$S_0 - \bar{S}_f$	$1 - \overline{\mathrm{Fr}}^2$	Δy, m	$\dfrac{1 - \overline{\mathrm{Fr}}^2}{S_0 - \bar{S}_f}$	Δx, m	x, m
(1)	(2)	(3)	(4)	(5)	(5)	(7)	(8)	(9)	(10)	(11)	(12)	(13)	(14)	(15)
0.82	2.46	4.64	0.53	2.85	1.01	0.0042								0
0.90	2.70	4.80	0.56	2.59	0.76	0.0033	0.88	0.0038	–0.0028	0.12	–0.08	–42.2	3.4	3.4
0.95	2.85	4.90	0.58	2.46	0.65	0.0028	0.7	0.003	–0.002	0.3	–0.05	–145.9	7.3	10.7
1.00	3.00	5.00	0.60	2.33	0.55	0.0024	0.6	0.0026	–0.0016	0.4	–0.05	–248.0	12.4	23.1
1.05	3.15	5.10	0.62	2.22	0.48	0.0021	0.52	0.0023	–0.0013	0.48	–0.05	–381.2	19.1	42.1
1.10	3.30	5.20	0.63	2.12	0.42	0.0019	0.45	0.002	–0.001	0.55	–0.05	–560.6	28.0	70.2
1.15	3.45	5.30	0.65	2.03	0.36	0.0016	0.39	0.0017	–0.0007	0.61	–0.05	–813.0	40.7	110.8
1.20	3.60	5.40	0.67	1.94	0.32	0.0015	0.34	0.0016	–0.0006	0.66	–0.05	–1,191.6	59.6	170.4
1.25	3.75	5.50	0.68	1.87	0.28	0.0013	0.3	0.0014	–0.0004	0.7	–0.05	–1,818.0	90.9	261.3
1.30	3.90	5.60	0.70	1.79	0.25	0.0012	0.27	0.0012	–0.0002	0.73	–0.05	–3,044.2	152.2	413.5
1.35[a]	4.05	5.70	0.71	1.73	0.23	0.0011	0.24	0.0011	–0.0001	0.76	–0.05	–6,493.8	324.7	738.2

[a] At $y = 1.35$ m, S_f will become equal to S_o. $(S_f - S_o)$ in Column (10) will approach zero, and $\dfrac{1 - \overline{\mathrm{Fr}}^2}{S_o - \bar{S}_f}$ in Column (13) will approach infinity. For this reason, the calculations are stopped at $y = 1.35$ m (see step (5(a)).

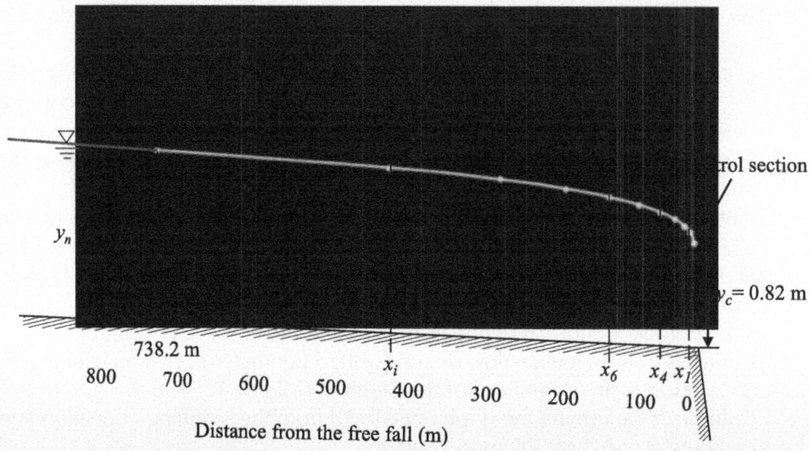

FIGURE 5.30 Water surface profile in the rectangular channel due to drawdown (Example 5.22).

i. **Column 14:** Calculate the distance step Δx from $(x_{i+1} - x_i) = \dfrac{1-\overline{\mathrm{Fr}}^2}{S_o - \bar{S}_f}(y_{i+1} - y_i)$, i.e., by multiplying Column 13 by Column 12.

j. **Column 15:** Calculate the cumulative distance x measured from the free fall. This is the distance from the free-fall structure where the flow depth y occurs. It is calculated by adding Δx from Column 14 to the x value for the previous depth.

6. Plot the water surface profile in the channel

   Use y and x values from Columns 1 and 15 to plot the water surface profile. The drawdown water surface profile is shown in Figure 5.30.

Example 5.23: Flow Parameters Downstream of a Hydraulic Jump

Water flows at the rate of 20.0 m³/s over a 5.0-m-wide spillway of a dam shown in Figure 5.28. A hydraulic jump is formed downstream of the spillway in a rectangular channel. If the depth of the water before the jump is 0.4 m, determine the depth y_2, velocity V_2, and Froude number Fr_1 and Fr_2 of the water before and after the hydraulic jump. Is the flow subcritical or supercritical after the jump?

Solution

1. Calculate the flow velocity before the jump

$$V_1 = \frac{Q}{A_1} = \frac{20.0\ \mathrm{m^3/s}}{0.4\ \mathrm{m} \times 5.0\ \mathrm{m}} = 10.0\ \mathrm{m/s}$$

2. Calculate the Froude number for the flow before the jump

   Apply Eq. (5.32). The hydraulic depth (D_h) of the rectangular spillway channel is equal to the flow depth y (Table 5.3).

$$\mathrm{Fr}_1 = \frac{V_1}{\sqrt{gD_{h1}}}$$

$$= \frac{10.0\ \mathrm{m/s}}{\sqrt{9.81\ \mathrm{m/s^2} \times 0.4\ \mathrm{m}}}$$

$$= 5.05$$

Since $\mathrm{Fr}_1 > 1$, the flow is supercritical upstream of the jump.

3. Determine y_2 from Eq. (5.44)

$$y_2 = \frac{y_1}{2}\left(\sqrt{1+8\mathrm{Fr}_1^2} - 1\right)$$

$$= \frac{0.4 \text{ m}}{2}\left(\sqrt{1+8\times(5.05)^2} - 1\right)$$

$$= 2.66 \text{ m}$$

4. Determine velocity V_2

$$V_2 = \frac{Q}{A_2}$$

$$A_2 = 2.66 \text{ m} \times 5.0 \text{ m} = 13.3 \text{ m}^2$$

$$V_2 = \frac{20.0 \text{ m}^3/\text{s}}{13.3 \text{ m}^2}$$

$$= 1.50 \text{ m/s}$$

5. Calculate the Froude number for the flow after the jump
   Apply Eq. (5.32)

$$\mathrm{Fr}_2 = \frac{V_2}{\sqrt{g y_2}}$$

$$= \frac{1.50 \text{ m}}{\sqrt{9.81 \text{ m/s}^2 \times 2.66 \text{ m}}}$$

$$= 0.29$$

Since $\mathrm{Fr}_2 < 1$, the flow is subcritical downstream of the jump.

5.6.5 Channel Transition

Changes in channel width, bottom elevation, shape, and slope result in changes in the flow velocity and depth. Transitions connect channels of different cross sections and sizes. In channels with smooth transitions, the energy loss is minimal. The energy equation can be used to determine the flow parameters throughout the transitional section and in the upstream and downstream sections. A rise in the bottom elevation or a decrease in the channel width at flow measurement structures are examples of smooth transitions. Consider the flow in the channel transition in Figure 5.31. The energy equation (5.7) can be written between sections (1) and (2) located upstream and downstream of the transition, and by neglecting the energy loss through the transition. The modification to the energy equation (5.7) is expressed below

$$y_1 + \frac{V_1^2}{2g} + z_1 = y_2 + \frac{V_2^2}{2g} + z_2$$

The sum of the depth (y) and the velocity head $\left(V^2/2g\right)$ is called *specific energy* (E). This equation is expressed by Eq. (5.49).

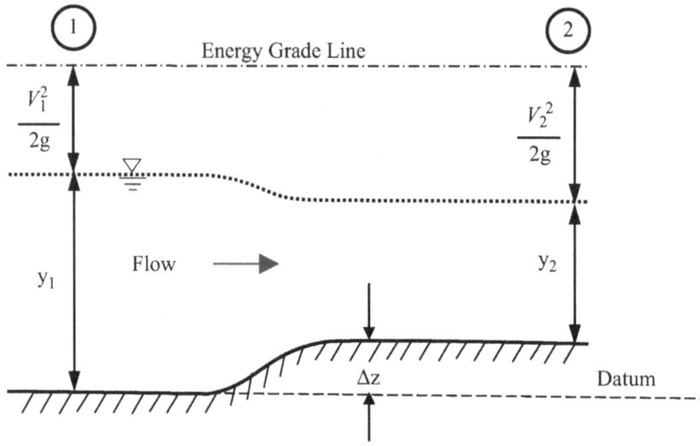

FIGURE 5.31 Longitudinal section of a channel with smooth transition and raised bottom elevation.

$$E = y + \frac{V^2}{2g} \tag{5.49}$$

Equation (5.49) is modified to Eq. (5.50) to represent specific energy, assuming no head loss (h_L).

$$E_1 = E_2 + \Delta z \tag{5.50}$$

where

E_1 = specific energy at section (1), m (ft)
E_2 = specific energy at section (2), m (ft)
Δz = deferential elevation in datum, m (ft)

Equation (5.51) shows that the specific energy is constant throughout channels with no head losses. A *specific energy diagram* shows the relationship between specific energy and flow depth for a specific value of discharge. A specific energy diagram is shown in Figure 5.32. This diagram can be used to analyze flow changes that occur through a channel transition.[2] If the upstream flow parameters are given, Eq. (5.50) can be solved along with the continuity equation to evaluate the flow depth and velocity downstream of the transition.

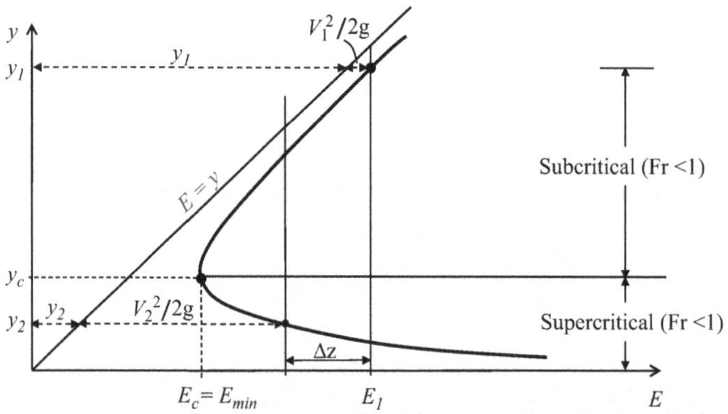

FIGURE 5.32 Specific energy diagram for a transition in a rectangular channel with a smooth and upward step in Figure 5.31.

Minor head losses occur in an open channel when turbulence is produced due to abrupt changes in the cross-section or flow direction. The energy loss due to abrupt transitions can be determined by applying a proper minor head loss coefficient (K) to the velocity head. This is similar to head losses associated with fittings in pressurized pipes. When the flow is subcritical, the head loss at contraction and expansion transitions is due to the difference in the velocity heads before and after the transition section. Equation (5.51) can be applied to calculate the head loss in expansion and contraction.[8]

$$\Delta h = K\frac{\left|V_1^2 - V_2^2\right|}{2g} \tag{5.51}$$

where

Δh = head loss at contraction and expansion transitions, m (ft)

V_1 and V_2 = flow velocity at sections upstream and downstream of the channel transition, m/s (ft/s)

$\left|V_1^2 - V_2^2\right|$ = absolute difference in velocities between cross-section upstream and downstream of transition, m/s (ft/s)

K = minor head loss coefficient, dimensionless. The typical values of K for contractions and expansions encountered in hydraulic calculations of wastewater treatment plants and conveyance systems are presented in Appendix C.

Example 5.24: Developing a Specific Energy Diagram for a Rectangular Channel

Water flows at a rate of 900 cfs in a 20.0-ft-wide rectangular channel. The channel bottom rises 1 ft through a smooth transition. Draw the specific energy diagram for this channel. Determine the specific energy if the water depth is 8.0 ft before the transition. Label flow depth and specific energy before the transition (y_1, E_1), critical depth and minimum energy (y_c, E_c), and depth and specific energy over the raised bottom (y_2, E_2).

Solution

1. Compute specific energy
   Apply Eq. (5.49)

$$E = y + \frac{V^2}{2g}$$

   Replace V with $\frac{Q}{A}$

$$E = y + E = \frac{Q^2}{2gA^2} = y + \frac{Q^2}{2gb^2y^2}$$

$$E = y + \frac{(900\ \text{cfs})^2}{2\times(32.2\ \text{ft/s}^2)(20.0\ \text{ft})^2\ y^2}$$

$$E = y + \frac{31.44\ \text{ft}^2}{y^2}$$

2. Calculate specific energy E for a range of flow depth y from 1.0 to 16.0 ft. The E values are listed in Table 5.6.
3. Plot the specific energy diagram. The diagram is shown in Figure 5.33 using E and y values in Table 5.6.

TABLE 5.6
Computation of Specific Energy at Different Depths in a Rectangular Channel (Example 5.24)

Flow depth, y (ft)	1.0	1.5	2.5	3.0	3.5	4.0	4.5	5.0	6.0	7.0	8.0	9.0	10.0	12.0	16.0
Specific energy, E (ft)	32.4	15.5	7.5	6.5	6.1	6.0	6.1	6.3	6.9	7.6	8.5	9.4	10.3	12.2	16.1

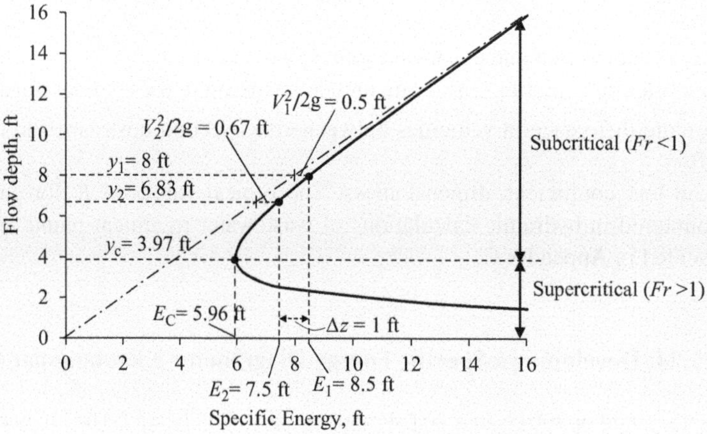

FIGURE 5.33 Specific energy diagram for a rectangular channel (Example 5.24).

4. Determine the specific energy in the channel if the flow depth is 8.0 m.

$$E = y + \frac{31.44\ \text{ft}^2}{y^2}$$

$$= 8.0\ \text{ft} + \frac{31.44\ \text{ft}^2}{(8.0\ \text{ft})^2}$$

$$= 8.5\ \text{ft}$$

Note: Same results can be obtained from Table 5.6 and Figure 5.33.

5. Show flow depth and specific energy before the transition, critical depth and minimum energy, and depth and specific energy over the raised bottom. These values are shown in Figure 5.33.

Example 5.25: Flow through a Channel Transition

Flow in a rectangular channel with a horizontal bottom goes through a transition section. The channel width before and after this section is 6.0 and 4.5 m. The mean flow velocity in the upstream section is 1.5 m/s. Calculate the depth and velocity in the contracted section (2) if the transition section is smooth. What would be the flow depth and velocity in the contracted section (section (2)) if the transition section is a sharp-cornered sudden contraction? Also, calculate the head loss because of the channel's abrupt contraction.

Solution

A. Determine the flow parameters for smooth channel transition
The smooth channel transition is shown in Figure 5.34a in solid lines
1. Calculate discharge

$$Q = V_1 A_1$$

$$A_1 = 3.0 \text{ m} \times 6.0 \text{ m} = 18.0 \text{ m}^2$$

$$Q = 1.5 \text{ m/s} \times 18.0 \text{ m}^2 = 27.0 \text{ m}^3/\text{s}$$

2. Calculate the Froude number for the flow before the transition (section 1)
Apply Eq. (5.32)

$$\text{Fr}_1 = \frac{V_1}{\sqrt{g y_1}}$$

$$= \frac{1.50 \text{ m}}{\sqrt{9.81 \text{ m/s}^2 \times 3.0 \text{ m}}}$$

$$= 0.27$$

Since $\text{Fr}_2 < 1$, the flow is subcritical upstream of the transition.
3. Determine the mean flow velocity V_2 at section 2
Apply continuity equation (Eqs. 5.3 and 5.4) between sections (1) and (2)

$$V_1 A_1 = V_2 A_2$$

$$27.0 \text{ m}^3/\text{s} = V_2 \left(4.5 \text{ m} \times y_2\right)$$

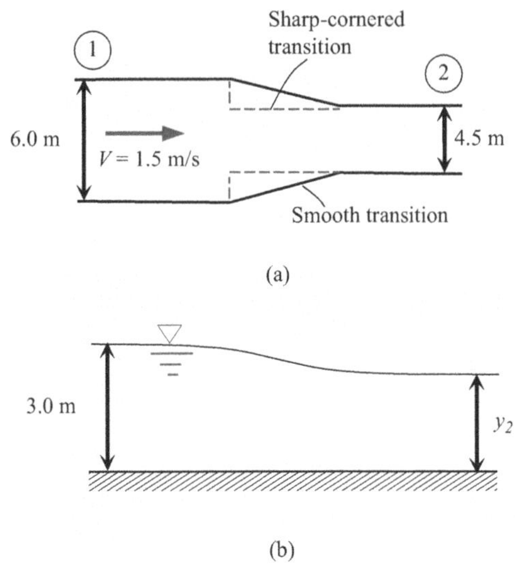

FIGURE 5.34 Flow through rectangular channels with smooth transition, and sharp-cornered sudden contraction: (a) plan view and (b) longitudinal section (Example 5.25).

$$V_2 = \frac{27.0\ \text{m}^3/\text{s}}{(4.5\ \text{m})\,y_2}$$

4. Apply specific energy equation (Eq. 5.51) between sections (1) and (2)
   Even though the transitions is smooth, the friction loss is not small and cannot be ignored.

$$E_1 = E_2 + \Delta z + \Delta h$$

Since the channel is horizontal, $\Delta z = 0$ and $\Delta h = K\dfrac{|V_1^2 - V_2^2|}{2g}$ (Eq. 5.51)

In channels with $V_2 > V_1$, Eq. (5.51) can be expressed as $\Delta h = K\dfrac{(V_2^2 - V_1^2)}{2g}$

$$E_1 = E_2 + \Delta h \quad \text{or} \quad y_1 + \frac{V_1^2}{2g} = y_2 + \frac{V_2^2}{2g} + K\frac{(V_2^2 - V_1^2)}{2g}$$

$$y_1 + \frac{V_1^2}{2g} = y_2 + \frac{V_2^2}{2g} + K\frac{V_2^2}{2g} - K\frac{V_1^2}{2g}$$

$$y_1 + \frac{V_1^2}{2g}(1+K) = y_2 + \frac{V_2^2}{2g}(1+K)$$

5. Calculate y_2
   For a smooth transition, the head loss coefficient K=0.25 (Appendix C)

$$y_1 + \frac{V_1^2}{2g}(1+0.25) = y_2 + \frac{V_2^2}{2g}(1+0.25)$$

$$3.0\ \text{m} + \frac{(1.5\ \text{m/s})^2}{2\times 9.81\ \text{m/s}^2}\times 1.25 = y_2 + \frac{V_2^2}{2\times 9.81\ \text{m/s}^2}\times 1.25$$

Replace $V_2 = \dfrac{27.0\ \text{m}^3/\text{s}}{(4.5\ \text{m})\,y_2}$ from step 2

$$3.0\ \text{m} + \frac{(1.5\ \text{m/s})^2}{2\times 9.81\ \text{m/s}^2}\times 1.25 = y_2 + \frac{\left(\dfrac{27.0\ \text{m}^3/\text{s}}{(4.5\ \text{m})\,y_2}\right)^2}{2\times 9.81\ \text{m/s}^2}\times 1.25$$

$$3.0\ \text{m} + 0.144\ \text{m} = y_2 + \frac{2.294\ \text{m}^3}{y_2^2}$$

$$3.144\ \text{m}\times y_2^2 = y_2^3 + 2.294\ \text{m}$$

$$y_2^3 - 3.144\ \text{m}\times y_2^2 + 2.294\ \text{m} = 0$$

This equation can be solved by trial and error or by using an equation solver. There are two positive and one negative solutions. The three values of y are 2.87, 1.05, and −0.77 m. The negative flow depth does not have a physical meaning. Since the flow is subcritical before the transition section (step A.2), the depth corresponding to the subcritical flow is the solution.[4] The depth after the transition section is

$$y_2 = 2.87 \text{ m}$$

6. Calculate the mean flow velocity

$$V_2 = \frac{27.0 \text{ m}^3/\text{s}}{(4.5 \text{ m})\, y_2} = \frac{27.0 \text{ m}^3/\text{s}}{4.5 \text{ m} \times 2.87 \text{ m}} = 2.09 \text{ m/s}$$

7. Calculate head loss at the contraction section from Eq. (5.51)

$$\Delta h = K \frac{\left(V_2^2 - V_1^2\right)}{2g}$$

$$= \frac{0.25\left(\left(2.09 \text{ m/s}^2\right)^2 - \left(1.5 \text{ m/s}^2\right)^2\right)}{2 \times 9.81\, E^2} = 0.027 \text{ m}$$

B. Determine flow parameters when the transition section is a sharp-cornered sudden contraction

The sharp-cornered sudden contraction is shown in Figure 5.34 in dotted lines

1. Apply specific energy equation (Eq. 5.50) between sections (1) and (2) and include the energy loss

$$E_1 = E_2 + \Delta z + \Delta h$$

Δh is calculated from Eq. (5.51)

In channel with contraction $V_2 > V_1$

$$\Delta h = K \frac{\left(V_2^2 - V_1^2\right)}{2g}$$

Since the channel is horizontal: $\Delta z = 0$

Apply the last equation from step A3

$$y_1 + \frac{V_1^2}{2g}(1 + K) = y_2 + \frac{V_2^2}{2g}(1 + K)$$

2. Calculate y_2

For sudden contraction, the head loss coefficient $K = 0.5$ (Appendix C)

$$y_1 + \frac{V_1^2}{2g}(1 + 0.5) = y_2 + \frac{V_2^2}{2g}(1 + 0.5)$$

$$3.0 \text{ m} + \frac{(1.5 \text{ m/s})^2}{2 \times (9.81 \text{ m/s}^2)} \times 1.5 = y_2 + \frac{\left(\dfrac{27.0 \text{ m}^3/\text{s}}{(4.5 \text{ m})\, y_2}\right)^2}{2 \times (9.81 \text{ m/s}^2)} \times 1.5$$

$$3.0\text{ m}+0.172\text{ m}=y_2+\frac{1.835\text{ m}^3}{y_2^2}\times 1.5$$

$$3.172\text{ m}\times y_2^2=y_2^3+2.753\text{ m}$$

$$y_2^3-3.172\text{ m}\times y_2^2+2.753\text{ m}=0$$

Once solved by trial and error, or by using an equation solver, this equation has two positive and one negative solutions. The three values of y are 2.83, 1.17, and −0.83 m. The negative flow depth does not have a physical meaning. Since the flow is subcritical before the transition section, the depth corresponding to the subcritical flow is the solution.[4] The depth after the transition section is

$$y_2=2.83\text{ m}$$

The mean flow velocity is

$$V_2=\frac{27.0\text{ m}^3/\text{s}}{4.5\text{ m}\times 2.83\text{ m}}=2.12\text{ m/s}$$

3. Calculate head loss at the contraction section from Eq. (5.51)
   In channels with $V_2>V_1$, Eq. (5.51) can be expressed as $\Delta h=K\frac{(V_2^2-V_1^2)}{2g}$

$$\Delta h=K\frac{|V_1^2-V_2^2|}{2g}$$

$$=(0.5)\frac{\left((2.12\text{ m/s}^2)^2-(1.5\text{ m/s}^2)^2\right)}{2\times 9.81\text{ m/s}^2}$$

$$=0.057\text{ m}$$

5.7 REVIEW PROBLEMS AND DISCUSSION TOPICS

5.1 Define the term "steady-uniform flow" in a pressurized pipe.

5.2 A venturimeter is installed to measure the discharge of water in a pipeline. An upstream valve controls the flow rate through the pipe system. The valve is being gradually closed to shut down the flow. Classify the flow through the converging and diverging sections of the venturimeter as uniform/non-uniform, and steady/unsteady. Is the flow through the throat of the venturimeter (the section with the smallest diameter) laminar or turbulent? The throat diameter is 0.25 m and the discharge through the pipe is 0.15 m³/s. Assume water temperature 20°C. The venturimeter installation is shown below.

5.3 Water flows at a rate of $0.06\,m^3/s$ through a 300-mm diameter pipe. This pipe is connected to a 150-mm diameter pipe through a reducer fitting. Calculate the average velocities in these two pipes.

5.4 Water is discharged from a large tank to the atmosphere. Determine the flow rate if the diameter of the pipe is 12 in. Ignore the energy loss in the short outlet pipe. The centerline of the outlet pipe is located 15 ft below the water surface in the reservoir. The tank installation is shown below.

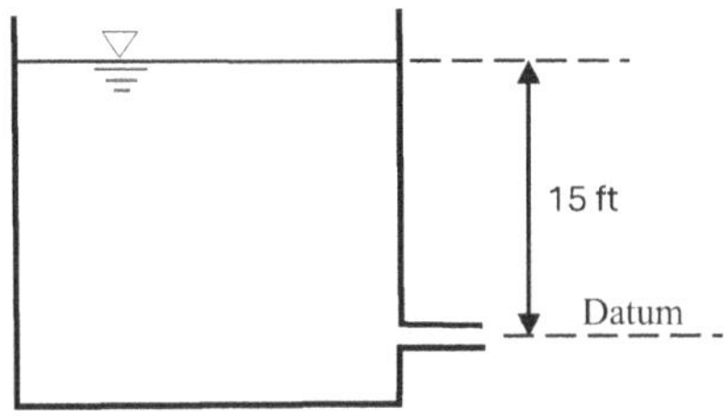

5.5 Water flows through a horizontal elbow that is inclined at 30° from the horizontal. The flow rate through the elbow is 0.3 cubic feet per second (cfs). The diameter of the elbow reduces from 5 in. at the inlet to 2 in. at the outlet. The elbow discharges water into the atmosphere. Compute the magnitude of the anchoring force needed to hold the elbow in place. The water temperature is 70°F. The elbow installation is shown below. Assume that the head loss through the elbow is negligible.

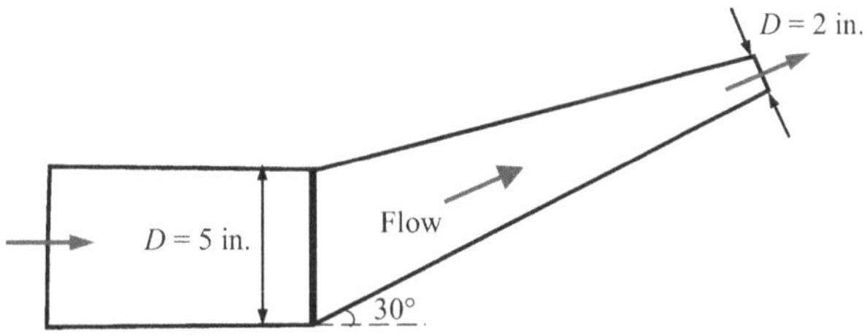

5.6 Reservoirs A and B are connected by a new cast iron pipe. Calculate the flow rate and velocity in the pipe. The connecting pipe section is 1,000 m long and 50 cm in diameter. The pipeline has a flushed end with reservoir A. There are two 90° regular elbows and a fully opened gate valve. The water surface elevation in reservoir B is 45 m below that of reservoir A. The water temperature is 15°C. The reservoirs and piping arrangement are shown below.

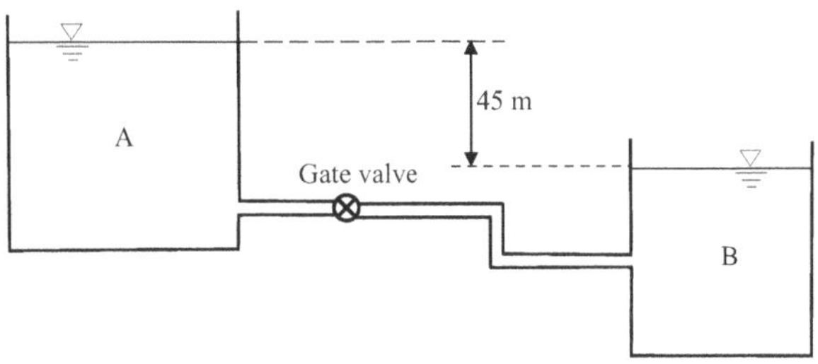

5.7 Water at 20°C is delivered from reservoir A to reservoir B. The connecting pipeline has two pipes connected in series and a gate valve. Calculate the flow rate and velocity through each pipe. The Hazen–Williams friction coefficient C for both pipes is 135. Draw the EGL and HGL. The reservoirs and piping arrangement are shown below.

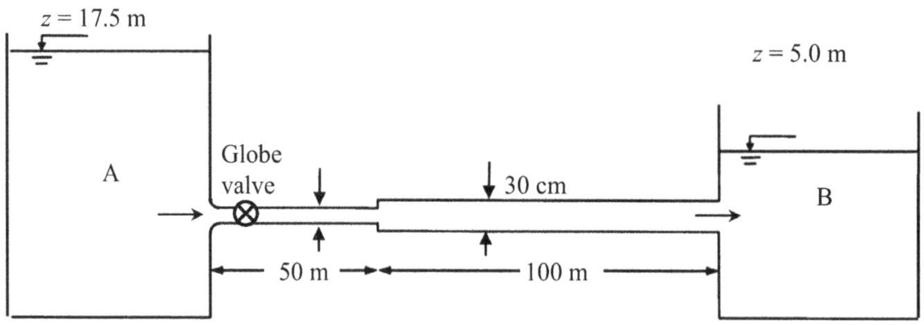

5.8 Three cast iron pipes are connected in parallel. The pipe configuration and pipe dimensions are given in the following figure. The total inflow at point A is 0.1 m^3/s. Determine (a) the flow rate in each pipe and (b) the pressure drop between points A and B. Water temperature is 20°C. Use the Darcy–Weisbach equation with $f = 0.025$.

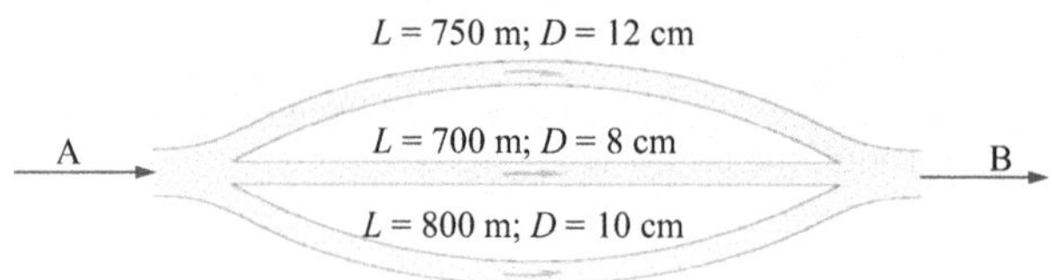

5.9 A 12-in. concrete sewer line is laid at a slope of 0.5%. The flow rate through the sewer is 0.35 ft^3/s. Calculate the flow depth and velocity.

5.10 A hydraulic jump is formed downstream of a spillway. The width of the channel remains constant. The flow depth and velocity upstream of the jump are 0.50 m and 13 m/s, respectively. Show that the flow is supercritical before the jump. Calculate the depth of the flow and velocity after the jump. The flow profile over the spillway is shown below.

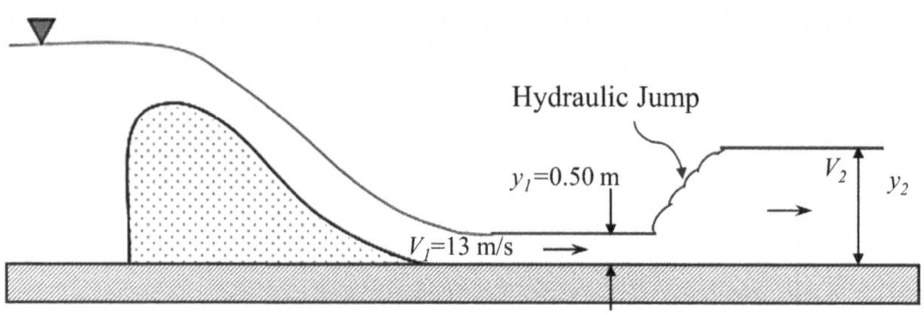

5.11 Water is flowing in a rectangular open channel. The depth of flow and velocity in the channel are 6 ft and 8 ft/s, respectively. There is a smooth contraction in the width of the channel from 10 to 9 ft. The bottom of the channel remains horizontal. Energy loss through the smooth contraction is negligible. Find the depth of flow and velocity in the contracted section. What would be the flow depth in the contracted section and head loss through the transition section if this section is a round-cornered sudden contraction?

REFERENCES

1. Chaudhry, M. H., *Open-Channel Flow*. Springer Science & Business Media, New York, 2007.
2. Chow, V. T., *Open Channel Hydraulics*, McGraw-Hill Book Co., New York, 1959.
3. Streeter, V. L., and E. B. Wylie, *Fluid Mechanics*, 6th ed., McGraw-Hill, New York, 1975.
4. Sturm, T. W., *Open Channel Hydraulics*, McGraw-Hill, New York, 2010.
5. Casey, T. J., *Water and Wastewater Engineering Hydraulics*, Oxford University Press, Oxford, 1992.
6. Gupta, R. S., *Hydrology & Hydraulic Systems*, 4th ed., Waveland Press Inc., Prospect Heights, IL, 2016.
7. Larock, B. E., R. W. Jeppson, and G. Z. Watters, *Hydraulics of Pipeline Systems*, CRC Press, Boca Raton, FL, 1999.
8. Metcalf and Eddy, Inc., *Wastewater Engineering: Collection and Pumping of Wastewater*, McGraw-Hill Book Co., New York, 1981.
9. Qasim, S. R., *Wastewater Treatment Plants: Planning, Design and Operation*, 2nd ed., CRC Press, Boca Raton, FL, 1999.
10. ShammasA, N. K. and L. K. Wang, *Water Supply and Wastewater Removal*, 3rd ed., John Wiley and Sons. Inc., Hoboken, NJ, 2011.
11. Ranga Raju, K. G., *Flow through Open Channels,* Tata McGraw-Hill, New Delhi, India, 1981.
12. Fair, G. M., J. C. Geyer, and D. A. Okun, *Water and Wastewater Engineering,* vol. I, *Water Supply and Wastewater Removal*, John Wiley & Sons, Inc., New York, 1966.
13. Steel, E. W., and T. J. McGhee, *Water Supply and Sewerage*, 6th ed., McGraw-Hill, Inc., New York, 1991.
14. McJunkin, F. E., and P. A. Vesilind, *Practical Hydraulics for the Public Works Engineer*, Public Works Magazine, 1968, 23 pages.
15. Joint Committee of the American Society of Civil Engineers and the Water Pollution Control Federation, *Design and Construction of Sanitary and Storm Sewers*, MOP 9, Water Environment Federation, Alexandria, VA, 1995.

6 Wastewater Conveyance, Watershed Management, and Nonpoint Source Pollution

Objectives and Design Considerations

6.1 CHAPTER OBJECTIVES

This is an introductory chapter to provide prospective on wastewater conveyance, watershed management, and nonpoint source pollution, including objectives and design considerations. The following topics are covered:

- Definition of wastewater conveyance, watershed management, and nonpoint source pollution
- Service area, objectives, and regulatory requirements
- Characteristics of wastewater and degree of treatment
- Basic design considerations to include current and future trends in population growth, design period, and design population of the service area
- Types and components of collection and conveyance systems
- Site selection of facility components and equipment, and hydraulics
- Energy and resource requirements, project economics, and environmental impact assessment
- General considerations in project planning, design plans and specifications, cost estimates and contract bid information, and construction management

6.2 DEFINITIONS OF WASTEWATER CONVEYANCE, WATERSHED MANAGEMENT, AND NONPOINT SOURCE POLLUTION

The wastewater conveyance systems carry point source discharges, while nonpoint sources (NPSs) result from surface runoffs. The definitions of wastewater conveyance systems, watershed management, and point and nonpoint source pollution are given below.

6.2.1 Wastewater Conveyance System

The wastewater conveyance infrastructures carry point source discharges. The term "point source discharges" means any discernible, confined, and discrete conveyance, including but not limited to any pipe, ditch, channel, tunnel, conduit, well, container, concentrated animal feeding operation, or vessel or other floating craft from which pollutants may be discharged. This term does not include agricultural stormwater discharges and return flows from irrigated agriculture.[1–3]

The wastewater included in the conveyance of point source discharges are: (1) sanitary sewage, (2) urban storm water, (3) combined sewer overflows (CSOs), and (4) infiltration and inflows. These flows are transported from generation sources to a central location for treatment, disposal,

DOI: 10.1201/9781003629399-6

and reuse. The sources and flow rates, and quality of wastewater for conveyance are presented in Chapter 4. The conveyance system contains a network of underground gravity sewers of different sizes, drainage channels, pumping stations and force mains, flow meters, and appurtenances. These invaluable assets are fully integrated to transport the flows generated under dry and wet weather conditions.[4,5]

6.2.2 Watershed Management

Every waterbody (e.g., rivers, lakes, ponds, streams, and estuaries) has a watershed. The watershed is the area of land that drains or sheds water into a specific receiving waterbody, such as a lake or a river. As rainwater or melted snow runs downhill in the watershed, it collects and transports sediment and other pollutants, and deposits them into the receiving water body.

Watershed management is a comprehensive planning process that results in a plan or a blueprint of how to best protect and improve the water quality and other natural resources in a watershed. Watershed boundaries very often extend over political boundaries such as adjacent municipalities and/or states. Runoff from rainwater or snowmelt can contribute significant amounts of pollution to lakes or rivers. Watershed management helps to control pollution of water and other natural resources in the watershed by identifying different pollution sources present in the watershed, how these pollutants are transported, and recommending ways to reduce or eliminate these pollution sources. Common activities that occur within a watershed that somehow affect the quality of watershed are new land development, runoff from already-developed areas, agricultural activities, confined animal feedlots, landfilling, and household activities such as gardening/lawn care, septic system use/maintenance, water diversion, and auto maintenance. Watershed management planning comprehensively identifies these activities and makes recommendations to reduce or eliminate the adverse impacts. Watershed management and comprehensive planning process are also important because they involve all affected municipalities and parties because all partners have a stake in the health of the watershed. This partnership is essential to the successful management of the land and water resources in the watershed.

Because watershed boundaries do not coincide with political boundaries, the actions of adjacent municipalities upstream can have as much of an impact on the downstream municipality's land and water resources as those actions carried out locally. Impacts from upstream sources can sometimes undermine the efforts of downstream municipalities to control pollution. Comprehensive planning for the resources within the entire watershed, with participation and commitment from all stakeholders in the watershed is critical to protecting the health of the watershed's resources.

Comprehensive watershed plans should first identify the characteristics of the watershed and inventory the watershed's natural resources. It is important to establish a baseline of the overall nature and quality of the watershed in order to plan properly for the improvement of the resources, and to actually measure those improvements. The important steps in watershed management planning are: (1) delineate and map the watershed's boundaries and the smaller drainage basins within the watershed; (2) inventory and map the resources in the watershed; (3) inventory and map the natural and manmade drainage systems in the watershed; (4) inventory and map land use and land cover; (5) inventory and map soils; (6) identify areas of erosion, including stream banks and construction sites; (7) identify the quality of water resources in the watershed as a baseline; and (8) inventory and map pollution sources, both point sources (such as industrial discharge pipes) and NPSs (such as municipal stormwater systems, failing septic systems, illicit discharges, feedlots, and landfills). Much of this information may already be compiled and available through the Natural Resources Conservation Service of the U.S. Department of Agriculture.

6.2.3 Nonpoint Source Pollution

NPS pollution generally results from land runoff, precipitation, atmospheric deposition, drainage, seepage, or hydrologic modification. NPS pollution, unlike pollution from municipal and industrial discharge, comes from many diffuse sources and is caused by rainfall or snowmelt moving over and through the ground. As the runoff moves, it picks up and carries away natural and human-made pollutants, and finally deposits them into lakes, rivers, wetlands, coastal waters, and ground waters. NPS pollution include: (1) excess fertilizers, herbicides, and insecticides from agricultural lands and residential areas; (2) oil, grease, and toxic chemicals from urban runoff; (3) sediment from improperly managed construction sites, crop and forest lands, and eroding streambanks; (4) salt from irrigation practices, streets, and drainage from abandoned mines; and (5) bacteria and nutrients from livestock, pet wastes and faulty septic systems.[3]

Most states report that NPS pollution is the leading cause of water quality problems. The effects of NPS pollutants on specific waters vary and may not always be fully assessed. However, we know that these pollutants have harmful effects on drinking water supplies, recreation, fisheries, and wildlife.

6.3 SERVICE AREA, OBJECTIVES, AND REGULATORY REQUIREMENTS

The definition of service area, objectives of collection and conveyance systems, and regulatory requirements are briefly presented in the subsequent sections.

6.3.1 Service Area

Service area (also called sewer district) is defined as the total land area that will be eventually served by the proposed wastewater collection and conveyance systems. The area may be based on natural drainage, political boundaries, or both. The service area may be located in one or several watersheds. Site visits and engineering data on topography, geology, hydrology, climate, ecological elements, and social and economic conditions, and land use or zoning plans should be studied. These factors may affect both developed and undeveloped lands. Such efforts should be carefully coordinated with the state, regional, and local planning agencies, and should be in conformance with the development and implementation of the area-wide waste management plan.[4,5]

6.3.2 Objectives of Collection and Conveyance Systems of Point and Nonpoint Source Pollution

The objectives of a wastewater collection and conveyance, and NPS pollution control are to meet the treatment and discharge quality standards established by the federal, state, and regional regulatory authorities. They also include preventing many adverse environmental conditions that may develop from the discharge of untreated wastewater and surface runoffs. Some of these adverse environmental conditions are: (1) unsightliness, nuisance and obnoxious odors in the service area, and at the disposal site; (2) contamination of water supply sources; (3) destruction of fish, shellfish and other aquatic life; (4) impairment of beneficial uses of receiving waters; and (5) decline in land values causing restriction in community growth and development. Ideally, a wastewater conveyance and treatment, and NPS pollution control should encourage the beneficial uses of effluent and residuals, protect the critical habitats for endangered or threatened species of flora or fauna, and enhance community growth and development.[4,5]

6.3.3 Regulatory Requirements

The passage of the Federal Water Pollution Control Amendments of 1972 (PL 92–500), and subsequent amendments brought a significant change in water pollution control philosophy. The specific national water quality goals and regulations are to eliminate the discharge of pollutants into all surface waters, and restore and maintain the physical, chemical, and biological integrity of the Nation's waters.[6,7] The highlights of these regulations are presented below:

1. Under the National Pollutant Discharge Elimination System (NPDES) permitting program, effluent limitations are established for conventional, nonconventional, toxic, and hazardous pollutants based upon their levels of concentration in the effluent.[6]

2. Industrial pretreatment standards are developed for all pollutants that "interfere with, pass through, concentrate in sludge, or are otherwise incompatible" with publicly owned treatment works.[6]
3. U.S. Environmental Protection Agency (U.S. EPA) published its definition of secondary treatment. The minimum national standards for secondary treatment, as defined by U.S. EPA, are given in Table 6.1.[8] The secondary treatment includes bar screen, grit removal, primary sedimentation, biological treatment, and disinfection. The process diagram of a secondary wastewater treatment plant is shown in Figure 6.1.
4. Under the Clean Water Act (CWA), each state agency is required to develop policies and program requirements for water quality planning and management, including water quality standards, water quality monitoring, water quality management plan, and water quality report.[9–11] Bodies of state waters are classified as either *effluent-limited* or *water quality-limited*. The state agency established total maximum daily load waste allocation for all surface waters throughout the state. The water body is classified as effluent-limited if the state's water quality standards are met by the discharge of secondary effluent. Water-quality-limited bodies of water require discharge of higher effluent quality. The revised total maximum daily load (TMDL) waste allocation includes (1) point sources, (2) nonpoint sources, (3) natural background levels, and (4) a margin of safety.[9–12]
5. NPS water pollution is regulated at the federal, state, and local levels. Yet, the pollution persists and remains a significant problem. There are several reasons why current regulations have failed to stem the flow of pollution. NPS water pollution comes from numerous and diverse sources, and control measures are expensive to implement.[13] There are three primary federal laws affecting NPS water pollution: (1) the CWA, (2) the Clean Water Action Plan, and (3) the Coastal Zone Management Act (CZMA).[1–3]
6. Under the CWA, Congress created Sections 208, 319, and 404. Section 208 was Congress's first attempt to regulate NPS water pollution, and it directed states and local governments to create management plans that identified future waste treatment needs and identified and controlled NPS water pollution.[1,14] The section notes that the area with management plan must discuss how to identify *agriculturally and silviculturally related NPSs of pollution*, and *runoff from manure disposal areas, and from land used for livestock and crop production*.[1,15] Section 208, in essence, is a federal funding mechanism for state programs that attempt to control NPS pollution.

TABLE 6.1
Minimum National Standards for Secondary Treatment as Defined by the U.S. EPA

Effluent Parameter	30-Day Average: Maximum Concentration, mg/L	30-Day Average: Minimum Removal, %	Seven-Day Average, Maximum Concentration, mg/L
BOD_5	30	85	45
$CBOD_5$[a]	25	85	40
TSS[b]	30	85	45
pH[c]		Between 6.0 and 9.0	

Note: Special considerations may apply to combined sewers, certain industrial categories, waste stabilization ponds, and less concentrated influent wastewater for separate sewers.

[a] CBOD = carbonaceous biochemical oxygen demand. $CBOD_5$ may be substituted for BOD_5 by the permitting authority.

[b] TSS = total suspended solids.

[c] The pH limit may be exceeded if publicly owned treatment works demonstrates that (1) inorganic chemicals are not added to waste stream as part of treatment process and (2) contributions from industrial sources do not cause the pH of the effluent to be less than 6.0 and greater than 9.0.

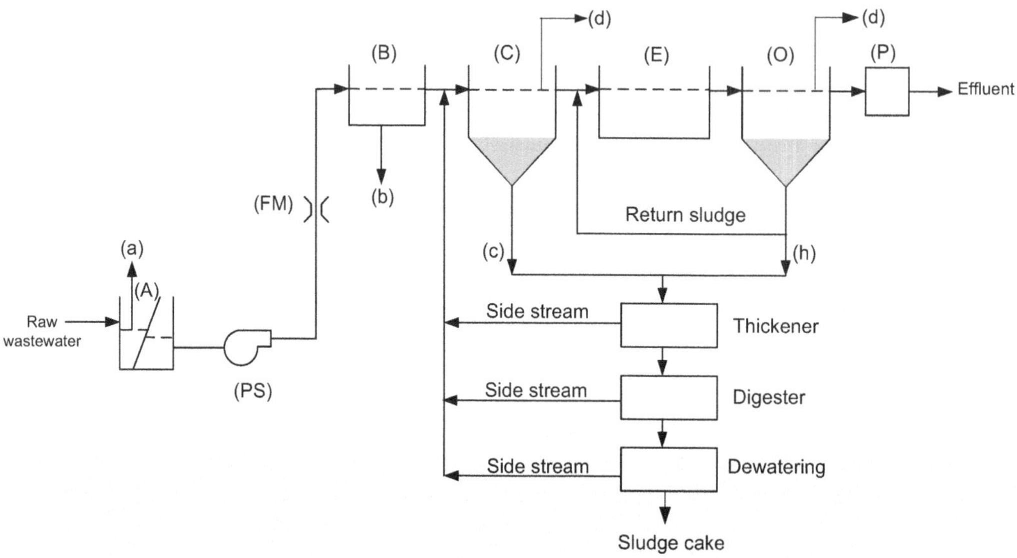

Bar screen, (a) Screenings, (PS) Pump station, (FM) Flow measurement, (B) Grit removal, (b) grit, (C) Primary sedimentation, (c) Primary sludge, (d) Scum, (E) Aeration basin, (O) Final clarifier, (h) Waste activated sludge, and (P) Disinfection

FIGURE 6.1 Process diagram of a conventional secondary wastewater treatment plant.

In 1987, Congress passed the Water Quality Act and created the new Section 319 to address the problem of NPSs. The section requires states to identify waterbodies that cannot meet water quality standards without control of NPSs.[1,16] The states must then identify best management practices and measures for those impaired sources, along with an implementation plan.[1,16] The EPA approves these plans, and if a state fails to develop a plan, the EPA must do so for the state.

Section 404 of the CWA requires that a permit be obtained from the U.S. Army Corps of Engineers for the *discharge of dredged and fill material into the navigable waters at specified disposal sites*.[1,17] The 404 permit process is frequently used to protect wetlands, which are essential to the control of nonpoint pollution because they slow the rate of surface water runoff and remove sediment and other pollutants before they reach lakes and streams.[1,18]

7. There is much concern regarding detection, monitoring, and health risks of many contaminants. Chemicals of greatest concern are gasoline additives, solvents, phenolic compounds, pesticides, herbicides, antibiotics, and the pharmaceuticals and personal care products. Many of these compounds are *endocrine disruptors*, meaning they may enhance hormone production, exaggerate the response, or block the effect of hormones on the body. Some reported findings include problems with development, behavior, and reproduction of many aquatic animals. Decline in male sperm count has also been reported. Increases in certain cancers have been blamed on endocrine-disrupting chemicals.[19–21] Some of these contaminants are released from underground storage tanks, agricultural runoff, and discharges from confined animal feeding operations. It is important that these contaminants are contained and removed from the NPSs.
8. Concentrated Animal Feeding Operations (CAFOs) may generate significant amounts of manure. One method to remove the manure is to apply it to land for fertilization. Due to excess amounts of manure applied to land areas, significant amounts of manure may then be washed away by rain into surface waters. While in general agricultural stormwater

runoff from CAFOs is a NPS, CAFOs may end up requiring a NPDES permit under the CWA.[22] Also, point source category regulatory requirements are established for (1) airports' discharge of pollutants from deicing operation,[23] (2) construction activities,[24] and (3) control of CSO.[25,26]

9. There have been two Clean Water Action Plans proposed by federal authorities that would affect NPS water pollution. They are both executive orders, not pieces of legislation from Congress. Both proposals provide funding and guidance to address NPS water pollution.[27,28]
10. The CZMA was passed in 1972 and provides for the management of the nation's coastal lands and the Great Lakes.[29] When Congress reauthorized the CZMA in 1990, it identified NPS water pollution as a significant factor in the degradation of coastal waters.[30] To address NPS water pollution in coastal areas, the 1990 amendments to the CZMA created the Coastal Zone Enhancement Grants Program.[31] The states must submit information on their programs to the Secretary of Commerce and the EPA Administrator, who are in charge of approving the plan. The plan is to be implemented in conjunction with the state's NPS water pollution plan under Section 319 of the CWA and through changes in the overall coastal zone management program.[32,33] If a state does not submit an approved program, the state may lose a percentage of the grant money provided under the CZMA and under Section 319 of the CWA.[34] The EPA is also to provide some guidance to states in developing their NPS coastal management plans through the publication of national guidelines on management measures.[35] These management measures must be economically achievable for new and existing NPS water pollution and must reduce pollution to the greatest extent achievable through the current best management practices.[36]

Example 6.1: Secondary Effluent Compliance

Grab samples were collected daily for seven days from the secondary effluent pipe. The BOD_5 and TSS concentrations are tabulated below. Determine 7-day average BOD_5 and TSS values. Compare the effluent quality with national standards for secondary treatment.

	Concentration, mg/L						
Parameter	**Mon.**	**Tue.**	**Wed.**	**Thu.**	**Fri.**	**Sat.**	**Sun.**
BOD_5	20	15	15	50	54	40	30
TSS	28	25	30	65	68	45	25

Solution

1. Determine 7-day average BOD_5 and TSS.

$$BOD_5 = (20+15+15+50+54+40+30)\text{mg/L}/7 = \left(224\text{ mg/L}\right)/7 = 32\text{ mg/L}$$

$$TSS = (28+25+30+65+68+45+25)\text{mg/L}/7 = \left(286\text{ mg/L}\right)/7 = 41\text{ mg/L}$$

2. Compare the effluent quality with the minimum national standards for secondary treatment in Table 6.1.

The Minimum National Secondary Treatment Standards for 7-day average BOD_5 and TSS are 45 mg/L each. Therefore, the effluent meets the secondary effluent quality standards.

Example 6.2: Effluent Quality from a Secondary Wastewater Treatment Plant

Municipal wastewater is treated at a secondary wastewater treatment plant. The raw wastewater has the following characteristics: BOD_5=210 mg/L, COD=400 mg/L, TSS=230 mg/L, total phosphorus (TP)=6 mg/L, organic nitrogen (ON)=15 mg/L, and ammonia nitrogen (AN)=30 mg/L. The performance of various units in terms of the removal efficiency of key constituents is given below. Estimate the effluent quality.

	Percent Removal, %					
Unit	**BOD_5**	**COD**	**TSS**	**TP**	**ON**	**AN**
Bar screen	0	0	0	0	0	0
Grit removal	0	0	0	0	0	0
Primary sedimentation	35	35	60	15	25	0
Activated sludge	90	90	85	18	73	12
Disinfection	0	0	0	0	0	0

Solution

1. Calculate the effluent quality in terms of BOD_5

   The calculation procedure for estimating BOD_5 in the effluent is given below as an example:

$$\text{Primary sedimentation BOD}_5 \text{ removal} = 35\%$$

$$\text{BOD}_5 \text{ in the effluent from primary sedimentation facility}$$

$$= (1 - 0.35) \times 210 \text{ mg/ L} = 0.65 \times 210 \text{ mg/L} = 137 \text{ mg/L}$$

$$\text{BOD}_5 \text{ in the effluent from activated sludge facility}$$

$$= (1 - 0.9) \times 137 \text{ mg/ L} = 0.1 \times 137 \text{ mg/ L} = 14 \text{ mg/L}$$

2. Calculate the effluent quality in terms of other parameters

   The effluent quality, including all other processes, is summarized below:

$$\text{BOD}_5 = 14 \text{ mg/L}$$

$$\text{COD} = (1 - 0.35)(1 - 0.90)\ 400 \text{ mg/L} = 0.065\ 400 \text{ mg/L} = 26 \text{ mg/L}$$

$$\text{TSS} = (1 - 0.60)(1 - 0.85)\ 230 \text{ mg/L} = 0.06\ 230 \text{ mg/L} = 14 \text{ mg/L}$$

$$\text{TP} = (1 - 0.15)(1 - 0.18)\ 6 \text{ mg/L} = 0.70\ 6 \text{ mg/L} = 4 \text{ mg/L}$$

$$\text{ON} = (1 - 0.25)(1 - 0.73)\ 15 \text{ mg/L} = 0.20\ 15 \text{ mg/L} = 3 \text{ mg/L}$$

$$\text{AN} = (1 - 0)(1 - 0.12)\ 30 \text{ mg/L} = 0.88\ 30 \text{ mg/L} = 26 \text{ mg/L}$$

The effluent quality from this plant meets the minimum national standard for secondary treatment.

Example 6.3: Secondary Treatment with Effluent Polishing

A secondary wastewater treatment plant is designed for effluent polishing for industrial reuse. Gravity filtration and a carbon adsorption column are provided to polish the effluent from the secondary plant. Estimate the effluent quality after post-filtration, carbon adsorption, and disinfection. The raw wastewater quality and removal rates for all constituents in conventional secondary treatment processes are the same as those given in Example 6.2. The BOD_5, COD, TSS, TP, ON, and AN removal for gravity filtration are 45%, 25%, 70%, 25%, 60%, and 0%, respectively. The BOD_5, COD, TSS, TP, ON, and AN removal for the carbon adsorption column are, respectively, 65%, 68%, 60%, 20%, 40%, and 0%.

Solution

1. State the procedure for calculating effluent quality from the process train using conventional secondary treatment processes, gravity filtration, and carbon adsorption column.

   The polishing gravity filter and carbon adsorption column are arranged in series. Therefore, their removal efficiencies are included with the process train.
2. Determine the effluent quality.

   The effluent quality is established from the removal efficiency of each unit.

$$BOD_5 = (1-0.35)\times(1-0.90)\times(1-0.45)\times(1-0.65)\times 210 \text{ mg/L} = 0.013\times 210 \text{ mg/L} = 3 \text{ mg/L}$$

$$COD = (1-0.35)\times(1-0.90)\times(1-0.25)\times(1-0.68)\times 400 \text{ mg/L} = 0.016\times 400 \text{ mg/L} = 6 \text{ mg/L}$$

$$TSS = (1-0.60)\times(1-0.85)\times(1-0.70)\times(1-0.60)\times 230 \text{ mg/L} = 0.007\times 230 \text{ mg/L} = 2 \text{ mg/L}$$

$$TP = (1-0.15)\times(1-0.18)\times(1-0.25)\times(1-0.20)\times 6 \text{ mg/L} = 0.42\times 6 \text{ mg/L} = 3 \text{ mg/L}$$

$$ON = (1-0.25)\times(1-0.73)\times(1-0.60)\times(1-0.40)\times 15 \text{ mg/L} = 0.049\times 15 \text{ mg/L} = 1 \text{ mg/L}$$

$$AN = (1-0)\times(1-0.12)\times(1-0)(1-0)\times 30 \text{ mg/L} = 0.88\times 30 \text{ mg/L} = 26 \text{ mg/L}$$

6.3.4 Characteristics of Wastewater and Degree of Treatment

Reliable estimates of wastewater characteristics from point and nonpoint sources are important to design the treatment facilities. The characteristics of wastewater are developed in terms of flow conditions, including physical, chemical, and biological quality. The wastewater characteristics data are needed for the initial year and for the design year. The data include minimum, average, and maximum dry weather flows; peak wet weather flows; sustained maximum flows; and chemical parameters such as BOD_5, TSS, pH, TDS, nitrogen, phosphorus, and toxic chemicals. Chapter 4 is devoted exclusively to provide the characteristics of municipal wastewater, stormwater, and surface runoffs.

The degree of treatment is based on the influent characteristics to the plant and the effluent quality required. If the effluent is discharged into the natural waters, it should comply with the NPDES permit requirements. If used for irrigation, the plant effluent must also satisfy the health regulations governing the types of crops that are irrigated. Other effluent uses, such as recreational lakes, agriculture, industrial, and municipal, may dictate their own effluent quality and thus the specific treatment requirements.

6.4 BASIC DESIGN CONSIDERATIONS

Many important design factors must be considered during the initial planning and design stages of a wastewater collection and conveyance system, and control of NPS pollution projects. Basic design factors are:[5]

1. Initial and design years, and design population
2. Types of proposed collection and conveyance systems for point and nonpoint source pollution control
3. Site selection of facility components and equipment
4. Hydraulic considerations for collection and conveyance systems
5. Energy and resource requirements and project economics
6. Environmental impact assessment

Some of these factors are covered in great detail in other chapters; the information presented below is an introduction to planning and design of wastewater conveyance and NPS pollution control projects.

Project planning is an essential part of pollution control from point and nonpoint source pollution. It takes several years to plan, design, and construct facilities for wastewater conveyance and NPS pollution control. For this reason, initial and design years and population served in the service area are important elements of the project planning. These topics are covered in the following sections.

6.4.1 Initial and Design Years and Design Population

Initial and design years: Facility components and size of service area required are dependent upon the future population served and planned activity. It generally takes several years to plan, design, and construct a wastewater collection and conveyance system, and control of NPS pollution. Accordingly, most of the project components are made large enough to satisfy the future community needs for several years in the future. The initial year is the year when the construction is completed, and the initial operation begins. The design or planning year is the year when the facility is expected to reach its full designed capacity. Selecting the design year requires sound judgment and skills in developing future population growth estimates from the past social and economic trends of a community. Design periods are generally chosen with the following factors in mind:

1. Useful life of the collection and conveyance system
2. Ease or difficulty in expansion
3. Performance of the facility during the initial years
4. Future growth in the population, commercial and industrial developments, activity in the watershed such as agricultural and feedlot operations, and animal production activities; housing developments and others; water demand; and wastewater and surface runoff characteristics
5. Cost of present and future construction, and availability of funds

The design periods of different components of a collection and conveyance facility may also vary. As an example, main conduits, channels, and appurtenances that cannot be expanded readily are designed for periods up to 50 years in the future. On the other hand, process equipment and pumps are constructed for shorter periods to avoid the construction of oversized units. In such cases, adequate space is left at the facility site for expansion at different staging periods. According to the guidelines of the construction grants program, the design period may be divided into several staging periods (10, 15, and 20 years).

Design population: The volume of point and nonpoint source wastewater generated in a community depends on the population and per capita contribution of wastewater. It is therefore important to estimate the population to be served at the design year. Accurate population prediction is quite difficult because many factors influence the growth of a service area. Among the important factors are industrial growth; state of development of the surrounding area; location with regard to transportation sources; availability of raw material, land, and water resources; local taxes and government activities; migration trends; and so on.

Sources of population data: The population data can be obtained from several sources. The U.S. Bureau of Census (Department of Commerce) publishes decennial census data. For the interim periods, reliable data can usually be obtained from local census bureaus; the state, county, or local planning commissions; the chambers of commerce; voter registration lists; the post office; newspapers; and public utilities (telephone, electric, gas, water, etc.). It is important that the design engineer becomes familiar with the population data sources and the type of information that can be obtained from these sources.[4,5]

Method of population forecasting: Many mathematical and graphical methods are used to project past and present population data to the design year. Widely employed methods are:[4,5,37]

1. Arithmetic growth
2. Geometric growth
3. Decreasing rate of increase
4. Mathematical or logistic curve fitting
5. Graphical comparison with similar cities
6. Ratio method
7. Employment forecast
8. Birth cohort

All these methods utilize different assumptions and therefore give different results. Selection of any method depends on the amount and type of data available and whether the projections are made for the short or long term. The arithmetic, geometric, decreasing rate of increase, and logistic curve-fitting methods are summarized in Table 6.2 (Eqs. 6.1–6.10). The remaining methods are presented in Table 6.3.

Population density: The average population density for the entire city rarely exceeds 7,500–10,000 per km^2 (30–40 per acre). Often, it is important to know the population density in different parts of the city in order to estimate the wastewater flows and to design the collection network. Density varies widely within a city depending on the land use. The average population densities based on land use characteristics are summarized in Table 6.4.

TABLE 6.2
Population Projections by Using Arithmetic, Geometric, Decreasing Rate of Increase, and Mathematical Curve Fitting

Method	Description	Basic Equations
Arithmetic method	Population is assumed to increase at a constant rate. The method is used for short-term estimates (1–5 years).	$\frac{dY}{dt} = K_a; \quad Y_t = Y_2 + K_a(T - T_2)$ (6.1) $K_a = \frac{Y_2 - Y_1}{T_2 - T_1}$ (6.2)
Geometric method	Population is assumed to increase in proportion to the number present. The method is commonly used for short-term estimates (1–5 years).	$\frac{dY}{dt} = K_p Y; \quad \ln Y_t = \ln Y_2 + K_p(T - T_2)$ (6.3) $K_p = \frac{\ln Y_2 - \ln Y_1}{T_2 - T_1}$ (6.4)

(Continued)

TABLE 6.2 (*Continued*)
Population Projections by Using Arithmetic, Geometric, Decreasing Rate of Increase, and Mathematical Curve Fitting

Method	Description	Basic Equations	
Decreasing rate of increase method	Population is assumed to reach some limiting value or saturation point. The method is normally applied for long-term estimates.	$\frac{dY}{dt} = K_d(Z-Y);$ $Y_t = Y_2 + (Z-Y_2)\left(1-e^{-K_d(T-T_2)}\right)$	(6.5)
		$Z = \frac{2Y_0Y_1Y_2 - Y_1^2(Y_0+Y_2)}{Y_0Y_2 - Y_1^2}$	(6.6)
		$K_d = -\frac{1}{T_2-T_1}\ln\left(\frac{Z-Y_2}{Z-Y_1}\right)$	(6.7)
Mathematical or logistic curve fitting	It is assumed that the population growth follows a logistic mathematical relationship. The most common relationship is an *S*-shaped curve.	$Y_t = \frac{Z}{1+ae^{b(T-T_0)}}$	(6.8)
		$a = \frac{Z-Y_0}{Y_0}$	(6.9)
		$b = \frac{1}{n}\ln\left(\frac{Y_0(Z-Y_1)}{Y_1(Z-Y_0)}\right)$	(6.10)

Source: Adapted in part from Refs. 4,5,37

$\frac{dY}{dt}$ = rate of changes in population with respect to time.

Y_0, Y_1, and Y_2 = populations at time T_0, T_1, and T_2.

Y_t = estimated population of the year of interest.

Z = saturation population.

K_a, K_p, and K_d = population constants for arithmetic, geometric and decreasing rate of increase.

a and b = constants.

n = constant interval between T_0, T_1 and T_2 (generally 10 years).

TABLE 6.3
Population Projections by Using Graphical Comparison, Ratio Method, Employment Forecast, and Birth Cohort

Method	Description
Graphical comparison	The procedure involves the graphical projection of the past population data for the city being studied. The population data of other similar but larger cities are also plotted in such a manner that all the curves are coincident at the present population of the city being studied. These curves are used as a guide in future population projections.
Ratio and correlation	In this method, the population of the city in question is assumed to follow the same trends as that of the region, county, or state. From the population records of a series of census years, the ratio is plotted and then projected to the year of interest. From the available estimated population of the region, county, or state, and the projected ratio, the population of the concerned city is obtained.

(*Continued*)

TABLE 6.3 (*Continued*)
Population Projections by Using Graphical Comparison, Ratio Method, Employment Forecast, and Birth Cohort

Method	Description
Employment forecast or other utility connections forecast	The population is estimated using the employment forecast. From the past data of population and employment, the ratio is plotted, and the population is obtained from the projected employment forecast. Procedure is similar to that of the ratio method. Similar procedure can be utilized for the forecast of various utility service connections such as telephone, electric, gas, water, and sewers. Utility companies conduct studies and develop reliable forecasts on future connections. Forecasts of postal and newspaper service points have also been used in population estimates.
Birth cohort	A birth cohort is defined by demographers as a group of people born in a given year or period.[a] The existing populations of males and females in different age groups are determined from the past records. From birth and death rates of each group and population migration data, the net increase in each group is calculated. The population data are then shifted from one group to the other until the design period is reached.

Source: Adapted in part from Refs. 4,5,37

[a] Demography is that branch of anthropology that deals with the statistical study of the characteristics of human population with reference to total size, density, number of deaths, births, migration, etc.

TABLE 6.4
Range of Population Densities in Various Sections of a City

	Population Range	
Land Use	**Persons/km²**	**Persons/Acre**
Residential areas		
Single-family dwellings, large lots	1,250–3,700	5–15
Single-family dwellings, small lots	3,700–8,700	15–35
Multiple-family dwellings, small lots	8,700–25,000	35–100
Apartment or tenement houses	25,000–250,000	100–1,000 or more
Commercial areas	3,700–7,500	15–30
Industrial areas	1,250–3,700	5–15
Total, exclusive of parks, playgrounds, and cemeteries	2,500–12,500	10–50

Source: Adapted in part from Refs. 4,5,37

Note: $km^2 = acre \times 247.1$.

Example 6.4: Population Estimates from Arithmetic, Geometric, Decreasing Rate of Increase, and Logistic Curve-Fitting Methods

Estimate the population of a city using arithmetic, geometric, decreasing rate of increase, and logistic curve-fitting methods. The population data from three census years for the city are given below: The design year for expansion and completion of point and nonpoint source collection and conveyance system is 2030.

Year	Population, Person
2000	10,000
2010	15,000
2020	18,000

Solution

1. Apply the following data from the table.

$$T_0 = 2000,\ Y_0 = 10{,}000 \text{ persons};\ T_1 = 2{,}010,\ Y_1 = 15{,}000 \text{ persons};$$

$$T_2 = 2020,\ Y_2 = 18{,}000 \text{ persons};\ T = 2{,}030, \text{ and } n = 10 \text{ years}.$$

2. Estimate population using arithmetic growth from Eqs. (6.1) and (6.2).
The most recent census data from 2020 is used for population projections.

$$K_a = \frac{(18{,}000 - 15{,}000) \text{ persons}}{(2020 - 2010) \text{ year}} = 300 \text{ persons/year}$$

$$Y_{2030} = 18{,}000 \text{ persons} + 300 \text{ persons/year} \times (2030 - 2020) \text{ years} = 21{,}000 \text{ persons}$$

3. Estimate population using geometric growth from Eqs. (6.3) and (6.4).

$$K_p = \frac{(\ln 18{,}000 - \ln 15{,}000)}{(2020 - 2010) \text{ year}} = \frac{9.798 - 9.616}{10 \text{ year}} = 0.0182 \text{ year}^{-1}$$

$$\ln Y_{2030} = \ln 18{,}000 + 0.0182 \text{ year}^{-1} \times (2030 - 2020) \text{ year} = 9.798 + 0.182 = 9.98$$

$$Y_{2030} = e^{9.98} = 21{,}600 \text{ persons}$$

4. Estimate population using the decreasing rate of increase method from Eqs. (6.5) through (6.7).

$$Z = \frac{2 \times (10{,}000 \times 15{,}000 \times 18{,}000) \text{ persons}^3 - (15{,}000 \text{ persons})^2 \times (10{,}000 + 18{,}000) \text{ persons}}{(10{,}000 \times 18{,}000) \text{ persons}^2 - (15{,}000 \text{ persons})^2}$$

$$= 20{,}000 \text{ persons}$$

$$K_d = -\frac{1}{(2020 - 2010) \text{ years}} \times \ln\left(\frac{(20{,}000 - 18{,}000) \text{ persons}}{(20{,}000 - 15{,}000) \text{ persons}}\right)$$

$$= \left(-0.1 \text{ year}^{-1}\right) \times (-0.916) = 0.0916 \text{ year}^{-1}$$

$$Y_{2030} = 18{,}000 \text{ persons} + (20{,}000 - 18{,}000) \text{ persons} \times (1 - e^{-0.0916 \text{ year}^{-1} \times (2030-2020) \text{ year}})$$

$$= 18{,}000 \text{ persons} + 1{,}200 \text{ persons} = 19{,}200 \text{ persons}$$

5. Estimate population using logistic curve fitting from Eqs. (6.8) through (6.10).

$$Z = 20{,}000 \text{ persons} \left(\text{from step 4}\right)$$

$$a = \frac{(20{,}000 - 10{,}000) \text{ persons}}{10{,}000 \text{ persons}} = 1$$

$$b = \frac{1}{10 \text{ years}} \times \ln\left(\frac{10{,}000 \text{ persons} \times (20{,}000 - 15{,}000) \text{ persons}^2}{15{,}000 \text{ persons} \times (20{,}000 - 10{,}000) \text{ persons}^2}\right)$$

$$= \frac{1}{10 \text{ years}} \times \ln 0.333 = -0.1099 \text{ year}^{-1}$$

$$Y_{2030} = \frac{20{,}000 \text{ persons}}{1 + 1 \times e^{-0.1099 \text{ year}^{-1} \times (2030-2000) \text{ year}}} = \frac{20{,}000 \text{ persons}}{1 + 0.0370} = 19{,}290 \text{ persons}$$

Example 6.5: Graphical Comparison Method

Estimate the population of City A by using graphical comparison with populations of Cities B and C. The design year is 2040.

	Population, 10^3 Persons		
Year	**City A**	**City B**	**City C**
1990	8.0	18.0	16.0
2000	11.0	20.3	20.0
2010	14.2	22.0	23.5
2020	18.0	23.2	26.0

Solution

1. Determine the years for Cities B and C so that their curves are coincident at the population of City A in the year 2020.

   City A population in 2020 = 18,000 persons

   City B population in 1990 = 18,000 persons

   City C reached a population of 18,000 persons between the years 1990 and 2000.

   Using the arithmetic rate of increase method (Eqs. 6.1 and 6.2), calculate the year T when the City C reaches a population of 18,000 persons.

$$K_a = \frac{(20{,}000 - 16{,}000) \text{ persons}}{(2000 - 1990) \text{ year}} = 400 \text{ persons/year}$$

$$18{,}000 \text{ persons} = 16{,}000 \text{ persons} + \frac{400 \text{ persons}}{\text{year}}(T - 1990) \text{ years}$$

$$T = 1990 + \frac{18{,}000 \text{ persons} - 16{,}000 \text{ persons}}{400 \text{ persons/year}} = 1995$$

2. Plot the census populations of the City A.
   The population of City A is plotted from 1990 to 2020
3. Plot the population data of City B
   Shift the year scale for City B so that the population curves start from the year 2020 or Point P. The population of City A in that year is 18,000 persons. The population of City B on a shifted scale (1990) is also 18,000.
4. Plot the estimated population of City C
   Shift the estimated population of City C to Point P. The estimated population of 18,000 for City C was reached in 1995. Plot the census population of City C for the years 1990, 2000, and 2010 on the shifted year scale.
   The graphical presentation of the populations of three cities is shown in Figure 6.2.
5. Project the population of City A using shifted population curves of Cities B and C as a guide.
6. The graphically estimated population of City A in the year 2040=23,800 persons.

Example 6.6: Ratio Method

Estimate the population of City A using the ratio method. The design year is 2040. The population data of the city and region are given below. The estimated population of the region in the year 2040 is obtained from the State Planning Commission. It is 988,000 persons.

	Population, 10^3 Persons	
Year	**City A**	**Region**
1990	50	455
2000	61	623
2010	72	766
2020	77	850

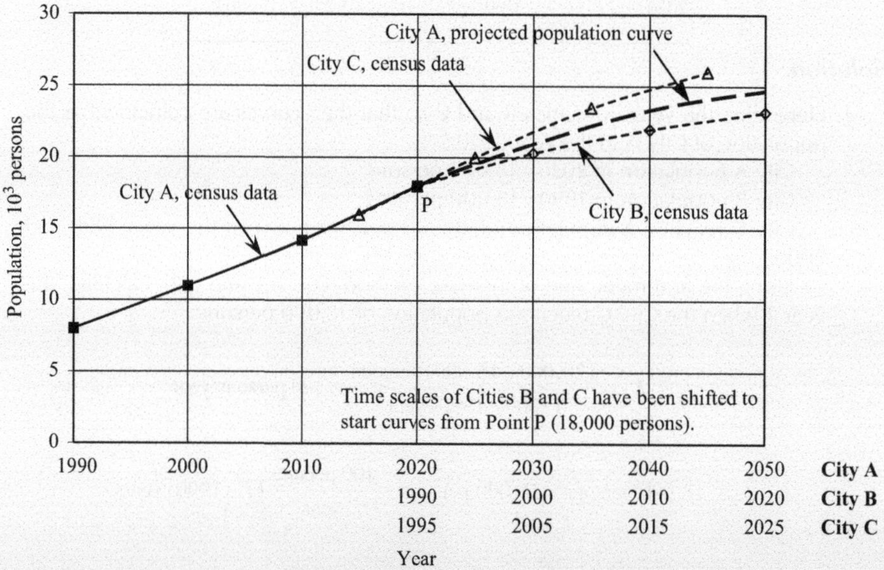

FIGURE 6.2 Population estimate by graphical comparison (Example 6.5).

Solution

1. Develop the ratio of city population and region population.

Year	City A Population, 10^3 Persons	Region Population, 10^3 Persons	Ratio
1990	50	455	0.110
2000	61	623	0.098
2010	72	766	0.094
2020	77	850	0.091
2040	—	988 (estimated)	—

2. Plot the population ratio by year.
   The population ratio is plotted in Figure 6.3.
3. Estimate the population ratio for the year 2040 and the population of City A in the year 2040.
   Extend the population ratio graph by a dashed line to the year 2040.
   Population ratio = 0.089
   The estimated population of City A in the year 2040 = 0.089 × 988,000 persons = 88,000 persons

Example 6.7: Employment Forecast

Estimate the population of a city using employment forecast. The design year is 2040. Use the following data. The employment forecast for the city is obtained from the Regional Planning Commission. For the year 2040, the estimated employment is 21,300.

	Population and Employment, 10^3 Persons	
Year	Population of the City	Employment
1990	20	6.80
2000	30	10.79
2010	39	14.77
2020	46	17.83
2040	—	21.30

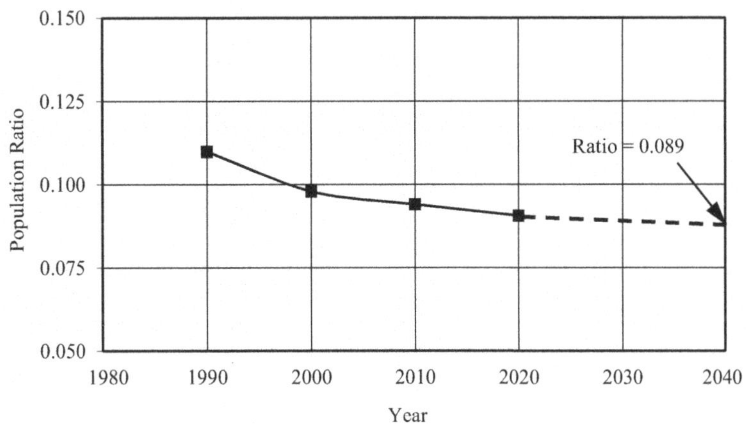

FIGURE 6.3 Population estimate by ratio method (Example 6.6).

Solution

1. Calculate the ratio of population to employment for the census years.

Year	Population of the City, 10^3 Persons	Employment in the City, 10^3 Persons	Ratio of Population to Employment
1990	20	6.80	2.94
2000	30	10.79	2.78
2010	39	14.77	2.64
2020	46	17.83	2.58
2040	—	21.30	—

2. Plot the ratios with respect to years.
   The ratios of the population to employment are shown in Figure 6.4.
3. Estimate the ratio of population to employment for the year 2040 and the population of the city.
   Extend the dotted line of the ratios of the population to employment to the year 2040.
   The ratio of population to employment for the year 2040 = 2.54
   Population of the city in the year 2040 = 2.54 × 21,300 persons = 54,100 persons

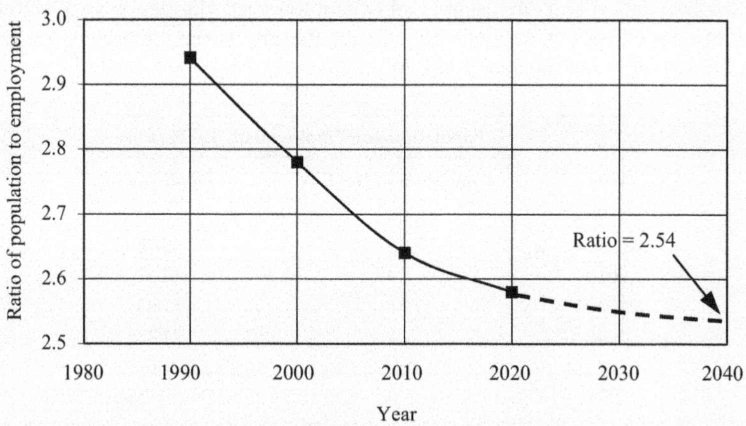

FIGURE 6.4 Population estimate by employment forecast (Example 6.7).

Example 6.8: Population Estimate Based on Population Density

A city is rapidly growing and is expected to reach its saturation population in the year 2040. The land use plan for the year 2040 was developed by the City Planning Commission. The areas for various sections of the city and the land use plan are provided below. The average ultimate population densities for different land uses are given below. Estimate (1) the saturation population of the city, and (2) the average population density for the entire city.

Type of Land Use	Single Family Large Lots	Single Family Small Lots	Multiple Family Small Lots	Apartments	Commercial	Industrial	Parks, Playgrounds, and Cemeteries
Area, acres	380	250	120	20	560	875	680
Population density, persons/acre	10	25	68	550	23	10	30

Solution

1. Tabulate the expected land use and population in the year 2040.
   The expected land use, average population densities, and population in the year 2040 are tabulated below.

Type of Land Use	Area, Acre	Population Density, Persons/Acre	Population, Persons
Single family, large lots	380	10	3,800
Single family, small lots	250	25	6,250
Multiple family, small lots	120	68	8,160
Apartments	20	550	11,000
Commercial	560	23	12,880
Industrial	875	10	8,750
Parks, playgrounds, and cemeteries	680	30	20,400
Total	2,885	—	71,240

2. Determine the saturation population of the city and the average population density of the entire city in 2040.
   a. Saturation population of the city=71,240 persons
   b. Average population density of the entire city=71,240 persons÷2,885 acres=25 persons/acre

6.4.2 Types of Proposed Collection and Conveyance Systems for Point and Nonpoint Source Pollution Control

The community wastewater collection and conveyance system consists of building sewers or connections, laterals or branch sewers, collectors or sub-collectors, trunk sewers, and intercepting sewers. The small sewers connect with larger sewers to build up a collection network. Finally, the wastewater is conveyed by an interceptor to a central location for treatment. All sewers follow the natural ground slope for the flow to occur under gravity. In service areas where the sewer lines are deep underground, a lift station may be needed to raise the flow so that gravity flow may be resumed. The wastewater collection and conveyance system may have many connections and appurtenances such as elbows, junction boxes, manholes, terminal cleanouts, collection and division boxes, pumping stations, pressure lines, and flow meters. The storm sewer collection system has inlets, gutters on the street curb line, storm sewer lines, and outfall structures. Chapter 7 is devoted to flow measurement devices. Detailed design of sanitary and storm sewers, pumping stations, and pressure lines is provided in Chapters 8 and 9.

The collection and conveyance system for NPS drainage includes surface drains, open channels, partially flowing concrete sewers and corrugated pipes, syphons, culverts, detention basins, and pumping stations. The combined sewers carry sanitary sewage, infiltration and inflow, and storm water. The collection network and conveyance system of combined sewers utilizes a considerably larger pipe network, flow meters, appurtenances, and inlet structures. The outlet structures of combined sewers have overflow and diversion structures to divert the excess flow to a detention facility. The examples of overflow and diversion structures are side weir, baffled side weir, transfer weir, leaping weir, orifice, and relief syphons. The management of CSOs is covered in detail in Chapter 11.

6.4.3 Site Selection of Facility Components and Equipment

The site selection of the common components of a wastewater conveyance system, such as pumping and lift stations, storage basin, collection and diversion units, and pretreatment facilities, should be based on careful consideration of the land use and development patterns in the region, as well as social, environmental, and engineering constraints. It is important to remember that the selection of a site for these facilities will have long-lasting social, economic, and political repercussions on the affected community and neighborhood. Therefore, public involvement in decision-making is crucial. The basic principles that must be considered during site evaluation are: (1) low elevations of pumping and lift stations, storage basin, and pretreatment and disposal facilities are necessary to permit gravity flow, (2) isolated site from presently build-up areas, and future developments, (3) large land area to maintain isolation (buffer land), (4) above flood zone; in low-lying areas proper flood protection measures must be taken, (5) year round, all-weather access roads, (6) alternate source of electric power, (7) effluent reuse and disposal potential, (8) geology to permit foundation stability, and (9) absence of archaeological and historical sites, and critical habitats for endangered or threatened species of flora or fauna.

The project planning team must investigate topography, drainage, surface and groundwater, soil type, prevailing winds, temperature, precipitation, seasonal solar angles, wildlife habitats, ecosystems, regional and local land use and zoning, transportation, archaeological and historical features, and other factors.

6.4.4 Energy and Resource Requirements and Project Economics

Primary energy is the energy used for operation, while secondary energy is needed to manufacture chemicals, other consumable materials, equipment, and construction materials. Resource requirements include land, equipment, instrumentation, and labor. The project economics is therefore not only based on the initial construction costs but also on annual operation and maintenance (O&M) costs. Under the Clean Water Act of 1977, it is required that the designers utilize project construction and operation alternatives that substantially conserve energy.[6,7,26] Additionally, a cost-benefit analysis is necessary to ensure that the construction and O&M costs are reasonable and appropriate for the planned level of collection and conveyance facilities.

6.4.5 Environmental Impact Assessment

The National Environmental Policy Act of 1969 was enacted to ensure that federal agencies consider environmental factors in the decision-making process and utilize an interdisciplinary approach in evaluating these issues.[38–40] The environmental impact assessment must evaluate all impacts-beneficial and adverse, and primary and secondary, that may result from the construction of collection and conveyance facilities. The primary impacts are those directly associated with the construction and operation of the facilities. For example, improvements in water quality of the receiving water and odors resulting from the operation of a wastewater treatment plant are the primary impacts. The secondary impacts result from the growth or change in land use induced or facilitated by the construction of the associated collection and conveyance systems. To address these federal environmental considerations, the design engineer should work closely with the federal, state, and local regulatory agencies that have responsibilities for planning, design, and operation of the collection and conveyance facilities.

Example 6.9: Facility Economics

A point and nonpoint source collection and conveyance facility is serving a design population of 80,000. Average wastewater flow collected and transported is 38,000 m^3/day (10 MGD). The total current capital cost of the project is 34.296 million dollars. The current total annual O&M cost is 1.895 million dollars. The interest rate is 7.652%, and the planning period is 15 years. Determine the following: (1) present worth, (2) equivalent annual cost, (3) unit cost $/m³ and $/1,000 gal, and (4) charges per family per month. Assume 3.5 members per family.

Solution

1. Describe the present worth.
   The present worth may be thought of as the sum which, if invested now at a given rate of interest, would provide exactly the funds required at the design period while making all necessary expenditures during the planning period.
2. Calculate the present worth of the annual O&M cost.
   The value of an annuity is expressed by Eq. (6.11).

$$\text{Value of annuity} = \frac{\left[1-\left(1+i\right)^{-n}\right]}{i} \tag{6.11}$$

where,
i = interest rate ratio, percent/100
n = planning period, years

$$\text{Value of annuity} = \frac{[1-(1+0.07652)^{-15}]}{0.07652} = 8.744$$

$$\text{Present worth of annual O\&M cost} = \$1.895\times10^{6}\times8.744\times\frac{1}{10^{6}} = 16.570 \text{ million dollars}$$

$$\begin{aligned}\text{Present worth of capital and O\&M cost} &= (34.296+16.570) \text{ million dollars}\\ &= 50.866 \text{ million dollars}\end{aligned}$$

3. Describe equivalent annual cost.
   The equivalent annual capital cost is the expression of non-uniform series of expenditures used as a uniform annual amount to simplify calculations of present worth.
4. Calculate the capital recovery factor.
   The capital recovery factor is expressed by Eq. (6.12).

$$\begin{aligned}\text{Capital recovery factor} &= \frac{i(1+i)^{n}}{\left[(1+i)^{n}-1\right]}\\ &= \frac{0.07652\times(1+0.07652)^{15}}{[(1+0.07652)^{15}-1]} = 0.1144/\text{year}\end{aligned} \tag{6.12}$$

5. Calculate the equivalent annual cost.
   The equivalent annual cost is calculated two ways.

a. The equivalent capital annual cost=34.296 million dollars×0.1144/year=3.924 million dollars/year

It may be noted that an annual payment of 3.924 million dollars over a 15-year period will pay back the loan principal and the interest on the loan.

$$\text{Total equivalent annual cost} = \left(3.924 + 1.895(\text{annual O\&M cost})\right)\text{ million dollars/year}$$

$$= 5.819 \text{ million dollars/year}$$

b. Total equivalent annual cost=50.866×0.1144=5.819 million dollars/year

An annual payment of 5.819 million dollars per year for 15 years will pay back the loan principal, interest, and O&M costs.

6. Calculate unit collection and conveyance cost.

Total equivalent annual cost=\$5.819×10⁶/year

$$\text{Average volume of watewater collected and conveyed in a year}$$

$$= 38{,}000 \text{ m}^3/\text{day} \times 365 \text{ day/year} = 13.87 \times 10^6 \text{ m}^3/\text{year}(3{,}650 \times 10^6 \text{ gallon/year})$$

$$\text{Unit cost of conveyance (SI unit)} = \frac{\$5.819 \times 10^6/\text{year}}{13.87 \times 10^6 \text{ m}^3/\text{year}} \times 1{,}000$$

$$= \$0.42/\text{m}^3 \text{ or } \$0.42 \text{ per m}^3$$

$$\text{Unit cost of conveyance (US customary unit)} = \frac{\$5.819 \times 10^6/\text{year}}{3{,}650 \times 10^6 \text{ gallon/year}} \times 1{,}000$$

$$= \$1.59/10^3 \text{ gallon or } \$1.59/1{,}000 \text{ gallons.}$$

7. Calculate charges per family.

$$\text{Number of families} = \frac{80{,}000 \text{ persons}}{3.5 \text{ persons/family}} = 22{,}900 \text{ families}$$

$$\text{Monthly charge per family} = \frac{\$5.819 \times 10^6/\text{year}}{22{,}900 \text{ families} \times 12 \text{ month/year}} = \$21.18/\text{month}$$

Example 6.10: Construction Cost Adjusted to a Different Time Period

The estimated construction cost of a wastewater treatment plant in 2018 is 27.70 million dollars. Estimate the construction cost in 2025 dollars.

Solution

The capital costs are adjusted to any period by using the Engineering News-Record Construction Cost Index (ENRCCI), or the U.S. EPA Cost Index. The project construction cost is estimated from Eq. (6.13).

Projected construction cost

$$\text{Projected construction cost}$$

$$= \frac{\text{Projected future value of index}}{\text{Value of index at time of estimate}} \times \text{Estimated cost at the time of estimate} \tag{6.13}$$

1. Determine the ENRCCI for 2018 and 2025 from the ENR website.

$$\text{ENRCCI for } 2018 = 11{,}062$$

$$\text{ENRCCI for } 2025 = 13{,}893$$

2. Determine the construction cost for 2025.

$$\text{Construction cost for } 2025 = \frac{13{,}893}{11{,}062} \times 27.70 \text{ million dollars} = 34.79 \text{ million dollars}$$

Example 6.11: Operation and Maintenance Cost

The annual O&M costs are important factors in the evaluation of alternative treatment processes. The principal elements of O&M costs are labor, energy, chemicals, and materials and supplies. The total O&M cost of a 10 MGD diffused aeration basin for 2018 was \$409,000. Various cost components and unit costs for aeration basin are provided below. Estimate the total annual O&M costs for 2025.

	O&M Cost, % of Total	Unit Energy Rate, \$/kWh		ENR Index	
Category	Cost	2018	2025	2018	2025
Labor (BCI)	20	—	—	6,105	8,578
Energy	62	\$0.129	\$0.176	—	—
Material and supplies (BCI)	18	—	—	6,019	8,500[a]
Total	100	—	—	—	—

[a] Estimated.

Solution

1. Calculate the costs of labor from the Engineering News Record Building Cost Index (ENR BCI).

$$\text{The annual cost of labor in 2025} = (\$409{,}000/\text{year} \times 0.20) \times \frac{8{,}578}{6{,}105}$$

$$= \$114{,}940/\text{year}$$

2. Calculate the cost of energy from unit energy rate.

$$\text{The annual cost of energy in 2025} = (\$409{,}000/\text{year} \times 0.62) \times \frac{\$0.176/\text{kWh}}{\$0.129/\text{kWh}}$$

$$= \$345{,}970/\text{year}$$

3. Calculate the cost of materials and supplies from ENR BCI Index.

$$\text{The annual cost of material and supply in 2025} = (\$409{,}000/\text{year} \times 0.18) \times \frac{8{,}500}{6{,}019}$$

$$= \$103{,}970/\text{year}$$

4. Calculate the estimated total O&M cost in 2025.

$$\text{Total annual O\&M cost in 2025} = (\$114{,}940 + \$345{,}970 + 103{,}970)/\text{year} = \$564{,}880/\text{year}$$

Example 6.12: Environmental Impact Assessment

A wastewater treatment project involves the expansion of sanitary and storm sewer, interceptors, and expansion of existing wastewater treatment and disposal facilities. List the major primary and secondary beneficial and adverse impacts of the project.

Solution

1. List the beneficial impacts.

   The beneficial impacts are:

   a. Encourage growth and increase in land use and property value induced by the expansion of sanitary and storm sewers and interceptor, and plant expansion.
   b. Provide compliance with effluent standards.
   c. Enhance surface water quality in the receiving water.
   d. Encourage many beneficial uses of effluent and receiving waters (agriculture, recreation, commerce, industry, and water supply).
   e. Eliminate groundwater pollution due to septic tanks and absorption fields.
   f. Reduce health hazards due to enhanced effluent quality.
   g. Provide opportunity for biosolids reuse.
   h. Improve shoreline and waterfront for recreation.
2. List the adverse impacts.

   Listed below are possible adverse impacts that may be minimized by modern design techniques and efficient plant operation.

   a. Traffic disruptions and higher noise levels during construction.
   b. Disrupt the environment and cause inconveniences to the citizens during the construction of sewers and interceptors.
   c. Possible occasional odors emanating from the treatment facility, and lower air quality.
   d. It may reduce the value of the adjacent property around the treatment plant.
   e. Require energy and chemicals for plant operation.
   f. Loss of construction materials and land permanently for the treatment facility and conveyance system.
   g. It is possible that some adverse impacts may occur on plant and animal communities, ecosystem, scenic views and aesthetics, and community growth patterns and land use around the treatment plant.

Example 6.13: Reviewing Agencies of Environmental Impact Assessment Report

Many federal, state, and local agencies, and private and public groups may review and provide comments on an environmental impact assessment report. List some of these potential agencies.

Solution

The federal and state agencies that review and comment on environmental impact assessment reports are: U.S. EPA, U.S. Fish and Wildlife Service, U.S. Army Corps of Engineers, State Environmental Quality Department, State Air Control Board, State Parks and Wildlife Department, State Water Development Board, State Historic Preservation Office, and concerned state and city departments. These agencies should be notified, and the information should be posted on their websites.

Various appointed environmental advisory committees and concerned citizens groups should also be notified. Some of the citizen groups are: League of Women Voters, Audubon Society, Conservation Societies, Historic Preservation Society, Sierra Club, and others.

Relevant documents should be placed in the designated depository and posted on their websites for public review and comments. The fact sheet to highlight major elements of the project and notices of public hearing should be posted on websites and sent to local newspapers, posted in libraries and city hall notice boards, and sent to the mailing list of concerned parties.

6.5 FACILITY PLANNING, DESIGN PLANS, AND SPECIFICATIONS

The task of planning and designing a collection and conveyance system for point and nonpoint source wastewater involves an understanding of the service area, sources and characteristics of wastewater, legal issues, and regulatory constraints. For most projects, the engineering services are performed in three steps: (1) engineering report or facility plan, (2) preparation of design plans, specifications, cost estimates, and contractual documents, and (3) construction and construction management.

6.5.1 Facility Planning

A facility plan is prepared to identify the water pollution problems in a specific area, evaluate alternatives, and recommend a solution. Through the facility plan, the consultants make many decisions that are subsequently used in preparation of the detailed plans and specifications for the point and nonpoint source collection and conveyance systems.

Example 6.14: Major Topics Covered in a Facility Plan

List the major topics covered and the final outcome of a facility plan.

Solution

1. The major topics covered in a facility plan are:
   a. Project description, need, and service area
   b. Design period
   c. Project requirements
   d. Existing and future conditions
   e. Population projections, industrial growths, and unit flow rates
   f. Wastewater characteristics: minimum, average, and peak flow rates; I/I; and sustained flow conditions
   g. Forecast, and evaluation and selection of design flow rates
   h. Point and nonpoint source collection and conveyance alternatives and evaluation
   i. Description of selected alternative
   j. Elements of conceptual design
   k. O&M of the facility and emergency operation
   l. Preliminary costs and financial status
   m. Environmental impact assessment report and public participation
   n. Implementation plan including financial arrangements, and time schedule for design and construction, and project milestones
2. Describe the outcome of the facility plan.

   At the completion of the facility plan, the project elements are fully defined so that preparation of detailed plans and specifications can proceed expeditiously. The completed facility plan in the prescribed format is submitted to the regional or state clearing house, EPA, and regulatory agency for review and comments, and public hearings. The plan is then revised or amended as necessary. The outcome of a facility plan, therefore, is a well-defined, cost-effective, and environmentally sound project that can be approved, and is capable of being implemented and being acceptable to the taxpayers, regulatory authorities, and general public.

6.5.2 Design Plans, Specifications, Cost Estimates, and Support Documents

Once a facility plan and project application of a community is approved by the state water quality agency, the design phase of the project is initiated. This phase of the project deals with the preparation of detailed engineering design plans, specifications, and cost estimates. The plans and specifications become the official document on which contractors base their bids for the construction of the facilities. Based on the project milestones established in this phase, the construction managers or administrators may hold the contractor responsible for completion of the project.

Example 6.15: Preliminary Information and Documents

Many documents and some basic information are needed before the task of developing the design plans, specifications, and cost estimates can be started. Provide a list of the needed important documents and preliminary information.

Solution

Following is a list of the preliminary information and important documents needed before starting the design plans, specifications, cost estimates, and support documents.

a. Application to prepare design plans and specifications is necessary on specified forms
b. Statement regarding sources of local share of project cost (general taxes, sewer revenue funds, etc.)
c. A copy of resolution authorizing the official representative (mayor, council member, or others) to act on behalf of the applicant
d. Statement regarding availability and right-of-way of the proposed site
e. Proposed contracts or explanation for selection of consulting engineers
f. User's charge or resolution that a user's charge system will be developed
g. Letter of agreement from industries participating in cost recovery system
h. Copy of existing sewer use ordinance or intent to develop one
i. Assurance of compliance with the Civil Rights Act, Uniform Relocation of Land Acquisition Policy Act, equal employment opportunity, and others

Example 6.16: Preparation of Design Plans and Specifications

The most common approach for preparation of the design plans and specifications consists of the following tasks: (1) conceptual design, (2) preliminary design, (3) special studies, and (4) final design and specifications. Describe each task.

Solution

The successful completion of design and construction phase of a project largely depends upon the quality of information developed during the early stages of the project.

1. List the purpose and basic steps conducted for conceptual design.

   The conceptual design is conducted to develop framework for completion of the preliminary design. The tasks accomplished are:

   a. Make the principal engineering decisions.
   b. Finalize the preliminary design criteria used in the facility plan.
   c. Select equipment and establish facility layout.
   d. Define the necessary field investigations, topographic surveys, geological and hydrological studies, soil borings, etc.

   e. Prepare the routing and alignment of collection and conveyance system, connecting piping, topography, layout, and hydraulic profile.
   f. Define operation and control strategies.
2. List the purpose and basic steps conducted for preliminary design.

   The preliminary design stage is an expansion of the conceptual design. Major design decisions are completed in this phase. The tasks accomplished are:
   a. The site plan is finalized, and equipment and piping arrangements are completed.
   b. Mitigation measures are utilized to reduce or lessen unavoidable environmental impacts.
   c. Space and utilities requirements and piping routes are finalized.
   d. Architectural concepts are developed.
   e. Preliminary cost estimates are developed.
   f. For large projects, the value engineering (VE) should be well along at this stage.
3. List the information developed during the special studies.

   Special studies are often needed to develop specialized information and to refine the design criteria. The specialized information developed is:
   a. Testing of equipment, piping, and conveyance systems, conducted to refine the equipment selection criteria
   b. Dispersion studies for effluent outfall siting
   c. Environmental studies to develop baseline data
   d. These special studies must be completed before the start of the final design plans and specifications to avoid costly design changes.
4. List the information that is developed for the final design.

   The detailed design plans consist of plan views, elevations, sections, and supplementary views that, together with the specifications and general layouts, provide the working information for the contract and construction facilities. The design plans specifically contain the following information:
   a. Dimensions, location, and type of equipment, location routing and size of piping, and appurtenances, and elevations.
   b. Final layout of piping, appurtenances, flow meters, equipment, buildings and roads, elevation of all units and structures, and ground elevations.
   c. Hydraulic profiles at maximum, average and peak hourly flows, and elevation of high and low water levels in the receiving waters.
   d. Technical specifications for the construction of sewers, channels, pumping stations, treatment units, and appurtenances must accompany the design plans.
   e. The specification shall include all construction information not shown on the drawings. Information such as quality of material, workmanship, fabrication requirements for the equipment, and complete requirements for all mechanical and electrical equipment – including machinery, valves, piping and jointing of pipes, electrical apparatus, wiring, instrumentation, and meters – should be included.
   f. The consultant is to prepare the construction cost estimates based on the scope of work covered in the project plans and specifications. This estimate is used to judge the reasonableness of the bids received.
   g. For the construction contracts, the specifications must also include construction schedule and project milestones, bid bond, performance bond and payment bond, fire and extended coverage, workmen's compensation, public liability and property damage, and "all risk" insurance as required by local and state law, and flood insurance as required during and after construction.
   h. Well-prepared design plans and specifications provide the following information: (1) allow the contractor to submit bids with small allowances for

unknowns, (2) utilize high-quality material for construction, (3) complete work in timely manner, (4) integrate new facilities with existing units under operation, and (5) require minimum changes during construction.

Example 6.17: Value Engineering

What is value engineering (VE)? Describe the purpose and procedure to conduct the VE.

Solution

VE is an intensive review of the project elements to utilize specialized cost control techniques. VE is conducted by a review team. The purpose is to obtain the best project without sacrificing quality or reliability. U. S. EPA has mandated that all projects having a total construction cost over $10 million and receiving federal funding are subject to VE analysis. The procedure is as follows:

a. The VE team members are senior professionals who are not involved with the design of the project.
b. The number of VE teams and the number of review sessions depend on the size and complexity of the project. It may vary from one team and one review session to multiple teams and multiple sessions.
c. For medium to large projects, one team and two review sessions lasting one week each are needed. The first session is conducted at approximately 20%–30% and the second session at 60%–70% completion of the design.

6.5.3 Construction and Construction Management

After the construction bid is approved and contractual agreements are completed, the construction phase of the project is initiated. The construction contract must clearly define integration of new construction with existing facility so that there is no violation of permit requirements, and construction does not create any safety hazards to the project personnel, during construction and operation phases.

Example 6.18: Construction Management Tasks

The construction management techniques are used for the timely construction of the project in accordance with the design plans and specifications. List the major construction management tasks.

Solution

The major construction and management tasks are:

a. Verify that the contractor has adequate technical and equipment resources on-site that are compatible with the project needs before construction begins.
b. Review the contractor's operation to ensure that the requirements of design plans and specifications are fully implemented.
c. Review and verify the construction schedule and progress in meeting the project milestones.
d. Control change orders and possible construction claims.

6.6 REVIEW PROBLEMS AND DISCUSSION TOPICS

6.1 Effluent sampling data was collected over a 30-day period at a wastewater treatment plant that has a primary sedimentation basin and a biological treatment facility. The frequency distributions of 30-day BOD_5 results are given below. Determine if 30-day average BOD_5 data is in compliance with the Minimum National Standards for Secondary Treatment as defined by the U.S. EPA.

BOD_5 concentration, mg/L	20	22	28	29	31	34	40
Frequency (number of days)	2	4	5	6	7	4	2

6.2 The initial and design years of a wastewater treatment facility are the years 2013 and 2038. The estimated populations of these years are 35,000 and 76,000, respectively. The expected flow rates during initial and design years are 550 and 650 Lpcd, respectively. Determine the flow rates during initial and design years, and the average increase per year.

6.3 Estimate the design year population of a community by using arithmetic, geometric, decreasing rate of increase, and logistic curve-fitting methods. The design year is 2030. Use the following census data.

Year	Population, Thousands
2000	31.6
2010	36.9
2020	42.3

6.4 The population and employment data of a city are given below. Estimate the population in year 2040 if the employment projection for the year 2040 is 9,200.

Year	Population	Employment
2000	20,000	7,500
2010	21,000	8,000
2020	23,000	8,800

6.5 The population of City A is estimated by graphical comparison. The population of City B and City C are used for comparison. These cities are larger and have reached the year 2020 population of City A in the past. Estimate the population of City A in the year 2040. Use the following census data.

	Population		
Year	City A	City B	City C
1990	52,000	90,000	85,000
2000	70,000	120,000	95,000
2010	80,000	128,000	105,000
2020	90,000	140,000	115,000

6.6 Draw a generalized process diagram of a secondary wastewater treatment plant using the activated sludge process. The process diagram should include processes for both liquid and sludge streams.

6.7 A wastewater treatment plant has the following process train: bar screen, grit chamber, primary sedimentation, extended aeration, final clarifier, gravity filtration, and chlorination/dechlorination. The average percent removal efficiencies of various treatment units of conventional secondary treatment plant and the gravity filter are given in Examples 6.2 and 6.3. On an average, 40% and 20% NH_3-N are nitrified in aeration basin and gravity filter, respectively. Estimate the effluent quality in terms of BOD_5, COD, TSS, TP, ON, NH_3-N, and NO_3-N. The influent quality after mixing with the return flows from the sludge-processing areas is as follows: $BOD_5=220$, COD=450, TSS=225, TP=9, ON=25, NH_3-N=21, and NO_3-N=0. All units are in mg/L.

6.8 The effluent discharge permit of a state requires effluent BOD_5, TSS, total nitrogen (TN), and TP less than 5, 5, 10, and 2 mg/L, respectively. The average influent BOD_5, TSS, TN, and TP are, respectively, 200, 240, 44, and 8 mg/L (assume TN=50% ON+50% NH_3-N). The selected process train uses a bar screen, grit removal, primary sedimentation, a biological nutrient removal (BNR) facility, gravity filters, and chlorination/dechlorination. The selected processes and percent removal efficiencies of these processes are given below. Assume no removal through bar screen and chlorination/dechlorination processes, and 93% NH_3-N removed in BNR facility (65% NH_3-N is denitrified (lost as N_2) and 80% nitrified to NO_3-N). Determine the effluent quality from each process and the overall plant effluent quality. Also, comment on whether the state effluent discharge permit is met.

	Removal Efficiency, %			
Parameter	Grit Chamber	Primary Sedimentation	BNR Facility	Gravity Filters
BOD_5	0	35	95	45
TSS	0	58	95	70
ON	0	30	83	60
NH_3-N	0	0	93	0
TP	0	15	80	25

6.9 The liquid elevations in a wet well and influent channel of grit chamber are 147.48 and 158.23 m, respectively. The elevation of centerline of the pump impeller is 150.00 m. Determine (1) static suction head, (2) static discharge head, and (3) total static head. Review portions of Chapters 5 and 9.

6.10 A submersible pump is used for pumping wastewater at a treatment plant. The diameter of the suction bell is 40 cm, and $K=0.05$. Calculate the components of head losses and total head loss in the force main. The diameter and length of a force main are 25 cm and 150 m, respectively. There is one enlarger from 15 to 25 cm, $K=0.2$, one check valve $K=2.5$, one plug valve $K=1.0$, two gate valves $K=0.19$ each, five 45° elbows $K=0.2$ each, and two 90° elbows $K=0.3$ each. The pump discharge rate is 0.10 m^3/s and Darcy–Weisbach coefficient of roughness $f=0.02$ – review portions of Chapters 5 and 9.

6.11 A pump is delivering a flow of 1,850 gpm at a total dynamic head (TDH) of 60 ft. Calculate the water power or electrical power output of the pump – review portions of Chapters 5 and 9.

6.12 An intercepting sewer is 500 mm in diameter and is serving a population of 6,500 residents. The average wastewater flow is 450 Lpcd. Calculate the depth of flow and velocity at minimum flow. The slope of the line is 0.0025. Assume $n=0.015$ and the minimum flow is one-third of the average flow – review portions of Chapters 5 and 8.

6.13 A trunk sewer has a diameter of 21 in. It has a slope of 0.001 and is flowing 40% full. Determine the depth, velocity, and discharge at partial flow conditions using hydraulic element equations. Assume $n=0.015$ – review portions of Chapters 5 and 8.

6.14 An intercepting sewer is carrying a flow of 0.12 m^3/s. It has a diameter of 61 cm and a slope of 0.001. The sewer enters a manhole that has an outlet sewer diameter of 61 cm at a slope of 0.0012. Draw the hydraulic profile and determine the invert elevation of the incoming sewer. Assume $n=0.013$ and the coefficients of exit and entrance are 0.42 and 0.14, respectively. Review portions of Chapters 5 and 8.

6.15 A wastewater conveyance system is designed to serve a population of 20,000 residents. The designed flow is 7,600 m^3/day (2.0 mgd). The construction cost of the facility is $8.0 million. The annual O&M cost is $300,000. Calculate: (1) present worth, (2) equivalent annual cost, (3) the unit treatment costs in $/$m^3$ and in $/1,000 gallons, and (4) $/family per year. Amortize the construction cost over 20 years of useful life at an interest rate of 7%.

6.16 The construction cost of municipal wastewater collection and pumping station in 2018 was 7.8 million dollars. What will be the estimated construction cost of the project in the year 2025? Use ENR construction cost indices.

6.17 What is the purpose of an environmental impact assessment report?

6.18 Describe the purpose of a facility plan. What are the major outcomes of a facility plan?

6.19 Several tasks are completed during the preparation phase of the design plans and specifications. List the information that is specifically developed for the completion of final design plans and specifications.

6.20 The construction management techniques are used for the timely construction of the project in accordance with the design plans and specifications. List the major construction and management tasks.

REFERENCES

1. Wikipedia, the free encyclopedia, *Nonpoint Source Pollution,* https://en.wikipedia.org/wiki/Nonpoint_source_pollution, 2019.
2. U.S. Environmental Protection Agency, *Summary of Clean Water Act, Laws and Regulations*, https://www.epa.gov/laws-regulations/summary-clean-water-act, 2019.
3. U.S. Environmental Protection Agency, *Polluted Runoff: Nonpoint Source (NPS) Pollution*, https://www.epa.gov/nps/basic-information-about-nonpoint-source-nps-pollution, 2019.
4. Qasim, S. R., and G. Zhu, *Wastewater Treatment and Reuse: Theory and Design Examples*, vol. 1, *Principles and Basic Treatment*, CRC Press, Taylor & Francis Group, Baca Raton, FL, 2019.
5. Qasim, S. R., *Wastewater Treatment Plants: Planning Design and Operation*, 2nd ed., CRC Press, Boca Raton, FL, 1999.
6. *Federal Water Pollution Control Act Amendments of 1972* (PL 92–500), 92nd Congress, October 18, 1972.
7. *Clean Water Act* (PL 92–217), 95th Congress, December 27, 1977.
8. U.S. Environmental Protection Agency, 40 CFR Part 133 – *Secondary Treatment Regulations 49 FR* 37006, September 20, 1982, and as amended later.
9. U.S. Environmental Protection Agency, 40 CFR Part 130 – *The Water Quality Planning and Management*, Sec. 303 of the CWA, 1985.
10. U.S. Environmental Protection Agency, 40 CFR Part 131 – *The Water Quality Standards*, Sec. 303 (c) of the CWA, 1983.
11. U.S. Environmental Protection Agency, 40 CFR Part 136 – *Guidelines Establishing Test Procedures for the Analysis of Pollutants*, 1973.
12. U.S. Environmental Protection Agency, *Total Maximum Daily Load (TDML)*, F-00-009, Washington, DC, 2000.
13. Mandelker, D. R., Controlling Nonpoint Source Water Pollution Can it be Done? *Chicago-Kent Law Review*, vol. 65, 1989, pp. 479–502.

14. Szalay, E., Breathing Life into the Dead Zone: Can the Federal Common Law of Nuisance be Used to Control Nonpoint Source Water Pollution? *Tulane Law Review*, vol. 85, November 2010, 215–246.
15. Title 33 of the United States Code, § 1288(b)(2)(F), *Areawide Waste Treatment Management.*
16. Title 33 of the United States Code, § 1329, *Nonpoint Source Management Program.*
17. Title 33 of the United States Code, § 1344(a), *Permit for Dredged or Fill Material.*
18. Olson, R., Evaluating the Role of Created and Natural Wetlands in Controlling Nonpoint Source Pollution, *Ecological Engineering*, vol. 1, Issue 1–2, March 1992, pp xi–xv.
19. U.S. Environmental Protection Agency, *Pharmaceuticals and Personal Care Products as Pollutants (PPCPs)*, https://web.archive.org/web/20150924133259/http://www.epa.gov/ppcp/
20. Daughton, C. G., and T. L. Jones-Lepp, (eds), *Pharmaceuticals and Personal Care Products in the Environment: Scientific and Regulatory Issues*, ACS, Washington, DC, June 2001.
21. Dietrich, D., S. Webb, and T. Petry, *Hot Spot Pollutants: Pharmaceuticals in the Environment*, Elsevier Science Ltd., Amsterdam, September 2003.
22. U.S. Environmental Protection Agency, 40 CFR Part 412 – *Concentrated Animal Feeding Operation (CAFO) Point Source Category*, 68 FR 7269, February 12, 2003, and as amended later.
23. U.S. Environmental Protection Agency, 40 CFR Part 449 – *Airport Deicing Source Category*, 77FR 29203, May 16, 2012, and as amended later.
24. U.S. Environmental Protection Agency, 40 CFR Part 450 – *Construction and Development Point Source Category*, 74 FR 63057, December 1, 2009, and as amended later.
25. U.S. Environmental Protection Agency, Combined Sewer Overflow (CSO) Control Policy, *59 FR, 18*688, vol. 59, no. 75, April 19, 1994, and as amended later.
26. U.S. Environmental Protection Agency, *Construction Grants 1985 (CG-85)*, EPA 430/9–84-004, Office of Water Program Operations (WH-546), Washington, DC, 1984.
27. U.S. Environmental Protection Agency, *Clean Water Action Plan 1998*, Retrieved 25 April 2011.
28. U.S. Environmental Protection Agency, *Clean Water Act Action Plan* 2009, Retrieved 25 April 2011.
29. NOAA Office of Ocean and Coastal Resource Management, *Congressional Action to Help Manage Our Nation's Coasts*, Retrieved 22 April 2011.
30. U.S. Environmental Protection Agency, *Summary of Coastal Zone Management Act and Amendments*, Retrieved 22 April 2011.
31. Title 16 of the United States Code § 1456b, *Coastal Zone Enhancement Grant.*
32. Title 33 of the United States Code § 1328, Aquaculture.
33. Title 16 of the United States Code § 1455b, *Protecting Coastal Waters.*
34. Title 16 of the United States Code § 1455a, *Coastal Resource Improvement Program.*
35. Holliday, E., and R. J. Gast, Bacteria in Beach Sands, an Emerging Challenge in Protecting Coastal Water Quality Bather Health, *Environmental Science and Technology*, vol. 45, no 2, pp. 370–379. December 2010. DOI: 10.1021/es102747s
36. Title 16 of the United States Code Chapter 13, *Coastal Zone Management.*
37. Qasim, S. R., E. Motley, and G. Zhu, *Water Works Engineering: Planning, Design and Operation.* Prentice Hall PTR, Upper Saddle River, NJ, 2000.
38. *The National Environmental Policy Act of 1964*, Public Law 91–190, 42U.S. Congress 4321–4347, January,1970.
39. Ross, W. A., Evaluating Environmental Impact Statements, *Journal of Environmental Management*, vol. 25, 1987, pp. 137–147.
40. Canter, L. W., *Environmental Impact Assessment*, 2nd ed., McGraw-Hill, Inc., New York, 1996.

7 Flow Measurement

7.1 CHAPTER OBJECTIVES

Flow measurement at a wastewater treatment plant and conveyance system is essential for process operation and control. Flow records are invaluable for future reference, particularly when the facility expansion is needed. Also, flow records at key locations of conveyance system provide valuable information for future expansion of pumping station, interceptors, and division channels. Most regulatory agencies require that the wastewater treatment plants maintain daily flow records at strategic locations in conveyance system and treatment plant.

The objectives of this chapter are to present:

- Purpose, location, and elements of flow measurement
- Types and principles of flow measurement devices
- Common flow measurement devices for wastewater applications
- Design and applications of flow measurement devices

7.2 PURPOSE OF FLOW MEASUREMENT

Flow data at a wastewater treatment plant and conveyance system are used to estimate the following:

1. Population and industries served
2. Infiltration/inflow reaching the wastewater collection system
3. Conduit and channel sizing
4. Pump and pumping station sizing
5. Application of chemical dose and air supply rate in the aeration basin
6. Sludge return and solids withdrawal rates

7.3 LOCATION AND TYPES OF FLOW MEASUREMENT DEVICES

There are many locations where a flow measuring device can be installed. These locations are (1) surface drainage pipe, (2) nonpoint source discharge, (3) wastewater force main or the interceptor, (4) at the head of the plant, (5) any location within the plant, or (6) before the plant outfall of treated effluent and storm sewer. Each location has advantages and disadvantages, and will serve some specific needs. Basic considerations for selection of a flow measurement device at any location are solids and debris in the flow, settleable solids, flow fluctuations, accuracy needed, accessibility for maintenance, hydraulic head available and head-loss constraints, type and characteristics of liquid streams, and cost.[1]

7.4 ELEMENTS OF FLOW MEASUREMENT SYSTEMS

Flow measurement systems are composed of primary and secondary elements. Primary elements produce a head, pressure, electrical current, or other measurable parameters that are proportional to the flow. The secondary element measures the parameter produced by the primary element and provides an indication of the flow rate. These indicators may be visual such as an analog or digital readout, or a telemetry signal. Secondary elements are an integral part of the control system. In general, the primary flow measurement systems are applicable to two conditions: (1) pressure pipes and (2) open channel. Some systems, however, are applicable to both. Various types of primary

DOI: 10.1201/9781003629399-7

flow elements and their principle of flow measurement are provided in Table 7.1.[1–3] The applicable range of flow, accuracy, repeatability, and many selection criteria are summarized in Table 7.2.[2–4]

7.4.1 Venturimeter

A Venturimeter utilizes the principle of differential pressure. It consists of (1) a converging section (called approach), (2) a throat, and (3) a diverging or recovery section. Due to the converging section, the velocity at the throat is increased; as a result, the piezometric head is decreased.

TABLE 7.1
Types of Flow Measurement Devices Available for Determining Liquid Discharge

Flow Measurement Devices	Principle of Flow Measurement
1. For Pressure Pipes	
a. Venturimeter[a]	The differential pressure is measured
b. Flow nozzle meter[a]	The differential pressure is measured
c. Orifice meter[a]	The differential pressure is measured
d. Pitot tube	The differential pressure is measured
e. Electromagnetic meter[a]	Magnetic field is induced, and voltage is measured
f. Rotameter	The rise of float in a tapered tube is measured
g. Turbine meter[a]	A velocity-driven rotational element (turbine, vane, wheel) is used
h. Ultrasonic velocity[a]	The ultrasonic transducers send and receive ultrasonic pressure pulses
i. Ultrasonic doppler[a]	The transducers transmit beams that are reflected to a receiver by suspended solids or gas bubbles
j. Elbow meter	The differential pressure is measured around a bend
2. For Open Channels	
a. Flumes (Parshall, Palmer–Bowlus)[a]	Critical depth is measured at the flume
b. Weirs[a]	Head is measured over a weir
c. Current meter	Rotational element is used to measure velocity
d. Pitot tube	The differential pressure is measured
e. Depth measurement[a]	Float is used to obtain the depth of flow
f. Sonic level meter[a]	The transducer emits and receives the beam reflected from the liquid surface
3. Computing Flow from Freely Discharging Pipes	
a. Pipes flowing full	
i. Nozzles and orifices	The water jet data are recorded
ii. Vertical open-end flow	The vertical height of water jet is recorded
b. Pipe partly flowing full	
Open-flow nozzle[a] (Kennison nozzle or California pipe method)	The depth of flow at free-falling end is determined
4. Miscellaneous Methods	
a. Dilution method	The concentration of a constant flow of a dye tracer is measured
b. Bucket and stopwatch	A calibrated bucket is used and time to fill it is noted
c. Measuring level change in a tank	Change in level in a given time is obtained
d. Calculation from water meter readings	Water meter readings over a given time period give average wastewater flow
e. Pumping rate	Constant pumping rate and pumping duration

Source: Adapted in part from Refs. 1–3

[a] Commonly used devices for wastewater flow measurement.

TABLE 7.2
Evaluation of Various Types of Devices Commonly Used for Wastewater Flow Measurement

	Application		Flow Range and Accuracy		Repeatability, Percent of Full Scale	Effects of Solids in Wastewater	Head Loss	Power Requirement	Simplicity and Reliability	Maintenance Requirement	Ease of Calibration	Cost	Portability
	Pressure Flow	Open Channel	Range	Accuracy Max. Flow, (%)									
Venturimeter	Y	N	10:1	±0.5[a]	±0.5	H[b]	L	L	G	M	G	H	N
Flow nozzle meter	Y	N	4:1	±0.3	±0.5	H	M	L	G	L	G	M	N
Orifice meter	Y	N	4:1	±1	±1	H	H	L	G	H	G	L	Y
Electromagnetic meter	Y	N	10:1	±1–2	±0.5	S	L	M	F	M	G	H	N
Turbine meter[c]	Y	N	15:1	±0.25	±0.05	H	M	L	F	H	G	H	N
Ultrasonic velocity	Y	N	10:1	±1–2	±1	M	L	M	F	M	G	H	N
Ultrasonic doppler	Y	N	10:1	±1–2	±1	M	L	M	F	M	G	H	N
Parshall flume	N	Y	20:1	±5	±0.5	S	L	L	G	L	G	M	Y
Palmer–Bowlus flume	N	Y	20:1	±10	±0.5	S	L	L	G	L	G	L	Y
Weirs	N	Y	20:1	±0.5	±0.5	H	H	L	G	M	G	L	Y
Depth measurement	N	Y	10:1	±50		M	L	L	G	L	P	L	Y
Open-flow nozzle	N	Y	20:1	±1	±0.5	S	H	L	G	M	F	L	Y

Source: Adopted in part from Refs. 2–4

Note: Estimate F=fair, G=good, H=high, M=medium, N=no, P=poor, S=slight, Y=yes.

[a]Based on full-scale reading.

[b]Effect of solids is substantially smaller if solids-bearing or continuous flushing-type Venturimeter is used.

[c]Positive displacement type.

The piezometric heads at the beginning of the approach and at the throat are measured. The difference in these two heads is analyzed by electrical or electromechanical instruments. The flow is a function of the difference in head. The Venturimeters may have full-body flanged ends or may be of insert type. Venturimeters provide a high degree of accuracy, but the pressure ports of piezometer tube are subject to plugging.[1,5]

Example 7.1: Venturimeter Equations and Calibration

A standard Venturimeter is used for flow measurement in a force main. The diameter of the pressure pipe is 92 cm (36 in.). The tube beta ratio (diameter of the throat/diameter of the force main) is 0.5. The discharge coefficient C_d of the Venturimeter is 0.985. Develop the calibration equation and calculate discharges when differential pressure heads are 1, 2, and 3 m of water. Assume that the force main and Venturimeter are horizontal.

Solution

1. Write Bernoulli's equation

   Bernoulli's equation (Eq. 5.6) has been discussed in Section 5.4.2. Apply Eq. (5.6) at sections 1 and 2 (above approach section and at the middle of the throat, Figure 7.1).

$$\frac{P_1}{\gamma} + Z_1 + \frac{V_1^2}{2g} = \frac{P_2}{\gamma} + Z_2 + \frac{V_2^2}{2g} + h_L$$

2. Simplify the equation by applying the given condition.

   The force main is horizontal, therefore $Z_1 = Z_2$

   Since the differential pressure in the Venturimeter is measured between small lengths of the pipe (above approach section and at the middle of the throat), the head loss (h_L) is assumed negligible. The simplified and rearranged equation is given by Eq. (7.1)

$$\frac{P_1}{\gamma} - \frac{P_2}{\gamma} = \frac{V_2^2}{2g} - \frac{V_1^2}{2g} \tag{7.1}$$

3. Use continuity equation and differential piezometric head values to develop a theoretical discharge equation

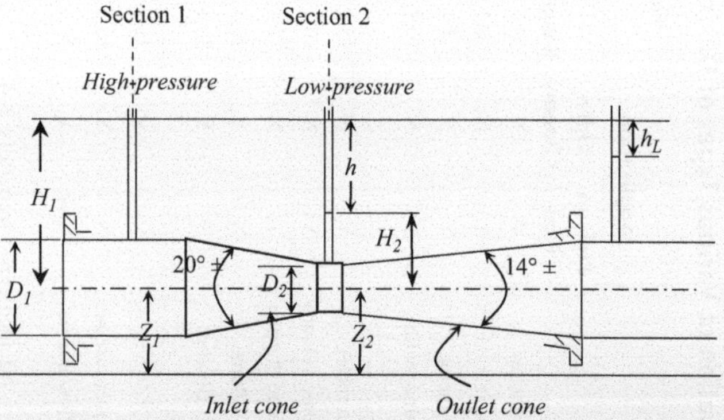

FIGURE 7.1 Typical Venturimeter used for flow measurement (Example 7.1).

Replace $\frac{P_1}{\gamma_1}$ and $\frac{P_2}{\gamma_2}$ by H_1 and H_2 in Eq. (7.1), and use continuity equation ($Q = V_1A_1 = V_2A_2$). The theoretical discharge is expressed by Eq. (7.2)

$$Q = \frac{A_1A_2\sqrt{2g(H_1 - H_2)}}{\sqrt{A_1^2 - A_2^2}} = \frac{A_1A_2\sqrt{2gh}}{\sqrt{A_1^2 - A_2^2}} = \frac{A_2\sqrt{2gh}}{\sqrt{1 - \left(\frac{A_2}{A_1}\right)^2}} = \frac{A_2\sqrt{2gh}}{\sqrt{1 - \left(\frac{D_2}{D_1}\right)^4}} \quad (7.2)$$

where

Q=theoretical pipe flow, m³/s

H_1, H_2=upstream and throat piezometric head, m

h=differential piezometric head (H_1-H_2), m

A_1, A_2=force main and throat areas, m

4. Develop flow equation for standard Venturi tube under actual operating condition

Under actual operating conditions, allowance for friction and discharge conditions is made. This is expressed by Eq. (7.3)

$$Q = C_1K_1A_2\sqrt{2g}\sqrt{h} \quad (7.3)$$

where

Q=actual pipe flow, m³/s

C_1=velocity, friction, or discharge coefficient (dimensionless). The value of C_1 is in the range of 0.97–0.99

K_1=Venturimeter coefficient, dimensionless (Eq. 7.4)

$$K_1 = \frac{1}{\sqrt{1 - \left(\frac{A_2}{A_1}\right)^2}} = \frac{1}{\sqrt{1 - \left(\frac{D_2}{D_1}\right)^4}} \quad (7.4)$$

The value of K_1 lies in the range of 1.0062–1.0328

D_1, D_2=diameter of pipe and throat, m

5. Determine the value of K_1 for the given Venturimeter (beta ratio $= \left(\frac{\text{throat diam}}{\text{pipe diam}}\right) = 0.5$)

$$K_1 = \frac{1}{\sqrt{1 - \left(\frac{0.5D_1}{D_1}\right)^4}} = 1.0328$$

6. Determine the value of C_1

Manufacturers of Venturi tubes, based on experimental tests, provide the value of C_1 in the equipment catalog. Assume that the value of C_1=0.985

7. Develop the calibration equation

The calibration equation is obtained by substituting different values in Eq. (7.3). This is given by Eq. (7.5)

$$Q = 0.985 \times 1.0328 \times \frac{\pi}{4}(0.46\text{ m})^2 \times \sqrt{2 \times 9.81\text{ m/s}^2} \times \sqrt{h} = 0.7489\sqrt{h} \quad (7.5)$$

$$Q = 0.7489\sqrt{h}$$

8. Calculate the discharge for a given h of 1, 2, and 3 m

The discharge values are calculated from Eq. (7.5) and are tabulated below.

h, m	$\sqrt{h}$	Q, m³/s (Eq. 7.5)
1	1	0.7489
2	1.4142	1.0591
3	1.7321	1.2972

Example 7.2: Venturimeter Head Loss and Head-Loss Coefficient

A horizontal Venturimeter is 30 cm×20 cm. The differential gauge is deflected 5.52 cm at a flow of 0.04 m³/s. The specific gravity of the gauge liquid is 1.25. Determine the meter coefficient K_1 and the discharge coefficient C_1. If the head in the Venturimeter is proportional to the differential pressure, calculate the head loss and the head-loss coefficient.

Solution

1. Apply Bernoulli's equation for pipe (D_1, Z_1, P_1, V_1) and throat (D_2, Z_2, P_2, V_2)

$$Z_1+\frac{P_1}{\gamma}+\frac{V_1^2}{2g}=Z_2+\frac{P_2}{\gamma}+\frac{V_2^2}{2g}+h_L$$

2. Rearrange and simplify the equation, $Z_1=Z_2$ and $h_L=\alpha\left(\frac{P_1}{\gamma}-\frac{P_2}{\gamma}\right)$ where α=head loss coefficient

$$\frac{V_2^2}{2g}-\frac{V_1^2}{2g}=\frac{P_1}{\gamma}-\frac{P_2}{\gamma}-h_L$$

$$\frac{V_2^2}{2g}-\frac{V_1^2}{2g}=\left(\frac{P_1}{\gamma}-\frac{P_2}{\gamma}\right)-\alpha\left(\frac{P_1}{\gamma}-\frac{P_2}{\gamma}\right)=(1-\alpha)\left(\frac{P_1}{\gamma}-\frac{P_2}{\gamma}\right)$$

3. Apply the gauge reading h_{guage} in the equation

$$\left(\frac{P_1}{\gamma}-\frac{P_2}{\gamma}\right)=\left(\frac{\gamma_{gauge}}{\gamma}\right)h_{gauge}=(\text{sp.gr})_{gauge}\times h_{gauge}$$

$$\frac{V_2^2}{2g}-\frac{V_1^2}{2g}=(1-\alpha)(\text{sp.gr})_{gauge}\times h_{gauge}$$

$$\frac{1}{2g}\left(\frac{Q}{\frac{\pi}{4}D_2^2}\right)^2-\frac{1}{2g}\left(\frac{Q}{\frac{\pi}{4}D_1^2}\right)^2=(1-\alpha)(\text{sp. gr.})_{gauge}\times h_{gauge}$$

4. Rearrange the equation

$$Q=\sqrt{(1-\alpha)}\left(\frac{1}{\sqrt{1-D_2^4/D_1^4}}\right)^2\left(\frac{\pi}{4}D_2^2\right)\sqrt{2g(\text{sp. gr.})_{guage}\times h_{liquid}}$$

5. Substitute the following values and simplify the equation

$$C_1 = \sqrt{(1-\alpha)}$$

$$K_1 = \frac{1}{\sqrt{1 - D_2^4 / D_1^4}}$$

$$Q = C_1 K_1 \left(\frac{\pi}{4} D_2^2\right) \sqrt{2g(\text{sp. gr.})_{\text{guage}} \times h_{\text{liquid}}}$$

6. Compute K_1

$$K_1 = \frac{1}{\sqrt{1 - D_2^4/D_1^4}} = \frac{1}{\sqrt{1-(20\ \text{cm})^4/(30\ \text{cm})^4}} = 1.116$$

7. Substitute K_1 and solve for C_1

$$0.04\ \text{m}^3/\text{s} = C_1 \times 1.116 \left(\frac{\pi}{4}\left(\frac{20\ \text{cm}}{100\ \text{cm/m}}\right)^2\right)\sqrt{2 \times (9.81\ \text{m/s}^2)(1.25)\left(\frac{5.52\ \text{cm}}{100\ \text{cm/m}}\right)}$$

$$C_1 = 0.98$$

8. Compute head-loss coefficient α

$$C_1 = \sqrt{(1-\alpha)}$$

$$C_1^2 = (1-\alpha) \quad \text{or} \quad (0.98)^2 = (1-\alpha)$$

$$\alpha = (1 - 0.9604) = 0.0396$$

9. Compute h_L

$$h_L = \alpha\left(\frac{P_1}{\gamma} - \frac{P_2}{\gamma}\right)$$

$$= 0.0396(\text{sp. gr.})_{\text{gauge}} \times h_{gauge} = 0.0396(1.25) \times 5.52\ \text{cm} = 0.27\ \text{cm}$$

Example 7.3: Manometer Reading in a Venturimeter

Two pressure points are connected by a manometer that contains mercury (Figure 7.2). The liquid above mercury is water. Determine the pressure at point B in lb/in^2 and feet of water. The pressure at point $A = 180$ psi.

Solution

1. Express pressure as a fluid height
   Eq. (7.6) expresses the pressure as a fluid height

$$P = S\gamma_0 H \tag{7.6}$$

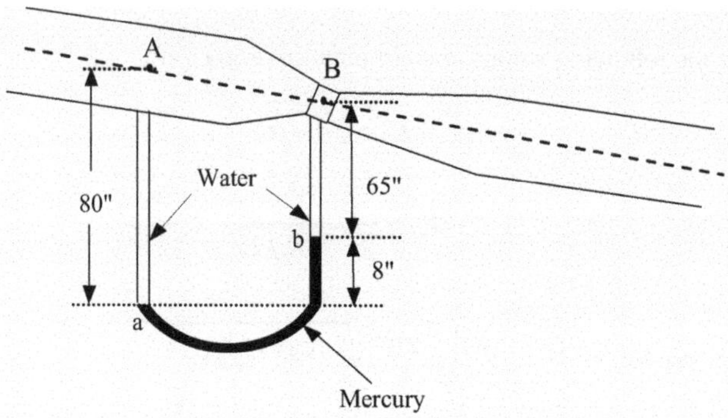

FIGURE 7.2 Manometer in a Venturimeter (Example 7.3).

where

P=pressure, N/m³ (lb/ft²)

S=specific gravity of fluid

γ_0=reference fluid specific weight, kg/m³ (lb/ft³)

H=height of fluid, m (ft)

2. Calculate pressure at points **a** and **b**

$$P = S\gamma_0 H$$

$$= P_A + S\gamma_0 h$$

$$= 180\,\frac{\text{lb}}{\text{in.}^2} + 1.0 \times 62.4\,\frac{\text{lb}}{\text{ft}^3} \times \frac{80\text{ in.}}{1{,}728\text{ in.}^3/\text{ft}^3} = 180\,\frac{\text{lb}}{\text{in.}^2} + 2.9\,\frac{\text{lb}}{\text{in.}^2} = 182.9\,\frac{\text{lb}}{\text{in.}^2}$$

$$P_b = P_a - S\gamma_0 h$$

$$= 182.9\,\frac{\text{lb}}{\text{in.}^2} - 13.5(\text{sp. gr. of mercury}) \times 62.4\,\frac{\text{lb}}{\text{ft}^3} \times \frac{8\text{ in.}}{1{,}728\text{ in.}^3/\text{ft}^3}$$

$$= 182.9\,\frac{\text{lb}}{\text{in.}^2} - 3.9\,\frac{\text{lb}}{\text{in.}^2} = 179.0\,\frac{\text{lb}}{\text{in.}^2}$$

3. Calculate pressure at point B

$$P_B = P_b - S\gamma_0 h$$

$$= 179.0\,\frac{\text{lb}}{\text{in.}^2} - 1.0 \times 62.4\,\frac{\text{lb}}{\text{ft}^3} \times \frac{65\text{ in.}}{1{,}728\text{ in.}^3/\text{ft}^3}$$

$$= 179.0\,\frac{\text{lb}}{\text{in.}^2} - 2.4\,\frac{\text{lb}}{\text{in.}^2} = 176.6\,\frac{\text{lb}}{\text{in.}^2}$$

$$= 176.6\,\frac{\text{lb}}{\text{in.}^2} \times 144\,\frac{\text{in.}^2}{\text{ft}^2} \times \frac{1}{62.4\text{ lb/ft}^3} = 407.5\text{ ft of water}$$

Example 7.4: Venturimeter in an Inclined Pipe

A Venturimeter with a 10-cm throat is installed in a pipe that is inclined upward at an angle of 45° to the horizontal (Figure 7.3). The diameter of the pipe is 20 cm. The distance

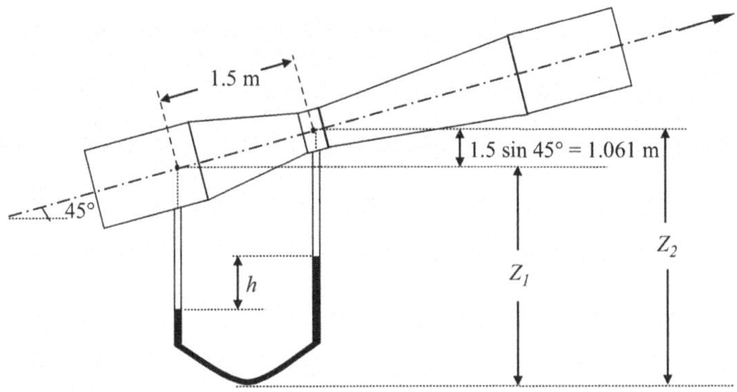

FIGURE 7.3 Venturimeter in an inclined plane (Example 7.4).

between the pressure taps along the pipe is 1.5 m, and the differential pressure is 20 kPa. Estimate the flow of water in the pipe $C_1 = 0.985$. Assume that the head loss is small.

Solution

1. Apply Bernoulli's equation (Eq. 5.6) at the center of pipe (D_1, Z_1, P_1, V_1), and throat (D_2, Z_2, P_2, V_2)

$$Z_1 + \frac{P_1}{\gamma} + \frac{V_1^2}{2g} = Z_2 + \frac{P_2}{\gamma} + \frac{V_2^2}{2g} + h_L\,(\text{small})$$

$$\frac{V_2^2}{2g} - \frac{V_1^2}{2g} = (Z_1 - Z_2) + \left(\frac{P_1}{\gamma} - \frac{P_2}{\gamma}\right)$$

2. Substitute the values of $(Z_1 - Z_2)$ and ΔP (differential pressure)

$$\frac{V_2^2}{2g} - \frac{V_1^2}{2g} = -1.5\ \text{m}\sin 45° + \frac{20\ \text{kN/m}^2}{9.81\ \text{kN/m}^3}$$

$$h = -1.061\ \text{m} + 2.039\ \text{m} = 0.978\ \text{m}$$

3. Calculate Q from Eq. (7.3)

$$Q = C_1 K_1 A_2 \sqrt{2g}\sqrt{h}$$

$$K_1 = \frac{1}{\sqrt{1-\left(\dfrac{D_2}{D_1}\right)^4}} = \frac{1}{\sqrt{1-\left(\dfrac{10}{20}\right)^4}} = 1.0328$$

$$Q = 0.985 \times 1.0328 \times \frac{\pi}{4}\left(\frac{10\ \text{cm}}{100\ \text{cm/m}}\right)^2 \sqrt{2\times 9.81\ \text{m/s}^2} \times \sqrt{0.978\ \text{m}}$$

$$= 0.985 \times 1.0328 \times 7.854 \times 10^{-3}\ \text{m}^2 \times 4.43\,\frac{\sqrt{\text{m}}}{\text{s}} \times 0.989\,\sqrt{\text{m}} = 0.035\ \text{m}^3/\text{s}$$

Example 7.5: Venturimeter Head Loss in an Inclined Pipe

In Example 7.4, the head loss through the Venturimeter is assumed small. Recalculate the flow through the meter if the head loss is 10% of the differential piezometric pressure head (h).

Solution

1. Apply Bernoulli's equation (Eq. 5.6), rearrange, and substitute values

$$\frac{V_2^2}{2g}-\frac{V_1^2}{2g}=(Z_1-Z_2)+\left(\frac{P_1}{\gamma}-\frac{P_2}{\gamma}\right)+h_L$$

$$\frac{V_2^2}{2g}-\frac{V_1^2}{2g}=(Z_1-Z_2)+\frac{\Delta P}{\gamma}+h_L$$

$$\frac{V_2^2}{2g}-\frac{V_1^2}{2g}=(Z_1-Z_2)+\frac{\Delta P}{\gamma}+h_L$$

$$h=-1.5\sin 45^\circ+\frac{20\ \text{kN/m}^2}{9.81\ \text{kN/m}^3}+0.1h$$

$$0.9h=-1.061\ \text{m}+2.039\ \text{m}$$

$$h=1.087\ \text{m}$$

2. Calculate Q

$$Q=C_1K_1A_2\sqrt{2g}\sqrt{h}$$

Substitute the values of K_1 and C_1 from Example 7.4

$$Q=0.985\times 1.0328\times\frac{\pi}{4}(0.1\ \text{m})^2\sqrt{2\times 9.81\ \text{m/s}^2}\sqrt{1.087\ \text{m}}=0.037\ \text{m}^3/\text{s}$$

Example 7.6: Venturimeter Head Loss as a Function of Velocity Head

A Venturimeter has 30-cm throat diameter. The head loss is proportional to the velocity head in the throat. The proportionality constant K=0.14. Calculate the head loss when discharge through the pipe is 0.1 m^3/s.

Solution

1. Calculate the velocity through the throat at 0.1 m^3/s

$$V=\frac{Q}{\frac{\pi}{4}D_2^2}=\frac{0.1\ \text{m}^3/\text{s}}{\frac{\pi}{4}(0.3\ \text{m})^2}=1.42\ \text{m/s}$$

2. Calculate the head loss

$$\text{Head loss } h_L=K\frac{V^2}{2g}=\frac{0.14(1.42\ \text{m/s})^2}{2\times 9.81\ \text{m/s}^2}=0.012\ \text{m}$$

Example 7.7: Venturimeter Head Loss as a Function of Differential

The head loss through a Venturimeter is typically 10% of the differential value. Manufacturers provide the standard differential value for their instrument at a unit flow (1 m^3/s). Calculate the head loss through a 107-cm Venturimeter with 48-cm throat at a flow of 1.314 m^3/s. The manufacturer's value of differential at 1 m^3/s is 1.52 m.

Solution

The head loss through a Venturimeter can be calculated using the manufacturer's standard differential values. Two alternative methods are used: (1) by calculating the differential at the required flow and (2) by calculating the velocity head coefficient at 1 m^3/s.

1. Calculate the head loss using differential at the required flow
   a. Set up the equation for differential head (Eq. 7.7)

$$\frac{h_1}{h_2} = \left(\frac{Q_1}{Q_2}\right)^2 = \left(\frac{V_1}{V_2}\right)^2 \tag{7.7}$$

   where

   Q_1, Q_2 = two flow rates through the meter for two conditions, m^3/s
   V_1, V_2 = velocity through the throat for two conditions, m/s
   h_1, h_2 = differentials corresponding to Q_1 and Q_2, m

   b. Calculate differential h_2 at the flow of 1.314 m^3/s
   Differential h_1 at 1 m^3/s = 1.52 m

$$\frac{1.52 \text{ m}}{h_2} = \left(\frac{1 \text{ m}^3/\text{s}}{1.314 \text{ m}^3/\text{s}}\right)^2 = 0.579$$

$$h_2 = \frac{1.52 \text{ m}}{0.579} = 2.625 \text{ m}$$

   c. Calculate the head loss at Q = 1.314 m^3/s

$$\text{Head loss} = 10\% \text{ of differential at } Q = 1.314 \text{ m}^3/\text{s}$$

$$= 2.625 \text{ m} \times 0.1 = 0.263 \text{ m}$$

2. Calculate the head loss using velocity head coefficient at 1 m^3/s
   a. Calculate the velocity through the throat at a flow of 1 m^3/s $= \dfrac{1.0 \text{ m}^3/\text{s}}{\frac{\pi}{4}(0.48 \text{ m})^2} = 5.526 \text{ m/s}$
   b. Head loss through Venturimeter at a flow of 1 m^3/s = 1.52 m×0.1 = 0.152 m
   c. Calculate velocity head coefficient at 1 m^3/s

$$\text{Velocity head equation is } h_L = K\frac{V^2}{2g}$$

   At 1 m^3/s, the velocity head coefficient is calculated

$$0.152 \text{ m} = K \frac{(5.526 \text{ m})^2}{2 \times 9.81 \text{ m/s}^2}$$

$$K = 0.098$$

d. Calculate head loss at the given flow

$$\text{Velocity through throat at the given flow} = \frac{1.314 \text{ m}^3/\text{s}}{\frac{\pi}{4}(0.48 \text{ m})^2} = 7.261 \text{ m/s}$$

$$\text{Head loss through the meter at the given flow} = \frac{0.098 \text{ m} \times (7.261 \text{ m/s})^2}{2 \times 9.81 \text{ m/s}^2} = 0.263 \text{ m}$$

7.4.2 Orifices

A standard orifice is one in which the water jet leaving the orifice does not again touch the walls of the orifice. The velocity of flow through an orifice is equal to the velocity acquired by a freely falling body in a space corresponding to the head over the orifice (Figure 7.4). The discharge and velocity through an orifice are expressed by Eqs. (7.8a) and (7.8b).

$$Q = C_d AV \tag{7.8a}$$

$$V = \sqrt{2gH} \tag{7.8b}$$

where

Q = discharge, m^3/s (cfs)
A = area of orifices, m^2 (ft^2)
V = velocity, m/s (ft/s)
H = head over the center of the orifice, m (ft)
g = acceleration due to gravity, 9.81 m/s^2 (32.2 ft/s^2)
C_d = orifice constant, 0.6–0.95 depending upon the character of the orifice (sharp to bell mouth)

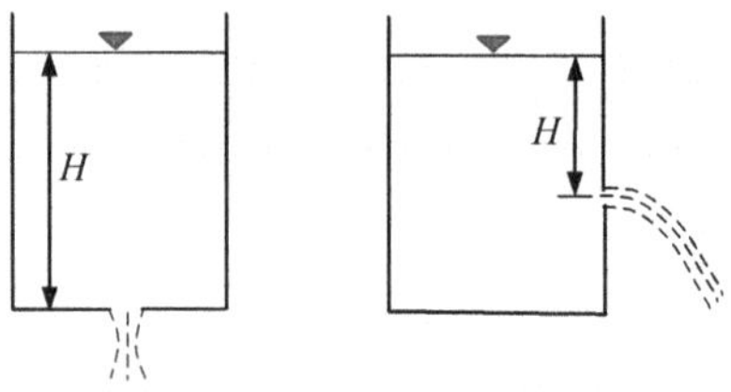

FIGURE 7.4 Flow through an orifice.

Example 7.8: Flow through a Sharp-Crested Orifice

The diameter of a sharp-crested orifice is 4 cm. The constant head over the center of orifice is 3 m. The orifice constant is 0.63. Calculate the discharge.

Solution

1. Calculate the area of the orifice.

$$A = \frac{\pi}{4}d^2 = \frac{\pi}{4}\left(4\text{ cm} \times 10^{-2}\text{ m/cm}\right)^2 = 1.257 \times 10^{-3}\text{ m}^2$$

2. Calculate the discharge velocity.

$$V = \sqrt{2gH} = \sqrt{2 \times 9.81\text{ m/s}^2 \times 3\text{ m}} = 7.62\text{ m/s}$$

3. Calculate the discharge.

$$Q = C_d AV = 0.63 \times 1.257 \times 10^{-3}\text{ m}^2 \times 7.672\text{ m/s} = 6.076 \times 10^{-3}\text{ m}^3\text{/s}$$

Example 7.9: Head Loss through Orifices in an Influent Channel

The influent channel of a sedimentation basin has eight circular orifices of 10-cm diameter each. The arrangement is shown in Figure 7.5. The flow of 0.126 m^3/s enters the channel and is then distributed equally through these orifices along the width of the basin. The coefficient of discharge is 0.85. Calculate the head loss across the orifice.

Solution

1. Calculate the area of each orifice

$$a = \frac{\pi}{4}\left(\frac{10\text{ cm}}{100\,\frac{\text{cm}}{\text{m}}}\right)^2 = 0.00785\text{ m}^2$$

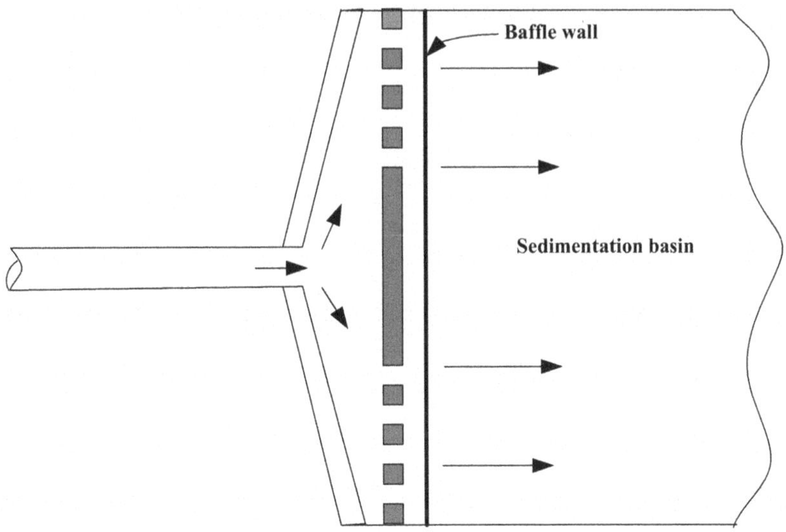

FIGURE 7.5 Influent structure of a sedimentation basin (Example 7.9).

2. Calculate flow per orifice

$$q = \frac{0.126\ \text{m}^3/\text{s}}{8} = 0.0158\ \text{m}^3/\text{s}$$

3. Apply the discharge equation (Eq. 7.8) and calculate head loss across the orifices

$$q = C_d a \sqrt{2gh_L}$$

$$0.0158\ \text{m}^3/\text{s} = 0.85 \times 0.00785\ \text{m}^2 \sqrt{2 \times 9.81\ \text{m/s}^2}\ \sqrt{h_L}$$

$$0.0158\ \text{m}^3/\text{s} = 0.0296\sqrt{h_L}$$

Head loss across the orifice $h_L = 0.28\ m$

7.4.3 Weirs

Weirs are an accurate method of measuring flow, provided the coefficient of discharge is accurately determined. The weirs are rectangular, triangular, and trapezoidal. The weirs may also be sharp-crested or broad-crested. They may be free-falling or submerged. The disadvantages of weirs for wastewater flow measurement are: (1) they encounter significant head loss, (2) solids settle upstream of the weir, and (3) free fall may cause the release of odorous gases. The benefit of a weir is that it maintains a desired elevation in the upstream channel.

Rectangular weir: Rectangular weir has vertical sides. It is located in the middle, on the side, or across the width of an open channel (Figure 7.6).

The most commonly used weir formula is expressed by Eqs. (7.9a)–(7.9c)

$$Q = \frac{2}{3} C_d L' H \sqrt{2gH} \tag{7.9a}$$

$$Q = 2.95 C_d (L - 0.1nH) H^{3/2} \quad \text{(SI units)} \tag{7.9b}$$

$$Q = 5.35 C_d (L - 0.1nH)\ H^{3/2} \quad \text{(U.S. customary units)} \tag{7.9c}$$

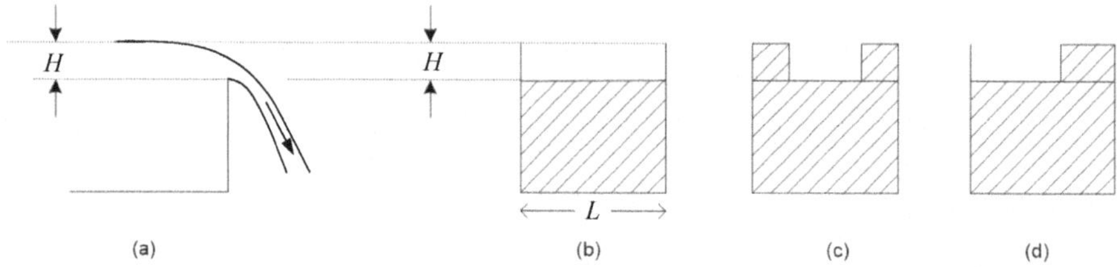

FIGURE 7.6 Rectangular weir: (a) sectional view, (b) across the width of a channel, (c) in the middle, and (d) on the side.

where

Q=flow, m^3/s (cfs)

C_d=weir coefficient. The value of C_d may vary from 0.6 to 0.95, dimensionless

g=acceleration due to gravity, 9.81 m/s^2 (32.2 ft/s^2)

$L'=(L-0.1nH)$

L=length of the weir, m (ft)

n=number of end contractions; $n=0$ for the weir that stretches across the width of the channel (Figure 7.6b), $n=1$ for weir that is flushed on one side of the channel (Figure 7.6d), and $n=2$ for weir that is away from both ends (Figure 7.6c)

H=observed head over the weir corrected to include the effect of velocity of approach, m (ft)

Francis equation: The Francis equation is most commonly used to estimate the flow over a rectangular weir. The Francis equation is based on $C_d=0.622$ and applies to a sharp-crested weir that stretches across the entire width of the channel. The Francis equation is expressed by Eqs. (7.9d) and (7.9e).

$$Q = 1.84LH^{3/2} \quad (\text{SI units}) \tag{7.9d}$$

$$Q = 3.33LH^{3/2} \quad (\text{U.S. customary units}) \tag{7.9e}$$

Triangular weir: Triangular weir or V-notch is used extensively for flow measurement (Figure 7.7). The basic expression for discharge is presented by Eq. (7.10).

$$Q = \frac{8}{15} C_d \sqrt{2g} \tan\frac{\theta}{2} H^{5/2} \tag{7.10}$$

where

Q=discharge, m^3/s (cfs)

H=head over V-notch, m (ft)

θ=angle of the notch, degree

C_d=weir coefficient, 0.6–0.95

Trapezoidal weir: The sides of a trapezoidal weir are inclined (Figure 7.8). The general discharge equations developed for rectangular and triangular weirs may be applied with individually calibrated discharge coefficients. The general equation of the trapezoidal weir is given by Eq. (7.11)

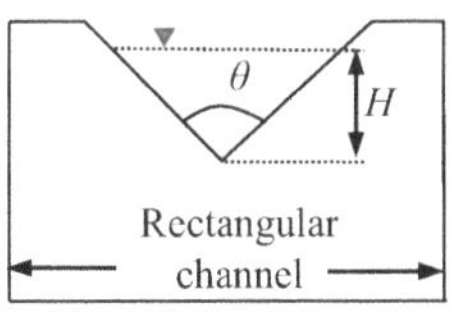

FIGURE 7.7 Triangular weir or V-notch in a rectangular channel.

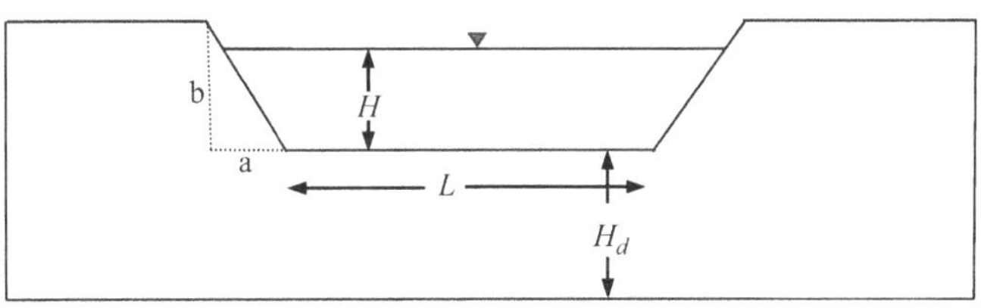

FIGURE 7.8 A trapezoidal weir in a rectangular channel.

$$Q = \frac{2}{3} C_{d1} \sqrt{2g} L H^{3/2} + \frac{8}{15} C_{d2} Z \sqrt{2g} H^{5/2} \tag{7.11}$$

where Z = the ratio of horizontal projection to vertical projection $\left(\frac{a}{b}\right)$

C_{d1} and C_{d2} = discharge coefficients through rectangular and triangular sections. Most commonly, they are assumed equal and expressed by the coefficient C.

The Cipolletti weir has $Z = \frac{1}{4}$, and Eq. (7.11) is reduced to Eq. (7.12)

$$Q = C_w L h^{3/2} \tag{7.12a}$$

$$Q = 1.859 L h^{3/2} \quad \text{(SI units)} \tag{7.12b}$$

$$Q = 3.367 L h^{3/2} \quad \text{(U.S. customary units)} \tag{7.12c}$$

where C_w is a coefficient which, in addition to other factors, accounts for the velocity of approach. The characteristics of the flow pattern affects C_w, which depends upon H and H_d/H. Values of C_w may vary from 3.20 to 4.18 in U.S. customary units. The standard value of C_w in the Cipolletti weir equation is 3.367. At this value of C_w, the coefficient of discharge C_d in Eq. (7.12c) is 0.629[6] (see Example 7.11).

Broad-crested weir: A broad-crested weir is usually built of concrete and has a wide crest. The critical depth occurs on the crest, as shown in Figure 7.9. The exact location of the critical depth depends on the crest profile, roughness, and shape. The discharge and critical velocity are expressed by Eq. (7.13)

$$Q = C_w L \sqrt{g} y_c^{3/2} \quad \text{or} \quad Q = C_w L \sqrt{g y_c}\, y_c \tag{7.13a}$$

$$v_c = \sqrt{g y_c} \tag{7.13b}$$

where

Q = discharge, m^3/s (cfs)
L = length of the weir across the channel width, m (ft)
v_c = critical velocity, m/s (ft/s)
C_w = weir coefficient, dimensionless
y_c = critical depth, m (ft). The discharge is maximum under a given condition

$$\text{when } y_c = \frac{2}{3} H \tag{7.13c}$$

Submerged weirs: A weir is submerged when the water surface in the channel downstream of the weir is higher than the crest. The flow profile is shown in Figure 7.10. The flow from a submerged weir is calculated from Eq. (7.14)

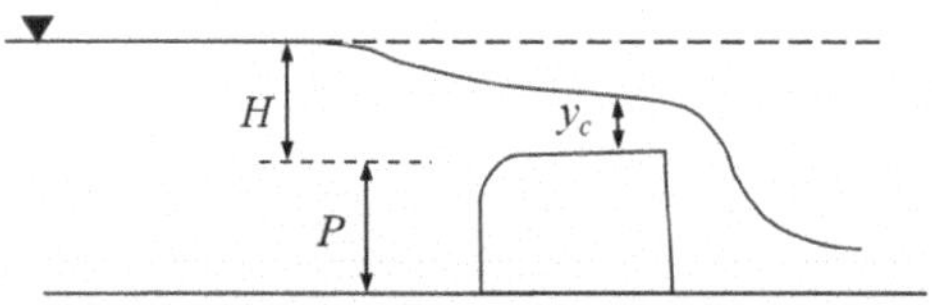

FIGURE 7.9 Water profile over a broad-crested weir.

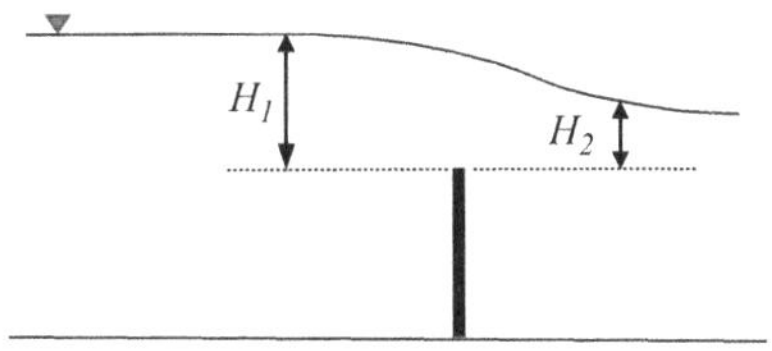

FIGURE 7.10 Flow profile over a submerged weir.

$$\frac{Q}{Q_1} = \left[1 - \left(\frac{H_2}{H_1}\right)^{n_1}\right]^{0.385} \tag{7.14}$$

where

Q, Q_1 = submerged flow discharge, and free-flow discharge, m^3/s (cfs)

H_1, H_2 = upstream and downstream heads, m (ft)

n_1 = exponent in free discharge equation of a particular weir. For example, $n_1 = 3/2$ for rectangular and Cipolletti weirs, and 5/2 for a V-notch

Example 7.10: Head over a Rectangular Weir

A sharp-crested weir is placed in the middle of a rectangular channel that is 2 m wide. The length of the weir is 0.8 m. The coefficient of discharge is 0.85, and the discharge in the channel is 0.25 m^3/s. Calculate the head over the weir.

Solution

1. Select the weir equation
   Use Eq. (7.9b), $n = 2$

   $$Q = 2.95C_d(L - 0.1nH)H^{3/2}$$

2. Calculate the head over the weir

   $$0.25 \text{ m}^3/\text{s} = 2.95 \times 0.85 \times (0.8 - 0.1 \times 2H)H^{3/2}$$

   $$0.0997 = (0.8 - 0.1 \times 2H)H^{3/2}$$

   Use trial and error solution

   $H = 0.261$ m

Example 7.11: Theoretical Discharge Equation of a Rectangular Weir

Develop the theoretical equation for discharge over a rectangular weir. Convert theoretical rectangular weir equation into the Francis equation.

Solution

1. Draw the definition sketch (Figure 7.11)
2. Calculate flow through the differential area

$$\Delta Q = (C_d)(\text{area})(\text{velocity})$$

$$= (C_d)(L)(dh)\left(\sqrt{2gh}\right)$$

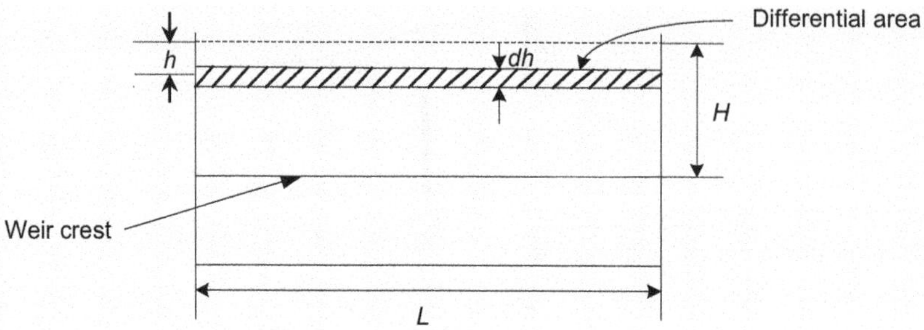

FIGURE 7.11 Rectangular weir (Example 7.11).

3. Develop the equation for flow over the rectangular weir (Eq. (7.9a) with $n=0$)

$$Q = C_d L\sqrt{2g}\int_0^H \sqrt{h}dh$$

$$Q = \frac{2}{3}C_d L\sqrt{2g}H^{3/2} \quad \text{or} \quad Q = \frac{2}{3}C_d L'H\sqrt{2gH} \tag{7.15}$$

This is same as Eq. (7.9a)

4. Convert Eq. (7.9a) (with $n=0$) into the Francis equation

Substitute $C_d=0.622$, $g=9.81$ m/s^2 (32.2 ft/s^2). The Francis equation in SI and U.S. customary units is given below (Eqs. (7.9d) and (7.9e)

$$Q = 1.84LH^{3/2} \quad (\text{SI units}) \quad Q = 3.33LH^{3/2} \quad (\text{U.S. Customary units})$$

Example 7.12: Water Depth in a Channel Upstream of a Weir

The theoretical weir equations (Eqs. 7.9c and 7.9d) are generally used to calculate the discharge over a rectangular weir. A sharp-crested rectangular weir is used to measure the flow in a rectangular channel that is 3 m wide. The weir crest is 2 m above the floor and extends across the channel width. Using theoretical weir equations and the Francis equation, calculate the water depth in the channel upstream of the weir if the discharge in the channel is 12.0 m^3/s. Weir coefficient is 0.622.

Solution

1. Calculate the head over the weir and water depth in the channel using the theoretical weir equation (Eq. 7.9a)

$$n = 0$$

$$12.0 \text{ m}^3/\text{s} = \frac{2}{3}(0.622)\times 3\sqrt{2\times 9.81}H^{3/2}$$

$$H = 1.68 \text{ m}$$

Calculate the depth in the channel upstream of the weir

$$\text{Depth} = 2.0 \text{ m} + 1.68 \text{ m} = 3.68 \text{ m}$$

2. Calculate head over the weir and depth in the channel using the Francis equation (Eq. 7.9e)

$$12.0\ \text{m}^3/\text{s} = 1.84 \times H^{3/2}$$

$$H = 1.68\ \text{m}$$

Calculate the depth in the channel

$$\text{Depth in the channel} = 2.0\ \text{m} + 1.68\ \text{m} = 3.68\ \text{m}$$

Example 7.13: Triangular Weir or V-Notch

An effluent weir plate contains 25, 90° V-notches. Calculate the discharge if the head over the weir notches is 20 cm. Assume C_d=0.6.

Solution

1. Calculate discharge through each V-notch (Eq. 7.12)

$$q = \frac{8}{15} \times 0.6\sqrt{2 \times (9.81\ \text{m/s}^2)}\tan\left(\frac{90°}{2}\right)\left(\frac{20\ \text{cm}}{100\ \text{cm / m}}\right)^{5/2} = 0.0254\ \text{m}^3/\text{s}$$

2. Calculate total flow through the effluent weir plate

$$Q_{\text{total}} = 25 \times 0.0254\ \text{m}^3/\text{s} = 0.64\ \text{m}^3/\text{s}$$

Example 7.14: Head over Cipolletti Weir

The discharge through a Cipolletti weir is 3.5 m³/s, and the head over the weir is 0.5 m. What should be the constructed length of the weir?

Solution

Use Eq. (7.12b) to calculate the weir length

$$3.5\ \text{m}^3/\text{s} = 1.859L(0.5)^{3/2}$$

$$L = 5.32\ \text{m}$$

Example 7.15: Discharge through a Trapezoidal Weir

A trapezoidal weir is a combination of a rectangular weir and a V-notch. The sides are inclined three vertical to one horizontal. The head over the weir is 2 m, and the length of the rectangular portion is 5 m. Calculate the discharge through the trapezoidal weir. Apply (1) V-notch and rectangular weir equations and (2) trapezoidal weir equation (Eq. 7.11). Assume C_d=0.7 for both portions.

Solution

1. Calculate the flow from the rectangular weir and V-notch equations.

   Flow from rectangular portion (Eq. 7.9a), $q_1 = \frac{2}{3}(0.7) \times 5\ \text{m}\sqrt{2g}(2)^{3/2} = 29.23\ \text{m}^3/\text{s}$

   Flow from V-notch (Eq. 7.10), $q_2 = \frac{8}{15} \times 0.7\sqrt{2g}\tan\frac{\theta}{2} H^{5/2}$

   Determine $\frac{\theta}{2}$

$$\tan\frac{\theta}{2} = 0.333$$

$$\frac{\theta}{2} = \tan^{-1}\left(\frac{1}{3}\right) = 18.43$$

$$q_2 = \frac{8}{15} \times 0.7 \times 4.4295 \times 0.333 \times (2)^{5/2}$$

$$= 3.12\ \text{m}^3/\text{s}$$

$$\text{Total } Q = (29.23 + 3.12)\ \text{m}^3/\text{s} = 32.35\ \text{m}^3/\text{s}$$

2. Calculate flow from the trapezoid equation (Eq. 7.11)

$$Q = \frac{2}{3}(0.7)\sqrt{2 \times 9.81} \times 5\ \text{m} \times (2)^{3/2} + \frac{8}{15} \times 0.7 \times \frac{1}{3}\sqrt{2g}(2)^{5/2}$$

$$= 29.23 + 3.12 = 32.35\ \text{m}^3/\text{s}$$

Example 7.16: Comparison of Flows from Trapezoidal and Cipolletti Weir Equations

A trapezoidal channel has side slope of four vertical to one horizontal. The length of the weir crest is 2 m, and head over the weir is 0.5 m. Compare the flows from trapezoidal and Cipolletti equations. Assume C_d=0.629.

Solution

1. Calculate flow from trapezoidal weir (Eq. 7.11)

$$Q = \frac{2}{3} \times 0.629\sqrt{2g} \times 2\ \text{m}\ (0.5\ \text{m})^{3/2} + \frac{8}{15} \times 0.629 \times \frac{1}{4}\sqrt{2g}(0.5)^{5/2}$$

$$= 1.29\ \text{m}^3/\text{s} + 0.06\ \text{m}^3/\text{s} = 1.35\ \text{m}^3/\text{s}$$

2. Calculate the flow from Cipolletti weir (Eq. 7.12b)

$$Q = 1.859 \times 2\ \text{m}(0.5)^{3/2}$$

$$= 1.32\ \text{m}^3/\text{s}$$

Example 7.17: Flow over a Broad Crested Weir

Calculate the discharge over a broad-crested weir. The crest is 2 m long and spans the entire width of the rectangular channel. The critical depth is 0.3 m, and the depth of flow over the weir crest upstream of the weir is 0.80 m, and has a free fall. Assume C_w=0.96.

Solution

The discharge over a broad-crested weir can be calculated from (1) critical depth and (2) critical velocity:

1. Calculate discharge from critical depth
   a. Calculate critical depth from Eq. (7.13c)

$$y_c = \frac{2}{3}H$$

$$= \frac{2}{3} \times 0.8 = 0.53 \text{ m}$$

   b. Calculate the discharge from Eq. (7.13a)

$$Q = 0.96 \times 2 \text{ m} \times \sqrt{9.81 \text{ m/s}^2}\left(0.53 \text{ m}\right)^{3/2}$$

$$= 0.96 \times 2 \text{ m} \times 3.132 \frac{\sqrt{\text{m}}}{\text{s}}\left(0.39 \text{ m}\sqrt{\text{m}}\right) = 2.35 \text{ m}^3/\text{s}$$

2. Calculate the discharge from critical velocity
   a. Calculate critical velocity from Eq. (7.13b)

$$v_c = \sqrt{9.81 \text{ m/s}^2 \times 0.53 \text{ m}} = 2.28 \text{ m/s}$$

   b. Calculate the discharge from Eq. (7.13a)

$$Q = 0.96 \times 2 \text{ m} \times 0.53 \text{ m} \times 2.28 \text{ m/s} = 2.32 \text{ m}^3/\text{s}$$

Example 7.18: Flow over a Submerged Weir

A submerged weir is across the width of a channel. The channel width is 1.5 m. The upstream and downstream heads over the weir are 0.5 and 0.2 m, respectively. Calculate the discharge. C_d=0.8 and n_1=3/2.

Solution

1. Calculate the discharge over a free-flowing weir (Eq. 7.9a)

$$Q_1 = \frac{2}{3}C_d\sqrt{2g}\left(L - 0.1nH_1\right) \times H^{3/2}$$

$$n = 0; \quad L = 1.5 \text{ m}; \quad H_1 = 0.5 \text{ m}$$

$$Q_1 = \frac{2}{3} \times 0.8 \times 1.5 \text{ m} \times \sqrt{2 \times 9.81 \text{ m/s}^2} \times \left(0.5 \text{ m}\right)^{3/2} = 1.252 \text{ m}^3/\text{s}$$

2. Calculate Q/Q_1 from Eq. (7.14)

$$\frac{Q}{Q_1} = \left[1 - \left(\frac{H_2}{H_1}\right)^{n_1}\right]^{0.385}$$

$$\frac{Q}{1.252\ \text{m}^3/\text{s}} = \left[1 - \left(\frac{0.2\ \text{m}}{0.5\ \text{m}}\right)^{3/2}\right]^{0.385} = [0.747]^{0.385} = 0.894$$

3. Calculate flow over submerged weir

$$Q = 0.894 \times 1.252\ \text{m}^3/\text{s} = 1.12\ \text{m}^3/\text{s}$$

7.4.4 Flumes

Flumes are widely used to measure flow in open channels. The geometry of a flume is such that critical depth occurs near the converging section and produces a backwater curve. The depth measured at the converging section is related to the flow. Two types of flumes are commonly used: (1) Parshall flume and (2) Palmer–Bowlus flume.

Parshall flume: Parshall flume consists of a converging section, a throat section, and a diverging section. Critical depth occurs in the throat section H_b, and the depth of the backwater curve H_a is measured in the converging section. The depth H_a can be related to the discharge. The design details and dimensions of standard flume are given in Figure 7.12 and Table 7.3.[7,8]

The discharge equations between gauge reading and throat width of various sizes, and free-flow flume capacity are given in Table 7.4.[7–10] In these equations, discharge Q is in cfs, the throat width W and head H_a are in feet. These equations are developed strictly for U.S. Customary units and for the standard dimensions of the Parshall flume. These equations do not have an equivalent version for the metric system.

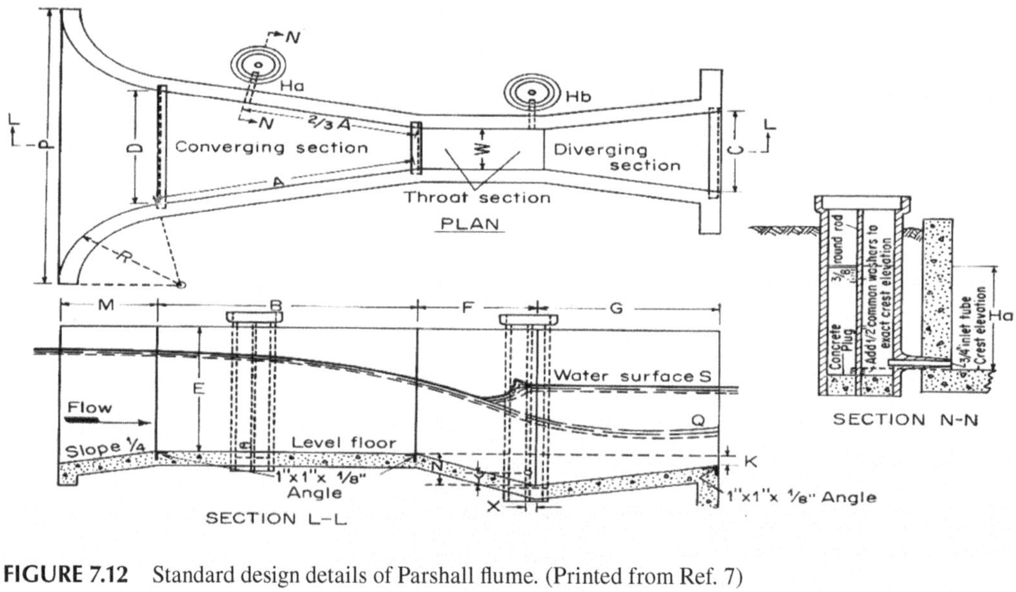

FIGURE 7.12 Standard design details of Parshall flume. (Printed from Ref. 7)

TABLE 7.3
Standard Dimensions of Parshall Flume[7]

Variables and Dimensions	Throat Size W in ft, and Dimensions of Variables in Feet and Inches									
	0.5		1		2		4		6	
	ft	in.	ft	in.	ft	in.	ft	in.	ft	in.
A	2	$\frac{7}{16}$	4	6	5	0	6	0	7	0
2/3 A	1	$4\frac{5}{16}$	3	0	3	4	4	0	4	8
B	2	0	4	$4\frac{7}{8}$	4	$10\frac{7}{8}$	5	$10\frac{5}{8}$	6	$10\frac{3}{8}$
C	1	$3\frac{1}{2}$	2	0	3	0	5	0	7	0
D	1	$3\frac{5}{8}$	2	$9\frac{1}{4}$	3	$11\frac{1}{2}$	6	$4\frac{1}{4}$	8	9
E	2	0	3	0	3	0	3	0	3	0
F	1	0	2	0	2	0	2	0	2	0
G	2	0	3	0	3	0	3	0	3	0
K	0	3	0	3	0	3	0	3	0	3
N	0	$4\frac{1}{4}$	0	9	0	9	0	9	0	9
R	1	4	1	8	1	8	2	0	2	0
M	0	0	1	3	1	3	1	6	1	6
P	2	$11\frac{1}{2}$	4	$10\frac{3}{4}$	6	1	8	$10\frac{3}{4}$	11	$3\frac{1}{2}$
X	0	2	0	2	0	2	0	2	0	2
Y	0	3	0	3	0	3	0	3	0	3
Minimum capacity, ft^3/s	0.05		0.11		0.42		1.3		2.6	
Maximum capacity, ft^3/s	3.9		16.1		33.1		67.9		103.5	

Source: Adapted in part from Refs. 7,8

TABLE 7.4
Discharge Relationship of Parshall Flume for Various Throat Widths

Throat Width	Discharge Equation	Free-Flow Capacity, cfs	Equation No.
3 in.	$Q = 0.992H_a^{1.547}$	0.03–1.9	(7.16)
6 in.	$Q = 2.06H_a^{1.58}$	0.05–3.9	(7.17)
9 in.	$Q = 3.07H_a^{1.53}$	0.09–8.9	(7.18)
1–8 ft	$Q = 4WH_a^{1.522W^{0.026}}$	Up to 140	(7.19)
10–50 ft	$Q = (3.6875W + 2.5)\ H_a^{1.6}$	Up to 2,000	(7.20)

Source: Adapted in part from Refs. 7–10

The ratio of H_b/H_a defines the submergence of the flume. Ideally, a flume should be set so that the degree of submergence does not exceed 0.6 for flumes having a throat less than 9 in. (0.23 m), and 0.7 for flumes with throat greater than 9 in. In all cases, the maximum submergence is 0.95. Above this submergence, the flume will not measure flow accurately. The submergence-correction curves, submergence-discharge, and head-loss curves for different sizes of flumes may be found in many textbooks.[1,3,7–10] The flow rate correction factors for Parshall flume of 1-ft throat width at different submergence ratios may be obtained from Figure 7.13a. The flow correction factors for Parshall flumes of throat width 1–8 ft (0.3–2.5 m) are provided in Table 7.5.

The flow correction procedure for Parshall flumes of throat width 1–8 ft (0.3–2.5 m) is shown in Examples 7.19–7.21. The head-loss curve for flume of different throat widths and percent of submergence at different discharges are shown in Figure 7.13b. The head-loss calculations are illustrated in Example 7.21.

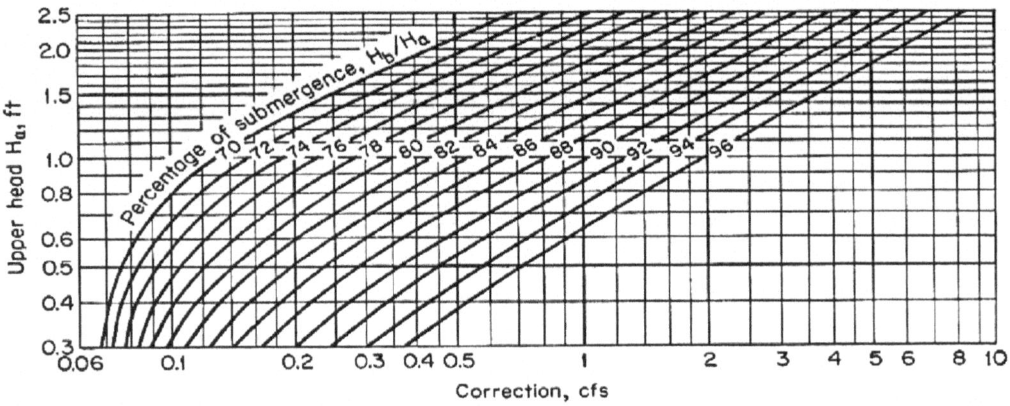

(a)

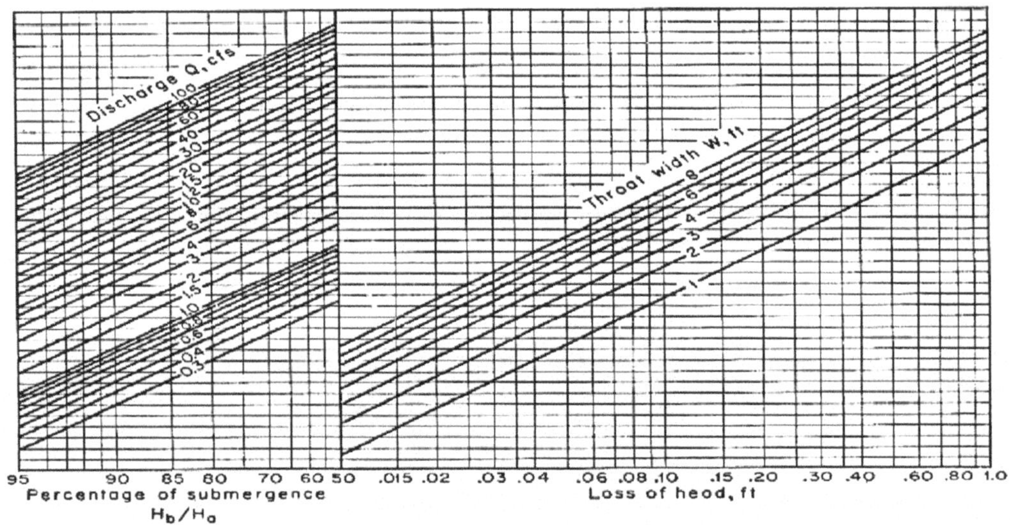

(b)

FIGURE 7.13 Flow rate correction and head loss for submerged Parshall flume: (a) correction factors at different submergence ratios for Parshall flumes of 1–8 ft throat width and (b) head loss through Parshall flume. (Printed from Ref. 7)

TABLE 7.5
Flow Correction Factors for Parshall Flume with a Throat Width of 1–8 ft (0.3–2.5 m)

Flume size	1.0	2.0	3.0	4.0	6.0	8.0
Correction factor	1.0	1.8	2.4	3.1	4.3	5.4

Source: Adapted in part from Refs. 7–10

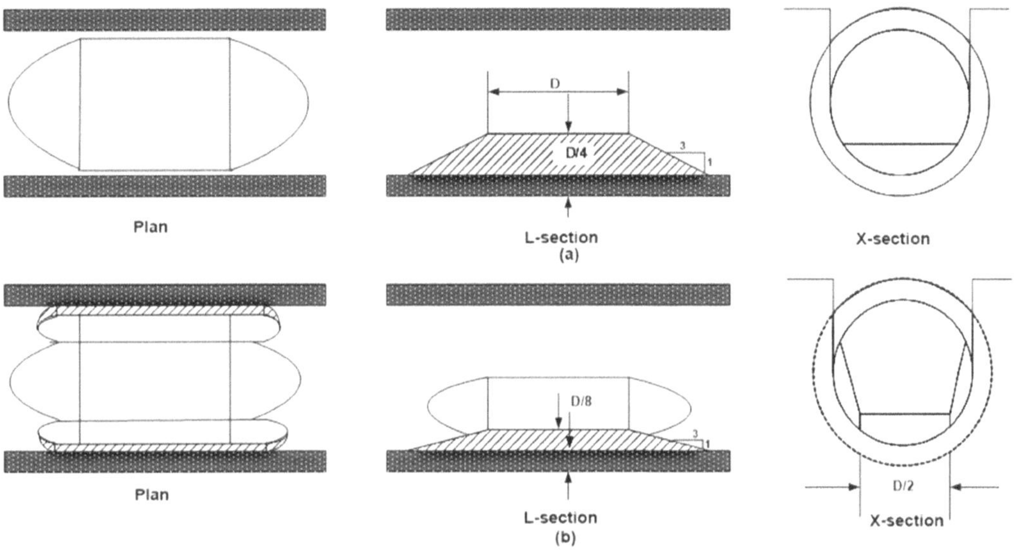

FIGURE 7.14 Plan and sectional views of Palmer–Bowlus flumes: (a) rectangular section and (b) trapezoidal section. (Source: Refs. 1, 11–13)

Palmer–Bowlus flume: The principle of operation of Palmer–Bowlus flume is similar to that of Venturimeter or Parshall flume. The flume is usually placed in an existing channel, sewer, or manhole. The flow is backed up above the control section such that the discharge is related to the upstream depth. Thus, by measuring the upstream depth, the discharge can be read from a calibration curve that is usually supplied with each unit. The head loss through the flume is small, and the flume is self-cleaning. The plan and sectional views of Palmer–Bowlus flume are shown in Figure 7.14. The theory and design of the Palmer–Bowlus flume may be found in Refs.[1,11–13]

Example 7.19: Discharge through a Free-Flowing Parshall Flume

Calculate the discharge through a Parshall flume that has a throat width of 1 ft. The gauge readings H_a and H_b are 1.2 and 0.8 ft, respectively.

Solution

1. Calculate discharge through the flume
   Use Eq. (7.19) for a throat width of 1 ft and $H_a = 1.2$ ft.

$$Q = 4WH_a^{1.522(W)^{0.026}} = 4(1)(1.2)^{1.522} = 5.28 \text{ cfs}$$

2. Check the submergence

$$\frac{H_b}{H_a} = \frac{0.8 \text{ ft}}{1.2 \text{ ft}} = 0.67$$

3. Determine the actual flow

Since the submergence is less than 0.7, no correction is needed. The actual flow is 5.28 cfs (0.150 m^3/s)

Example 7.20: Discharge through a Parshall Flume with Submergence

Calculate the actual flow through a Parshall flume that has a throat width of 1 ft. The upstream gauge reading is 1.5 ft, and the downstream reading H_b is 1.2 ft.

Solution

1. Calculate the discharge through the flume (Eq. 7.19)

$$Q = 4WH_a^{1.522(W)^{0.026}} = 4(1)(1.5)^{1.522} = 7.41 \text{ cfs}$$

2. Calculate submergence

Submergence, $\dfrac{H_b}{H_a} = \dfrac{1.2}{1.5} = 0.8$

Since submergence is greater than 0.7 correction is needed

3. Determine the correction

Using H_a = 1.5 ft, and $\dfrac{H_b}{H_a} = 0.8$, obtain the correction in cfs from Figure 7.13a

Correction = 0.70 cfs

4. Determine the actual flow

The actual flow = 7.41 cfs − 0.70 cfs = 6.71 cfs

Example 7.21: Flow and Head Loss in a Parshall Flume under Submergence

Calculate the actual flow and head loss through a Parshall flume that has a throat width of 2 ft. The submergence is 0.8 and H_a = 1.5 ft.

Solution

1. Calculate the free-flow discharge from a 2 ft flume (Eq. 7.19)

$$Q = 4WH_a^{1.522(W)^{0.026}} = 4(2)1.5^{1.522(2)^{0.026}} = 15.0 \text{ cfs}$$

2. Determine the correction factor

For a 1-ft wide Parshall flume at H_a = 1.5 ft and submergence of 0.8, the correction factor = 0.70 cfs (Figure 7.13a). The correction factor for 2 ft wide flume (Table 7.5) = 1.8 × 0.70 cfs = 1.26 cfs (Figure 7.13a)

3. Determine the actual flow

The actual flow = (15.0 − 1.26) cfs = 13.74 cfs

4. Determine the head loss

The head loss is determined from Figure 7.13b

For submergence = 80%, discharge = 13.74 cfs, W = 2 ft, the head loss = 0.37 ft

Example 7.22: Parshall Flume Design

A Parshall flume is to be installed in a channel that has a flow in the range of 10–25 cfs (0.280–0.708 m^3/s) and has a moderate slope. The depth of flow in the channel, both upstream and downstream of the flume, at 25 cfs, is 3.20 ft (0.98 m). Design the Parshall flume.

Solution

1. Select the Parshall flume
   Flume of several sizes can be selected. For the sake of economy, select the smallest practical size that is applicable to the flow in the expected range. A 4-ft flume will measure the flow in the expected range of 1.3–67.9 ft^3/s (Table 7.3). Select a 4-ft flume with $H_b/H_a = 0.7$
2. Calculate H_a and H_b, at 25 cfs flow (Eq. 7.19)

$$25\text{ cfs} = 4(4)\ H_a^{1.522(4)^{0.026}}$$

$$1.563 = H_a^{1.58}$$

$$H_a = 1.33\text{ ft}$$

$$H_b = 1.33\text{ ft} \times 0.7 = 0.93\text{ ft}$$

3. Determine the elevation of floor at the throat

$$\text{Floor elevation} = 3.20\text{ ft} - 1.33\text{ ft} = 1.87\text{ ft}$$

4. Calculate the head loss
   The head loss in a 4-ft flume carrying 25 cfs is determined from Figure 7.13b

$$\text{Head loss } h_L = 0.49\text{ ft}$$

5. Determine water depth above the floor of throat downstream

$$\text{Water depth} = 1.33\text{ ft} - 0.49\text{ ft}$$

$$= 0.84\text{ ft}$$

6. Determine the water surface level in the downstream channel
   At 70% submergence, the water surface in the throat at H_b gauge is essentially level with the water surface at the lower end of the flume.
7. Determine the bottom level of the channel downstream of the flume
   The water depth in the downstream channel is 3.20 ft
   The bottom of the channel below the bottom of the throat = 3.2 ft − H_b = 3.2 ft − 0.84 ft = 2.36 ft
8. Draw the sketch
   The water surface elevation and the flume and channel bottom elevation are shown in Figure 7.15. All dimensions are selected from Table 7.3.

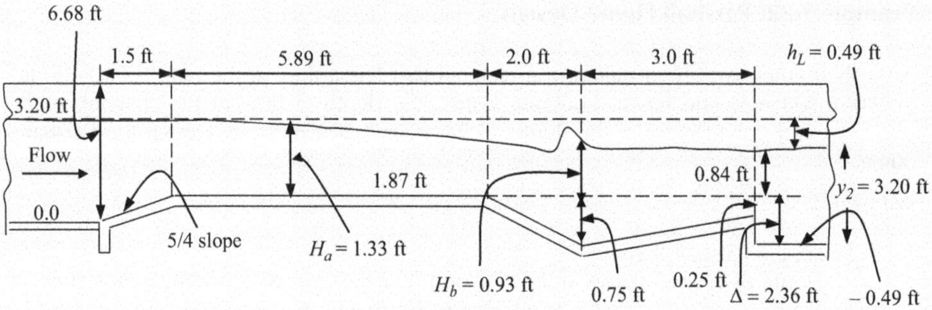

FIGURE 7.15 Design details of Parshall flume (Example 7.22).

Example 7.23: Rating Curve of a Palmer–Bowlus Flume

A Palmer–Bowlus flume is installed in a manhole. The rating curve is developed experimentally and is shown in Figure 7.16. Determine the discharge through the flume. The depth of flow upstream of the flume is 27 cm.

Solution

Determine the discharge through the flume
The upstream depth = 27 cm
The discharge from rating curve (Figure 7.16) = 0.03 m^3/s

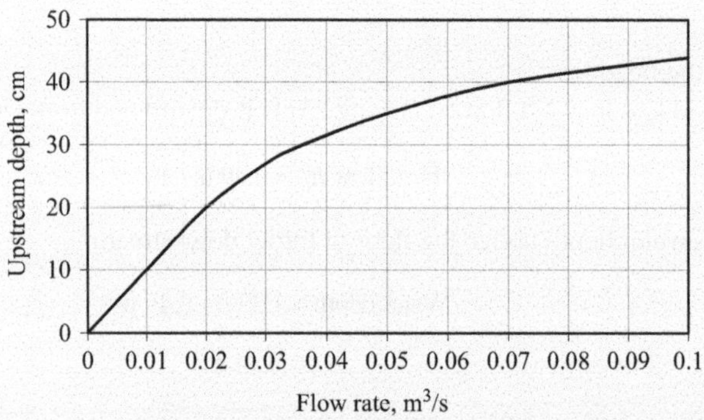

FIGURE 7.16 Palmer–Bowlus flume rating curve (Example 7.23).

7.4.5 Free Discharge from Open-End Pipe

At an elevated outfall, it is possible to estimate the flow falling from a filled or partly filled pipe. The procedure requires measuring the coordinates of the falling stream after it has left the pipe and is freely discharging into the air. Several methods have been developed to utilize this principle. These are horizontal or sloped open-end pipes. The examples are the California pipe method, and open channel flow nozzles such as Kennison nozzle and parabolic flume.

Horizontal or sloped open-end pipe: The flow from a horizontal freely discharging pipe is measured from the coordinates of the falling stream. This method gives an approximate flow. The procedure is given in Example 7.24.

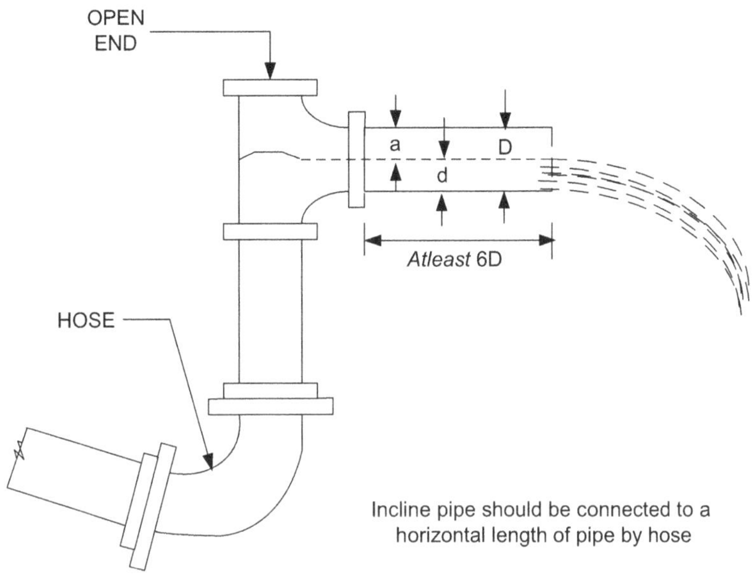

FIGURE 7.17 Installation of a California pipe flow measurement device. (Printed from Ref. 14)

California pipe method: The California pipe method is used to measure the flow in a partly filled horizontal pipe having free discharge. If the pipe is flowing almost full, an air vent should be provided. The horizontal part of the pipe should be at least six times the diameter. If the pipe is not horizontal, a horizontal section should be added (Figure 7.17). The outfall depth in the pipe is related to the critical depth. The flow relationships are given by Eq. (7.21).[14]

$$Q = TW \tag{7.21a}$$

$$T = 3,900\left(1 - \frac{a}{D}\right)^{1.88} \tag{7.21b}$$

$$W = D^{2.48} \tag{7.21c}$$

where

Q = flow, gpm
D = pipe diameter, ft
d = depth of flow, ft
$a = (D - d)$, ft

Open-flow nozzle: For accurate and continuous measurement of flow, two types of open-flow nozzles are commonly used. These are Kennison nozzle or parabolic flume.[14] These nozzles are inserted in a pipe or channel. Each nozzle is rated according to the relationship between the water level in the nozzle and the flow. These devices can register the flow continuously by placing a float in the nozzle, which may be connected to a recorder. The manufacturers provide the calibration curve of these devices. Figure 7.18 shows the details of the Kennison nozzle and the open-flow nozzle with an in-stream transmitter.[14]

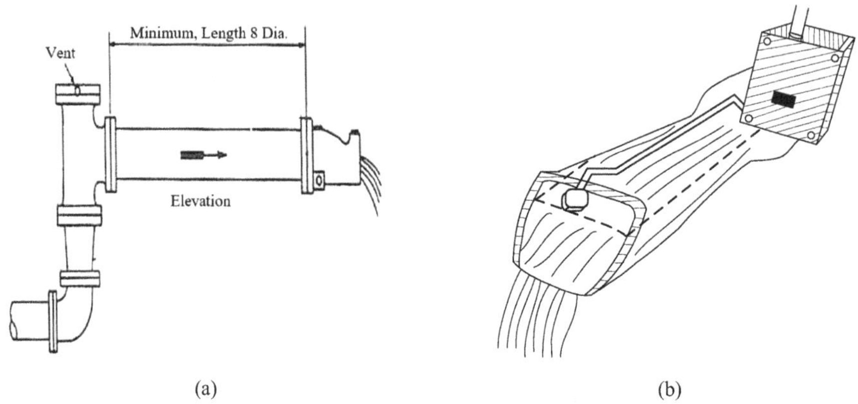

FIGURE 7.18 Open-flow nozzles: (a) Kennison nozzle and (b) open-flow nozzle with in-stream transmitter. (Printed from Ref. 14)

Example 7.24: Discharge from Horizontal Open-End Sewer

A nearly horizontal 12-inch outfall sewer is discharging freely into air. The characteristics of outfall are shown in Figure 7.19. Determine the flow in m^3/s.

Solution

1. Select the flow equation

The most commonly used relationship for flow estimation is expressed by Eq. (7.22)

$$Q = \frac{1{,}800 \times a \times X}{\sqrt{Y}} \tag{7.22}$$

where

Q = discharge, gallons/min

a = wetted cross-sectional area of the pipe section, ft^2

b = vertical distance from sewer invert to the intersection of falling stream with X coordinate, ft

X = distance between end of the pipe and the center of falling stream measured parallel to the length of pipe, ft

Y = vertical distance from water surface at the depth of flow in the pipe and the intersection of falling stream with X coordinate, ft

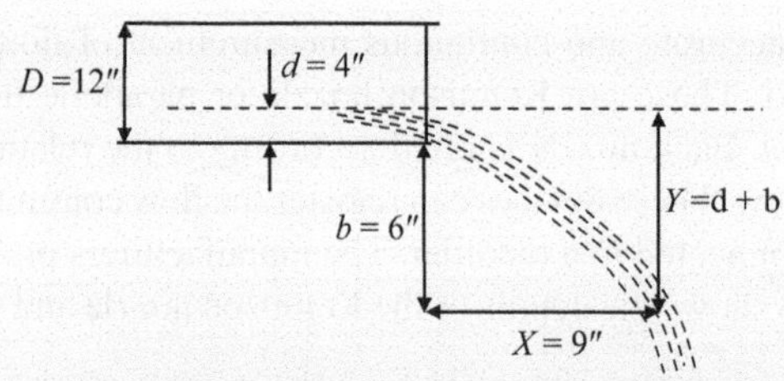

FIGURE 7.19 Free-falling discharge from an outfall pipe (Example 7.24).

2. Calculate the wetted area
   $d = 4$ in., $D = 12$ in., $d/D = 4$ in./12 in. $= 0.333$
   Use Eq. (5.33) to calculate θ

$$\cos\frac{1}{2}\theta = \left(1 - \frac{2d}{D}\right) = (1 - (2 \times 0.333)) = 0.334$$

$$\theta = 2 \times 70.5 = 141°$$

3. Calculate the partial flow area of the circular pipe
   The geometric elements equations of circular pipe are given in Table 5.3

$$a = \frac{1}{8}\left(\frac{\pi\theta}{180} - \sin\theta\right)D^2$$

$$= \frac{1}{8}\left(\frac{\pi 141}{180} - \sin 141°\right)\left(\frac{12 \text{ in.}}{12 \text{ in./ft}}\right)^2$$

$$= \frac{1}{8}(2.461 - 0.629)1 \text{ ft}^2$$

$$= 0.229 \text{ ft}^2$$

4. Calculate discharge

$$X = 9 \text{ in.} = 0.75 \text{ ft}$$

$$b = \frac{6 \text{ in.}}{12 \text{ in./ft}} = 0.5 \text{ ft}$$

$$Y = (d + b) = \frac{4 \text{ in.}}{12 \text{ in./ft}} + 0.5 \text{ ft} = 0.833 \text{ ft}$$

$$Q = \frac{1{,}800 \times a \times X}{\sqrt{Y}} = \frac{1{,}800 \times 0.229 \text{ ft} \times 0.75 \text{ ft}}{\sqrt{0.833 \text{ ft}}} = 339 \text{ gpm}$$

$$Q = \frac{339 \text{ gallon}}{\text{min}} \times \frac{3.8 \text{ L}}{\text{gallon}} \times \frac{10^{-3} \text{ m}^3/\text{L}}{60 \text{ s/min}} = 0.022 \text{ m}^3/\text{s}$$

Example 7.25: Discharge Measurement from a California Pipe

Calculate the flow in a freely discharging 8-in sewer line. Use California pipe method. Depth of flow $d = 3$ in.

Solution

1. Calculate T from Eq. (7.21a)

$$a = D - d$$

$$= (8 \text{ in.} - 3 \text{ in.}) = 5 \text{ in.}$$

$$\frac{a}{D} = \frac{5 \text{ in.}}{8 \text{ in.}} = 0.625$$

$$T = 3{,}900(1 - 0.625)^{1.88} = 616.94$$

2. Calculate W from Eq. (7.21c)

$$W = D^{2.48} = \left(\frac{8 \text{ in.}}{12 \text{ in.}}\right)^{2.48} = 0.366$$

3. Calculate the discharge from Eq. (7.21a)

$$Q = T \times W$$

$$= 616.94 \times 0.366 = 226 \text{ gpm}\left(0.0143 \text{ m}^3/\text{s}, 0.503 \text{ cfs}\right)$$

Note: 1 cfs = 448.9 gpm, and 1 m^3/s = 1,580 gpm

Example 7.26: Discharge Measurement from an Open-Flow Nozzle

A standard open-flow nozzle is inserted in a channel that is freely discharging. The manufacturer's rating curve is given in Figure 7.20. Determine the discharge through the channel if the in-stream transmitter reading is 30 cm.

Solution

Determine the discharge through the channel
The channel in-stream transmitter reading = 30 cm
Discharge through the channel = 0.1 m^3/s

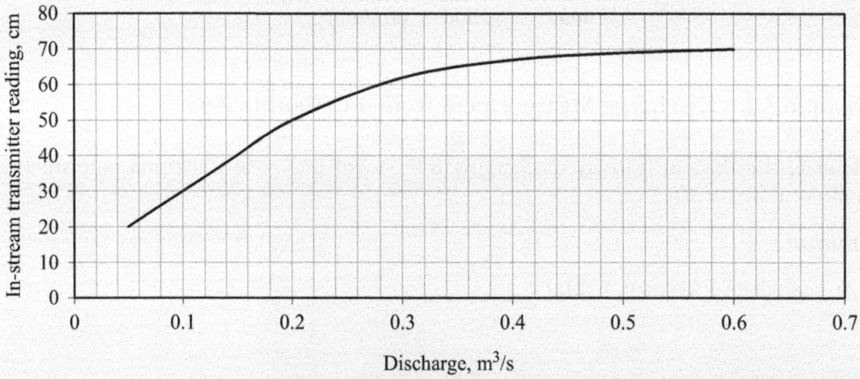

FIGURE 7.20 Manufacturer's rating curve of the open-flow nozzle.

7.4.6 Depth of Flow Measurement

The depth measurement in a channel is frequently used to calculate the flow. The slope of the channel and the coefficient of roughness must be known to calculate the flow. The surface elevation in a sewer, manhole, or a channel is measured by a float, bubbler tube, or by ultrasonic liquid indicator.[14,15] Installation of these devices is shown in Figure 7.21. A rating curve can be developed from the surface elevation and the channel discharge.

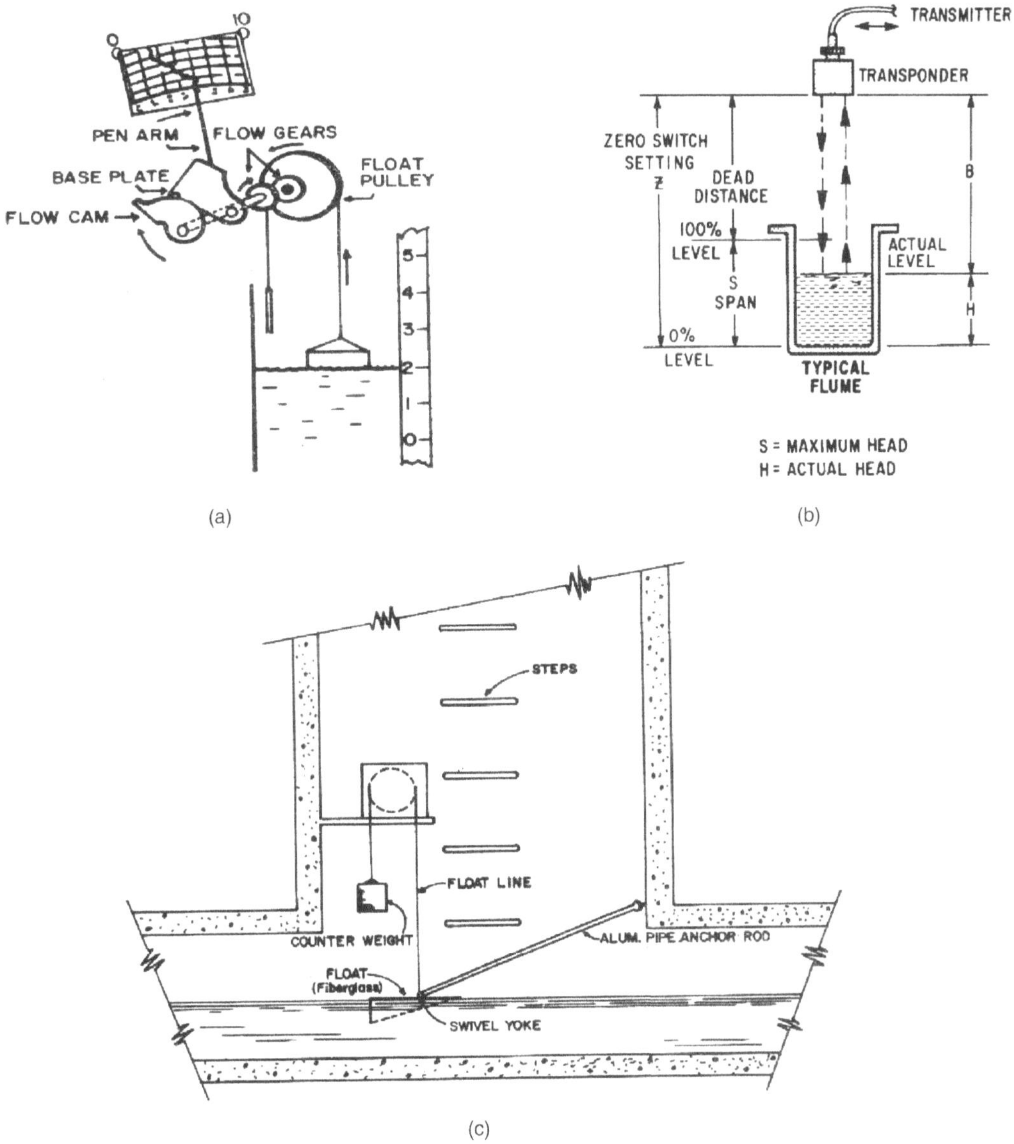

FIGURE 7.21 Devices used for measurement of depth and surface elevations in sewers and open channels: (a) float activated water surface elevation measuring device, (b) ultrasonic open channel liquid level indicator, and (c) recorder and scow float used in a sewer manhole. (Printed from Ref. 14)

Example 7.27: Flow Calculations Based on Depth of Flow in a Sewer Line

The measured depth in a 91 cm sewer is 42 cm. The sewer slope is 0.00046 m/m, and Manning's coefficient *n* is 0.013. Determine the flow.

Solution

1. Calculate the central angle θ under partial flow

$$\frac{d}{D} = \frac{42\text{ cm}}{91\text{ cm}} = 0.46$$

Use Eq. (5.33) to calculate θ

$$\cos\frac{1}{2}\theta = \left(1 - \frac{2d}{D}\right) = 0.08$$

$$\frac{1}{2}\theta = \cos^{-1}(0.08) = 85.4°$$

$$\theta = 170.8°$$

2. Determine the hydraulic elements *p*, *a*, and *r* of circular pipe.
   The equations of hydraulic elements of circular sewers under partial flow are given in Table 5.3.

$$a = \frac{1}{8}\left(\frac{\pi\theta}{180} - \sin\theta\right)D^2$$

$$= \frac{1}{8}\left(\frac{\pi \times 170.8}{180} - \sin 170.8°\right)(0.91\text{ cm})^2$$

$$= \frac{1}{8}(2.981 - 0.160)(0.91\text{ cm})^2$$

$$= 0.292\text{ m}^2$$

$$p = \frac{\pi D\theta}{360} = \frac{\pi \times 0.91\text{ m} \times 170.8}{360} = 1.36\text{ m}$$

$$r = \frac{a}{p} = \frac{0.292\text{ m}^2}{1.36\text{ m}} = 0.214$$

3. Calculate velocity from Manning's equation (Eq. 5.34)

$$v = \frac{1}{n}r^{2/3}S^{1/2} = \frac{1}{0.013}(0.215)^{2/3}(0.00046)^{1/2} = 0.59\text{ m/s}$$

4. Calculate discharge

$$Q = av = 0.292\text{ m}^2 \times 0.59\text{ m/s} = 0.172\text{ m}^3\text{/s}$$

TABLE 7.6
Ratios of Hydraulic Elements of Circular Sewer Flowing Partially Full (Example 7.28)

$\frac{d}{D}$	$\frac{a}{A}$	$\frac{q}{Q}$	$\frac{v}{V}$
0.1	0.05	0.02	0.33
0.2	0.14	0.08	0.56
0.3	0.25	0.18	0.74
0.4	0.37	0.33	0.87
0.5	0.50	0.44	0.98
0.6	0.63	0.67	1.07
0.7	0.75	0.84	1.13
0.8	0.86	0.98	1.16
0.9	0.95	1.07	1.13
1.0	1.0	1.0	1.0

Note: The lower- and uppercase notations are, respectively, for partially and full flow conditions.

Example 7.28: Hydraulic Elements of a Partially Flowing Sewer

Tabulate the ratios of hydraulic elements of a circular sewer flowing partially full.

Solution

The hydraulic elements of a circular sewer are given in Figure 5.26. The ratios of hydraulic elements for sewer flowing partially full are calculated from Eq. (5.33) and the hydraulic element equations given in Table 5.3. The procedure to calculate the hydraulic elements of circular sewer is shown in Example 7.27. The ratios *d/D*, *a/A*, *q/Q*, and *v/V* are summarized in Table 7.6.

Example 7.29: Calibration or Rating Curve of Depth of Flow with Respect to Sewer Discharge

An intercepting sewer is 76 cm in diameter and has a slope of 0.002 m/m. Develop the calibration or rating curve between the depth of flow and discharge. Assume $n=0.015$. Use the ratios of *d/D* and *q/Q* developed in Table 7.6.

Solution

1. Calculate the flow velocity when sewer is flowing full.
   Apply Manning's equation (Eq. 5.34)

$$V = \frac{1}{n} \times R^{2/3} \times S^{1/2}$$

$$V = \frac{1}{0.015} \times \left(\frac{0.76}{4}\right)^{2/3} \times (0.002)^{1/2} = 0.985 \text{ m/s}$$

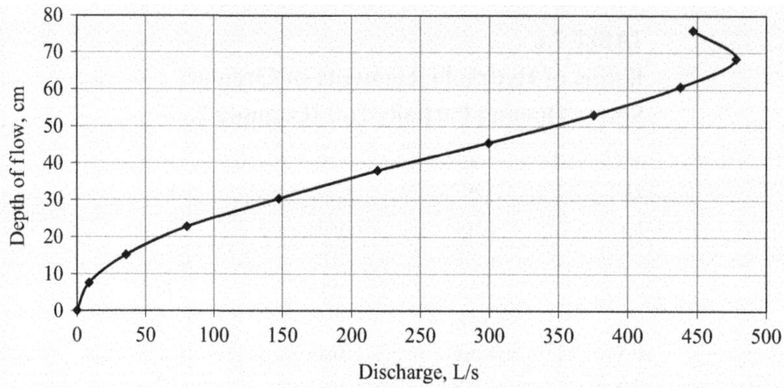

FIGURE 7.22 Calibration or rating curve of 76 cm diameter interceptor at a slope of 0.002 m/m (Example 7.28).

2. Calculate discharge when sewer is flowing full.

$$A = \frac{\pi}{4} \times (0.76)^2 = 0.454 \text{ m}^2$$

$$Q = V \times A = 0.985 \text{ m/s} \times 0.454 \text{ m}^2$$

$$= 0.447 \text{ m}^3/\text{s or } 447 \text{ L/s}$$

3. Develop the partial flow relationship from Table 7.6 with respect to depth of flow.
   The ratios of d/D and q/Q are given in Table 7.6. These ratios are used to calculate d and q. The calculated values of d and q for different assumed ratios are summarized below.

$\frac{d}{D}$	0.1	0.2	0.3	0.4	0.5	0.6	0.7	0.8	0.9	1.0
d, cm	7.6[a]	15.2	22.8	30.4	38.0	45.6	53.2	60.8	68.4	76.0
$\frac{q}{Q}$	0.02	0.08	0.18	0.33	0.49	0.67	0.84	0.98	1.07	1.00
q, L/s	8.9[b]	35.8	80.5	147.3	219.0	299.5	375.5	438.1	478.3	447.0

[a] 0.1×76 cm=7.6 cm.
[b] 0.02×447 L/s=8.9 L/s

4. Draw the sewer rating or calibration curve
   The discharge versus depth of flow relationship for 76 cm intercepting sewer at a slope of 0.002 m/m is shown in Figure 7.22

7.4.7 Dilution Method

Chemical or radioactive tracers are often used to determine the flow in a river, stream, or sewer system. A constant rate of a chemical or a radioactive tracer is continuously added to the stream and its concentration is measured at a distance downstream. The tracer should be conservative (i.e., nonreactive and does not decay with time). A mass balance analysis is performed. Eq. (7.23) is used to calculate the flow in the stream.[1,2,14]

$$Q_s = \frac{Q_t(C_t - C)}{(C - C_0)} \quad (7.23)$$

where

Q_s = stream discharge, m³/s
Q_t = tracer discharge, m³/s
C_t = concentration of tracer in tracer feed, mg/L
C = concentration of tracer in stream including background concentration, mg/L
C_0 = background concentration of tracer in the stream

Example 7.30: Flow Measurement by a Tracer

A salt solution (NaCl) was continuously discharged into a turbulent creek. The concentrations of salt and feed rate of the salt in the tracer solution were 10,000 mg/L and 150 L/s, respectively. At a distance of 200 m downstream, an equilibrium concentration of 205 mg/L of the salt was reached. The background salt concentration in the stream was 21 mg/L. Determine the flow in the creek.

Solution

Calculate the flow in the creek.
Flow rate of the tracer = 150 L/s × 10^{-3} m³/L = 0.15 m³/s
Discharge in the creek from Eq. (7.23)

$$Q_s = \frac{Q_t(C_t - C)}{(C - C_0)} = \frac{0.15\ \text{m}^3/\text{s}\ (10{,}000\ \text{mg/L} - 205\ \text{mg/L})}{(205\ \text{mg/L} - 21\ \text{mg/L})}$$

$$= \frac{0.15\ \text{m}^3/\text{s} \times 9{,}795\ \text{mg/L}}{184\ \text{mg/L}} = 7.99\ \text{m}^3/\text{s}$$

7.5 REVIEW PROBLEMS AND DISCUSSION TOPICS

7.1 A Venturimeter is installed in a force main that is pumping raw wastewater. The diameter of the force main is 82 cm (32 in.). The tube beta ratio is 0.4 and C_d = 0.985. The force main and Venturimeter are horizontal. Develop the calibration equation and also calculate flows when differential pressure heads are 0.5, 1.0, and 3.0 m.

7.2 A horizontal Venturimeter 35 cm × 25 cm is installed to measure the discharge in a water transmission line. The differential gauge reading at a flow of 0.066 m³/s is 5.8 cm. The specific gravity of the gauge liquid is 1.25. Determine the meter coefficient K_1 and the discharge coefficient C_1. Also, calculate the head loss and the head-loss coefficient if the head loss in the Venturimeter is proportional to the differential pressure.

7.3 Two pressure points in a Venturimeter are connected by a manometer that contains mercury. The liquid above mercury is water, and the pressure at point A is 200 psi. Determine the pressure at point B in lb/in².

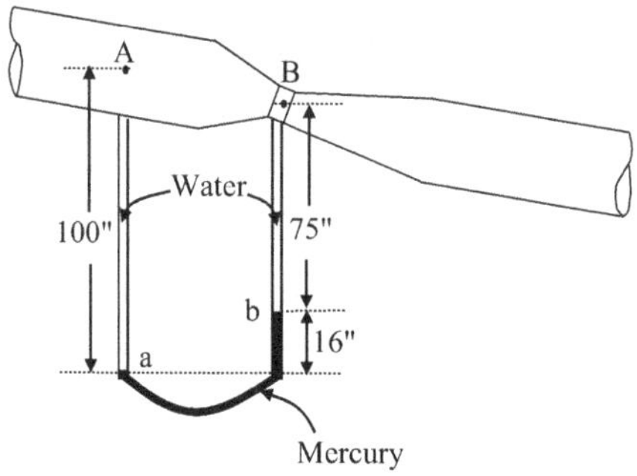

7.4 A Venturimeter is inclined upward at an angle of 45° to the horizontal. The diameters of the pipe and throat are 30 and 15 cm, respectively. The distance between the pressure taps along the pipe is 1.75 m. Assume that the head loss in the Venturimeter is small and the differential pressure is 25 kPa. Determine the discharge in the pipe. Also, recalculate the discharge if the head loss in the Venturimeter is 15% of differential pressure head. $C_1=0.985$.

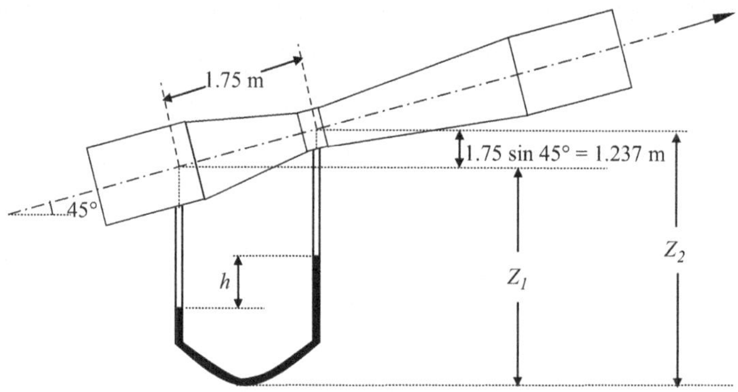

7.5 The throat diameter of a Venturimeter is 25 cm. The discharge through the pipe is $0.15\,m^3/s$, and head loss is proportional to the velocity head. Compute the head loss. $K=0.14$.

7.6 A Venturimeter of 54-cm throat is installed in a 120 cm diameter common header. The manufacturer's standard value of differential at $1\,m^3/s$ is 1.84 m, and head loss is 10% of the differential. Calculate the head loss at a flow of $1.523\,m^3/s$ by two alternative methods: (1) from the differential at the required flow and (2) from the velocity head coefficient at $1\,m^3/s$.

7.7 A sharp-crested orifice is installed in a tank. The diameter of the orifice is 8 cm, and head over the orifice is 500 cm. Calculate the discharge from the orifice. $C_d=0.63$.

7.8 A straight weir is placed at the center of a rectangular channel. The channel is 2-m wide and weir length is 60 cm. Calculate the flow in the channel if the head over the weir is 0.314 m. $C_d=0.80$.

7.9 An effluent weir plate contains 20, 90° V-notches. Calculate the effluent flow if the head over each notch is 0.15 m. $C_d=0.75$.

7.10 A trapezoidal weir has a side slope of 4 vertical to 1 horizontal. The bottom width of the weir is 19.0 ft. The head over the weir is 1.5 ft. Calculate the flow by using (1) Trapezoidal weir equation, $C_d=0.62$, and (2) Cipolletti weir equation.

7.11 The sides of a trapezoidal weir are inclined at 3 vertical to 1 horizontal. The length of the rectangular portion is 6.5 m, and the head over weir is 2.22 m. Develop the equation for the flow using (1) V-notch and rectangular weir equations, and (2) Trapezoidal equation. Assume C_d of 0.85 in both rectangular and V-notch sections. The variables are shown below.

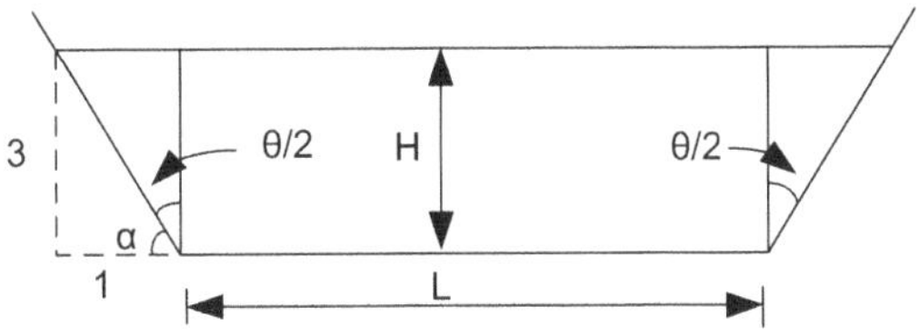

7.12 A broad-crested weir has a 2-m long crest across the width of a rectangular channel. The discharge over the weir is 2.8 m³/s. Calculate the critical depth and critical velocity. C_w=0.96.

7.13 The discharge of a submerged weir is 1.314 m³/s. The channel width and upstream head are 1.62 and 0.68 m, respectively. Calculate (1) free discharge, and (b) downstream head. C_d=0.85 and n_1=3/2.

7.14 The upstream and downstream gauge readings of a Parshall flume are 2.0 and 1.56 ft, respectively. The throat width is 2.0 ft. Calculate the actual discharge under submergence.

7.15 Calculate the actual flow and head loss through a Parshall flume that has a throat width of 1.5 ft. The submergence is 0.7 and H_a=1.5 ft.

7.16 The California pipe method is used to calculate the free discharge from a partially flowing pipe. A 16-in diameter outfall sewer is discharging freely. Determine the free discharge by the California pipe method. a=8 in.

7.17 A sanitary sewer is serving a subdivision. The diameter and slope of the sewer are 87 cm and 0.00046 m/m, respectively. Calculate the flow if the depth of flow is 35 cm. n=0.013.

7.18 An intercepting sewer is 92 cm in diameter and has a slope of 0.001 m/m. Develop the calibration curve between the depth of flow and discharge. Assume n=0.013.

7.19 Dilution method is used to determine the flow in a turbulent creek. A salt tracer solution (NaCl) was continuously discharged into the creek. The concentration and feed rate of the salt solution were 12,000 mg/L and 175 L/s, respectively. At a distance of 230 m downstream, an equilibrium concentration of 275 mg/L of salt was reached. The background salt concentration in the stream was 19 mg/L. Determine the flow in the creek.

REFERENCES

1. Metcalf and Eddy, Inc., *Wastewater Engineering: Collection and Pumping of Wastewater*, McGraw-Hill Book Co., New York, 1981.
2. Metcalf and Eddy, Inc., *Wastewater Engineering: Treatment, and Reuse*, 4th ed., McGraw-Hill Inc., New York, 2003.
3. Qasim, S. R., *Wastewater Treatment Plants Planning, Design and Operation*, 2nd ed., CRC Press, Boca Raton, FL, 1999.
4. Qasim, S., E. M. Motley, and G. Zhu, *Water Works Engineering: Planning, Design and Operation*, Prentice-Hall, PTR, Upper Saddle River, NJ, 2000.
5. Badger Meter Inc., *Badger Venturi Tube Fabricated Flow Element*, No. 941081, Tulsa, OK, February 1997.
6. Linsley, R. K., J. B. Franzini, D. Freyberg, and G. Tchobanoglous. *Water Resources Engineering*, 4th ed., McGraw-Hill Book Co., New York, 1992.
7. U.S. Department of the Interior, Bureau of Reclamation, *Water Measurement Manual*, U.S. Government Printing Office, Washington, DC, Stock No. 2403–00086, 1974.

8. Parshall, R. L., and C. Rohwer, *The Venturi Flume*, Colorado Agricultural Experimental Station Bulletin No. 265, 1921.
9. Parshall, R. L., The Imported Venturi Flume, *Transactions of the American Society of Civil* Engineers, vol. 89, 1926, pp. 841–851.
10. Chow, V. T., *Open-Channel Hydraulics*, McGraw-Hill Book Co., New York, 1959.
11. Ludwig, J. H., and R. G. Ludwig, Design of Palmer-Bowlus Flume, *Sewage and Industrial Wastes*, vol. 23, 1951, p. 1096.
12. Palmer, H. K., and F. D. Bowlus, *Adaptation of Venturi Flume to Flow Measurements in Conduits, Transactions of American Society of Civil Engineers*, vol. 101, 1936, p. 1195.
13. Joint Committee of the Water Pollution Control Federation and American Society of Civil Engineers, *Design and Construction of the Sanitary and Storm Sewers*, WPCF Manual of Practice No. 9, Water Pollution Control Federation, Alexandria, VA, 1970.
14. U.S. Environment Protection Agency, *Handbook for Monitoring Industrial Wastewater*, Technology Transfer, August 1973.
15. Shelley, P. E., and G. A. Kirkpatrick, *Sewer Flow Measurement: A State of the Art Assessment*, U.S. Environmental Protection Agency, EPA-600/2-75-027, November 1975.

8 Sanitary and Storm Sewer Design

8.1 CHAPTER OBJECTIVES

Sewers are underground conduits for conveying the wastewater from the point of origin to a treatment plant or to a point of disposal. In this chapter, the theory, design, and construction procedures of sanitary and storm sewers are provided. The specific objectives of this chapter are to present

- Sanitary sewerage system
- Hydraulics of full and partially full conduits
- Theory and design of sanitary sewers
- Theory and design of storm sewers
- Construction of community sewerage system

8.2 SANITARY SEWERAGE SYSTEM

8.2.1 Types of Municipal Sewerage System

There are three types of sewers: sanitary, storm, and combined. Sanitary sewers are designed to carry wastewater from residential, commercial, and industrial areas, and a certain amount of infiltration/inflow. Storm sewers are exclusively designed to carry the surface runoff. Combined sewers carry both the sanitary and storm flows. These days, combined sewers are seldom designed and constructed in the United States.[1,2]

8.2.2 Sewer Size and Distribution

A community sewerage system consists of (1) building sewers (also called building connections), (2) laterals or branch sewers, (3) mains, collectors, or sub-collectors, (4) trunk sewers, and (5) intercepting sewers.[2,3] All sewers except the building sewers are constructed in streets, special easements, or available rights-of-way. They follow the natural ground slope. The recommended minimum sewer size is 20 cm (8 in.), although many states allow 15 cm (6 in.) lateral sewers. Each type of sewer is briefly described in Table 8.1. The total average length of the sewerage system is 4.3–11.5 m (14–38 ft) per capita. The typical distribution of pipe sizes is summarized in Table 8.2.[2,4] A typical layout of a municipal sewerage system is shown in Figure 8.1.

8.2.3 Appurtenances

Sewer appurtenances include manholes, building connections, junction chambers or boxes, and terminal cleanouts. General information about these appurtenances is summarized in Table 8.3.

DOI: 10.1201/9781003629399-8

TABLE 8.1
Types of Sewers in a Typical Collection System

Type of Sewer	Purpose
Building	Building sewers connect to the building plumbing and convey wastewater from the building to the lateral or branch sewer. Building sewers begin outside the building and connect to the line in the street or special easements. Building sewers are 10–15 cm (4–6 in.) in diameter and constructed on a slope of 0.02 m/m.
Lateral or branch	Lateral sewers form the first element of a central wastewater collection system, and are usually in streets or special easements. They are used to collect wastewater from one or more building sewers and convey it to a main sewer.
Main	Main sewers are used to convey wastewater from one or more lateral sewers to trunk sewers, or to intercepting sewers.
Trunk	Trunk sewers are large sewers that are used to convey wastewater from main sewers to treatment plant, or to a large intercepting sewer.
Intercepting	Intercepting sewers are larger sewers that are used to intercept a number of main or trunk sewers, and convey the wastewater to a treatment plant.

Source: Adapted in part from Refs. 2–4

TABLE 8.2
Typical Distribution of Sewer Sizes

Sewer Diameter [cm (in.)]	Distribution (%)
10, 12.5, 15, (4, 5, 6)	3.6
20 (8)	73.1
25, 30, 35, 37.5, 40 (10, 12, 14, 15, 16)	13.4
45, 50, 53 (18, 20, 21)	3.8
60 (24)	1.7
67.5 (27) and above	4.4
Total	100.0

Source: Adapted in part from Refs. 2–4

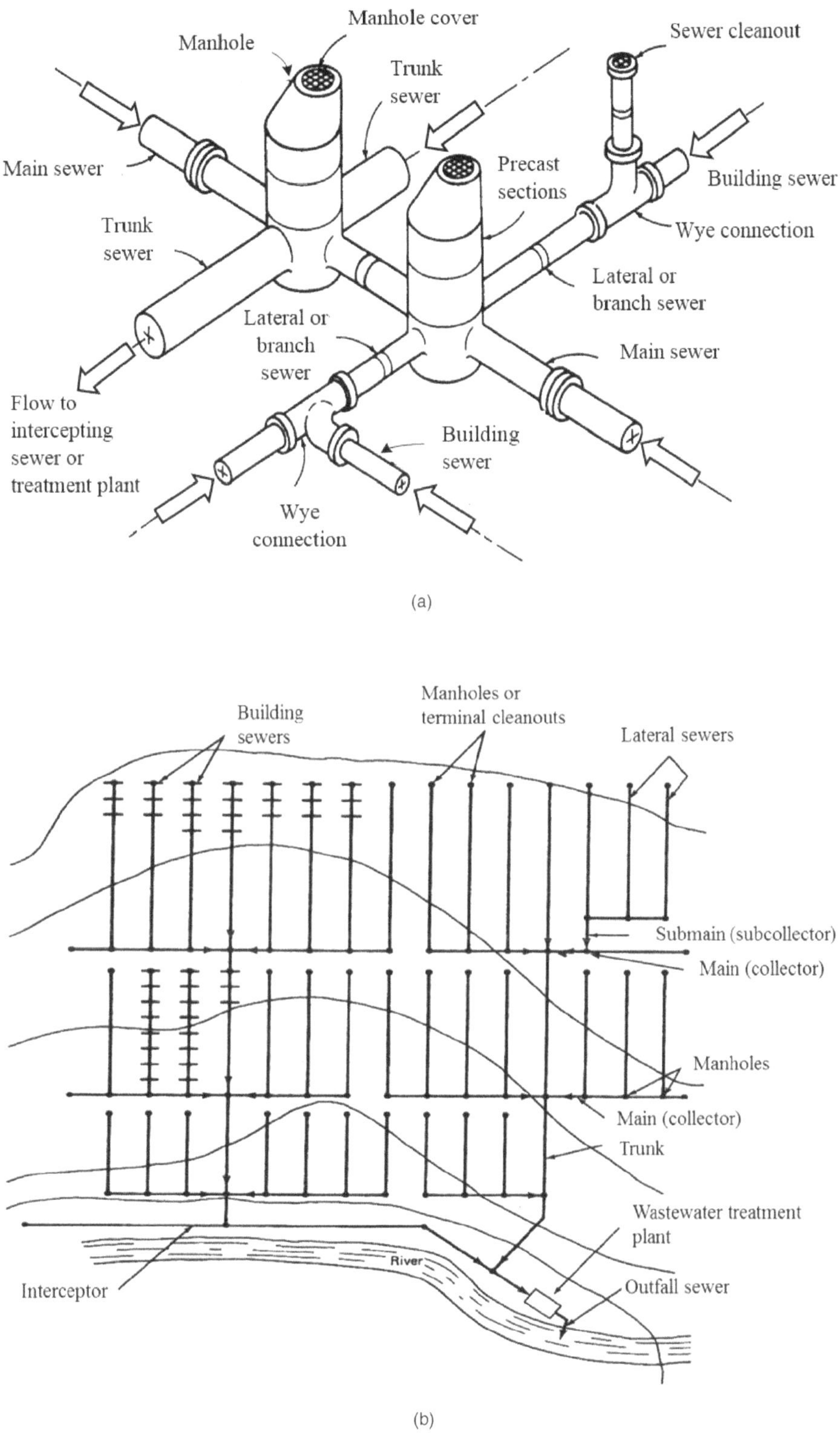

FIGURE 8.1 Sanitary sewerage system: (a) definition sketch for types of sanitary sewers used in collection system and (b) sanitary sewer layout.

TABLE 8.3
General Information about Sewer Appurtenances

Appurtenance	General Information
Manhole	Located at the (a) end of the line (terminal cleanout), (b) intersection of sewers, (c) change in grade, (d) change in alignment except in curved sewers, and (e) maximum spacing of 90–180 m (300–600 ft) depending upon available length of sewer cleaning equipment. Manholes for small sewers are generally 1.2 m (4 ft) in diameter. Manholes should not be located in low areas unless watertight covers are provided.
Drop manholes	Located if the elevation difference between incoming and outgoing sewers is greater than 0.5 m (1.5 ft). They reduce turbulence and stripping of odorous gases.
Terminal cleanout	Located at the end of the line. They are often curved upward as a continuation of the line.
Building connections	Building connections are also called house connections, service connections, or service laterals. They are provided during sewer construction by providing wyes or tees that are plugged until service connections are made.

Source: Adapted in part from Refs. 1–3

Example 8.1: Length and Size Distribution of Sewerage System

A city of 40,000 residents has single- and multiple-family housing, apartments, and commercial and industrial areas. Determine the total length of the collection system and the lengths of various pipe sizes. Based on the population density, it is estimated that the average length of sewer is 27 ft/capita. Also, calculate the equivalent sewer diameter. Use the information summarized in Table 8.2.

Solution

1. Calculate total length of sewerage system

$$\text{Total length} = \frac{27\text{ ft}}{\text{person}} \times 40{,}000 \text{ people} = 1{,}080{,}000 \text{ ft}$$

$$= 1{,}080{,}000 \text{ ft} \times \frac{1}{5{,}280 \text{ ft/mile}} = 205 \text{ miles}$$

2. Determine sewer size distribution and length
   The typical sewer size distribution is obtained from Table 8.2 and is provided below.

Sewer Diameter, in.	Distribution, %	Length, Mile
6 and below	3.6	7.4
8	73.1	149.8
10–16	13.4	27.5
18–21	3.8	7.8
24	1.7	3.5
27 and above	4.4	9.0
Total	100	205.0

3. Determine equivalent sewer diameter.
   The equivalent sewer diameter is the weighted average of all sewer diameter

$$\text{Equivalent dia.} = \frac{(6\times3.6+8\times73.1+13\times13.4+19.5\times3.8+24\times1.7+27\times4.4)\text{ in.}}{100}$$

$$= 10.2\text{ in.}$$

8.2.4 Sewer Layout and Positioning in Street

The layout of the sewer system will depend on local topography. The best grades and best locations for sewers are obtained when they follow the natural drainage channels. The following rules may apply for different conditions.[5]

1. If the sewer district is along a stream with terrain sloping gently to it, a trunk sewer alignment close to the stream may be the best. An example of such a layout is shown in Figure 8.2.
2. If the district is on different levels, the sewers may be laid in different directions, but eventually lead to a common trunk line (Figure 8.3).
3. If the district is located on opposite sides of a ridge, a special design may be needed to provide the most economical layout (Figure 8.4).

Procedure for establishing layout: Proper sewer layout plan and profiles must be completed before design flows can be established. The following is a list of basic rules that must be followed in developing a sewer layout plan and profile.[2,5]

1. Select the site for wastewater treatment plant. For gravity system, the best site is generally the lowest elevation of the entire drainage area.
2. The preliminary layout of the sewers is made from the topographic maps. In general, sewers are located on streets or on available rights-of-way and are sloped in the same direction as the slope of the natural ground surface.

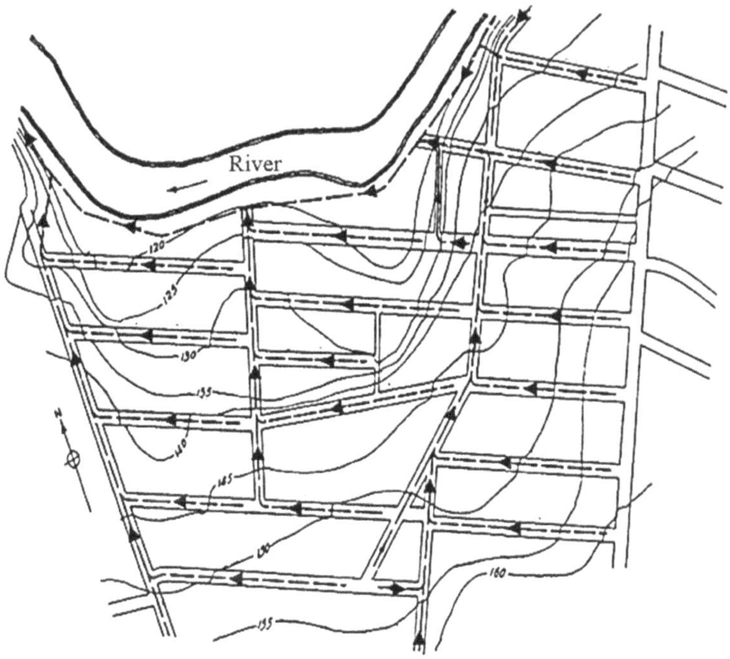

FIGURE 8.2 Sewer layout that has trunk alignment close to a stream.

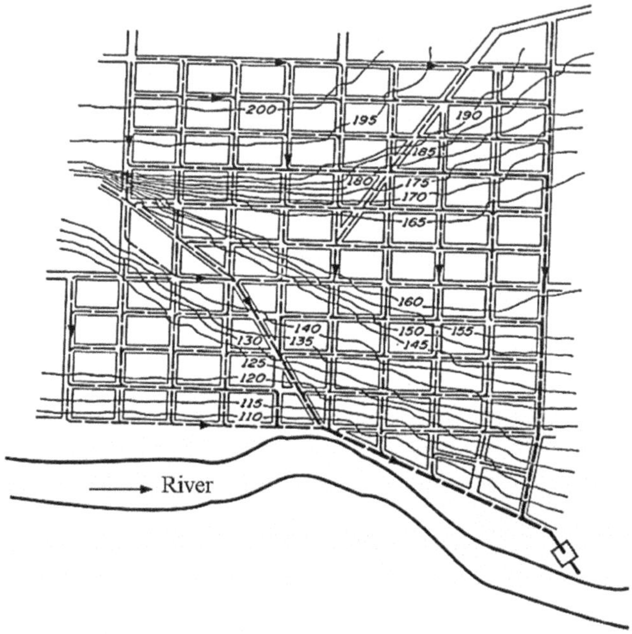

FIGURE 8.3 Sewer layout for districts at different levels.

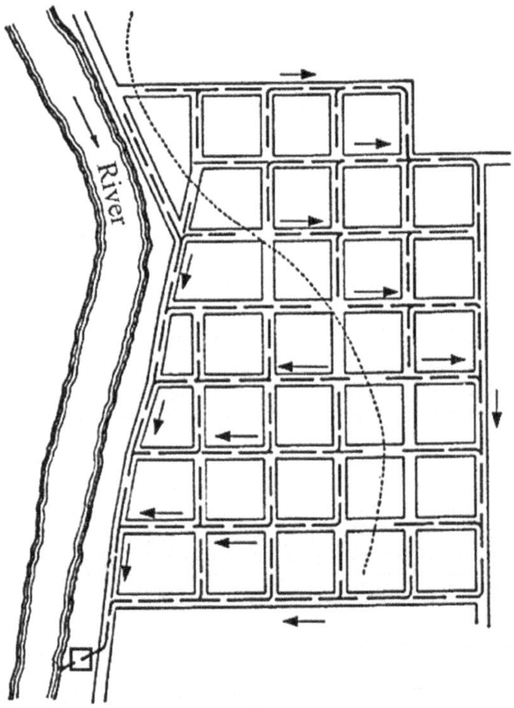

FIGURE 8.4 Sewer layout in district separated by a ridge.

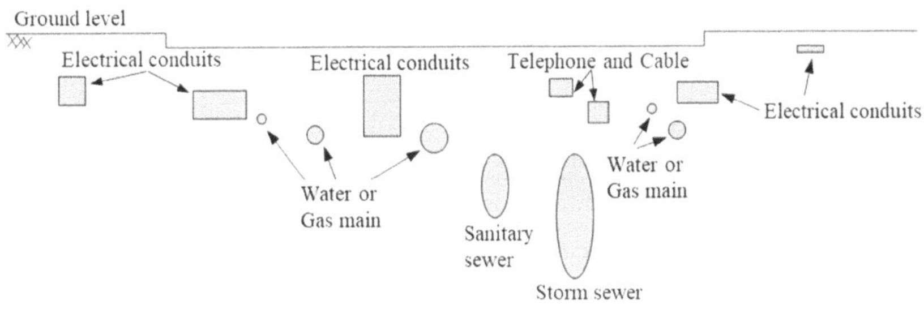

FIGURE 8.5 Subsurface utilities commonly found in streets.

3. The trunk sewers are commonly located in valleys. Each line starts from the intercepting sewer and extends uphill until the edge of the drainage area is reached, and further extension is not possible without working downhill.
4. Main sewers start from the trunk line and extend uphill, intercepting the laterals.
5. All laterals or branch lines are located in the same manner as the main sewers. Buildings' sewers are directly connected to the laterals.
6. Preliminary layout and routing of wastewater flow is done by considering several feasible alternatives. In each alternative, factors such as total length of sewers and cost of construction of laying deeper lines versus cost of construction, operation, and maintenance of lift station should be evaluated to arrive at a cost-effective sewerage system.
7. Sewers should not be located near water mains. State and local regulations must be consulted for the appropriate separation distance between the sewers and water lines.
8. After the preliminary sewer layout plan is prepared, the street profiles are drawn. These profiles should show the street elevations, existing sewer lines, and manholes. These profiles are used to design proposed lines.

Sewer depths and ground cover: The depth of the sewer should be such that all house sewers will drain by gravity. Normally, the sewer lines are constructed 1–2 m (3–6 ft) below the ground surface. In sewer districts with basements, the depth is normally 2–3 m. Construction cost of sewers increases rapidly with depth. It is uneconomical if the sewer depth exceeds 3.6 m. The depth of ground cover depends upon (1) topography, (2) depth of water table, (3) lowest point to be served, (4) freeze depth, and (5) load distribution.

Sewer location: Where sewers are located on both sides of the street, the most desirable location is the center of the street. However, on wide streets with a high cost of surfacing, sewers on both sides may be economical. Other locations, such as sidewalk, parking strip, or near the curb, may also be considered.

Subsurface information: Both surface and subsurface conditions will affect the construction cost. This includes the character of soil, presence of rocks, type of pavement, position of curbs, location of manhole, and subsurface structures and utility connections. Some of the usual subsurface structures found underneath a street are shown in Figure 8.5.

8.2.5 Basic Design Considerations

Many factors are important for sewer design. These factors are design period, design flow, minimum size, velocity, slope, and depth. Many of these factors are briefly described below.[6,7] Additional information may be found in Refs. 1,2,5–7.

Design period and design flow: The ultimate population that will be served by the sewer at the end of the design period should be used to estimate the design flow. Design periods of 25–50 years or more are not uncommon.

Design flow: Sanitary sewers should be designed to carry peak wet weather flow. The following design flows are recommended.

Laterals and sub-mains	1,500 Lpcd (400 gpcd)
Mains and trunk sewers	1,330 Lpcd (350 gpcd)
Intercepting sewers	1,000 Lpcd (260 gpcd)

Industrial contributions and allowance for infiltration/inflow should be made to all these values. The average wastewater flow, daily variation, sustained flow, infiltration/inflow allowance, peak wet weather flow, and 2-h design flows are presented in Section 4.5. Readers should refer to this section to develop a design flow for a conveyance system.

Minimum sewer size, velocity, slope, and liquid depth: Recommended minimum lateral sewer size is 20 cm (8 in.), although many states allow 15 cm (6 in.) lateral sewers.

Velocity: Velocity above 0.3 m/s (1 ft/s) under low flow is recommended. Under peak dry weather conditions, a velocity of 0.6 m/s (2 ft/s) or greater must be attained to flush the lines each day. Velocity greater than 3 m/s (10 ft/s) should be avoided as erosion and damage to sewer and manhole may occur.

Slope: Flat slopes encourage solids deposition. Minimum recommended slope should be such that a velocity of 0.6 m/s (2 ft/s) is reached when flowing full and $n = 0.013$. Most state-design criteria provide a table of minimum slope (see Example 8.3). If sewers of slopes less than the recommended value are provided, the state agency may require depth and velocity computations at minimum, average, and peak design flow conditions.

Liquid depth: The liquid depth under low flow is generally very small. A good design practice is to design a sewer that can carry peak wet weather flow at maximum depth of 80%–90% of the sewer diameter.

Example 8.2: Design Capacity of a Sewer

A 15-in. interceptor is 8,500 ft long. The average infiltration/inflow allowance is 1,500 gal/day/in./mile, and the peaking factor is 1.8. It serves a design population of 4,000 residents and receives an industrial contribution of 300,000 gpd. Determine the design capacity of the interceptor.

Solution

1. Calculate peak dry weather flow

$$\text{Peak dry weather flow} = 260\,\frac{\text{gallons}}{\text{person/day}} \times 4{,}000 \times 10^{-6}\,\frac{\text{million gallons}}{\text{gallon}} = 1.04\ \text{mgd}$$

2. Calculate average infiltration/inflow allowance

$$\text{Average I/I allowance} = \frac{1{,}500\ \text{gal}}{\text{day/in./mile}} \times \frac{8{,}500\ \text{ft}}{5{,}280\ \text{ft/mile}} \times 15\ \text{in.} = 36{,}220\ \text{gal/day}$$

$$\text{Peak I/I allowance} = 36{,}220\ \text{gal/day} \times 1.8 \times 10^{-6}\ \text{million gallon/gal} = 0.07\ \text{mgd}$$

3. Industrial contribution = 0.30 mgd
4. The design capacity of the interceptor = (1.04 + 0.07 + 0.3) mgd = 1.41 mgd

8.2.6 Design Equations

Circular sewer flowing full: Many design equations have been proposed for designing sewers and drains.[1,8–10] The Manning equation has received the most widespread application. The Manning equation in various forms is presented in Chapter 5. Readers should review Eqs. (5.34) through (5.37). These equations are listed below.

$$V = \frac{1}{n} R^{2/3} S^{1/2} \quad \text{(SI unit)}$$

$$Q = \frac{0.312}{n} D^{8/3} S^{1/2} \quad \text{(Circular pipe flowing full, SI unit)}$$

$$V = \frac{1.486}{n} R^{2/3} S^{1/2} \quad \text{(U.S. customary unit)}$$

$$Q = \frac{0.464}{n} D^{8/3} S^{1/2} \quad \text{(Circular pipe flowing full, U.S. customary unit)}$$

Manning's coefficient of roughness n depends on the material and age of the conduit. Commonly used values of *n* for different materials may be found in Table 5.4.

Various types of nomographs have also been developed for solution of problems involving sewers flowing full. Nomographs also based on the Manning equation for circular pipe flowing full and variable n values are provided in Figure 5.25. Similar nomographs may be found in Refs. 1,2,6,10,11.

Circular sewer flowing partially full: Sanitary sewers are primarily designed to flow partially full. The hydraulic element equations for circular pipe flowing under part-full flow conditions are given by Eq. (5.33), and several other equations are summarized in Table 5.3. The central angle θ is in degrees. It has been found experimentally that the value of n decreases with the depth of flow. However, in most designs, n is assumed constant for all flow depths. The hydraulic elements of circular sewers for partial flow area are also graphically presented in Figure 5.26. Use of the hydraulic element equations and nomographs is shown below in several examples. The decrease in n with depth is shown in Figure 5.26. It is a common practice to use d, v, q, a, and p (lowercase) notations for depth of flow, velocity, discharge, area, and wetted perimeter under partial flow conditions, while D, V, Q, A, and P (uppercase) notations are used for sewer flowing full. A sewer line that is flowing 80% full has a d/D ratio of 0.8. Equation (5.33) and geometric elements of circular pipes obtained from Table 5.3 are listed below.

$$\cos\frac{1}{2}\theta = \left(1 - \frac{2d}{D}\right) \quad \text{or} \quad \theta = 2\cos^{-1}\left(1 - \frac{2d}{D}\right)$$

$$a = \frac{D^2}{8}\left(\frac{\pi\theta}{180} - \sin\theta\right)$$

$$p = \frac{\pi D\theta}{360}$$

$$r = \frac{D}{4}\left(1 - \frac{180\sin\theta}{\pi\theta}\right)$$

Example 8.3: Minimum Slope and Velocity of a Sanitary Sewer

Most states' design criteria provide a table of minimum slope for different sewer sizes. The table is based on Manning $n = 0.013$ and a velocity of 2 ft/s (0.6 m/s) when flowing full. Develop this table for standard sewer sizes 6–54 in. (150–1,370 mm). The minimum wall thickness of the standard concrete sewers is also included in this table.

Solution

1. Rearrange the Manning equation (Eq. 5.36)

$$V = \frac{1.486}{n} R^{2/3} S^{1/2}$$

$$S = \frac{\left(\frac{V \times n}{1.486}\right)^2}{R^{\frac{4}{3}}}$$

2. Substitute the values $n = 0.013$, $V = 2$ ft/s and $R = D/4$

$$S = 0.001943\left(D^{-1.3}\right)$$

3. Substitute the diameter D in ft and solve for S. The slope values for standard pipe sizes are provided in Table 8.4.
4. The minimum wall thickness of standard sewers is obtained from Concrete Pipe Design Manual.[12]

TABLE 8.4
Minimum Recommended Slopes of Sanitary Sewers that Gives 2 ft/s (0.6 m/s) Velocity When Flowing Full, and Minimum Wall Thickness (Example 8.3)

Standard Internal Diameter		Slope	Min. Wall Thickness	
in.	mm	ft/ft (m/m)	in.	mm
6	150	0.0049	5/8	15.88
8	200	0.0033	3/4	19.05
10	250	0.0025	7/8	22.23
12	310	0.0019	1¾	44.45
15	380	0.0014	1⅞	47.63
18	460	0.0011	2	50.80
21	530	0.00092	2¼	57.15
24	610	0.00077	2½	63.50
27	690	0.00066	2⅝	66.68
30	760	0.00056	2¾	69.85
36	910	0.00045	3	76.20
42	1,050	0.00037	3½	88.90
48	1,200	0.00031	4	101.60
54	1,370	0.00026	4½	114.30

Example 8.4: Diameter and Velocity in Sewer Flowing Full

A sewer is carrying a peak flow of 18 mgd when full. It has a slope of 0.00022 ft/ft and n = 0.011. Determine the diameter and velocity using equations.

Solution

1. Calculate the diameter of the sewer line from Eq. (5.38)

$$Q = \frac{0.312}{n} D^{\frac{8}{3}} S^{\frac{1}{2}}$$

$$Q = \frac{0.312}{0.01} D^{\frac{8}{3}} (0.001)^{\frac{1}{2}}$$

$$1.0 = 0.987 D^{8/3}$$

$$D = \left(\frac{1.0}{0.987}\right)^{3/8} = (1.013)^{0.375}$$

$$D = 1.0 \text{ m} (39 \text{ in.})$$

2. Calculate area of the sewer line

$$\text{Area} = \frac{\pi}{4}(1.0 \text{ m})^2 = 0.785 \text{ m}^2$$

3. Calculate velocity

$$\text{Velocity} = \frac{Q}{\text{area}} = \frac{1.0 \text{ m}^3/\text{s}}{0.785 \text{ m}^2} = 1.26 \text{ m/s}$$

4. Calculate velocity from Eq. (5.34)

$$V = \frac{1}{0.01}\left(\frac{1.0 \text{ m}}{4}\right)^{\frac{2}{3}} (0.001)^{\frac{1}{2}} = 1.26 \text{ m/s}$$

Example 8.5: Partial Flow Velocity and Depth, Constant *n*

A 60 cm (24 in.) sewer is designed to carry a peak design flow of 0.01 m^3/s at a slope of 0.0008 m/m. Calculate the actual depth of flow and velocity. Assume constant value of n = 0.015.

Solution

1. Calculate the flow of Q and velocity V when sewer is full – use Eq. (5.38)

$$Q = \frac{0.312}{0.015}(0.6 \text{ m})^{\frac{8}{3}} (0.0008)^{\frac{1}{2}}$$

$$Q = 0.15 \text{ m}^3/\text{s}$$

$$V = \frac{1}{0.015}\left(\frac{0.6\text{ m}}{4}\right)^{\frac{2}{3}}(0.0008)^{\frac{1}{2}}$$

$$V = 0.53\text{ m/s}$$

2. Calculate the partial flow and full flow ratio $\left(\frac{q}{Q}\right)$

$$\left(\frac{q}{Q}\right) = \frac{0.01\text{ m}^3/\text{s}}{0.15\text{ m}^3/\text{s}} = 0.067$$

3. Read the *d/D* and *v/V* ratios from Figure 5.26b for known *q/Q* ratio of 0.067. Use the curves with dotted lines for constant *n*

$$\left(\frac{d}{D}\right) = 0.21$$

$$\left(\frac{v}{V}\right) = 0.64$$

4. Calculate the depth of flow and velocity under partial flow

$$\text{Depth of flow} = 0.21 \times 0.6\text{ m} = 0.13\text{ m}$$

$$\text{Velocity} = 0.64 \times 0.53\text{ m/s} = 0.34\text{ m/s}$$

Example 8.6: Partial Flow Velocity and Depth, Variable *n*

It has been found experimentally that Manning's *n* varies as depth of flow changes. Repeat Example 8.5 for variable value of *n*. It is given that $n = 0.015$ when the sewer line is flowing full.

Solution

1. Determine *Q* and *V* when sewer is flowing full (Example 8.5)
   For diameter = 0.6 m, $q = 0.01\text{ m}^3/\text{s}$, slope = 0.0008 m/m and $n = 0.015$

$$V = 0.53\text{ m/s} \quad \text{and} \quad Q = 0.15\text{ m}^3/\text{s}$$

$$\frac{q}{Q} = \left(\frac{0.01\text{ m}^3/\text{s}}{0.15\text{ m}^3/\text{s}}\right) = 0.067$$

2. Determine *d/D* and *v/V* for variable *n*
   From Figure 5.26b read *d/D* and *v/V* using the curves with solid lines for variable *n*

$$\left(\frac{d}{D}\right) = 0.23$$

$$\left(\frac{v}{V}\right) = 0.52$$

3. Determine d and v

$$d = 0.23 \times 0.60 \text{ m} = 0.14 \text{ m}$$

$$v = 0.52 \times 0.53 \text{ m/s} = 0.28 \text{ m/s}$$

Example 8.7: Diameter and Slope of a Sewer Line under Partial Flow

Calculate the diameter and slope of a trunk sewer to carry 50 L/s when flowing 60% full. The velocity in the line at this flow is 0.6 m/s (n = 0.013)

Solution

1. Determine q/Q and v/V at 60% full
   At 60% full, d/D = 0.6
   From Figure 5.26b determine q/Q and v/V

$$q/Q = 0.67$$

$$v/V = 1.08$$

2. Determine Q and V (sewer line full)

$$Q = 50 \times 10^{-3} \text{ m}^3/\text{s} \times \frac{1}{0.67} = 0.075 \text{ m}^3/\text{s}$$

$$V = \frac{0.6 \text{ m/s}}{1.08} = 0.56 \text{ m/s}$$

3. Determine diameter D and slope S

   Use Eq. (5.38), $0.075 \text{ m/s} = \dfrac{0.312}{n} D^{8/3} S^{1/2}$

   Use Eq. (5.34), $0.56 \text{ m/s} = \dfrac{1}{n}\left(\dfrac{D}{4}\right)^{2/3} S^{1/2}$

   Divide one equation by the other

$$\frac{0.56 \text{ m/s}}{0.075 \text{ m}^3/\text{s}} = \frac{(D/4)^{2/3} S^{1/2}}{0.312\ (D)^{8/3} S^{1/2}} = \frac{1}{0.312}\left(\frac{1}{4}\right)^{2/3} D^{-2}$$

$$D = 0.414 \text{ m}$$

   From Eq. (5.34), $0.56 \text{ m/s} = \dfrac{1}{0.013}\left(\dfrac{0.414 \text{ m}}{4}\right)^{2/3} \times S^{1/2}$

   $S = 0.0011 \text{ m/m}$

   Select standard sewer diameter of 460 mm or 18 in. (Table 8.4)

Example 8.8: Depth and Velocity at a Given Minimum Flow

An intercepting sewer is 400 mm in diameter and is serving a population of 5,000 residents. The average wastewater flow is 450 Lpcd. Calculate the depth of flow and velocity at minimum flow. The slope of the line is 0.003. Assume $n = 0.013$ and minimum flow is one-third of the average flow.

Solution

1. Calculate the minimum flow

$$\text{Average flow} = 5{,}000 \text{ person} \times \frac{450 \text{ L}}{\text{person} \cdot \text{day}} \times 10^{-3} \text{ m}^3/\text{L} \times \frac{\text{day}}{86{,}400 \text{ s}}$$

$$= 0.026 \text{ m}^3/\text{s}$$

$$\text{Minimum flow} = \frac{1}{3} \times 0.026 \text{ m}^3/\text{s} = 0.0087 \text{ m}^3/\text{s}$$

2. Calculate the velocity and discharge when flowing full Use Eqs. (5.34) and (5.38)

$$V = \frac{1}{0.013}\left(\frac{0.4 \text{ m}}{4}\right)^{2/3} \times (0.003)^{1/2} = 0.91 \text{ m/s}$$

$$Q = \frac{0.312}{0.013}(0.4 \text{ m})^{8/3} \times (0.003)^{1/2} = 0.114 \text{ m}^3/\text{s}$$

3. Determine $\frac{q_{\min}}{Q}$

$$\frac{q_{\min}}{Q} = \frac{0.0087 \text{ m}^3/\text{s}}{0.114 \text{ m}^3/\text{s}} = 0.076$$

4. Determine *d*/*D* and *v*/*V* ratios from Figure 5.26b

$$\left(\frac{d}{D}\right) = 0.22$$

$$\left(\frac{v}{V}\right) = 0.65$$

5. Determine *d* and *v* at minimum flow.

$$d = 0.22 \times 0.4 \text{ m} = 0.088 \text{ m}$$

$$v = 0.65 \times 0.91 \text{ m/s} = 0.59 \text{ m/s}$$

Example 8.9: Depth and Velocity Calculated from Hydraulic Elements Equations and Nomograph

A sewer is flowing 35% full. The sewer diameter and slope are 18 in. and 0.0018. Determine depth, velocity, and discharge under partial full conditions using hydraulic elements equations $n = 0.013$. Also, check partial flow velocity and discharge using the nomograph.

Solution

A. Calculations based on hydraulic elements equations

1. Determine the velocity and discharge from Eq. (5.36) when sewer is flowing full.

$$V = \frac{1.486}{0.013}\left(\frac{1.5\text{ ft}}{4}\right)^{2/3}(0.0018)^{1/2} = 2.52\text{ ft/s}$$

$$Q = V.A = 2.52\text{ ft/s} \times \frac{\pi}{4}(1.5\text{ ft})^2 = 4.453\text{ ft}^3/\text{s}$$

2. Calculate the hydraulic elements of the sewer running partially full
   Use Eq. (5.33) and equations listed in Table 5.3

$$d/D = 0.35$$

$$\cos\frac{1}{2}\theta = \left(1 - \frac{2d}{D}\right) = 0.3$$

$$\theta = 145.1°$$

$$a = \frac{(1.5\text{ ft})^2}{8}\left(\frac{\pi \times 145.1}{180} - \sin 145.1°\right) = 0.551\text{ ft}^2$$

$$p = \frac{\pi \times 1.5\text{ ft} \times 145.1}{360} = 1.9\text{ ft}$$

$$r = \frac{1.5\text{ ft}}{4}\left(1 - \frac{180\sin 145.1}{\pi \times 145.1}\right) = 0.29\text{ ft}$$

3. Calculate the velocity and discharge From Eq. (5.36),

$$v = \frac{1.486}{0.013}(0.29\text{ ft})^{\frac{2}{3}}(0.0018)^{\frac{1}{2}} = 2.1\text{ ft/s}$$

$$q = v \times a = 2.1\text{ ft/s} \times 0.551\text{ ft}^2$$

$$= 1.16\text{ ft}^3/\text{s}$$

B. Calculations based on nomograph

1. Determine v and q from Figure 5.26b
   Read the ratios v/V and q/Q using given $d/D = 0.35$

$$\left(\frac{v}{V}\right) = 0.82 \quad \text{and} \quad v = 2.1\text{ ft/s}$$

$$\left(\frac{q}{Q}\right) = 0.26 \quad \text{and} \quad q = 1.16\text{ ft}^3/\text{s}$$

8.2.7 Hydraulic Profile

The water surface in a sewer line follows the slope of the energy grade line or invert slope. At a manhole or junction box, many head losses are encountered. The head losses are due to (1) exit loss, (2) change in direction, (3) contractions and expansions, (4) transitions, and (5) entrance loss.[1,2,13] All these minor losses may be conveniently expressed as a function of velocity head. Minor head loss equation was presented in Chapter 5. Equation (5.27) expresses the minor head loss relationship and is given below.

$$h_L = K\frac{v^2}{2g}$$

The values of K for different conditions of open channel flow and sewer appurtenances are summarized in Appendix C. This is a simplified procedure and gives conservative value if higher values of K are selected from Appendix C. Other alternative methods used are to: (1) apply the energy equation at two sections and then solve for velocity or depth of flow by a trial-and-error procedure. It is a tedious and time-consuming procedure. Details and examples of this method are covered in Example 5.3, (2) match the crown elevations of all sewers meeting at a manhole. Since the outgoing sewer is always the largest, a drop in manhole invert is necessary to align with the outgoing sewer, and (3) drop the manhole invert by about 0.03 m below the invert of the lowest incoming sewer.[2]

Example 8.10: Hydraulic Profile

An intercepting sewer is 61 cm in diameter and has a slope of 0.0013. It carries a flow of 113.3 L/s. The sewer enters a standard manhole that is 76 cm in diameter. The outlet sewer is also 61 cm in diameter and has a slope of 0.0014. Using $n = 0.015$, draw the hydraulic profile and determine the invert elevation of the incoming sewer. Assume that the minor loss coefficients of exit and entrance are 0.4 and 0.1, respectively, and the manhole invert is in line with that of the outlet sewer.

Solution

1. Calculate the depth of flow and velocity in the upstream sewer.
   Use Eqs. (5.34) and (5.38)

$$V_1 = \frac{1}{0.015}\left(\frac{0.61\text{ m}}{4}\right)^{2/3}(0.0013)^{1/2} = 0.686\text{ m/s}$$

$$Q_1 = \frac{0.312}{0.015}(0.61\text{ m})^{8/3}(0.0013)^{1/2} = 0.201\text{ m}^3/\text{s}$$

$$\frac{Q}{Q_1} = \frac{113.3\times10^{-3}\text{ m}^3/\text{s}}{0.201\text{ m}^3/\text{s}} = 0.56$$

From Figure 5.26b

$$\frac{d_1}{D_1} = 0.54;\quad \frac{v_1}{V_1} = 1.04$$

$$d_1 = 0.54\times0.61\text{ m} = 0.33\text{ m}$$

$$v_1 = 1.04\times0.686\text{ m/s} = 0.713\text{ m/s}$$

2. Calculate the depth of flow and velocity in the downstream sewer line. Use Eqs. (5.34) and (5.38)

$$V_2 = \frac{1}{0.015}\left(\frac{0.61\ \text{m}}{4}\right)^{2/3}(0.0014)^{1/2} = 0.712\ \text{m/s}$$

$$Q_2 = \frac{0.312}{0.015}(0.61\ \text{m})^{8/3}(0.0014)^{1/2} = 0.208\ \text{m}^3/\text{s}$$

$$\frac{Q}{Q_2} = \frac{113.13\times10^{-3}\ \text{m}^3/\text{s}}{0.208\ \text{m}^3/\text{s}} = 0.55$$

From Figure 5.26b

$$\frac{d_2}{D_2} = 0.53; \quad \frac{v_2}{V_2} = 1.03$$

$$d_2 = 0.53\times0.61\ \text{m} = 0.323\ \text{m}$$

$$v_2 = 1.03\times0.712\ \text{m/s} = 0.733\ \text{m/s}$$

3. Calculate head loss at entrance and exit of the manhole.
   Apply minor head loss equation (Eq. 5.27)

$$H_{L,\text{entrance}} = K\frac{v_2^2}{2g} = \frac{0.1\times(0.733\ \text{m/s})^2}{2\times9.81\ \text{m/s}^2} = 0.003\ \text{m}$$

$$h_{L,\text{exit}} = K\frac{v_1^2}{2g} = \frac{0.4\times(0.713\ \text{m/s})^2}{2\times9.81\ \text{m/s}^2} = 0.001\ \text{m}$$

4. Compute the hydraulic profile
   Water surface elevation in the downstream sewer above invert at the entrance point = 0.323 m
   Entrance loss in the outlet sewer at the manhole = 0.003 m
   Water surface elevation in the manhole = 0.323 m + 0.003 m = 0.326 m
   Water surface elevation in the incoming sewer line = 0.326 m + 0.010 m = 0.336 m
   Invert elevation of the incoming sewer line above the invert of the manhole = 0.336 m – depth of flow in upstream sewer = 0.336 m – 0.330 m = 0.006 m
5. Draw the hydraulic profile

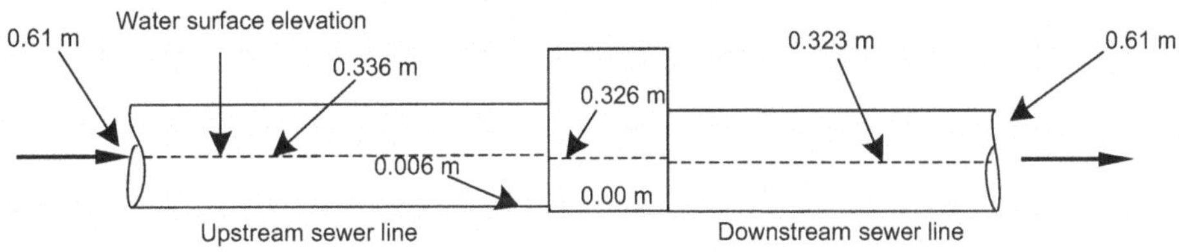

Example 8.11: Invert Elevation and Hydraulic Profile

Two influent sewer lines *A* and *B* are connected to a manhole at 90°. The exit line C is 12-in. diameter, and the depth of flow is 7 in. The sewer lines *A* and *B* are 10 and 8 in. in diameter and have flow depths of 6 and 4 in., respectively. The invert of line *B* is 5 in. higher than the invert of line *A*. Draw the hydraulic profile (1) by matching the crowns of line *A* and *C* and (2) by dropping 3 in. the invert of manhole from the invert of line *A*.

Solution

A. Draw the layout plan of sewers.

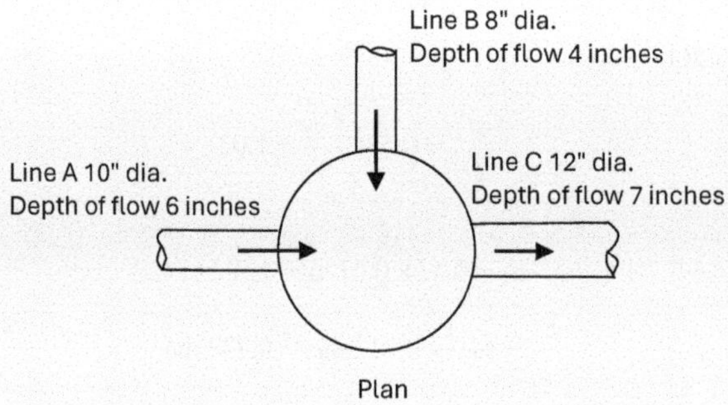

B. Draw the L-section by matching the crowns of Lines *A* and *C*

The crown is the top inside elevation of a sewer line. The invert is the inside bottom elevation of sewer or manhole. The crowns of sewers *A* and *C* are matched. The hydraulic profile is shown in Figure 8.6.

C. Draw the L-section by dropping 3 in. the invert of the manhole from the invert of Line *A*

The invert of the manhole is dropped 3 in. below the invert of line *A*. The hydraulic profile is shown in Figure 8.7.

Example 8.12: Sizing and Profiling a Trunk Sewer

Two proposed trunk sewers *AC* and *BC* will be connected to an existing trunk sewer at manhole *C*. The existing trunk sewer *CD* is 18-in. diameter and has a slope of 0.0018 from manhole *C* to *D*. Its design capacity is 2.20 mgd. The invert elevation of manhole *C* is 97.05 ft. The design capacities of trunk sewers *AC* and *BC* are 0.80 and 0.70 mgd, respectively.

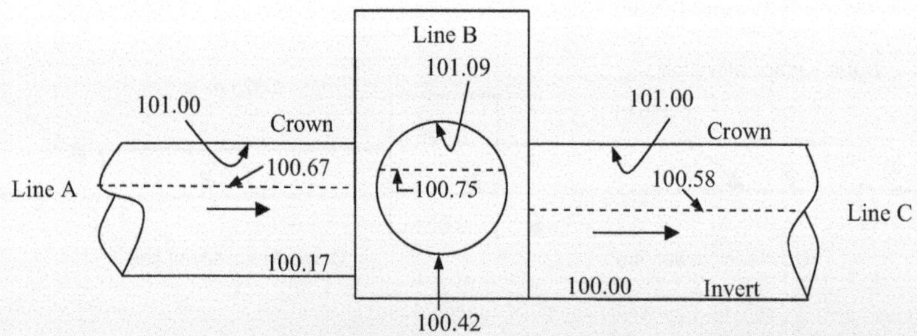

FIGURE 8.6 The hydraulic profile of sewer lines A and C with matching crown elevations.

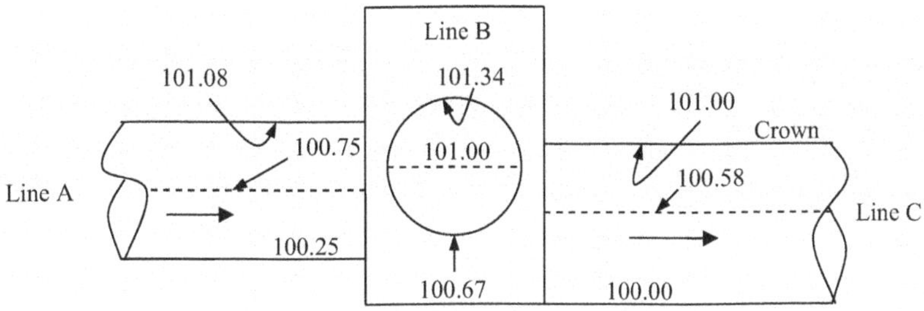

FIGURE 8.7 The hydraulic profile of sewer lines A and C with manhole invert elevation dropped by 3 in.

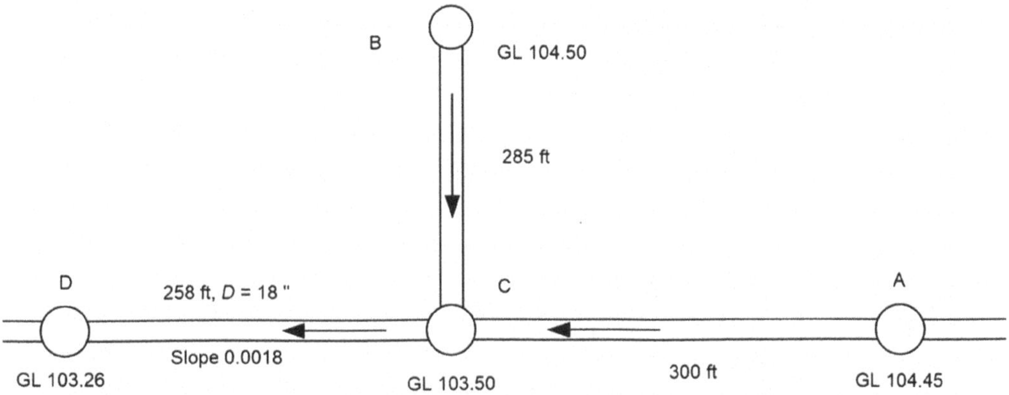

FIGURE 8.8 Plan of existing and proposed trunk sewers.

The ground surface elevations at manholes *A*, *B*, *C*, and *D* and the lengths of trunk sewers are given in Figure 8.8. Design interceptors *AC* and *BC*, and draw the profiles. The required minimum ground cover is around 4.75 ft. ($n = 0.013$).

Solution

1. Develop the hydraulic characteristics and profile of existing trunk sewer *CD*

$$V \text{ when full (Eq. (5.36))} = \frac{1.486}{0.013} \times \left(\frac{1.50 \text{ ft}}{4}\right)^{\frac{2}{3}} \times (0.0018)^{1/2} = 2.52 \text{ ft/s}$$

$$Q \text{ when full} = V \times A$$

$$= 2.52 \times \frac{\pi}{4} \times (1.50 \text{ ft})^2 = 4.45 \text{ cfs}$$

$$= 2.88 \text{ mgd}$$

$$\frac{q}{Q} = \frac{2.20 \text{ mgd}}{2.88 \text{ mgd}} = 0.76$$

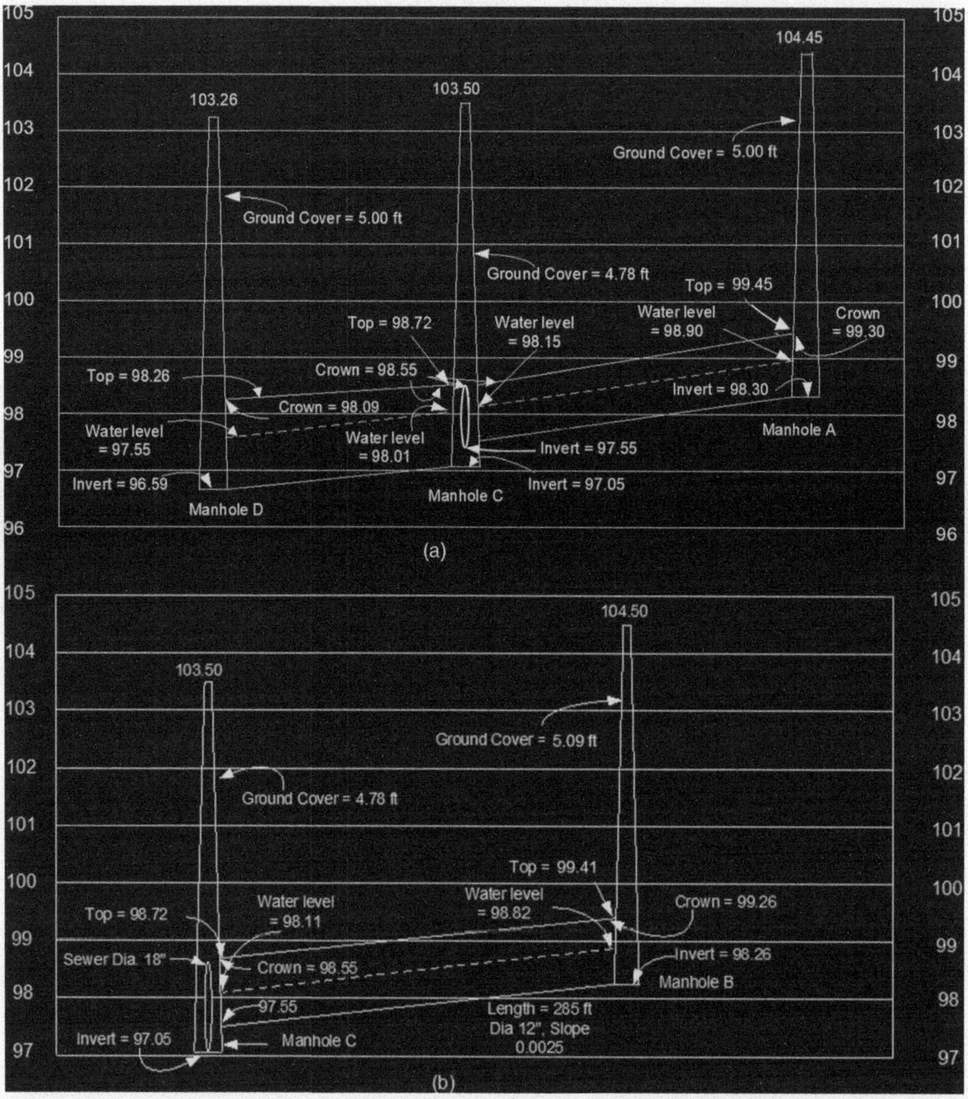

FIGURE 8.9 The hydraulic profiles and L-sections of trunks: (a) Trunks AC and CD, and (b) Trunk BC.

From Figure 5.26b $\frac{d}{D} = 0.64$ and $\frac{v}{V} = 1.16$

$$d = 0.96 \text{ ft} \quad \text{and} \quad v = 2.92 \text{ ft/s}$$

2. Draw the sewer profile *AC* and *CD*

   The hydraulic profile and the L-section of the trunk lines *AC* and *CD* are shown in Figure 8.9. The following information at manhole *C* may be noted.

   The invert elevation of manhole *C* and trunk *CD* are given = 97.05 ft

   Water surface elevation of trunk *CD* = 98.01 ft

   Crown elevation of trunk *CD* = 98.55 ft

   Top elevation of the trunk *CD* (wall thickness from Table 8.4) = 98.72 ft

Ground elevation = 103.50 ft
Ground cover over manhole C = 4.78 ft

3. Design and layout trunk AC

Assume a trunk diameter of 12 in. and a slope of 0.0025 (the sewer slope is selected to utilize the available ground slope and keep the ground cover around 5 ft)

$$\text{Velocity when full } V\,(\text{Eq. (5.36)}) = \frac{1.486}{0.013} \times \left(\frac{1.0\text{ ft}}{4}\right)^{2/3} \times (0.0025)^{1/2} = 2.30\text{ ft/s}$$

$$Q \text{ when full} = V \times A$$

$$= 2.30\text{ ft/s} \times \frac{\pi}{4} \times (1.0\text{ ft})^2 = 1.81\text{ cfs}$$

From Figure 5.26b $\frac{q}{Q} = \frac{0.80\text{ mgd}}{1.17\text{ mgd}} = 0.68$

$$\frac{d}{D} = 0.60 \quad \text{and} \quad \frac{v}{V} = 1.08$$

$$d = 0.6\text{ ft} \quad \text{and} \quad v = 2.48\text{ ft/s}$$

4. Draw the hydraulic profile and L-section of trunk AC

Match the crown of trunks AC and CD at manhole C.
Crown elevation of trunk AC at manhole C = 98.55 ft.
The invert elevation of trunk AC at manhole C = 97.55 ft.

The hydraulic profile and L-section of trunk AC are shown in Figure 8.9. The key elevations at manhole A are given below:

Invert elevation = 98.30 ft
Water surface elevation = 98.90 ft
Crown elevation = 99.30 ft
Top of trunk elevation = 99.45 ft
Ground elevation = 104.45 ft
Ground cover = 5.00 ft

5. Design and layout trunk BC

Assume trunk diameter of 12 in. and slope of 0.0025 (the sewer slope is selected to utilize the available ground slope, and keep the ground cover around 5 ft)

$$\text{Velocity when full } V\,(\text{Eq. (5.36)}) = \frac{1.486}{0.013} \times \left(\frac{1.0\text{ ft}}{4}\right)^{2/3} \times (0.0025)^{1/2} = 2.30\text{ ft/s}$$

$$Q \text{ when full} = V \times A$$

$$= 2.30 \times \frac{\pi}{4} \times (1.0\text{ ft})^2 = 1.81\text{ cfs} = 1.17\text{ mgd}(1\text{ mgd} = 1.547\text{ cfs})$$

$$\frac{q}{Q} = \frac{0.70\text{ mgd}}{1.17\text{ mgd}} = 0.60$$

From Figure 5.26b $\frac{d}{D} = 0.56$ and $\frac{v}{V} = 1.06$

$$d = 0.56\text{ ft} \quad \text{and} \quad v = 2.44\text{ ft/s}$$

6. Draw hydraulic profile BC
   Match the crown of trunk *BC* with *AC* and *CD* at manhole *C*
   Crown elevation of trunk *CD* at manhole *C* = 98.55 ft
   The invert elevation of trunk *BC* at manhole *C* = 97.55 ft
   The key elevations at manhole *B* are given below:
   Invert elevation = 98.26 ft
   Water surface elevation = 98.82 ft
   Crown elevation = 99.26 ft
   Top of trunk elevation = 99.41 ft
   Ground cover = 5.09 ft

8.2.8 Design Considerations for a Community Sewerage System

Designing a community sewerage system requires considerable experience. A great deal of information is needed to make proper decisions concerning the layout, sizing, and construction of a sewer network that is efficient and cost-effective. The design engineer needs to develop information about the service area, topographic maps, preliminary sewer layout plans, and design and construction considerations.[1,2,5,6,14]

Service area: Service area is the ultimate area that will eventually be served by the sewerage system. It should include land-use plan, zoning regulations, population density, and social and economic conditions.

Topographic maps: Reliable topographic maps with existing and proposed streets, utility lines, watershed boundaries, and contours provide the most important information for preliminary flow routing.

Layout plan: A number of tentative layout plans may be established, with most attention being given to the elevations of the various controlling points. The procedure for establishing layout is given in Section 8.2.4. The first sewers to be located are the mains, including the outfall lines and the interceptor. The sub-mains are then located, followed by laterals. As far as possible, each sewer should be sloped in the same direction as the surface of the street. A typical sewer layout plan and profile are shown in Figure 8.10.[14]

Design and construction considerations: Many design and construction factors must be investigated before sewer design can be completed. Factors such as design period; peak, average, and minimum flows; peaking factors; I/I allowance; sewer slopes, minimum velocities, and design equations; sewer material; joints and connections; appurtenances; and sewer installation are all important elements in developing the sewer design.

8.2.9 Design Calculations and Organization of Computation Table

After the preliminary sewer layout plan and profile are prepared, the design computations are accomplished. Design computations for sewers are repetitive and therefore are best performed in a tabular format. Most textbooks provide a computational table that can be adapted for a desired application. References 1,3,6,8,10,13 may be consulted on this subject.

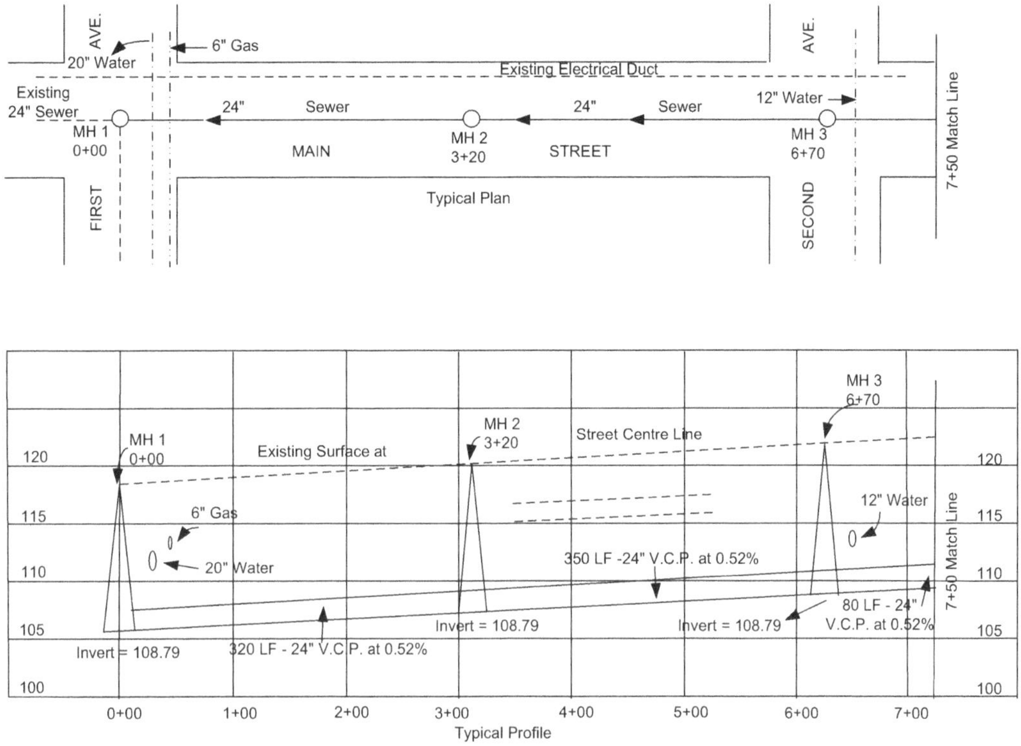

FIGURE 8.10 Typical plan and profile drawings for sanitary sewer: (a) plan and (b) profile. (Printed from Ref. 14 with permission of the National Clay Pipe Institute.) 1 ft = 0.3048 m.

Example 8.13: Estimation of Design Flows in Sanitary Sewers

A sanitary sewer system is designed for a new development that has single- and multi-family residential areas, commercial zones (including shopping areas), industrial park, and a community college. The data on expected saturation population densities and wastewater flow rates for various types of housing and developments were derived from the actual records published in the zoning plans and planning reports of the area. The sewer layout plan, types of developments and area served, and other information are given in the location map (Figure 8.11). The planning and design information of the sewer district, such as peaking factors of commercial, industrial, and institutional areas, saturation population densities, average unit flow rates, and ground surface elevations, are provided in Tables 8.5 and 8.6. The peak infiltration/inflow (I/I) allowance for newly developed residential areas is 14 m^3/ha.day up to 50 ha, then the allowance decreases as the area increases. This relationship is shown in Figure 4.3. The total length of sewer lines in the commercial, industrial, and institutional areas is nearly half that of the residential areas. Therefore, the peak I/I allowance in these areas is assumed to be half that of residential areas. Estimate the design peak flow rates for each sanitary sewer.

Solution

1. Describe briefly the process of laying out sanitary sewers

   Establish the ultimate limits of the drainage areas and the sources of wastewater contribution. Mark the drainage areas and the sub-areas of each sewer line. Layout

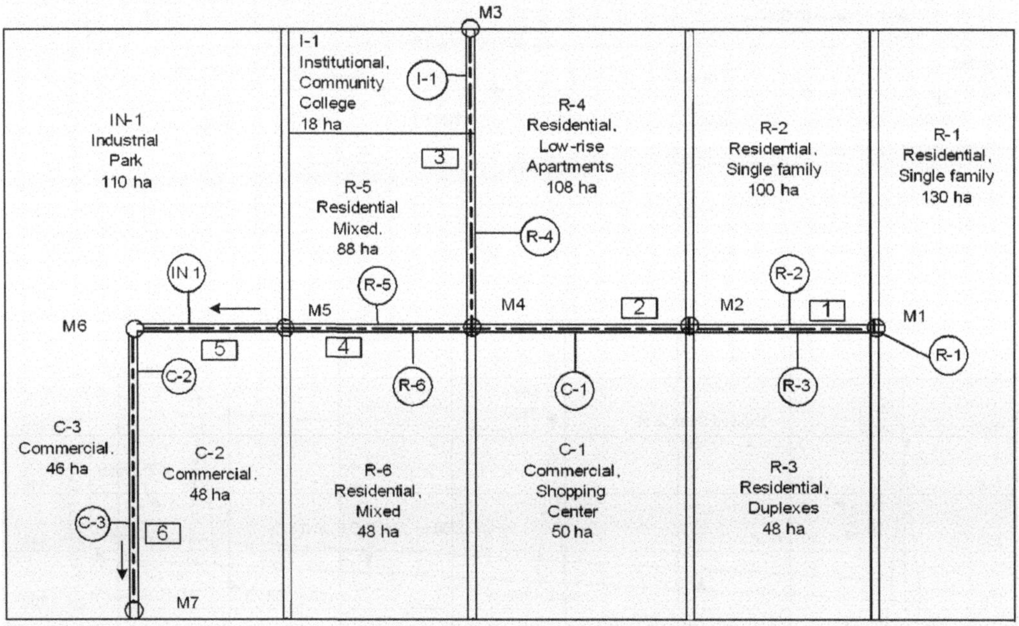

FIGURE 8.11 Location map and sewer layout of the planned subdivision for design of sanitary sewer (Example 8.13). C, commercial; I, institutional; IN, industrial; R, residential.

TABLE 8.5
Planning and Design Information for Sanitary Sewer (Example 8.13)

Sewer			Upper Manhole		Lower Manhole	
No.	Length, m	Service Area	No.	Ground Surface Elevation, m	No.	Ground Surface Elevation, m
1	120	Residential, R-1, R-2, R-3	M1	204.50	M2	204.25
2	130	Commercial, C-1	M2	204.25	M4	204.10
3	160	Residential, R-4, Institutional, I-1	M3	205.00	M4	204.10
4	125	Residential, R-5, R-6	M4	204.10	M5	203.94
5	110	Industrial, IN-1	M5	203.94	M6	203.85
6	160	Commercial, C-2, C-3	M6	203.85	M7	203.70

the trunk lines and number them. Follow the procedure given in Section 8.2.4 for establishing the desired layout. Locate the manholes and number them.

2. Describe the process of flow generation

   Establish the saturation population densities of the residential areas. Identify residential, commercial, institutional, and industrial zones. Determine average unit wastewater generation rate and peaking factors for each area. Calculate peak dry weather flows from residential, commercial, and institutional areas, including peak I/I. Finally, establish the combined peak design wet weather flows. Procedure for calculating each component of the flows is given below.

3. Prepare a computation table to determine the peak flows from the residential areas.

   The residential areas are located in service areas R-1 through R-6 and are served by lines 1, 3, and 4. The cumulative flows are carried through all downstream lines. The calculation steps are summarized in Table 8.7.

TABLE 8.6
Planning and Design Information for Sewer District (Example 8.13)

Area Designation	Size, ha	Description	Population Density, People/ha	Flow Rate Lpcd	Flow Rate m³/ha.day	Peaking Factor
R-1	130	Residential, single-family housing, high income	30	350		Eq. (4.4)
R-2	100	Residential, single-family housing, medium income	38	300		Eq. (4.4)
R-3	48	Residential, duplexes	50	260		Eq. (4.4)
R-4	108	Residential, low-rise apartments	124	220		Eq. (4.4)
C-1	50	Commercial, shopping center	—		15	1.8
R-5	88	Residential, mixed	70	255		Eq. (4.4)
I-1	18	Institutional, community college (day population)	3,000 (total)	70		4.0
R-6	48	Residential mixed	100	250		Eq. (4.4)
C-2	48	Commercial	—		10	1.8
C-3	46	Commercial	—		10	1.8
IN-1	110	Industrial	—		30	2.1

Note: The flow rates (m^3/ha.day) for commercial and industrial zones are estimated from density and types of establishments.

4. Prepare a computation table to determine the peak flows from institutional, commercial, and industrial or institutional areas

   The commercial contributions reach sewer lines 2, 5, and 6, and institutional and industrial contributions reach from lines 3 and 5, respectively. The calculation steps for flow generation from commercial, industrial, and institutional areas are summarized in Table 8.8.
5. Prepare a computational table to determine peak I/I allowance from all areas.

   The peak I/I allowances from residential, commercial, industrial, and institutional areas are calculated and summarized in Table 8.9.
6. Summarize the design cumulative peak residential, commercial, institutional, and industrial flows, and peak I/I flows

   The cumulative peak residential, commercial, industrial, and institutional flows are developed in Table 8.8. Also, peak I/I allowances are developed and summarized in Table 8.9. Finally, the peak flows from residential, commercial, industrial, and institutional areas, as well as peak I/I allowances from all these areas, are added to obtain peak design flow for each sewer line. These calculations are summarized in Table 8.10.

Example 8.14: Design and Computational Tables for Design of Sanitary Sewers in Example 8.13

The newly developed city subdivision is shown in Figure 8.11. The peak design flows from residential, commercial, industrial, and institutional areas, and peak I/I allowance from all these areas were calculated in Example 8.13 and are summarized in Tables 8.6–8.9. The cumulative peak design flows for each sewer line are provided in Table 8.10. Design each sewer line and draw the typical profile of lines 1, 2, 4, and 5. Match the crowns of these lines. Show the invert elevations of lines 3 and 6. Minimum ground cover established by

TABLE 8.7
Peak Design Flow Estimates from Residential Areas (Example 8.13)

Sewer Line, Upper and Lower Manholes (1)	Area ID and Size, ha (2)	Population Density, Persons/ha (3)	Population, Persons (4) = (2) × (3)	Cumulative Population Served by Line, Person (5)	Unit Flow, Lpcd (6)	Average Flow, m³/day (7) = (4) × (6)	Cumulative Flow for Line, m³/day (8)	Peaking Factor, M (9)	Total Peak Design Residential Flow, m³/day (10) = (8) × (9) Except for Line 3
Line 1	R-1, 130	30	3,900	3,900	350	1,365	1,365	2.95[a]	9,231[b]
M1–M2	R-2, 100	38	3,800	7,700	300	1,140	2,505		
	R-3, 48	50	2,400	10,100	260	624	3,129		
Line 2	No direct residential	—	—	10,100	—	—	3,129	2.95[a]	9,231[c]
M2–M4	sub-areas served								
Line 3	R-4,108	124	13,392	13,392	124	1,661	1,661	2.83[d]	4,701[e]
M3–M4									
Line 4	R-5, 88	70	6,160	29,652	255	1,571	1,571	2.42[f]	18,298[g]
M4–M5	R-6, 48	100	4,800	34,452	250	1,200	2,771		
Line 5	No direct residential	—	—	34,452	—	—	7,561	2.42[f]	18,298[h]
M5–M6	sub-areas served								
Line 6	No direct residential	—	—	34,452	—	—	7,561	2.42[f]	18,278[i]
M6–M7	sub-areas served								

[a] Equation (4.4) $M = 1 + \frac{14}{4+\sqrt{P}} = 1 + \frac{14}{4+\sqrt{10.1}} = 2.95$.

[b] Line 1 receives residential flows from areas R-1, R-2, and R-3. $Q = 3{,}129\ \text{m}^3/\text{day} \times 2.95 = 9{,}231\ \text{m}^3/\text{day}$.

[c] Line 2 does not serve residential sub-areas, but it receives residential flows from line 1 = 9,231 m³/day.

[d] $M = 1 + \frac{14}{4+\sqrt{13{,}392}} = 2.83$.

[e] Line 3 receives residential flow from area R-4. $Q = 1{,}661\ \text{m}^3/\text{day} \times 2.82 = 4{,}701\ \text{m}^3/\text{day}$.

[f] $M = 1 + \frac{14}{4+\sqrt{34{,}452}} = 2.42$.

[g] Line 4 carries residential flow from areas R-5 and R-6, and residential flows from lines 2 and 3. $Q = (2{,}771 + 1{,}661 + 3{,}129)\ \text{m}^3/\text{day} \times 2.42 = 18{,}298\ \text{m}^3/\text{day}$.

[h] Line 5 does not serve residential sub-areas, but receives residential flows from line 4 = 18,298 m³/day.

[i] Line 6 does not serve the residential sub-areas, but receives residential flow from line 5 = 18,298 m³/day.

TABLE 8.8
Peak Design Flow Estimation from Commercial, Industrial, and Institutional Areas (Example 8.13)

Sewer Line Upper and Lower Manhole	Types of Area Served	Area ID and Size, ha	Unit Flow, m^3/ha.day	Population Served, Person	Unit Flow, Lpcd	Average Flow, m^3/day	Peaking Factor, M	Peak Flow, m^3/day	Cumulative Peak Flow, m^3/day
Line 1 M1–M2	Only residential areas served. See Table 8.7	—		—	—	—	—	—	—
Line 2 M2–M4	Commercial	C-1, 50	15	—	—	750	1.8	1,350	1,350
Line 3 M3–M4	Institutional and residential areas served. See Table 8.7 for residential areas	I-1, 18	—	3,000	70	210	4.0	840	840
Line 4 M4–M5	Only residential areas served. See Table 8.7	—	—	—	—	—	—	—	—
Line 5 M5–M6	Industrial	IN-1, 110	30	—	—	3,300	2.1	6,930	6,930
Line 6 M6–M7	Commercial	C-2, 48	10	—	—	480	1.8	864	
		C-3, 46	10	—	—	460	1.8	828	1,692

TABLE 8.9
Peak I/I Allowance from Residential, Commercial, Industrial, and Institutional, and Cumulative Peak I/I in Each Line (Example 8.13)

Service Line, and Upper and Lower Manholes	Type of Area Served	Area ID	Area Size, ha	Peak I/I Rate,[a] m^3/ha.day	Peak I/I Rate Applied, m^3/ha.day	Peak I/I Reaching Line, m^3/day	Cumulative Peak I/I in Each Line, m^3/day
Line 1	Residential	R-1	130	10.1	10.1	1,313	1,313
M1–M2	Residential	R-2	100	12.0	12.0	1,200	1,313 + 1,200 = 2,513
	Residential	R-3	48	14.0	14.0	672	672 + 2,513 = 3,185
Line 2 M2–M4	Commercial	C-1	50	13.9	6.95[b]	348	348 + 3,185 = 3,533
Line 3	Institutional	I-1	18	14.0	7.00[b]	126	126
M3–M4	Residential	R-4	108	11.8	11.8	1,274	126 + 1,274 = 1,400
Line 4	Residential	R-5	88	11.9	11.9	1,047	1,047
	Residential	R-6	48	14.0	14.0	672	672 + 1,047 + 1,400 + 3,533 = 6,652
Line 5 M5–M6	Industrial	IN-1	110	11.7	5.85[b]	644	644 + 6,652 = 7,296
Line 6	Commercial	C-2	48	14.0	7.00[b]	336	
M6–M7	Commercial	C-3	46	14.0	7.00[b]	322	336 + 322 + 7,296 = 7,954

[a] Peak I/I allowance is obtained from Figure 4.5.
[b] Peak I/I allowance in commercial, industrial, and institutional areas is half of residential areas because the total length of sewer lines in these areas is assumed half that of residential areas.

the local community building code is 1.0 m. The minimum recommended slope of sanitary sewers that gives 2 ft/s (0.6 m/s) velocity when flowing full between the upper and lower manholes is given in Table 8.4.

Solution

The procedure to design a sewerage system involves tedious and repetitive calculations carried out in a tabular format. The following steps describe the procedure

1. Select the preliminary diameter and slope
   The preliminary diameter and slope of all sewer lines are selected. The slopes are set to follow the natural ground slope. The velocity in the sewer at full capacity should be greater than 2 ft/s (0.60 m/s). Velocity above 2 ft/s prevents the deposition of solids at low flows.
2. Determine the ground surface elevations at the manholes.
   The ground surface elevations at manholes are obtained by interpolation from the topographical map. The information is provided in Tables 8.5 and 8.11.
3. Prepare the computational table to arrange the preliminary design data
   The computational table contains sewer number, upper and lower manhole numbers, sewer length, peak design flow, sewer diameter and slope, flow and

TABLE 8.10
Peak Design Flow for Each Sewer Line (*Example 8.13*)

Sewer Line		Upper Manhole		Lower Manhole		Peak Residential Flow from Table 8.7, m^3/Day	Peak Commercial Flow from Table 8.8, m^3/Day	Peak Industrial Flow from Table 8.8, m^3/Day	Peak Institutional Flow from Table 8.8, m^3/Day	Peak I/I Allowance from Table 8.9, m^3/Day	Total Peak Design Flow	
No.	Length	No.	Ground Surface Elevation, m	No.	Ground Surface Elevation, m						m^3/Day	m^3/s
1	120	M1	204.50	M2	204.25	9,231	—	—	—	3,185	12,416	0.14
2	130	M2	204.25	M4	204.10	9,231	1,350	—	—	3,533	14,114	0.16
3	160	M3	205.00	M4	204.10	4,701	—	—	840	1,400	6,941	0.08
4	125	M4	204.10	M5	203.94	18,298	1,350	—	840	6,652	27,140	0.31
5	110	M5	203.94	M6	203.85	18,298	1,350	6,930	840	7,296	34,714	0.40
6	160	M6	203.85	M7	203.70	18,298	3,042	6,930	840	7,954	37,064	0.43

The peak design flow includes peak residential, commercial, industrial, and I/I allowance from all areas (Example 8.13).

TABLE 8.11
Computation Table for Design of Sanitary Sewer (Example 8.14)

A

						Full Flow Conditions			Partial Flow Conditions				
Sewer Line (1)	Upper Manhole (2)	Lower Manhole (3)	Length, m(4)	Peak Design Flow q, m^3/s (5)	Sewer Dia. D, m (in.) (6)	Slope (7)	Q, m^3/s (8)	V, m/s (9)	q/Q (10)	d/D (11)	v/V (12)	v, m/s (13)	Depth of flow d, m (14)
1	M1	M2	120	0.14	0.46 (18)	0.00250	0.15	0.91[a]	0.93	0.75	1.15	1.05	0.35
2	M2	M4	130	0.16	0.53 (21)[b]	0.00150	0.17[c]	0.77	0.94	0.76	1.16	0.90	0.40
3	M3	M4	160	0.08	0.38 (15)	0.00300	0.09	0.85	0.93	0.75	1.15	0.97	0.27
4	M4	M5	125	0.31	0.91 (36)	0.00085	0.54	0.84	0.57	0.54	0.98	0.82	0.49
5	M5	M6	110	0.40	0.91 (36)[d]	0.00095	0.57	0.88	0.70	0.62	1.14	1.00	0.56
6	M6	M7	160	0.43	0.91 (36)[d]	0.00100	0.59	0.91	0.73	0.63	1.15	1.05	0.57

B

Ground Elevation Upper End, m (15)	Invert Elevation Upper End, m (16)	Cover, m (17)	Fall of Sewer, m (18)	Invert Elevation Lower End, m (19)	Ground Elevation Lower End, m (20)	Cover, m (21)
204.50	202.49	1.50	0.300	202.19	204.25	1.55
204.25	202.11	1.55	0.200	201.91	204.10	1.60
205.00	202.55	2.03	0.480	202.07	204.10	1.60
204.10	201.51	1.60	0.110	201.40	203.94	1.55
203.94	201.32	1.55	0.110	201.21	203.85	1.65
203.85	201.13	1.65	0.160	201.97	203.70	1.74

[a] (Eq. 5.34) $= V = \dfrac{1}{0.013}\left(\dfrac{0.46\text{ m}}{4}\right)^{2/3} \times (0.0025)^{1/2} = 0.91\text{ m/s}$.

[b] The outgoing sewer has larger diameter than incoming sewer.

[c] Q (continuity equation) $= \pi/4\ (0.53\text{ m})^2 \times 0.77\text{ m/s} = 0.17\text{ m}^3/\text{s}$.

[d] The diameters of incoming and outgoing sewers are the same. The invert elevation of outgoing sewer is lowered by about 2–3 in. (0.05–0.08 m).

velocity when running full, ratios of q/Q, d/D and v/V under partial flow condition, ground surface and invert elevations, and ground cover. The sewer number, upper and lower manhole numbers, sewer length, peak design flows, and ground surface elevations are obtained from Figure 8.11 and Tables 8.5 and 8.10. The sewer slope and diameter are selected to follow the natural ground slope, ground cover, and minimum velocity requirement. The discharge Q and velocity V when flowing full are calculated from continuity and Manning's equations. The partial flow ratio d/D and v/V are obtained from Figure 5.26b. Finally, d and v are calculated from these ratios. These entries are provided in the computation table (Table 8.11).

4. Establish the invert elevations of sewers

   The invert elevations of sewers are established to meet ground slope, ground cover, and velocity requirements. The following steps are applied:

   a. Locate the invert elevation of the upper end of the sewer. The invert elevation of the upper end of the sewer is selected such that the minimum ground cover requirement is satisfied. The upper and invert elevation is calculated as follows:

$$\text{Upper invert elevation} = (\text{ground surface elevation}) - (\text{ground cover}) - (\text{sewer wall thickness}) - (\text{sewer diameter})$$

The wall thickness of standard sewers is listed in Table 8.4. The standard diameters of sewers used in this example are 0.38 m (15 in.), 0.46 m (18 in.), 0.53 m (21 in.), and 0.91 m (36 in.). The wall thicknesses of these sewers are, respectively, 47.63 mm (1⅞ in.), 50.80 mm (2 in.), 57.15 mm (2¼ in.), and 76.20 mm (3 in.).

The sample calculations for upper end of sewer 1 at manhole *M*1 are given below.

Invert deviation at M1 = 204.50 m (ground level elevation) − 1.50 m (ground cover) − 0.05 m (wall thickness) − 0.46 m (sewer diameter) = 202.49 m

b. Proper sewer alignment is necessary to achieve smooth flow transition through the manhole and other appurtenances. The crowns of all sewers of straight path, such as sewers 1, 2, 4, and 5 in the Example problem, should be matched. The diameter of the outgoing sewer is normally the largest; therefore, invert elevation will be the lowest, and there will be a natural drop at the manhole. If the sewer size of incoming and outgoing sewers remains the same, then the invert of the outgoing sewer is often lowered by about 7.6 cm (3 in.). The invert of incoming sewer from the cross direction is kept at least 7.6 cm (3 in.) higher than the manhole invert. A drop manhole may be necessary if the elevation of the incoming sewer is higher than 0.6 m (2 ft) from the manhole invert. The outgoing sewer in the cross direction should have the lowest invert elevation. Adapting these invert elevations will keep smooth transition through manhole without backflow wastewater into the smaller pipe. This topic is covered in Section 8.2.7. Also check the invert elevations at manholes *M*1, *M*2, *M*4, *M*5, and *M*6 in Table 8.11.

c. The lower end invert elevation is computed by subtracting the drop fall of sewer in its entire length.

Lower end invert elevation of sewer = (upper end invert elevation) − sewer slope × length of sewer)

The sample calculation for Sewer Line 1 is given below.

$$\text{Lower end sewer invert elevation} = 202.49\text{ m} - 0.00250\text{ m/m} \times 120\text{ m}$$

$$= 202.49\text{ m} - 0.30\text{ m} = 202.19\text{ m}$$

5. Prepare the sewer profile worksheet

The design calculations and data of the sewerage system are developed and summarized in columns (1) through (21) of Table 8.11. The sewer profile is prepared by plotting elevations on *Y*-axis, and sewer length or coordinate on *X*-axis. The preliminary profile of the sanitary sewers 1, 2, 4, and 5 is shown in Figure 8.12. The sectional view of sewers 3 and 6 is also shown at manholes *M*4 and *M*6, respectively.

Example 8.15: Design of a Sewerage System

The sewerage system of a city subdivision is extended to accommodate future growth. The street layout is shown in Figure 8.13. The proposed sewer layout is marked by dotted lines. The lateral sewers run north-south and discharge into the sub-main, which runs along the River Drive. The average municipal wastewater flow is 100 gpcd. This sewerage system is designed for peak design flow. The peak wet weather flow is 3 times the average flow and includes the peak I/I contribution. Also, an industrial plant discharges liquid waste at a maximum flow rate of 1.27 mgd. This plant is located on the north of Center Street and between

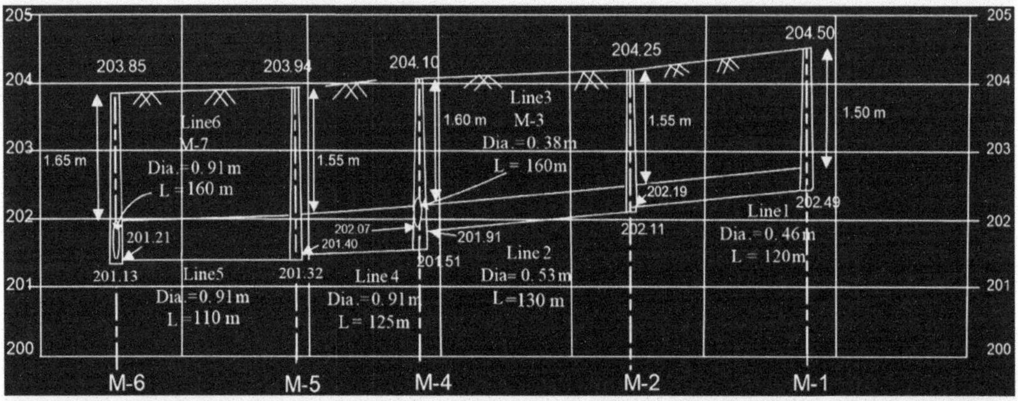

FIGURE 8.12 Sewer profile of inline sewers 1, 2, 4, and 5, and cross-section of sewers 3 and 6 (Example 8.14).

Sixth and Eighth Streets. The information obtained from the city survey includes the length of each block and the elevation of pavement surface at each street intersection. The invert elevation of the sewer line shall be at least 5.5 ft below the pavement surface. The proposed sub-main will connect to an existing sub-main that runs along the River Drive, and starts from 11th Street. The invert elevation of manhole 11 (MH 11) at the intersection of the 11th Street and River Drive is 374.52 ft. The existing sewers and sub-main are shown by solid lines. Design the sewage system. The minimum velocity at full flow must not be less than 2 ft/s. Match the crowns of the incoming and outgoing sewer lines on straight paths. Minimum sewer diameter is 8 in. and $n = 0.013$.

Note: The purpose of matching the sewer crown of straight path is discussed in Section 8.2.7.

Solution

Describe the procedure for sewer alignment: The sewer alignment is started from the lower end at the existing MH 11. This manhole has a fixed invert elevation at 374.52 ft. The upward sewer lines 20 and 19 connect to MH 10 and MH *j*. These sewer lines should have slopes greater than the minimum slope to maintain a velocity higher than 2 ft/s when flowing full, and ground cover at manholes greater than 6 ft. These requirements must be met with every selected sewer line and manhole.

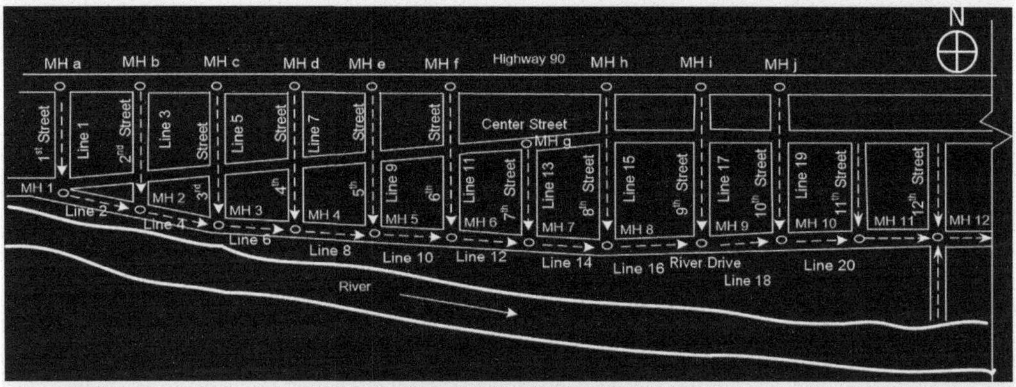

FIGURE 8.13 Layout of the proposed sanitary sewer system (Example 8.15).

1. **Generate the peak design flow for each sewer line**: The peak design flow is generated in each sewer line using the area, population density, average and peak wet weather flow data, and industrial flow. This information is summarized in columns (1) through (13) of Table 8.12.
2. **Design the sewerage system**: The design calculations of sewer lines and their alignments are started from MH 11 since the invert elevation of this manhole is fixed at 374.52 ft. The calculations are carried out upward. Select the diameter *D* and slope *S* of sewer lines 20 and 19. The slope should be compatible with the natural ground slope, ground cover, and minimum velocity requirements. Calculate the full flow velocity *V* and discharge *Q* from Manning's and continuity equations. The velocity *V* should be larger than 2 ft/s. The peak design flows of sewer lines 20, 19, and other sewers are listed in Table 8.12. Calculate *q/Q* ratios and determine *d/D* and *v/V* ratios from Figure 5.26b. This information is summarized in columns (1) through (14) of Table 8.13. From these ratios and selected *D* and *V* data, calculate the depth *d* and velocity *V* under partial flow conditions. Similarly, develop the information for all sewers. The *d* and *V* values of all sewers are listed in columns (15) and (16) of Table 8.13. The crown elevations of all sewers in straight path are matched. The invert elevations of upper and lower manholes and ground surface elevation of all sewers are given in columns (17) through (21) of Table 8.13.
3. **Draw the sewer alignment along the River Drive**: The generalized sewer alignment of all sewers in the straight path along the River Drive is shown in Figure 8.14. In this alignment, the invert elevations of all cross sewers meeting at the manhole are given.

The generalized sewer alignment is shown in Figure 8.14.

8.3 STORM SEWERS

Storm sewers are designed and built to collect surface runoff from streets, lawns, roofs, and other sources. Sanitary wastewater is totally excluded. The flow of surface runoff for the design of storm sewer is difficult to estimate. It depends upon the intensity of rainfall and surface and subsurface conditions. Both factors are highly variable and difficult to predict. The design of storm drainage system requires study of (1) boundary of drainage or tributary area and characteristics, (2) frequency and intensity of rainfall, (3) time of concentration, (4) stormwater runoff, and (5) design of storm sewer.

8.3.1 Boundary of Drainage Area and Characteristics

The boundary of area tributary and its characteristics must be fully investigated. This requires a preliminary layout of the system and a tentative location of inlet points. Therefore, consideration must be given to the street network, the number of homes and establishments that will be built over the design period, related factors when estimating the imperviousness of the area, and the time of concentration. A survey map is prepared to show all existing and proposed streets, buildings, contour lines, and elevations at all important points. The runoff coefficients from different tributary areas are given in Table 8.14.

TABLE 8.12
Generate the Peak Design Flow for Each Sewer Line (Example 8.15)

Line No.	Location	Manhole (MH)		Length, ft	Area, Acre		Population, Persons		Wastewater, mgd		Industrial Peak Flow, mgd	Total Peak Design Flow, mgd
		From	To		Increment	Cumulative	Per acre	Total	Avg.	Peak		
(1)	(2)	(3)	(4)	(5)	(6)	(7)	(8)	(9)	(10)	(11)	(12)	(11) + (12) = (13)
1	First Street	MH a	MH 1	600	4		50	200	0.020	0.060[a]	—	0.060
2	River Dr.	MH 1	MH 2	400	3	7	50	350	0.035	0.105	—	0.105
3	Second Street	MH b	MH 2	700	5		50	250	0.025	0.075	—	0.075
4	River Dr.	MH 2	MH 3	400	4	16	50	800	0.080	0.240	—	0.240
5	Third Street	MH c	MH 3	800	6		50	300	0.030	0.090	—	0.090
6	River Dr.	MH 3	MH 4	400	4	26	50	1,300	0.130	0.390	—	0.390
7	Fourth Street	MH d	MH 4	840	6		50	300	0.030	0.090	—	0.090
8	River Dr.	MH 4	MH 5	400	4	36	50	1,800	0.180	0.540	—	0.540
9	Fifth Street	MH e	MH 5	880	7		50	350	0.035	0.105	—	0.105
10	River Dr.	MH 5	MH 6	400	4	47	50	2,350	0.235	0.705	—	0.705
11	Sixth Street	MH f	MH 6	920	7		50	350	0.035	0.105	—	0.105
12	River Dr.	MH 6	MH 7	400	4	58	50	2,900	0.290	0.870	—	0.870
13	Seventh Street	MH g	MH 7	600	5		50	250	0.025	0.075	1.27	1.345[b]
14	River Dr.	MH 7	MH 8	400	4	67	50	3,350	0.335	1.005	—	2.350
15	Eighth Street	MH h	MH 8	1,000	8		50	400	0.040	0.120	—	0.120
16	River Dr.	MH 8	MH 9	400	4	79	50	3,950	0.395	1.185	—	2.455
17	Ninth Street	MH i	MH 9	980	8		50	400	0.040	0.120	—	0.120
18	River Dr.	MH 9	MH 10	400	4	91	50	4,550	0.455	1.365	—	2.635
19	Tenth Street	MH j	MH 10	960	7		50	350	0.035	0.105	—	0.105
20	River Dr.	MH 10	MH 11	400	4	102	50	5,100	0.510	1.530	—	2.800

[a] 4 acre × 50 person/acre × 100 gpcd × 3 (peaking factor) × 10^{-6} MG/gal = 0.060 mgd.

[b] 0.075 mgd + 1.27 mgd = 1.345 mgd.

TABLE 8.13
Tabulation of Data for Design of Sanitary Sewers for a City Subdivision (Example 8.15)

A

Line		Manhole		Length	Total Peak Flow		Dia. D	Slope S	Vel. V	Q	q/Q	d/D	v/V
No.	Location	From	To	ft	mgd	cfs	in.		ft/s	cfs			
(1)	(2)	(3)	(4)	(5)	(6)	(7)	(8)	(9)	(10)	(11)	(12)	(13)	(14)
1	First Street	MH a	MH 1	600	0.060	0.09	8	0.0033	2.00	0.70	0.13	0.26	0.67
2	River Dr.	MH 1	MH 2	400	0.105	0.16	8	0.0033	2.00	0.70	0.23	0.35	0.80
3	Second Street	MH b	MH 2	700	0.075	0.12	8	0.0033	2.00	0.70	0.17	0.28	0.73
4	River Dr.	MH 2	MH 3	400	0.240	0.37	8	0.0033	2.00	0.70	0.53	0.53	1.03
5	Third Street	MH c	MH 3	800	0.090	0.14	8	0.0033	2.00	0.70	0.20	0.32	0.74
6	River Dr.	MH 3	MH 4	400	0.390	0.60	8	0.0033	2.00	0.70	0.86	0.72	1.14
7	Fourth Street	MH d	MH 4	840	0.090	0.14	8	0.0033	2.00	0.70	0.20	0.32	0.74
8	River Dr.	MH 4	MH 5	400	0.540	0.84	10	0.0025	2.00	1.36	0.62	0.56	1.05
9	Fifth Street	MH e	MH 5	880	0.105	0.16	8	0.0033	2.00	0.70	0.23	0.34	0.78
10	River Dr.	MH 5	MH 6	400	0.705	1.09	12	0.0027	2.40	1.89	0.58	0.55	1.05
11	Sixth Street	MH f	MH 6	920	0.105	0.16	8	0.0033	2.00	0.87	0.18	0.32	0.74
12	River Dr.	MH 6	MH 7	400	0.870	1.35	12	0.0019	2.00	1.57	0.86	0.72	1.14
13	Seventh Street	MH g	MH 7	600	1.345	2.08	15	0.0015	2.00	2.45	0.85	0.70	1.13
14	River Dr.	MH 7	MH 8	400	2.350	3.64	21	0.0033	3.58	8.63	0.42	0.46	0.95
15	Eighth Street	MH h	MH 8	1,000	0.120	0.19	8	0.0043	2.27	0.79	0.24	0.35	0.80
16	River Dr.	MH 8	MH 9	400	2.455	3.80	21	0.00092	2.00	4.81	0.79	0.67	1.13
17	Ninth Street	MH i	MH 9	980	0.120	0.19	8	0.00330	2.00	0.70	0.27	0.35	0.83
18	River Dr.	MH 9	MH 10	400	2.635	4.08	24	0.00077	2.00	6.28	0.65	0.58	1.07
19	Tenth Street	MH j	MH 10	960	0.105	0.16	8	0.00330	2.00	0.70	0.23	0.34	0.78
20	River Dr.	MH 10	MH 11	400	2.800	4.33	24	0.00077	2.00	6.28	0.69	0.62	1.09

(Continued)

TABLE 8.13 (*Continued*)
Tabulation of Data for Design of Sanitary Sewers for a City Subdivision (Example 8.15)

B

Line No. (1)	Location (2)	Manhole From (3)	Manhole To (4)	Length ft (5)	Total Peak Flow mgd (6)	Total Peak Flow cfs (7)	Dia. *D* in. (8)	Slope *S* ft/ft (9)	*d* in. (15)	*v* ft/s (16)	Ground Elev. Upper End ft (17)	Invert Elev. Upper End ft (18)	Fall of Sewer ft (19)	Invert Elev. Lower End ft (20)	Ground Elev. Lower End ft (21)
1	1st Street	MH a	MH 1	600	0.060	0.09	8	0.0033	2.10	1.34	394.43	387.48	1.98	385.50	392.45
2	River Dr.	MH 1	MH 2	400	0.105	0.16	8	0.0033	2.80	1.60	392.45	385.07	1.32	383.75	390.42
3	2nd Street	MH b	MH 2	700	0.075	0.12	8	0.0033	2.24	1.46	394.21	386.51	2.31	384.20	390.24
4	River Dr.	MH 2	MH 3	400	0.240	0.37	8	0.0033	4.24	2.06	390.24	383.75	1.32	382.43	388.77
5	3rd Street	MH c	MH 3	800	0.090	0.14	8	0.0033	2.56	1.48	394.10	385.64	2.64	383.00	388.97
6	River Dr.	MH 3	MH 4	400	0.390	0.60	8	0.0033	5.76	2.28	388.77	382.18[a]	1.32	380.86	388.51
7	4th Street	MH d	MH 4	840	0.090	0.14	8	0.0033	2.56	1.48	393.61	383.92	2.77	381.15	388.51
8	River Dr.	MH 4	MH 5	400	0.540	0.84	10	0.0025	5.60	2.10	388.51	380.86	1.00	379.86	387.79
9	5th Street	MH e	MH 5	880	0.105	0.16	8	0.0033	2.72	1.56	393.03	383.15	2.90	380.25	387.78
10	River Dr.	MH 5	MH 6	400	0.705	1.09	12	0.0027	6.96	2.62	387.79	379.69[b]	1.10	378.59	387.60
11	6th Street	MH f	MH 6	920	0.105	0.16	8	0.0033	2.56	1.48	392.47	382.04	3.04	379.00	387.60
12	River Dr.	MH 6	MH 7	400	0.870	1.35	12	0.0019	8.64	2.28	387.60	378.59	0.76	377.83	387.01
13	7th Street	MH g	MH 7	600	1.345	2.08	15	0.0015	10.50	2.70	391.23	378.90	0.90	378.00	387.01
14	River Dr.	MH 7	MH 8	400	2.350	3.64	21	0.0033	9.66	3.40	387.01	377.08[b]	1.32	375.76	385.89
15	8th Street	MH h	MH 8	1,000	0.120	0.19	8	0.0043	2.80	1.82	391.19	380.39	4.63	376.00	385.89
16	River Dr.	MH 8	MH 9	400	2.455	3.80	21	0.00092	14.07	2.26	385.89	375.76	0.37	375.39	385.31
17	9th Street	MH i	MH 9	980	0.120	0.19	8	0.00330	2.96	1.66	392.97	378.73	3.23	375.50	385.89
18	River Dr.	MH 9	MH 10	400	2.635	4.08	24	0.00077	13.92	2.14	385.31	375.14[b]	0.31	374.83	383.92
19	10th Street	MH j	MH 10	960	0.105	0.16	8	0.00330	2.72	1.56	392.50	378.00	3.17	374.83	383.92
20	River Dr.	MH 10	MH 11	400	2.800	4.33	24	0.00077	14.88	2.18	383.92	374.83	0.31	374.52	383.35

[a] 3-in. drop in manhole invert elevation.
[b] Drop in manhole invert elevation due to increase in the diameter of outgoing sewer.

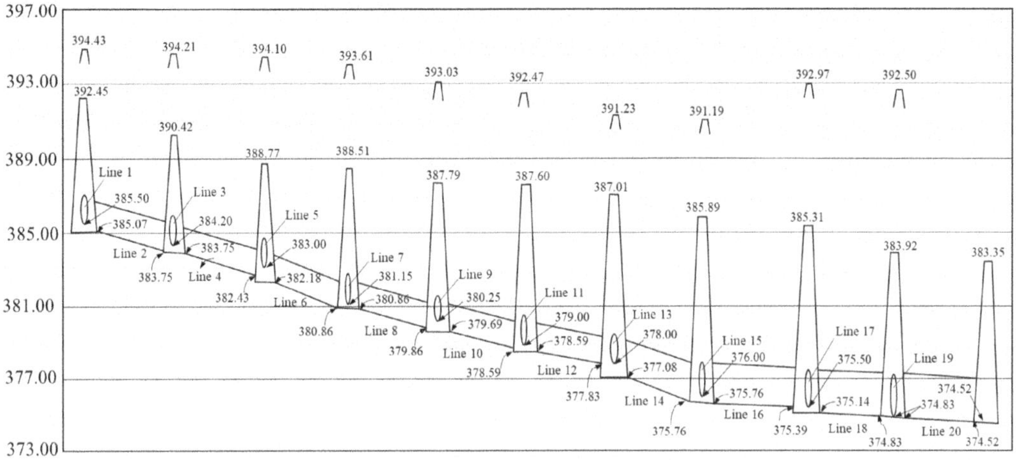

FIGURE 8.14 Longitudinal section of sewer lines on straight path on River Drive, and sectional view of cross sewer lines meeting the longitudinal sewer lines (Example 8.15). Note: Ground cover = Ground surface elevation – (invert elevation + pipe wall thickness). Sewer diameter and wall thickness are given in Table 8.4.

TABLE 8.14
Coefficient of Runoff for Various Types of Areas

Type of Development	Coefficient of Runoff
Residential	
Single family	0.30–0.50
Multi-units	0.40–0.70
Apartments	0.50–0.70
Business	
Downtown	0.70–0.95
Commercial	0.85–0.90
Industrial	0.50–0.90
Streets	
Asphalt and concrete	0.70–0.95
Brick	0.70–0.85
Other Areas	
Roofs	0.75–0.95
Lawns, Sandy Soil	
Flat, 2% or less	0.05–0.10
Average, 2%–7%	0.10–0.15
Steep, 7% or greater	0.15–0.20
Lawns, Heavy Soil	
Flat, 2% or less	0.13–0.17
Average, 2%–7%	0.18–0.22
Steep, 7% or greater	0.25–0.35

Source: Adapted in part from Refs. 1,2,6,7

Example 8.16: Average Coefficient of Runoff

A residential district has the following land-use and runoff coefficients. Determine the average coefficient of runoff for the entire district.

Roofs, 20%, $C = 0.90$
Pavement, 25%, $C = 0.85$
Lawns, gardens, vacant lots, 50%, $C = 0.10$
Wooded, 5%, $C = 0.05$

Solution

Calculate weighted average of the coefficient of runoff for the entire district

Assume district area = A

$$C_{avg} = \frac{A_1C_1 + A_2C_2 + A_3C_3 + \cdots}{A_1 + A_2 + A_3}$$

$$C_{avg} = \frac{0.20A \times 0.90 + 0.25A \times 0.85 + 0.50A \times 0.10 + 0.05A \times 0.05}{0.20A + 0.25A + 0.50A + 0.05A}$$

$$= \frac{0.18 + 0.213 + 0.05 + 0.003}{1} = 0.446$$

The weighted average coefficient of runoff for the district = 0.45

8.3.2 Frequency and Intensity of Rainfall

The rainfall intensity–duration–frequency (IDF) data are derived from rain-gauge measurement records. These records are reasonably numerous for the large cities, but not for small communities. The U.S. National Weather Service maintains a large network of stations where such data are collected and published. The average frequency of rainfall occurrence used for sewer design is based on the degree of protection provided by a given storm sewer system. The range of frequencies used in engineering offices is:

Residential areas 2–15 years, with 5 years most common
Commercial and high-value districts 10–50 years
Flood protection works for 50 years or more.

Rainfall–intensity–frequency curves: The intensity of rainfall is dependent on the duration of rainfall. Short-duration and less frequent storms are more intense. The typical IDF curves are shown in Figure 8.15.

Empirical relations: In the absence of long-term gauge measurements and other field data, empirical relations have been developed to express the rainfall intensity. Some of these relations are given by Eqs. (8.1) and (8.2).

$$i = \frac{A}{t + B} \tag{8.1}$$

$$i = \frac{C}{t^k} \tag{8.2}$$

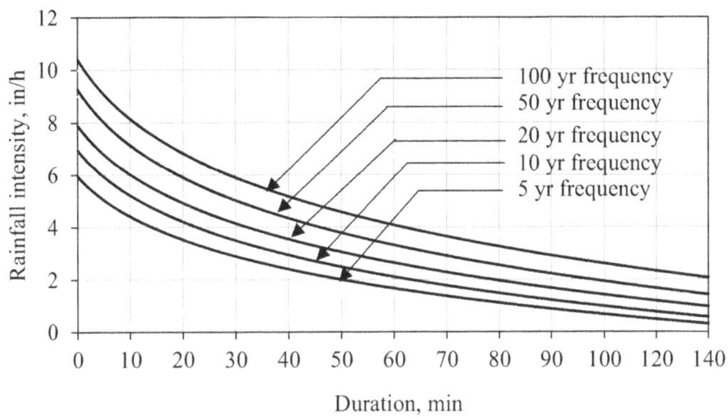

FIGURE 8.15 Typical intensity–duration–frequency curves of northeastern region of the United States.

where

i = intensity of rainfall, in/h
t = storm duration, min
A, B, and C = constants that depend on the frequency and climatic conditions
k = constant, the value depends on local conditions

For records covering a wide range of conditions, the values of constants A and B in Eq. (8.1) have been derived for design purposes. Some of these values are:

a. Northeastern states for storms involving ordinary heavy rainfall, $A = 105$ and $B = 15$
b. Southern states for storms of 5-year frequency, $A = 135$ and $B = 18$

In the exponential formula (Eq. 8.2), the values of the constant C and k have also been derived. These values are:

a. Northern states for 10-year frequency storm, $C = 15$ and $k = 0.5$
b. Western states for 10-year frequency storm, $C = 7.1$ and $k = 0.54$

The values of these constants for different parts of the United States can be found in Refs. 3,5,6,13.

Example 8.17: Rainfall Intensity from Intensity–Frequency Curves

Determine the rainfall intensity for a storm that has 20 min duration and 10-year frequency. Use the typical IDF curves given in Figure 8.15.

Solution

Identify the rainfall intensity curve.
Select 10-year frequency curve given in Figure 8.15.
Draw a vertical line from 20-min duration to intercept the 10-year frequency curve.
The intensity of rainfall i for the storm is 4.35 in./h

Example 8.18: Preparation of Intensity–Duration–Frequency Curve

Develop the rainfall IDF curves from long-term rain-gauge data. The IDF analysis was carried out from the recording of rain-gauge data collected over a long period of time. The recording rain-gauge data are based on accumulated amount of precipitation with respect to time. From the record of the rain-gauge readings, the specific rainfall durations of 5, 10, 15, 20, 30, and 60 mins were selected. The precipitation values were arranged in descending order. The results are summarized in Table 8.15. Prepare IDF curves for 10-year and 5-year frequencies.

Solution

1. Arrange the precipitation values in descending order.
   The precipitation data for 5, 10, 15, 20, 30, and 60 min duration are obtained and arranged in descending order. The highest value of precipitation for a given intensity is assigned the lowest rank of 1, and the lowest value is assigned the highest rank of 24. These values are arranged in Table 8.15.
2. Determine the return period for each value.
   The return period is obtained from Eq. (8.3)

$$T = \frac{(n+1)}{m} \tag{8.3}$$

TABLE 8.15
Analysis of Frequency for Different Durations of Precipitation Depths (Example 8.18)

	Precipitation, Inches for the Given Duration						Return Period, Year
Rank, m	5 min	10 min	15 min	20 min	30 min	60 min	$T = (n + 1)/m$
(1)	(2)	(3)	(4)	(5)	(6)	(7)	(8)
1	0.45	0.70	0.92	1.10	1.53	2.17	25.0[a]
2	0.43	0.68	0.86	1.01	1.35	1.94	12.5
3	0.41	0.66	0.83	0.95	1.31	1.50	8.3
4	0.41	0.65	0.80	0.90	1.10	0.93	6.3
5	0.40	0.65	0.76	0.89	0.88	0.80	5.0
6	0.38	0.63	0.75	0.80	0.87		4.2
7	0.37	0.60	0.74	0.80	0.83		3.6
8	0.35	0.56	0.67	0.74	0.82		3.1
9	0.34	0.53	0.60	0.70	0.72		2.8
11	0.31	0.47	0.55	0.60	0.64		2.3
12	0.29	0.42	0.51	0.57	0.59		2.1
.							
.							
.							
.							
24	0.17	0.27	0.36	0.44	0.44	0.47	1.04

[a] (24 + 1)/1 = 25 year.

TABLE 8.16
Computed Values for 10-Year and 5-Year Precipitation Intensities (Example 8.18)

Return Period	Intensity (in./h) of Duration of					
(Years)	5 min	10 min	15 min	20 min	30 min	60 min
10	5.04[a]	4.02[c]	3.38	2.94	2.66	1.72
5	4.80[b]	3.90	3.04	2.67	1.76	0.80

[a] Interpolated value of precipitation in inches for 5-min duration = (0.43 + 0.41)/2 = 0.42 in. Intensity of rainfall for 5-min duration and 10-year frequency = (0.42 in./ 5 min) × 60 min/h = 5.04 in./h.

[b] 0.4 in./5 min × 60 min = 4.8 in./h.

[c] 10 min duration $= \frac{(6.8+0.66)\text{ min}}{2\times 10\text{ min}} \times 60\text{ min} = 4.02\text{ min}$.

where
T = return period, year
n = total number of years of data in the record
m = rank of the data

3. Determine the intensity of rainfall for different durations corresponding to 10-year and 5-year frequency. Interpolate from Table 8.15 for each duration, the depth of rainfall in inches for 10-year and 5-year frequency. The interpolation range for 10-year frequency is 12.5 and 8.3 years ($\frac{(12.5+8.3)\text{ year}}{2} = 10.4$ year or 10 year). For 5-year frequency the year is 5.0. The precipitation depths are converted to intensity by dividing the depth of precipitation by duration and converting to inches per hour. For example, the 0.40-in. precipitation in 5 min is 0.4 in./5 min × 60 min/h = 4.8 in./h. Similarly, the interpolated precipitations for all durations are converted to intensity data. These values are summarized in Table 8.16.
4. Prepare the IDF curve.
   The data generated in Table 8.16 are used to prepare the IDF curves. These curves for 10- and 5-year return periods are shown in Figure 8.16.

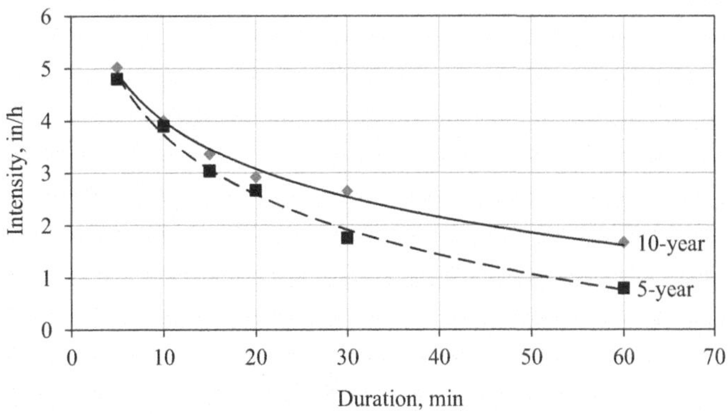

FIGURE 8.16 Intensity–duration–frequency curve (Example 8.18).

Example 8.19: Rainfall Intensity from Eq. (8.1)

In the northeastern states of the United States, the maximum rainfall intensity may be calculated from Eq. (8.1). For storms involving ordinary heavy rainfall at a 5-year frequency, the values of constants A and B are 105 and 15. Calculate the maximum rate of rainfall likely to occur during a storm lasting 20 min.

Solution

1. Substitute the values of constants in Eq. (8.1)

$$i = \frac{105}{t + 15}$$

2. Calculate the maximum rainfall intensity for a storm that has duration of 20 min.

$$i = \frac{105}{20+15} = 2.9 \text{ in./h}$$

Example 8.20: Rainfall Intensity from Eq. (8.2)

The exponential formula of the type illustrated by Eq. (8.2) can be used to determine the maximum rainfall intensity for a given duration. The values of constants C and k are determined from the gaging record over a long period. For a study area, the values of constants C and k are respectively 15 and 0.55 for a 10-year frequency storm. Determine the maximum rainfall intensity of a storm of 20-min duration.

Solution

1. Substitute the values of the constant C and k in Eq. (8.2)

$$i = \frac{15}{t^{0.55}}$$

2. Calculate the maximum rainfall intensity for a storm that has a duration and frequency of 20 min and 10-year

$$i = \frac{15}{(20)^{0.55}} = 2.9 \text{ in./h}$$

8.3.3 Time of Concentration

The time of concentration t_c is the time required for runoff from the hydraulically most remote part of the drainage area to reach the point of reference. Since the rainfall intensity decreases with the increase in storm duration, the duration should be as short as possible. On the other hand, if the rainfall duration is less than t_c, then only a part of the drainage area will be contributing to the runoff. For the entire area to contribute, the shortest storm duration should be equal to t_c. There are many ways to estimate t_c. It is the sum of overland flow time, also called inlet time, and channel flow time. Many empirical formulas are also used. Equations (8.4)–(8.6) express the Kerby, Izzard, and Bransby-Williams formulas:[13,15–19]

$$\text{Kerby Formula} \quad t_i = 0.828 \times \left(\frac{r \times L}{S^{0.5}}\right)^{0.467} \tag{8.4}$$

$$\text{Izzard Formula} \quad t_i = \frac{41.025(0.007i + K)L^{0.33}}{S^{0.333} i^{0.667}} \tag{8.5}$$

$$\text{Brasby-Williams Formula} \quad t_i = \frac{0.00765 \times L}{S^{0.2} \times A^{0.1}} \tag{8.6}$$

where

t_i = overland flow time, also called inlet time, min
L = length of flow path, ft
S = land slope, ft/ft
A = drainage area, acres
r = coefficient for surface; r = 0.02 smooth pavement, 0.1 bare packed soil, 0.3 rough bare or poor grass, 0.4 average grass, 0.8 dense grass and timber
i = intensity of rainfall, in./h
K = coefficient for surface; K = 0.007 smooth asphalt, 0.012 concrete pavement, 0.017 tar and gravel pavement, 0.046 closely clipped sod, and 0.060 dense bluegrass turf

Izzard formula requires a trial-and-error solution as the intensity of rainfall must be known and has an applicable range of $(i \times L) \leq 500$.[13,15] The Kerby formula has an applicable range $L < 1{,}300$ ft. Observations in the field during a rain may help. In general, the reasonable starting assumption for the time of concentrations for different conditions is summarized in Table 8.17.

The inlet time normally varies from 5 to 10 min for high-density areas with closely spaced street inlets. For well-developed areas, it may vary from 10 to 20 min, and for residential areas with widely spaced streets, the inlet time may be 20–30 min.

TABLE 8.17
Approximate Values of Inlet Times for Different Land Uses

Condition and Type of Development	Time of Conc., min
Streets and commercial areas, coefficient of runoff 0.85–0.90	
Gutter grade greater than 3%	5
Gutter grade lesser than 3%	7
Apartment houses, and suburban and business buildings, coefficient of runoff 0.70–0.75	
Gutter grade greater than 3%	6
Gutter grade lesser than 3%	8
Urban and built-up residential areas, coefficient of runoff 0.50–0.65	
Gutter grade greater than 3%	7
Gutter grade lesser than 3%	10
Suburban areas, coefficient of runoff 0.30–0.50	
Gutter grade greater than 3%	10
Gutter grade lesser than 3%	12

Source: Adopted in part from Refs. 15–19

Example 8.21: Time of Concentration

An urbanized residential watershed is given below. Estimate the time of concentration at point *C*. The velocity in the storm drain is 3.9 ft/s.

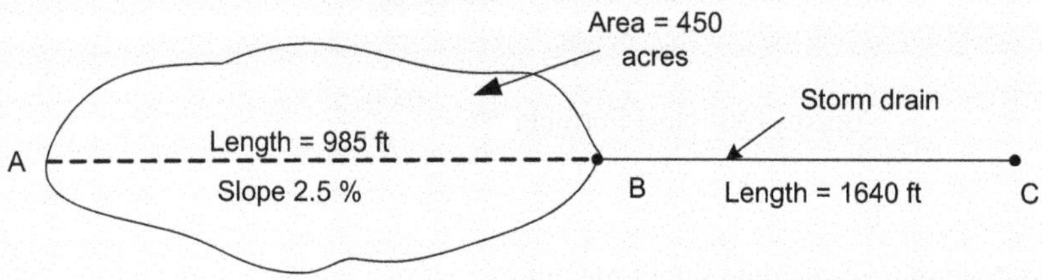

Solution

1. Estimate the inlet time t_i
   a. Land-use and slope information is provided in Table 8.18.
   The inlet time at point *B* in urban and built-up residential areas may vary depending upon the gutter grade. For grades greater than 2% and less than 3% the estimated inlet times may be from 7 to 10 min. For this problem, assume an average value of 8.5 min.
   b. Apply the Kerby formula (Eq. 8.4)[13]
   Assume $r = 0.02$ for smooth pavement and $S = 2.5\%$

$$T_i = 0.828 \times \left(\frac{0.02 \times 985 \text{ ft}}{0.025^{0.5}} \right)^{0.467} = 7.9 \text{ min (overland flow time or inlet time)}$$

   c. Apply Izzard formula (Eq. 8.5)
   The intensity of rainfall is calculated from Eq. (8.1)
   Select $A = 105$, $B = 18$, and $t = 15$ min for 5-year frequency curve (Figure 8.16)

$$i = \frac{105}{15+18} = 3.18 \text{ in./h}$$

$$i \times L = 3.18 \times 985$$

$$= 3{,}132 > 500; \text{ therefore, the formula is not applicable.}$$

   d. Apply the Bransby-Williams formula (Eq. 8.6), $L < 3{,}000$ ft

$$t_i = \frac{0.00765 \times 985 \text{ ft}}{0.025^{0.2} \times (450 \text{ acres})^{0.1}} = 8.6 \text{ min (overland flow time or inlet time)}$$

2. Determine channel time

$$\text{Channel flow time} = \frac{1{,}640 \text{ ft}}{3.9 \text{ ft/s}} = 421 \text{ s} = 7.0 \text{ min}$$

3. Determine the time of concentration t_c

$$t_c = \text{inlet time} + \text{channel time} = 7.9 \text{ min} + 7.0 \text{ min} = 14.9 \text{ min}$$

   Also, $t_c = 8.6$ min + 7.0 min = 15.6 m
   The range of t_c in this example is 14.9–15.6 min. Select $t_c = 15.2$ min.

8.3.4 Stormwater Runoff

The quantity of stormwater runoff may be calculated by any of the following methods: (1) the rational method, (2) empirical formulas, (3) rainfall-runoff correlation studies, (4) hydrograph methods, (5) the inlet method, and (6) computer models. The method selected depends on the size of drainage area, local geographic and hydrologic conditions, past rainfall and runoff data, and degree of protection needed against flooding. The most commonly used method for storm sewer design is the rational method. Readers should refer to Chapter 10 and other sources for estimating runoff by other methods.[16–21] The basic relationships of rational method are expressed by Eq. (8.7)

$$Q = 0.00278 \times C \times i \times A \quad \text{(SI units)} \tag{8.7a}$$

$$Q = C \times i \times A \quad \text{(U.S. customary units)} \tag{8.7b}$$

$$Q = i \times \sum C_j \times a_j \, \text{(U.S. customary units)} \tag{8.7c}$$

where

Q = maximum rate of runoff, m³/s (ft³/s)
C = average runoff coefficient
C_j = runoff coefficient of sub-drainage area a_j
i = average rainfall intensity, mm/h (in./h)
A = drainage area tributary to the point of reference, ha (acre)
a_j = total sub-drainage area, acre

Equation (8.7b) is dimensionally homogeneous (i.e., if A is in ft², i is in ft/s, and Q is in cfs). However, this equation also yields a correct value of Q in cfs when i is in in./h, and A is in acres.

Example 8.22: Surface Runoff

A drainage area has three sub-areas a_1, a_2, and a_3. The surface runoff coefficients of these sub-drainage areas are given in Table 8.18. The rainfall intensity is expressed by equation $i = 105/(t+15)$. The 5-year frequency maximum rainfall intensity storm has duration of 10 min. Determine the maximum rate of runoff from the area.

Solution

1. Determine the 10-min duration rainfall intensity of 5-year frequency storm

$$i = \frac{105}{10+15} = 4.2 \text{ in./h}$$

2. Calculate the maximum surface runoff

$$\text{Average } C = \frac{(2\times 0.6)+(5\times 0.2)+(3\times 0.4)}{10} = \frac{3.4}{10} = 0.34$$

a. Maximum surface runoff from Eq. (8.7b)

$$Q = 0.34 \times 10 \text{ acres} \times 4.2 \text{ in./h} = 14.28 \text{ cfs}$$

b. Maximum surface runoff from Eq. (8.7c)

$$Q = 4.2 \text{ in./h} \times \lfloor (2 \text{ acres} \times 0.6) + (5 \text{ acres} \times 0.2) + (3 \text{ acres} \times 0.4) \rfloor = 14.28 \text{ cfs}$$

TABLE 8.18
Drainage Area Details (Example 8.22)

Sub-drainage Rate Designation	Area, Acres	Coefficient of Runoff, C_j
a_1	2	0.6
a_2	5	0.2
a_3	3	0.4

Example 8.23: Design Flow for Storm Sewer Design

A storm sewer system consists of four drainage areas. The details are shown in Figure 8.17. The 5-year storm intensity is calculated from equation $i = 108/(t_c + 18)$. The characteristics of the drainage system are provided in Table 8.19. Determine the 5-year design flow for each section of the sewer line.

Solution

1. Determine 5-year peak design sewer flow in the sewer line 1–2
   Inlet time for drainage area A = 8 min
   Channel time = 0

$$t_c = 8 \text{ min}$$

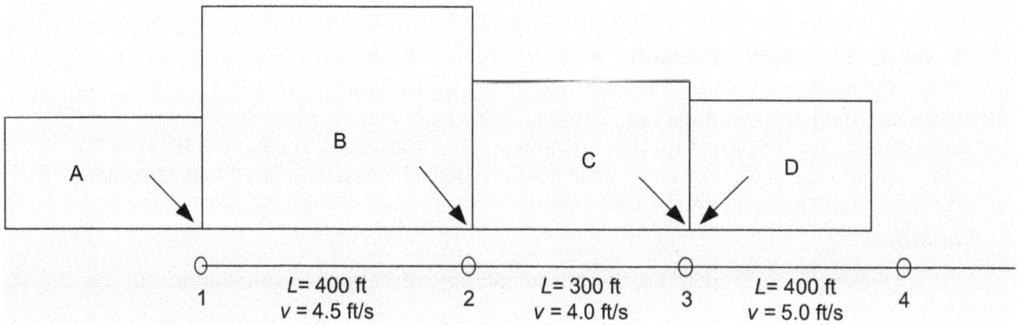

FIGURE 8.17 Storm drainage system (Example 8.23).

TABLE 8.19
Drainage Area Details (Example 8.23)

Identification	Area, Acres	Runoff Coefficient	Inlet Time, min
A	5	0.9	8
B	12	0.8	12
C	8	0.6	10
D	6	0.5	6

$$i = \frac{108}{8+18} = 4.2 \text{ in./h}$$

$$\text{Design peak flow,} \quad Q_{1-2} = i \times \Sigma C_j \times A_j$$

$$= 4.2 \text{ in./h} \times (0.9 \times 5) = 18.9 \text{ cfs}$$

2. Determine 5-year peak design flow in sewer line 2–3

$$t_c \text{ for drainage area } A = 8 \text{ min} + \text{channel time}$$

$$= 8 \text{ min} + \frac{400 \text{ ft}}{4.5 \text{ ft/ s}} \times \frac{1}{60 \text{ s/min}}$$

$$= 8 \text{ min} + 1.5 \text{ min} = 9.5 \text{ min}$$

t_c for drainage area B = inlet time 12 min + channel time 0 min = 12 min

Use the larger of two t_c values.
The t_c value of 12 min is used for areas (A + B)

$$i = \frac{108}{12+18} = 3.6 \text{ in./h}$$

Design peak flow Q_{2-3} = 3.6 in./h[(0.9 × 5) + (0.8 × 12)] acres = 50.8 cfs

3. Determine 5-year peak design flow in sewer line 3–4

$$t_c \text{ from drainage area } (A+B) = 12 \text{ min inlet time} + \text{channel time}$$

$$= 12 \text{ min} + \frac{300 \text{ ft}}{4.0 \text{ ft/s}} \times \frac{1}{60 \text{ s/min}}$$

$$= 12 \text{ min} + 1.3 \text{ min} = 13.3 \text{ min}$$

$$t_c \text{ from drainage area } C = 10 \text{ min}$$

$$t_c \text{ from drainage area } D = 8 \text{ min}$$

Use the largest of the three t_c values of 13.3 min for areas (A + B)

$$i = \frac{108}{13.3+18} = 3.45 \text{ in./h}$$

$$\text{Design peak flow } Q_{3-4} = 3.45 \text{ in./h}\left[(0.9 \times 5) + (0.8 \times 12) + (0.6 \times 8) + (0.5 \times 6)\right] \text{acres}$$

$$= 75.5 \text{ cfs}$$

4. Summarize the design flows calculated for three sewer lines
The design flow calculated for three sewer lines is given below.

Sewer line between MH	Design flow, cfs
1–2	18.9
2–3	50.8
3–4	75.8

8.3.5 Design of Storm Sewers

The design of storm sewers is similar to the design of sanitary sewer system. It involves three basic steps: (1) develop sewer layout plan and determine the characteristics of the drainage area, (2) determine the quantity of storm water and establish sewer design capacity, and (3) develop the sewer design and profile.

Sewer layout and characteristics of drainage area: The storm sewers are laid to reach the outfall point in a suitable water course. The proposed drainage area is surveyed to establish its characteristics at the end of the design period (which may be 20 years or more). Consideration must be given to the street network, the probable number of homes that will be built, and related factors when estimating the imperviousness of the area as developed, and the time of concentration. From the results of the survey, a map is prepared that shows all existing and proposed streets, existing and proposed buildings, finished contour lines, and elevations of all important points.

Quantity of storm water and sewer design capacity: Develop the rainfall intensity, duration, and frequency relations, and the runoff coefficients expected when the district is fully developed. Estimate the probable inlet time, which is the time required for water to flow over the surface of the ground to the first inlet point. By adding the channel time, the time of concentration is established at each entry point. Finally, the peak design capacity of the sewer line is established.

Sewer design and profile: Storm sewers are designed to flow full. A minimum velocity of 0.9 m/s (3 ft/s) is maintained when flowing full in order to produce a minimum non-depositing slope for sand or grit. To achieve this velocity, the slope of the line is often increased in excess of surface grade. Special circumstances may require higher minimum velocities. The upper limit on velocity is about 5 m/s (15 ft/s) because of scour considerations. The minimum size of a storm sewer should be 300–350 mm (12–15 in.). The building storm connections are preferably at least 150 mm (6 in.) in diameter.

Design procedure: The design procedure of storm sewer involves the following steps:

a. Develop sewer plan.
b. Determine the size and characteristics of the area contributing the runoff to each inlet point over the design period.
c. Develop the rainfall IDF relations to be used.
d. Calculate peak design flow.
e. Assume the sewer diameter and a minimum velocity of 0.9 m/s (3 ft/s) when flowing full.
f. Determine the slope of the sewer line from Manning equation (Eq. 5.34 or 5.36), and compare this slope with the available natural ground slope.
g. Establish the sewer invert elevation, and determine the excavation needed and ground cover.

For large subdivisions with network of storm sewers, the design computations become repetitious. These calculations are best performed in a tabular format similar to that of sanitary sewers. Most textbooks provide computational tables that can be adapted.

Example 8.24: Peak Design Flow Capacity of a Sewer Line

The inside diameter of a circular concrete storm sewer is 18 in., $n = 0.013$ and slope is 0.0055. Determine the peak design capacity when flowing full.

Solution

1. Determine the velocity from Manning equation (Eq. 5.34)

$$V = \frac{1.486}{0.013} \times \left(\frac{1.5\ \text{ft}}{4}\right)^{2/3} \times (0.0055)^{1/2} = 4.4\ \text{ft/s}$$

2. Determine peak design flow when flowing full

$$Q_{\text{peak}} = \frac{\pi}{4} \times (1.5\ \text{ft})^2 \times 4.4\ \text{ft/s} = 7.78\ \text{cfs}$$

Example 8.25: Sewer Diameter, Ground Cover, and Sewer Profile

A storm sewer for multi-family development is to serve a drainage area of 10 acres. A circular concrete pipe will be used. A minimum velocity of 3 ft/s shall be reached when flowing full. Determine the sewer diameter and slope. Draw the sewer profile between the manholes. Use the following information

Runoff coefficient, C	= 0.40
Time of concentration, t_c	= 10 min
Rainfall intensity, i	= 110/(t + 18)
Storm duration, t	= 20 min
Roughness coefficient, n	= 0.015
Ground elevations	
Upper manhole	= 338.25 ft
Lower manhole	= 337.82 ft
Sewer length	= 350 ft
Minimum ground cover	= 3 ft

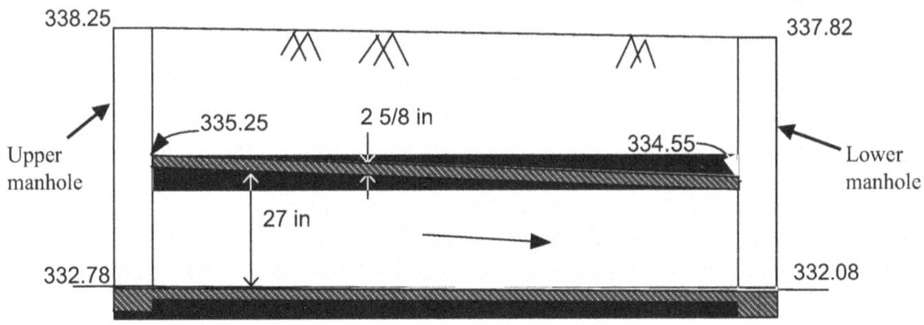

Solution

1. Determine peak design flow

$$i = \frac{110}{20+18} = 2.9\ \text{in./h}$$

$$Q = C \times i \times A$$

$$= 0.4\ 2.9 \text{ in./h} \times 10 \text{ acres} = 11.6 \text{ ft}^3/\text{s}$$

2. Calculate sewer diameter

$$11.6 \text{ ft}^3/\text{s} = \frac{\pi}{4} \times D^2 \times 3 \text{ ft/s}$$

$$D = 2.2 \text{ ft} \quad \text{or} \quad 27 \text{ in.}$$

3. Determine the slope (Eq. 5.36)

$$3 \text{ ft/s} = \frac{1.486}{0.015} \times \left(\frac{27 \text{ ft}}{12 \times 4}\right)^{2/3} \times S^{1/2} \quad \text{or} \quad S = 0.002 \text{ ft/ft}$$

4. Determine the invert elevation at the upper end

$$\text{Invert elevation} = (338.25 \text{ ft} - 3.0 \text{ ft (cover)}) - \frac{27 \text{ in.}}{12 \text{ in./ft}}(\text{sewer diameter})$$

$$-\frac{2\,\tfrac{5}{8} \text{ in.}}{12 \text{ in./ft}}(\text{wall thickness}) = 332.78 \text{ ft}$$

5. Determine the invert elevation at the lower end

$$\text{Invert elevation} = 332.78 \text{ ft} - (0.002 \times 350 \text{ ft}) = 332.08 \text{ ft}$$

6. Draw the sewer profile

Example 8.26: Storm Sewer Design

Design a storm sewer system for a city subdivision. The subdivision was surveyed for surface type, rainfall pattern, and street slopes, and was divided into several sections. The boundaries were established based on the position of inlet gutters. The subdivision plan and streets, and sewer layout are shown in Figure 8.18. The standard inlet time of 15 min for inlet gutter grades less than 3% is used for determination of runoff. The coefficients of runoff and drainage areas were estimated from field inspections of street layout and type of surface. The intensity of rainfall is obtained from Eq. (8.2). $k = 0.61$ and $C = 518.2$, $i =$ mm/h, and $t =$ min. The Manning's n is 0.013, and minimum velocity in the storm sewers is 3 ft/s. Draw the longitudinal section of straight path of storm sewers 3, 4, and 5.

Solution

1. Determine the peak design flow for all sewer lines.
   The peak design flow calculations are conducted in a tabular form using the service area characteristics given in Figure 8.18. The channel times in the table are calculated from the sewer lengths and minimum velocity of 3 ft/s in the storm sewer. The calculation procedure for developing the tabular data is given below.
2. Develop the peak design flow for storm sewer design
   The procedure for developing the peak design flow for sewers 1 through 5 is given below

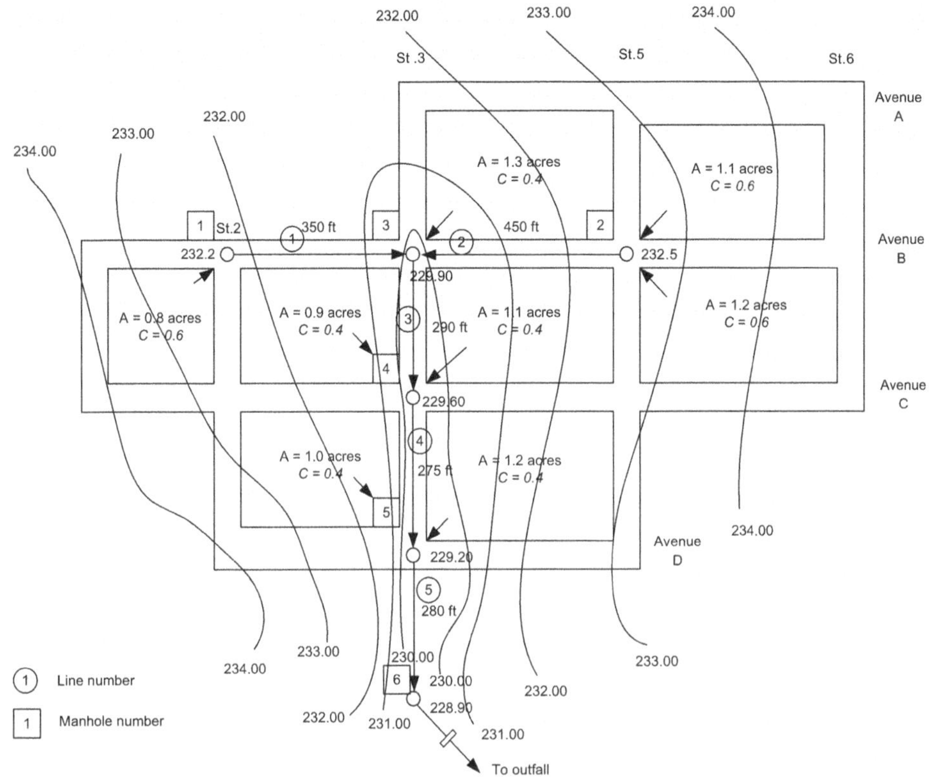

FIGURE 8.18 Layout of the proposed storm sewer system (Example 8.26).

a. Sewer 1 receives storm flow from one land are $A = 0.8$ acre
At gutter 1 inlet time = 15 min and channel time = 0
Average rainfall intensity i is calculated from Eq. (8.2)

$$i = \frac{c}{t^k} = \frac{518.2}{(15)^{0.61}} = 99.33\,\frac{\text{mm}}{\text{h}} \times 10^{-1}\,\frac{\text{cm}}{\text{mm}} \times \frac{1}{2.54\ \text{cm/in.}}$$

$$= 3.91\ \text{in./h}$$

$$\sum (CA) = 0.6 \times 0.8\ \text{acre} = 0.48\ \text{acre}$$

$$Q = \sum (CA)i = 0.48\ \text{acre} \times (3.91\ \text{in./h}) = 1.88\ \text{cfs}$$

b. Sewer 2 receives storm flow from two land areas $A = (1.1 + 1.2)$ acres = 2.3 acres.
At gutter 2, inlet time = 15 min, channel time = 0 and $i = 3.91$ in./h

$$\sum (CA) = [(0.6 \times 1.1\ \text{acre}) + 0.6 \times 1.2\ \text{acre}] = 1.38\ \text{acre}$$

$$Q = \sum (CA) \times i = 1.38\ \text{acre} \times 3.91\ \text{in./h} = 5.4\ \text{cfs}$$

c. Sewer 3 receives storm flow from one land area A = 1.3 acres, and two sewers 1 and 2. A total of four land areas are served: (0.8 + 1.1 + 1.2 +1.3) acres = 4.4 acres

$$\sum(CA) = [(0.6\times 0.8\text{ acre}) + (0.6\times 1.1\text{ acre}) + (0.6\times 1.2\text{ acre}) + (0.4\times 1.3\text{ acre})]$$

$$= (0.48\text{ acre} + 0.66\text{ acre} + 0.72\text{ acre} + 0.52\text{ acre})$$

$$= 2.38\text{ acre}$$

$$\text{Channel time in Sewer 1} = \frac{350\text{ ft}}{3\text{ ft/s}} \times \frac{1}{60\text{ s/min}} = 1.94\text{ min}$$

$$\text{Channel time in Sewer 2} = \frac{450\text{ ft}}{3\text{ ft/s}} \times \frac{1}{60\text{ s/min}} = 2.50\text{ min}$$

$$t_c = \text{inlet time} + \text{larger channel time} = 15\text{ min} + 2.50\text{ min (larger value is used)}$$

$$= 17.5\text{ min}$$

$$\text{Average rainfall intensity } i = \frac{518.2}{(17.5)^{0.61}} = 90.42\frac{\text{mm}}{\text{h}} \times 10^{-1}\frac{\text{cm}}{\text{mm}} \times \frac{1}{2.54\text{ cm/in.}}$$

$$= 3.56\text{ in./h}$$

$$Q = \sum(CA) \times i = 2.38\text{ acre} \times 3.56\text{ in./h} = 8.47\text{ cfs}$$

d. Sewer 4 receives storm flow from two land areas A = 0.9 and A = 1.1 acres, and Sewer 3

A total of six areas are served (0.8 + 1.1 + 1.2 + 1.3 + 0.9 + 1.1) acres = 6.4 acres

$$\sum(CA) = [(0.6\times 0.8) + (0.6\times 1.1) + (0.6\times 1.2) + (0.4\times 1.3) + (0.4\times 0.9) + (0.4\times 1.1)]\text{ acres}$$

$$= 3.18\text{ acres}$$

$$\text{Channel time in Sewer 3} = \frac{290\text{ ft}}{3\text{ ft/s}} \times \frac{1}{60\text{ s/min}} = 1.61\text{ min}$$

$$t_c = 17.50\text{ min} + 1.61\text{ min} = 19.11\text{ min}$$

$$\text{Average rainfall intensity } i = \frac{518.2}{(1{,}911)^{0.61}} = 85.69\text{ mm/h} \times 10^{-1}\frac{\text{cm}}{\text{mm}} \times \frac{1}{2.54\text{ cm/in.}}$$

$$= 3.37\text{ in./h}$$

$$Q = \sum(CA) \times i = 3.18\text{ acres} \times 3.37\text{ in./h} = 10.72\text{ cfs}$$

e. Sewer 5 receives storm flow from two land areas A = 1.1 acres and A = 1.2 acres, and Sewer 4

A total of eight land areas are served: (0.8 + 1.1 + 1.2 + 1.3 + 0.9 + 1.1 + 1.0 + 1.2) acres = 8.6 acres

$$\sum(CA) = [(0.6\times0.8)+(0.6\times1.1)+(0.6\times1.2)+(0.4\times1.3)+(0.4\times0.9)+(0.4\times1.0)+(0.4\times1.2)\ \text{acres}]$$

$$= (0.48+0.66+0.72+0.52+0.36+0.44+0.40+0.48)\ \text{acres}$$

$$= 4.06\ \text{acres}$$

$$\text{Channel time in Sewer 4} = \frac{275\ \text{ft}}{3\ \text{ft/s}}\times\frac{1}{60\ \text{s/min}} = 1.53\ \text{min}$$

$$t_c = 19.11\ \text{min} + 1.53\ \text{min} = 20.64\ \text{min}$$

$$\text{Average rainfall intensity } i = \frac{518.2}{(20.64)^{0.61}}$$

$$= 81.76\ \text{mm/h}\times10^{-1}\ \frac{\text{cm}}{\text{mm}}\times\frac{1}{2.54\ \text{cm/in.}}$$

$$= 3.22\ \frac{\text{in.}}{\text{h}}$$

$$Q = \sum(CA)\times i = 4.06\ \text{acres}\times3.22\ \text{in./h}$$

$$= 13.07\ \text{cfs}$$

3. Summarize the design flows in storm sewer.

   The design flows are summarized below.

Sewer Line	Design Flow, cfs
1	1.88
2	5.40
3	8.47
4	10.72
5	13.07

4. Arrange the peak design storm flow data in a tabular form.

   The calculation steps for peak design flow for storm sewer design are summarized in Table 8.20.
5. Develop and tabulate sewer design data

   The design data, such as sewer diameter, grade velocity, invert and ground surface elevations, and ground cover, are calculated and summarized in columns 1 through 16 of Table 8.21.
6. Draw the longitudinal section of straight path sewers 3, 4, and 5 (Figure 8.19).

TABLE 8.20
Peak Design Flow for Each Sewer Line (Example 8.26)

Sewer Line Number (1)	Location (2)	Manhole From (3)	Manhole To (4)	Length of Sewer, ft (5)	Area *A*, Acres (6)	Coefficient of Runoff, C (7)	*A* × *C*, Acres (8)	Σ*A* × *C*, Acres (9)	Local Inlet Time, min (10)	Channel Time, min (11)	Time of Conc, min (12)	Rainfall Intensity, in./h (13)	Total Design Flow, cfs (14)
1	Avenue B	1	3	350	0.8	0.6	0.48	0.48	15.00	0	15.00	3.91	1.88
2	Avenue B	2	3	450	1.1	0.6	0.66	0.66	15.00	0	15.00	3.91	
					1.2	0.6	0.72	1.38	15.00	0	15.00	3.91	5.40
3	St. 3	3	4	290	1.3	0.4	0.52	2.38	15.00	0	15.00		
									15.00	1.94	16.94		
									15.00	2.50	**17.50**	3.56	8.47
4	St. 3	4	5	275	0.9	0.4	0.36	2.74	15.00	0	15.00		
					1.1	0.4	0.44	3.18	15.00	0	15.00		
									17.50	1.61	**19.11**	3.37	10.72
5	St. 3	5	6	280	1.0	0.4	0.40	3.58	15.00	0			
					1.2	0.4	0.48	4.06	15.00	0	15.00		
					8.6	4.06			**19.11**	1.53	**20.64**	3.22	13.07

Note: In step 2, calculations are shown for the data entered in this table.
The values are made bold as they are the total time of concentration of each sewer line.

TABLE 8.21
Tabulation of Data for Design of Storm Sewers (Example 8.26)

A

Line (1)	Manhole From (2)	Manhole To (3)	Design Flow, cfs (4)	Length of Sewer, ft (5)	Max. Diameter for Velocity of 3 ft/s,[a] in. (6)	Design Parameters: Standard Sewer Diameter,[b] in. (7)	Design Parameters: Velocity Full,[c] ft/s (8)	Design Parameters: Sewer Grade,[d] ft/ft (9)
1	1	3	1.88	350	10.71	10	3.44	0.0073
2	2	3	5.40	450	18.16	15	4.40	0.0070
3	3	4	8.47	290	22.75	21	3.52	0.0028
4	4	5	10.73	275	25.60	24	3.41	0.0022
5	5	6	13.07	280	28.26	27	3.29	0.0018

B

Line (1)	Manhole From (2)	Manhole To (3)	Ground Elevation Upper End, ft (10)	Invert Elevation Upper End, ft (11)	Ground Cover, ft (12)	Fall of Sewer, ft (13)	Invert Elevation Lower End, ft (14)	Ground Elevation Lower End, ft (15)	Ground Cover, ft (16)
1	1	3	232.20	227.79	3.51[e]	2.55	225.24	229.90	3.76
2	2	3	232.50	227.97	3.12	3.15	224.82	229.90	3.67
3	3	4	229.90	224.21	4.75	0.812	223.39	229.60	4.27
4	4	5	229.60	223.12	4.27	0.605	222.51	229.20	4.47
5	5	6	229.20	222.26	4.47	0.504	221.76	228.90	4.67

[a] Diameter D is calculated from continuity equation ($Q = VA$), $V = 3$ ft/s.

[b] Standard sewer diameter estimated from Table 8.4.

[c] Actual velocity calculated from continuity equation ($Q = VA$).

[d] Slope or sewer grade calculated from Manning's equation (Eq. 5.36).

[e]
$$232.20\text{ ft} - \left[\begin{array}{l} 227.79\text{ ft (invert elev.)} + \dfrac{10\text{ in.}}{12\text{ in./ft}}\text{(sewer diam.)} \\ + \dfrac{7/8\text{ in.}}{12\text{ in./ft}}\text{(wall thickness from Table 8.4)} \end{array}\right]$$

$$= 3.51\text{(ground cover)}$$

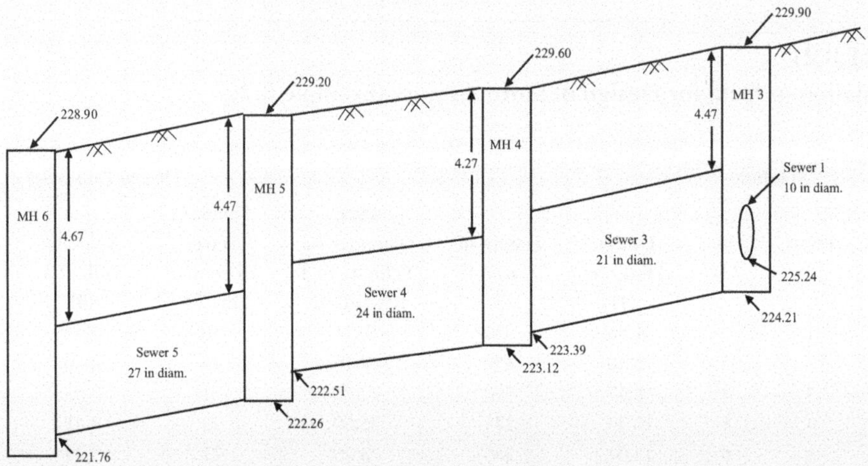

FIGURE 8.19 Longitudinal section of straight path of storm sewers 3, 4, and 5 (Example 8.26).

8.4 COMPUTER SIMULATION AND MODELING OF SEWERS

These days, computer modeling of sewers has become an essential part of any wastewater collection infrastructure project. The traditional sewer design procedure uses detailed design tables that contain the necessary information to develop design flows and complete the design calculations. These tables have been developed in Examples 8.15 and 8.16. The process is complex and time-consuming. A desirable approach is to use modeling software to perform a comprehensive hydraulic analysis that enables engineers to optimally analyze, design and size sanitary, storm, and combined sewer systems. A wide variety of computer software are available from numerous venders. Information about these venders and computer simulation programs may be found on the internet. These software packages have built-in hydraulic and hydrologic capabilities. They are universally used by the design engineer to optimally check the design assumptions, constraints, and design requirements both for existing and proposed systems. Also, these software packages may check the selected size and slopes of sanitary and storm sewer systems, arbitrary arrays of storages, pumps, and force mains. They also address the increasing population demands, new and aging and overloaded infrastructures, perform overflow remediation analysis, suggest cost-effective system expansions and rehabilitation strategies, and decrease the design risks. The U.S. Environmental Protection Agency (USEPA) has also developed innovative software models, tools, and technologies for communities. These software and tools are used widely for large-scale planning, analysis, and design related to sanitary, storm, and combined sewers, stormwater runoff management, and drainage systems in urban and nonurban areas.[22]

As a point of caution, it is stated that these computer software packages can only serve as a tool in assisting a design engineer in developing an optimum and cost-effective design. They do not replace the basic knowledge of sewer design presented in this chapter, and the design calculations summarized in many tables included in Examples 8.13–8.16.

8.5 SURFACE RUNOFF QUALITY AND TREATMENT

8.5.1 Quality Assessment

Storm sewers carry surface runoffs from paved streets, parking lots, building rooftops, and lawn and grassy areas during rainfall. Such discharges often contain pollutants in quantities that can adversely affect water quality. Pollutants commonly found in stormwater runoffs include debris, sediments, BOD, nutrients, and toxic compounds. The stormwater management program includes the following activities: (1) separation of municipal wastewater from stormwater; (2) separation of industrial wastewater from storm water; (3) surface water quality improvement from watershed (drainage area) by the best manage practices such as pools, infiltration basins and channels, bioretention swales, filter strips, vegetated channels, and the like; and (4) the best construction management practices. The stormwater program is divided into two phases. Phase I was developed by the U.S. EPA in the year 1990 in response to the 1987 Amendments to the Clean Water Act (CWA). In Phase I, the permit is required from (1) medium and large municipal separate storm water sewer systems located in incorporated places or counties with populations of 100,000 or more, and (2) 11 categories of industrial activity, including construction activity that disturbs five or more acres of land. **Phase II** was developed in December 1999, and it requires permit coverage from (1) small municipal separate storm water systems (population less than 100,000), and (2) small construction activities (between 1 and 5 acres of land). Drainage from Confined Animal Feeding Operations (CAFOs) may contain manure, pesticides, herbicides, antibiotics, priority pollutants, and nutrients. While in general agricultural stormwater runoff from CAFOs is a nonpoint source, these discharges may end up requiring the National Pollutant Discharge Elimination System (NPDES) permit under Clean Water Act.[23] The following are the requirements of the NPDES program: (1) wet weather sampling data are needed to characterize the quantity and quality of stormwater, (2) estimation of pollutant concentrations and pollutant loads, (3) Best Management Practices, and (4) schedule of activities.

EPA has developed benchmark concentration values for NPDES stormwater permit for industrial activities. These values are used to determine if a facility is successfully implementing the Storm Water Pollution Prevention Plan.

8.5.2 Treatment

Stormwater treatment involves screens that remove trash and debris and suspended solids at high flows. Other applicable technologies are sedimentation basins, hydrocyclone, and filters to remove suspended solids. Currently available technologies for stormwater treatment are proprietary devices available from many equipment manufacturers and suppliers. An in-depth coverage of stormwater treatment and proprietary devices may be found in Section 10.10.

Example 8.27: Stormwater Treatment Devices

A number of proprietary devices are available for stormwater management. List 6 stormwater equipment suppliers and their websites.

Solution

List the stormwater treatment equipment suppliers.

Stormwater treatment devices utilize sedimentation screening, filtration, hydrocyclone, and more. Equipment suppliers provide information about their products. Six selected equipment manufacturers and their websites are listed in Table 8.22.

TABLE 8.22
List of Stormwater Treatment Suppliers (Example 8.27)

No.	Name of the Supplier	Website
1	Advanced Drainage Systems	https://www.ads-pipe.com/
2	BaySaver Technologies, Inc.	https://baysaver.com/
3	CDS Technologies	www.cdstech.com
4	CULTEC, Inc.	https://cultec.com/
5	Hancor, Inc.	www.hancor.com
6	Westech Engineering, Inc.	https://www.westech-inc.com/

8.6 CONSTRUCTION OF COMMUNITY SEWERAGE SYSTEM

Construction of a community sewerage system involves (1) selection of manufactured sewers and appurtenances, (2) structural requirements of buried pipes, (3) construction inspection, and (4) acceptance tests.

8.6.1 Selection of Sewers and Appurtenances

The sewers and manholes are manufactured from various basic materials in accordance with nationally recognized product specifications.

Sewers: Sewers are made from concrete, reinforced concrete, vitrified clay, asbestos-cement, brick masonry, cast iron, corrugated steel, sheet steel, and plastic. Important factors in selection of sewer material include (1) chemical characteristics of wastewater and degree of resistance to corrosion against acid, gases, and solvents, (2) resistance to scour and flow (friction coefficient), (3) external forces and internal pressures, (4) soil conditions, (5) type of backfill and bedding material, (6) useful life, (7) strength and water tightness of joints required and effective control of infiltration and inflow, (8) availability in diameter, length, and ease of installation, and (9) cost of construction and maintenance. Basic properties of sewers of different materials are summarized in Table 8.23.

Manholes: Manholes are generally constructed from (1) brick masonry, (2) precast concrete, (3) cast-in-place concrete, (4) concrete block, and (5) fiberglass. Manhole covers are made of cast iron and should be tightly fitted to eliminate inflow, uncovering, and rattling. They should also have provisions to vent gases and admit air. Manhole steps should be placed at intervals of 30–50 cm (12–16 in.), and be designed to prevent corrosion and slipping.

Sewer joints: Sewer joints are designed to make sewers watertight, root-resistant, flexible, and durable. Joints are of many types and are selected to serve the intended purpose (see Example 8.28).

Example 8.28: Sewer Joints

Draw the construction details of the common types of sewer joints. Also, indicate their purpose and application.

Solution

Concrete pipe joints are covered by National Standards. There are American Society for Testing and Materials (ASTM) and American Association of State Highway and Transportation Officials standards for concrete products' joints. The various standards are discussed in their publications. The versatility of joints allows the designer to match project requirements with the appropriate joint and/or seal. Note that joints for most pipe products are specified in separate standards than the pipe and precast manufacturing and design standards. The different types of joints made in the concrete pipe are (1) rigid spigot and socket joint, (2) rigid collar joint, (3) semi-flexible spigot and socket joint, (4) semi-flexible collar joint, (5) internal flush joint, and (6) external flush joint. However, the most commonly used joint for concrete pipe is rigid spigot and socket joint, as this joint does not require machinery and is easy to work with.

a. **Rigid spigot and socket joint:** This is a commonly preferred type of joints in the concrete pipes, as this is easy to install. The spigot of the pipe is slipped inside the socket of the previously laid pipe and adjusted in the correct position, as shown in Figure 8.20a. The joint is filled with a 1:2 cement mortar mixture of stiff consistency, which is rammed with a caulking tool. The excess cement mortar shall be removed, and the joint shall be cured.

TABLE 8.23
Characteristics of Sewers Made from Common Construction Materials

Construction Material	Available Sizes (Dia.)		Available Section Length		Reference Standard	Subject to Corrosion or Erosion	Strength	Type of Joint	Application
	cm	in.	m	ft					
Concrete	10–61	4–24	1.2–2.4	4–8	ASTM C14[a]				
Plain	30–305	12–120	1.2–2.4	4–8	ASTM C76	Yes	Good	Bell and spigot, or tongue and	Used for sewers, force mains,
Plain reinforced	30–366	12–144	1.2–7.3	4–24	ASTM	Yes	Good	groove with rubber gasket,	submerged outfalls, inverted
Reinforced concrete	30–366	12–144	1.2–7.3	4–24	C300–303	Yes	Good	hot or cold-poured	siphons. Low cost, short
cylinder					ASTM	Yes	Good	bituminous material, cement	lengths, many joints
Mortar lined (and coated) steel					C200–210			mortar	
Vitrified clay	10–122	4–48	0.9–1.8	3–6	ASTM C700	No	Brittle	Mortar, rubber gasket, compression	Used for sewers. Short lengths, many joints, susceptible to infiltration. Not a common material for new pipes.
Ductile iron (optional-cement mortar lined. Optional-epoxy coated)	8–163	3–64	5.5–6.1	18–20	ANSI[b]/ AWWA[c] C151/821.51	Yes	Excellent	Bell and spigot, flanged, mechanical or grooved coupled, rubber or neoprene ring-type push on joint, restrained joint or ball and socket	Used for force mains, yard piping, etc. Long laying lengths, tight joints, high-pressure applications. Susceptible to soil corrosion
Corrugated and plain steel	5–792	2–312	6.4–12.8	21–42	AWWA C200	Yes	Good	Band bolt, coating or lining or coupling	Used for sewers and drainage culverts

(Continued)

TABLE 8.23 (*Continued*)
Characteristics of Sewers Made from Common Construction Materials

Construction Material	Available Sizes (Dia.) cm	Available Sizes (Dia.) in.	Available Section Length m	Available Section Length ft	Reference Standard	Subject to Corrosion or Erosion	Strength	Type of Joint	Application
Plastic		4–15	4.0–6.1	13–20	ASTM D3034	No	Fair	Rubber cement, chemical weld sleeves, compression gasket	Light weight, tight joints, long laying lengths, thin walls, susceptible to sunlight and temperature which affect shape and strength. Susceptible to long-term deflections.
Gravity	10–38	18–27	4.0–6.1	13–20	ASTM F679				
Pressure	46–69	21–48	4.0–6.1	13–20	ASTM F794				
	53–122	1.5–12	4.0–6.1	13–20	ASTM D2241	No	Fair		
	4–30	4–12	4.0–6.1	13–20	AWWA C900				
	10–30	14–48	4.0–6.1	13–20	AWWA C905				
	36–122								
Cast Iron[d]	5–120	2–48	Long length	Long length	AWWA C100	Yes	Excellent	Bell and spigot, flanged, mechanical or groove coupled, rubber or neoprene ring-type push on joint, or ball and socket.	Used for force mains, water mains, yard piping, etc. Long laying lengths, tight joints, withstand high pressures. Susceptible to soil corrosion.
Asbestos-cement[d]	10–120	4–42	—	—	AWWA C400	Yes	Good	Collar with rubber ring	Used for force mains, water distribution lines, sewer. Light weight, easy to transport and handle.
Brick-masonry[d]	Greater than 91	Greater than 36	Any length	Any length	None	Yes	Good	Mortar	Uncommon these days except for lining of concrete sewers

Source: Adopted in part from Ref. 2

[a] ASTM, American Society for Testing and Materials.

[b] ANSI, American National Standards Institute.

[c] AWWA, American Water Works Association.

[d] These are not in common use in the United States.

b. **Rigid collar joint**: The rigid collar joint is made by joining two adjacent pipes with butt to butt position and adjusted in the correct position, as shown in Figure 8.20b. A collar is used to cover the joint equally on both pipes. The void space in between the pipe and collar shall be fixed with a 1:2 cement mortar and rammed with a caulking tool. The excess cement mortar shall be removed, and the joint shall be cured.
c. **Semi-flexible spigot and socket joint:** The mechanism involved in the joining of pipes is the same as that of rigid spigot and socket joint. The difference is that a rubber ring is placed on the spigot, which shall be slipped into the socket of the adjacent pipe, as shown in Figure 8.20c. The remaining gap is filled with a stiff mixture of cement mortar 1:2 and rammed with a caulking tool.
d. **Semi-flexible collar joint:** The ends of both pipes are fitted with a rubber ring, which is further compressed in between the spigot and the collar which seals the joint, as shown in Figure 8.20d. The remaining gap is filled with a stiff mixture of cement mortar 1:2 and rammed with a caulking tool.

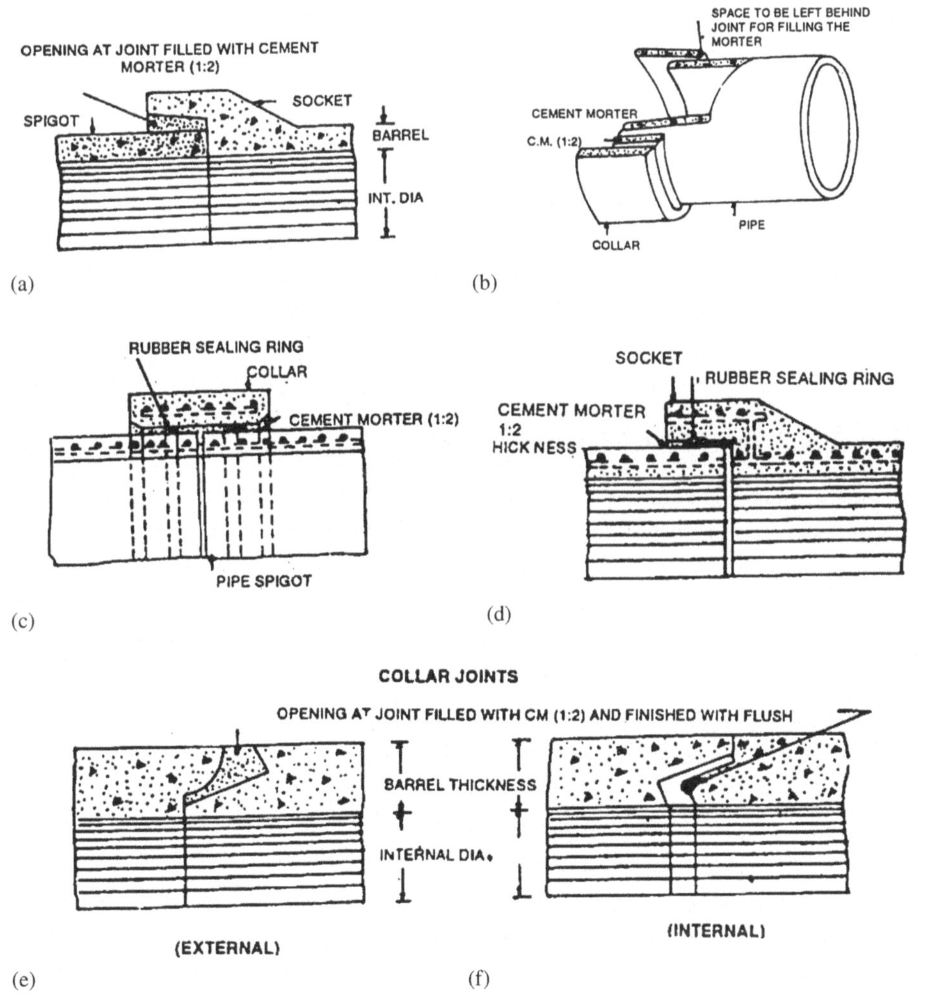

FIGURE 8.20 Construction details of common types of sewer joints: (a) rigid spigot and socket joint, (b) rigid collar joint, (c) semi-flexible spigot and socket joint, (d) semi-flexible collar joint, (e) external flush collar joint, and (f) internal flush collar joint (Example 8.28).

e. **External flush joint:** This type of joints is preferred for pipes that are too small for jointing from inside. The ends of both pipes are butt-joined against each other and adjusted in the correct position, as shown in Figure 8.20e. The joint space is then filled with 1:2 cement mortar, compacted using trowel or rammer, and cured.
f. **Internal flush joint:** As the name suggests, the joint is flushed with both inside and outside of the pipe wall. The ends of both pipes are specially shaped to form a self-centering joint with an internal jointing space 1.3 cm wide, as shown in Figure 8.20f. The joint space is then filled with 1:2 cement mortar, compacted using trowel or rammer, and cured. This joint is generally used for a culvert pipe of 60 cm diameter and over.

8.6.2 Structural Requirements of Buried Pipes

The constructed sewer must be able to withstand the forces under which it must function. The structural design requires that the supporting strength of the conduits, as installed, divided by a suitable factor of safety, must equal or exceed the loads imposed on it by the weight of the earth and any superimposed load. The supporting strength of a buried conduit is a function of installation conditions, as well as the strength of the pipe. The three most common conditions of installation are (1) trench in a natural ground, (2) embankment, and (3) tunnel. Loading conditions due to gravity and earth forces for each installation are different. The loadings involve weight for the prism of earth directly over it, frictional shearing forces, cohesive forces, forces induced by backfill and settlement, and others. Martson's equation and many modifications for different soils and installation conditions are used to check the structural integrity of the sewer line.[7,24] Many types of nomographs have been developed to facilitate the calculations. Readers are referred to several sources to obtain procedures, nomographs, and solved examples for determining the structural requirements of sewer installation.[6,7,12,14,24]

Example 8.29: Structural Requirements of Buried Pipe

An underground sewer line is subjected to a concentrated H-20 wheel load vertically centered on the pipe. The diameter of pipe is 15 in., wall thickness is 1.5 in., and its length is 5 ft. The pipe is under a soil core of 5 ft. The supporting strength of the conduit as installed is 900 lbs/ft. Calculate the safety factor.

Solution

1. Select the equation for load calculation

   The Holl's and Newmark's integration of Boussinesq equation for calculation of vertical stresses is used. This is expressed by Eq. (8.8)[14]

$$W_{sc} = \frac{C_s \times P \times F}{L} \tag{8.8}$$

where

W_{sc} = the load exerted on the buried conduit, lb/ft of length

P = concentrated live load in lb. The H-20 wheel load is standard load used for highway and bridge design. It refers to loading resulting from the passage of trucks having a gross weight of 20 tons. Each rear wheel carries a load of 8 tons without impact (Figure 8.21)

F = impact factor. For highway F = 1.5

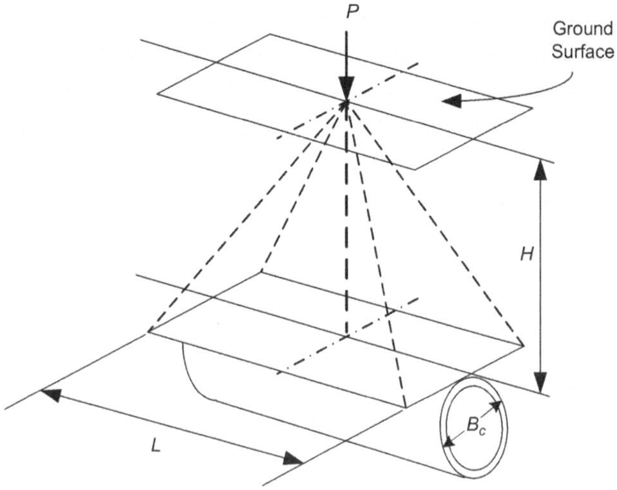

FIGURE 8.21 Superimposed concentrated load vertically centered over (Example 8.29). (Printed from Ref. 14 with permission of National Clay Pipe Institute.)

L = effective length of pipe, ft. L = 3 ft for pipes greater than 3 ft length.
C_s = load coefficient, a function of $B_c/2H$ and $L/2H$. The C_s values for different conditions are given in Table 8.24
H = height of fill from top of conduit to ground surface, ft
B_c = outside diameter of pipe, ft

2. Determine the values of variable in Eq. (8.8)
   P = 16,000 lb
   F = 1.5 (highway) for H-20 wheel load
   L = 3.0 ft (for pipes greater than 3 ft in length)
   H = 5 ft

TABLE 8.24
Values of Load Coefficients, C_s for Concentrated and Distributed Superimposed Loads Vertically Centered over Conduit (Example 8.29)

$\frac{D}{2H}$ or $\frac{B_c}{2H}$	$\frac{M}{2H}$ or $\frac{L}{2H}$													
	0.1	0.2	0.3	0.4	0.5	0.6	0.7	0.8	0.9	1.0	1.2	1.5	2.0	5.0
0.1	0.019	0.037	0.053	0.067	0.079	0.089	0.097	0.103	0.108	0.112	0.117	0.121	0.124	0.128
0.2	0.037	0.072	0.103	0.131	0.155	0.174	0.189	0.202	0.211	0.219	0.229	0.238	0.244	0.248
0.3	0.053	0.103	0.149	0.190	0.224	0.252	0.274	0.292	0.306	0.318	0.333	0.345	0.355	0.630
0.4	0.067	0.131	0.190	0.241	0.284	0.320	0.349	0.373	0.391	0.405	0.425	0.440	0.454	0.460
0.5	0.079	0.155	0.224	0.284	0.336	0.379	0.414	0.441	0.463	0.481	0.505	0.525	0.540	0.548
0.6	0.089	0.174	0.252	0.320	0.379	0.428	0.467	0.499	0.524	0.544	0.572	0.596	0.613	0.624
0.7	0.097	0.189	0.274	0.349	0.414	0.467	0.511	0.546	0.584	0.597	0.628	0.650	0.674	0.688
0.8	0.103	0.202	0.292	0.373	0.441	0.499	0.546	0.584	0.615	0.639	0.674	0.703	0.725	0.740
0.9	0.108	0.211	0.306	0.391	0.463	0.524	0.574	0.615	0.647	0.673	0.711	0.742	0.766	0.784
1.0	0.112	0.219	0.318	0.405	0.481	0.544	0.597	0.639	0673	0.701	0.740	0.774	0.800	0.816
1.3	0.117	0.229	0.333	0.425	0.505	0.572	0.628	0.674	0.711	0.740	0.783	0.820	0.849	0.868
1.5	0.121	0.238	0.345	0.440	0.525	0.596	0.650	0.703	0.742	0.774	0.820	0.861	0.894	0.916
2.0	0.124	0.244	0.355	0.454	0.540	0.613	0.674	0.725	0.766	0.800	0.849	0.894	0.930	0.956

B_c = 15 in. diameter + 2 × 1.5 in. wall thickness = 1.5 ft
$B_c/2H$ = 1.5 ft/10 ft = 0.15
$L/2H$ = 3 ft/10 ft = 0.30
C_s = 0.078 (from Table 8.24 by interpolation)

3. Calculate the load on the conduit

$$W_{sc} = \frac{0.078 \times 16{,}000 \text{ lb} \times 1.5}{3.0 \text{ ft}}$$

$$= 624 \text{ lb/ft}$$

4. Calculate the safety factor

$$\text{Safety factor} = \frac{900 \text{ lb/ft}}{624 \text{ lb/ft}} = 1.44$$

8.6.3 Sewer Construction and Inspection

Sewer construction involves excavation, sheeting and bracing of trenches, sewer installation, backfilling, joints and infiltration, and acceptance tests for new sewers. These construction features are discussed briefly below.

Excavation: After the sewer alignment is marked on the ground, the trench excavation begins. Machinery such as backhoe, clamshell, dragline, front-end loader, or other specialized equipment is used. Hand excavation may be possible only for short distances. Hard rocks may be broken by drilling; explosives may also be used where situations permit.

Sheeting and bracing: Trenches in unstable soil conditions require sheeting and bracing to prevent caving. Sheeting is placing planks in actual contact with the trench sides. Bracing is placing crosspieces extending from one side of the trench to the other. Sheeting and bracing may be of various types depending on the depth and width of the trenches and type of soils supported. Common types are stay bracing, poling boards, box sheeting, vertical sheeting, and skeleton (open) sheeting. In many situations, pumping may be necessary to dewater the trenches.

Sewer installations: After the trench is completed, the bottom of the trench is checked for elevation and slope. In firm cohesive soils, the trench bottom is shaped to fit the pipe barrel and projecting collars. Often, granular materials such as crushed stones, slag, gravel, and sand are used to provide uniform bedding for pipe.

The pipes are inspected and lowered, with particular attention being given to the joints. The pipe lengths are placed on line and grade with joints pressing together with a lever or winch. The joints are then filled per specifications. Typical pipe installations are shown in Figure 8.22. These installations are trenches in natural ground, embankment, and tunnel.

Backfilling: The trenches are filled soon after the pipes are laid. The fill should be carefully compacted in layers of 15 cm (6 in.) deep around, under, and over the pipe. After completion of the filling, the surface is finished.

Joints and infiltration: Joints should be designed to make sewers watertight, root-resistant, flexible, and durable. The leakage (infiltration and exfiltration) shall not exceed 0.2 m^3/day/cm of pipe diameter per kilometer length (200 gpd/in. of pipe diameter per mile). It has been demonstrated that joints made from rubber gasket and hot-poured bituminous material produced almost no infiltration, whereas cement mortar joints cause excessive infiltration. The air test shall, as a minimum, conform to the test procedure described in ASTM for clay and concrete pipes.

FIGURE 8.22 Pipe installations for trench in (a) tunnel, (b) natural ground, and (c) embankment.

8.6.4 Acceptance Test for New Sanitary Sewers

Newly built sanitary sewer lines should be tested for water tightness. Testing should be done in sections not more than 300 m (1,000 ft). The testing methods are water testing and low-pressure air testing. The testing procedure is summarized below.[1]

Water testing for infiltration: During wet weather, water testing for infiltration is done when the groundwater table is at least 0.3–0.6 m (1–2 ft) above the crown of the sewer. The upstream end of sewer is plugged. A V-notch weir is installed in the manhole at lower end. Sufficient readings are taken to determine the infiltration rate.

Water testing for exfiltration: When groundwater table is low, leakage may be determined by measuring exfiltration. The sewer line is plugged at the lower end. The pipe section is filled with water at the upper manhole to a desired level. The rate of leakage is computed on the basis of the observed drop in water level or by metering the water needed to maintain the original level.

Low-pressure air testing: A section of sewer between manholes is plugged. Air is introduced into the sewer in the range of 20–35 kN/m^2 (3–5 lb/in.^{2}). Most common pressure for municipal sewer is 24 kN/m^2 (35 lb$_f$/in.^{2}). It has been established that a section of sewer is performing satisfactorily if the time required (in seconds) for pressure to drop 6.9 kN/m^2 (1 lb$_f$/in.^{2}), that is from 24.0 to 17.1 kN/m^2 (3.5–2.5 lb$_f$/in.^{2}) is greater than or equal to the least of the value computed from Eqs. (8.9) and (8.10).

$$t_Q = \frac{0.032}{Q} \times \left(d_1^2 \times L_1 + d_2^2 \times L_2 + \cdots + d_n^2 \times L_n\right) \quad \text{(SI units)} \tag{8.9a}$$

$$t_Q = \frac{0.022}{Q} \times \left(d_1^2 \times L_1 + d_2^2 \times L_2 + \cdots + d_n^2 \times L_n\right) \quad \text{(U.S. customary units)} \tag{8.9b}$$

$$t_q = \frac{1.0184}{q} \times \left(\frac{d_1^2 \times L_1 + d_2^2 \times L_2 + \cdots + d_n^2 \times L_n}{d_1 \times L_1 + d_2 \times L_2 + \cdots + d_n \times L_n}\right) \quad \text{(SI units)} \tag{8.10a}$$

$$t_q = \frac{0.085}{q} \times \left(\frac{d_1^2 \times L_1 + d_2^2 \times L_2 + \cdots + d_n^2 \times L_n}{d_1 \times L_1 + d_2 \times L_2 + \cdots + d_n \times L_n}\right) \quad \text{(U.S. customary units)} \tag{8.10b}$$

where

t_Q, t_q = allowable time for decrease in pressure from 24.0 to 17.1 kN/m^2 (3.5–2.5 lb$_f$/in.^{2}), s
Q = 56.7 m^3/min (2.0 ft^3/min) of air loss
q = 0.913 L/min/m^2 of internal pipe surface (0.0030 ft^3/min/ft^2) of internal pipe surface of air loss
(d_1, L_1), (d_2, L_2), … = diameter, mm (in.), and length, m (ft) of sections tested

For safety reasons, air supply valve should be set so that internal pressure does not exceed 55 kN/m^2 (8 lb$_f$ /in.^{2}).

Example 8.30: Infiltration Water Testing

A newly constructed 48-in. interceptor is tested for infiltration. A 980 ft section of sewer between two manholes was tested under the wet weather condition. The groundwater table was about 2 ft above the crown. The head over the 90° V-notch constructed in the downstream manhole was 0.5 in. Determine the infiltration in gallons per inch of diameter per mile per day. C_d = 0.75

Solution

1. Calculate the flow over the 90° V-notch (Eq. 7.10)

$$q = \frac{8}{15} \times C_d \times \sqrt{2g} \times \tan\frac{\theta}{2} \times H^{\frac{5}{2}}$$

$$q = \frac{8}{15} \times 0.75 \times \sqrt{2 \times 32.2 \frac{\text{m}}{\text{s}^2}} \times 1 \times \left(\frac{0.5 \text{ in.}}{12 \frac{\text{in.}}{\text{ft}}}\right)^{5/2} = 0.001133 \text{ cfs}$$

$$= 0.001133 \frac{\text{ft}^3}{\text{s}} \times 7.48 \frac{\text{gal}}{\text{ft}^3} \times 24 \times 60 \times 60 \frac{\text{s}}{\text{day}} = 732 \text{ gal/day}$$

$$\text{Infiltration} = 732 \frac{\text{gal}}{\text{day}} \times \frac{1}{48 \text{ in.}} \times \frac{1}{980 \text{ ft}} \times 5{,}280 \frac{\text{ft}}{\text{mile}}$$

$$= 82 \text{ gal/in.dia.mile.day}$$

Since the permissible minimum rate of infiltration is 100 gal/in. dia/mile/day, this interceptor meets the acceptable requirement for infiltration (see Section 4.5.2).

Example 8.31: Exfiltration Water Testing

A leakage test was conducted on a newly built interceptor section between two manholes. The diameter and length of the interceptor were 910 mm and 280 m, respectively. There was no rainfall in the area, and the groundwater table was low. The testing was done by measuring the exfiltration. All incoming sewer lines were blocked at upper manholes, and the interceptor was blocked at the lower end. The interceptor was filled with water, and a head of 0.6 m above the top of the pipe was maintained in the upper manhole. The testing was conducted for 24 h. Over this period, the volume of water required to maintain the original level was 260 L/h. Determine the exfiltration rate in cubic meter per centimeter of pipe per kilometer per day.

Solution

1. Determine the exfiltration rate

$$\text{The total exfiltration per day} = 260\,\frac{\text{L}}{\text{h}} \times \frac{24\text{ h}}{\text{day}} \times 10^{-3}\,\frac{\text{m}^3}{\text{L}}$$

$$= 6.24\text{ m}^3/\text{day}$$

$$\text{The exfiltration rate} = 6.24\text{ m}^3/\text{day} \times \frac{1}{91\text{ cm}} \times \frac{1}{280\text{ m}} \times 1{,}000\,\frac{\text{m}}{\text{km}}$$

$$= 0.25\text{ m}^3/\text{cm.km.day}$$

2. Compare the exfiltration rate with permissible infiltration rate

   The permissible minimum infiltration rate = 0.09 m^3/cm.km.day (see Section 4.5.2)

   The exfiltration of 0.25 m^3/cm.km.day exceeds the permissible value of minimum infiltration.

   The interceptor does not meet the acceptance requirements for exfiltration.

Example 8.32: Evaluation of Air Testing Results

An air test was conducted on a newly constructed section of a sewer sub-main that is 300 mm (12 in.) in diameter and 100 m (328 ft) in length. In this test, 200 m (656 ft) of 150 mm (6 in.) building connections were also included. The air pressure was equalized at 27.5 kN/m^2 (4 lb_f/in.^{2}) for 6 min. The drop in pressure from 24 to 17.1 kN/m^2 (3.5–2.5 lb_f/in.^{2}) occurred in 4.7 min. Check if the constructed sewer system meets the acceptance requirements.

Solution

1. Determine from Eq. (8.9a) the acceptance time t_Q

$$t_Q = \frac{0.032}{Q} \times \left(d_1^2 \times L_1 + d_2^2 \times L_2\right)$$

$$t_Q = \frac{0.032}{56.7} \times \left((300 \text{ m})^2 \times 100 \text{ m} + (150 \text{ m})^2 \times 200 \text{ m} \right)$$

2. $t_Q = 7{,}619$ s or 127 min Determine from Eq. (8.10a) the acceptance time t_q

$$t_q = \frac{1.0184}{q} \times \left(\frac{d_1^2 \times L_1 + d_2^2 \times L_2}{d_1 \times L_1 + d_2 \times L_2} \right)$$

$$t_q = \frac{1.0184}{0.913} \times \left(\frac{(300 \text{ m})^2 \times 100 \text{ m} + (150 \text{ m})^2 \times 200 \text{ m})}{300 \text{ m} \times 100 \text{ m} + 150 \text{ m} \times 200 \text{ m}} \right)$$

$$t_q = 251 \text{ s} \quad \text{or} \quad 4.2 \text{ min}$$

3. Compare the results to establish acceptability
   The observed time is 4.7 min to drop 6.9 kN/m^2 (1 lb/in.^{2}) pressure (from 24 to 17.1 kN/m^2 (3.5–2.5 lb$_f$/in.^{2})). The least acceptable time calculated from Eqs. (8.9a) and (8.10a) is 4.2 min. Since the observed time of 4.7 min is greater than the least acceptable time of 4.2 min, the sewer system meets the acceptance requirement.

Example 8.33: Exfiltration Test of Manhole

A manhole is tested for leakage. The acceptable leakage value is 12.4 L/vertical m.day (1.0 gal/vertical ft.day) when tested with no external groundwater. The testing was done by blocking all incoming and outgoing sewer lines, then filling the manhole with water to the top of the cone section. The top of the cone section is 3.2 m from the manhole invert level. There was no external groundwater. The testing was continued for 12 h, and it took 1.5 L/h to maintain the original level. Determine the leakage rate.

Solution

1. Determine the manhole leakage rate

$$\text{The leakage rate} = 1.5 \frac{\text{L}}{\text{h}} \times 24 \frac{\text{h}}{\text{d}} \times \frac{1}{3.2 \text{ m}}$$

$$= 11.3 \text{ L/vertical m.day}$$

2. Compare the results
   The observed value of 11.3 L/vertical m.day is less than the controlling value of 12.4 L/vertical m.day.
   Therefore, the manhole meets the acceptance requirements.

8.7 REVIEW PROBLEMS AND DISCUSSION TOPICS

8.1 A city has a population of 55,000 residents. The land-use plan shows single- and multiple-family housing and apartments, and commercial and industrial areas. The estimated average length of sewer is 28 ft/capita. Determine the length of the collection system and lengths of various pipe sizes. Also, calculate the equivalent sewer diameter. The typical size distribution is summarized below.

Diameter, in.	6 and below	8	12	18	24	27 and above
% Distribution	3.6	73.1	13.4	3.8	1.7	4.4

8.2 A trunk sewer is 18-in. diameter and has a length of 9,240 ft. The average design flow is 120 gpcd. Determine the design capacity. Use the following information:

I/I allowance = 1,600 gal/day. in. mile
Dry weather peaking factor = 2.0
Population = 5,200 residents
Industrial contribution = 50,000 gpd

8.3 A sanitary sewer is designed to carry a peak design flow of 0.1 m^3/s when full. The slope of the line is 0.0015 and $n = 0.013$. Determine its diameter and velocity using equations and nomograph.

8.4 A trunk sewer is designed to carry a peak design flow of 0.01 m^3/s. The slope of the line is 0.00095 and its diameter is 53 cm (21 in.). Calculate the actual depth of flow and velocity for:
(a) constant value of $n = 0.013$ and (b) variable value of n ($n = 0.013$ when sewer line is running full).

8.5 Calculate the diameter and slope of a sewer line that carries a flow of 75 L/s when flowing at 60% full. The velocity in the line at this flow is 0.65 m/s ($n = 0.015$).

8.6 An intercepting sewer is 500-mm diameter and is serving a population of 6,500 residents. The average wastewater flow is 450 Lpcd. Calculate the depth of flow and velocity at minimum flow. The slope of the line is 0.0025. Assume $n = 0.015$ and the minimum flow is one-third of the average flow.

8.7 A trunk sewer has a diameter of 21 in. It has a slope of 0.001 and is flowing 40% full. Determine the depth, velocity, and discharge at a partial flow condition using hydraulic element equations, $n = 0.015$.

8.8 An intercepting sewer is carrying a flow 0.12 m^3/s. It has a diameter of 61 cm and a slope of 0.001. The sewer enters a manhole that has an outlet sewer diameter of 61 cm at a slope of 0.0012. Draw the hydraulic profile and determine the invert elevation of the incoming sewer. Assume $n = 0.013$ and coefficient of exit and entrance are 0.42 and 0.14, respectively.

8.9 Three proposed trunk sewers *AC*, *BC*, and *EC* will be connected to an existing trunk sewer at manhole *C*. The sewer plan is shown in Figure 8.23. The existing trunk sewer is 21-in. diameter and has a slope of 0.0020 from manhole *C* to *D*. Its design capacity is 3.20 mgd. The invert elevation of manhole *C* is 315.68 ft. The design capacities of trunk sewers *AC*, *BC*, and *EC* are 0.85, 0.80, and 1.0 mgd, respectively. The ground surface elevations at manholes *A*, *B*, *C*, *D*, and *E* and lengths of trunk sewers are shown in Figure 8.23. Design interceptors *AC*, *BC*, and *EC*, and draw the profiles. The required ground cover is around 5 ft. ($n = 0.013$)

8.10 Design the sewerage system for a city section shown in Figure 8.24. The average municipal wastewater flow is 120 gpcd. This sewerage system is designed for peak design flow. The peak wet weather flow is 2.7 times the average flow, including the peak I/I contribution. Also, an industrial plant discharges liquid waste in lines 10 and 11 at a maximum flow rate of 1.35 mgd in each line. The information regarding the sewer layout, sewer and manhole numbers, and elevation of pavement surface at each street intersection is shown in Figure 8.24. The sewer length and area served by each sewer are provided in Table 8.25. The average population density is 55 persons per acre. The invert elevation of the manhole 6 is fixed at 192.47 ft. The velocity at full pipe flow

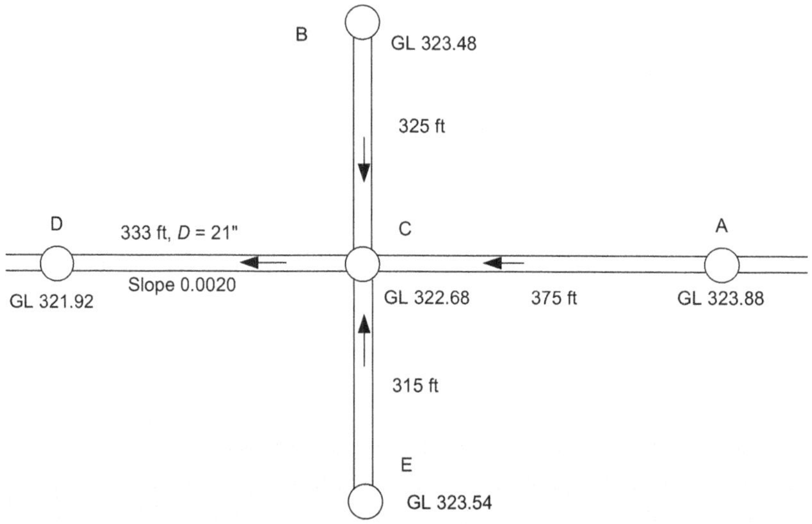

FIGURE 8.23 Plan of existing and proposed trunk sewers (Problem 8.9).

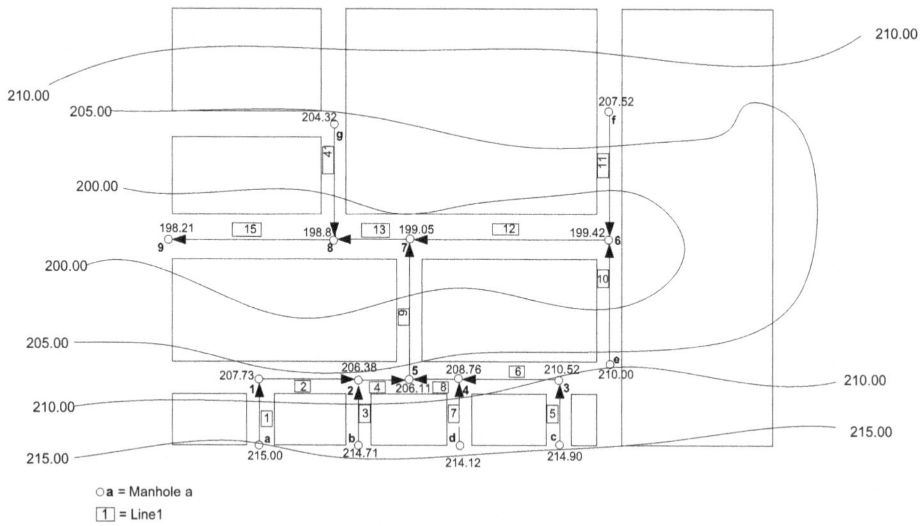

FIGURE 8.24 Layout of the sewerage system of the city (Problem 8.10).

must not be less than 2 ft/s and not more than 4.5 ft/s. Match the crowns of incoming and outgoing sewer lines on straight paths. Minimum sewer diameter must be 8 in., and minimum ground cover required is 4 ft. $n = 0.013$.

8.11 A residential district has the following land-use and runoff coefficients. Determine the average coefficient of runoff for the entire district.

Roof 22%, $C = 0.90$; Pavement 28%, $C = 0.88$; Lawn, garden, and vacant lots 43%, $C = 0.10$; Wooded land 7%, $C = 0.05$

8.12 A storm sewer is designed to carry the surface runoff from a 5-year storm. The rainfall intensity is expressed by equation $i = 135/(t + 18)$. Calculate the maximum rainfall intensity from a storm that has a duration of 30 min.

8.13 Determine the maximum rainfall intensity to estimate the surface runoff. The storm duration is 20 min. Use the exponential formula (Eq. 8.2). The values of constant C and k for 10-year design storm frequency are 15 and 0.5, respectively.

8.14 A watershed of an urban built-up residential subdivision is shown in Figure 8.25. The stormwater flows in a storm drain with a velocity of 1.25 m/s. Estimate the time of concentration at point C. Apply the inlet time from the Kerby and Bransby-Williams formulas. Assume $r = 0.025$.

8.15 A storm sewer serves six drainage areas. The layout of the drainage area is shown in Figure 8.26. The rainfall intensity in the drainage areas for 5-year frequency storm is expressed by equation $i = 120/(t + 15)$.
Determine 5-year peak design flow for each section of the sewer line. The drainage characteristics for the areas are summarized in Table 8.26.

8.16 A storm sewer is designed to serve a multi-family development. A circular concrete storm sewer having a minimum velocity of 3 ft/s when flowing full is provided. Determine the sewer diameter and slope. Also, draw the longitudinal section. Use the following information:
Runoff coefficient $C = 0.42$
Time of concentration $t_c = 35$ min
Drainage area = 38 acres
Roughness coefficient, $n = 0.015$
Ground elevations
Upper manhole = 261.38 ft
Lower manhole = 260.63 ft
Sewer length = 375 ft
Minimum ground cover needed = 3 ft
Use the typical 5-year IDF curve in Figure 8.15.

8.17 A city subdivision is shown in Figure 8.27. Apply the following conditions to design the storm sewers.
1. The tributary areas and coefficient of runoff are given on the layout plan (Figure 8.27).
2. The inlet time for each area is 20 min at each gutter.

TABLE 8.25
Summary Data Table (Problem 8.10)

Line No.	1	2	3	4	5	6	7	8	9	10	11	12	13	14	15
Length, ft	300	500	300	250	300	500	300	250	750	680	680	1,000	350	600	750
Local Area, Acres	5	10	5	10	3	7	50	11	12	12.5	12.5	25	15	20	10

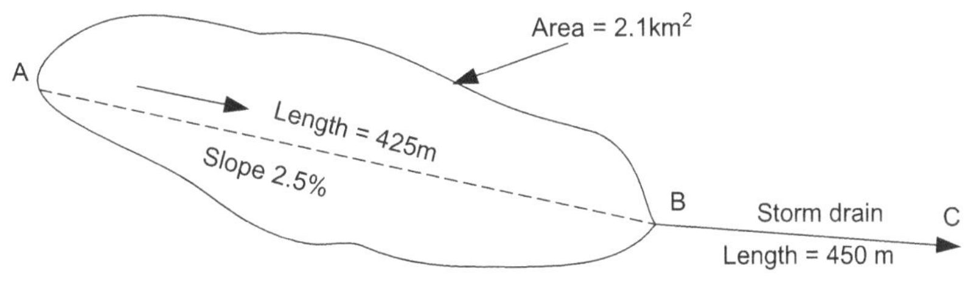

FIGURE 8.25 Urban watershed area (Problem 8.14).

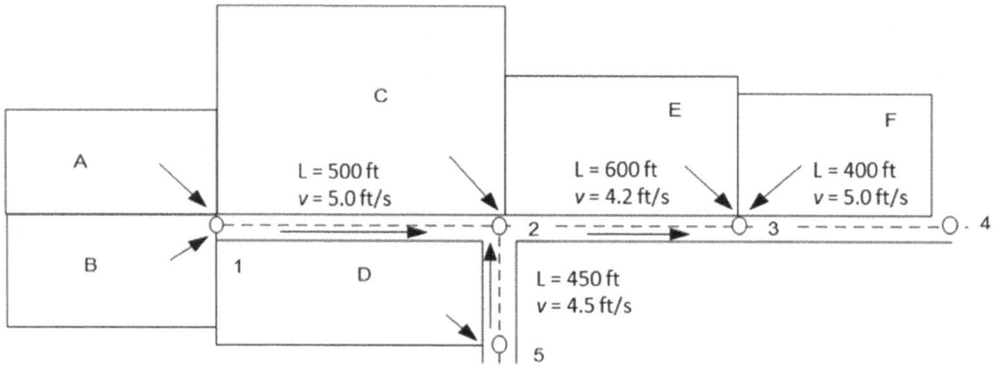

FIGURE 8.26 Storm drainage system (Problem 8.15).

TABLE 8.26
Drainage Area Details (Problem 8.15)

Identification	Areas, Acres	Runoff Coefficient	Inlet Time, min
A	5	0.80	8
B	6	0.90	8.5
C	10	0.85	12
D	12	0.60	14
E	8	0.55	7.5
F	6	0.70	6.5

3. The rainfall intensity in the drainage area for 5-year frequency is calculated from the formula $i = 135/(t + 18)$.
4. Manning's constant $n = 0.013$.
5. The invert elevation at Manhole 1 is 95.19 ft.
6. The minimum velocity in the storm sewer shall remain between 3 and 4 ft/s when flowing full.
7. Match the crowns of all sewers meeting at a manhole.
8. The minimum ground cover is 1.5 ft.

8.18 An underground sewer line is subjected to a concentrated H-20 wheel load vertically centered on the pipe. The diameter of the pipe is 21 in. and its length is 4 ft. The pipe is under a soil cover of 5 ft. Determine the load exerted over the buried pipe.

8.19 A leakage test was conducted on a newly built section of an interceptor. The diameter and length of the interceptor are 910 mm and 320 m, respectively. There was no rainfall in the area, and the groundwater table was low. The testing between two manholes was done by measuring the exfiltration. All incoming sewer lines were blocked on the upper manholes, and the interceptor was blocked on the lower end. A water head of 0.45 m above the top of the pipe was maintained at the upper manhole. The testing was conducted for 36 h. Over this period, the volume of water required to maintain the original level was 450 L/h. Determine the exfiltration rate in cubic meter per day per centimeter of pipe diameter per km.

8.20 A newly constructed section of a sewer sub-main is 375-mm (15 in.) diameter and 125 m (410 ft) long. The standard air test was conducted on this line to determine its acceptability requirements. In this test, 220 m (721.6 ft) of 150 mm (6 in.) building connections

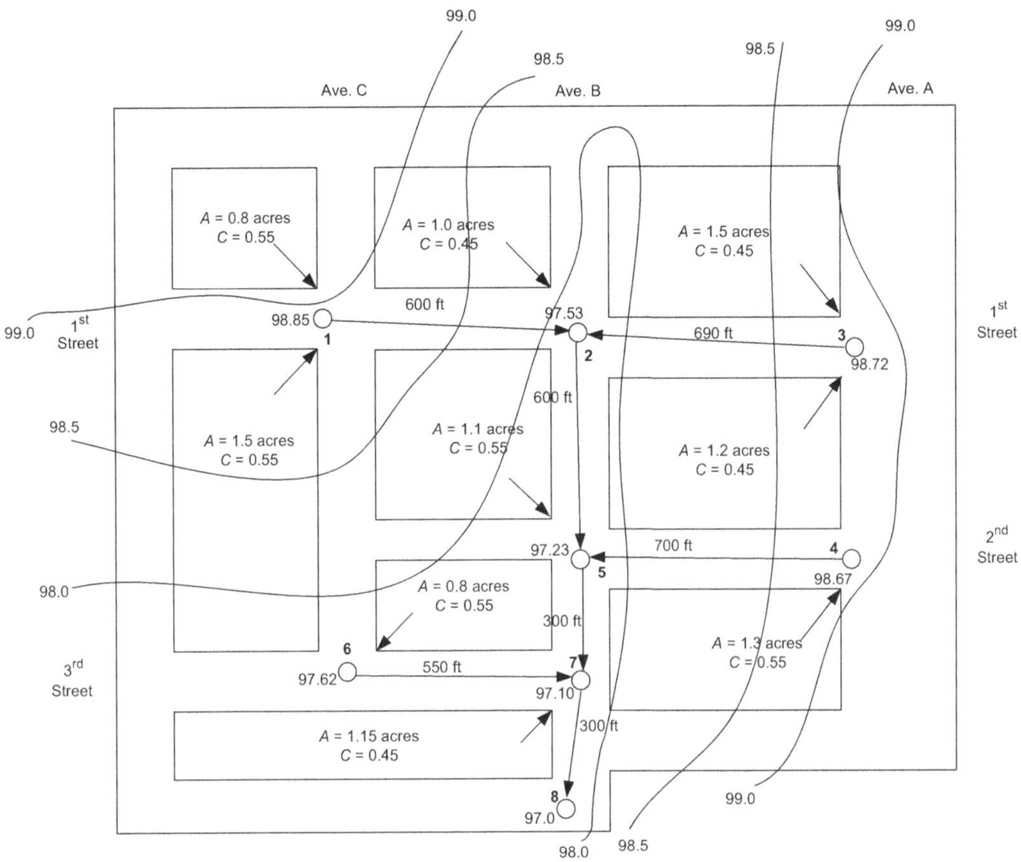

FIGURE 8.27 Layout of a storm sewer system for a city section (Problem 8.17).

were also included. The air pressure was equalized at 27.5 kN/m² (4 lb_f/in.²) for 6 min. A drop in air pressure from 24.0 to 17.1 kN/m² (3.5 to 2.5 lb/in.²) occurred in 5.3 min. Check if the constructed sewer system meets the acceptance requirements.

REFERENCES

1. Metcalf and Eddy, Inc., *Wastewater Engineering: Collection and Pumping of Wastewater*, McGraw-Hill Book Co., New York, 1981.
2. Qasim, S. R., *Wastewater Treatment Plants: Planning, Design and Operation*, 2nd ed., CRC Press, Boca Raton, FL, 1999.
3. Peavy, H. S., D. R. Rowe, and G. Tchobanoglous, *Environmental Engineering*, McGraw-Hill Book Co., New York, 1985.
4. Metcalf and Eddy, Inc., *Water Pollution Abatement Technology Capabilities and Costs for Publicly Owned Treatment Works*, Report Prepared for National Commission on Water Quality, PB-250 690, National Technical Information Service, March 1976.
5. Hardenbergh, W. A., and E. B. Rodie, *Water Supply and Waste Disposal*, International Textbook Company, Scranton, PA, 1966.
6. Joint Committee of the American Society of Civil Engineers and the Water Pollution Control Federation, *Design and Construction of Sanitary and Storm Sewers*, MOP 9, Water Environment Federation, Alexandria, VA, 1995.

7. The Joint Task Force of the American Society of Civil Engineers and the Water Pollution Control Federation, *Gravity Sanitary Sewer Design and Constructions*, MOP FD-5, Water Environment Federation, Alexandria, VA, 1995.
8. Fair, G. M., J. C. Geyer, and D. A. Okun, *Water and Wastewater Engineering, vol.* I, *Water Supply and Wastewater Removal*, John Wiley & Sons, Inc., New York, 1966.
9. Chow, V. T., *Open Channel Hydraulics*, McGraw-Hill Book Co., New York, 1959.
10. Steel, E. W., and T. J. McGhee, *Water Supply and Sewerage*, 6th ed., McGraw-Hill, Inc., New York, 1991.
11. Practical Hydraulics for the Public Works Engineer, Public Works Magazine, 1968.
12. American Concrete Pipe Association, *Concrete Pipe Design Manual*, American Concrete Pipe Association, Irving, TX, 2002.
13. Gupta, R. S., *Hydrology & Hydraulics Systems*, 2nd ed., Waveland Press Inc., Prospect Heights, IL, 2001.
14. National Clay Pipe Institute, *Clay Pipe Engineering Manual*, National Clay Pipe Institute, Lake General, WI, 1995.
15. Izzard, C. F., *Hydraulics of Runoff from Developed Surfaces*, Proceedings of the 26th Annual Meeting, Division of Engineering and Industrial Research, National Research Council, Highway Research Board, vol. 26, Washington, DC, December 5–8, 1946, pp. 129–150.
16. Maidment, D. R. (Editor), *Handbook of Hydrology*, McGraw-Hill, New Tork, 1993.
17. Viessman, W., and G. L. Lewis, *Introduction to Hydrology*, 5th ed., Prentice Hall, Pearson Education, Inc., Upper Saddle River, NJ, 2003.
18. Ward, A. D., and W. J. Elliot, *Environmental Hydrology*, CRC Press, Boca Raton, FL, 1995.
19. Jha, R., V. P. Singh, V. Singh, L. B. Roy, and R. Thandiyath, *Hydrological Modeling: Hydraulics, Water Resources, and Coastal Engineering*, Springer, 2022.
20. Nix, S. G., *Urban Stormwater Modeling and Simulation*, CRC Press, Boca Raton, FL., 1994.
21. Kibler, D. F., M. E. Jennings, G. L. Lewis, B. A. Tschantz, and S. G. Walesh, ASCE Task Committee on Urban Stormwater Software, Microcomputer Software in Urban Hydrology, HYDATA, *American Water Resources Association*, vol. 10, no. 5, September 1991, pp. 27–32.
22. U.S. Environmental Protection Agency, *Storm Water Management Model (SWMM)*, U.S. EPA Research, Office of Science Information Management, Durham, NC, Updated June 20, 2024. https://www.epa.gov/water-research/storm-water-management-model-swmm
23. U.S. Environmental Protection Agency, 40 CFR Part 412-Concentrated Animal Feeding Operation (CAFO) Point Source Category, 68 FR 7269, February 12, 2003, as amended later.
24. Martson, A., and A. O. Anderson, *The Theory of Loads on Pipes in Ditches, and Test of Cement and Clay Drain Tile and Sewer Pipe*, Iowa Engineering Experiment Station, Bulletin No. 31, 1913.

9 Pumps and Pumping Stations

9.1 CHAPTER OBJECTIVES

Pumps are needed to transport water and wastewater, chemicals, and residuals. Pumping involves imparting energy to the liquid to raise its head, causing it to flow from a lower elevation to a higher elevation. The objectives of this chapter are to present:

- Types of pumps and pumping stations
- Pump applications at wastewater treatment plants
- Hydraulic terms and definitions
- Principles of centrifugal pumps, characteristics, design, and selection
- Pumping station design calculation, including operation and maintenance

9.2 PUMP TYPES AND APPLICATIONS

According to the Hydraulic Institute, all pumps may be classified as *kinetic energy* or *positive displacement*.[1] Brief descriptions and applications of many types of pumps in these two classes are provided in Table 9.1.[2] Examples of kinetic and positive displacement pumps are shown in Figure 9.1.[2,3] Because kinetic energy pumps are most commonly used in wastewater applications, an in-depth discussion on these pumps is provided in this chapter.

TABLE 9.1
Pump Types and Major Applications in Wastewater

Major Classifications	Pump Type	Brief Description	Major Pumping Applications
Kinetic	Centrifugal	Consists of an impeller enclosed in a casing with inlet and discharge connections. The head is developed principally by centrifugal force.	Raw wastewater, secondary sludge, return and waste sludge, settled primary and thickened sludge, and effluent
	Peripheral (torque flow or vortex)	Consists of a recessed impeller on the side of the casing entirely out of the flow stream. A pumping vortex is set up by viscous drag.	Scum, grit, sludge, and raw wastewater
Positive displacement	Rotary	Consists of a fixed casing containing gears, vanes, pistons, cams, screws, etc., operating with minimum clearance. The rotating element pushes the liquid around the enclosed casing into the discharge pipe (Figure 9.1a)	Lubricating oils, chemical solutions, small flows of water, and wastewater
	Screw	Uses a spiral screw operating in an enclosed casing (Figure 9.1b)	Grit, settled primary and secondary sludge, thickened sludge, raw wastewater
	Diaphragm	Uses flexible diaphragm or disk fastened over edges of a cylinder (Figure 9.1c)	Chemical solutions

(*Continued*)

DOI: 10.1201/9781003629399-9

TABLE 9.1 (*Continued*)
Pump Types and Major Applications in Wastewater

Major Classifications	Pump Type	Brief Description	Major Pumping Applications
	Plunger	Uses a piston or plunger that operates in a cylinder. The pump discharges liquid during piston or plunger movement through each stroke (Figure 9.1d)	Scum and primary, secondary, and settled sludge. Chemical solutions.
	Airlift	Air is bubbled into a vertical tube partly submerged in water. The air bubbles reduce the unit weight of the fluid in the tube. The higher unit weight fluid displaces the low unit weight fluid, forcing it up into the tube (Figure 9.1e)	Secondary sludge return and circulation.
	Pneumatic ejector	Air is forced into the receiving chamber which ejects the wastewater from the receiving chamber (Figure 9.1f)	Raw wastewater at small installations (100–600 L/min)

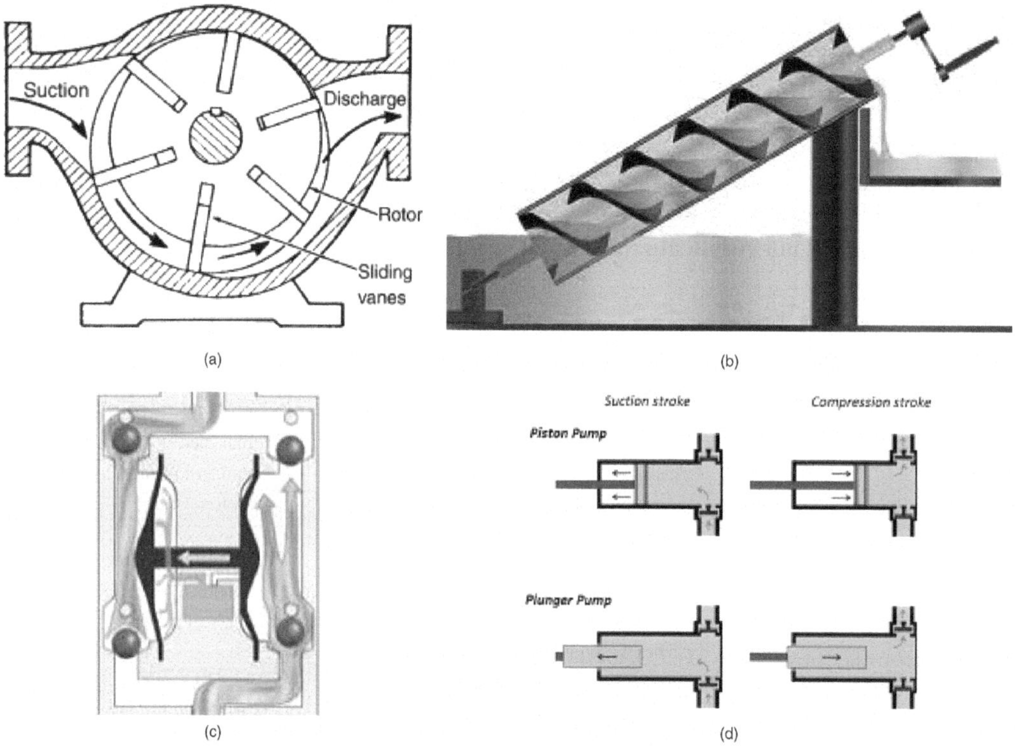

FIGURE 9.1 Types of pumps: (a) Operation of rotary pump, (b) operation of screw pump, (c) operation of a diaphragm pump. Reciprocating action shows simultaneous compression (pumping) from left diaphragm and suction from right diaphragm, and (d) operation of piston and plunger pumps. Left and right strokes are suction and compression (pumping) cycles, (e) air lift pump, and (f) pneumatic ejector.

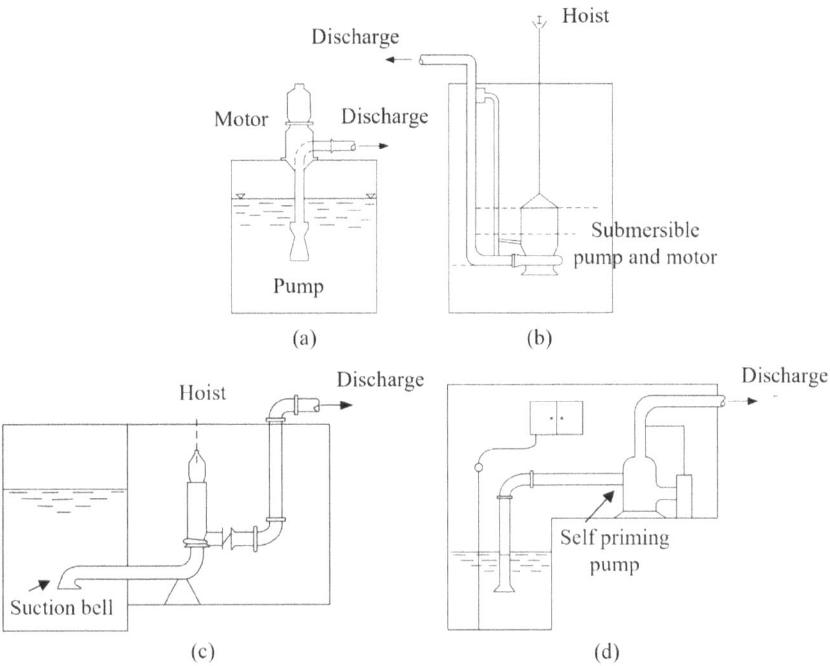

FIGURE 9.2 Wet and dry-well pumping stations: (a) wet-well suspended, (b) wet-well submersible, (c) dry-well flooded suction, and (d) dry-well self-priming.

9.3 TYPES OF PUMPING STATIONS

The pumping stations are classified as (1) wet well and (2) dry well. Both types of pumping stations are described below. Detailed discussion on pumping stations may be found in Refs. 4–9.

9.3.1 Wet-Well Pumping Stations

The wet-well stations employ either suspended or submersible pumps. Suspended pumps have the motor mounted above the liquid level in the wet well, while the pump remains submerged. Submersible pumps have integral motors with special seals suitable for operation below liquid level. The suspended and submersible pumping arrangements in wet well are illustrated in Figure 9.2a and b.

9.3.2 Dry-Well Pumping Stations

The dry-well stations employ either dry-well or self-priming centrifugal pumps. The dry-well centrifugal pumps operate within a dry-well adjacent to the wet well. The volute of the pump is positioned below the low water level in the wet well to ensure a positive suction or prime (Figure 9.2c). In the other case, a recessed dry section may be located within a wet well (Figure 9.2d). The pumps are specially designed to provide a suction lift and are self-priming.

9.4 HYDRAULIC TERMS AND DEFINITIONS

Common terms used in pumping are (1) head; (2) capacity, discharge, or flow rate; and (3) work, power, and efficiency. These terms are described below.

9.4.1 HEAD

Head describes hydraulic energy expressed as the height of a column of liquid above a datum. Many head terms are applied to the analysis of pumps and pumping stations. These include static head, static discharge head, total static head, friction head, velocity head, minor head losses, and total dynamic head (TDH). These head terms are described below.

Static suction head and lift: The static suction head (Figure 9.3) is the difference in elevation between the liquid level in the suction reservoir and the center line of the pump impeller. If the level in the suction reservoir is higher than the center line of the pump impeller, the pump has flooded suction, and the head is called static suction head. If the suction liquid level is below the center line of the impeller, the head is a static suction lift. Pumps with static suction lift need priming system, and should be avoided for raw wastewater pumping.

Static discharge head: The static discharge head is the difference in elevation between the discharge liquid level and the center line of the pump impeller (Figure 9.3).

Total static head: The total static head is the difference in elevation between the discharge liquid level and the free suction liquid level (Figure 9.3).

Friction head: The friction head consists of loss of head in piping (suction and discharge). It is calculated from the Darcy–Weisbach or Hazen–Williams equations. These equations have been discussed and presented in Chapter 5. The Darcy–Weisbach equation is expressed by Eq. (5.13a), and Hazen–Williams equation in different form is expressed by Eqs. (5.18)–(5.23). Readers should carefully review Section 5.5.2 and Examples 5.7–5.11 on this subject. For ready reference, Eqs. (5.13a) and (5.18)–(5.21) are listed below.

$$h_f = \frac{fLV^2}{2gD} \quad \text{(Darcy-Weisbach)}$$

$$h_f = 6.82\left(\frac{V}{C}\right)^{1.85} \times \frac{L}{D^{1.167}} \quad \text{(Hazen-Williams SI units)}$$

$$h_f = 3.02\left(\frac{V}{C}\right)^{1.85} \times \frac{1}{D^{1.167}} \quad \text{(Hazen-Williams U.S. customary units)}$$

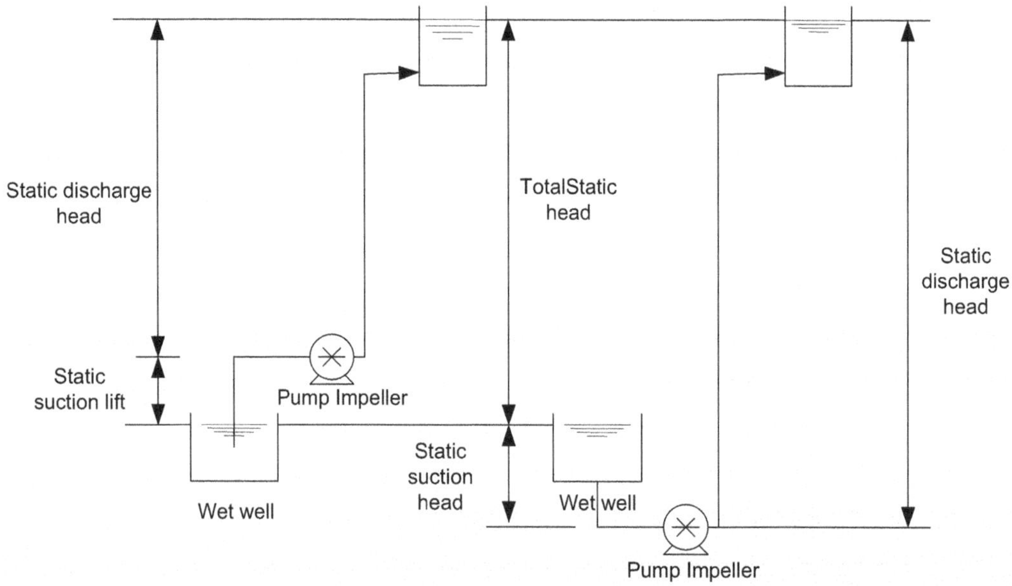

FIGURE 9.3 Head terms used in pumping.

$$V = 0.355CD^{0.63}\left(\frac{h_f}{L}\right)^{0.54} \quad \text{(Hazen-Williams SI units)}$$

$$V = 0.550CD^{0.63}\left(\frac{h_f}{L}\right)^{0.54} \quad \text{(Hazen-Williams U.S. customary units)}$$

$$Q = 0.278CD^{2.63}\left(\frac{h_f}{L}\right)^{0.54} \quad \text{(Hazen-Williams SI units)}$$

$$Q = 0.432CD^{2.63}\left(\frac{h_f}{L}\right)^{0.54} \quad \text{(Hazen-Williams U.S. customary units)}$$

All variables in these equations have been defined in Section 5.5.2.

Velocity head: The velocity head is the kinetic energy contained in the flowing liquid. In determining the head at any point in a piping system, the velocity head must be added to the gauge reading. The exit loss from a pipe into the atmosphere is the velocity head. The velocity head or exit loss is given by Eq. (5.27). For ready reference, it is given below.

$$h_v = \frac{V^2}{2g}$$

where

h_v = velocity head or exit loss, m (ft)

Minor head loss: The minor head losses are produced because of fittings, valves, bends, entrances, exits, etc., and are normally calculated as a function of velocity head. The minor losses have been discussed in Chapter 5 and are expressed by Eq. (5.27). This equation is given below for ready reference.

$$h_m = K\frac{V^2}{2g}$$

where

h_m = minor losses, m (ft)

K = head loss coefficient (see Appendix C)

The velocity head or exit loss is often added to minor head losses.

Total dynamic head: The TDH is head against which the pump works to move the liquid. It is the total energy barrier that must be overcome before the water can be lifted. It is determined by considering the static head, friction head, and minor head losses in suction and discharge pipings, and velocity head in the discharge pipe. Eq. (9.1) is used to determine TDH

$$\text{TDH} = H_{\text{stat}} + h_{\text{fs}} + \Sigma h_{\text{ms}} + h_{\text{fd}} + \Sigma h_{\text{md}} + \frac{V_d^2}{2g} \tag{9.1}$$

where

H_{stat} = static head, m (ft)

h_{fs}, h_{fd} = friction head loss in suction and discharge pipings, m (ft)

Σh_{ms}, Σh_{md} = sum of minor loss in fittings and valves in suction and discharge pipings. Entrance loss in the suction bell is included in the minor losses in suction pipings, m (ft)

V_d = velocity in discharge pipe, m/s (ft/s)

Bernoulli's energy equation (Eq. 5.5) has been discussed in Chapter 5. It is also applied to determine TDH. The energy equation (Eq. 9.2a) is written between the suction bell and the discharge nozzle of the pump. Distances above and below the datum are considered positive and negative, respectively. The datum is normally referenced to the centerline of the pump impeller. The TDH is expressed by Eq. (9.2)

$$Z_d + \frac{P_d}{\gamma} + \frac{V_d^2}{2g} = Z_s + \frac{P_s}{\gamma} + \frac{V_s^2}{2g} + \text{TDH} \tag{9.2a}$$

$$\text{TDH} = \left(\frac{P_d}{\gamma} + \frac{V_d^2}{2g} + Z_d\right) - \left(\frac{P_s}{\gamma} + \frac{V_s^2}{2g} + Z_s\right) \tag{9.2b}$$

where

P_d and P_s = gauge pressures on discharge and suction nozzles of the pump, N/m² (lb/ft²)

γ = specific weight of water, N/m³ (lb/ft³)

V_d and V_s = velocity in discharge and suction nozzles of the pump, m/s (ft/s)

Z_d and Z_s = elevation of gauges on discharge and suction nozzles of the pump above datum, m (ft)

The components of head, head loss, and velocity on the suction and discharge sides of a pump are shown in the definition sketch (Figure 9.4).

Example 9.1: Definition Equations of Total Dynamic Head (TDH)

Develop TDH equations (Eq. 9.1) using the definition sketch given in Figure 9.4.

Solution

1. Apply Eq. (9.2b)

$$\text{TDH} = (\text{discharge energy} - \text{suction energy})$$

$$= \left(\frac{P_d}{\gamma} + \frac{V_d^2}{2g} + Z_d\right) - \left(\frac{P_s}{\gamma} + \frac{V_s^2}{2g} + Z_s\right)$$

2. Develop Eq. (9.1)

Referring to Figure 9.4, the pressure heads at the pump nozzles are represented by terms HD and HS in Eqs. (9.3a) and (9.3b).

$$\frac{P_d}{\gamma} = H_D \quad \text{and} \quad \frac{P_s}{\gamma} = H_S \tag{9.3a}$$

$$\text{TDH} = H_D - H_S + \frac{V_d^2}{2g} - \frac{V_s^2}{2g} \tag{9.3b}$$

where

H_D and H_S = discharge and suction heads measured at the discharge and suction nozzles of the pump referenced to the centerline of the pump impeller, m (ft)

H_D and H_S can be expressed by static heads, and friction and minor losses in Eqs. (9.3c) and (9.3d)

$$H_D = h_d + h_{fd} + \Sigma h_{md} \tag{9.3c}$$

$$H_S = h_s - h_{fs} - \Sigma h_{ms} - \frac{V_s^2}{2g} \tag{9.3d}$$

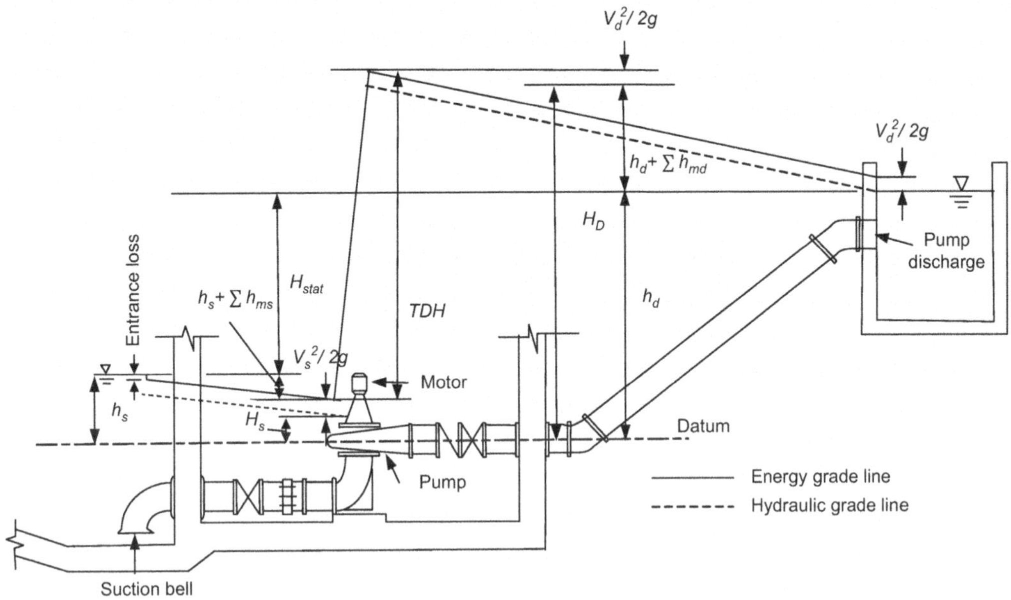

FIGURE 9.4 Definition sketch for the head on a pump (Example 9.1).

where

h_d and h_s=static discharge and suction heads with respect to the centerline of pump impeller, m (ft)

Substitute these values in Eq. (9.3b)

$$\text{TDH} = h_d + h_{fd} + \Sigma h_{md} - h_s + h_{fs} + \Sigma h_{ms} + \frac{V_s^2}{2g} + \frac{V_d^2}{2g} - \frac{V_s^2}{2g}$$

$$H_{\text{stat}} = h_d - h_s$$

These substitutions provide Eq. (9.1).

$$\text{TDH} = H_{\text{stat}} + h_{fs} + \Sigma h_{ms} + h_{fd} + \Sigma h_{md} + \frac{V_d^2}{2g}$$

9.4.2 Capacity, Discharge, or Flow Rate

The capacity, discharge, or flow rate of a pump is the volume of liquid pumped per unit of time. It is expressed as cubic meters per second (m^3/s), liters per second (L/s), gallons per minute (gpm), cubic feet per second (ft^3/s), etc.

9.4.3 Work Power and Efficiency

The work done by a pump is proportional to the product of the specific weight of the fluid being discharged and the total head against which the flow is moved. The pump efficiency is the ratio of the useful pump output power to the input power. Pump power output, and pump and motor efficiencies are expressed by Eqs. (9.4)–(9.6)

$$P_w = K'Q(\text{TDH})\gamma \tag{9.4}$$

$$E_p = \frac{P_w}{P_p} \tag{9.5}$$

$$E_e = \frac{P_p}{P_m} \tag{9.6}$$

where

P_w = power output of the pump (water power), kW [hp (horsepower)]
P_p = power input to the pump (brake power), kW (hp)
P_m = power input to the motor (electrical energy or wire power), kW (hp)
Q = capacity, discharge, or flow rate, m^3/s (ft^3/s)
TDH = total dynamic head, m (ft)
γ = specific weight of the liquid pumped, kN/m^3 (lb/ft^3)
E_p = pump efficiency, usually 70%–90%
E_e = motor efficiency, usually 90%–98%
K' = constant depending on the units of expression
(TDH = m, $Q = m^3/s$, $\gamma = 9.81\ kN/m^3$, P_w = kW, $K' = 1$ kW/(kN.m/s))

(TDH = ft, $Q = ft^3/s$, $\gamma = 62.4\ lb/ft^3$, P_w = hp, $K' = \dfrac{1\ \text{hp}}{550\ \text{ft.lb/s}}$

Note: Power = kN.m/s or kW (SI unit)
Power = ft.lb/s or one hp = 550 ft.lb/s (U.S. customary unit)

Example 9.2: Gauge Pressure and Head

The pressure gauge reading at the pump discharge nozzle is 120 kPa (kN/m^2). Convert the pressure into (a) m of water, (b) lb_f/ft^2, (c) ft of water, (d) atm, and (e) inches of mercury.

Solution

a. Head in $m = \dfrac{P}{\gamma} = \dfrac{120\ \text{kN/m}^2}{9.81\ \text{kN/m}^3} = 12.2\ \text{m}$

b. Pressure in $\text{lb}_f/\text{ft}^2 = 120\ \text{kN/m}^2 \times \dfrac{20.88\ \text{lb}_f/\text{ft}^2}{\text{kN/m}^2} = 2{,}506\ \text{lb}_f/\text{ft}^2$

c. Head in ft of water $= 2{,}506\ \text{lb}_f/\text{ft}^2 \times \dfrac{1}{62.4\ \text{lb}_f/\text{ft}^3} = 40.2\ \text{ft}$

d. Pressure. in atm $= 120\ \text{kN/m}^2 \times \dfrac{\text{atm}}{101.325\ \text{kN/m}^2} = 1.18\ \text{atm}$

e. Head in inches of mercury $= 120\ \dfrac{\text{kN}}{\text{m}^2} \times \dfrac{0.2961\ \text{in. of mercury}}{\text{kN/m}^2} = 35.5\ \text{in. of mercury}$

Also, head in inches of mercury $= \dfrac{40.2\ \text{ft of water} \times 12\ \text{in./ft}}{13.585\ \text{in. of water/in. of mercury}} = 35.5\ \text{in. of mercury}$

Example 9.3: Static Heads

The liquid elevations in the wet well and discharge well are 98.21 and 108.67 m, respectively. The elevation of the centerline of the pump impeller is 100.00 m. Determine: (1) static suction head, (2) static discharge head, and (3) total static head.

Solution

Calculate suction head

a. Static suction head = 98.21 m − 100.00 m = −1.79 m (suction lift)

b. Static discharge head = 108.67 m − 100.00 m = 8.67 m

c. Total static head = 108.67 m − 98.21 m = 10.46 m

Also, total static head = 8.67 m − (−1.79 m) = 10.46 m

Example 9.4: Head Losses in a Discharge Pipe

A submersible pump is pumping primary settled wastewater from a sump to an aeration basin. There is no suction piping. The diameter of the suction bell is 30 cm, and $K = 0.04$. Calculate the components of head losses in the force main. The diameter and length of force main are 20 cm and 120 m, respectively. There is one enlarger $K = 0.25$, one check valve $K = 2.5$, one plug valve $K = 1.0$, one gate valve $K = 0.19$, three 45° elbows $K = 0.2$, and two 90° elbows $K = 0.3$. The pump discharge is 0.08 m³/s and the Darcy–Weisbach coefficient of roughness $f = 0.02$.

Solution

1. Calculate velocity head in the force main

$$\text{Area } A \text{ of the force main} = \frac{\pi}{4} \times (0.2\text{ m})^2 = 0.0314\text{ m}^2$$

$$\text{Velocity } V \text{ in the force main } = \frac{Q}{A} = \frac{0.08\text{ m}^3/\text{s}}{0.0314\text{ m}^2} = 2.55\text{ m/s}$$

$$\text{Velocity head, } h_v = \frac{V^2}{2g} = \frac{(2.55\text{ m/s})^2}{2 \times 9.81\text{ m/s}^2} = 0.33\text{ m}$$

2. Calculate the total friction head loss in the force main
   From Eq. (5.13a)

$$h_f = \frac{0.02 \times 120\text{ m}}{0.2\text{ m}} \times \frac{V^2}{2g} = 12.0 \times 0.33\text{ m} = 3.96\text{ m}$$

3. Calculate minor losses

$$\text{Velocity through suction bell} = \frac{0.08\text{ m}^3/\text{s}}{\pi/4\,(0.3\text{ m})^2} = 1.13\text{ m/s}$$

$$\text{Head loss through suction bell} = \frac{0.04 \times (1.13\text{ m/s})^2}{2 \times 9.81\text{ m/s}^2} = 0.003\text{ m}$$

$$\text{Head loss through enlarger} = 0.25 \times 0.33 \text{ m} = 0.08 \text{ m}$$

$$\text{Head loss through check valve} = 2.5 \times 0.33 \text{ m} = 0.83 \text{ m}$$

$$\text{Head loss through gate valve} = 0.19 \times 0.33 \text{ m} = 0.06 \text{ m}$$

$$\text{Head loss through plug valve} = 1.0 \times 0.33 \text{ m} = 0.33 \text{ m}$$

$$\text{Head loss through } 45^\circ \text{ elbows} = 3 \times 0.2 \times 0.33 \text{ m} = 0.20 \text{ m}$$

$$\text{Head loss through } 90^\circ \text{ elbows} = 2 \times 0.3 \times 0.33 \text{ m} = 0.20 \text{ m}$$

$$\Sigma h_m = 1.70 \text{ m}$$

4. Calculate total head loss

$$\text{Total head loss } h_L = h_f + \Sigma h_m + h_v = 3.96 \text{ m} + 1.70 \text{ m} + 0.33 \text{ m} = 5.99 \text{ m}$$

Example 9.5: Hydraulic and Energy Grade Lines

A horizontal force main is carrying a flow of 0.20 m^3/s. At a point *A*, the diameter of the force main is 30 cm, and the pressure gauge reading is 180 kN/m^2. At a point *B*, 50 m from *A*, there is a gate valve (K=2.5). At a point *C*, 80 m from *B*, the pipe diameter is reduced to 25 cm (K=0.25). A pressure gauge is located at a point *D*, 150 m from point *C*. What will be the pressure gauge reading in kN/m^2? Draw the hydraulic and energy grade lines. Assume C=110.

Solution

1. Compute pressure head, velocity head, and friction head loss in pipe *AB*

$$\text{Pressure head at point } A = 180 \frac{\text{kN}}{\text{m}^2} \times \frac{1}{9.81 \text{ kN/m}^2 \cdot \text{m}}$$

$$\text{Velocity in the force main} = \frac{0.20 \text{ m}^3/\text{s}}{\pi/4 \left(0.3 \text{ m}\right)^2} = 2.83 \text{ m/s}$$

$$\text{Velocity head} = \frac{\left(2.83 \text{ m/s}\right)^2}{2 \times 9.81 \text{ m/s}^2} = 0.41 \text{ m}$$

$$\text{Friction head loss in line } AB \; h_{f\text{AB}} \left(\text{Eq. (5.18)}\right) = 6.82 \left(\frac{2.83}{110}\right)^{1.85} \times \frac{50 \text{ m}}{\left(0.3 \text{ m}\right)^{1.167}}$$

$$= 1.59 \text{ m}$$

2. Compute head loss through gate valve at *B* and friction head loss in pipe *BC*
   Head loss due to gate valve = 2.5 × 0.41 m = 1.03 m

$$h_{f\text{BC}} \left(\text{Eq. (5.18)}\right) = 6.82 \left(\frac{2.83}{110}\right)^{1.85} \times \frac{80 \text{ m}}{(0.3)^{1.167}} = 2.55 \text{ m}$$

3. Compute velocity head and friction head loss in pipe *CD*, and head loss due to reducer at point *C*

$$\text{Velocity in pipe } CD = \frac{0.2\ \text{m}^3/\text{s}}{\pi/4(0.25\ \text{m})^2} = 4.07\ \text{m/s}$$

$$\text{Velocity head} = \frac{(4.07\ \text{m/s})^2}{2\times 9.81\ \text{m/s}^2} = 0.84\ \text{m}$$

$$h_{f\text{CD}}(\text{Eq. (5.18)}) = 6.82\left(\frac{4.07}{110}\right)^{1.85} \times \frac{150\ \text{m}}{(0.25)^{1.167}} = 11.58\ \text{m}$$

$$\text{Head loss due to pipe reduction} = 0.25\times 0.84\ \text{m} = 0.21\ \text{m}$$

4. Compute pressure gauge reading at point *D*

$$\text{Total head at point } D = 18.76\ \text{m} - 1.59\ \text{m} - 1.03\ \text{m} - 2.55\ \text{m} - 0.21\ \text{m} - 11.58\ \text{m}$$
$$= 18.76\ \text{m} - 16.96\ \text{m} = 1.80\ \text{m}$$

$$\text{Pressure head at point 9} = 1.80\ \text{m} - 0.84\ \text{m} = 0.96\ \text{m}$$

$$\text{Pressure gauge reading at point 9} = 0.96\ \text{m}\times 9.81\ \frac{\text{kN}}{\text{m}^2\cdot\text{m}} = 9.42\ \text{kN/m}^2$$

5. Draw the hydraulic and energy grade lines

The hydraulic grade line is drawn by subtracting the head losses in the direction of flow. Energy grade line is drawn by adding the velocity head to the hydraulic grade line. The hydraulic grade and energy grade lines are shown in Figure 9.5.

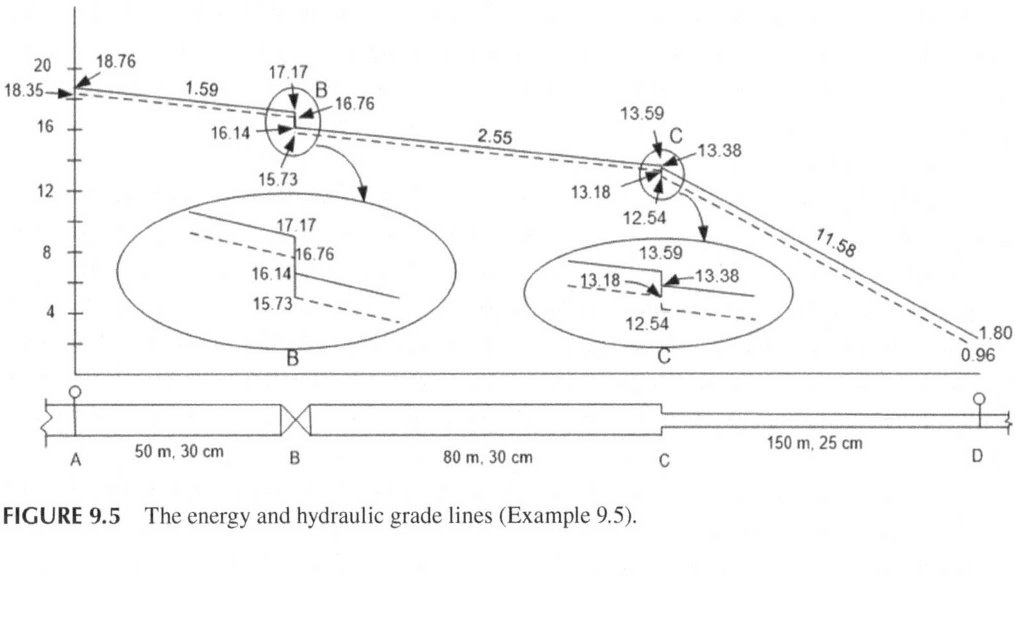

FIGURE 9.5 The energy and hydraulic grade lines (Example 9.5).

Example 9.6: Calculations for Total Dynamic Head (TDH)

Wastewater is pumped from a sump into an equalization basin. The piping schematic is shown in Figure 9.6a. Calculate the components of head losses in suction and discharge pipings and TDH. The pump discharge is 0.10 m³/s. Draw the profile of energy grade and hydraulic grade lines. The datum is the center line of the pump impeller. The elevation of center of pump impeller is 150.00 m. The reading on the discharge gauge located at the centerline of the pump is 346.78 kN/m².

Valves and fitting schedule:

A = Suction bell, 28 cm, *K* = 0.04
B = 45° elbow, 20 cm, *K* = 0.20

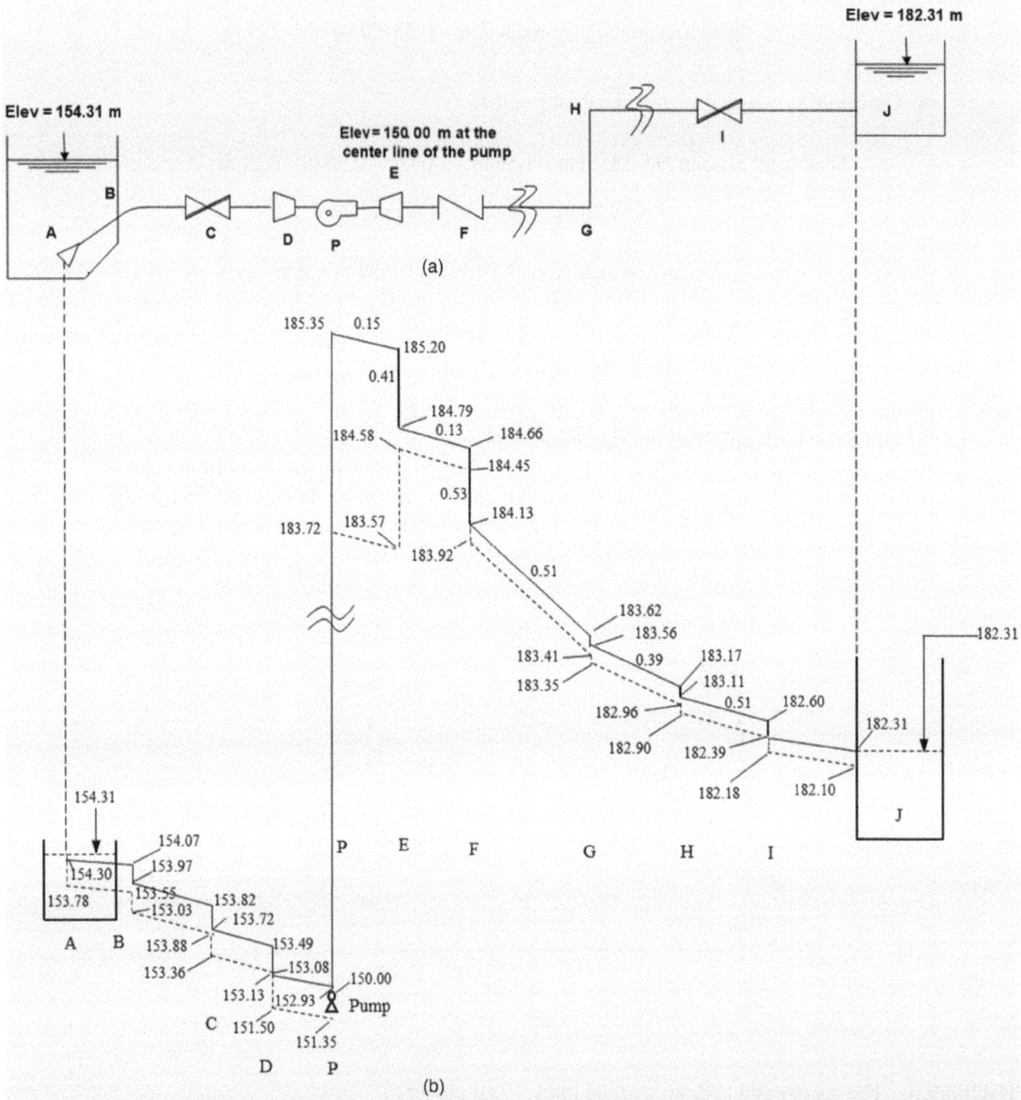

FIGURE 9.6 The piping schematic, and energy and hydraulic grade lines: (a) piping schematic, and (b) energy and hydraulic grade lines (Example 9.6).

C = Gate valve, 20 cm, $K=0.20$
D = Reducer, 20 × 15 cm, $K=0.25$
E = Enlarger, 15 × 25 cm, $K=0.25$
F = Check valve, 25 cm, $K=2.50$
G, H = 90° elbows, 25 cm, $K=0.30$
I = Plug valve, 25 cm, $K=1.00$

Piping schedule:

AB = 20 cm dia, and $L=3.0$ m; BC = 20 cm dia, and $L=2$ m; CD = 20 cm, and $L=3.0$ m; DP and PE = 15 cm dia, and $L=0.5$ m; EF = 25 cm dia, and $L=5$ m; FG = 25 cm, and $L=20$ m; GH = 25 cm dia, and $L=15$ m; HI = 25 cm dia, and $L=20$ m; IJ = 25 cm dia, and $L=3$ m. Hazen–Williams C for all pipe sections is 100.

Note: Pump impeller center line is the datum at an elevation of 150.00 m. It is the reference point for both suction and discharge lines.

Solution

1. Determine the pressure head generated at the pump centerline

$$h_p = \frac{P_d}{\gamma} = \frac{346.79 \text{ kN/m}^2}{9.81 \text{ kN/m}^3} = 35.5 \text{ m}$$

Elevation with respect to pump center line

$$150.00 \text{ m} + 35.35 \text{ m} = 185.35 \text{ m}$$

2. Calculate minor and friction head losses in suction and discharge piping

The minor head losses and friction head losses in valves and fittings, and pipings of suction and discharge pipes are summarized in Table 9.2.

3. Calculate TDH

The exit loss or head loss due to pipe discharge, or velocity head, is calculated from Eq. (5.27).

$$\frac{V_d^2}{2g} = \frac{(2.04 \text{ m})^2}{2 \times 9.81 \text{ m/s}^2} = 0.21 \text{ m}$$

$$H_{\text{stat}} = 182.31 \text{ m} - 154.31 \text{ m} = 28.00 \text{ m}$$

TDH is calculated from Eq. (9.1)

$$\text{TDH} = H_{\text{stat}} + h_{\text{fs}} + \Sigma h_{\text{ms}} + h_{\text{fd}} + \Sigma h_{\text{md}} + \frac{V_d^2}{2g}$$

$$= 28.00 \text{ m} + 0.76 \text{ m} + 0.62 \text{ m} + 1.77 \text{ m} + 1.27 \text{ m} + 0.21 \text{ m} = 32.63 \text{ m}$$

4. Draw the pipe schematic and hydraulic grade and energy grade lines

The pipe schematic, and hydraulic grade and energy grade lines are drawn in Figure 9.6a and b, respectively.

TABLE 9.2
Minor Head Losses in Valves and Fittings, and Friction Losses in Pipes of Suction and Discharge Pipings (Example 9.6)

Valves, Fittings, and Pipings	Velocity V, m/s	Velocity Head $V^2/2g$, m	K or C	Head Loss h_L, m
Suction Piping				
Valves and fittings (Eq. 5.27)				
Suction bell A, 28 cm	1.62	0.13	0.04	0.01[a]
45° elbow B, 20 cm	3.18	0.52	0.20	0.10
Gate valve C, 20 cm	3.18	0.52	0.20	0.10
Reducer D, 15 cm	5.66	1.63	0.25	0.41
Σh_{ms}				0.62
Straight pipe section (Eq. 5.18)				
AB, 20 cm, 3 m	3.18	0.52	100	0.23[b]
BC, 20 cm, 2 m	3.18	0.52	100	0.15
CD, 20 cm, 3 m	3.18	0.52	100	0.23
DP, 15 cm, 0.5 m	5.66	1.63	100	0.15
Σh_{fs}				0.76
Discharge Piping				
Valves and fittings (Eq. 5.27)				
Enlarger E, 15 cm	5.66	1.63	0.25	0.41
Check valve F, 25 cm	2.04	0.21	2.50	0.53
Two 90° elbows G and H, 25 cm	2.04	0.21	0.30	0.12
Plug valve I, 25 cm	2.04	0.21	1.00	0.21
Σh_{md}				1.27
Straight pipe sections (Eq. 5.18)				
PE, 15 cm, 0.5 m	5.66	1.63	100	0.15
EF, 25 cm, 5 m	2.04	0.21	100	0.13
FG, 25 cm, 20 m	2.04	0.21	100	0.51
GH, 25 cm, 15 m	2.04	0.21	100	0.39
HI, 25 cm, 20 m	2.04	0.21	100	0.51
IJ, 25 cm, 3 m	2.04	0.21	100	0.08
Σh_{fd}		0.21		1.77

[a] $h_{ms} = KV^2/2g = 0.04 \times 0.13 = 0.01$

[b] h_f from Eq. (5.18) $= 6.82\left(\frac{V}{C}\right)^{1.85} \times \frac{L}{D^{1.167}} = 6.82\left(\frac{3.18}{100}\right)^{1.85} \times \frac{3\text{ m}}{(0.2)^{1.167}} = 0.23.$

Example 9.7: Determination of TDH from Bernoulli's Equation

Wastewater is pumped from a wet well to a primary sedimentation basin. The pump discharge is 0.5 m³/s. The diameter of the discharge pipe is 35 cm, and the reading on the discharge gauge located at the centerline of the pump is 125 kN/m². The diameter of the suction pipe is 40 cm, and reading on the suction gauge located 0.6 m below the centerline of the pump is 10 kN/m². Determine the head on the pump using Bernoulli's equation.

Solution

1. Determine the velocity in discharge and suction pipings

$$V_d = \frac{0.5\text{ m}^3/\text{s}}{\pi/4(0.35\text{ m})^2} = 5.20\text{ m/s}$$

$$V_s = \frac{0.5\ \text{m}^3/\text{s}}{\pi/4(0.4\ \text{m})^2} = 3.98\ \text{m/s}$$

2. Determine pressure head in discharge and suction pipings

$$h_p = \frac{P_d}{\gamma} = \frac{125\ \text{kN/m}^2}{9.81\ \text{kN/m}^3} = 12.74\ \text{m}$$

$$h_s = \frac{P_s}{\gamma} = \frac{10\ \text{kN/m}^2}{9.81\ \text{kN/m}^3} = 1.02\ \text{m}$$

3. Apply Bernoulli's equation (Eq. 5.5)

$$\left(\frac{P_s}{\gamma} + \frac{V_s^2}{2g} + Z_s\right) + \text{TDH} = \left(\frac{P_d}{\gamma} + \frac{V_d^2}{2g} + Z_d\right)$$

$$\text{TDH} = \left[12.74\ \text{m} + \frac{(5.20\ \text{m/s})^2}{2 \times 9.81\ \text{m/s}^2} + 0\right] - \left[1.02 + \frac{(3.98\ \text{m/s})^2}{2 \times 9.81\ \text{m/s}^2} + (-0.6\ \text{m})\right]$$

$$= (12.74 + 1.38 + 0 - 1.02 - 0.81 + 0.6)\ \text{m} = 12.89\ \text{m}$$

Example 9.8: Wire Power and Efficiency

A pump is delivering 2,000 gpm at a TDH of 64 ft. The pump efficiency is 85% and the motor efficiency is 92%. Calculate the wire power or electrical energy input to the motor.

Solution

1. Convert pump discharge into cfs

$$Q = 2{,}000\ \text{gpm} \times \frac{1}{60\ \text{s/min}} \times \frac{1}{7.48\ \text{gal/ft}^3} = 4.456\ \text{cfs}$$

2. Calculate water power or power output of pump (Eq. 9.4)

$$P_w = K'Q(\text{TDH})\gamma$$

$$P_w = \frac{1\ \text{hp}}{550\ \text{ft.lb/s}} \times 4.456\ \frac{\text{ft}^3}{\text{s}} \times 64\ \text{ft} \times 62.4\ \frac{\text{lb}}{\text{ft}^3}$$

$$P_w = 32.35\ \text{hp}$$

3. Calculate power input to the pump (Eq. 9.5)

$$P_p = \frac{P_w}{E_p}$$

$$P_p = \frac{32.35}{0.85} = 38\ \text{hp}$$

4. Calculate wire power or power input to motor (Eq. 9.6)

$$P_m = \frac{P_p}{E_m} = \frac{38\ \text{hp}}{0.92} = 41.3\ \text{hp}$$

9.5 CENTRIFUGAL PUMPS

9.5.1 Pump Elements

Centrifugal pumps are most commonly used in water and wastewater applications. These pumps have two principal parts: (1) a rotating element called the impeller and (2) pump casing. The impeller rotates and forces the liquid being pumped into a rotary motion. The liquid leaves the impeller with much higher pressure and velocity than when it entered. The pump casing is designed to direct the liquid to the impeller and lead it to the discharge nozzle. Basic design features may be found in Refs. 4–6. The basic features of centrifugal pumps are summarized in Table 9.3.

9.5.2 Specific Speed and Pump Categories

Specific Speed: The pump performance parameters are: (1) rotational speed, (2) pumping head (TDH), (3) discharge capacity, (4) power applied, and (5) efficiency. These parameters have a certain relationship to each other that varies from pump to pump. The pressure and discharge of a pump vary with speed. These three main parameters – speed, discharge, and head – are combined into a single term known as the *specific speed.* It is defined as the rotational speed at which the pump discharges a unit flow at a unit head and at the maximum efficiency. Under any similar operating conditions of head and capacity, the specific speed is the same for all pumps of geometrically

TABLE 9.3
Basic Features of Centrifugal Pumps

Types or Variations	Description
Suction	
Single-suction	A pump equipped with one or more single-suction impellers.
Double-suction	A pump equipped with one or more double-suction impellers.
Number or Stages	
Single-stage	A pump in which the total head is developed by one impeller.
Multi-stage	A pump having two or more impellers acting in series in one casing.
Position of the Shaft	
Horizontal	A pump with the shaft normally in horizontal position.
Vertical	
Dry-pit type	A pump with vertical shaft located in a dry well.
Submerged type	A pump with vertical shaft located in a wet well.
Casing	
Volute	A pump having casing made in the form of a spiral or volute. The velocity is transformed into head by gradually increasing area of water passage. Dry-well pumps generally have volute case.
Circular	A pump having a casing of constant cross-section, and concentric with the impeller. Curved vanes are used to transform velocity into head. Wet-well pumps have circular casing.
Diffusion (turbine)	A pump that transforms velocity into head by diffusion vanes. Diffusion casing is used with wet-well pumps.
Impeller	
Enclosed (front and back shroud)	A pump with enclosed impeller and shroud is commonly used for pumping sanitary sewage.
Open or semi-enclosed	A pump constructed with the vanes attached to an upper or inbound shroud is best suited for pumping large volumes with intermittent service. Common applications include stormwater pumping.

similar designs.[3] The specific speed is used to group the pumps with respect to the similarity of their design, and to compare the performance of the pumps of different designs. As a general rule, (1) the specific speed of any pump changes as its operating condition changes, (2) pumps with high specific speeds are more efficient than those with low, (3) high specific speed is associated with high discharge at low head, (4) the efficiency of a pump starts to drop at specific speed below 20 (1,000 in U.S. customary units), (5) the pump efficiency also starts to drop at specific speed over 100 (5,000 U.S. customary units), and (6) at all specific speeds, pumps of smaller capacities have lower efficiency than pumps of higher capacities. The specific speed is generally expressed by Eq. (9.7).[2]

$$N_s = \frac{NQ^{1/2}}{(\text{TDH})^{3/4}} \tag{9.7}$$

where

N_s = pump specific speed[1]

N = pump rotational speed, rev/min

Q = flow at optimum efficiency, gpm (m³/s)

Pump categories: Centrifugal pumps are classified in many ways. These classifications or types are generally based on the manner in which liquid enters and leaves the casing, and the manner in which the impeller or vanes impart the energy to the water to transform the velocity into the head. Many of these variations are presented in Table 9.3. Based on the impeller configuration and specific speed, pumps are traditionally divided into three major classes: radial flow, mixed flow, and axial flow. Basic characteristics of the radial-flow, mixed-flow, and axial-flow pumps are given below. Major differences in pump classes are presented in Table 9.4. Details of typical radial-flow, mixed-flow, and axial-flow pumps are shown in Figure 9.7.

9.5.2.1 Radial-Flow Pumps

1. The pressure head is developed basically by the action of centrifugal force.
2. The liquid normally enters the impeller at the hub and flows radially to the periphery.
3. The impellers used may be single- or double-suction.
4. The impeller may have vanes that may be straight or double curvature, such as the Francis–Vane.[2]

TABLE 9.4
The Major Differences in Three Classes of Centrifugal Pumps

Category	Radial Flow	Mixed Flow	Axial Flow
Specific speed	Less than 80 for single-suction (4,200 U.S. customary units) and 115 for double-suction (6,000 U.S. customary units)	80–120 (4,200–6,250 U.S. customary units)	More than 200 (10,000 U.S. customary units)
Discharge direction	Radial to the axis of rotation	Partially radial and partially parallel to the axis of rotation	Parallel to the axis of rotation
Pumping action	Centrifugal force and lifting action	Centrifugal force	Lifting action
Head ranges	High to medium (more than 16 m)	Medium to low (10–16 m)	Low (less than 10 m)
Capacity	Low to medium (below 0.06 m³/s (1,000 gpm))	Medium to high (0.06–0.3 m³/s) (1,000–5,000 gpm)	High (over 0.3 m³/s (5,000 gpm))

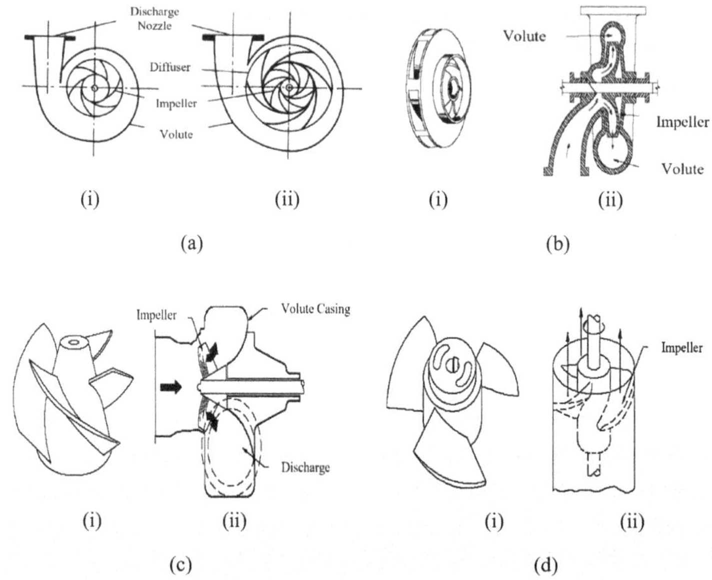

FIGURE 9.7 Details of typical centrifugal pumps: (a) pump casings (i) volute and (ii) diffusion; (b) radial-flow pump (i) impeller and (ii) volute and impeller; (c) mixed-flow pump (i) impeller and (ii) axial entry, and axial and radial discharge; and (d) axial-flow pump (i) impeller and (ii) axial discharge.

5. Pump shaft may be horizontal or vertical.
6. The pump has specific speeds below 80 (4,200 in U.S. customary units) with single impeller and 115 (6,000 in U.S. customary units) with double-suction impeller.
7. For raw wastewater pumping, the non-clog type is used.
8. These pumps are customarily used for low to medium flows and medium to high heads.

9.5.2.2 Mixed-Flow Pumps

1. Head is developed partly by centrifugal force and partly by the lift of the vanes on the liquid.
2. The liquid enters axially and discharges in an axial and radial direction.
3. The casing may be either the volute or the diffusion type. This gives two variations of mixed-flow volute and mixed propeller pumps.
4. The pump may have single- or double-suction and may be single- or multi-stage.
5. The specific speed of the pump is in the range of 80–120 (4,200–6,250 U.S. customary units).
6. As specific speed increases, mixed-flow pumps become more like axial-flow pumps.
7. The mixed-flow pumps are applicable for medium-head applications (7–16) and for medium to large capacities.
8. The pumps require positive submergence, although with proper selection of rotative speeds, they may be used for limited suction lift applications.
9. Common applications are for pumping raw wastewater and stormwater flow. For raw wastewater pumping, a non-clog type pump must be used.

9.5.2.3 Axial-Flow Pump

1. The pumps are also called propeller pumps. They develop most of the head by a propelling or lifting action of the vanes on the liquid.
2. They have a single inlet impeller with the flow entering axially and discharging axially.

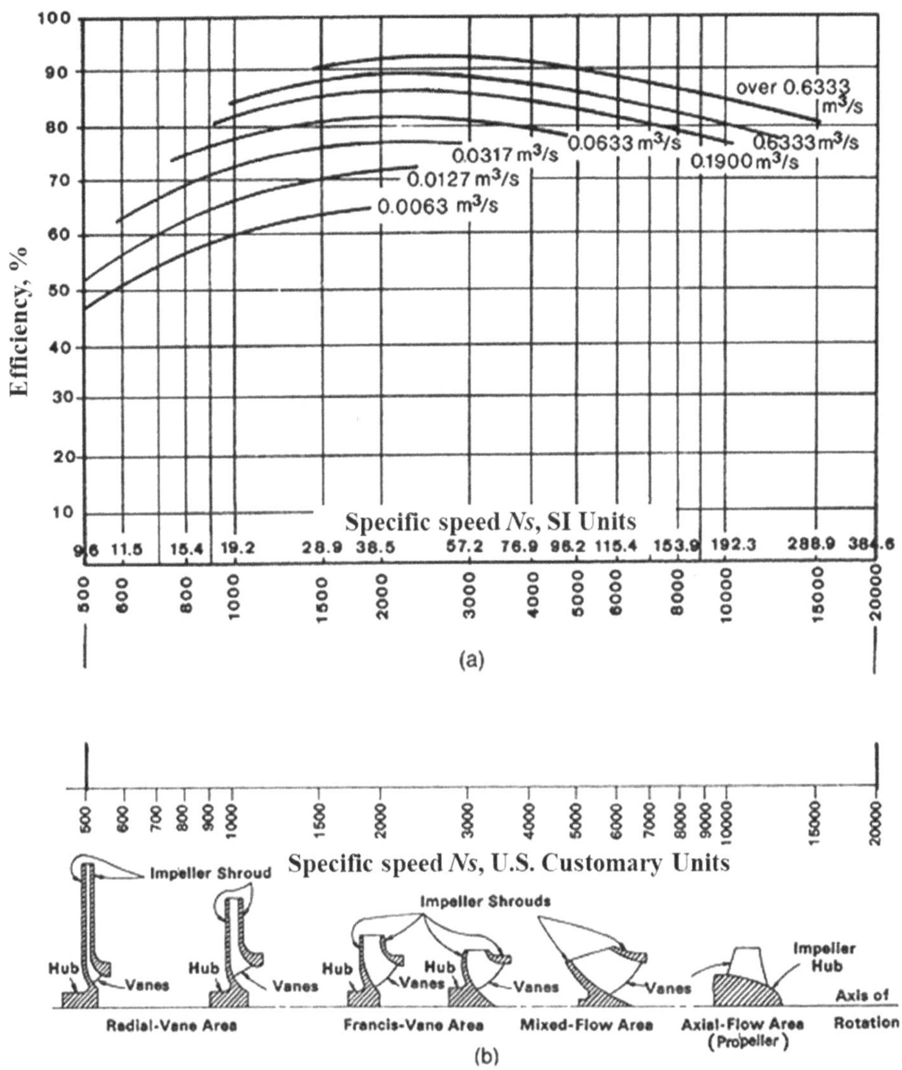

FIGURE 9.8 Comparison of pump capacity, efficiency, and impeller designs with specific speed: (a) pump efficiency versus specific speed and (b) profile of several pump impeller designs placed according to where each design fits on the specific-speed scale.

3. Specific speed is usually above 200 (10,000 U.S. customary units).
4. These pumps are customarily used for large flows (over 600 L/s) and low head (below 10 m) installations. Best applications include irrigation work and stormwater pumping. They should not be used for pumping raw wastewater, sludge, or unscreened stormwater.

9.5.2.4 Comparison of Specificity of Pump Class

Pump comparison based on the impeller configuration specific speed, head, capacity, and speed is best illustrated in Figure 9.8 and by use of Examples 9.9–9.11.[1–3]

Example 9.9: Specific Speed and Pump Selection

A designer needs to select a pump type for the pumping of raw wastewater for a total head of 30 m. The maximum anticipated capacity is 0.25 m³/s at an operating drive of 1,200 rpm. Calculate the specific speed and select the pump and anticipated efficiency.

Solution

1. Calculate the specific speed from Eq. (9.7)

$$N_s = \frac{1{,}200 \times (0.25)^{1/2}}{(30)^{3/4}} = 46.81(2{,}415 \text{ in U.S. customary units})$$

2. Select the pump
   The pump has low specific speed, medium to high capacity, and high to medium head (Table 9.4). This application falls within the range of a radial-flow centrifugal pump.
3. Determine anticipated efficiency
   Referring to Figure 9.8, the application falls within the radial-flow, Francis–Vane area, with an anticipated pump efficiency of 87%.

Example 9.10: Specific Speed of a Pump for Different Operating Conditions

A pumping station has a wet well to pump the screened wastewater to the treatment plant. The pumping starts when high levels in the wet well are reached. At a high wet well level, the pumping rate and TDH are 0.063 m³/s and 28.0 m. The pump speed is 1,400 rpm. Calculate the specific speed under operating conditions. Also, determine the pump efficiency and type of pump impeller.

Solution

1. Calculate the specific speed at high wet well level

$$N_s = \frac{1{,}400 \times (0.063)^{1/2}}{(28)^{3/4}} = 28.9(1{,}500 \text{ in U.S. customary units})$$

2. Determine the pump efficiency
   The pump efficiency from Figure 9.8a = 81%
3. Comment on pump impeller design (Figure 9.8b)
   The pump impeller design falls in the Francis–Vane area.

Example 9.11: Identification of Pump Type

The head, capacity, and specific speed of different pumps are listed below. Identify the values that match axial-flow, mixed-flow, and radial-flow pumps.

1. Specific speed: (a) 20 (SI units), (b) 12,000 (U.S. customary units), and (c) 115 (SI units)
2. Pump head: (d) 9 m and (e) 14 m
3. Pump capacity (f) 800 gpm and (g) 6,000 gpm
4. Speed, head, and capacity: (h) 1,500 rpm, 8 m, and 0.5 m³/s, and (i) 1,100 rpm, 80 ft, and 1,200 gpm

Solution

Identify the pump types (radial, mixed, and axial flow) based on specific speed, head, and capacity. For speed, head, and capacity data in (4), calculate the specific speed from Eq. (9.7). These pumps are identified below for different operating characteristics.

Pump Types	Radial Flow	Mixed Flow	Axial Flow
Operating characteristics	(a), (f), (i)[a]	(c), (e)	(b), (d), (g), (h)[b]

[a] $N_s = \left(1,100 \times \sqrt{1,200}/80^{3/4}\right) = 1,427$ (U.S. customary units).

[b] Based on specific speed $N_s = \dfrac{1,500 \times (0.5)^{1/2}}{(8)^{3/4}} = 223$ (SI units).

9.5.3 Relationships for Geometrically Similar Pumps

In a series of geometrically similar pumps, the flow patterns are similar. By applying the principles of dimensional analysis, several relationships can be developed. These relationships are useful in predicting the performance of a geometrically similar pump from pumping tests on a similar model pump. These relationships are expressed by Eq. (9.8)[3,4]

$$\frac{Q_2}{Q_1} = \frac{N_2}{N_1}\left(\frac{D_2}{D_1}\right)^3 \tag{9.8a}$$

$$\frac{H_2}{H_1} = \left(\frac{N_2}{N_1}\right)^2\left(\frac{D_2}{D_1}\right)^2 \tag{9.8b}$$

$$\frac{P_2}{P_1} = \left(\frac{N_2}{N_1}\right)^3\left(\frac{D_2}{D_1}\right)^5 \tag{9.8c}$$

where

Q = capacity, m³/s (cfs)
H = head, m (ft)
P = power, hp
N = speed, rpm
D = impeller diameter, m (ft)

The subscript 1 refers to the parameters for which the values are known, and the subscript 2 refers to the parameters of the pump for which the values are to be predicted.

Example 9.12: Relationship for Geometrically Similar Pumps

A model pump is designed and built. The model pump has an impeller diameter of 12 cm. When tested at a speed of 800 rpm at a total head of 76 cm, it developed water power of 0.187 kW. A geometrically similar pump of 38 cm diameter is to operate at the same efficiency at a head of 15 m. Determine the expected speed and power of this pump. Also, compare the specific speeds of both pumps.

Solution

1. Determine the expected speed of the geometrically similar pump (Eq. 9.8b)

$$\frac{H_2}{H_1}=\left(\frac{N_2}{N_1}\right)^2\left(\frac{D_2}{D_1}\right)^2$$

$$\frac{15\ \text{m}}{0.76\ \text{m}}=\left(\frac{N_2}{800}\right)^2\left(\frac{38\ \text{cm}}{12\ \text{cm}}\right)^2$$

$$N_2=1{,}122\ \text{rpm}$$

2. Determine the expected power of the geometrically similar pump (Eq. 9.8c)

$$\frac{P_2}{P_1}=\left(\frac{N_2}{N_1}\right)^3\left(\frac{D_2}{D_1}\right)^5$$

$$\frac{P_2}{0.187\ \text{kW}}=\left(\frac{1122}{800}\right)^3\left(\frac{38\ \text{cm}}{12\ \text{cm}}\right)^5$$

$$P_2=164\ \text{kW}$$

3. Determine the capacity of both pumps
   The capacity of model pump is calculated from Eq. (9.4)

$$P_1=K'Q_1\times H_1\times\gamma$$

$$0.187\ \text{kW}=1\times Q_1\times 0.76\ \text{m}\times 9.81\ \text{kN/m}^3$$

$$Q_1=0.025\ \text{m}^3/\text{s}$$

   The capacity of geometrically similar pump is calculated from Eq. (9.8a)

$$\frac{Q_2}{Q_1}=\frac{N_2}{N_1}\left(\frac{D_2}{D_1}\right)^3$$

$$\frac{Q_2}{0.025\ \text{m}^3/\text{s}}=\frac{1{,}122}{800}\left(\frac{38\ \text{cm}}{12\ \text{cm}}\right)^3$$

$$Q_2=1.113\ \text{m}^3/\text{s}$$

4. Determine the specific speed of both pumps
   The specific speed of the model pump (Eq. 9.7)

$$=\frac{800\times(0.025\ \text{m}^3/\text{s})^{1/2}}{(0.76\ \text{m})^{3/4}}=155\,(\text{SI units})$$

   The specific speed of geometrically similar pump (Eq. 9.7)

$$=\frac{1{,}122\times(1.113\ \text{m}^3/\text{s})^{1/2}}{(15\ \text{m})^{3/4}}=155\,(\text{SI units})$$

5. Compare the specific speed of both pumps
   The geometrically similar pumps have same specific speed.

9.5.4 Rotational Speed, Capacity, Head, and Power Relationships for Same Pump (Affinity Laws)

Effect of speed: Relationship exists between discharge, head, and power for the same pump operating at different rotational speeds. These relationships are called affinity laws and are expressed by Eq. (9.9)[3,10]

$$\frac{Q_2}{Q_1} = \frac{N_2}{N_1} \tag{9.9a}$$

$$\frac{H_2}{H_1} = \left(\frac{N_2}{N_1}\right)^2 \tag{9.9b}$$

$$\frac{P_2}{P_1} = \left(\frac{N_2}{N_1}\right)^3 \tag{9.9c}$$

Effect of impeller geometry: The shape and size of the impeller has an effect upon pump performance. These relationships for head, capacity, and power are expressed by Eq. (9.10)[3,10]

$$\frac{Q_2}{Q_1} = \frac{D_2}{D_1} \tag{9.10a}$$

$$\frac{H_2}{H_1} = \left(\frac{D_2}{D_1}\right)^2 \tag{9.10b}$$

$$\frac{P_2}{P_1} = \left(\frac{D_2}{D_1}\right)^3 \tag{9.10c}$$

The terms Q, H, D, N, and P, and subscripts 1 and 2 are defined in Eq. (9.8).

Example 9.13: Effect of Change in Impeller Speed and Diameter

A centrifugal pump was tested at an impeller speed of 1,600 rpm. The pump gave the following results: $Q_1 = 4{,}000$ gpm, $H_1 = 157$ ft, $D_1 = 15$ in., $P_1 = 190$ hp. Calculate capacity, head, and power if: (1) impeller speed is changed from 1,600 to 1,200 rpm and (2) both impeller speed and diameter are changed from 1,600 to 1,200 rpm, and 15 to 13.5 in., respectively.

Solution

1. Determine the pump characteristics due to a change in impeller speed from 1,600 to 1,200 rpm, Pump $D_1 = 15$ in.

From Eq. (9.9a) $\frac{N_2}{N_1} = \frac{1{,}200 \text{ rpm}}{1{,}600 \text{ rpm}} = 0.75$

$$\frac{Q_2}{4{,}000 \text{ gpm}} = 0.75, \quad Q_2 = 3{,}000 \text{ gpm}$$

From Eq. (9.9b) $\frac{H_2}{157 \text{ ft}} = (0.75)^2, H_2 = 88 \text{ ft}$

From Eq. (9.9c) $\frac{P_2}{190\text{ hp}} = (0.75)^3, P_2 = 80\text{ hp}$

2. Determine the characteristics of the pump due to a change in impeller speed from 1,600 to 1,200 rpm, and diameter, from 15 to 13.5 in.

From Eq. (9.10a) $\frac{D_2}{D_1} = \frac{13.5\text{ in.}}{15\text{ in.}} = 0.90$

$$\frac{Q_2}{4{,}000\text{ gpm} \times 0.75} = 0.90$$

$$Q_2 = 4{,}000 \times 0.90 \times 0.75 = 2{,}700\text{ gpm}$$

From Eq. (9.10b) $\frac{H_2}{157\text{ ft} \times (0.75)^2} = (0.90)^2$

$$H_2 = 157 \times (0.90)^2 \times (0.75)^2 = 72\text{ ft}$$

From Eq. (9.10c) $\frac{P_2}{190\text{ hp} \times (0.75)^3} = (0.90)^3$

$$P_2 = 191 \times (0.90)^3 \times (0.75)^3 = 58\text{ hp}$$

3. Compare the characteristics

Condition	Q, gpm	H, ft	Speed, rpm	Power, hp	D, in.
Test condition	4,000	157	1,600	190	15
RPM changed from 1,600 to 1,200, and D=15 in.	3,000	88	1,200	80	15
RPM changed from 1,600 to 1,200, and diameter changed from 15 to 13.5 in.	2,700	72	1,200	58	13.5

9.5.5 Pump Characteristic Curves

The pump manufacturers conduct tests on their pumps and provide information in the form of characteristic curves (also called pump curves). These curves are prepared at a constant speed, by plotting on Y-axis: (1) TDH (meters or feet), (2) the efficiencies (percent), and (3) the power input (kW or hp), against the pump capacity (m^3/s, gpm, or mgd) on X axis. These curves and the best efficiency point (bep) are shown in Figure 9.9a. Since a pump casing can accommodate impellers of several sizes, and impellers can operate at different speeds, the manufacturers also present pump data in the form shown in Figure 9.9b and c. This feature is helpful because the capacity of an existing pump can be expanded by replacing the impeller and motor.[3,9] The pump characteristic curves are discussed below.

Head-capacity curves: The pump tests are conducted by measuring the energy head at the discharge and suction nozzles. The capacity of the pump is varied by throttling a valve at the discharge pipe, and the corresponding head is measured. The TDH is calculated from Eq. (9.1b), and calculations are shown in Examples 9.1 and 9.7. The data points are plotted to give head-capacity

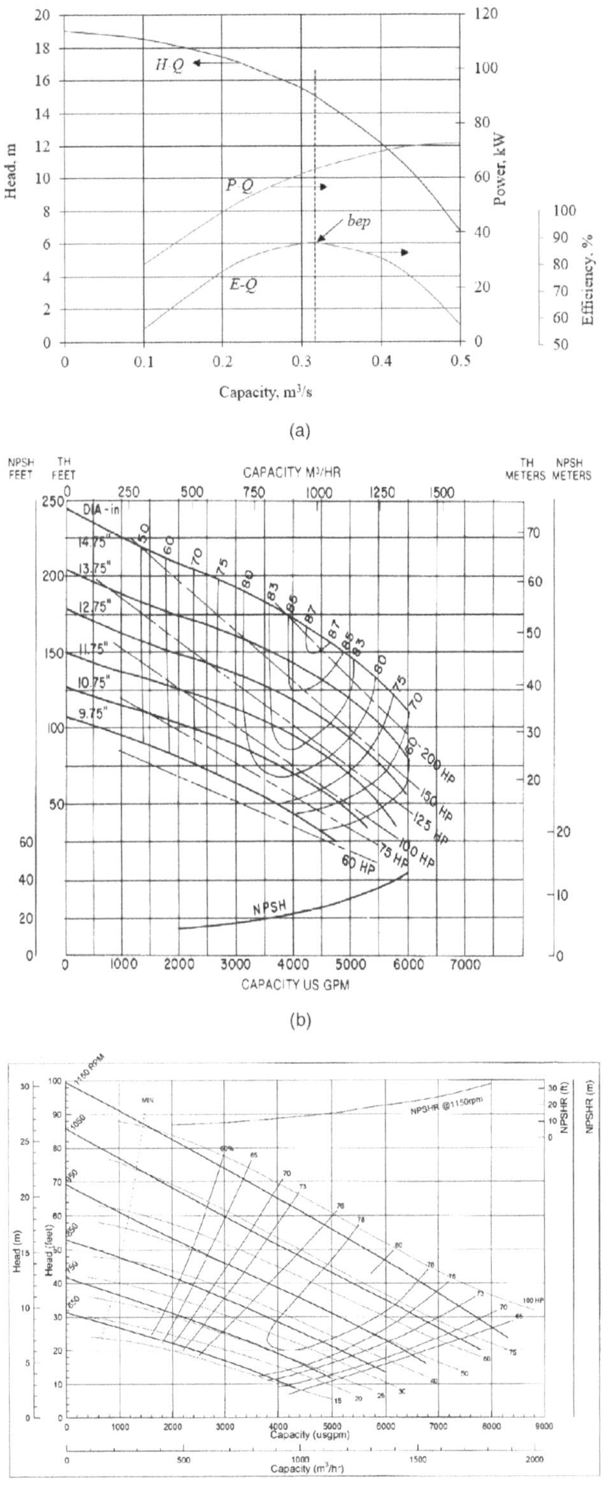

FIGURE 9.9 Manufacturer's characteristic curves of centrifugal pumps: (a) Head-capacity, power, and efficiency curves; (b) variable impeller size 25.4–38.1 cm (10–15 in.) and 1,175 rpm speed; and (c) variable speed 650–1,150 rpm and impeller diameter 30.5 cm (12 in.).

TABLE 9.5
Operating Characteristics of Radial-Flow, Mixed-Flow, and Axial-Flow Pumps

	Centrifugal Pumps		
Characteristics	Radial Flow	Mixed Flow	Axial Flow
1. Flow	Uniform[a]	Uniform	Uniform
2. Effect of increasing head on:			
(a) Capacity	Decrease	Decrease	Decrease
(b) Power required	Decrease	Small decrease to large increase	Large increase
3. Effect of decreasing head on:			
(a) Capacity	Increase	Increase	Increase
(b) Power required	Increase	Slight increase to decrease	Decrease
4. Effect of closing discharge valve on:			
(a) Pressure	Up to 30% increase	Considerable increase	Large increase
(b) Power demand	Decrease 50%–60%	10% decrease to 80% increase	Increase 80%–150%

[a] No pulsation (even discharge)

curve of the pump at the given speed (Figure 9.9a). The shape of the head-capacity curve is important in selecting a pump for a specific application. The shape of head-capacity curve may be rising head, flat, steep, or drooping head. Such curves are a function of specific speed and pump design.

Power-capacity curves: During the pump test, the power input is measured at different discharge values. These values are plotted with respect to capacity on the same diagram (Figure 9.9a). The power curve demonstrates the power required to operate the pump at a given discharge. Typically, radial-flow pumps have a power curve that climbs from left to right. Therefore, these pumps are often started against a closed valve to reduce the starting torque. The valve is slowly opened after the pump starts. This reduces the effects of a surge or water hammer phenomenon. Mixed-flow and axial-flow pumps have power curves that drop from left to right, with a sharp increase near the Y-axis. These pumps should not be started against a closed valve. Additional coverage on this topic may be found in Section 9.5.10. The characteristic curves of radial, mixed, and axial-flow pumps are also covered in this section. The operating characteristics of radial-flow, mixed-flow, and axial-flow pumps are summarized in Table 9.5.

Efficiency curves: The power output of the pump and pump efficiency are calculated from Eqs. (9.11) and (9.12). These values are plotted with respect to the capacity in the same diagram as head-capacity and power-capacity curves. The efficiency curve demonstrates the performance of a pump at various discharge rates. Typically, this curve has a hump or peak (Figure 9.9a). Ideally, the pump should be selected to operate near its peak or *bep*.

Example 9.14: Pump Efficiency

In a manufacturer's pumping test, the pump is delivering a flow of 0.3 m^3/s at a TDH of 15 m. The power input to the pump is 54 kW. Calculate the pump efficiency.

Solution

1. Determine water power delivered by the pump

$$Q = 0.3\ \text{m}^3/\text{s} \quad \text{and} \quad \text{TDH} = 15\ \text{m}$$

Use Eq. (9.4)

$$P_w = K'Q \times (\text{TDH}) \times \gamma$$

$$= 1 \times 0.3\ \text{m}^3/\text{s} \times 15\ \text{m} \times 9.81\ \text{kN/m}^3 = 44.2\ \text{kN.m/s}$$

$$= 44.2\ \text{kW}$$

$$\text{Pump efficiency (Eq. 9.5)} = \frac{44.2\ \text{kW}}{54\ \text{kW}} \times 100 = 82\%$$

Example 9.15: Manufacturers' Characteristic Curves of Multiple Diameter Impeller to a Single-Impeller Characteristic Curve

The pump manufacturers provide the pump characteristic curves in different forms depending upon the variability of the impeller diameter, and impeller speed. The manufacturers' curves for variable impeller diameter at 1,175 rpm speed are given in Figure 9.9b. Draw the characteristic curves of a 13.75-in. impeller diameter pump operating at the manufacturer's speed of 1,175 rpm.

Solution

The head-capacity, power, and efficiency curves of a pump 13.75-in. diameter at an impeller speed of 1,175 rpm are given in Figure 9.9b. These curves are directly convertible to the form shown in Figure 9.9a.

1. Tabulate the head, efficiency, power, and net-positive suction head (NPSH) required values for a 13.75-in. impeller pump with respect to pump capacity and the impeller speed of 1,175 rpm.

   The head-capacity, efficiency, and power curves data are derived from Figure 9.9b for 13.75-in. impeller pump. These values are summarized in Table 9.6.

2. Plot the characteristic curves for the 13.75-in. impeller pump, and 1,175 rpm speed. The capacity, head, efficiency, power, and NPSH required are plotted in Figure 9.10.

TABLE 9.6
Capacity, Head, Efficiency, Power, and NPSH Required Values Obtained from the Manufacturer's Characteristics Curves for Pump 13.75-in. Impeller Diameter and 1,175 rpm Speed (Example 9.15)

Capacity, gpm	Head, ft	Efficiency, %	Power, bhp	NPSH Required, ft
0	202	—	—	—
1,000	188	—	125	—
2,000	175	64	137	14
3,000	160	77	147	18
4,000	142	85.3	160	23
5,000	130	83.4	172	29
6,000	119	80	188	34
7,000	107	77.8	195	38

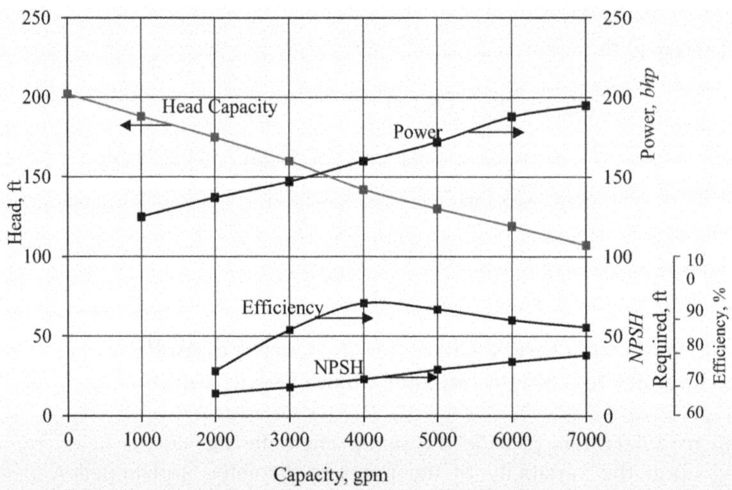

FIGURE 9.10 Pump characteristic curves for a pump 13.75-in. diameter and 1,175 rpm derived from the manufacturer's curve of multi-diameter impeller given in Figure 9.9b (Example 9.15).

Example 9.16: Head-Capacity Curves at Different Speeds

A centrifugal pump has the following head-capacity test data at an impeller speed of 1,200 rpm. Develop the head-capacity curves at operating speeds of 850 and 700 rpm.

Capacity, m³/s	0	0.1	0.2	0.3	0.4	0.5
Head, m	40.0	39.0	36.6	34.4	30.5	23.0

Solution

1. Develop the head-capacity relations at operating speeds of 850 and 700 rpm.
   Use affinity laws Eqs. (9.9a) and (9.9b)
   Head-capacity relations at 850 rpm

$$Q_2 = Q_1 \times \frac{850 \text{ rpm}}{1,200 \text{ rmp}} = 0.708Q_1$$

$$H_2 = H_1 \times (0.708)^2 = 0.501H_1$$

   Head-capacity relations at 700 rpm

$$Q_2' = Q_1 \times \frac{700 \text{ rpm}}{1,200 \text{ rmp}} = 0.583Q_1$$

$$H_2' = H_1 \times (0.583)^2 = 0.340H_1$$

2. Develop the head-capacity data at operating speeds of 850 and 700 rpm.

Substitute different values of Q_1 and H_1 from the test data at 1,200 rpm. These values for speeds 1,200, 850, and 700 rpm are tabulated below.

Test data $N_1 = 1{,}200$ rpm		Projected data $N_2 = 850$ rpm		Projected data $N'_2 = 850$ rpm	
Capacity Q_1, m³/s	**Head H_1, m**	**Capacity Q_2, m³/s**	**Head H_2, m**	**Capacity, m³/s**	**Head H'_2, m**
0.0	40.0	0.0	20.0	0.0	13.6
0.1	39.0	0.071	19.5	0.058	13.3
0.2	36.6	0.142	18.3	0.117	12.7
0.3	34.4	0.212	17.2	0.175	11.7
0.4	30.5	0.283	15.3	0.233	10.1
0.5	23.0	0.354	11.5	0.292	7.8

3. Plot the head-capacity curve at 1,200, 850, and 700 rpm (Figure 9.11).

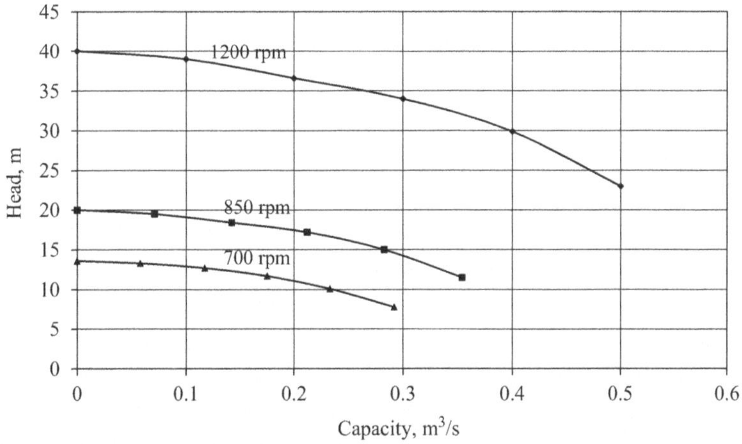

FIGURE 9.11 Head-capacity curves of a centrifugal pump at impeller speeds of 1,200, 850, and 700 rpm (Example 9.16).

9.5.6 Multiple Pump Operation

Wastewater pumping stations utilize more than one pump. These pumps may be operating in parallel or series. Both parallel and series operations are discussed below.[10]

Parallel operation: Parallel operation is used to increase the flow. When two or more pumps operate in parallel, the combined head-capacity curve is found by adding the discharge of all pumps at the same head. The relations of parallel pump combination are expressed by Eq. (9.11).

$$H = H_A = H_B = \cdots \tag{9.11a}$$

$$Q = Q_A + Q_B + \cdots \tag{9.11b}$$

$$E = \frac{Q_A + Q_B + \cdots}{\left(\dfrac{Q_A}{E_A} + \dfrac{Q_B}{E_B} + \cdots\right)} \tag{9.11c}$$

$$P = \gamma \times H\left(\frac{Q_A + Q_B + \cdots}{E}\right) \quad \text{(SI unit)} \tag{9.11d}$$

$$\text{or} \quad P = \gamma \times H\left(\frac{Q_A + Q_B + \cdots}{550E}\right) \quad \text{(U.S. customary unit)} \tag{9.11e}$$

where

H_A, H_B, Q_A, Q_B, E_A, and E_B = head, capacity, and efficiency of pumps A and B
H, Q, E, and P = head, capacity, efficiency, and power output of two pumps in parallel combination.

Series operation: Series operation is used to increase the station head. The booster pumps may be installed in the suction line or in force main. The total heads for all pumps are added up for a given discharge.

The relations of series pump combinations are expressed by Eq. (9.12)

$$H = H_A + H_B + \cdots \tag{9.12a}$$

$$Q = Q_A = Q_B = \cdots \tag{9.12b}$$

$$E = \frac{H_A + H_B + \cdots}{\left(\frac{H_A}{E_A} + \frac{H_B}{E_B} + \cdots\right)} \tag{9.12c}$$

$$P = \gamma \times Q\left(\frac{H_A + H_B + \cdots}{E}\right) \quad \text{(SI unit)} \tag{9.12d}$$

$$\text{or} \quad P = \gamma \times Q\left(\frac{H_A + H_B + \cdots}{550E}\right) \quad \text{(U.S. customary unit)} \tag{9.12e}$$

Example 9.17: Parallel Combination of Two Pumps of Different Characteristics

The head-capacity and efficiency data of two different pumps are given below. Develop the combined head-capacity and efficiency curves of these two pumps arranged in parallel. Calculate the break power of two pumps at bep.

	Pump *A*		Pump *B*	
Head, m	**Capacity, m³/s**	**Efficiency, %**	**Capacity, m³/s**	**Efficiency, %**
35	0.025	50	0.056	60
30	0.043	65	0.071	70
25	0.057	85	0.082	80
20	0.067	65	0.092	65
15	0.075	50	0.100	50
10	0.082	40	0.110	30

Solution

1. Add the discharge of two pump at each head to obtain the combined discharge.

Tabulate the results.

Head, m	Combined Flow, m³/s	Combined Efficiency, %
35	0.081	56.51
30	0.114	68.03
25	0.139	81.98
20	0.159	65.00
15	0.175	50.00
10	0.192	33.59

2. Calculate the combined efficiency of two pumps from Eq. (9.11c)

$$E = \frac{0.081\ \text{m}^3/\text{s}}{\left(\dfrac{0.025\ \text{m}^3/\text{s}}{0.50}\right) + \left(\dfrac{0.056\ \text{m}^3/\text{s}}{0.60}\right)} = 0.5651 \quad \text{or} \quad 56.51\%$$

Similarly, calculate the combined efficiency for all points and tabulate the result in the above table.

3. Plot the head-capacity curves of pumps *A* and *B*, and the combined head-capacity and efficiency curves of two pumps operating in parallel (Figure 9.12)
4. Obtain the operating head and capacity at bep.

   At bep, the operating head and capacity of two pumps in parallel are

   Head = 25.2 m

   Capacity = 0.135 m³/s
5. Calculate the power input at bep from (Eq. 9.11d)

$$P = \frac{25.2\ \text{m} \times 9.81\ \text{kN/m}^3 \times 0.135\ \text{m}^3/\text{s}}{0.81} = 41.2\ \text{kW}$$

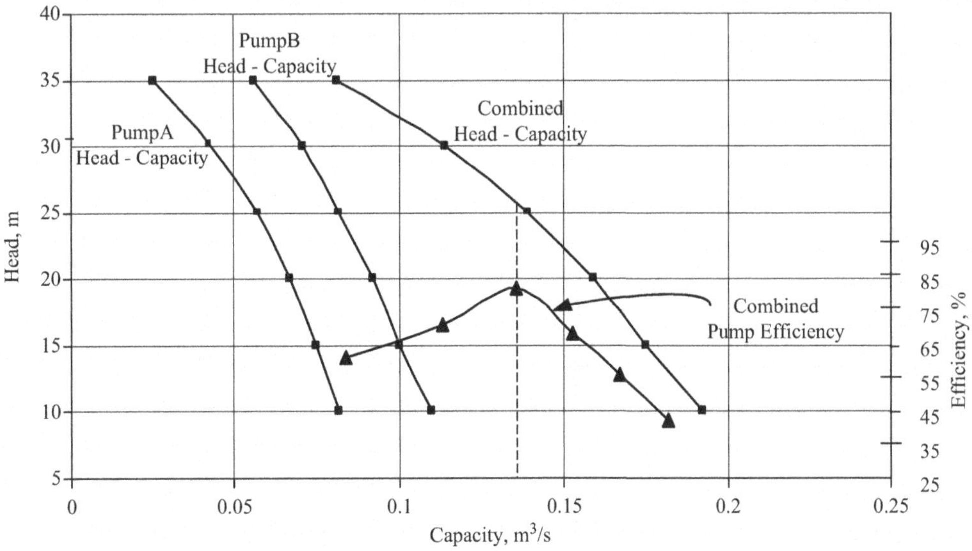

FIGURE 9.12 Head-capacity curves of pumps *A* and *B*, and combined head-capacity and efficiency curves of two centrifugal pumps arranged in parallel (Example 9.17).

Note: The bep is the peak efficiency point of the combined pump efficiency curve. The peak efficiency is 81% of the pump combination.

Example 9.18: Parallel Combinations of Two, Three, and Four Identical Pumps

The manufacturer's head-capacity data of a centrifugal pump are given below. Develop the head-capacity curves of one pump; and the combined head-capacity curves of two, three, and four pumps in parallel.

Capacity, gpm	0	750	1,750	2,500	3,000	3,500
Head, ft	75	70	60	50	40	30

Solution

1. Arrange the parallel pump combination data.

   Tabulate pump-operating heads and capacity of one, two, three, and four pumps with respect to the head.

	Capacity, gpm			
Head, ft	**One Pump**	**Two Pumps**	**Three Pumps**	**Four Pumps**
75	0	0	0	0
70	750	1,500	2,250	3,000
60	1,750	3,500	5,250	7,000
50	2,500	5,000	7,500	10,000
40	3,000	6,000	9,000	12,000
30	3,400	6,800	10,200	13,600

2. Plot the curves

   The *H-Q* curve of one pump developed from manufacturer's data, and parallel combinations of two, three, and four pumps are shown in Figure 9.13.

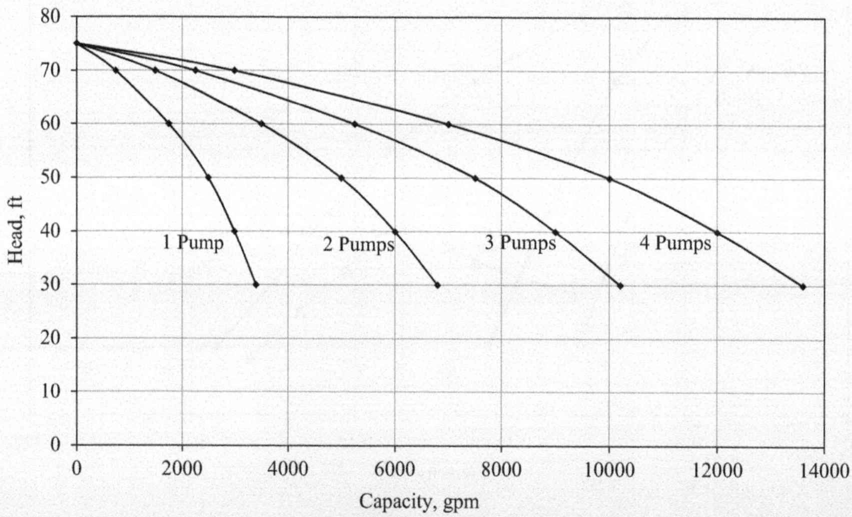

FIGURE 9.13 Head-capacity curves of one pump, and combined head-capacity curves of two, three, and four identical pumps in parallel combination (Example 9.18).

Example 9.19: Series Combination of Two Pumps of Different Characteristics

The manufacturer's head-capacity and efficiency data of two different pumps are given below. Develop the head-capacity and efficiency curves of these two pumps arranged in series combination. Also, calculate the break power of two pumps at bep.

	Pump A		Pump B	
Capacity, m³/s	Head, m	Efficiency, %	Head, m	Efficiency, %
0	40	0	45	0
0.02	36	60	42	32
0.04	31	84	39	63
0.06	23	63	34	86
0.08	12	32	27	64

Solution

1. Determine the combined head of two pumps at different flows

   Add the head of two pumps at different capacities to obtain the combined head. Tabulate the data below.

Capacity, m³/s	Combined Head, m	Combined Efficiency, %
0	85	0
0.02	78	41
0.04	70	71
0.06	57	75
0.08	39	49

2. Calculate the combined efficiency of two pumps from Eq. (9.12c)

$$E = \frac{78 \text{ m}}{(36 \text{ m}/0.6) + (42 \text{ m}/0.32)} = 0.41 \quad \text{or} \quad 41\%$$

   Similarly, calculate the combined efficiency for all points, and tabulate the results in the above table.
3. Plot the head-capacity curves of pumps *A* and *B*, and the combined head-capacity and efficiency curves of two pumps operating in series (Figure 9.14).
4. Obtain the operating head and capacity at bep.

   Best efficiency from Figure 9.14 = 79.5%

   At the bep, the operating head and capacity of the two pumps in series are

   Head = 62 m

   Capacity = 0.0525 m³/s
5. Calculate the combined power input from Eq. (9.12d)

$$P = \frac{62 \text{ m} \times 9.81 \text{ kN/m}^3 \times 0.0525 \text{ m}^3/\text{s}}{0.795} = 40.2 \text{ kW}$$

Example 9.20: Series Combinations of Two, Three, and Four Identical Pumps

The manufacturer's head-capacity data of a centrifugal pump are given below. Develop the head-capacity curves of one pump, and two, three, and four pumps in series combination

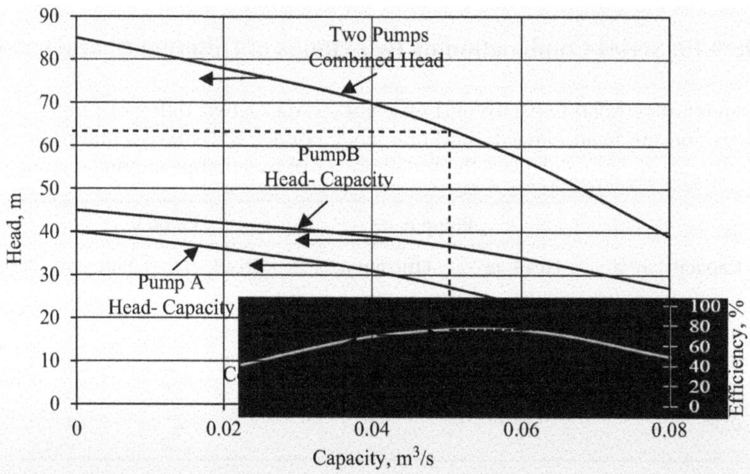

FIGURE 9.14 Head-capacity curves of pumps *A* and *B*, and combined head-capacity and efficiency curves of two pumps arranged in series (Example 9.19).

Capacity, gpm	0	1,000	2,000	3,000	4,000	5,000	6,000
Head, ft	42	38	36	31	27	20	10

Solution

1. Develop the combined head-capacity curves.
   Add the heads of two, three, and four pumps at the same capacity. The series combination data are summarized below.

	Head, ft			
Capacity, gpm	**One Pump**	**Two Pumps**	**Three Pumps**	**Four Pumps**
0	42	84	126	168
1,000	38	76	114	152
2,000	36	72	108	144
3,000	31	62	93	124
4,000	27	52	81	108
5,000	20	40	60	80
6,000	10	20	30	40

2. Plot the head-capacity curves of one, two, three, and four pumps in series (Figure 9.15).

9.5.7 System Head-Capacity Curve

The system head-capacity curve is prepared by the design engineer for the actual suction and discharge piping arrangements. This curve is prepared by plotting TDH over a range of flows from zero to the maximum expected flow (Figure 9.16a). The system head-capacity curve is superimposed over the head-capacity curve of the pump system (Figure 9.16b). The intersection point of the pump head-capacity and system-head capacity (*P*) gives the operating point, that is, the operating head and capacity of the pump for the selected piping system. For good selection, this point should be as close as possible to the peak efficiency. If the efficiency is too low, another pump should be selected. Several important considerations regarding the system head-capacity curve warrant discussion.

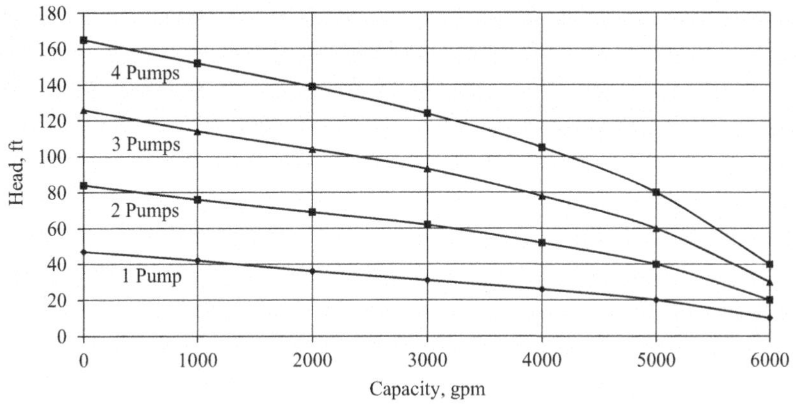

FIGURE 9.15 Combined head-capacity curves of one pump, and two, three, and four pumps arranged in series (Example 9.20).

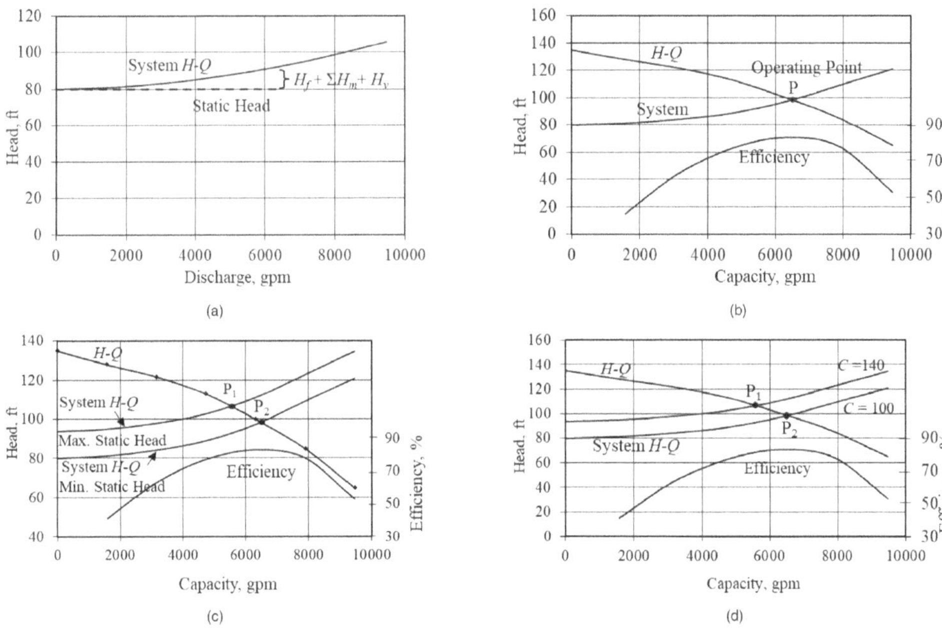

FIGURE 9.16 The system head-capacity curves for one pump operating under different conditions: (a) constant static head, (b) operating-point at a constant static head, (c) operating-points at two extreme static heads, and (d) operating-points of new and old pumping systems.

1. Some pumping stations may have variable static head conditions. Maximum- and minimum-level wet well and discharge reservoirs are examples of such condition. In this situation, the system head curve is prepared for both extreme head conditions that yield a band of possible operating range between two extreme static heads. For the best pump selection, the peak efficiency should lie between two operating points P_1 and P_2 (Figure 9.16c).
2. If two or more pumps are arranged in parallel, only losses common to all pumps from the start of the common discharge header to the delivery point are included in the system head curve. Detailed discussion with an example is covered in Section 9.5.8.

3. As the discharge pipe *ages* over time, the friction factor also changes. Therefore, the system head-capacity curve should be developed for minimum and maximum values of friction factors to obtain the pump operating-point. The Hazen–Williams friction coefficient C for new and old cast iron pipes are, respectively, 140 and 100 (Table 5.2). The pump operating-points of new and old pumping systems are P_1 and P_2, respectively, are shown in Figure 9.16d.

Example 9.21: Pump Operating-Point

The manufacturer's characteristic curve data of a pump are given below. This pump is used to transfer water from reservoir *A* to reservoir *B*. The piping system is shown in Figure 9.17. Develop the system head-capacity curve and determine the pump-operating head, capacity, efficiency, power input, and annual power cost. The valves, fittings, and specials and their *K* values are listed below. The diameter and length of piping are 24 in. and 2,000 ft. C=100. Unit power cost is 5 cents/kWh. Assume continuous pump operation.

Q, gpm	0	1,580	3,160	4,740	6,320	7,900	9,480
Head, ft	130	128	121	113	100	76	50
Efficiency, %		41	65	77	84	80	65

One entrance EN (24 in.) K=0.4
Four 90° elbow (24 in.) K=0.3
One reducer R and one enlarger E (14 in.×24 in.) K=0.3
One gate valve G (24 in.) K=0.2
One check valve CK (24 in.) K=2.5
One exit EX (24 in.) K=1.0

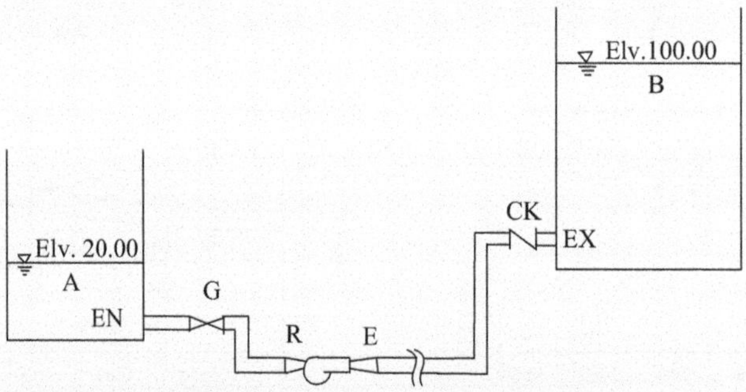

FIGURE 9.17 Piping and water elevations in reservoir (Example 9.21).

Solution

1. Compute system losses for valves, fittings, and specials.

The minor head losses are calculated for six assumed flow conditions. The calculation steps are summarized below.

Item No.	Valves, Fittings, and Specials	No.	K	1,580 gpm (3.52 cfs) Vel., ft/s	H_m, ft	3,160 gpm (7.04 cfs) Vel., ft/s	H_m, ft	4,740 gpm (10.56 cfs) Vel., ft/s	H_m, ft
(1)	(2)	(3)	(4)	(5)	(6)	(7)	(8)	(9)	(10)
1	Entrance	1	0.4	1.12[a]	0.007	2.24	0.031	3.38	0.071
2	90° elbow	4	0.3	1.12	0.023	2.24	0.094	3.38	0.213
3	Reducer	2[b]	0.3	3.29[c]	0.101	6.59	0.405	9.88	0.909
4	Gate valve	1	0.2	1.12	0.004	2.24	0.016	3.38	0.036
5	Check valve	1	2.5	1.12	0.049	2.24	0.195	3.38	0.444
6	Exit	1	1.0	1.12	0.020	2.24	0.078	3.38	0.177
	ΣH_m				0.20		0.82		1.85

Item No.	Valves, Fittings, and Specials	No.	K	6,320 gpm (14.08 cfs) Vel., ft/s	H_m, ft	7,900 gpm (17.60 cfs) Vel., ft/s	H_m, ft	9,480 gpm (21.12 cfs) Vel., ft/s	H_m, ft
(1)	(2)	(3)	(4)	(11)	(12)	(13)	(14)	(15)	(16)
1	Entrance	1	0.4	4.48	0.125	5.60	0.195	6.72	0.280
2	90° elbow	4	0.3	4.48	0.374	5.60	0.584	6.72	0.842
3	Reducer	2[b]	0.3	13.18	1.618	16.47	2.527	19.77	3.641
4	Gate valve	1	0.2	4.48	0.062	5.60	0.097	6.72	0.140
5	Check valve	1	2.5	4.48	0.779	5.60	1.218	6.72	1.753
6	Exit	1	1.0	4.48	0.312	5.60	0.487	6.72	0.701
	ΣH_m				3.27		5.11		7.36

[a] $Q = \dfrac{1{,}540\text{ gal}}{\text{min}} \times \dfrac{1}{(60\text{ s/min})} \times \dfrac{1}{7.48\text{ gal/ft}^3} = \dfrac{3.52\text{ ft}^3}{\text{s}}$; vel. $= \dfrac{3.52\text{ ft}^3/\text{s}}{(\pi/4)\times(2.0\text{ ft})^2} = 1.12$ ft/s.

[b] 14 in.×24 in. reducer and enlarger have similar head loss characteristics, therefore total number is 2.

[c] vel. $= (3.52\text{ ft}^3/\text{s})/\pi/4\times(14\text{ in.}/12\text{ in./ft})^2 = 3.29$ ft/s

2. Calculate H_f, $(H_f+\Sigma H_m)$ and TDH at different flows.

Use Eq. (5.19) to calculate frictional head loss. The static head is (100.00−20.00)=80 ft. The TDH values for different pump discharges are calculated below.

Q, gpm	Velocity, ft/s	H_f, ft	$(H_f+\Sigma H_m)$, ft	Static Head, ft	TDH, ft
0	0	0	0	80	80.0
1,580	1.12	0.67[a]	0.87	80	80.9
3,160	2.24	2.39	3.21	80	83.2
4,740	3.38	5.11	6.96	80	87.0
6,320	4.48	8.61	11.88	80	91.9
7,900	5.60	13.02	18.13	80	98.1
9,480	6.72	18.24	25.60	80	105.6

[a] $1.12=0.550\times100\times(2.0)^{0.63}(H_f/2{,}000)^{0.54}$; $H_f=0.67$ ft.

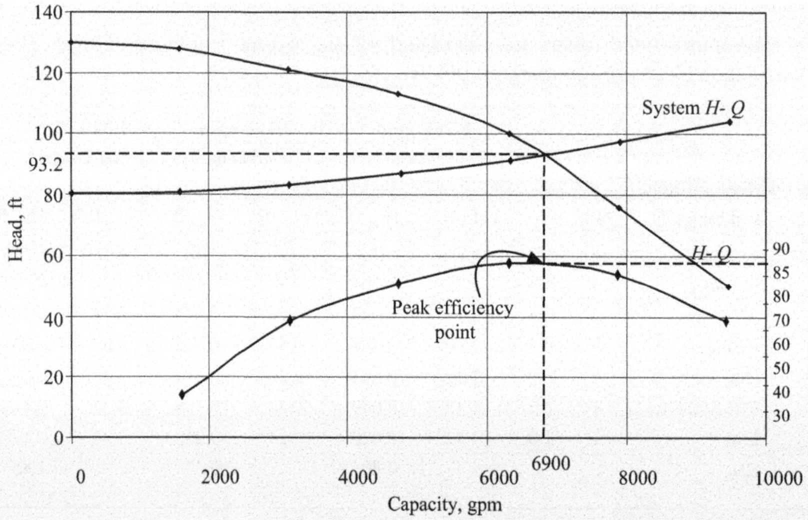

FIGURE 9.18 Pump characteristics curves, system curve, and pump operating-points (Example 9.21).

3. Draw the *H-Q*, Eff-*Q*, and system *H-Q* curves (Figure 9.18)
4. Determine the operating head and discharge of the pump, pump efficiency, and power input.

   The pump-operating head and capacity point are at the intersection point of *H-Q* and system head-capacity curve.

$$Q = 6{,}900 \text{ gpm}, \quad H = 93.2 \text{ ft}$$

$$\text{Pump efficiency} = 85\%, \text{ and given motor efficiency} = 92\% \text{ (assumed)}$$

$$\text{Power input to pump} = \frac{\gamma \times Q \times H}{550 \times \text{Efficiency of pump motor}}$$

$$= \frac{62.4 \text{ lb/ft}^3 \times \left(6{,}900 \text{ gpm}/449 \left(\text{gpm/cfs}\right)\right) \times 93.2 \text{ ft}}{550 \text{ ft lb/s.hp} \times 0.85 \times 0.92}$$

$$= 191.2 \text{ hp}$$

$$\text{Wire power input} = 191.2 \text{ hp} \times \frac{1}{0.92} = 208 \text{ hp}$$

$$= 208 \text{ hp} \times 0.76 \text{ kW/hp} = 158.8 \text{ kW}$$

5. Determine annual power cost

$$\text{Annual power cost} = 158.8 \text{ kW} \times 24 \frac{\text{h}}{\text{day}} \times 365 \frac{\text{day}}{\text{year}} \times \frac{0.05}{\text{kWh}} = \$69{,}554/\text{year}$$

Example 9.22: Pump-Operating Points for Variable Static Head

A wastewater treatment plant uses a pumping station to raise screened wastewater from a wet well to a grit chamber. The wet well has a bubbler control system to start and stop the pumping operation. The pumping starts at high wet-well elevation. As soon as the minimum wet-well elevation is reached, the pumping stops. The piping arrangements are

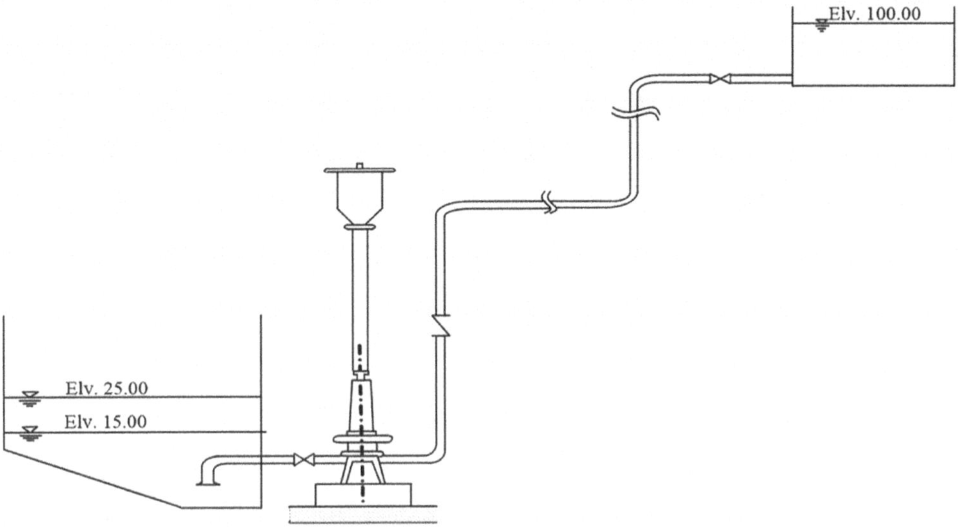

FIGURE 9.19 Pump station and piping arrangement (Example 9.22).

shown in Figure 9.19. The pump is identical to that used in Example 9.21, and the manufacturer's characteristic curves are shown in Figure 9.18. Draw the system head-capacity curves. Obtain the operating points and pump efficiency, and power input for two extreme situations. The valves, fittings, and specials include:

One suction bell $K=0.1$
Five 90° elbows $K=0.3$
One check valve $K=2.5$
Two gate valves $K=0.2$

The suction and discharge pipings are 24 in. diameter, and the total length is 2,000 ft. $C=100$. Motor efficiency is 90%.

Solution

1. Compute system losses for valves, fittings, and specials.
   The minor head losses are calculated for six assumed flow conditions. The calculation steps are tabulated below.

				1,580 gpm		3,160 gpm		4,740 gpm	
Item No. (1)	Valves, Fittings, and Specials (2)	No. (3)	K (4)	Vel., ft/s (5)	H_m, ft (6)	Vel., ft/s (7)	H_m, ft (8)	Vel., ft/s (9)	H_m, ft (10)
1	Suction Inlet	1	0.1	1.12[a]	0.002	2.24	0.008	3.38	0.018
2	90° elbow	5	0.3	1.12	0.029	2.24	0.117	3.38	0.266
3	Gate valve	2	0.2	1.12	0.008	2.24	0.031	3.38	0.071
4	Check valve	1	2.5	1.12	0.049	2.24	0.195	3.38	0.443
	ΣH_m				0.088		0.351		0.798

				6,320 gpm		7,900 gpm		9,480 gpm	
Item No. (1)	Valves, Fittings, and Specials (2)	No. (3)	K (4)	Vel., ft/s (11)	H_m, ft (12)	Vel., ft/s (13)	H_m, ft (14)	Vel., ft/s (15)	H_m, ft (16)
1	Suction Bell	1	0.1	4.48	0.031	5.60	0.049	6.72	0.070
2	90° elbow	5	0.3	4.48	0.467	5.60	0.730	6.72	1.052
3	Gate valve	2	0.2	4.48	0.125	5.60	0.195	6.72	0.280
4	Check valve	1	2.5	4.48	0.779	5.60	1.217	6.72	1.753
	ΣH_m				1.402		2.191		3.155

[a] $Q = 1{,}580\,\frac{\text{gal}}{\text{min}} \times \frac{1}{60\text{ s/min}} \times \frac{1}{7.48\text{ gal/ft}^3} = 3.52\text{ cfs};\ \text{vel.} = \frac{3.52\text{ ft}^3/\text{s}}{\pi/4 \times (2\text{ ft})^2} = 1.12\text{ ft/s}.$

2. Calculate H_f, $(H_f + \Sigma H_m)$ and TDH at different flows.

   Use Eq. (5.19) to calculate frictional head loss. The static heads for high and low wet-well elevations are, respectively, (100.00−25.00) ft=75 ft and (100.00−15.00) ft=85 ft. The TDH at maximum and minimum wet-well elevations for different pump discharges are calculated below.

				At Maximum Wet-Well Elevation		At Minimum Wet-Well Elevation	
Q, gpm	Velocity, ft/s	H_f, ft	ΣH_m, ft	Static Head, ft	TDH, ft	Static Head, ft	TDH, ft
0	0	0	0	75	75.00	85	85.00
1,580	1.12	0.67[a]	0.09	75	75.76	85	85.76
3,160	2.24	2.39	0.35	75	77.74	85	87.74
4,740	3.38	5.11	0.80	75	80.91	85	90.91
6,320	4.48	8.61	1.40	75	85.01	85	95.01
7,900	5.60	13.02	2.19	75	90.21	85	100.21
9,480	6.72	18.24	3.16	75	96.40	85	106.40

[a] $1.12 = 0.550 \times 100 \times (2.0)^{0.63} \times \left(\frac{H_f}{2{,}000}\right)^{0.54};\ H_f = 0.67\text{ ft}.$

3. Draw the H-Q and Eff.-Q curves from Figure 9.18

   These curves are shown in Figure 9.20.

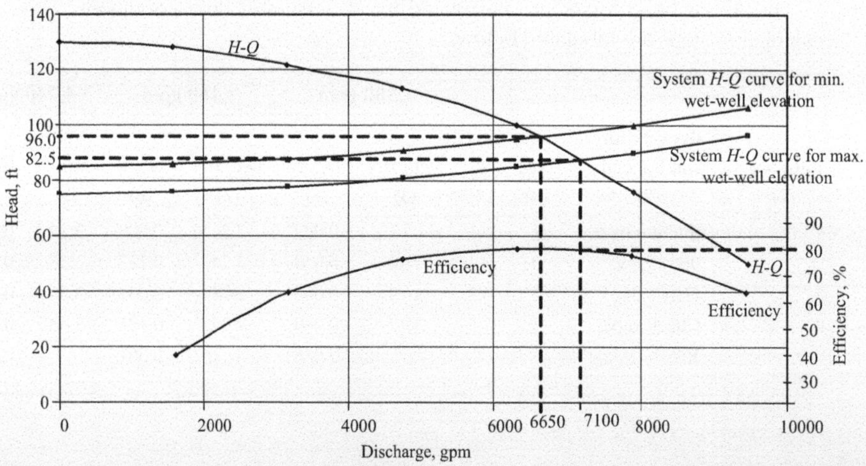

FIGURE 9.20 The H-Q and system head-capacity curves, operating-points, and efficiency of a pump at two extreme static head conditions (Example 9.22).

4. Draw system H-Q curves.
   These curves are shown in Figure 9.20.
5. Determine operating head and discharge of the pump, pump efficiency, and power input at minimum wet-well elevation.
   The operating head and capacity of the pump are at the intersection point of H-Q and system curve.

$$Q = 6{,}650 \text{ gpm}, H = 96.0 \text{ ft and pump efficiency} = 81\%, \text{ motor efficiency} = 90\%$$

$$\text{Power} = \frac{\gamma \times Q \times H}{550 \times \text{Efficiency}} = \frac{62.4 \text{ lb/ft}^3 \times \left(6{,}650 \text{ gpm}/449 \left(\text{gpm/cfs}\right)\right) \times 96.0 \text{ ft}}{550 \text{ ft lb/s.hp} \times 0.81 \times 0.90}$$

$$= 221 \text{ hp}$$

6. Determine operating head and discharge of the pump, pump efficiency, and power input at maximum wet-well elevation. The operating head and capacity of the pump at the intersection point of H-Q and system curve are

$$Q = 7{,}100 \text{ gpm}, \quad H = 82.5 \text{ ft}$$

$$\text{Pump efficiency} = 80\%, \text{ and motor efficiency} = 90\% (\text{given})$$

$$\text{Power} = \frac{\gamma \times Q \times H}{550 \times \text{Efficiency}} = \frac{62.4 \text{ lb/ft}^3 \times \left(7{,}100 \text{ gpm}/449 \left(\text{gpm/cfs}\right)\right) \times 82.5 \text{ ft}}{550 \text{ ft lb/s.hp} \times 0.80 \times 0.90}$$

$$= 206 \text{ hp}$$

Example 9.23: Pump Operating-Points for New and Old Piping Systems

As transmission lines age over many years of operation, the head loss increases, and pump operating-points may change from new to old piping systems. In the Hazen–Williams equation, the C values for new and old cast iron pipes may normally change from 140 to 100, respectively. Draw the system head-capacity curves for piping system and pump in Example 9.22 for new and old pipes. Draw the system head-capacity curve for:

1. Max. wet-well elevation $C=100$
2. Min. wet-well elevation $C=100$
3. Max. wet-well elevation $C=140$
4. Min. wet-well elevation $C=140$

Obtain the operating-points and pump efficiency for new and old piping for two extreme wet-well conditions.

Solution

1. Develop system head-capacity curves
   a. **$C=100$**: In Example 9.22, the system head-capacity curves were developed at max and min wet-well elevations. To achieve this, the velocity in pipes, H_f, ΣH_m and TDH values were calculated at six assumed flows. The Hazen–Williams coefficient $C=100$ for older pipes was used

b. **$C=140$**: The system head-capacity curve for new pipes ($C=140$) is calculated using the procedure identical to that for $C=100$. The final values for $C=100$ and 140 are tabulated below.

Q, gpm	Velocity, ft/s	H_f, ft $C=100$	H_f, ft $C=140$	ΣH_m, ft $C=100$ or 140	TDH at Max. Wet-Well Elev. Static Head 75.0 ft $C=100$	TDH at Max. Wet-Well Elev. Static Head 75.0 ft $C=140$	TDH at Minimum Wet-Well Elev. Static Head 85.0 ft $C=100$	TDH at Minimum Wet-Well Elev. Static Head 85.0 ft $C=140$
0	0	0	0	0	75.00	75.00	85.00	85.00
1,580	1.12	0.67[a]	0.35[b]	0.09	75.76	75.44	85.76	85.44
3,160	2.24	2.39	1.27	0.35	77.74	76.62	87.74	86.62
4,740	2.38	5.11	2.73	0.80	80.91	78.53	90.91	88.53
6,320	4.48	8.61	4.60	1.40	85.01	81.00	95.01	91.00
7,900	5.60	13.02	6.95	2.19	90.21	84.14	100.21	94.14
9,480	6.72	18.24	9.74	3.16	96.40	87.90	106.40	97.90

[a] $1.12=0.550\times100\times(2.0)^{0.63}\left(\frac{H_f}{2{,}000}\right)^{0.54};H_f=0.67\text{ ft.}$

[b] $1.12=0.550\times140\times(2.0)^{0.63}\left(\frac{H_f}{2{,}000}\right)^{0.54};H_f=0.35\text{ ft.}$

2. Draw the H-Q, and system head-capacity curves (Figure 9.21)
Note: The pump characteristics curves in Examples 9.21–9.23 are for the identical pumps. The system curves for old pipe ($C=100$) in Examples 9.22 and 9.23 are developed from the same data.

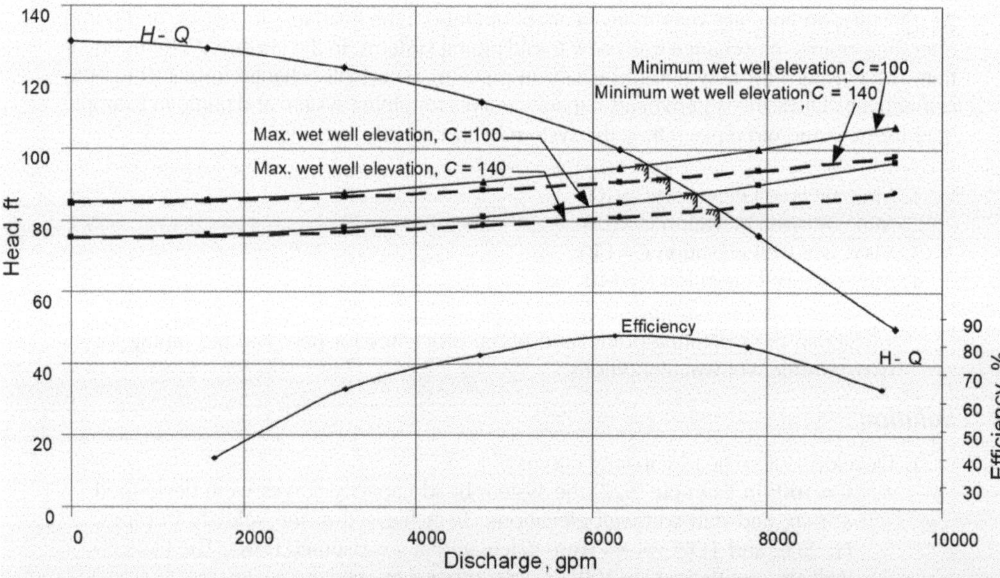

FIGURE 9.21 *H-Q* and efficiency curves, and system head-capacity curves and operating-points for two extreme static head conditions, for new and old piping (Example 9.23).

3. Obtain the pump operating-points at the minimum wet-well elevation

   At minimum wet-well elevation, the static head is 85 ft, and the operating *H*, *Q*, *E*, and power input (bhp) values for old and new force mains with $C=100$ and 140 are summarized below.

	C=100	C=140
H	96 ft	90 ft
Q	6,650 gpm	6,950 gpm
E	81%	80%
P	221 hp[a]	220 hp

[a] $\text{Power} = \dfrac{62.4\ \text{lb/ft}^3 \times (6{,}650\ \text{gpm}/449\ \text{gpm/cfs}) \times 96.0\ \text{ft}}{550\ \text{ft.lb/s.hp} \times 0.81 \times 0.9} = 221\ \text{hp}.$

4. Obtain the pump operating-points at maximum wet-well elevation (static head = 75 ft)

   At maximum wet-well elevation *H*, *Q*, *E*, bhp

	C=100	C=140
H	82.5 ft	81.5 ft
Q	7,100 gpm	7,400 gpm
E	80%	79%
P[b]	206 hp[b]	214 hp

[b] $\text{Power} = \dfrac{62.4\ \text{lb/ft}^3 \times (7{,}100\ \text{gpm}/449\ \text{gpm/cfs}) \times 82.5\ \text{ft}}{550\ \text{ft.lb/s.hp} \times 0.80 \times 0.9} = 206\ \text{hp}.$

5. Select the pump motor

   Select a standard pump motor that is larger and closest to 221 hp.

9.5.8 Modified Pump Head-Capacity Curve, Station Head and Capacity, and Pump Operating-Point

When two or more pumps operate in parallel, these pumps will discharge into a common discharge header or force main. A modified head-capacity curve must be developed to determine the pump operating-point. The procedure is as follows:

1. Subtract from the manufacturer's head-capacity curve all head losses in the individual pump pipings from the suction bell to the tee connecting the discharge pipe to the common header. These losses are called station losses.
2. Using the modified head-capacity curve, prepare the combined head-capacity curve for pumps in parallel combination.
3. Prepare the system head-capacity curve for the common header (discharge piping common to all pumps).
4. The point of intersection of the head-capacity curve of parallel combination and system head-capacity curve of the common header gives the station head and capacity.
5. The head and capacity delivered by each pump and its efficiency are found by reverting to the manufacturer's head-capacity curve. The procedure is shown in Example 9.24.

Example 9.24: Station Losses and Preparation of Modified Head-Capacity Curve

The manufacturer's head-capacity curve of a pump is given below. Calculate the station losses, and draw the modified head-capacity curve of one pump. The piping arrangement is shown in Figure 9.22.

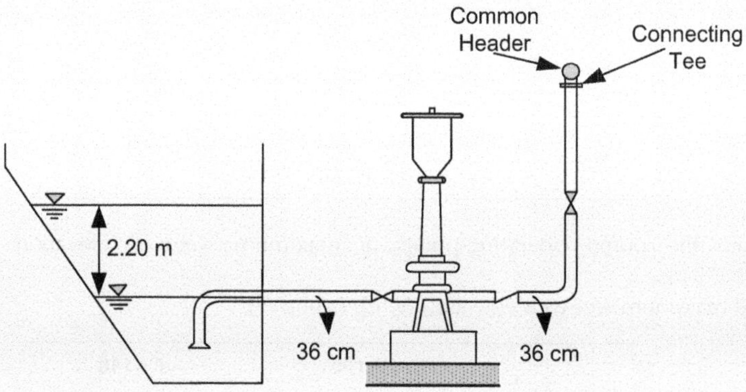

FIGURE 9.22 The piping system for preparation of modified head-capacity curve arranged in parallel combination (Example 9.24).

Q, m³/s	0	0.05	0.10	0.15	0.20	0.25	0.30
Head, m	20	19.4	18.5	17.0	16.0	14.2	10.0
Efficiency, %		51	65	77	84	80	65

The minor head loss coefficient K for valves, fittings, and specials are:

One suction bell, 36 cm $K=0.08$
Two 90° elbows, 36 cm $K=0.3$
Two gate valves, 36 cm $K=0.2$
One check valve, 36 cm $K=2.5$
One tee, 36 cm $K=1.8$

The suction and discharge pipings connecting to common header are 36 cm in diameter, and total length is 8 m. $C=100$.

Solution

1. Determine the minor and friction head losses from suction bell to common pipe header at several assumed flows.

Item No. (1)	Valves, Fittings, and Specials (2)	No. (3)	K (4)	0.05 m³/s		0.10 m³/s		0.15 m³/s	
				Vel., m/s (5)	H_m, m (6)	Vel., m/s (7)	H_m, m (8)	Vel., m/s (9)	H_m, m (10)
1	Suction	1	0.08	0.49[a]	0.001[b]	1.00	0.004	1.47	0.009
2	Gate valve	2	0.20	0.49	0.005	1.00	0.020	1.47	0.044
3	Check valve	1	2.50	0.49	0.031	1.00	0.127	1.47	0.275
4	90° elbow	2	0.30	0.49	0.007	1.00	0.031	1.47	0.066

5	Tee	1	1.80	0.49	0.022	1.00	0.092	1.47	0.198
	ΣH_m				0.066		0.274		0.593
Friction head loss[c]					0.010		0.036		0.073
	$\Sigma H_m + H_f$				0.076		0.310		0.666

Item No. (1)	Valves, Fittings and, Specials (2)	No. (3)	K (4)	0.20 m³/s		0.25 m³/s		0.30 m³/s	
				Vel., m/s (11)	H_m, m (12)	Vel., m/s (13)	H_m, m (14)	Vel., m/s (15)	H_m, m (16)
1	Suction	1	0.08	2.00	0.016	2.46	0.025	3.00	0.037
2	Gate valve	2	0.20	2.00	0.082	2.46	0.123	3.00	0.183
3	Check valve	1	2.50	2.00	0.510	2.46	0.771	3.00	1.147
4	90° elbow	2	0.30	2.00	0.122	2.46	0.185	3.00	0.275
5	Tee	1	1.80	2.00	0.367	2.46	0.555	3.00	0.826
	ΣH_m				1.097		1.659		2.468
Friction head loss[c]					0.129		0.189		0.272
	$\Sigma H_m + H_f$				1.226		1.848		2.740

[a] $V = \dfrac{0.05\text{ m}^3/\text{s}}{\pi/4 \times (0.36\text{ m})^2} = 0.49\text{ m/s}$.

[b] $H_m(\text{Eq. (5.27)}) = 0.08 \times \dfrac{(0.49\text{ m/s})^2}{2 \times 9.81\text{ m/s}^2} = 0.001\text{ m}$.

[c] $H_f(\text{Eq. (5.18)}) = 6.82\left(\dfrac{0.49}{100}\right)^{1.85} \times \dfrac{8\text{ m}}{(0.36)^{1.167}} = 0.0095\text{ m} \approx 0.01\text{ m}$.

2. Draw the station losses, manufacturer's, and modified head-capacity curves.
   The plotting points of manufacturer's and modified head-capacity curves are given below.

Q, m³/s	0	0.05	0.10	0.15	0.20	0.25	0.30
Head, m	20	19.40	18.50	17.00	16.00	14.20	10.00
$(H_f + \Sigma H_m)$, m	0	0.076	0.310	0.666	1.226	1.848	2.740
Modified head, m	20	19.32[a]	18.19	16.33	14.77	12.35	7.26

[a] (19.40 − 0.076) m = 19.324 m or 19.32 m.

The manufacturer's head capacity, station losses, modified head capacity, and efficiency curves are shown in Figure 9.23.

Example 9.25: Station Head and Capacity, Operating-Points, and Efficiency

The manufacturer's characteristic curves of a pump are given in Example 9.24. The modified head-capacity curve of this pump, based on actual piping, valves, fittings, and specials, is also developed in Example 9.24. This modified head-capacity curve also applies to the pump and piping in Example 9.25. Develop the system head-capacity curve using the piping, valves, fittings, and specials in the common header given below.

Two 90° elbows, diam. 45 cm $K = 0.3$
Two 45° elbows, diam. 45 cm $K = 0.2$

One Venturimeter, diam. 45 cm $K=0.2$
One plug valve, diam. 45 cm $K=1.0$
Discharge to atmosphere, $K=1.0$

The common header is 45 cm in diameter and 120 m long. The static head is 12.5 m and $C=100$. Determine the station head and capacity when one pump, and two pumps are operating in parallel. Also, determine the power input to each pump when one or two pumps are operating. Motor efficiency = 92%.

Solution

1. Draw the modified head-capacity curve

   The manufacturer's head-capacity curve and the modified head-capacity curve for the given piping system were developed in Example 9.24 and drawn in Figure 9.23. These curves are redrawn in Figure 9.24 for Example 9.25.

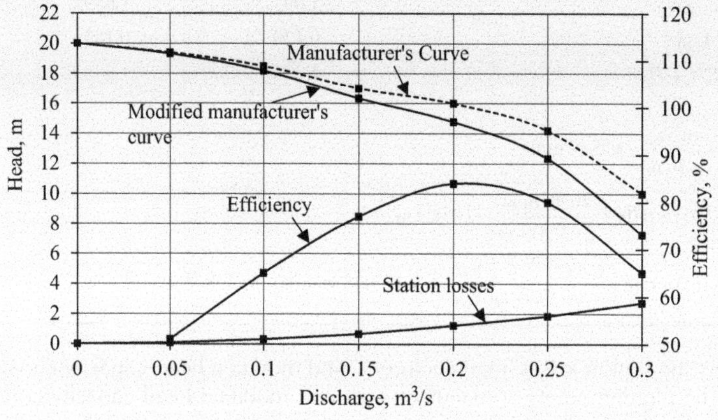

FIGURE 9.23 Station losses, and manufacturer's and modified head-capacity curves (Example 9.24).

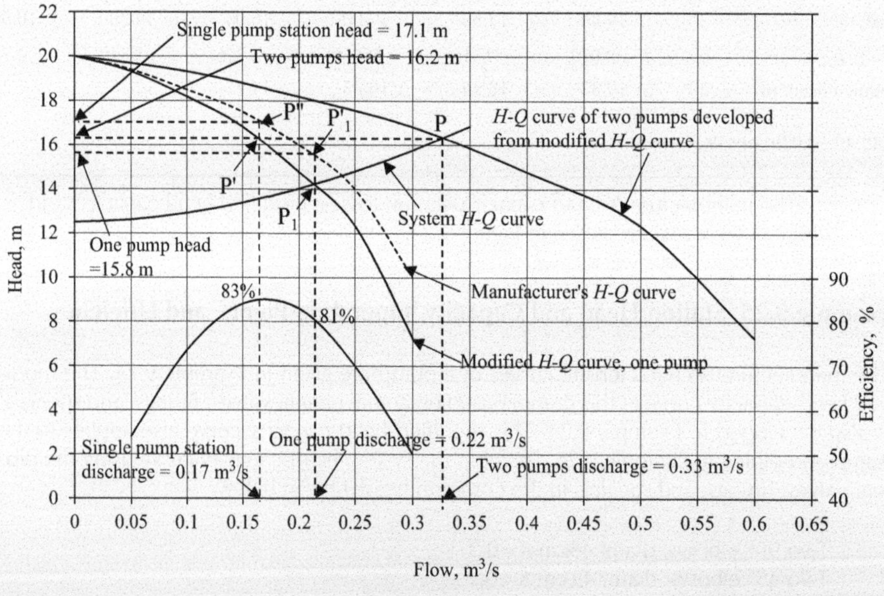

FIGURE 9.24 Station losses, and manufacturer's and modified head-capacity curves (Example 9.25).

2. Draw the head-capacity curve of two pumps in parallel combination.

   The *H-Q* curve of parallel combination is developed by adding the discharge of two pumps, and plotting them against the modified *H-Q* curve for different heads. The plotting values are given below.

Head, m	*Q* for One Pump, m³/s	*Q* for Two Pumps, m³/s
19.32	0.05	0.10
18.19	0.10	0.20
16.33	0.15	0.30
14.77	0.20	0.40
12.35	0.25	0.50
7.26	0.30	0.60

   The *H-Q* curves of one and two pumps in parallel are shown in Figure 9.24.

3. Develop the system head-capacity curve

   a. Determine minor and friction losses in the common header.

      The minor losses due to valves, fittings, and specials, and friction loss in the common header are tabulated below.

Item No. (1)	Valves, Fittings, and Specials (2)	No. (3)	*K* (4)	0.05 m³/s		0.10 m³/s		0.15 m³/s	
				Vel., m/s (5)	H_m, m (6)	Vel., m/s (7)	H_m, m (8)	Vel., m/s (9)	H_m, m (10)
1	90° elbow	2	0.3	0.31[a]	0.003[b]	0.63	0.012	0.94	0.027
2	45° elbow	2	0.2	0.31	0.002	0.63	0.008	0.94	0.018
3	Venturimeter	1	0.2	0.31	0.001	0.63	0.004	0.94	0.009
4	Plug valve	1	1	0.31	0.005	0.63	0.020	0.94	0.045
5	Discharge to atmosphere	1	1	0.31	0.005	0.63	0.020	0.94	0.045
	ΣH_m				0.016		0.065		0.144
Friction head loss					0.047[c]		0.175		0.368

Item No. (1)	Valves, Fittings, and Specials (2)	No. (3)	*K* (4)	0.20 m³/s		0.25 m³/s		0.30 m³/s	
				Vel., m/s (11)	H_m, m (12)	Vel., m/s (13)	H_m, m (14)	Vel., m/s (15)	H_m, m (16)
1	90° elbow	2	0.3	1.26	0.049	1.57	0.075	1.89	0.109
2	40° elbow	2	0.2	1.26	0.032	1.57	0.050	1.89	0.073
3	Venturimeter	1	0.2	1.26	0.016	1.57	0.025	1.89	0.036
4	Plug valve	1	1	1.26	0.081	1.57	0.126	1.89	0.182
5	Discharge to atmosphere	1	1	1.26	0.081	1.57	0.126	1.89	0.182
	ΣH_m				0.259		0.402		0.583
Friction head loss					0.633		0.95		1.339

[a] *V*=0.05 m3/s/π/4×(0.45 m)2=0.31 m/s.

[b] *hm* (Eq. 5.27)=0.6 × (0.31 m/s)2/(2×9.81 m/s2)=0.003 m.

[c] $h_f\,(\text{Eq. (5.18)}) = 6.82\left(\frac{0.31\ \text{m/s}}{100}\right)^{1.85} \times \frac{120}{(0.45\ \text{m})^{1.167}} = 0.047\ \text{m}.$

b. Calculate TDH of common header.
The minor and friction losses, the static head, and TDH values for different pumps discharge are summarized below.

Q, m³/s	Velocity, m/s	H_f, m	ΣH_m, m	Static Head, m	TDH, m
0	0	0	0	12.5	12.50
0.05	0.31	0.047	0.016	12.5	12.56
0.10	0.63	0.175	0.065	12.5	12.74
0.15	0.94	0.368	0.144	12.5	13.01
0.20	1.26	0.633	0.259	12.5	13.39
0.25	1.57	0.950	0.402	12.5	13.85
0.30	1.89	1.339	0.583	12.5	14.42

c. Draw the system H-Q of common header.
The system H-Q is drawn in Figure 9.24

4. Define points P, P', P'', P_1, and P_1' in Figure 9.24.

P is at the intersection of H-Q curve of two pumps in parallel and the system H-Q curve.

P' is at the intersection of modified H-Q curve of one pump and a horizontal line from P.

P'' is vertically above P' at the intersection of manufacturer's H-Q curve.

P_1 is at the intersection of modified H-Q curve of one pump and the system H-Q curve.

P_1' is vertically above P_1 at the intersection of manufacturer's H-Q curve.

5. Determine the operating head and capacity for different conditions.

The operating head and capacity of the pump are at the intersection point of H-Q and system H-Q curves. The head capacity of two pumps in parallel and single pump in operation, and efficiencies are tabulated below.

Pump-Operating Condition	Head, m	Capacity, m³/s	Efficiency, %
Two pumps operating in parallel, P	16.2	0.33	
Each pump operating in parallel, P''	17.1	0.17	81
Only one pump operating, P_1'	15.8	0.22	83

6. Calculate the power input

The power input to the pumps is obtained from the head and capacity, and pump efficiency under different operating conditions.

a. Power input to each pump operating in parallel (Eq. 9.4)

$$Q = 0.17\ \text{m}^3/\text{s},\quad H = 17.1\ \text{m},\quad \text{Pump efficiency} = 81\%$$

$$\text{Power} = \frac{\gamma \times Q \times H}{\text{Efficiency}} = \frac{9.81\ \text{kN/m}^3 \times 0.17\ \text{m}^3/\text{s} \times 17.1\ \text{m}}{0.81} = 35.2\ \text{kW}$$

$$\text{Power input to both pumps} = 2 \times 35.2 = 70.4\ \text{kW}$$

b. Power input to single pump in operation (Eq. 9.4)

$$Q = 0.22\ \text{m}^3/\text{s},\quad H = 15.8\ \text{m},\quad \text{Pump efficiency} = 83\%$$

$$\text{Power input to each pump} = \frac{\gamma \times Q \times H}{\text{Efficiency}}$$

$$= \frac{9.81\ \text{kN/m}^3 \times 0.22\ \text{m}^3/\text{s} \times 15.8\ \text{m}}{0.83}$$

$$= 41.1\ \text{kW}$$

7. Select pump motor

   Motor efficiency is 92%. Power input to motor = 41.1 kW/0.92 = 44.7 kW.

   Select standard motor for each pump that is larger and closest to 44.7 kW.

   Select 45 kw motor for each pump.

9.5.9 Cavitation and Net-Positive Suction Head

Cavitation: Cavitation reduces pump capacity and efficiency, and can damage the pump. It occurs in pumps when the absolute pressure at the inlet drops below the vapor pressure of the fluid being pumped. Under cavitation, vapors form bubbles at the inlet, and are carried into a zone of higher pressure where they collapse abruptly. The surrounding fluid rushes to fill the void with such a force that hammering action occurs with distinctive rattling or pinging noise. It may sound as if rocks are being pumped. If this condition is allowed to continue, severe pitting of metallic surfaces may occur.[3,6]

Net-positive suction head: The NPSH is the absolute pressure of water at the pump center line or impeller eye as water enters the pump suction. Two values of NPSH are important in pump design: (1) NPSH required ($NPSH_{rq}$), and (2) NPSH available ($NPSH_{av}$). The $NPSH_{rq}$ is the minimum value required at the pump to prevent cavitation. It is determined by the pump manufacturer based on an extensive testing program. The $NPSH_{av}$ is the absolute pressure of the liquid as it enters the pump suction. It depends upon the suction head and pump suction piping, and is calculated by the design engineer. Equation (9.20) is used to calculate $NPSH_{av}$.[2,7,8]

$$NPSH_{av} = H_{abs} + h_s - H_{LS} - H_{vp} \tag{9.13}$$

where

$NPSH_{av}$ = available net-positive suction head, m (ft)

H_{abs} = absolute pressure at the surface of the liquid in the suction well, m (ft). This is the barometric pressure at the given altitude (see Appendix C). For safety, the barometric pressure is reduced by 0.4 m of water to account for any storm activity.

h_s = the static head of the liquid above the center line of the pump, m (ft). If the liquid level is below, the center line h_s is negative.

H_{LS} = total suction head loss due to friction and minor losses ($h_{fs} + h_{vs} + \Sigma h_{ms}$), m (ft)

H_{vp} = absolute vapor pressure of the liquid at the pump operating-temperature, m (ft). H_{vp} at various temperatures are given in Table B.6.

The $NPSH_{av}$ should be evaluated over the entire head and capacity range for which the pump is expected to operate. It is recommended that the $NPSH_{av}$ should exceed the $NPSH_{rq}$ by one meter or more. If the $NPSH_{av}$ is less than the $NPSH_{rq}$ or these values are too close to each other, serious operating problems may result. Following adjustments are recommended.

- Raise water surface elevation in the suction well
- Lower pump centerline elevation
- Increase the suction-piping size to reduce head losses

- Decrease the pump speed, which will decrease the pump discharge
- Check if suction line has any obstruction
- Check the specific speed of the pump

Very high specific speed for optimum design parameters may be detrimental to the pump. The cavitation constant is calculated from Eq. (9.14) to check the specific speed

$$\sigma = \frac{\text{NPSH}_{\text{rq}}}{\text{TDH}} \tag{9.14a}$$

$$\sigma = \frac{K \times N_s^{4/3}}{10^6} \tag{9.14b}$$

where

σ=cavitation constant. The range of the cavitation constant is 0.05 for a specific speed of 1,000, to 1.0 for specific speed of 8,000 (U.S. customary units).

K=1,210 when N_s is in SI units (m^3/s and m), and K=6.3 when N_s is in U.S. customary units (gpm and ft).

Example 9.26: Net-Positive Suction Head (NPSH)

Calculate the NPSH_{av} for the pumping system shown in Figure 9.22 (suction bell to pump center line). The static suction head=2.20 m. At the maximum station capacity, the pump discharge is 0.25 m^3/s. The pump station is located 1,050 m above the sea level, and operating-temperature is 25°C. The suction pipe is 36 cm in diameter and 4-m long from the suction end to pump centerline; C=100. The NPSH_{rq} from the manufacturer's data is 6.5 m. Determine whether the pump may have cavitation problem.

Solution

1. Calculate the velocity and the velocity head in the suction pipe up to pump impeller (see Figure 9.22)

$$Q = 0.25 \text{ m}^3/\text{s}, \quad \text{and pipe diameter} = 36 \text{ cm}$$

$$V = \frac{0.25 \text{ m}^3/\text{s}}{\pi/4 \times (0.36 \text{ m})^2} = 2.46 \text{ m/s}$$

$$\text{Velocity head } V^2/2g = \frac{(2.46 \text{ m/s})^2}{2 \times 9.81 \text{ m/s}^2} = 0.31 \text{ m}$$

2. Calculate total minor losses from Eq. (5.27), and velocity head
   Suction end h_m=0.08×0.31 m=0.04 m
   90° elbow h_m=0.30×0.31 m=0.09 m
   Gate valve h_m=2.50×0.31 m=0.78 m
   Total minor losses Σh_m=0.91 m
   Velocity head $V^2/2g$=1.00×0.31 m=0.31 m
3. Calculate friction head loss in the suction pipe from Eq. (5.18)

$$2.46 \text{ m/s} = 0.355 \times 100 \times (0.36)^{0.63} \times \left(\frac{h_f}{4}\right)^{0.54}$$

$$h_f = 0.09 \text{ m}$$

4. Calculate total suction pipe head losses
   Total head losses in suction piping H_{LS}=0.91 m+0.31 m+0.09 m=1.31 m
5. Determine the values of H_{abs} and H_{vp}
   The absolute barometric pressure P at sea level from Table B.6 is 10.3 m of water.
   The absolute barometric pressure is reduced at a rate of 0.096 m per 100 m of altitude above sea level.

$$\text{Absolute barometric pressure at pump station} = 10.33 - \frac{0.096 \text{ m}}{100 \text{ m}} \times 1{,}050 \text{ m}$$

$$= 9.33 \text{ m}$$

   As a safety factor, the barometric pressure is reduced by 0.4 m of water to account for the storm activity.

$$H_{abs} = 9.33 - 0.4 \text{ m} = 8.93 \text{ m}$$

   Vapor pressure at 25°C (Table B.2)=23.76 mm Hg (mercury)

$$H_{vp} = \frac{23.76 \text{ mm Hg}}{1{,}000 \text{ mm/m}} \times 13.58 \left(\text{sp. gr. of Hg}\right)$$

$$= 0.32 \text{ m (water)}$$

6. Calculate $NPSH_{av}$ (Eq. 9.13)

$$NPSH_{av} = H_{abs} + h_s - H_{LS} - H_{vp}$$

$$= 9.33 \text{ m} + 2.20 \text{ m} - 1.31 \text{ m} - 0.32 \text{ m}$$

$$= 9.90 \text{ m}$$

7. Check the NPSH

$$NPSH_{rq} = 6.50 \text{ m}$$

$$NPSH_{av} = 9.90 \text{ m}$$

   Since the $NPSH_{av}$ is greater than $NPSH_{rq}$ by more than 1 m, there will not be any cavitation problem.

Example 9.27: Allowable Suction Head Based on Cavitation Constant

A pump station is located at sea level, and the operating-temperature is 30°C. The TDH is 42 m. Determine the maximum allowable suction head for a pump with a cavitation constant equal to 0.3. The $NPSH_{av}$ is equal to $NPSH_{rq}$ +1. Also, calculate the specific speed, and check if the specific speed is within the acceptable range.

Solution

1. Calculate $NPSH_{av}$
   Equation (9.14a) is used to calculate $NPSH_{rq}$,

$$\sigma = 0.3 = \frac{\text{NPSH}_{\text{rq}}}{42\text{ m}}$$

$$\text{NPSH}_{\text{rq}} = 12.6\text{ m}$$

$$\text{NPSH}_{\text{av}} = 13.6\text{ m}$$

2. Substitute variables in Eq. (9.13) to calculate NPSH_{rq}

$$\text{NPSH}_{\text{rq}} = H_{\text{abs}} + h_s - H_{\text{Ls}} - H_{\text{vp}}$$

$$H_{\text{Ls}} = h_{\text{fs}} + \Sigma h_{\text{ms}} + \frac{V_s^2}{2g}$$

$$H_s = h_s - \left(h_{fs} + \Sigma h_{\text{ms}} + \frac{V_s^2}{2g} \right)$$

$$H_s = h_s - H_{\text{Ls}}$$

where H_s is maximum allowable suction head (Figure 9.4)

3. Determine H_{abs} and H_{vp}
   H_{abs} at sea level (Table B.6) = 10.3 m

$$H_{\text{vp}} \text{ at } 30°\text{C (Table B.2)} = \frac{4.24\text{ kN/m}^2}{9.81\text{ kN/m}^3} = 0.43\text{ m}$$

4. Calculate H_s from Eq. (9.13)

$$\text{NPSH}_{\text{av}} = H_{\text{abs}} + H_s - H_{\text{vp}}$$

$$13.6\text{ m} = 10.3\text{ m} + H_s - 0.43\text{ m}$$

$$H_s = 3.73\text{ m}$$

The head in the suction well should be at least 3.73 m above the centerline of the pump impeller.

5. Calculate specific speed
   Equation (9.14b) is used to calculate the specific speed N_s

$$0.3 = \frac{1{,}210 \times N_s^{4/3}}{10^6}$$

$$N_s = \left(\frac{0.3 \times 10^6}{1{,}210} \right)^{3/4} = 63(\text{SI units})$$

The specific speed in U.S. customary unit is 63 × 52 = 3,280. The specific speed of 3,280 is nearly within the acceptable range for cavitation constant of 0.30.

9.5.10 Pump Selection and Design Considerations

Pump application chart: Wastewater treatment plants utilize many types of pumps for different applications. Common applications are pumping of raw wastewater, various types of sludge, grit, and chemical solutions. Most pumps fall into kinetic and positive displacement pumps (Tables 9.1 and 9.3). A typical pump application chart is shown in Figure 9.25.

Pump rated point: The pump rated point is one of the most important values used for pump selection. The pump rated point corresponds to the required head and discharge provided by a single pump that is operating in parallel combination to develop the total station capacity. For the best pump selection, this point should correspond to peak efficiency.

To obtain the pump rated point for pump selection, the following information is needed: (1) develop system head-capacity curve for the common header, (2) establish the maximum station capacity with all pumps in operation, (3) select total number of pumps to be arranged in parallel combination, (4) obtain the operating-point of a single pump. This point must correspond to the peak efficiency of the selected pump. The procedure is illustrated in Example 9.29.

Suction head and NPSH: For continued efficient operation over the design period, the pump must be free of cavitation. The $NPSH_{av}$ for the selected pumping system should be at least 1 m more than the $NPSH_{rq}$. If the installation requires suction lift, the pump must be self-priming.

Pump efficiency, bep, and pump operating range: The pump must operate near peak efficiency or *bep*. If pumping occurs over a band of static heads, the peak efficiency point must lie between two extreme operating conditions.

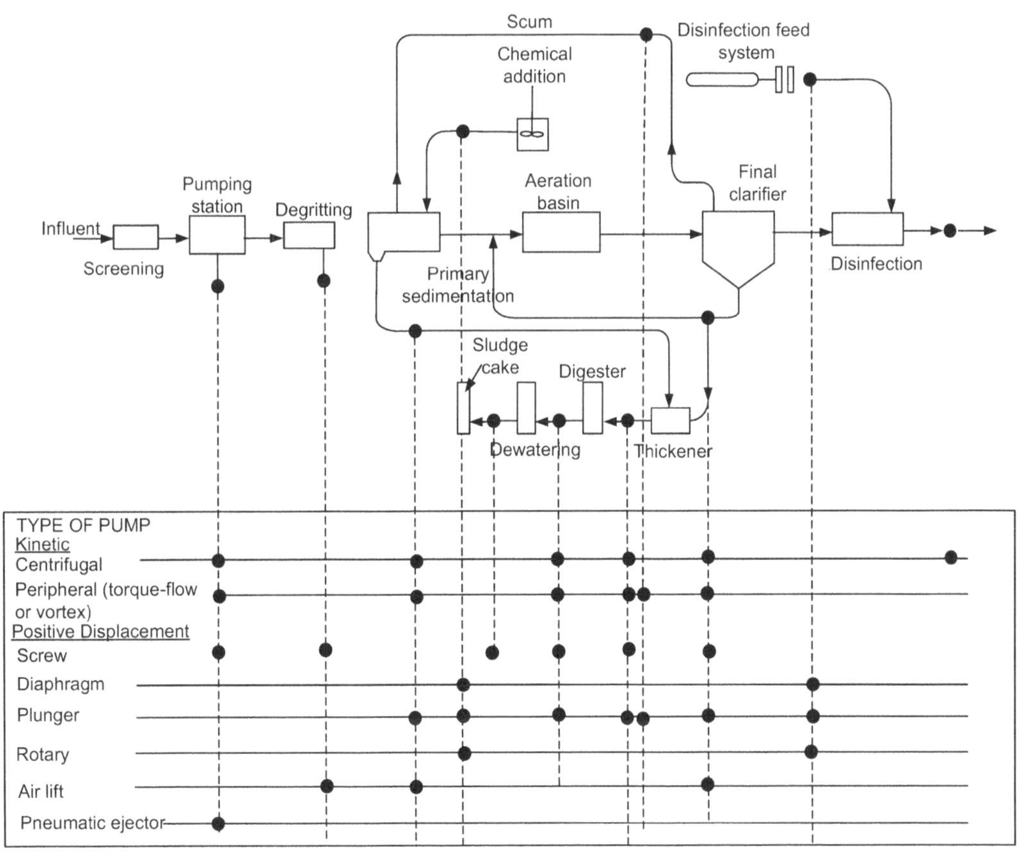

FIGURE 9.25 Pump application points and types of pumps at a wastewater treatment plant.

At *bep*, the radial loads on bearings are at a minimum. If the pump operates beyond the *bep*, the cavitation, vibration, and overheating may result. For these reasons, the operating range of the pump is limited between 60% and 120% of the *bep*. The pump-operating envelope at different pump speeds is illustrated in Figure 9.26a.3 Quite often, the head, power input, and efficiency are presented with respect to discharge as a percent of *bep*. An example of such a representation for a radial-flow pump is shown in Figure 9.26b.3

Shut-off and run-out: Pumps should be selected to operate near the middle of their head-capacity curve. Operations too near the left-hand side of the curve (near shut-off point) result in low flows. Continued operation in this condition will result in damage to the pump from overheating. Axial-flow and mixed-flow pumps usually have a sharp rise in discharge head and require more power to operate near the left-hand side of the curve. Higher discharge head can damage pipes and valves from excessive pressure, and higher power requirements can overload the pump driver.

Radial-flow pumps, however, can be started against a closed valve to take the advantage of lower starting torques. If this is done, the valve should be opened soon after the pump accelerates to full speed to prevent heat damage. Operation of pumps too far to the right-hand side (run-out) of the curve results in discharge rates in excess of the suction capacity. Continued operation in this

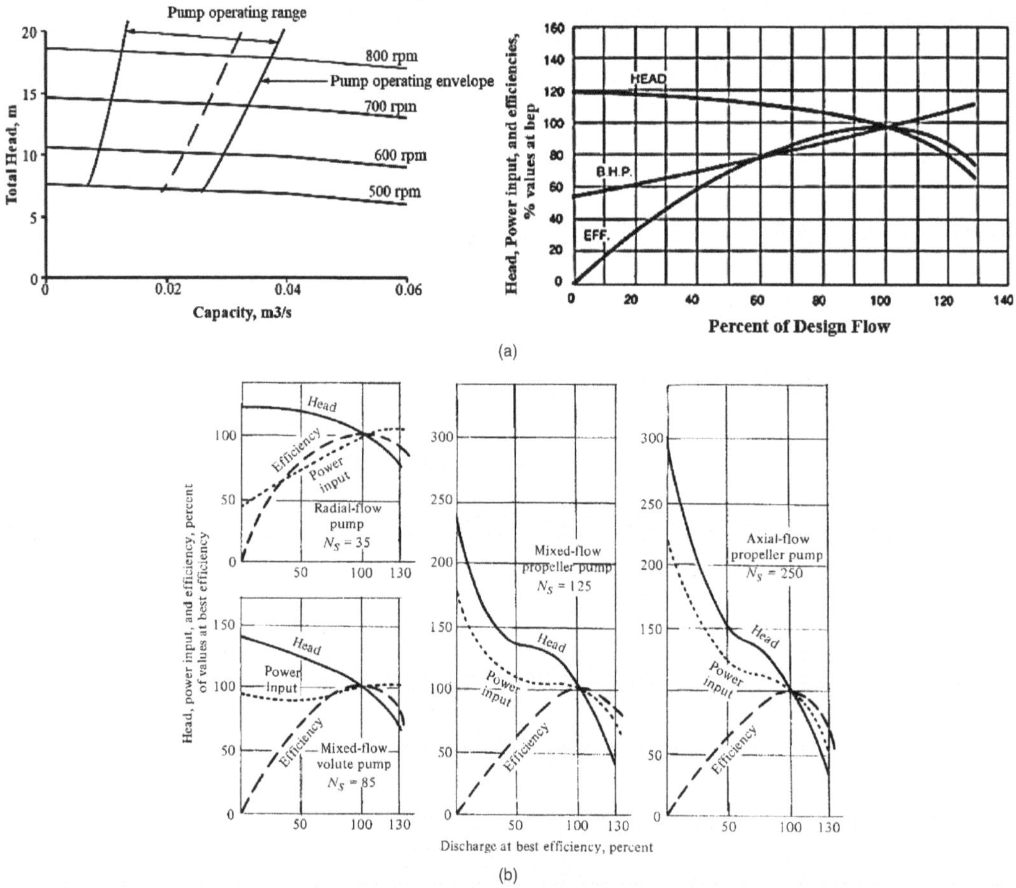

FIGURE 9.26 The operating characteristics of typical centrifugal pump with respect to *bep*: (a) pump operating envelop at different pump speeds, (b) characteristic curves of radial-flow pump, as percent of discharge and *bep*, and (c) comparison of typical characteristics of radial-flow, mixed-flow, and axial-flow pumps.

condition will result in damage to the pump from cavitation. The characteristic curves of typical radial-flow, mixed-flow, and axial-flow centrifugal pumps as a percent of *bep* are shown in Figure 9.26c.

Unstable operation: Some axial-flow and mixed-flow pumps have an inflection point or knee in the head-capacity curve. An example of such a curve is illustrated in Figure 9.27. Such pumps should always be operated at a head that is lower than the minimum head of the knee. Operation near the knee will result in oscillations or unstable flow and will ultimately damage the pump.

Impeller and motor selection: A pump casing is designed to accommodate impellers of several sizes; therefore, a particular pump casing may be used for different head and flow by replacing the impeller and motor. It is a good design practice to select a smaller impeller initially. As the need for large flow is identified, the pump impeller and motor are replaced to meet the situation.

Non-clog pump: Pumps should be selected to handle the largest solids that may be required to pass. Experience has shown that in pumping raw wastewater, such materials as rope, string, rags, sticks, cans, rubber goods, and grease present the greatest problem. Screens prior to pumping are recommended. The term non-clog pump has been used for radial-flow and mixed-flow pumps. They are designed with the following considerations: (1) pump size, impeller arrangement and smoothness, and contour of channels and impellers are such that they permit smooth passage of clogging materials, (2) provide open passage slightly smaller than the discharge pipe, (3) provide minimum number of vanes: two in smaller and not more than four in large pumps, and (4) provide recessed blades entirely into the curved shroud so that the flow does not pass through the impeller. The peripheral (torque flow or vortex) pumps are designed to offer this unique non-clog feature.

Constant-speed pumps: The selection of constant-speed pump should be based on the following considerations:

- These pumps are the simplest, cost-efficient, and very reliable.
- They operate on level controls. Therefore, the pumping may not be continuous. All pumps in parallel combination operate at maximum wet-well elevation. As the water surface

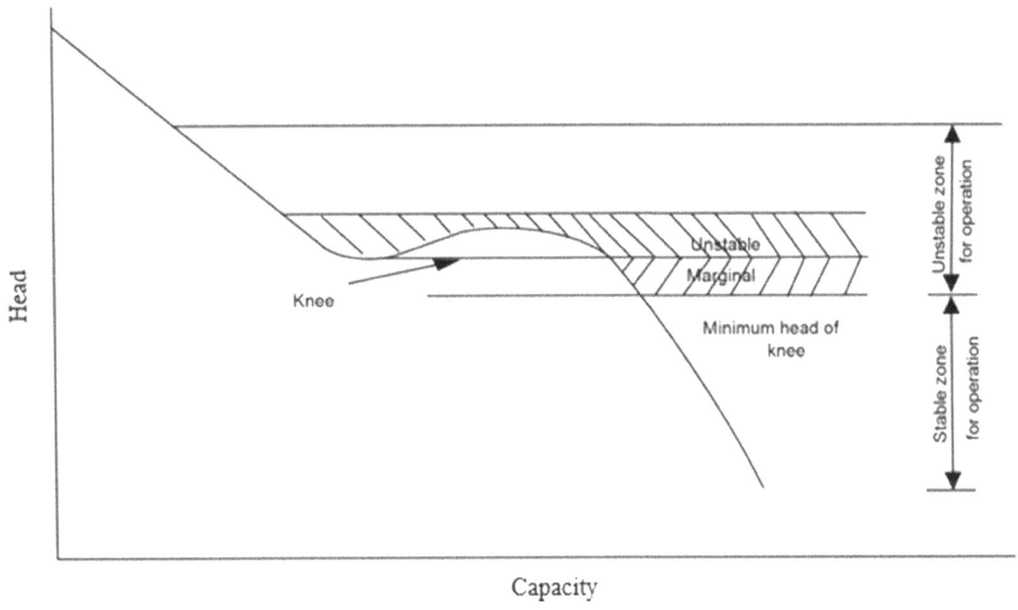

FIGURE 9.27 Example of a knee in a head-capacity curve.

elevation in the wet-well drops, different pumps at designated elevations are turned off until the lowest elevation is reached and pumping stops.

- Pumping station having large flow variations may require a large wet well and multiple pumps to maintain nearly uniform flow.
- Constant-drive pumps should operate near peak efficiency if properly designed.

Variable-speed pumps: The variable-speed pump drive offers many important features:

- Nearly continuous flow from the pump station can be obtained.
- The pump speed is regulated in proportion to the influent flow to maintain constant station capacity.
- The pumps reach the best efficiencies when operating at maximum speed. At lower speeds, the pump efficiency normally drops.
- The variable-speed pumps lead to a smaller wet-well size. Also, the water hammer is controlled by slowing down the pump gradually before shutdown.

Example 9.28: Pump Characteristic Curves as Percent of bep Values

The *bep* is shown in Figure 9.9a. The pump efficiency, head, capacity, and power at *bep* are 88.0%, 15.0 m, 0.32 m^3/s, and 63.0 kW. Develop head, power, and efficiency curves as a percent of *bep* values.

Solution

1. Determine the discharge, head, power, and efficiency corresponding to *bep* in Figure 9.9a
   From Figure 9.9a, these values at *bep* are obtained and given below.

$$Q = 0.32\ \text{m}^3/\text{s},\quad P = 63.0\ \text{kW},\quad H = 15.0\ \text{m},\quad \text{Eff.} = 88\%$$

2. Tabulate the capacity, head, efficiency, and power input values as a percent of values at bep.

Capacity		Head		Efficiency		Power Input	
m^3/s	% of *bep* Value	m	% of *bep* Value	%	% of *bep* Value	kW	% of *bep* Value
0	0	19	130				
0.1	31	18.5	123	56	64	34	54
0.2	63	17.3	115	78	89	48	76
0.32	100	15.0	100	88	100	62	100
0.4	125	12.0	80	81	92	70	111
0.5	156	7.0	47	57	65	74	117

3. Plot the percent head, efficiency, and power input curves with respect to percent capacity values of *bep*. This plot is shown in Figure 9.28.

Example 9.29: Pump Rated Point and Pump Selection

The system head-capacity curve of a common header was developed for a pumping station. There are four identical pumps and one standby pump arranged in parallel. The system

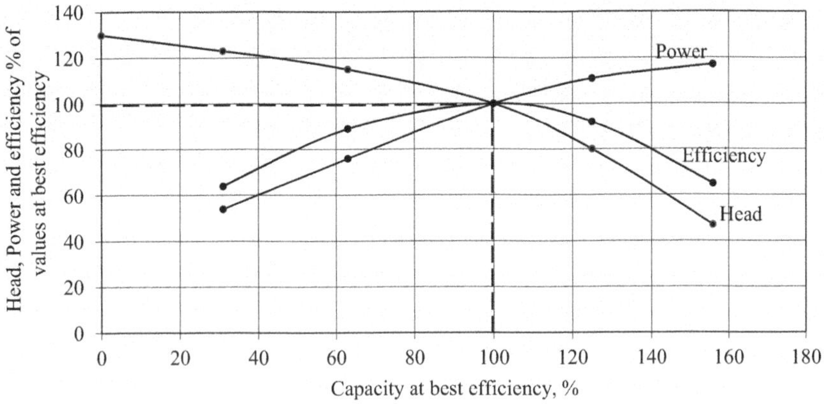

FIGURE 9.28 Head, power input, and efficiency as percent of *bep* (Example 9.28).

head-capacity curve and station capacity are shown in Figure 9.29. Maximum station capacity and station head when all four pumps are operating are 1.5 m^3/s (23.78×10^3 gpm) and 13.8 m (45.3 ft), respectively. Determine the pump operating-point. Also, make the preliminary pump selection for this application from the manufacturer's characteristic curves.

Solution

1. Determine the pump rated point

   The station capacity = 1.5 m^3/s

   The station head is obtained from the station capacity and system *H-Q* curve.

   Station head = 13.8 m

   $$\text{Discharge of each pump} = \left(1.5\ \text{m}^3/\text{s}\right) \times \frac{1}{4} = 0.38\ \text{m}^3/\text{s}\ \left(6{,}000\ \text{gpm}\right)$$

   The pump rated point corresponds to $H = 13.8$ m (45.3 ft), and $Q = 0.38\ m^3/s$ (6,000 gpm).
2. Select a pump

   The preliminary selection of a pump is made from the manufacturer's catalog. Select a pump to check the requirement. For the best selection, the pump rated point should correspond to peak efficiency. The manufacturer's characteristic curves of the selected pump are given in Figure 9.30. This pump has a variable

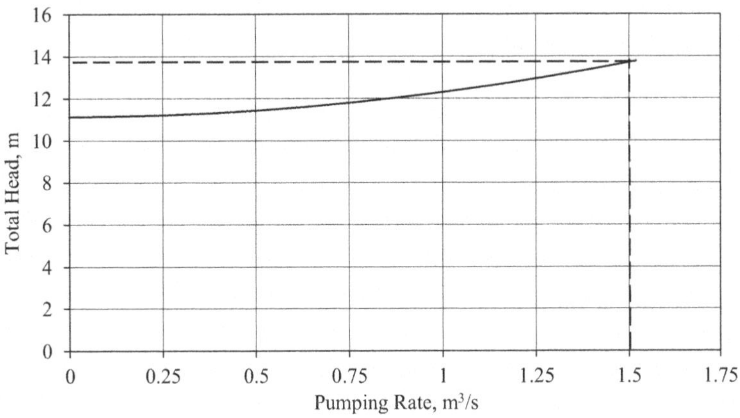

FIGURE 9.29 The system head-capacity curve and station head-capacity (Example 9.29).

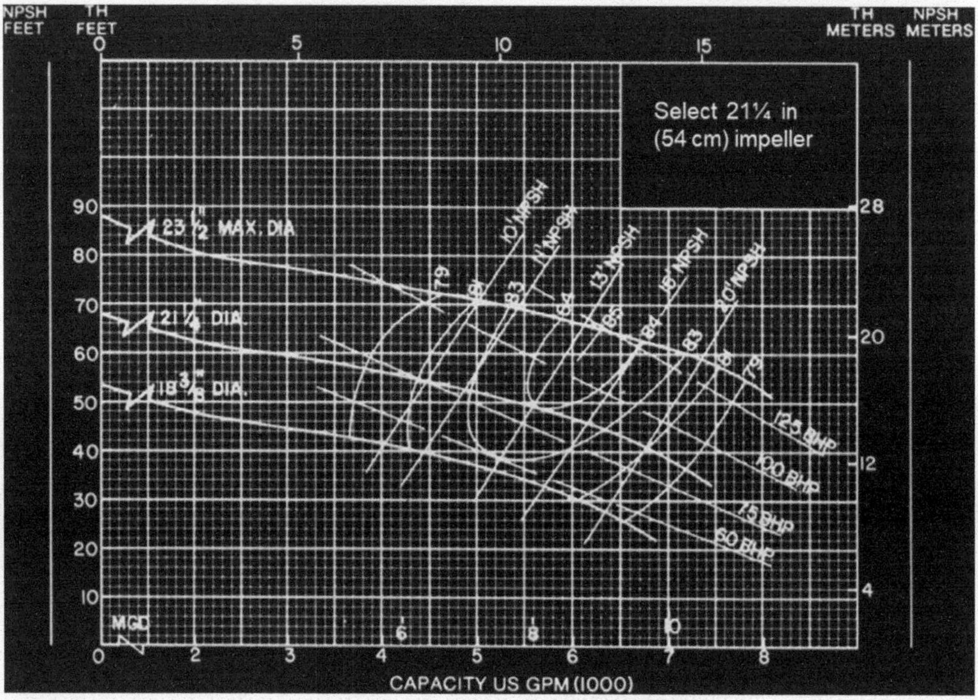

FIGURE 9.30 Manufacturer's characteristics curve for selected pump from rated point for pump station design (Example 9.29).

drive and accommodates three different sizes of impellers. The diameter of the selected impeller is 54 cm (21¼ in.).

3. Check the pump efficiency for rated points

   Corresponding to $H = 13.8$ m (45.3 ft) and $Q = 0.38\ m^3/s$ (6,000 gpm), the pump efficiency is 83.4% which is near the maximum efficiency of this pump. Therefore, based on preliminary selection using *bep*, this is a good choice, and detailed design is warranted. Detailed design of the pumping station using the selected pump is given in Example 9.30.

Example 9.30: Pump Station Design

Screened wastewater is pumped from a wet well to a grit chamber. The pump station has four identical pumps and one standby pump in parallel combination. The manufacturer's pump characteristic curves for the selected pump are given in Figure 9.30. The diameter of selected impeller is 54 cm (21¼ in.), and it operates at a variable speed. The piping schematic plan and profile are shown in Figure 9.31. The inventory of valves, fittings, and specials, and K values are listed in Table 9.7. The common header is a 92-cm force main, 129.0 m long, and $C = 100$. The minimum static head at maximum water surface elevation in the wet well is 9.91 m. Determine the following for the minimum static head condition: (1) maximum station capacity and operating head when all four pumps are operating; (2) operating head, capacity, efficiency, and power input to each pump in (1); and (3) pump operating-head, capacity, efficiency and power input when only one pump is operating. Draw the hydraulic and energy grade lines with respect to water surface elevation in the wet well.

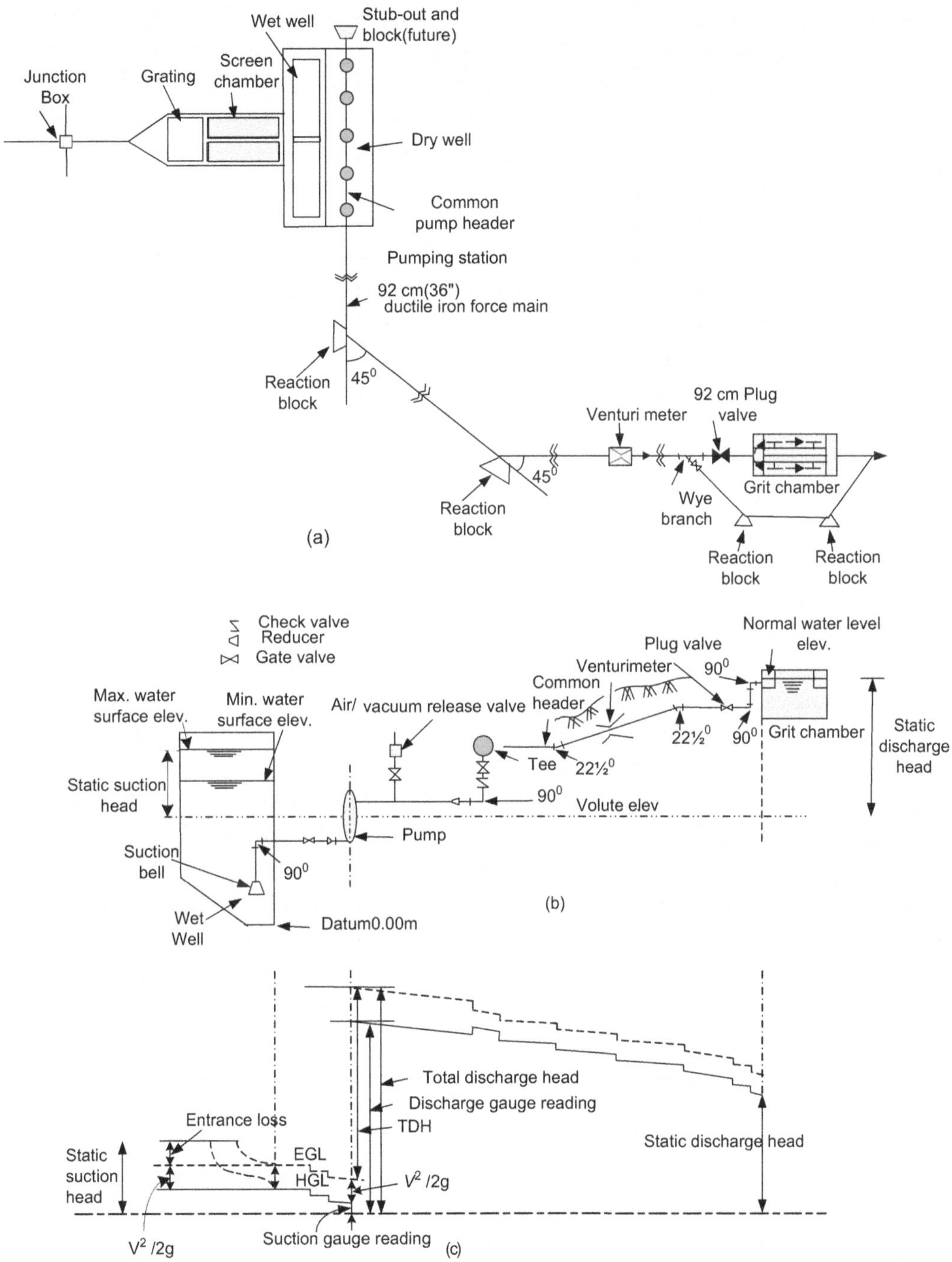

FIGURE 9.31 Pumping station layout, pipings, and hydraulic and energy grade lines: (a) piping schematic plan, (b) piping schematic vertical section, and (c) hydraulic and energy grade lines (Example 9.30).

TABLE 9.7
Inventory of Valves, Fittings, and Specials in Common Header, and in Piping of Individual Pump (Example 9.30)

Common Header, 92 cm Diameter Force Main				Individual Pump Piping			
Item No.	Valves, Fittings, and Specials	Numbers Provided	*K* Value	Item No.	18	Numbers Provided	*K* Value
1	90° elbow	2[a]	0.30	1	Entrance, suction bell 81 cm (32 in.)	1[a]	0.04
2	45° elbow	2[b]	0.20	2	90° elbow 61 cm (24 in.)	1[a]	0.30
3	22½° elbow	2[a]	0.15	3	Gate valve 61 cm (24 in.)	1[a]	0.19
4	Venturimeter	1[c]	0.14	4	Reducer 35.5 cm (14 in.)	2[a]	0.25
5	Wye branch	1[b]	1.00	5	Check valve 51 cm (20 in.)	1[a]	2.50
6	Plug valve	1[c]	1.00	6	90° elbow 51 cm (20 in.)	1[a]	0.30
7	Outlet (head velocity)	1[c]	1.00	7	Gate valve 51 cm (20 in.)	1[a]	0.19
				8	Tee 51 cm×51 cm (20 in.×20 in.)	1[d]	1.80
				9	The straight connecting pipings between fittings are 61, 51, and 35.5 cm in diameter and 2.0, 1.0, and 2.0 m long, respectively.		

[a] Shown in Figure 9.31b.
[b] Shown in Figure 9.31a.
[c] Shown in Figure 9.31a and b. The throat diameter is 46 cm.
[d] The tee connects the pump discharge pipe to the common station header. Location of the tee is shown in Figure 9.31b.

Note: It is customary to also determine the pump operating-conditions under maximum static head. This condition is normally reached at minimum wet-well elevation. See review Problems 9.17 and 9.18 for additional information.

Solution

1. Prepare system head-capacity curve
   a. Compute system losses in force main
      The system losses are in valves, fittings, and specials provided in the common header. These losses are calculated for six assumed flows and are summarized in Table 9.8. The sum of minor losses is given in row 8.
   b. Compute friction losses in the force main
      The friction losses are also calculated for six assumed flows, and are summarized in Table 9.8.
      $C=100$. The friction losses are given in row 9.
   c. Compute total losses in the force main
      Total losses are the sum of minor and friction losses in the force main.
      Total losses for six assumed flows are given in Table 9.8 (row 10).
   d. Compute TDH
      TDH is sum of total losses in the force main and static head. TDH values with respect to min static head (maximum wet well) are calculated in Table 9.8 (row 12).
2. Plot TDH (system head vs capacity)
   The TDH of force main is the sum of the total losses plus static head. The static head is 9.91 m. The TDH values for six assumed flows are provided in Table 9.8. The TDH values for static head (row 12) are plotted against the assumed pumping rate in Figure 9.32.
3. Prepare the modified head-capacity curve
   a. Draw the manufacturer's head-capacity curve for 54 cm (21¼ in.) diameter impeller.
      Derive the data from the manufacturer's curve in Figure 9.30. The procedure is similar to that in Example 9.15. The manufacturer's *H*-*Q* curve for one pump is shown in Figure 9.32 by the dotted line.
   b. Calculate the minor losses in the piping of the individual pumps
      The individual pump piping is from the suction bell to the tee that joins the piping of individual pump to the common header. The minor losses are due to fittings, valves, and specials. These losses are calculated for five assumed flows. The results are summarized in Table 9.9.
   c. Calculate frictional head losses due to straight connecting pipes
      There are three straight pipe sizes 61, 51, and 35.5 cm in diameter, and 2.0, 1.0, and 2.0 m long.
      The head losses are calculated from Eq. (5.18) and are summarized in Table 9.9.
   d. Plot the station losses curve and the modified head-capacity curve of one pump.
      The station losses and modified head-capacity curves are shown in Figure 9.32.
4. Prepare pump combination curves
   a. Calculate the capacity of two, three, and four identical pumps in parallel combination.
      The pump combination curves for two, three, and four pumps are prepared from modified head-capacity curves. The calculations are summarized in Table 9.10.
   b. Plot the parallel combination curves
      The modified head-capacity curve of the given pump and the combination curves of two, three, and four pumps, are shown in Figure 9.32.

TABLE 9.8
System Head Loss Calculations for Force Main from Common Header to Grit Chamber (Example 9.30)

				Assumed Flow 0.253 m^3/s (4,000 gpm)		Assumed Flow 0.507 m^3/s (8,000 gpm)		Assumed Flow 0.761 m^3/s (12,000 gpm)		Assumed Flow 1.013 m^3/s (16,000 gpm)		Assumed Flow 1.267 m^3/s (20,000 gpm)		Assumed flow 1.521 m^3/s (24,000 gpm)	
Item No.	Valves, Fittings, and Specials in 92 cm Diameter Force Main	Numbers Provided	*K* Value	Vel., m/s	Head Loss, m	Vel., m/s	Head Loss, m	Vel., m/s	Head Loss, m	Vel., m/s	Head Loss, m	Vel., m/s	Head Loss, m	Vel., m/s	Head Loss, m
1	90° elbow	2	0.30	0.38	0.004	0.76	0.018	1.14	0.04	1.52	0.071	1.91	0.112	2.29	0.16
2	45° elbow	2	0.20	0.38	0.003	0.76	0.012	1.14	0.026	1.52	0.047	1.91	0.074	2.29	0.107
3	22½° elbow	2	0.15	0.38	0.002	0.76	0.009	1.14	0.02	1.52	0.035	1.91	0.056	2.29	0.08
4	Venturimeter	1	0.14	1.52	0.017	3.05	0.066	4.58	0.15	6.1	0.266	7.62	0.414	9.15	0.597
5	Wye branch	1	1.00	0.38	0.007	0.76	0.03	1.14	0.066	1.52	0.118	1.91	0.186	2.29	0.267
6	Plug valve	1	1.00	0.38	0.007	0.76	0.03	1.14	0.066	1.52	0.118	1.91	0.186	2.29	0.267
7	Outlet (head velocity)	1	1.00	0.38	0.007	0.76	0.03	1.14	0.066	1.52	0.118	1.91	0.186	2.29	0.267
8	Subtotal	—	—	—	0.047	—	0.195	—	0.434	—	0.773	—	1.214	—	1.745
9	Friction head loss in 92 cm force main, 129 m long[a]				0.032		0.117		0.247		0.42		0.641		0.896
10	Total losses in force main $((\Sigma h_m)+h_v+h_f)$				0.079		0.312		0.681		1.193		1.855		2.641
11	Static head at maximum wet-well elevation				9.91		9.91		9.91		9.91		9.91		9.91
12	TDH with respect to maximum wet-well elevation				**9.99**		**10.22**		**10.59**		**11.1**		**11.77**		**12.55**

[a] Friction loss is calculated from Hazen–Williams Eq. (5.18).

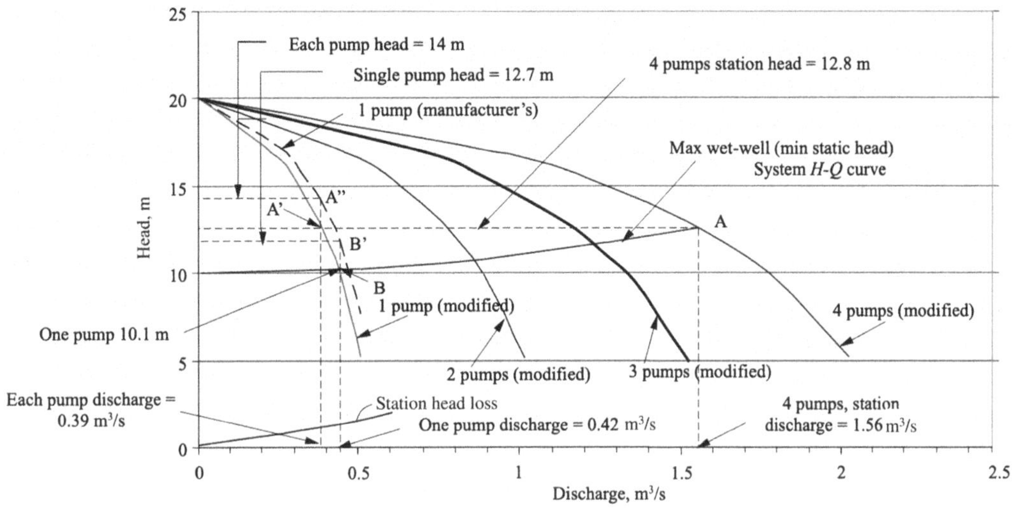

FIGURE 9.32 System head-capacity, pump-modified head capacity, and pumps parallel combination curves showing operating-heads and capacities (Example 9.30).

5. Determine the station head and capacity and individual pumps operating-points at minimum static head.
   a. The station head and capacity are found from the intersection point of system-head capacity, and the head-capacity of four pumps in parallel combination curves.
   b. The head and capacity of single pump when four pumps are operating in parallel combination are found as follows:
      i. Go horizontally from station head-capacity point to intersect the modified head-capacity curve.
      ii. Go vertically to the manufacturer's curve. The intersection point gives the head and capacity of each pump when four pumps are operating in parallel.
   c. The head and capacity when one pump is in operation is found from the intersection point of system head-capacity, and modified head-capacity (one pump), and manufacturer's head-capacity curves.
   d. The operating conditions of pumping station and individual pumps are shown in Figure 9.32.
      The values are summarized in Table 9.11.
6. Determine pump efficiency
   a. The operating efficiency of each pump at minimum station head corresponds to the point *A"*.

   Corresponding to the head and capacity at this point (Table 9.11) (head = 14.0 m and capacity = 0.39 m^3/s), the pump efficiency is read from Figure 9.30. The pump efficiency is 83.4%.

   b. The operating efficiency of one pump at head = 12.7 m and capacity = 0.42 m^3/s corresponds to the point *B'* is 81.7% (Figure 9.30).
7. Determine power input to the pump

   The power input to the pump is water power divided by pump efficiency. These calculations are summarized in Table 9.12.
8. Select motor for pumps

   The next higher standard or available motor over 73.3 kw is 82 kw.

   Provide 82 kW standard motor (110 hp) for each pump.
9. Draw hydraulic and energy grade lines

   The hydraulic and energy grade lines with respect to maximum water surface elevation in the wet well are shown in Figure 9.31c.

TABLE 9.9
Individual Pump Losses for Preparation of Modified Head-Capacity Curve (Example 9.30)

Item No.	Valves, Fittings, and Specials, and Numbers Provided	*K* Value	Assumed flow 0.253 m^3/s (4,000 gpm)		Assumed Flow 0.317 m^3/s (5,000 gpm)		Assumed Flow 0.380 m^3/s (6,000 gpm)		Assumed Flow 0.443 m^3/s (7,000 gpm)		Assumed Flow 0.507 m^3/s (8,000 gpm)	
			Vel., m/s	Head Loss, m	Vel., m/s	Head Loss, m	Vel., m/s	Head Loss, m	Vel., m/s	Head Loss, m	Vel., m/s	Head Loss, m
1	One entrance, suction bell (32 in.) 81 cm	0.04	0.49	0.001	0.615	0.001	0.737	0.001	0.860	0.002	0.984	0.002
2	One 90° elbow (24 in.) 61 cm	0.30	0.87	0.012	1.085	0.018	1.300	0.026	1.516	0.035	1.735	0.046
3	One gate valve (24 in.) 61 cm	0.19	0.87	0.007	1.085	0.011	1.300	0.016	1.516	0.022	1.735	0.029
4	Two reducer/enlarger (14 in.) 35.5 cm	0.25	2.56	0.167	3.203	0.262	3.839	0.376	4.476	0.511	5.122	0.669
5	One check valve (20 in.) 51 cm	2.50	1.24	0.196	1.552	0.307	1.860	0.441	2.168	0.599	2.482	0.785
6	One 90° elbow (20 in.) 51 cm	0.30	1.24	0.024	1.552	0.037	1.860	0.053	2.168	0.072	2.482	0.094
7	One gate valve (20 in.) 51 cm	0.19	1.24	0.015	1.552	0.023	1.860	0.034	2.168	0.046	2.482	0.060
8	One tee (20 in. × 20 in.) 51 cm × 51 cm	1.80	1.24	0.141	1.552	0.221	1.860	0.317	2.168	0.431	2.482	0.565
	Total minor losses	—	—	0.563	—	0.880	—	1.264	—	1.718	—	2.250
	Total friction head loss in connecting pipe[a]			0.060		0.090		0.126		0.168		0.216
	Total head losses			**0.623**		**0.970**		**1.390**		**1.883**		**2.466**

[a] The friction head losses in connecting pipings are calculated for Eq. (5.18). The connecting pipings are 61.0, 51.0, and 35.5 cm in diameter and are, respectively, 2.0, 1.0, and 2.0 m long.

TABLE 9.10
Calculations for Preparation of Modified Pump and Parallel Pump Combination Curves (Example 9.30)

Pump Characteristic Head-Capacity Curve		Individual Pump Head Loss,[b] m	Pump-Modified Head (One Pump), m	Pump Capacity, Parallel Combination		
Capacity, m^3/s	Total Head,[a] m			Two Pumps, m^3/s	Three Pumps, m^3/s	Four Pumps, m^3/s
0.253	17.22	0.62	16.60	0.506[c]	0.759[d]	1.012[e]
0.317	16.00	0.97	15.03	0.634	0.951	1.268
0.380	14.33	1.39	12.94	0.760	1.140	1.520
0.443	11.89	1.88	10.01	0.886	1.329	1.772
0.507	7.62	2.47	5.15	1.014	1.521	2.028

Note: It is customary not to extend the manufacturer's curve. Points beyond manufacturer's curve may be associated with excessive power demand and cavitation. A point beyond the pump capacity of 0.507 m^3/s (8,000 gpm) is outside the manufacturer's curve for 54 cm (21¼ in.) diameter impeller.

[a] Values taken from Figure 9.30 for 54 cm (21¼) in diameter impeller of the selected pump.
[b] Values taken from Table 9.9.
[c] $2\times0.253=0.506$.
[d] $3\times0.253=0.759$.
[e] $4\times0.253=1.012$.

TABLE 9.11
Summary of Pump Operating-Heads and Capacities at Max Wet-Well Elevation (Min. Static Head) (Example 9.30)

Condition and Description	Reference Point in Figure 9.32	Operating Head m	Operating Head ft	Operating Capacity m^3/s	Operating Capacity gpm	Remarks
Intersection of system curve and head-capacity curve of four pumps in parallel combination	*A*	12.8	42.0	1.56	24,630	Station head and capacity with four pumps in parallel operation
Project horizontally from point *A* to the individual pump's modified curve	*A′*	12.8	42.0	0.39	6,160	Operating condition of each pump when four pumps are operating in parallel
Project vertically from point *A′* to the manufacturer's head-capacity curve of individual pump.	*A″*	14.0	45.9	0.39	6,160	Line *A′ A″* represents station losses (1.2 m)[a]. Point *A″* corresponds to the manufacturer's characteristic curve
Intersection of system curve and modified head-capacity curve of one pump in operation	*B*	10.1	33.1	0.42	6,630	Operating condition of single pump
Project vertically from point *B* to pump characteristic curve	*B′*	12.7	41.7	0.42	6,630	Line *BB′* represents station losses (2.6 m)[a]. Point *B* corresponds to the manufacturer's characteristic curves

[a] Station losses are the head losses in the suction and discharge piping of each pump from suction bell up to the common header (Table 9.9).

TABLE 9.12
Summary Calculations of Pump Efficiency, Power Output of the Pump, Power Input to the Pump, and Motor Power (Example 9.32)

Operating Conditions for a Single Pump	Maximum Wet-Well Level for Each Pump (Minimum Static Head, Four Pumps Operating)	Maximum Wet-Well Level (Minimum Static Head, Single Pump Operating)
Pump operating-head	14.0 m (45.9 ft) (point *A″*)	12.7 m (41.7 ft) (point *B″*)
Pump operating-discharge	0.39 m^3/s (6,160 gpm) (point *A″*)	0.42 m^3/s (6,630 gpm) (point *B″*)
Pump operating-efficiency	83.4% (Figure 9.30)	81.7% (Figure 9.30)
Power output of pump, water power, Eq. (9.11)	53.7 kW (71.5 hp)[a]	52.7 kW (70.4 hp)[a]
Power input to the pump, brake power, Eq. (9.12)	64.4 kW (85.7 hp)[b]	64.0 kW (85.5 hp)[b]
Motor power, Eq. (9.13)	73.2 kW (97.4 hp)[c]	73.3 kW (97.9 hp)[c]

[a] $14.0 \times 0.39\ \text{m}^3/\text{s} \times 9.81\ \text{kN/m}^3 = 53.7\ \text{kW}; \dfrac{45.9\ \text{ft} \times 6{,}160\ \text{gpm} \times 62.4\ \text{lb/ft}^3}{7.48\ \text{gal/ft}^3 \times 60\ \text{s/min} \times 550\ \text{ft.lb/hp.s}} = 71.5\ \text{hp}.$

[b] 53.7 kW/0.834 = 64.4 kW (71.5 hp/0.834 = 85.7 hp).

[c] 64.4 kW/0.88 = 73.2 kW (85.7 hp/0.88 = 97.4 hp). It is assumed that the full-load motor efficiency is 88%.

9.5.11 Sludge Pumping

Pumps commonly used for sludge pumping are non-clog centrifugal, peripheral (torque flow or vortex), screw, and plunger type. Sludge is a non-Newtonian fluid and flow is turbulent. Accurate friction loss calculations for long-distance pumping are based on *thixotropic behavior, rheological concept*, and other flow properties of sludge.[11,12] For sludge pumping within a plant the head loss calculation is based on the Hazen–Williams equation (Eq. 5.18). The *C* value is decreased for different types of sludge. The following rule of thumb is applied:

1. Waste activated sludge (total solids up to 1%) $C=90$
2. Primary sludge (total solids up to 4.5%) $C=55$
3. Thickened or digested sludge (total solids up to 8%) $C=35$

The following operational considerations are given:

1. Provide sludge piping larger than 15 cm (6 in.) diameter
2. Operate pump on time cycle to achieve high velocity during pumping cycle.
3. Provide liberal number of cleanouts and hose gate for cleaning the stoppages.

Detailed discussion on sludge characteristics and pumping may be found in Refs. 2–5.

Example 9.31: TDH of Sludge Pump

Primary sludge is pumped from a primary sedimentation basin to a gravity thickener. The sludge pipe is 15 cm in diameter. The static head is 3 m, and pipe length is 40 m. Assume that minor losses are 10% of friction losses. The sludge pumping is 24 cycles/day. During sludge pumping cycle, the sludge flow is 0.016 m^3/s. The Hazen–William coefficient *C* is 50. Calculate TDH.

Solution

1. Calculate velocity in sludge pipe during pumping cycle

$$V = \frac{0.016\ \text{m}^3/\text{s}}{\frac{\pi}{4} \times (0.15)^2\ \text{m}^2} = 0.91\ \text{m/s}$$

2. Calculate friction head loss

$$h_f\,(\text{Eq. (5.18)}) = 6.82\left(\frac{0.91}{50}\right)^{1.85} \times \frac{40\ \text{m}}{(0.15)^{1.167}}$$

$$h_f = 6.82 \times 0.0006041 \times \frac{40}{0.1093} = 1.50\ \text{m}$$

3. Calculate TDH

$$\text{Minor losses,}\quad \Sigma h_m = 1.50\ \text{m} \times 0.1 = 0.15\ \text{m}$$

$$\text{Velocity head} = \frac{(0.91)^2}{2 \times 9.81\ \text{m/s}^2} = 0.04$$

$$\text{Static head} = 3.00\ \text{m}$$

$$\text{TDH} = 3.00\ \text{m} + 1.50\ \text{m} + 0.15\ \text{m} + 0.04\ \text{m} = 4.69\ \text{m}$$

Example 9.32: Cyclic Pump Discharge

Sludge is pumped from a primary sedimentation basin to sludge thickener. On average, 6800 kg of dry solids are removed daily. The solids content in the sludge is 4%, and the specific gravity is 1.02. The diameter of the sludge pipe is 15 cm, and the velocity in the pipe must not be less than 0.8 m/s. The sludge pump operates every hour. Calculate the pumping and idle times per cycle.

Solution

1. Calculate sludge volume pumped per cycle
   There are 24 cycles/day

$$\text{Sludge volume/day} = \frac{6{,}800\ \text{kg/day}}{0.04 \times 1{,}020\ \text{kg/m}^3} = 167\ \text{m}^3/\text{day}$$

2. Calculate pumping and idle times

$$\text{Pumping rate} = \frac{\pi}{4} \times (0.15\ \text{m})^2 \times 0.8\ \text{m/s} \times 60\ \frac{\text{s}}{\text{min}} = 0.85\ \text{m}^3/\text{min}$$

$$\text{Pumping time per cycle} = \frac{6.96\ \text{m}^3/\text{cycle}}{0.85\ \text{m}^3/\text{min}} = 8.2\ \text{min/cycle}$$

$$\text{Idle time per cycle} = 60\ \frac{\text{min}}{\text{cycle}} - 8.2\ \frac{\text{min}}{\text{cycle}} = 51.8\ \text{min/cycle}$$

The sludge pump runs for 8.2 min/cycle and pumps sludge at a rate of 0.85 m^3/min. The idle time per cycle is 51.8 min. Total cycle time is 60 min.

9.5.12 Pump Drives and Sizing of Pumping Stations

Pump drives convert electrical or thermal energy into the rotating kinetic energy to drive the pump. The *internal combustion engines*, *turbines* (gas and steam), and *electric motors* are common pump drives. The drive selection is based on (1) cost of energy, (2) availability and reliability of the energy resource, and (3) characteristics of the load demand, and power output of the drive.

Electric motors are the most widely used pump drives. They offer (1) high efficiency of energy conversion, (2) simple and reliable, and (3) widespread availability of reliable electric power. Electric motors are *primarily alternating current (a-c)*, although some *direct current (d-c)* motors are also used. The a-c motors in common use are *induction* and *synchronous*. The electric motors work on either constant speed or variable speed. The constant-speed motors may be either the *squirrel cage*, *synchronous*, or *wound rotor induction* type. Two most common variable-speed motors are *eddy current type* and *variable frequency controller*. The typical electric motor efficiencies based on their power ratings are summarized in Table 9.13.

Pump stations: Pump stations are the structures that house pumps, pipings, equipment, and chemicals. The pump stations are wet pit or dry pit. The wet-pit pump stations have one chamber, the suction bell, pump, pipings, and motor are all located in the wet well. Submersible and vertical pumps require wet-pit structure. The dry-pit stations utilize wet-well and dry-well compartments. The suction bell and pipings are located in the wet well, while the pumps, motors, and other controls are located in the dry pit that is separated by water and gas-tight wall. Separate entrances are provided to each well. The wet-pit, dry-pit, and combination of wet- and dry-pit pump stations are shown in Figure 9.2, and are described in Section 9.3.

Wet-well dimensions: Wet-wells contain toxic and flammable gases. The floor should be horizontal from the wall to a point 0.3–0.4 m (12–18 in.) beyond the outermost edges of the suction bell. The remaining section should be sloped upward to the opposite wall at a slope of 1:1 or steeper. The suction bell should be turned down to reduce vortex formation. The vertical height of the bell from the floor should not be more than $D/2$ and not less than $D/3$ (D=diameter of the suction bell). The velocity at the entrance must be less than 3 m/s. The minimum center-to-center spacing between the pumps and from the walls to the center of the suction bell, height above the floor and the submergence, and distance to the screen or trash rack are specified in the Hydraulic Institute Standards. Some of these recommendations are provided in Figurc 9.33.[1,2,4,9]

Wet-well volume and retention time: Many regulatory agencies limit maximum retention time at average design flow. This criterion is to minimize septic conditions and odor problems. The wet-well volume should be large enough to provide the required storage to allow sufficient time between successive start or stop cycles. The time between starts is a function of the pumping rate and the flow entering the station. Equation (9.15) is used to calculate the minimum time of one pumping cycle:

TABLE 9.13
Motor Efficiencies Based on Power Rating

Motor Power Rating (kW)	Typical Efficiency (%)
1–5	70–80
5–7.5	80–85
7.5–20.0	85–88
20 and above	88–92

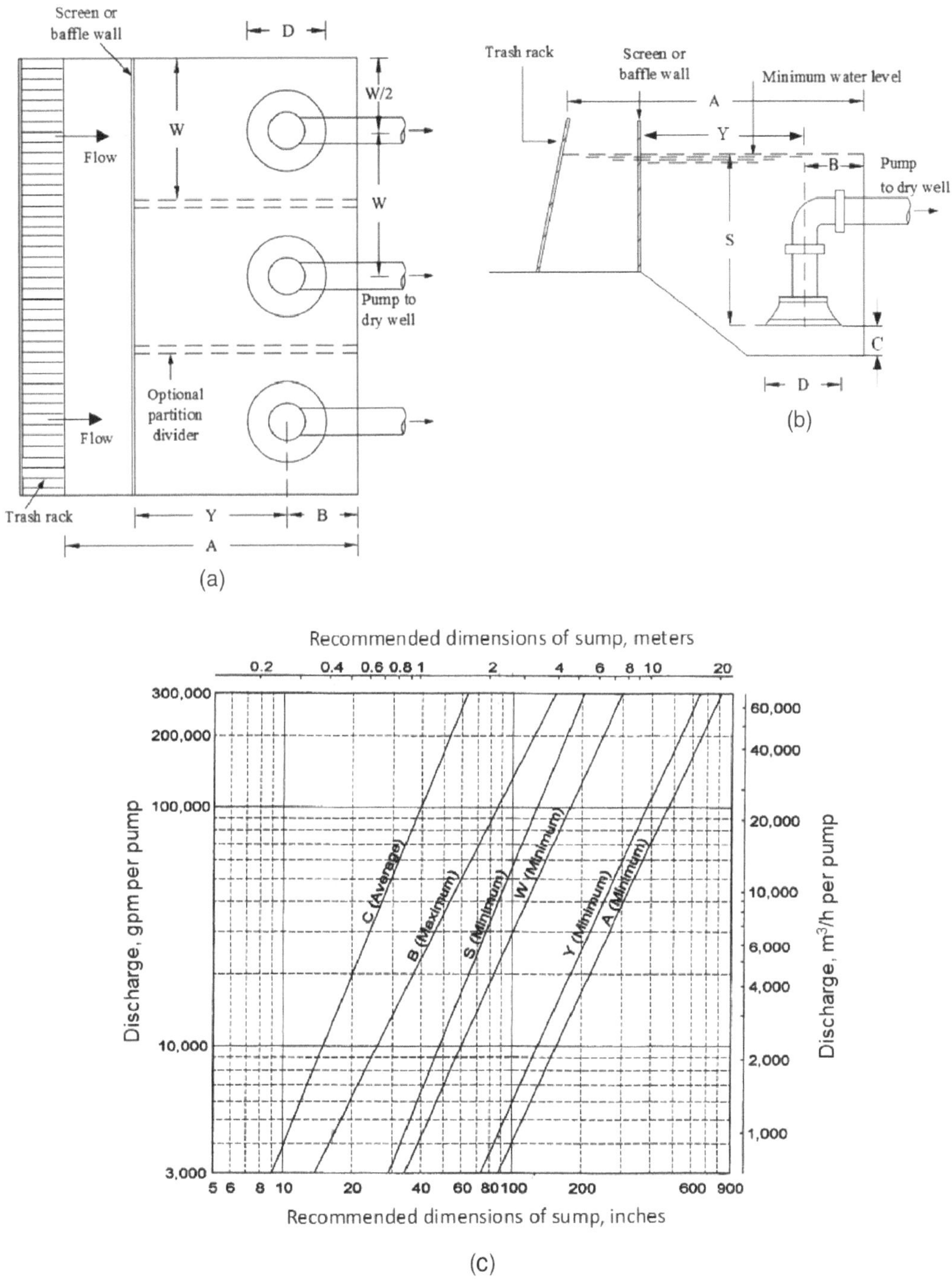

FIGURE 9.33 Wet-well design criteria: (a) wet-well variables-plan view, (b) wet-well variables-section view of wet-well pump with turned-down suction bell mouth elbow, and (c) sump variables and dimensions for different flows.

$$t = \frac{4V}{Q} \tag{9.15}$$

where

t=pumping cycle time, min

V=volume between maximum and minimum wet-well elevations, m^3

Q=pumping rate of one pump, m^3/min

Many design standards specify maximum pumping cycles per hour. If the number of cycles calculated from t exceeds the specified number of cycles per hour, then the wet-well volume should be increased.[2,3]

Baffle wall and flow distribution: Baffle wall is provided at the entrance of wet well to distribute the flow uniformly to all pumps. These walls have large number of orifices. For good flow distribution, a horizontal jet should be developed at the orifice, causing an appreciable velocity head.

Ventilation in dry well: The dry-well houses pumps, controls, and pipings. Ventilation should be provided for all dry wells below ground surface, and in all wet-wells housing equipment. As a general rule, ventilation equipment should provide at least six air changes per hour, or meet the minimum standard of the applicable building code. Pumping equipment generates a great deal of heat during operation. This heat is typically generated from the lost efficiencies of the pump drivers. If this heat is not removed from the pump station, high temperatures may damage the equipment. The heat load generated by electric motors is calculated from Eq. (9.16)[13,14]

$$H = PUL \times \left(\frac{1-E}{E}\right) \tag{9.16}$$

where

H=heat load, kW (Btu/h)

E=motor efficiency in decimal fraction

L=motor load factor, 1.0 or decimal fraction. The motor load factor is the fraction of the load being delivered under the estimated cooling conditions.

P=motor power rating, kW. The motor power rating is the power output of the motor.

U=motor use factor. It is 1.0 or a decimal fraction. The motor use factor may be applied when motor use is intermittent, with significant non-use during all hours of operation. For conventional application, its value is 1.0.

The pump stations are typically cooled by forced air ventilation. This method requires forcing a sufficient quantity of outside air through the pump station to remove the heat generated from the pumping operation. Equation (9.17) is typically used to estimate the heat removed by air flow.[13,14]

$$Q = \frac{H}{C_p \times \rho \times (T_i - T_o)} \tag{9.17}$$

where

Q=air flow rate required to remove heat, m^3/s (ft^3/s)

H=quantity of heat removed, kW (Btu/h)

C_p=specific heat of air; the specific heat of air at 77°C is 1.02 kW.s/kg.°C or 1.02 kJ/kg. °K (0.245 Btu/lb.°F)

ρ=density of air, 1.2 kg/m^3 (0.075 lb/ft^3)

T_i=temperature of indoor air, °C (°F)

T_o=maximum outdoor temperature, °C (°F)

Example 9.33: Pump and Piping, and Pump Station Design

A wet well has five pumps (including one standby pump) arranged in parallel. The wet well is 15.0 ft×50 ft. There is a baffle wall 5.0 ft from the free fall into the wet well. At full station capacity (four pumps operating), each pump delivers 6,160 gpm at a total head of 45.9 ft. The diameter of the suction bell is 32 in. Determine the wet-well dimensions and pump layout. The depth of water in the wet well is 8.0 ft. The dry well is 25.0 ft × 50 ft. Give the dimensions of pump station, pumps arrangement, common header, and valves and fittings.

Note: See the design procedure of four pumps in Example 9.30

Solution

1. Determine the minimum dimensions from the recommendations in Figure 9.33. At the single pump discharge of 6,160 gpm, the minimum dimensions of components *A*, *B*, *C*, *S*, *W*, and *Y* are obtained from Figure 9.33c. The recommended minimum and actual wet-well dimensions are summarized in Table 9.14. Pumps arrangement, common header, and valves are shown in Figure 9.34.
2. Show the dimensions of wet well and design details.

   The dimensions of wet well, pumps arrangement, and suction bell location are shown in Figure 9.34.

Example 9.34: Pump Cycle Time

The wet well of a pump station is 10 m×3 m. The pump station contains three pumps, including one standby pump. The pumping rate of one pump is 12 m³/min. The water depth between the maximum and minimum water surface elevations in the wet well is 1.8 m. The specified maximum pumping cycle is 4 cycles/h. Check if the specified pumping cycle condition is met.

Solution

1. Calculate the effective volume of the wet well

$$V = 10\text{ m} \times 3\text{ m} \times 1.8\text{ m} = 54\text{ m}^3$$

2. Calculate pump cycle time

$$t = \frac{4 \times 54\text{ m}^3}{12\text{ m}^3/\text{min}} = 18\text{ min}$$

$$\text{Number of cycles per hour} = \frac{60\text{ min/h}}{18\text{ min/cycle}} = 3.3\text{ cycle/h}$$

3. Compare result

   The specified maximum pumping cycle is four per hour.

   3.3 cycles/h meets the specified pumping cycle per hour requirement.

Example 9.35: Air Circulation Rate in Dry Well

In a dry well, proper air circulation is necessary to dissipate the heat generated. A dry pit has four identical pumps. The pump operation is controlled by liquid level elevation in the

TABLE 9.14
Minimum and Actual Dimensions of Different Components of Suction Well (Example 9.33)

Components of Suction Well	Minimum Dimensions Obtained from Figure 9.33c, in.	Actual Dimensions Provided in the Design (Figure 9.34), ft
A[a]	120	10.0
B	19	1.7
C	13	1.3
D	32	2.7
S	39	3.3
W	48	5.0
$W/2$	24	2.4
Y	100	8.3

[a] Dimension A is normally the distance from the trash rack to the common wall between the wet well and dry pit (Figure 9.33b). In this design, the trash rack is not in the wet well, and screened wastewater flow is toward the suction well. Therefore, it is assumed that the dimension A is from the common wall between wet well and dry well to the baffle wall, or $A = B + Y$.

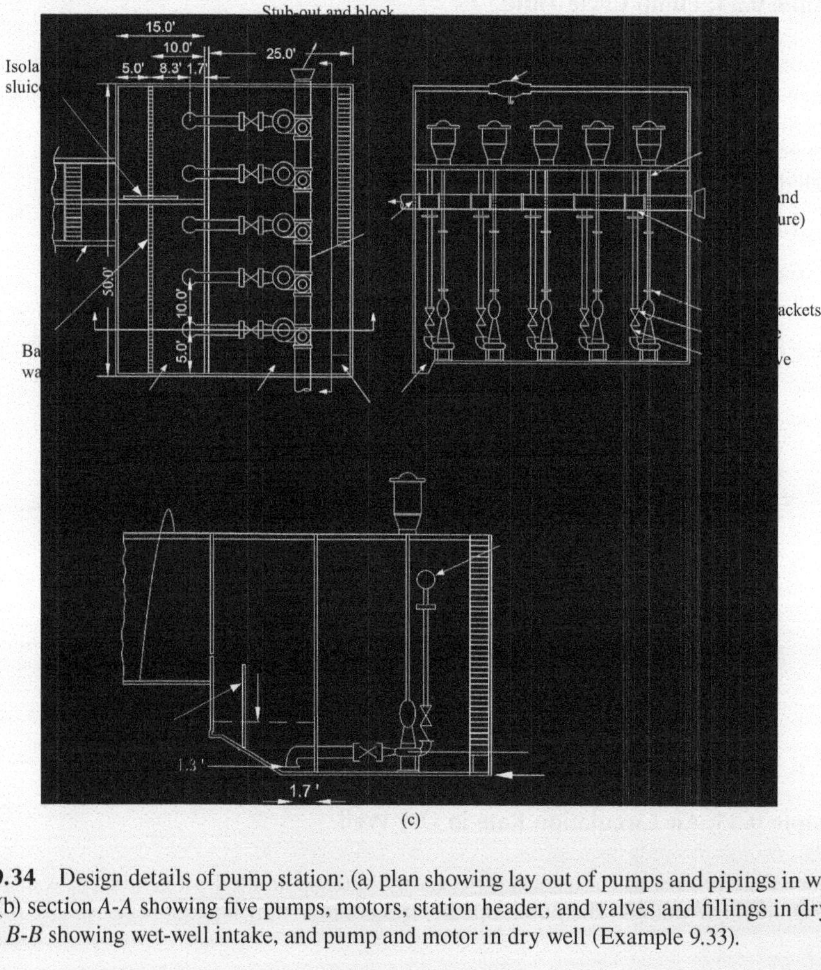

(c)

FIGURE 9.34 Design details of pump station: (a) plan showing lay out of pumps and pipings in wet well and dry well; (b) section *A-A* showing five pumps, motors, station header, and valves and fillings in dry well; and (c) section *B-B* showing wet-well intake, and pump and motor in dry well (Example 9.33).

wet well. At maximum wet-well elevation, all four pumps operate. Calculate the minimum air circulation rate and the number of air changes per hour. The air space in the dry well is 650 m^3. Each pump motor is rated at 82 kW, and the full-load efficiency and power factor are 88% and 75%. The maximum indoor and outdoor temperatures are 48°C and 37.8°C, respectively.

Solution

1. Identify equation to calculate the air circulation rate
   The air circulation rate is calculated from Eqs. (9.16) and (9.17). The following variables are substituted in these equations

$$P = 82\text{ kW},\quad E = 0.88,\quad U = 0.75,\quad L = 1.0,\quad C_p = 1.02\text{ kW.s/kg.°C}$$

$$\rho = 1.2\text{ kg/m}^3,\quad T_i = 48\text{°C},\quad T_o = 37.8\text{°C}$$

2. Calculate heat load

$$H = PUL \times \left(\frac{1-E}{E}\right)$$

$$= 82\text{ kW} \times 0.75 \times 1.0 \times \left(\frac{1-0.88}{0.88}\right)$$

$$= 8.4\text{ kW}$$

3. Calculate air circulation per pump

$$Q = \frac{8.4\text{ kW}}{1.02\text{ kW.s/kg°C} \times 1.2\text{ kg/m}^3 \times (48-37.8)\text{°C}}$$

$$= 0.67\text{ m}^3\text{/s/pump}$$

4. Calculate air circulation rate in the dry pit
   Maximum number of pumps that can operate at any time = 4
   Provide an intermittent air circulation system with a minimum air exhaust rate of 0.67 m^3/s per pump × 4 pump = 2.7 m^3/s

5. Calculate the number of air changes per hour

$$\text{Number or air changes} = \frac{2.7\text{ m}^3\text{/s }(60 \times 60)\text{ s/h}}{650\text{ m}^3\text{/change}} = 15\text{ changes/h}$$

   An intermittent exhaust fan operation is advisable. Use fan operation with 45-min on and 15-min off timer.

$$\text{Air supply required for 15 changes/h} = 2.7\text{ m}^3\text{/s} \times \frac{60\text{ min}}{45\text{ min}} = 3.6\text{ m}^3\text{/s}$$

Example 9.36: Design of Baffle Wall

A wet well is 10 m long and 5 m wide. The maximum and minimum liquid levels in the wet well are 5.0 and 2.5 m, respectively. The screened wastewater drops freely from the screening facility into the wet well. The cross-section of the wet well is shown in Figure 9.35a.

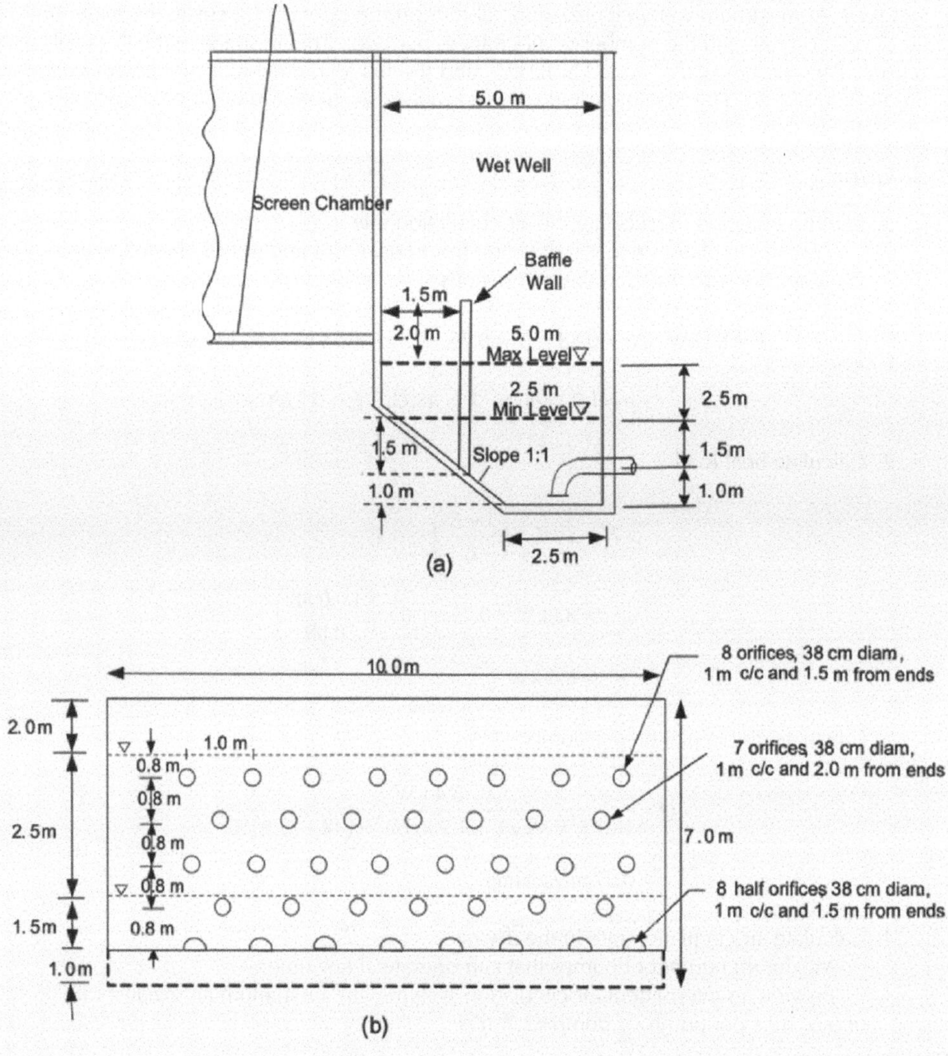

FIGURE 9.35 Baffle wall design: (a) cross-section of wet wall and (b) baffle wall details (Example 9.36).

The maximum pumping rate is 1.50 m^3/s when all four pumps are operating. A baffle wall is located 1.5 m from the free-falling wall. Design the baffle wall for proper flow distribution to all pumps.

Solution

For uniform flow distribution, the most important design consideration for a baffle wall is the size of the orifice openings that will create a head loss at least equal to the upstream kinetic energy. The thickness of the baffle wall should be at least equal to the diameter of the orifice to fully develop a horizontal jet. Also, large number of orifices should be provided.

1. Determine the dimensions of the baffle wall and area for location of orifices

   The height of the baffle wall with openings is from the maximum level in the wet well to the bottom of the baffle wall that is supported over 1:1 slope. The length of the baffle wall covers the entire length of the wet wall. The baffle dimensions are shown in Figure 9.35b.

The baffle wall area with orifices in the wet wall = 10.0 m × 4.0 m

2. Determine the velocity through the orifice

Assume a velocity head of 0.02 m across the baffle wall. Higher velocity head across the baffle wall will ensure proper flow distribution.

The velocity through the orifice is calculated from Eq. (7.8). This equation, in modified form, is expressed by Eq. (9.18)

$$V = C_d \times \sqrt{2g \times \Delta H} \quad (9.18)$$

where

V = velocity through the orifice, m/s (ft/s)

C_d = coefficient of discharge. Commonly used value is 0.60

ΔH = the of head loss across the orifice, m (ft)

$$V = 0.6 \times \sqrt{2 \times 9.81 \times 0.02} = 0.38 \text{ m/s}$$

3. Calculate the total area of all orifices

Total area of all orifices = (1.50 m^3/s)/(0.38 m/s) = 3.95 m^2

4. Calculate the total number of orifices

Provide 38 cm (15 in.) diameter openings in the baffle wall

Total number of orifices = (3.95 m^2/((π/4) × 0.38 m^2) = 34.8 or 34 orifices.

5. Arrange the orifices in the baffle wall

Provide 30 circular 38-cm-diameter, and eight half-circle orifice openings in the baffle wall. The design details are shown in Figure 9.35b.

Example 9.37: Liquid Level Bubbler Control

A liquid level bubbler control is provided to indicate high and low wet-well levels. It also controls the adjustable speed derive to stop or start the pump within the wet-well parameters. In a wet well, four identical pumps are arranged in parallel operation. The maximum and minimum wet-well elevations above the wet-well floor are 5.00 and 3.00 m. The centerline elevation of the pump volute is 2.50 m. Develop the following information (1) the major components of a bubble control system, and (2) elevations above the floor of the wet well to control the pump operation from the max. to min. wet-well elevations. The operating differential levels of the bubbler tube are 0.25 m between each change in pump operating-conditions.

Solution

1. Identify the major components of the bubbler tube control system

The major components of the bubbler control system include (1) compressor and air tank with gauge and pressure relief values of specified rated capacity; (2) drain valve, bleed valve, pressure maintenance switch, and pressure regulator gauge; (3) air filter with drain, air flow regulator of suitable range, level gauge, electronic pressure transducer, and needle valve; (4) power on-off switch, power-on indicating light, high- and low-level alarms, and pneumatic piping and fittings to connect to the bubbler tube; and (5) digital output of pressure signal from the bubbler tube that is directly proportional to the wet-well level above the bottom of the bubbler tube. The simplified control loop diagram of bubbler system is given in Figure 9.36.

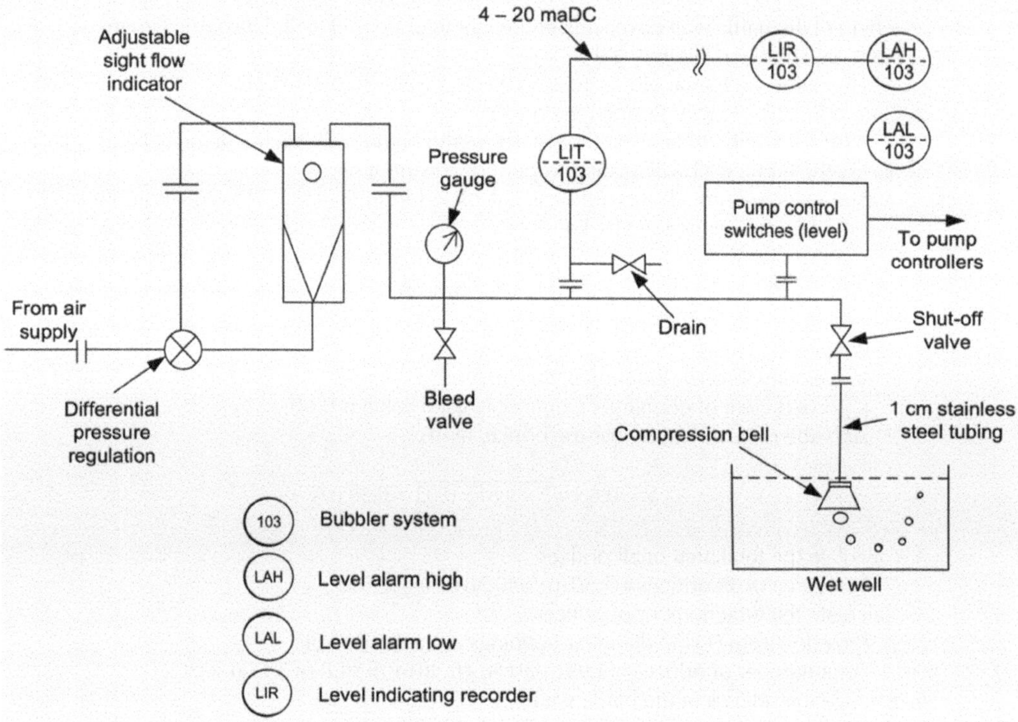

FIGURE 9.36 Simplified control loop diagram of bubble control system (Example 9.37).

2. Identify the pump operating-elevations

   The pump operating-elevations from maximum to minimum wet-well elevations are identified on the bubbler tube. The operating-conditions of the pumping stations are summarized in Table 9.15.

TABLE 9.15
Operating Conditions of the Pumping Station (Example 9.37)

Condition	Elev. above Floor of the Wet Well, m	Third Lag Pump			Second Lag Pump			First Lag Pump			Lead Pump		
		Max Speed[a]	Min Speed[b]	Stop	Max Speed	Min Speed	Stop	Max Speed	Min Speed	Stop	Max Speed	Min Speed	Stop
End of bubbler tube	5.50	×			×			×			×		
High-level alarm	5.25	×			×			×			×		
High level in the wet well	5.00	×			×			×			×		
Third lag pump minimum speed	4.75		×		×			×			×		
Third lag pump stopped	4.50			×	×			×			×		
Second lag pump minimum speed	4.25			×		×		×			×		
Second lag pump stopped	4.00			×			×	×			×		
First lag pump minimum speed	3.75			×			×		×		×		
First lag pump stopped	3.50			×			×			×	×		
Lead pump minimum speed	3.25			×			×			×		×	
Minimum level in the wet well	3.00			×			×			×		×	
Low-level alarm and shut down	2.75			×			×			×			×
End of bubbler tube	2.50												

Note: The operating elevations to start the pumps are generally higher than those to stop the pumps given in this table. The control system starts from minimum to maximum wet-well elevations.

[a] Maximum speed (rpm).

[b] Variable-speed drive pumps generally have minimum speed at approximately half the maximum speed.

9.6 REVIEW PROBLEMS

9.1. The pressure gauge reading at the pump discharge nozzle is 150 kPa (kN/m^2). Convert the pressure into (1) m of water, (2) lb$_f$/ft^2, (3) ft of water, (4) atm, and (5) inches of mercury.

9.2. The liquid elevations in wet well are 147.48 and 158.23 m, respectively. The elevation of centerline of the pump impeller is 150.00 m. Determine (1) static suction head, (2) static discharge head, and (3) total static head.

9.3. A submersible pump is used for pumping wastewater at a treatment plant. The diameter of the suction bell is 40 cm, and $K = 0.05$. Calculate the components of head losses and total head loss in the force main. The diameter and length of force main are 25 cm and 150 m, respectively. There is one enlarger from 15 to 25 cm $K = 0.25$, one check valve $K = 2.5$, one plug valve $K = 1.0$, two gate valves $K = 0.19$, five 45° elbows $K = 0.2$, and two 90° elbows $K = 0.3$. The pump discharge is 0.10 m^3/s and Darcy–Weisbach coefficient of roughness $f = 0.02$.

9.4. A horizontal force main is carrying a flow of 0.10 m^3/s. At a point A, the diameter of the force main is 25 cm, and the pressure gauge reading is 200 kN/m^2. At a point B 75 m from A, there is a gate valve ($K = 2.5$). At a point *C* 110 m from *B*, the pipe diameter is reduced to 20 cm ($K = 0.25$). A pressure gauge is located at a point *D*, 170 m from point *C*. What will be the pressure gauge reading in kN/m^2. Draw the hydraulic and energy grade lines. Assume $C = 110$.

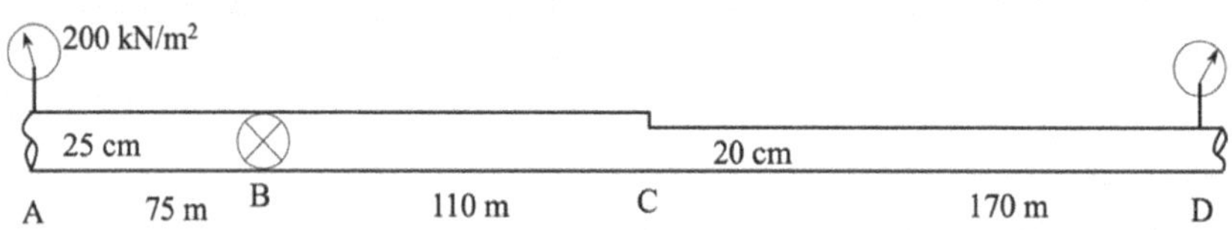

9.5. A pump discharges 0.75 m^3/s of wastewater from a wet well. The diameter of the discharge pipe is 40 cm, and the reading on the discharge gauge located at the centerline of the pump is 140 kN/m^2. The diameter of the suction pipe is 50 cm, and the reading on the suction gauge located 0.8 m below the centerline of the pump is 12 kN/m^2. Determine the head delivered by the pump. Use Bernoulli's equation.

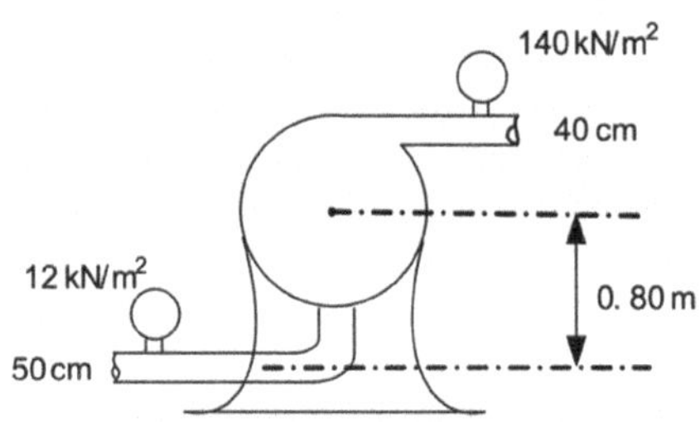

9.6. A pump is delivering 1,850 gpm at a TDH of 60 ft. The pump efficiency is 85% and motor efficiency is 92%. Calculate the wire power or electrical energy.

9.7. A pumping station has a wet well to pump the screened wastewater. The pumping starts and stops when the high and low levels in the wet well are reached. At a high wet-well level, the pumping rate and TDH are 0.094 m^3/s and 22 m. At a low wet-well level, the pumping rate and TDH are 0.072 m^3/s and 25 m, respectively. The pump speed is 1,250 rpm. Calculate the specific speed under both operating-conditions.

9.8. A model pump has an impeller diameter of 15 cm. The test results showed that the model pump developed a water power of 0.195 kW at a speed of 900 rpm and under a total head of 80 cm. A geometrically similar pump of 48 cm impeller diameter is to operate at the same efficiency at a head of 18 m. Determine the expected speed and power output of this pump. Also, compare the specific speeds of both pumps.

9.9. A centrifugal pump was tested at an impeller speed of 1,250 rpm. The pump gave the following results: $Q_1 = 3{,}250$ gpm, $H_1 = 145$ ft, and $P_1 = 190$ hp. Calculate capacity, head, and power if (1) impeller speed is changed from 1,250 to 1,000 rpm, and (2) both impeller speed and diameter are changed from 1,250 to 1,000 rpm, and 18 to 15 in., respectively.

9.10. A manufacturer's pumping test was conducted to determine the efficiency of a pump. The pump delivered 0.25 m^3/s at a TDH of 12 m. The power input to the pump was 57 kW. Determine the pump efficiency.

9.11. A centrifugal pump has the following head-capacity test data at an impeller speed of 1,100 rpm. Develop the head-capacity curves at operating-speeds of 800 and 725 rpm.

Capacity, m^3/s	0	0.1	0.2	0.3	0.4	0.5
Head, m	50.0	49.0	46.6	44.4	40.5	33.0

9.12. The head-capacity and efficiency data of two different pumps are given below. Develop the characteristic curves of these two pumps arranged in a parallel combination.

	Pump *A*		Pump *B*	
Head, ft	**Capacity, gpm**	**Efficiency, %**	**Capacity, gpm**	**Efficiency, %**
85	0	70	0	83
80	825	85	635	80
70	1,900	72	1,150	60
60	2,400	52	1,500	50
50	3,000	40	2,100	37
40	3,450	35	2,500	23

9.13. The head-capacity and efficiency data of two pumps are given below. Develop the head capacity, efficiency, and power input curves of two pumps arranged in series combination.

	Pump *A*		Pump *B*	
Capacity, m^3/s	**Head, m**	**Efficiency, %**	**Head, m**	**Efficiency, %**
0	45	0	51	0
0.02	40	60	50	32
0.04	35	84	44	63
0.06	21	63	38	86
0.08	10	32	31	64

9.14. A pump is used to transfer water from reservoir *A* to *B*. The pump head, capacity, and efficiency data are tabulated below. The pipe system is shown in the following Figure 9.37. The valves, fittings, and specials, and their *K* values are listed below. The diameter and length of piping are 30 in. and 3,500 ft; $C = 110$. Develop the system head-capacity curve and determine the pump operating-head, capacity, efficiency, and power input.

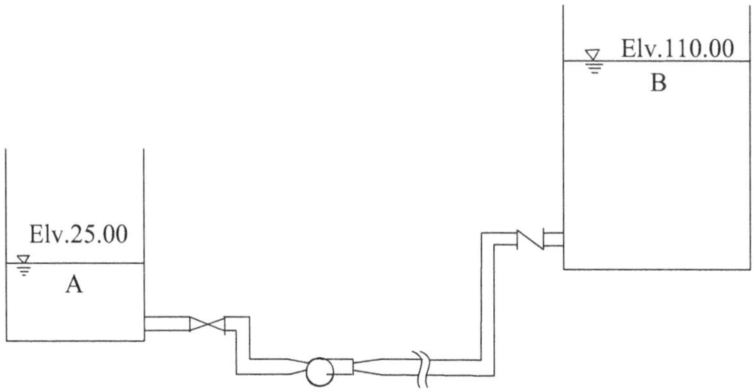

FIGURE 9.37 Piping and water elevations in reservoirs (Review Problem 9.14).

Q, gpm	0	1,500	3,000	4,750	6,300	7,900	9,500
Head, ft	141	135	129	122	102	85	70
Efficiency, %		41	65	77	84	80	65

One entrance $K=0.3$
Four 90° elbow $K=0.4$
One gate valve $K=0.15$
One check valve $K=2.4$
One exit $K=1.0$

9.15. The wet well of a pumping station has a bubbler control system to start and stop the pump operation. The pumping starts at high elevation and stops as soon as the minimum wet-well elevation is reached. The pump station and piping arrangements are given below (Figure 9.38). The manufacturer's characteristic curves data for the centrifugal

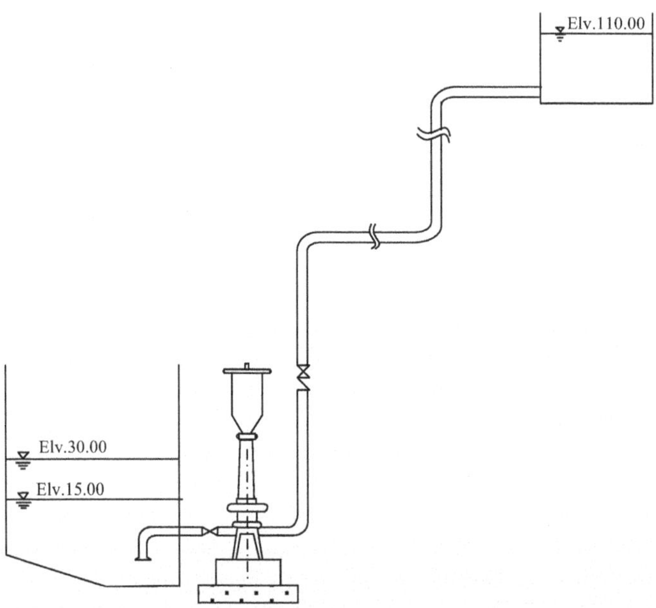

FIGURE 9.38 Pump station and piping arrangement (Review Problem 9.15).

pump are given in the Review Problem 9.14. Draw the system head-capacity curves and obtain the operating-points and pump efficiency for two extreme wet-well conditions. The valves, fittings, and specials include:

One suction bell $K = 0.20$
Five 90° elbows $K = 0.30$
One check valve $K = 2.50$
Two gate valves $K = 0.15$

The suction and discharge piping are 30 in. diameter, and the total length is 3,500 ft. $C = 110$.

9.16. The manufacturer's head-capacity and efficiency curves of a pump are given below. Calculate the station losses, and draw the modified head-capacity curve. The piping arrangement from the suction well to the common header is shown below. The suction piping is 45 cm in diameter, and total straight length is 8 m.

Q, gpm	0	0.05	0.10	0.15	0.20	0.25	0.30
Head, ft	20	19.4	18.5	17.0	16.0	14.2	10.0
Efficiency, %		41	65	77	84	80	65

The total number of valves, fittings, and specials in the suction piping and their minor head loss coefficients K are provided below:

One suction bell $K = 0.10$
Two 90° elbows $K = 0.30$
Two Gate valves $K = 0.25$
One Check valve $K = 2.50$
One Tee $K = 1.95$

The longitudinal section of pumping station from the suction well to the common header is shown in Figure 9.39.

The pumping station is designed to pump screened wastewater from the wet well to a grit chamber at a maximum static head of 14 m. Three identical pumps in parallel arrangements meet the maximum station capacity. The common header is 60 cm in diameter and 120 m long. There are two 45° ($K = 0.20$) and four 90° elbows ($K = 0.30$), one gate valve ($K = 0.25$), one check valve ($K = 2.50$), and one Venturimeter ($K = 0.38$). C for suction and discharge pipings is 100. Determine the following: (1) maximum station

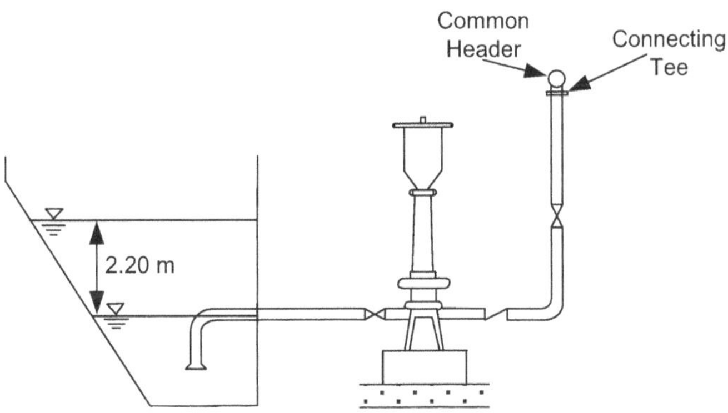

FIGURE 9.39 The longitudinal section of pumping station from suction well to common header (Review Problem 9.16).

head and capacity; (2) head and capacity of each pump, and (3) pump efficiency and power input to each pump.

9.17. This problem is a continuation of Example 9.30. Determine the following for maximum static head condition (minimum wet-well elevation): (1) maximum station capacity and operating-head when all four pumps are operating, (2) operating-head, capacity, and efficiency and power input to each pump in (1), and (3) pump operating-head, capacity, efficiency and power input if only one pump is operating. Draw the hydraulic and energy grade lines with respect to minimum water surface elevation in the wet well.

Hint: All calculations have been performed in Tables 9.8, 9.9, and 9.11. Develop system head-capacity curve for maximum static head of 11.3 m and develop a figure similar to Figure 9.32 to obtain the operating-head and capacities for different conditions. Organize the results in a table similar to Tables 9.8–9.11.

9.18. Calculate the $NPSH_{av}$ of a pumping system. The static suction head = 2.20 m. The diameter and total length of suction piping from suction bell to pump impeller are 36 cm and 4 m, respectively. $C = 100$. At the operating-point, the pump discharge is 0.25 m^3/s. The pump station is located 1,050 m above sea level, and operating-temperature is 25°C. The K values for valves, fittings, and special are: suction bell ($K = 0.08$), one 90° elbow ($K = 0.30$), and one gate valve ($K = 0.19$). The $NPSH_{rq}$ from the manufacturer's data is 7.5 m. Determine whether the system will have a cavitation problem.

NOTES

1 N_s in U.S. customary units = 51.6 × N_s in SI units.

2 The pump impeller with Francis-Vane has vanes of double curvature, may have open or closed impeller, and single or double suction.

REFERENCES

1. Hydraulic Institute, *American National Standards for Centrifugal Pumps*, 15th ed., Hydraulic Institute, Parsippany, NJ, 1994.
2. Qasim S. R., *Wastewater Treatment Plants, Planning, Design and Operation*, 2nd ed., CRC Press, Boca Raton, FL, 1999.
3. Metcalf & Eddy, Inc., *Wastewater Engineering: Collection, Pumping of Wastewater*, McGraw-Hill Book Co, New York, 1981.
4. American National Standard Institute, Inc., *American National Standards for Pump Intake Design*, ANSI/HI 9.8–1998, Hydraulic Institute, Parsippany, NJ, 1998.
5. Walker, R., *Pump Selection: A Consulting Engineers Manual*, Ann Arbor Science Publishers, MI, 1972.
6. Karassik, I. J., W. C. Krutzsch, W. H. Fraser, and J. P. Messina, *Pump Handbook*, 2nd ed., McGraw-Hill Book Co., New York, 1986.
7. Stewart, H. L., *Pumps*, 2nd ed., Macmillan Publishing Book Co., New York, 1986.
8. Bartlett, R. E., *Pumping Stations for Water and Sewage*, Applied Science Publishers, Ltd., London, 1977.
9. Jones, G. M., R. L. Sanks, G. Tchobanoglous, and B. E. Bosserman II (Editors), *Pumping Station Design*, 3rd ed, Butterworth-Heinemann, Elsevier, Inc., Burlington, MA, 2006.
10. Gupta, R. S., *Hydrology and Hydraulic Systems*, 2nd ed., Waveland Press, Inc, Prospect Heights, IL, 2001.
11. Carthew, G. A., C. A. Goehring, and J. E. Van Teylingen, Development of Dynamic Head Loss Criteria of Raw Sludge Pumping, *Journal of Water Pollution Control Federation*, vol. 55, no. 5, 1983, pp. 472–483.
12. Mulbarger, M. C., S. R. Copas, J. R. Kordic, and F. M. Cash, Pipeline Friction Losses for Wastewater Sludges, *Journal of Wastewater Pollution Control Federation*, vol. 53, no. 8, 1981, pp. 1303–1313.
13. Avallone, E. A., and T. Baumeister III., *Marks' Standard Handbook for Mechanical Engineers*, 9th ed., McGraw-Hill Book Co., New York, 1986.
14. American Society of Heating, Refrigeration and Air Conditioning Engineers. ASHRAE Handbook: 1993 Fundamentals (I-P Edition), Atlanta, GA, 1993.

10 Best Watershed Management Practices for Control of Nonpoint Source Pollution

10.1 CHAPTER OBJECTIVES

Managing the release of contaminants from stormwater runoff is one of the challenges in protecting water quality in many receiving waters. The stormwater may carry many types of contaminants, including pathogens, suspended and dissolved organic and inorganic solids, oxygen-demanding pollutants, nutrients, and floatable matter. These contaminants have many adverse effects on water quality and beneficial use of water resources. Yet contaminants are discharged from stormwater runoff from many towns in the United States. The objectives of this chapter are to present:

- Regulatory status of stormwater runoff
- Characteristics, impacts, and management of stormwater runoffs
- Hydrographs and unit hydrographs
- Stormwater runoff estimation
- Source control of best management practices (BMPs) for stormwater runoffs and watershed management
- Performance and effectiveness of source-control stormwater BMPs
- Urban runoff quality and flow
- Urban runoff treatment by source-control BMPs and proprietary systems

10.2 REGULATORY STATUS OF STORMWATER RUNOFF

It has been well established that government policies and regulations play a critical role in environmental improvements. Wastewater management laws govern the conveyance, storage, treatment, and disposal of wastewater generated from urban and rural areas. Stormwater runoffs create serious water quality impairments during wet weather conditions. Important regulations dealing with these sources of water pollution are discussed below.

The stormwater runoff from urban areas is a significant contributor to the surface water quality impairment of the nation's waters. Stormwater runoff from urban and industrial areas typically contains significant quantities of pollutants, similar to those found in wastewater and industrial discharges. Consequently, these pollutants cause similar impacts on water quality. Pollutants commonly found in stormwater runoff include nitrogen, phosphorus, sediments, heavy metals, biochemical oxygen demand (BOD), and synthetic organic compounds such as refractory organics, pesticides, and herbicides. In addition to many of these pollutants, the increased quantity of stormwater discharged from rapidly urbanizing areas also poses a threat of significant impact on aquatic ecosystems because of physical alterations in the characteristics of receiving waters.

To help improve the quality of stormwater discharges, Congress amended the Clean Water Act (CWA) in 1987, which directs the U.S. Environmental Protection Agency (EPA) to develop National Pollutant Discharge Elimination System (NPDES) permit application requirements for the following classes and types of stormwater discharges:[1–3] (1) discharges from municipal separate storm sewer systems serving a population of 100,000 or more, (2) stormwater discharges associated with

DOI: 10.1201/9781003629399-10

industrial activity, and (3) stormwater discharges that contributed to a violation of a water quality standard or are significant contributors of pollutants to surface waters. Thus, the discharge from municipal separate storm sewer systems will not necessarily require traditional end-of-pipe controls; rather, municipalities will be required to develop and implement site-specific stormwater management programs.

Under NPDES regulations, municipalities submit a two-part permit application for discharges from their stormwater systems. The permitting requirements will include a stormwater system inventory and baseline condition and description, followed by monitoring and planning for stormwater management. The monitoring will involve measures (1) to reduce pollutants in runoff from commercial and residential areas, (2) to direct and remove illicit connections and improper disposal into storm sewers, (3) to reduce pollutants in runoff from industrial sites, and (4) to reduce pollutants in runoff from construction sites.[4] Municipalities must develop stormwater management plans that include controls to reduce the discharge of pollutants to the maximum extent practicable. The management plans must also address the legal, administrative, and financial aspects of the municipality's stormwater control program.[5]

Example 10.1: Regulatory Status of Stormwater Discharges

Provide a listing of major regulations and actions taken by regulatory agencies to control stormwater discharges.

Solution

The regulatory status of stormwater discharges is listed below:

1. Under the Clean Water Act of 1987, Congress directed the U.S. EPA to develop NPDES permit application requirement for the following classes and types of stormwater discharges: (a) municipal separate stormwater systems serving a population of 100,000 or more, (b) industrial activity, and (c) stormwater discharges that contribute to violations of water quality standards or are significant contributors of pollutants to surface water.
2. Under NPDES regulations, municipalities are required to submit a two-part permit application for discharges from stormwater systems to conduct (a) an inventory and document baseline conditions, and (b) monitoring and planning for stormwater management.
3. The monitoring will involve measures to reduce pollution from (a) commercial and residential areas, (b) illicit connections and improper disposal into storm sewers, (c) industrial sites, and (d) construction sites.
4. Municipalities must develop stormwater management plans to control the discharge of pollutants to the maximum extent practicable. The management plan must also address the legal, administrative, and financial aspects of the municipality's stormwater control program.

10.2.1 Stormwater Runoff from Urban Areas

The stormwater runoffs from urban areas are a significant source of water quality impairment of the nation's water. The *Clean Watershed Needs Survey 2012 Report to Congress* was submitted in 2016. This report documented that a total capital investment of $271 billion is needed to ensure that the nation's water quality objectives of the CWA are met.[5–8] This investment also includes a new conveyance system and the correction of combined sewer overflows (CSOs), which causes serious water quality impairments in natural water systems.[8–10] EPA has also established a CSO control policy that provides a framework and lays out clear expectations for municipalities, NPDES permitting, and enforcement authorities for controlling CSOs.[10–13]

10.3 STORMWATER RUNOFF: CAUSES AND EFFECTS

The transformation of undeveloped land into urban areas interrupts the natural drainage in a watershed. As a result, the hydrology of a watershed is significantly altered due to urbanization. The storm runoff from urbanized areas is significantly larger than runoff from land with natural ground cover. The effect of urbanization on imperviousness, infiltration, and surface runoff is shown schematically in Figure 10.1.[14] Due to urbanization, the runoff volume and peak discharge may increase, and the time to peak flow may decrease.

The stormwater drainage systems are built to control and manage storm runoff by collecting water from urbanized areas quickly and transporting it to the receiving waters.[15,16] There are three types of sewer systems to transport municipal and stormwater runoff: *sanitary*, *storm*, and *combined sewer systems* (CSSs). The combined sewers still exist in older urban areas but are no longer built in the United States. In CSSs, stormwater is combined with sanitary sewage and conveyed through a single pipe to wastewater treatment plants. During severe rainfall events, the total volume of wastewater carried by CSSs (wastewater plus stormwater) may exceed the capacity of treatment facilities. As a result, the untreated wastewater is discharged directly into surface waterbodies such as lakes, rivers, estuaries, or coastal waters. These discharges are called CSOs. The CSOs can be a major source of localized pollution of the receiving waters.[15,17]

The storm sewers are designed to convey only stormwater. The collected stormwater runoff is usually discharged directly into receiving waters. The pollutants picked up and washed away by stormwater from the residential, commercial, and industrial areas contribute significantly to water quality impairment of receiving streams and other waterbodies. Stormwater runoff entering receiving streams is considered a *nonpoint source* (NPS) as opposed to *point sources* from municipal and industrial discharges. In the U.S., NPS pollutant is reported as major contributors to degraded water quality.[16,18,19] NPS is still the leading cause of water quality problems and a challenge from both regulatory and assessment perspectives.[17,19–21] In Chapter 8, the design procedure of sanitary and storm sewers is presented in detail. Readers are referred to this chapter to find design examples and solutions of sanitary and storm sewers. Chapter 11 is devoted exclusively to the management of CSO.

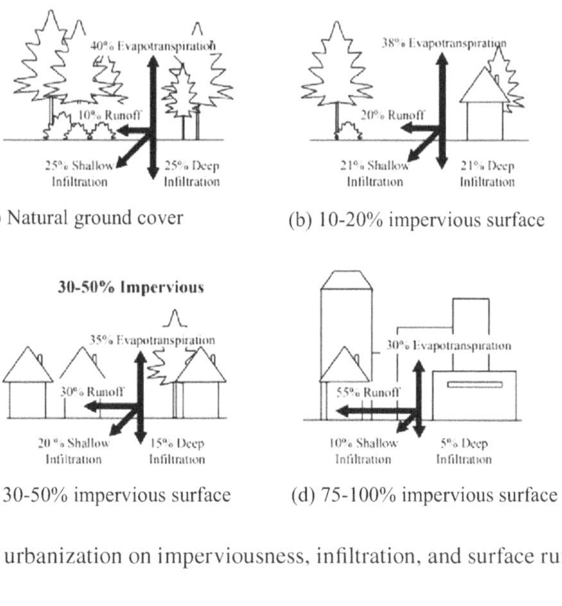

FIGURE 10.1 Effects of urbanization on imperviousness, infiltration, and surface runoff.

The imperviousness of a watershed increases as more and more landscape is converted from vegetation, trees, and agricultural lands to buildings, roads, rooftops, driveways, and parking lots. This limits the percolation of rainwater into the soil. The excess rainfall from impervious surfaces becomes stormwater runoff and drains rapidly into the nearby receiving water. Increased stormwater runoff from urbanized areas causes the following:[15,20–22]

1. Increased flooding occurs because of increased peak flow and a reduction in stormwater travel time;
2. Degradation of stream channels results from accelerated stream velocities, causing banks and channel erosion. Transport and deposition of eroded material in lakes and reservoirs may destroy food sources and habitats of aquatic life;
3. Large quantities of debris and contaminants such as organics, salt, sand and silt, heavy metals, oils, and pesticides are washed off into receiving waters. It has been estimated that runoff from the first hour of a moderate-to-heavy storm in a typical U.S. city will contribute a greater pollution load than would the city's untreated municipal wastewater during the same period;
4. Groundwater recharge is reduced due to an increase in the imperviousness of the watershed. The percolation of rainwater into the soil is reduced, thus reducing the groundwater storage. This also reduces the baseflow of the stream during low-flow conditions;
5. In older urban areas, the CSOs during wet weather events cause localized contamination of natural waters; and
6. Coastal waters are adversely impacted by pollutants from urban stormwater runoffs and CSOs. Beach closures, contaminated shellfish, and loss of commercial and recreational fisheries are common in coastal waters of large metropolitan areas.

The urbanization and increased imperviousness of the associated watersheds impact the physical characteristics of the receiving streams. Table 10.1 provides a check matrix showing the physical impacts on watershed due to increased imperviousness.[23] A comparison between surface water quality and relative levels of pollution in the U.S. streams in urban, agricultural, and undeveloped areas is presented in Table 10.2.[23,24] Only specific urban pollutants, including nutrients, metals, pesticides, and herbicides, are compared in this table. In many cases, the concentrations of some select water quality parameters in urban storm runoff are comparable to those found in untreated municipal wastewater. Table 10.3 presents a comparison between the concentrations of selected water quality parameters in urban storm runoff with the concentrations in municipal wastewater.[25]

TABLE 10.1
Environmental Impacts from Increased Watershed Imperviousness

	Resulting Impacts				
Increased Impervious Leads to:	**Flooding**	**Habitat Loss**	**Erosion**	**Channel Widening**	**Streambed Alteration**
Increased volume	✓	✓	✓	✓	✓
Increased peak flow	✓	✓	✓	✓	✓
Increased peak duration	✓	✓	✓	✓	✓
Increased stream temperature		✓			
Decreased baseflow		✓			
Changes in sediment loading	✓	✓	✓	✓	✓

Source: Adapted in part from Ref. 23

TABLE 10.2
Water Quality and Relative Levels of Pollution in the U.S. Streams in Urban, Agricultural, and Undeveloped Areas

Water Quality Parameter	Urban Areas	Agricultural Areas	Undeveloped Areas
Nitrogen	Medium	Medium-high	Low
Phosphorus	Medium-high	Medium-high	Low
Herbicides	Medium	Medium-high	Low
Pesticides	Medium-high	Low-medium	Very low
Metals	High	Medium	Very low
Toxic organics	High	Medium	Very low

Source: Adapted in part from Refs. 23,24

TABLE 10.3
Comparison of Water Quality Parameters in Urban Runoff with Domestic Wastewater (mg/L)

	Urban Runoff		Domestic Wastewater		
	Separate Sewers		Before Treatment		After Secondary
Constituent and Concentration	Range	Typical	Range	Typical	Typical
COD	200–275	75	250–1,000	500	80
TSS	20–2890	150	100–350	200	20
Total P	0.02–4.30	0.36	4–15	8	2
Total N	0.4–20.0	2	20–85	40	30
Lead	0.01–1.20	0.18	0.02–0.94	0.10	0.05
Copper	0.01–0.40	0.05	0.03–1.19	0.22	0.03
Zinc	0.01–2.90	0.02	0.02–7.68	0.28	0.08
Fecal Coliform per 100 mL	400–50,000		10^6–10^8		200

Source: Adapted in part from Refs. 12,25
Note: Concentrations are in mg/L, unless indicated.

Example 10.2: Water Balance Due to Change in Surface Imperviousness

A 20-ha land tract has been changed from natural ground cover to 30%–50% impervious due to housing development. Compare the evapotranspiration, surface runoff, and infiltration under both conditions. Use the information given in Figure 10.1. The average rainfall over the entire land tract is 5 cm.

Solution

1. Calculate the total rainfall over the entire land tract

$$Q_{\text{rainfall}} = 20 \text{ ha} \times 10{,}000 \text{ m}^2/\text{ha} \times 5 \text{ cm} \times 10^{-2} \text{ m/cm}$$

$$= 10{,}000 \text{ m}^3$$

2. Calculate from Figure 10.1 the total evapotranspiration, percolation, and runoff from natural land

$$\text{Total evapotranpiration} = 10{,}000\ \text{m}^3 \times 0.4 = 4{,}000\ \text{m}^3$$

$$\text{Total percolation} = 10{,}000\ \text{m}^3 \times 0.5 = 5{,}000\ \text{m}^3$$

$$\text{Total runoff} = 10{,}000\ \text{m}^3 \times 0.1 = \underline{1{,}000\ \text{m}^3}$$

$$10{,}000\ \text{m}^3$$

3. Calculate from Figure 10.1 the total evapotranspiration, percolation, and runoff from 30–50% impervious land

$$\text{Total evapotranpiration} = 10{,}000\ \text{m}^3 \times 0.35 = 3{,}500\ \text{m}^3$$

$$\text{Total percolation} = 10{,}000\ \text{m}^3 \times 0.35 = 3{,}500\ \text{m}^3$$

$$\text{Total runoff} = 10{,}000\ \text{m}^3 \times 0.30 = 3{,}000\ \text{m}^3$$

$$\overline{10{,}000\ \text{m}^3}$$

4. Compare the total evapotranspiration, percolation, and runoff from 30–50% imperviousness

$$\text{Reduction in evapotranpiration} = \frac{(3{,}500 - 4{,}000)\ \text{m}^3 \times 100}{4{,}000\ \text{m}^3} = -12.5\%$$

$$\text{Reduction in percolation} = \frac{(3{,}500 - 5{,}000)\ \text{m}^3 \times 100}{5{,}000\ \text{m}^3} = -30\%$$

$$\text{Increase in surface runoff} = \frac{(3{,}000 - 1{,}000)\ \text{m}^3 \times 100}{1{,}000\ \text{m}^3} = 200\%$$

Example 10.3: Change in Sediment Loading Due to Flooding

The average flow and total suspended solids (TSS) concentration in a creek are 30 cfs and 200 mg/L. During a flood condition, the flow and TSS concentrations reached 350 cfs and 2,500 mg/L. Calculate the increase in sediment load due to erosion.

Solution

1. Calculate the sediment load during average flow

$$\text{Sediments load} = 30\ \text{ft}^3/\text{s} \times 200\ \text{g/m}^3 \times \frac{\text{lb}}{453.6\ \text{g}} \times 1/(3.281\ \text{ft/m})^3$$

$$\times 86{,}400\ \frac{\text{s}}{\text{day}} \times \frac{1}{2{,}000\ \text{lb/T}} = 16.2\ \text{T/day}$$

2. Calculate the sediment load during flood conditions

$$\text{Sediments load} = 350\ \text{ft}^3/\text{s} \times 2{,}500\ \text{g/m}^3 \times \frac{\text{lb}}{453.6\ \text{g}} \times 1/(3.281\ \text{ft/m})^3$$

$$\times 86{,}400\ \frac{\text{s}}{\text{day}} \times \frac{1}{2{,}000\ \text{lb/T}} = 2359.4\ \text{T/day}$$

3. Calculated the increase in sediment load due to erosion

$$\text{Increase in sediment load} = \frac{(2359.4 - 16.2)\ \text{T/day}}{16.2\ \text{T/day}} \times 100$$

$$= 14{,}464\%$$

10.4 HYDROGRAPH AND UNIT HYDROGRAPH

Stormwater hydrograph, peak discharge, or volume of stormwater runoff are used in the design of stormwater drainage systems. The design of a stormwater drainage system requires estimation of runoff. The quantity of stormwater runoff and its temporal variations depend on the physical characteristics of rainfall and the drainage basin. The rainfall characteristics include its pattern, intensity, and duration. The shape and size of the area, slope, and soil type are the physiographic characteristics of a drainage basin. Unfortunately, the rainfall events are predictable only in a statistical sense, and whatever return period may be selected for design, some storms of higher intensity will certainly occur, which will overtax the system. The actual quantity of runoff, therefore, will be a function of the statistical distribution of frequency of the design storm, distribution of storm in time (the hyetograph), the antecedent conditions, the season of the year, and the physical layout of the service area. Several methods are used for the estimation of rainfall-runoff from a watershed. Most methods may require the use of a hydrograph and a unit hydrograph. Therefore, the definition of a *hydrograph* and a *unit hydrograph* is given before presenting several methods of estimation of rainfall-runoff from watersheds. After that, a detailed discussion on developing a hydrograph and a unit hydrograph is presented.

10.4.1 Hydrograph

A hydrograph is a graph showing the rate of flow (discharge) versus time past a specific point in a river, channel, or conduit carrying flow. The rate of flow is typically expressed in cubic meters or cubic feet per second. A typical hydrograph for a single storm is shown in Figure 10.2. Hydrograph characteristics include the *rising limb* of the hydrograph, also known as the concentration curve, that reflects a prolonged increase in discharge from a catchment area, typically in

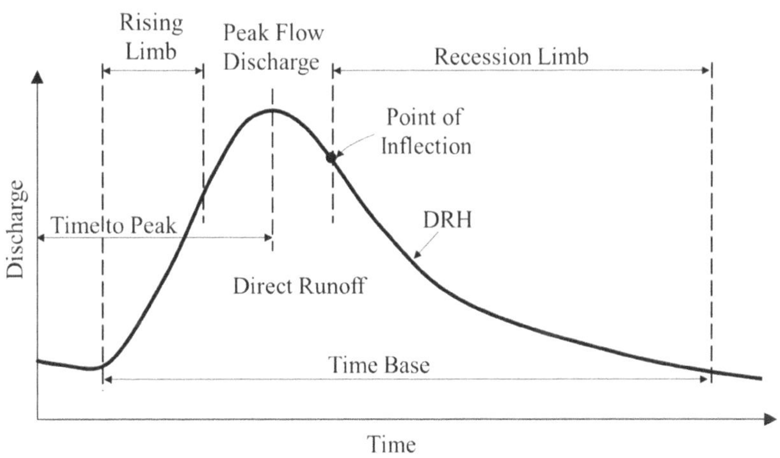

FIGURE 10.2 Example of a typical stormwater hydrograph.

response to a rainfall event. The *highest point* on the hydrograph is when the rate of discharge is the greatest. The *falling limb (or recession limb)* extends from the point of inflection at the end of the crest to the beginning of the natural groundwater flow. The shape of the recession limb depends upon basin characteristics only and is independent of the storm. The end of stormflow and the return to groundwater-derived *baseflow* is often taken as the point of inflection of the recession limb. The recession limb represents the withdrawal of water from the storage built up in the basin during the earlier phases of the hydrograph. The *lag time* is the interval from the maximum rainfall to the peak discharge. The *time to peak* is the time interval from the start of rainfall to the peak discharge. There are two types of hydrographs: (1) natural hydrograph and (2) synthetic hydrograph. The natural hydrographs are obtained from the flow measurements at streamflow gauges. In the absence of gauge stations in the study area, a synthetic hydrograph can be developed based on the basin parameters and storm characteristics. There are several methods to develop synthetic hydrographs. The most common methods are the Natural Resources Conservation Service (NRCS) and Snyder's methods. These methods are discussed in Sections 10.5.3 and 10.5.5.

10.4.2 Storm Runoff and Baseflow

The area beneath a stormwater hydrograph is the volume of the streamflow. This volume can be divided into storm runoff and baseflow. The baseflow is contributed by groundwater, and it is usually carried by the stream year-round. The storm runoff or *direct runoff* is produced by excess rainfall. The baseflow and direct runoff are shown in Figure 10.3. The first step in the estimation of stormwater runoff is to separate the baseflow from storm runoff. There are several methods for hydrograph separation. These methods are discussed in detail in Ref. 26 and many hydrology textbooks. A simple approach to separate the baseflow is presented here. In this approach, point A is the beginning of the direct runoff hydrograph (DRH). This point is connected with a straight line to the end of the DRH (point B). These points and the line separating the baseflow from direct runoff are shown in Figure 10.3. The area below the line AB is the baseflow, and the area above this line is direct runoff. The ordinates of the DRH are calculated by subtracting baseflow from the total discharge at each time step.

The stormwater runoff hydrograph can also be estimated from the basin unit hydrograph and the temporal rainfall pattern. The unit hydrograph and the development of the basin unit hydrograph are presented below.

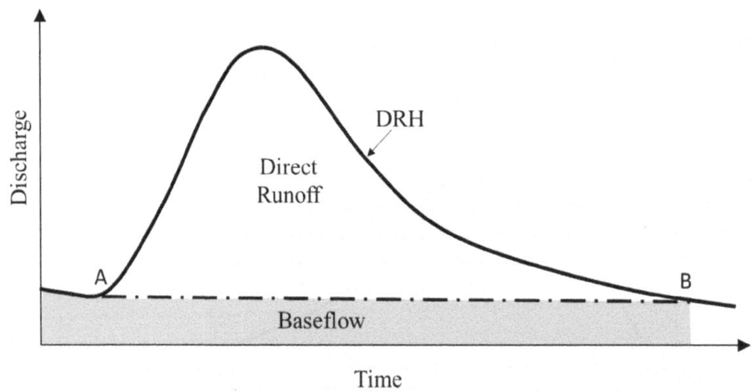

FIGURE 10.3 Baseflow separation from stormwater runoff hydrograph to obtain DRH.

10.4.3 Unit Hydrograph

The shape of the hydrograph at the basin's outlet or the study point reflects the physical characteristics of the basin such as shape, size, and slope, and storm characteristics, including duration, amount, and spatial pattern. Since the basin features may not change in the short term, similar storms are expected to have similar hydrographs. If the pattern and duration of storms are similar, the DRHs may have the same time base, but the ordinates of the hydrograph are proportional to the runoff volume of each storm. The *unit hydrograph* is a DRH observed at the study point in a basin due to one unit depth of rainfall excess of a storm of a specified duration. The unit depth of excess rainfall may be taken to be 1 mm or 1 in. over the entire drainage basin.[26,27] The storm duration may be considered a few hours or less, but it must be less than the basin time of concentration.[26] If a storm has the same duration but different amount of rainfall excess, the hydrograph of direct runoff will have the same shape and time base as the unit hydrograph and ordinates of flow proportional to the amount of rainfall excess. A storm that produces 2 in. of excess rainfall will produce a hydrograph with the same shape as the hydrograph of 1 in. of excess runoff with the same duration; however, all of the ordinates will be twice the unit hydrograph.

Unit hydrographs may be derived from streamflow data (natural hydrographs) and the temporal pattern of rainfall excess. For ungauged basins, a synthetic unit hydrograph can be developed using basin physiographic data. The development of the synthetic stormwater unit hydrograph is presented in Section 10.5.

10.4.4 Derivation of Unit Hydrograph from Streamflow Hydrograph

The derivation of the unit hydrograph from measured streamflow data includes the following steps:

1. Baseflow is separated from the streamflow hydrograph to obtain the DRH. The baseflow separation is described in Example 10.4.
2. The direct runoff volume is calculated from the area under the DRH, which is equal to the volume of excess rainfall.
3. The average rainfall excess depth is computed by dividing the volume of rainfall excess by the area of the basin using Eq. (10.1)

$$\bar{P} = \frac{V_{DR}}{A} \tag{10.1}$$

   where
   $\bar{P}$ = average rainfall excess depth, mm (in.)
   V_{DR} = direct runoff volume that is the volume under the hydrograph, m^3 (ft^3)
   A = drainage area of the basin, m^2 (ft^2)
4. The ordinates of the unit hydrograph are calculated by dividing each of the ordinates of the DRH by $\bar{P}$.

   The resulting hydrograph is a unit hydrograph of duration equal to the duration of the storm.

Example 10.4: Hydrograph Separation

The hourly streamflow data from a rainfall of 3-h duration for a river at a site are given in Table 10.4. Separate the baseflow from the DRH. Calculate the equivalent depth of the direct runoff. The drainage area is 300 acres.

TABLE 10.4
Streamflow Hydrograph (Example 10.4)

Time, h	*Q*, cfs	Time, h	*Q*, cfs
1	36	9	85
2	34	10	65
3	58	11	50
4	114	12	40
5	184	13	31
6	218	14	25.8
7	150	15	23
8	110	16	23

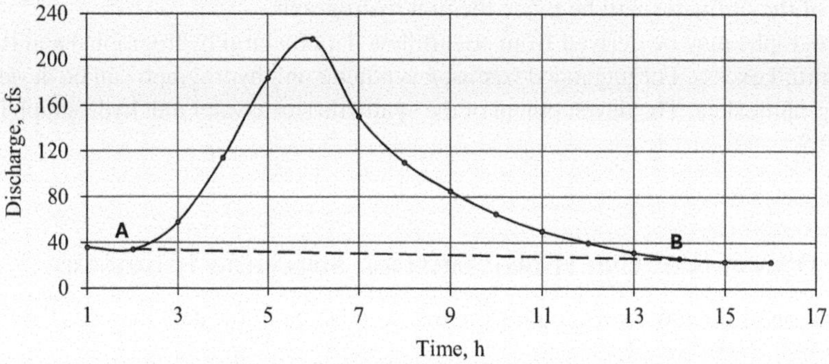

FIGURE 10.4 Baseflow separation and direct runoff calculations (Example 10.4).

Solution

1. Plot the streamflow hydrograph
   The hydrograph is plotted in Figure 10.4 from the data given in Table 10.4.
2. Separate the baseflow from the DRH
   a. Join point A (beginning of the direct runoff) to point B (end of the direct runoff) with a straight line. This line is shown in Figure 10.4. Note that point B is selected by judgment at time $t = 14$ h. Selecting point B a few hours before or after $t = 14$ h has no significant effect on the calculation of baseflow and DRH.
   b. Determine the baseflow ordinates at each hour from Figure 10.4. These values are listed in Column (3) of Table 10.5.
3. Determine the ordinates of the DRH
   The ordinates of direct runoff are obtained by subtracting the baseflow ordinates from the streamflow hydrograph ordinates at each hour. For example, the DRH at hour 3 = (58.0 cfs − 33.3 cfs) = 24.7 cfs. The ordinates of DRH are given in Column (4) of Table 10.5.
4. Calculate the direct average runoff
   The direct average runoff is obtained by graphically integrating the area bounded by the DRH. The average value of the discharge ordinate between two consecutive time intervals is multiplied by the time duration. For example, the direct average runoff between time intervals 2 and 3 h is (0.0 + 24.7) cfs/2.0 × 1 h = 12.4 cfs.h. The direct average runoffs are summarized in Column (5) of Table 10.5.
5. Calculate the sum of total direct average runoff × time
   The direct average runoff is the product of runoff and time interval. These values for 16 time intervals are listed in Column (7) of Table 10.5. The sum of 16

TABLE 10.5
Computation of Direct Runoff Hydrograph (Example 10.4)

Time, h	Total Flow, cfs	Baseflow, cfs	Direct Runoff, cfs	Average Runoff, cfs	Duration, h	Runoff × Time, cfs.h
(1)	(2)	(3)	(4)	(5)	(6)	(7)
1	36	36	0.0	0.0	1	0.0
2	34	34	0.0	12.4	1	12.4
3	58	33.3	24.7	53.1	1	53.1
4	114	32.6	81.4	116.7	1	116.7
5	184	32.0	152.0	169.4	1	169.4
6	218	31.3	186.7	153.1	1	153.1
7	150	30.5	119.5	99.8	1	99.8
8	110	30.0	80.0	67.9	1	67.9
9	85	29.2	55.8	46.2	1	46.2
10	65	28.5	36.5	29.3	1	29.3
11	50	27.9	22.1	17.5	1	17.5
12	40	27.2	12.8	8.7	1	8.7
13	31	26.5	4.5	2.3	1	2.3
14	25.8	25.8	0.0	0.0	1	0.0
15	23	23.0	0.0	0.0	1	0.0
16	23	23.0	0.0			
Total						**776.0**

calculated values of (Runoff×Time)=776.0 cfs×1 h. The sum of (Runoff×Time) is also calculated from the area of the curve bounded by DRH in Figure 10.4. For each time interval, the area under the DRH is calculated by multiplying the duration of the interval by the average value of two consecutive runoffs. The total runoff volume is the sum of the areas of all intervals.

Area under the DRH

$$= (1/2)\begin{bmatrix} (0+0)+(0+24.7)+(24.7+81.4)+(81.4+152) \\ +(152+186.7)+(186.7+119.5)+(119.5+80) \\ +(80+55.8)+(55.8+36.5)+(36.5+22.1)+(22.1+12.8) \\ +(12.8+4.5)+(4.5+0)+(0+0)+(0+0) \end{bmatrix} \times (1\text{ h})$$

$$= 776\text{ cfs.h}$$

These values are summarized in Column (7) of Table 10.5

6. Calculate runoff volume

$$\text{Runoff volume, } V_{DR} = 776\text{ cfs.h} \times (3{,}600\text{ s/h}) = 2{,}793{,}600\text{ ft}^3$$

7. Calculate the equivalent depth of the direct runoff
   Apply Eq. (10.1) to calculate the equivalent depth of direct runoff

$$\bar{P} = \frac{V_{DR}}{A}$$

$$= \frac{2{,}793{,}600\text{ ft}^3}{300\text{ acres}(43{,}560\text{ ft}^2/\text{acre})} = 0.21\text{ ft} \times 12\text{ (in/ft)} = 2.56\text{ in.}$$

Example 10.5: Derivation of Unit Hydrograph from Streamflow Hydrograph

The streamflow hydrograph in Example 10.4 was produced by a rainfall of 3-h duration. Compute the 3-h direct runoff and unit hydrographs for this basin.

Solution

1. Develop the unit hydrograph data from the streamflow hydrograph
   a. Separate the baseflow from the streamflow hydrograph to obtain the DRH. These values were calculated in Example 10.4 and are listed in Table 10.5 Column (4).
   b. Calculate direct runoff volume
      The direct runoff volume was calculated in step 6 of Example 10.4.

$$V_{DR} = 2{,}793{,}600 \text{ ft}^3$$

   c. Determine average rainfall excess depth
      The direct runoff volume was calculated in step 7 of Example 10.4.

$$\bar{P} = 2.56 \text{ in.}$$

   d. Calculate the ordinates of the unit hydrograph
      Divide each ordinate of the DRH by $\bar{P}$. For example, in Table 10.6 at hour 3, the unit hydrograph value of 24.7 cfs/2.56 in. = 9.6 cfs/in. The ordinates of the unit hydrograph are listed in Column (5) of Table 10.6.
   e. Determine the ordinates of total flow, baseflow, and DRHs
      The ordinates of the total flow, baseflow, and DRHs are obtained from Table 10.5 and are listed in columns (2), (3), and (4) of Table 10.6. These values are also plotted in Figure 10.5.

TABLE 10.6
Calculation Data for Unit Hydrograph (Example 10.5)

Time, h	Total Flow, cfs	Baseflow, cfs	Direct Runoff, cfs	Unit Hydrograph, cfs
(1)	(2)	(3)	(4)	(5)
1	36	36	0.0	0.0
2	34	34	0.0	0.0
3	58	33.3	24.7	9.6
4	114	32.6	81.4	31.8
5	184	32.0	152.0	59.4
6	218	31.3	186.7	72.9
7	150	30.5	119.5	46.7
8	110	30.0	80.0	31.3
9	85	29.2	55.8	21.8
10	65	28.5	36.5	14.3
11	50	27.9	22.1	8.6
12	40	27.2	12.8	5.0
13	31	26.5	4.5	1.8
14	25.8	25.8	0.0	0.0
15	23	23.0	0.0	0.0
16	23	23.0	0.0	0.0

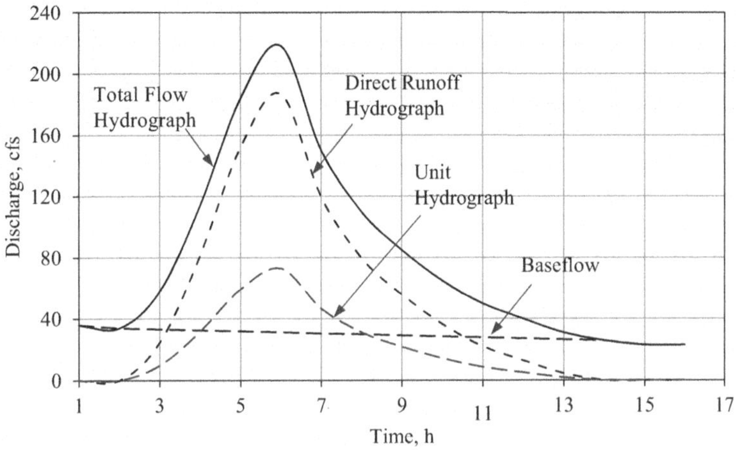

FIGURE 10.5 Total, direct, and unit hydrographs and baseflow (Example 10.5).

2. Draw the unit hydrograph
   The total flow, direct runoff and unit hydrographs, and baseflow are plotted in Figure 10.5.

10.5 STORMWATER RUNOFF ESTIMATION

The design of a stormwater drainage system requires the estimation of runoff. The quantity of stormwater runoff and its temporal variations depend on the physical characteristics of rainfall and the drainage basin. The rainfall characteristics include its pattern, intensity, and duration. The shape and size of the area, slope, and soil type are the physiographic characteristics of a drainage basin. There are several methods available to estimate rainfall-runoff characteristics. These methods are: (1) rational method, (2) modified rational method, (3) NRCS method, (4) rainfall-runoff regression equations (e.g., U.S. Geological Survey (USGS) method), (5) hydrograph methods (e.g., Soil Conservation Service (SCS) unit hydrograph method and Snyder's unit hydrograph method), (6) inlet method, and (7) computer simulation techniques. The rational method, the modified rational method, and the NRCS method have been most commonly used in the past, but now computer simulation techniques are extensively used to estimate runoff rate and volume.

10.5.1 Rational Method

The rational method uses an empirical formula to calculate the storm peak runoff. This method was originally introduced in the United States in 1889; however, it is still a very popular method for estimating stormwater runoff from small drainage basins with generally uniform surface cover and topography.[28] The rational method is used to estimate the stormwater runoff peak flows; therefore, it can be used to design stormwater gutters, drainage inlets, sewers, culverts, and small roadside ditches. However, the stormwater runoff volume and hydrographs are required for the design of storage facilities such as detention ponds. The modified rational method can be used for the design of such facilities. Nevertheless, using the runoff coefficient *C* in the rational method, which is derived for peak flows, may not be justified for estimating runoff volume and hydrograph. This deficiency has been improved, and the design procedure for detention facilities is presented in several stormwater management and design manuals.[29–31]

The stormwater runoff estimation using the rational method is discussed in Chapter 8. Readers are referred to Section 8.3 to find the application of the rational method in design examples for storm sewers. The rational formula to determine peak runoff from small watersheds is expressed in different forms by Eqs. (8.16a)–(8.16c). Additionally, many examples and solutions to illustrate their application to calculate the storm flow from small watersheds are also presented in Eqs. (8.7a)–(8.7c), which are listed below. All variables in these equations have been defined with these equations in Chapter 8.

$$Q = 0.00278 \times C \times i \times A \quad \text{(SI units)}$$

$$Q = C \times i \times A \quad \text{(U.S. customary units)}$$

$$Q = i \times \Sigma C_j \times a_j \quad \text{(U.S. customary units)}$$

The maximum rate of runoff is expressed by Eq. (10.2). This equation has a correlation factor C_a to account for the return period.

$$Q = C_a \times C \times i \times A \quad \text{(U.S. customary units)} \tag{10.2}$$

where

Q=maximum rate of runoff, m^3/s (ft^3/s)

A=drainage area contributing flow to the point of study, ha (acre)

C=runoff coefficient, assumed dimensionless. Typical values of C for storms of 5–10 years return periods are summarized in Table 10.7.

C_a=correlation factor to account for return period or antecedent conditions.

The value of C_a varies from 1 to 1.25, and there is no need to adjust the C coefficient for hydrological conditions. [12,32]

i=uniform rainfall intensity lasting for a critical period of time t_c, mm/h (in/h); t_c is time of concentration in min

The imperviousness of a watershed increases as more and more landscape is converted from vegetation, trees, and agricultural lands to buildings, roads, rooftops, driveways, and parking lots. The relationship of watershed imperviousness to runoff coefficient is shown in Figure 10.6.[22] Many relationships that express the intensity of rainfall and time of concentration as a function of physical characteristics of the drainage area have been developed for several parts of the country. Additionally, flow estimation and design examples of storm sewers are presented in Chapter 8.

TABLE 10.7
Typical Runoff Coefficients *C* for 5- to 10-Year Rainfall Frequency

Description of Area	Runoff Coefficients
Business	
Downtown areas	0.70–0.95
Neighborhood areas	0.50–0.70
Residential	
Single-family areas	0.30–0.50
Multifamily, detached	0.40–0.60
Multifamily, attached	0.60–0.75
Residential (suburban)	0.25–0.40

(Continued)

TABLE 10.7 (*Continued*)
Typical Runoff Coefficients *C* for 5- to 10-Year Rainfall Frequency

Description of Area	Runoff Coefficients
Apartment dwelling areas	0.50–0.70
Industrial	
Light areas	0.50–0.80
Heavy areas	0.60–0.90
Parks, cemeteries	0.10–0.25
Playgrounds	0.20–0.35
Railroad yard areas	0.20–0.40
Unimproved areas	0.10–0.30
Streets	
Asphaltic	0.70–0.95
Concrete	0.80–0.95
Brick	0.70–0.85
Drives and walks	0.75–0.85
Roofs	0.75–0.95
Lawns, Sandy Soil	
Flat, 2%	0.05–0.10
Average, 2%–7%	0.10–0.15
Steep, 7%	0.15–0.20
Lawns, Heavy Soil	
Flat, 2%	0.13–0.17
Average, 2%–7%	0.18–0.22
Steep, 7%	0.25–0.35

Source: Adapted in part from Refs. 33–36

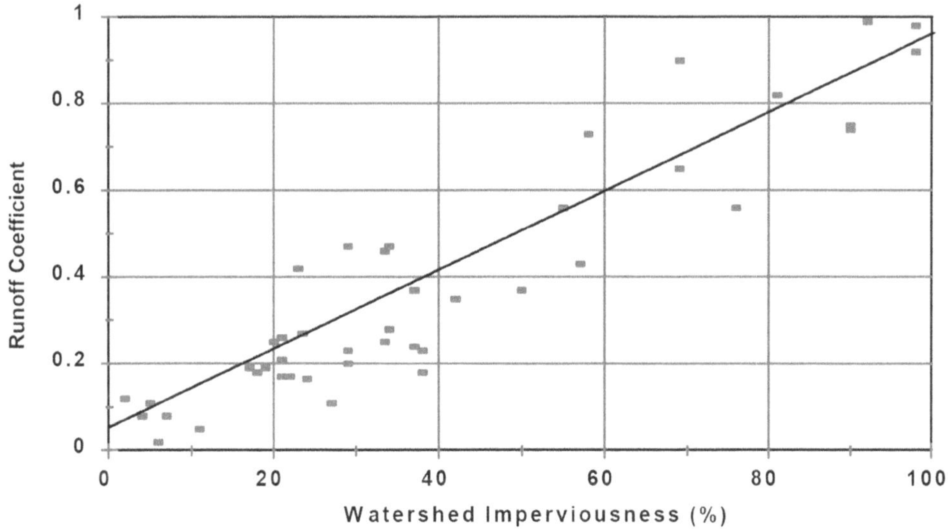

FIGURE 10.6 Relationship of watershed imperviousness to runoff coefficient *C*. (Printed from Ref. 22)

Example 10.6: Effect of Urbanization on Stormwater Runoff

A 10-acre rural watershed has been developed into a residential area. Due to urbanization, the coefficient of runoff of the sub-drainage areas a_1, a_2, and a_3 has changed from 0.6, 0.2, and 0.4 rural to 0.8, 0.7, and 0.5 urban, respectively. Determine the percentage increase in the maximum rate of stormwater runoff. The rainfall intensity of the 5-year frequency storm and duration are 4.2 in./h and 10 min. The areas of sub-drainage a_1, a_2, and a_3 are 2, 5, and 3 acres, respectively.

Solution

1. Calculate the weighted average of the coefficient of runoff

$$\text{Average } C \text{ for rural watershed} = \frac{(2 \text{ acres} \times 0.6) + (5 \text{ acres} \times 0.2) + (3 \text{ acres} \times 0.4)}{10 \text{ acres}}$$

$$= \frac{3.4 \text{ acres}}{10 \text{ acres}} = 0.34$$

$$\text{Average } C \text{ for developed watershed} = \frac{(2 \text{ acres} \times 0.8) + (5 \text{ acres} \times 0.7) + (3 \text{ acres} \times 0.5)}{10 \text{ acres}}$$

$$= \frac{6.6 \text{ acres}}{10 \text{ acres}} = 0.66$$

2. Calculate the maximum stormwater runoff
   Apply Eq. (8.7b)

$$Q = CiA$$

Storm runoff from rural watershed (before urbanization)

$$Q = 0.34 \times 4.2 \text{ in./h} \times 10 \text{ acres} = 14.28 \text{ cfs}$$

Storm runoff from developed watershed (after urbanization)

$$Q = 0.66 \times 4.2 \text{ in./h} \times 10 \text{ acres} = 27.72 \text{ cfs}$$

3. Calculate the percentage change in the maximum rate of stormwater runoff after urbanization

Themaximum rate of runoff from the areabefore urbanization = 14.28 cfs

Percent increase in maximum rate of stormwater runoff due to urbanization

$$= \frac{27.72 \text{ cfs} - 14.28 \text{ cfs}}{14.28 \text{ cfs}} \times 100 = 94\%$$

Example 10.7: Units of Expression of Rational Formula

The rational formula $Q = CiA$ in U.S. customary units is dimensionally homogeneous; i.e., A in acres, i in in./h gives Q in cfs. Apply unit conversion to prove this relationship.

Solution

Apply unit conversions in U.S. customary units to rational formula $Q = CiA$

$$Q = C \times \text{in./h} \times \frac{1}{12}\ \text{ft/in.} \times \frac{1}{3{,}600}\ \text{h/s} \times \text{acre} \times 43{,}560\ \text{ft}^2/\text{acre}$$

$$= C \times 1.008\ \text{ft}^3/\text{s} \quad \text{or} \quad C \times \text{cfs}$$

10.5.2 Modified Rational Method

The modified rational method is used to calculate peak stormwater runoff, volume, and hydrograph. In this method, the peak of the stormwater runoff is calculated from the rational method (Eqs. 8.7a–8.7c). The rainfall duration (t_d) should be considered equal to or longer than the time of concentration (t_c) to get the maximum peak discharge. If rainfall duration is equal to the time of concentration, the shape of the hydrograph is a triangle. If the rainfall duration is greater than the time of concentration, the hydrograph is a trapezoid. Figure 10.7 shows triangular and trapezoidal hydrographs. In both hydrographs, the rising and receding limbs express a duration equal to the time of concentration. A value of 1.5 t_c can be used for receding limb to better represent the actual storm and runoff. This approach is also similar to the SCS unit hydrograph with the receding limb that is longer than the rising limb.

The modified rational method can be used in engineering practices for preliminary sizing of detention ponds. The inflow hydrograph to the pond is assumed to be trapezoidal in shape. A trapezoidal stormwater runoff hydrograph developed from the modified rational method for estimation of storage volumes for detention ponds is shown in Figure 10.8. The rising and falling limbs of the inflow hydrograph have a duration equal to the time of concentration (t_c)

The required storage volume can be obtained from Eq. (10.3). [37]

$$V = 60\left(Q_p t_d - Q_a\left(\frac{t_d + t_c}{2}\right)\right) \tag{10.3}$$

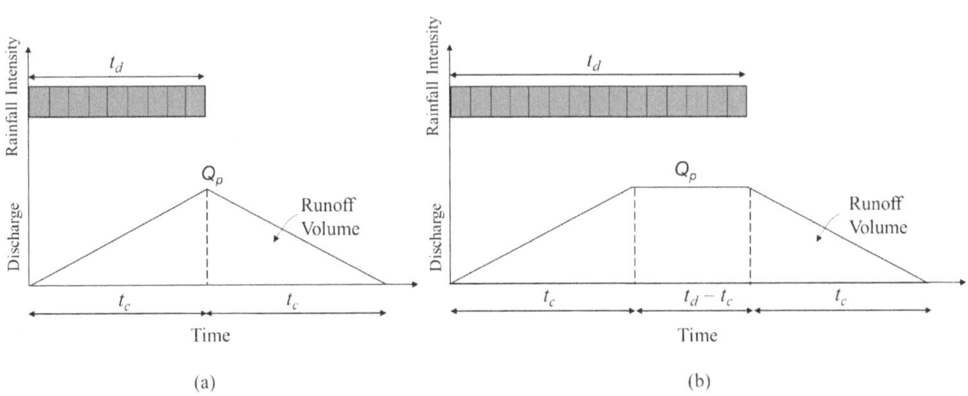

FIGURE 10.7 Stormwater runoff hydrograph using Modified Rational Method, (a) triangular hydrograph, (b) trapezoidal hydrograph. t_c is time of concentration, t_d is storm duration, and Q_p is peak discharge.

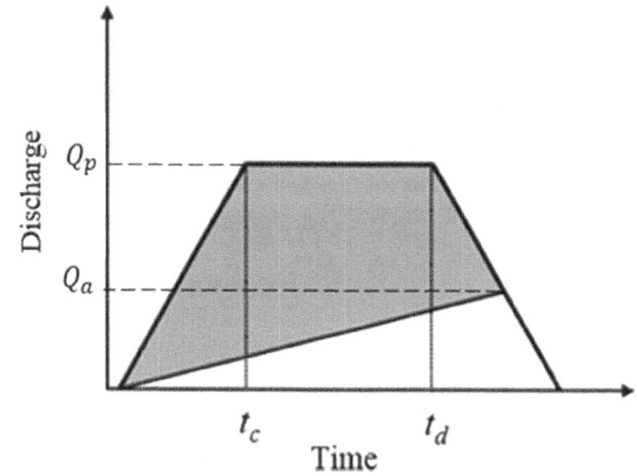

FIGURE 10.8 Trapezoidal stormwater runoff hydrograph for estimating storage volume of detention ponds using Modified Rational Method.

where

V = required detention volume, m^3 (ft^3)

Q_p = peak stormwater discharge (peak inflow rate for detention pond) or pre-development stormwater discharge, m^3/s (ft^3/s)

Q_a = allowable peak outflow from detention pond or pre-development maximum stormwater runoff, m^3/s (ft^3/s)

t_d = storm duration, min

t_c = time of concentration for post-development condition, min

The detention volume calculated from Eq. (10.3) should be used with caution and only for preliminary sizing of the pond. The modified rational method may result in undersized detention facilities in flat and sandy watersheds. A modification to the runoff coefficient (C) or time of concentration (t_c) should be considered in the design of these facilities. [31] This method should not be used to estimate the required detention volume for basins greater than 20–30 acres. [37]

Example 10.8: Stormwater Runoff Hydrograph and Volume from Modified Rational Method

A 15-acre residential area has a coefficient of runoff C of 0.75 and a time of concentration of 30 min. Construct a runoff hydrograph from the area for a 2 in./h storm. Apply the Modified Rational Method to calculate the storm runoff for two conditions: (1) the rainfall duration is 30 min, and (2) the rainfall duration is 2 h. Calculate the stormwater runoff for both storms.

Solution

A. Compute stormwater hydrograph for rainfall duration (t_d) of 30 min
   1. Calculate the peak stormwater runoff
      Apply Eq. (8.7b)

$$Q_p = CiA$$

$$= 0.75 \times 2 \text{ in./h} \times 15 \text{ acres} = 22.5 \text{ cfs}$$

2. Develop stormwater hydrograph

   Since the rainfall duration (t_d) is equal to the time of concentration (t_c), the hydrograph will be triangular. The rising and receding limbs have a 30-min duration, equal to t_c
   i. at time t=0 min, discharge Q=0 cfs
   ii. at time t=t_c = 30 min, discharge Q=Q_p=22.5 cfs
   iii. at time t=60 min, discharge Q=0 cfs
3. Plot stormwater hydrograph

   The stormwater hydrograph is plotted in Figure 10.9a using the hydrograph ordinates from step 2, t=0, 30, and 60 min.
4. Calculate stormwater volume

   Stormwater volume V=area beneath the triangular hydrograph in Figure 10.9a

$$= 1/2(\text{peak runoff rate} = 1/2(\text{peak runoff rate } Q_p \times \text{time base})$$

$$= 1/2(22.5 \text{ cfs} \times 60 \text{ min} \times (60 \text{ s/min})) = 40{,}500 \text{ ft}^3$$

B. Compute stormwater hydrograph for rainfall duration (t_d) of 2 h
1. Calculate the peak stormwater runoff

   Peak runoff Q_p from step A.2(ii)=22.5 cfs
2. Develop stormwater hydrograph

   Since the rainfall duration (t_d) is greater than the time of concentration (t_c), the hydrograph will be trapezoidal. The rising and receding limbs have a duration t_c equal to 30 min.
   i. at time t=0 min, discharge Q=0 cfs
   ii. at time t=t_c = 30 min, discharge Q=Q_p=22.5 cfs
   iii. at time t=120 min, discharge Q=Q_p=22.5 cfs
   iv. at time t=150 min, discharge Q=0 cfs
3. Plot stormwater hydrograph

   The stormwater hydrograph is plotted in Figure 10.9b using the hydrograph ordinates from step B.2, t=0, 30, 120, and 150 min.
4. Calculate stormwater volume

   Stormwater volume V=area of the trapezoidal hydrograph in Figure 10.9b

$$= 1/2(Q_p \times t_c) + (Q_p \times (t_d - t_c)) + 1/2(Q_p \times t_c) = (Q_p \times t_c) + (Q_p \times (t_d - t_c))$$

$$= (22.5 \text{ cfs} \times 30 \text{ min} \times (60 \text{ s/min})) + (22.5 \text{ cfs} \times (120 - 30) \text{ min} \times (60 \text{ s/min}))$$

$$= 40{,}500 \text{ ft}^3 + 121{,}500 \text{ ft}^3 = 162{,}000 \text{ ft}^3$$

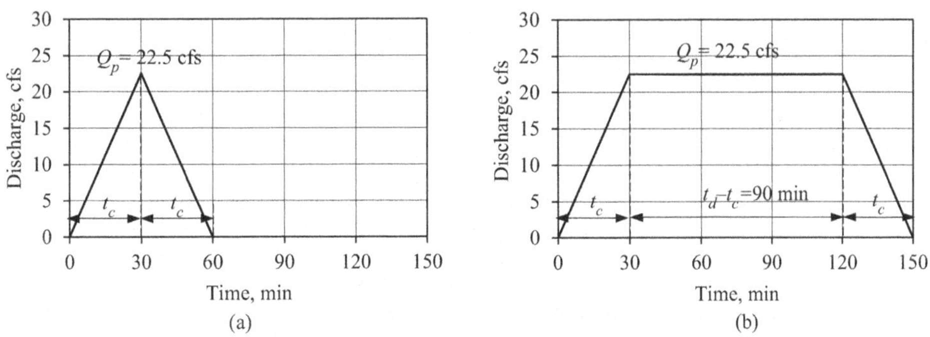

FIGURE 10.9 Stormwater runoff hydrograph using Modified Rational Method, (a) triangular hydrograph, and (b) trapezoidal hydrograph (Example 10.8).

Example 10.9: Storage Volume Requirement to Reduce Peak Discharge to Pre-development Period

A 10-acre watershed will be urbanized. A detention basin is needed for this watershed to reduce the peak discharge to the pre-urbanization period. The rainfall duration and intensity for a storm of a 5-year return period obtained from the meteorological data on record are given in Table 10.8. Determine the storage volume required for the detention basin. The estimated time of concentration for the post-development watershed is 7.5 min. The design storm has a pre-development 5-year frequency, and rainfall duration and intensity of 10 min and 4.2 in./h, respectively. The average coefficients of runoff C for unurbanized and urbanized watersheds are 0.34 and 0.66, respectively.

Solution

1. Calculate peak stormwater runoff from rural watershed (before urbanization)
   Apply Eq. (8.7b)
   For a 5-year storm, $C=0.34$, $i=4.2$ in./h and $A=10$ acres

$$Q_a = CiA$$

$$= 0.34 \times 4.2\ \text{(in./h)} \times 10\ \text{acres} = 14.28\ \text{cfs}$$

2. Calculate peak discharge for each storm duration and intensity pair in Table 10.8.
   $C=0.66$ and $A=10$ acres

$$Q_p = CiA$$

$$Q_p = 0.66 \times i\ \text{(in./h)} \times 10\ \text{acres} = 66i\ \text{cfs}$$

   For a 5-year frequency storm when $i=4.8$ in./h

$$Q_p = 6.6 \times 4.8 = 31.7\ \text{cfs}$$

   Similarly, substitute different values of i (in./h) from Table 10.8 in the above relationship
   The peak discharge for each storm duration and intensity pair is tabulated in Column (3) of Table 10.9.
3. Calculate the volume of detention basin required for each storm duration, intensity, and peak discharge
   Apply Eq. (10.3), time of concentration = 7.5 min

$$V = 60\left(Q_p t_d - Q_a\left(\frac{t_d + t_c}{2}\right)\right)$$

TABLE 10.8
Duration and Precipitation Intensity for a Storm of 5-Year Return Period (Example 10.9)

Duration (min) and Intensity (in./h) of the Design Storm					
5 min	**10 min**	**15 min**	**20 min**	**30 min**	**60 min**
4.80P in./h	3.90 in./h	3.10 in./h	2.60 in./h	1.85 in./h	1.25 in./h

TABLE 10.9
Volume of Detention Basin for Storm Duration, Precipitation Intensity for a Storm of 5-Year Return Period (Example 10.9)

t_d, min	i, in./h	Q_p, cfs	Volume, ft^3
(1)	(2)	(3)	(4)
5	4.8	31.7	4,155
10	3.9	25.7	7,923
15	3.1	20.5	8,811
20	2.6	17.2	8,859
30	1.85	12.2	5,895
60	1.25	8.3	963

$$V = 60\left(6.6\ i\ t_d - 14.28\ \text{cfs}\left(\frac{t_d + 7.5\ \text{min}}{2}\right)\right)$$

When $i = 4.8$ in./h and $t_d = 5$ min storm duration

$$V = 60\left(6.6 \times 4.8 \times 5 - 14.28\ \text{cfs}\left(\frac{5\ \text{min} + 7.5\ \text{min}}{2}\right)\right) = 4{,}155\ \text{ft}^3$$

Substitute different values of i and t_d pairs from Table 10.8 in the above equation

The volume of the detention basin required for each storm duration, intensity, and peak discharge is tabulated in Column (4) of Table 10.9.

4. Determine the storage volume required for the detention basin

The design storm duration is the one with maximum detention storage volume, $t_d = 20$ min

The maximum storage volume required for the detention basin = 8,859 ft^3

PA detention basin of theoretical volume of 8,859 ft^3 will reduce the stormwater runoff from the urbanized watershed to the pre-urbanization watershed.

10.5.3 Natural Resources Conservation Service (NRCS) Method

The NRCS method, formerly known as the SCS, is used to (1) estimate stormwater runoff, (2) peak discharge and volume, (3) calculate the time of concentration, and (4) develop stormwater hydrographs. This approach is more sophisticated than the rational method. It considers temporal rainfall distribution, the initial rainfall losses due to interception and depression storage, and a decreasing time-averaged infiltration rate during a storm. This method can be used in the design of stormwater control and management facilities such as storm drain systems, outlet structures, culverts, and others. The NRCS method uses the runoff curve number (CN) to estimate storm runoff. The stormwater runoff estimation using the NRCS method is discussed in several hydrology textbooks and stormwater management technical manuals (Refs. 26–28,31,37).

Stormwater runoff estimation: The NRCS method for estimating stormwater runoff is expressed by Eq. (10.4).[38]

$$Q = \frac{(P - I_a)^2}{(P - I_a) + S} \tag{10.4}$$

where

Q = accumulated runoff, in.
P = accumulated rainfall, in.
I_a = initial abstraction, in.
S = potential maximum soil moisture retention after runoff begins, in.

Initial abstraction (I_a) is the total rainfall losses before the runoff starts. It includes water retained by ground surface depressions, intercepted water by vegetation and other objects such as buildings, infiltration, and evaporation. I_a is estimated from an empirical relationship as a percentage of S. This relationship is given by Eq. (10.5)

$$I_a = 0.2S \quad (10.5)$$

Equation (10.6) is obtained by combining Eqs. (10.4) and (10.5). No runoff (Q) is produced if $P \leq 0.2S$.

$$Q = \frac{(P - 0.2S)^2}{P + 0.8S} \quad (10.6)$$

The value of S for a watershed is related to soil type and cover conditions. The NRCS method uses the runoff CN to estimate S from Eq. (10.7)

$$\text{CN} = \frac{1,000}{10 + S} \quad (10.7)$$

The CN resembles the runoff coefficient (C) used in the rational method and incorporates the hydrological effect of land use, soil type, and antecedent moisture in stormwater runoff calculations. The CN is a dimensionless number between 0 and 100. For natural soil, CN is smaller than 100, and for impervious material, CN =100.

The soil type and texture are classified as group *A*, *B*, *C*, and *D*. The soil group type, texture, and minimum infiltration rate of each soil type are listed in Table 10.10. The antecedent moisture condition of the soil may vary greatly. Based on moisture conditions, these soils are grouped into three categories: (I) dry conditions (low moisture), (II) average conditions, and (III) wetter conditions (high moisture, heavy rainfall over the preceding few days).

The NRCS has developed a system of CN*s* that are based on land use, soil type, and antecedent moisture. [38] Four generalized tables have been developed in Ref. 38 for the average antecedent moisture condition (Condition II). These tables provide CNs for (1) urban areas, (2) cultivated agricultural lands, (3) other agricultural lands, and (4) arid and semiarid rangelands. These tables are provided in Appendix B (Tables B.7a–B.7d). The adjustment of CNs for dry condition (I) and

TABLE 10.10
Hydrologic Soil Groups

Group	Soil Textures	Minimum Infiltration Rate (in./h)
A	Sand, loamy sand, or sandy loam	>0.3
B	Silt loam or loam	0.15–0.3
C	Sandy clay loam	0.05–0.15
D	Clay loam, silty clay loam, sandy clay, silty clay, or clay	0.0–0.05

Source: Adapted in part from Ref. 38

wet condition (III) is also tabulated in Appendix B (Table B.8). A composite CN is determined and used in stormwater runoff estimation from a watershed with several different land uses. Composite CNs can be calculated by using the weighted-average method. CNs for many combinations may be computed from Figures B.1 and B.2 (Appendix B). The runoff estimation using the NRCS method is discussed in detail in publications of U.S. Department of Agriculture (Refs. 38,39), and several textbooks and stormwater technical manuals (Refs. 30,31,40–44).

The effect of urbanization on the percentage of imperviousness and how surface runoff is conveyed from impervious areas to the drainage system should be incorporated in the calculation of CN. It is important to identify whether impervious areas connect directly to the drainage system, or they spread over lawns or other pervious areas where infiltration can occur. The *connected impervious area* is an area that is connected so that the runoff from it flows directly into the drainage system. It is also considered connected if runoff from it occurs as a concentrated shallow flow that runs over a pervious area and then into the drainage system. An *unconnected impervious area* is that where the runoff from the areas is spread over a pervious area as sheet flow. The procedure for calculating CNs for connected and unconnected areas is discussed in detail in Ref. 38. Composite CN for areas with connected impervious surfaces and areas with unconnected impervious surfaces, where the total impervious area is less than 30%, can be calculated using Figures B.1 and B.2.

Estimation of time of concentration and travel time: Time of concentration (t_c) is the time required for runoff from the hydraulically most remote part of the drainage area to reach the point of reference within the watershed. Travel time (t_t) is the time it takes runoff to travel from one point to another point within a watershed through the various components of the stormwater system. The time of concentration is calculated by adding the travel times for water as it moves through the watershed in (1) sheet flow, (2) shallow concentrated flow, (3) open channel flow, and (4) a combination of these. The time of concentration has been discussed previously in Section 8.3.3. Equations (8.4)–(8.6) express the time of concentration based on the characteristics of the drainage area. Readers are referred to Section 8.3.3 and Examples 8.21 and 8.23, and their solutions, for detailed discussion and calculations of the time of concentration.

Computation of stormwater peak discharge: The peak stormwater discharge can be estimated from Eq. (10.8). This equation can be used for both rural and urban areas.

$$Q_p = q_u AQF_p \tag{10.8}$$

where

Q_p = peak discharge, cfs

q_u = unit peak discharge, cfs/mi^2/in.

A = drainage area, mi^2

Q = runoff depth corresponding to 24-h rainfall (P) of a desired frequency is calculated from Eq. (10.6), in.

F_p = pond and swamp adjustment factor. The F_p is required if pond and swamp areas are spread throughout the watershed and are not considered in the calculation of time of concentration t_c. The values of F_p is given in Table 10.11.

The unit peak discharge (q_u) is determined from the graphs presented in Ref. 38. These graphs are developed for four synthetic 24-h rainfall distributions and cover four regions of the United States. These regions are Type I, Type IA, Type II, and Type III, and are listed in Table 10.12. These graphs are shown in Figures B.3–B.6. The time of concentration (t_c), I_a/P ratio, and rainfall distribution type are required to obtain q_u from these graphs. I_a Ris determined from Table 10.13 using the CN for the watershed. P is the accumulated rainfall depth in 24-h in inches for a selected frequency and location, and is applied in Eqs. (10.4) and (10.6).

NRCS synthetic unit hydrograph development: The NRCS method can be used to develop a synthetic stormwater hydrograph. An average dimensionless unit hydrograph is developed based on the analysis of a large number of watersheds of different sizes and physiography. The ordinate of

TABLE 10.11
Adjustment Factor (F_p) for Pond and Swamp Areas that are Spread Throughout the Watershed

Percentage of Pond and Swamp Areas	F_p
0.0	1.00
0.2	0.97
1.0	0.87
3.0	0.75
5.0	0.72

Source: Adapted from Ref. 38

TABLE 10.12
Synthetic 24-Hour Rainfall Distribution in Four Regions of the United States

Region Type	Components of the Region
Type I	Hawaii, Alaska
Type IA	Coastal side of Sierra Nevada and Cascade Mountains in southern and northern California, Oregon, and Washington
Type III	Gulf of Mexico and Atlantic coastal areas, where tropical storms bring large 24-hour rainfall amounts
Type II	Rest of the United States

Source: Adapted in part from Ref. 38

TABLE 10.13
Values of Initial Loss (I_a) for Different Curve Numbers

Curve Number	I_a, in.	Curve Number	I_a, in.
40	3.000	70	0.857
41	2.878	71	0.817
42	2.762	72	0.778
43	2.651	73	0.740
44	2.545	74	0.703
45	2.444	75	0.667
46	2.348	76	0.632
47	2.255	77	0.597
48	2.167	78	0.564
49	2.082	79	0.532
50	2.000	80	0.500
51	1.922	81	0.469
52	1.846	82	0.439
53	1.774	83	0.410

(Continued)

TABLE 10.13 (*Continued*)
Values of Initial Loss (I_a) for Different Curve Numbers

Curve Number	I_a, in.	Curve Number	I_a, in.
54	1.704	84	0.381
55	1.636	85	0.353
56	1.571	86	0.326
57	1.509	87	0.299
58	1.448	88	0.273
59	1.390	89	0.247
60	1.333	90	0.222
61	1.279	91	0.198
62	1.226	92	0.174
63	1.175	93	0.151
64	1.125	94	0.128
65	1.077	95	0.105
66	1.030	96	0.083
67	0.985	97	0.062
68	0.941	98	0.041
69	0.899		

Source: Adapted in part from Ref. 38

this hydrograph is Q/Q_p, which is the ratio of discharge Q to the peak discharge Q_p, and the abscissa is t/t_p, which is the ratio of time t and t_p, time to peak. The ratios of the NRCS dimensionless unit hydrograph are given in Table 10.14, and the plot is shown in Figure 10.10.

The peak discharge Q_p and time to peak t_p are computed from Eqs. (10.9) and (10.10)

TABLE 10.14
Ratios of NRCS Dimensionless Parameters t/t_p and Q/Q_p for Synthetic Hydrograph

t/t_p	Q/Q_p	t/t_p	Q/Q_p
0.0	0.000	1.4	0.780
0.1	0.030	1.5	0.680
0.2	0.100	1.6	0.560
0.3	0.190	1.8	0.390
0.4	0.310	2.0	0.280
0.5	0.470	2.2	0.207
0.6	0.660	2.4	0.147
0.7	0.820	2.6	0.107
0.8	0.930	2.8	0.077
0.9	0.990	3.0	0.055
1.0	1.000	3.5	0.025
1.1	0.990	4.0	0.011
1.2	0.930	4.5	0.005
1.3	0.860	5.0	0.000

Source: Adapted in part from Refs. 38,39

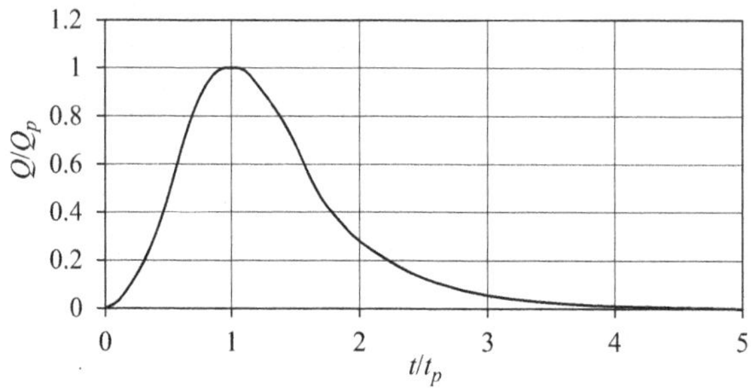

FIGURE 10.10 NRCS dimensionless unit hydrograph. (Printed from Refs. 38,39)

$$Q_p = \frac{484A}{t_p} \quad \text{(U.S. customary units)} \tag{10.9a}$$

$$Q_p = \frac{2.08A}{t_p} \quad \text{(SI units)} \tag{10.9b}$$

$$t_p = 0.67t_c \tag{10.10}$$

where

Q_p = peak discharge, cfs (m³/s)
A = drainage area, mi² (km²)
t_p = time to peak, h

Example 10.10: Stormwater Runoff Estimation Using NRCS Method

A storm event dropped an average depth of 4.3 in. of rain over a residential area. The residential area has an average lot size of ¼ acres, and the soils fall in hydrologic soil group A. Estimate stormwater runoff using the NRCS method.

Solution

1. Determine CN
   CN = 61 from Table B.7a for a residential area with an average lot size of ¼ acres and soil from hydrologic soil group A (Table 10.10).
2. Calculate S
   From Eq. (10.7), $\text{CN} = \frac{1{,}000}{10+S}$

$$61 = \frac{1{,}000}{10+S}$$

$$610 + 61S = 1{,}000$$

$$S = 6.19 \text{ in.}$$

3. Calculate runoff
   Apply Eq. (10.6)

$$Q = \frac{(P-0.2S)^2}{P+0.8S} = \frac{(4.3\text{ in.} - 0.2\times 6.19\text{ in.})^2}{4.3\text{ in.} + 0.8\times 6.19\text{ in.}} = \frac{9.38\text{ in.}^2}{9.25\text{ in.}}$$

$$Q = 1.01\text{ in.}$$

Example 10.11: Direct Runoff from a Watershed with Two Land Uses

The total area of a watershed is 500 acres. Area 1 is 350 acres in rotation meadow, managed with contouring and good rotation, and Area 2 is 150 acres in row crop, managed with contouring and good rotation. Soils in both areas are in the hydrologic soil group B. Estimate direct runoff from the watershed using the weighted-average CN. The rainfall over the entire watershed is 5.1 in.

Solution

1. Determine CN for both Area 1 and Area 2
   Area 1, $CN_1=69$ from Table B.7b for rotation meadow, contoured, good rotation, and soil group B (Table 10.10)
   Area 2, $CN_2=75$ from Table B.7b for row crop, contoured, good rotation, and soil group B (Table 10.10)
2. Calculate the weighted-average CN

$$\text{Weighted average CN} = \frac{\text{CN}_1 \times A_1 + \text{CN}_2 \times A_2}{A_1 + A_2}$$

$$= \frac{69 \times 350 \text{ acres} + 75 \times 150 \text{acres}}{350 \text{ acres} + 150 \text{ acres}} = 70.8$$

3. Calculate S
   From Eq. (10.7), $\text{CN} = \dfrac{1{,}000}{10 + S}$

$$70.8 = \frac{1{,}000}{10 + S}$$

$$708 + 70.8S = 1{,}000$$

$$70.8S = 292$$

$$S = 4.12 \text{ in.}$$

4. Calculate runoff
   From Eq. (10.6)

$$Q = \frac{(P - 0.2S)^2}{P + 0.8S} = \frac{(5.1 \text{ in.} - 0.2 \times 4.12 \text{ in.})^2}{5.1 \text{ in.} + 0.8 \times 4.12 \text{ in.}} = \frac{18.28 \text{ in.}^2}{8.40 \text{ in.}}$$

$$Q = 2.17 \text{ in.}$$

Example 10.12: Stormwater Runoff Estimation Using NRCS Method and Peak Discharge

A 250-acre fully developed urban area is located in Tennessee. 175 acres of the watershed consist of 1/4-acre residential lots. The rest of the watershed is open space in good hydrologic condition. Soil in both areas is in the hydrologic soil group C. Estimate stormwater runoff using the NRCS method and peak discharge from Eq. (10.8) for a 24-h storm with total rainfall of 3.5 in. The estimated time of concentration for the watershed is 1.5 h. There is no pond or swamp in this area.

Solution

A. Calculate stormwater runoff Q using the NRCS method

1. Determine CN for both 1/4-acre residential lots and open space area

   1/4-acre residential lots, $CN_1=83$ from Table B.7a for soil group C

   Open space area, $CN_2=74$ from Table B.7a for soil group C in good hydrologic condition

2. Calculate the weighted-average CN

$$\text{Weighted average CN} = \frac{CN_1 \times A_1 + CN_2 \times A_2}{A_1 + A_2}$$

$$= \frac{83 \times 175 \text{ acres } + 74 \times 75 \text{ acres}}{175 \text{ acres} + 75 \text{ acres}} = 81.3$$

3. Calculate S

   From Eq. (10.7), $CN = \dfrac{1{,}000}{10+S}$

$$81.3 = \frac{1{,}000}{10+S}$$

$$813 + 81.3S = 1{,}000$$

$$81.3S = 187$$

$$S = 2.3 \text{ in.}$$

4. Calculate runoff

   From Eq. (10.6)

$$Q = \frac{(P-0.2S)^2}{P+0.8S} = \frac{(3.5 \text{ in.} - 0.2 \times 2.3 \text{ in.})^2}{3.5 \text{ in.} + 0.8 \times 2.3 \text{ in.}} = \frac{9.24 \text{ in}^2}{5.34 \text{ in.}}$$

$$Q = 1.73 \text{ in.}$$

B. Compute peak discharge Q_p

1. Determine initial loss I_a

   $I_a=0.46$ in. from Table 10.13 for a watershed with CN=81.3

2. Calculate unit peak discharge q_u

   $I_a/P=(0.46 \text{ in.})/(3.5 \text{ in.})=0.13$

   Since the watershed is located in Type II region (Table 10.12), from Figure B.5, unit peak discharge $q_u=270$ cfs/mi²/in. for the time of concentration $t_c=1.5$ h and $I_a/P=0.13$

3. Determine adjustment factor (F_p) for pond and swamp areas

   Since there is no pond or swamp in the watershed, $F_p = 10$ (Table 10.11)

4. Calculate peak discharge

   From Eq. (10.8), $Q_p = q_u A Q F_p$

$$Q_p = (270 \text{ cfs/mi}^2\text{/in.})\left(250 \text{ acres} \times \frac{1 \text{ mi}^2}{640 \text{ acres}}\right)(1.73 \text{ in.}) \times 1.0$$

$$Q_p = 182.5 \text{ cfs}$$

Example 10.13: Surface Runoff from Several Successive Storms

Determine the direct runoff (in in./h) for each of the successive 20-min periods storm tabulated below. Assume CN=75 for the NRCS method.

Successive 20-min Storm Period	Rainfall Intensity, in./h
0–20	0.5
20–40	2.5
40–60	6.2
60–80	4.8
80–100	1.0
100–120	2.1
120–140	3.2

Solution

1. Calculate S

   From Eq. (10.7), $\mathrm{CN} = \dfrac{1{,}000}{10+S}$

$$75 = \frac{1{,}000}{10+S}$$

$$750 + 75S = 1{,}000$$

$$75\mathrm{S} = 250$$

$$S = 3.33 \text{ in.}$$

2. Develop an equation for runoff

   From Eq. (10.6), $Q = \dfrac{(P-0.2S)^2}{P+0.8S}$

$$= \frac{(P - 0.2 \times 3.33 \text{ in.})^2}{P + 0.8 \times 3.33 \text{ in.}} = \frac{(P-0.666)^2}{P+2.664}$$

   In this equation, runoff depth (Q) and rainfall depth (P) are accumulated values in inches.

3. Calculate the amount of rainfall during each successive storm
   i. Amount of rain during 0–20 min storm=0.5 in./h×20 min×h/60 min=0.17 in.
      Accumulated rainfall P=(0+0.17 in.)=0.17 in.
      Accumulated direct runoff Q=0 since accumulated rainfall P of 0.17 in. is less than the initial abstraction I_a from Eq. (10.5) (I_a=0.2×3.33 in.=0.666 in.)
      Runoff depth increment=0
      Runoff intensity increment=0
   ii. Amount of rain during 20–40 min storm=2.5 in./h×20 min×h/60 min=0.83 i n.
      Accumulated rainfall P=(0.17 in.+0.83 in.)=1.00 in.

$$\text{Accumulated direct runoff } Q = \frac{(1.00 \text{ in.} - 0.666 \text{ in.})^2}{(1.00 \text{ in.} + 2.664 \text{ in.})} = \frac{0.112 \text{ in.}^2}{3.664 \text{ in.}} = 0.03 \text{ in.}$$

$$\text{Runoff depth increment} = 0.03 \text{ in.} - 0 \text{ in.} = 0.03 \text{ in.}$$

$$\text{Runoff intensity increment} = \frac{0.03 \text{ in.}}{20 \text{ min}} \times \frac{60 \text{ min}}{\text{h}} = 0.09 \text{ in.}$$

iii. Amount of rain during 40–60 min storm = 6.2 in./h × 20 min × h/60 min = 2.07 in.
Accumulated rainfall P = (1.00 + 2.07 in.) = 3.07 in.

$$\text{Accumulated direct runoff } Q = \frac{(3.07\text{ in.} - 0.666\text{ in.})^2}{(3.07\text{ in.} + 2.664\text{ in.})} = \frac{5.779\text{ in.}^2}{5.734\text{ in.}} = 1.00\text{ in.}$$

$$\text{Runoff depth increment} = 1.00\text{ in.} - 0.03\text{ in.} = 0.97\text{ in.}$$

$$\text{Runoff intensity increment} = \frac{0.97\text{ in.}}{20\text{ min}} \times 60\frac{\text{min}}{\text{h}} = 2.91\text{ in./h}$$

Similarly, calculate the amount of rainfall, accumulated rainfall, accumulated direct runoff, runoff depth increment, and runoff intensity increment during each successive storm.

4. Summarize the results

The amount of rainfall, accumulated rainfall, accumulated direct runoff, runoff depth increment, and runoff intensity increments are summarized in seven Columns of Table 10.15.

5. Plot the rainfall and runoff data from the successive seven 20-min storm periods

The distribution of rainfall and runoff rates is shown in Figure 10.11 for seven successive 20-min storms. In Figure 10.11, the areas in solid plus hatched areas show rainfall, and the hatched areas show runoff. For example, for a storm during t = 40–60 min, the rainfall intensity is 6.2 in./h, and the rate of runoff produced is 2.91 in./h.

TABLE 10.15
Calculation of Runoff Using NRCS Method (Example 10.13)

Time, min	Rainfall Intensity, in./h	Amount of Rain, in.	Accumulated Rainfall, in.	Accumulated Direct Runoff, in.	Runoff Depth Increment, in.	Runoff Intensity Increment, in/h
(1)	(2)	(3)	(4)	(5)	(6)	(7)
0–20	0.5	0.17	0.17	0.00	0.00	0.00
20–40	2.5	0.83	1.00	0.03	0.03	0.09
40–60	6.2	2.07	3.07	1.00	0.97	2.91
60–80	4.8	1.60	4.67	2.18	1.18	3.54
80–100	1.0	0.33	5.00	2.45	0.27	0.81
100–120	2.1	0.70	5.70	3.03	0.58	1.74
120–140	3.2	1.07	6.77	3.94	0.91	2.73

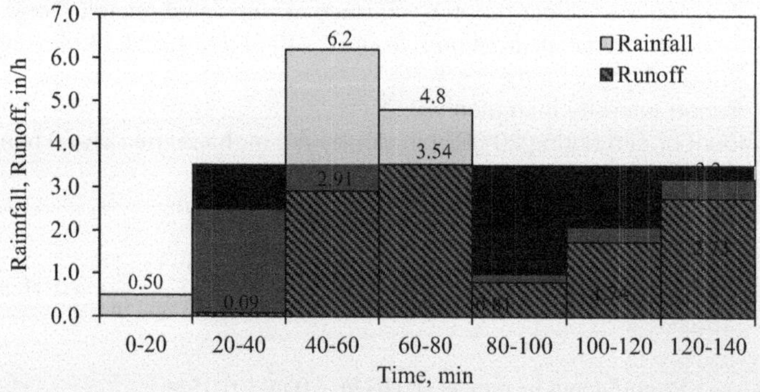

FIGURE 10.11 Rainfall and runoff intensity calculated using NRCS method for seven successive 20-min storms (Example 10.13).

Example 10.14: Peak Stormwater Discharge from NRCS Method

A watershed experienced a 24-h storm of 8.6 in. The watershed area is 350 acres and is located in a northeastern state. The time of concentration is 1.3 h, the CN is 70, and 20% of the watershed area is formed by ponds. Compute peak discharge runoff.

Solution

1. Determine initial loss I_a
   $I_a = 0.857$ in. from Table 10.13 for a watershed with CN = 70
2. Determine unit peak discharge q_u
   $I_a/P = (0.857 \text{ in.})/(8.6 \text{ in.}) = 0.1$
   From Figure B.4, unit peak discharge $q_u = 100$ cfs/mi²/in. for the time of concentration $t_c = 1.3$ h and $I_a/P = 0.10$
3. Estimate runoff depth Q
   a. Calculate S
      From Eq. (10.7), $\text{CN} = \dfrac{1{,}000}{10+S}$

$$70 = \frac{1{,}000}{10+S}$$

$$700 + 70S = 1{,}000$$

$$70S = 300$$

$$S = 4.29 \text{ in.}$$

   b. Calculate runoff depth

$$\text{From Eq. (10.6), } Q = \frac{(P-0.2S)^2}{P+0.8S} = \frac{(8.6 \text{ in.} - 0.2 \times 4.16 \text{ in.})^2}{8.6 \text{ in.} + 0.8 \times 4.16 \text{ in.}} = \frac{59.93 \text{ in.}^2}{12.03 \text{ in.}}$$

$$Q = 4.98 \text{ in.}$$

4. Compute peak discharge Q_p
   From Eq. (10.8), $Q_p = q_u AQF_p$
   Since 20% of the watershed area is formed by ponds, $F_p = 0.97$ (Table 10.11)

$$Q_p = (100 \text{ cfs/mi}^2/\text{in.})\left(350 \text{ acres} \times \frac{1 \text{ mi}^2}{640 \text{ acres}}\right)(4.98 \text{ in.}) \times 0.97$$

$$Q_p = 264 \text{ cfs}$$

Example 10.15: Synthetic Stormwater Hydrograph from NRCS Method

Compute the synthetic stormwater hydrograph for the watershed in Example 10.14.

Solution

1. Determine peak discharge Q_p
   a. Calculate time to peak t_p
      From Eq. (10.10), $t_p = 0.67t_c$ and $t_c = 1.3$ h from Example 10.14

$$t_p = 0.67 \times 1.3 \text{ h} = 0.871 \text{ h}$$

b. Calculate peak discharge Q_p

From Eq. (10.9a), $Q_p = \frac{484A}{t_p}$ and A from Example 10.14 = 350 acres

$$Q_p = \frac{484 \times \left(350 \text{ acres} \times \frac{1 \text{ mi}^2}{640 \text{ acres}}\right)}{0.871 \text{ h}} = 303 \text{ cfs}$$

2. Compute the stormwater hydrograph using ratios for the NRCS dimensionless unit hydrograph

   The ratio of NRCS dimensionless parameters t/t_p and Q/Q_p is used from Table 10.14.
3. Develop a hydrograph

   Calculate t and Q for different t/t_p and Q/Q_p ratios in Table 10.16.

$$t / t_p = 0.1$$

$$t = 0.1 \times t_p = 0.1 \times 0.871 \text{ h} = 0.0871 \text{ h} = 0.09 \text{ h}$$

$$Q / Q_p = 0.03$$

$$Q = 0.03 \times Q_p = 0.03 \times 303 \text{ cfs} = 9.09 \text{ cfs} = 9.1 \text{ cfs}$$

Similarly, calculate t and Q for all t/t_p and Q/Q_p ratios in Table 10.16 and arrange the corresponding values of t and Q in this table.

The hydrograph ordinates are given in Table 10.16 and plotted in Figure 10.12.

TABLE 10.16
Ordinates of Stormwater Hydrograph (Example 10.15)

t/t_p	Q/Q_p	t, h	Q, cfs	t/t_p	Q/Q_p	t, h	Q, cfs
0.0	0	0.00	0.0	1.7	0.46	1.48	139.4
0.1	0.03	0.09	9.1	1.8	0.39	1.57	118.2
0.2	0.10	0.17	30.3	1.9	0.33	1.65	100.0
0.3	0.19	0.26	57.6	2.0	0.28	1.74	84.8
0.4	0.31	0.35	93.9	2.2	0.207	1.92	62.7
0.5	0.47	0.44	142.4	2.4	0.147	2.09	44.5
0.6	0.66	0.52	200.0	2.6	0.107	2.26	32.4
0.7	0.82	0.61	248.5	2.8	0.077	2.44	23.3
0.8	0.93	0.70	281.8	3.0	0.055	2.61	16.7
0.9	0.99	0.78	300.0	3.2	0.04	2.79	12.1
1.0	1.00	0.87	303.0	3.4	0.029	2.96	8.8
1.1	0.99	0.96	300.0	3.6	0.021	3.14	6.4
1.2	0.93	1.05	281.8	3.8	0.015	3.31	4.5
1.3	0.86	1.13	260.6	4.0	0.011	3.48	3.3
1.4	0.78	1.22	236.3	4.5	0.005	3.92	1.5
1.5	0.68	1.31	206.0	5.0	0.000	4.36	0.0
1.6	0.56	1.39	169.7				

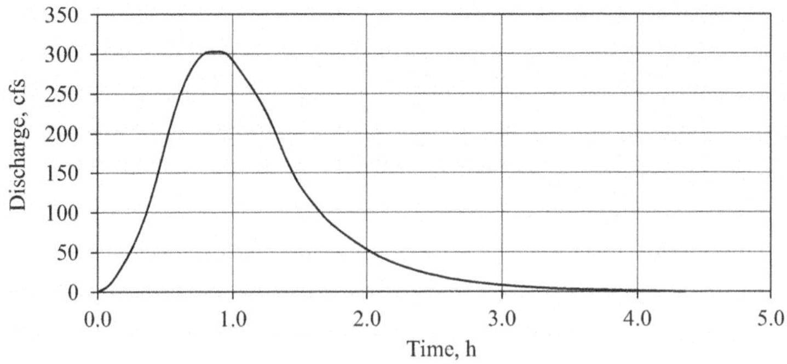

FIGURE 10.12 Synthetic stormwater hydrograph (Example 10.15).

Example 10.16: Effect of Urbanization on Synthetic Stormwater Hydrograph

A major development is being planned in a 100-acre forested watershed. A new residential and industrial complex will occupy 50% of the watershed. The maximum rate of stormwater runoff is expected to change from 100 to 200 cfs, and the time of concentration is expected to decrease from 25 to 11 min as a result of the urbanization. Develop the pre- and post-development synthetic stormwater hydrographs for the watershed. Plot the hydrographs and discuss the effect of urbanization on the stormwater runoff volume, peak discharge, and time to peak. Use the NRCS dimensionless ratios of unit hydrograph given in Table 10.14.

Solution

1. Calculate time to peak t_p for pre- and post-development stormwater hydrograph
   Time to peak (t_p) is the time from the beginning of the rainfall to the time of the stormwater peak discharge. t_p is calculated from Eq. (10.10)

$$t_p = 0.67t_c$$

   where
   t_c is time of concentration
   Time to peak (t_p) for pre-development $= 0.67 \times 25$ min $= 16.8$ min
   Time to peak (t_p) for post-development $= 0.67 \times 11$ min $= 7.4$ min
2. Determine the peak discharge Q_p for pre- and post-development
   Peak discharge (Q_p) for pre-development = 100 cfs (given data)
   Peak discharge (Q_p) for post-development = 200 cfs (given data)
3. Calculate the ordinates of the pre-and post-development hydrographs
   Apply the NRCS dimensionless synthetic hydrograph data given in Table 10.14. Also, calculate the time (min) and corresponding discharge (cfs) for both hydrographs. This information is summarized in Table 10.17.
   Use the procedure presented in Example 10.15
4. Plot pre- and post-development hydrographs in Figure 10.13.
5. Comment on both hydrographs
   The post-development hydrograph reaches the peak much sooner, and the peak discharge is significantly higher than that for the pre-development hydrograph. This clearly indicates that there is significantly increased discharge, and there is a possibility of local flash flooding.

TABLE 10.17
NRCS Pre- and Post-development Stormwater Hydrographs (Example 10.16)

Time Ratio	Discharge Ratio	Pre-development		Post-development		Time Ratio	Discharge Ratio	Pre-development		Post-development	
t/t_p	Q/Q_p	t, min	Q, ft³/s	t, min	Q, ft³/s	t/t_p	Q/Q_p	t, min	Q, ft³/s	t, min	Q, ft³/s
0.0	0.0	0.0	0.0	0.0	0.0	1.7	0.46	28.6	46	12.6	92
0.1	0.03	1.7	3.0	0.7	6.0	1.8	0.39	30.2	39	13.3	78
0.2	0.1	3.4	10	1.5	20	1.9	0.33	31.9	33	14.1	66
0.3	0.19	5.0	19	2.2	38	2	0.28	33.6	28	14.8	56
0.4	0.31	6.7	31	3.0	62	2.2	0.207	37	20.7	16.3	41.4
0.5	0.47	8.4	47	3.7	94	2.4	0.147	40.3	14.7	17.8	29.4
0.6	0.66	10.1	66	4.4	132	2.6	0.107	43.7	10.7	19.2	21.4
0.7	0.82	11.8	82	5.2	164	2.8	0.077	47	7.7	20.7	15.4
0.8	0.93	13.4	93	5.9	186	3.0	0.055	50.4	5.5	22.2	11
0.9	0.99	15.1	99	6.7	198	3.2	0.04	53.8	4.0	23.7	8.0
1.0	1.00	16.8	100	7.4	200	3.4	0.029	57.1	2.9	25.2	5.8
1.1	0.99	18.5	99	8.1	198	3.6	0.021	60.5	2.1	26.6	4.2
1.2	0.93	20.2	93	8.9	186	3.8	0.015	63.8	1.5	28.1	3.0
1.3	0.86	21.8	86	9.6	172	4.0	0.011	67.2	1.1	29.6	2.2
1.4	0.78	23.5	78	10.4	156	4.5	0.005	75.6	0.5	33.3	1.0
1.5	0.68	25.2	68	11.1	136	5.0	0.000	84	0.0	37	0.0
1.6	0.56	26.9	56	11.8	112						

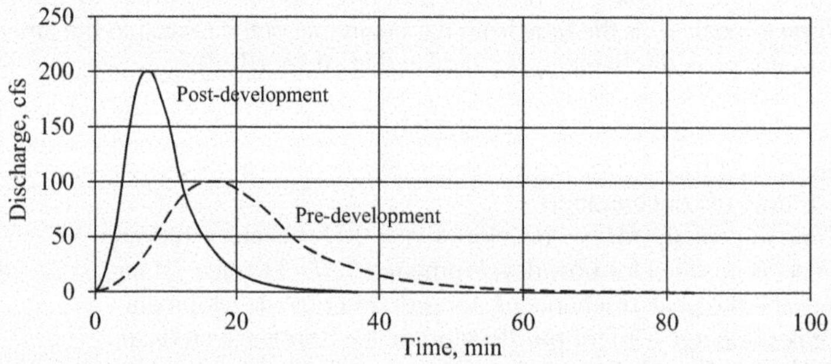

FIGURE 10.13 Synthetic pre- and post-development stormwater hydrographs (Example 10.16).

10.5.4 Rainfall-Runoff Regression Equations

The USGS has developed rainfall-runoff regression equations for ungauged urban and rural areas in the U.S. The equations may be used to estimate stormwater peak discharges with recurrence intervals (T) that range from 2 to 100 years. The equations for rural and urban areas are expressed by Eq. (10.11)[45]

$$Q_T = a(A)^b \quad \text{(rural area)} \tag{10.11a}$$

$$Q_T = a(A)^b c(\text{UI})^d \quad \text{(urban area)} \tag{10.11b}$$

where

Q_T = peak discharge with recurrence intervals (T), cf

A = drainage area, mi²

UI = urbanization index, which represents the percentage of the drainage area that is covered by impervious surfaces

a, b, c, d = regression coefficients. These coefficients are developed for different regions of the U.S. and are available on the USGS website.

In addition to the drainage area (A) and percentage of imperviousness (UI), other independent variables are considered for the development of the regression equation. These variables include basin characteristics such as stream length, stream slope, basin shape factor, and precipitation characteristics.

The USGS regression equations should not be used outside of their limits. These limitations may include the size of the drainage area, the percentage of imperviousness in urban areas, and basins with significant in-channel detention or basin storage.

Example 10.17: Peak Stormwater Peak Discharge from Rainfall-Runoff Regression Equations

The USGS has developed regression equations for different regions of the United States to estimate peak stormwater discharge.[45] The equations for Dallas-Fort Worth urban areas are given by Eqs. (10.12a)–(10.12f).[46]

$$Q_2 = 42.83(A)^{0.704}(\mathrm{UI})^{0.836} \tag{10.12a}$$

$$Q_5 = 82.92(A)^{0.724}(\mathrm{UI})^{0.751} \tag{10.12b}$$

$$Q_{10} = 120.7(A)^{0.73}(\mathrm{UI})^{0.697} \tag{10.12c}$$

$$Q_{25} = 184.8(A)^{0.745}(\mathrm{UI})^{0.632} \tag{10.12d}$$

$$Q_{50} = 246.4(A)^{0.752}(\mathrm{UI})^{0.587} \tag{10.12e}$$

$$Q_{100} = 362.1(A)^{0.752}(\mathrm{UI})^{0.510} \tag{10.12f}$$

where

Q_T = peak discharge with recurrence intervals (T), cfs

A = drainage area, mi²

UI = urbanization index, which represents the percentage of the drainage area that is covered by impervious surfaces

Calculate the peak discharge for a drainage area located in the Dallas-Fort Worth region for storms with 2-year and 100-year recurrence intervals. The drainage area is 40 mi² and it is 85% urbanized.

Solution

1. Calculate peak discharge for storms of a 2-year recurrence interval
   Apply Eq. (10.12a)

$$Q_2 = 42.83(A)^{0.704}\ (\mathrm{UI})^{0.836}$$

$$= 42.83(40)^{0.704}\ (0.85)^{0.836}$$

$$= 501.9 \text{ cfs}$$

2. Calculate peak discharge for storms of a 100-year recurrence interval
   Apply Eq. (10.12f)

$$Q_{100} = 362.1(A)^{0.752}\ (\mathrm{UI})^{0.510}$$

$$= 362.1(40)^{0.752}(0.85)^{0.510}$$

$$= 5{,}340.5 \text{ cfs}$$

10.5.5 Snyder's Hydrograph Method (Synthetic Hydrograph Methods)

The stormwater runoff volume and duration are required to design retention and detention ponds and similar storm drain structures. In these cases, stormwater hydrographs are needed rather than peak flows. If the basin is gauged, the stream hydrograph may be used to estimate the stormwater hydrograph. The hydrograph separation and calculation of direct runoff are discussed in Section 10.4.1. For basins where streamflow measurements are not available, *synthetic hydrographs* can be developed based on the physical characteristics of the watershed and storm characteristics. Two common methods for developing synthetic hydrographs are Snyder's method and the NRCS method. The *Snyder's hydrograph* method is given below. The NRCS method has been presented in Section 10.5.3.

The *Snyder's hydrograph* method is used to prepare a synthetic hydrograph. The peak stormwater flow and time parameters of the synthetic hydrographs may be calculated using Eqs. (10.13)–(10.18). These equations relate to characteristics such as physical geometry of the basin and storm duration.[26,28,37,40]

$$t_p = C_t(LL_C)^{0.3} \tag{10.13}$$

$$Q_p = \frac{C_P A}{t_p} \tag{10.14}$$

$$t_r = \frac{t_p}{5.5} \tag{10.15}$$

$$T = 3 + \frac{t_p}{8} \tag{10.16}$$

$$t_{pR} = t_p + 0.25(t_R - t_r) \tag{10.17}$$

$$Q_{pR} = Q_p \frac{t_p}{t_{pR}} \tag{10.18}$$

where

t_r = standard duration of rainfall excess, h

t_R = duration of rainfall excess other than the standard duration (t_r) adopted in a specific study, h

t_p = lag time from midpoint of rainfall excess duration (t_r) to the peak of the unit hydrograph, h

t_{pR} = lag time from midpoint of duration (t_R) to the peak of the unit hydrograph, h

T = time base of the hydrograph, d

Q_p = peak flow of unit hydrograph for the standard duration t_r, m³/s (cfs)

Q_{pR} = peak flow of unit hydrograph for duration t_R, m³/s (cfs)

L = distance along the stream from a study point to the upstream limit (upper end) of the basin, km (mi)

L_C = distance from a study point (lower end) to a point on the stream nearest the centroid of the basin area, km² (mi)

A = drainage area, km² (mi²)

C_t = regional coefficient representing the slope of the basin. Typical value of C_t range from 1.4 to 1.7 for distances (L and L_C) in kilometers and 1.8–2.2 for distances in miles. The larger values are associated with flatter basins.

C_p = regional coefficient indicating the basin storage capacity. The C_p varies between 0.15 and 0.19 for calculation in SI units and between 360 and 440 in U.S. customary units. The larger values represent steeper basins or highly urbanized areas with smaller storage capacity. [26,28,37,40]

In developing the synthetic hydrograph, the peak flow Q_p, lag time t_p, and duration of standard rainfall excess t_r are calculated from Eqs. (10.13) to (10.16). If the duration of storm t_R is different from the standard duration of rainfall t_r, the value of peak flow and lag time are adjusted using Eqs. (10.17) and (10.18).

The relationships developed by the U.S. Corps of Engineers (Eqs. 10.19 and 10.20) are used to plot the Snyder hydrograph using adjusted peak discharge (Q_{pR}), adjusted lag time (t_{pR}), time base of T, and two time parameters, W_{50} and W_{75}. W_{50} and W_{75} are the widths of the hydrograph at discharges of 50% and 75% of the peak discharge. The widths are distributed one-third before the peak discharge and two-third after. The Snyder hydrograph and its time parameters are shown schematically in Figure 10.14.

$$W_{50} = \frac{770A^{1.08}}{Q_{pR}^{1.08}} \quad \text{(U.S customary units)} \tag{10.19a}$$

$$W_{75} = \frac{440A^{1.08}}{Q_{pR}^{1.08}} \quad \text{(U.S. customary units)} \tag{10.19b}$$

$$W_{50} = \frac{0.23A^{1.08}}{Q_{pR}^{1.08}} \quad \text{(SI units)} \tag{10.20a}$$

$$W_{75} = \frac{0.13A^{1.08}}{Q_{pR}^{1.08}} \quad \text{(SI units)} \tag{10.20b}$$

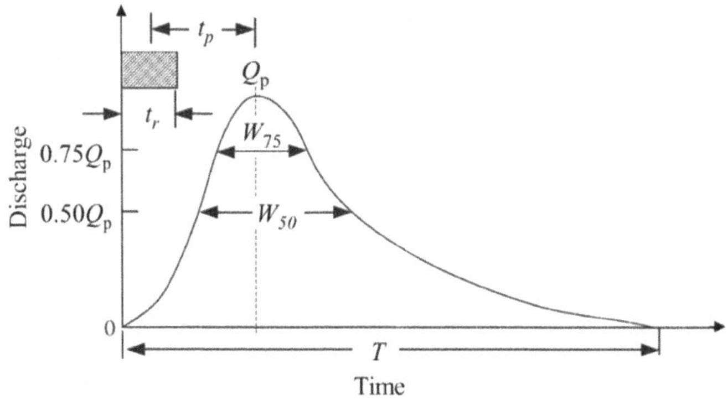

FIGURE 10.14 Snyder's hydrograph and its time parameters.

where

A = drainage area, km² (mi²)
Q_{pR} = adjusted peak flow of hydrograph for duration t_R, m³/s (cfs)
W_{50} = width of the hydrograph at discharges of 50% of Q_{pR}, h
W_{75} = width of the hydrograph at discharges of 75% of Q_{pR}, h

Example 10.18: Synthetic Hydrograph Developed from Snyder's Method

The basin shown below has a surface area (A) of 960 mi². The stream length from the outlet to the upstream boundary of the basin (L) is 70 miles, and the distance along the stream from the outlet to the closest point to the basin centroid (L_c) is 50 miles. (1) Derive a 6-h synthetic hydrograph using Snyder's method, (2) prove that the hydrograph developed by Snyder's method is a unit hydrograph, and (3) calculate the total runoff if the depth of water over the entire watershed is 2.25 in. Assume C_t=2.2 and C_p=360.

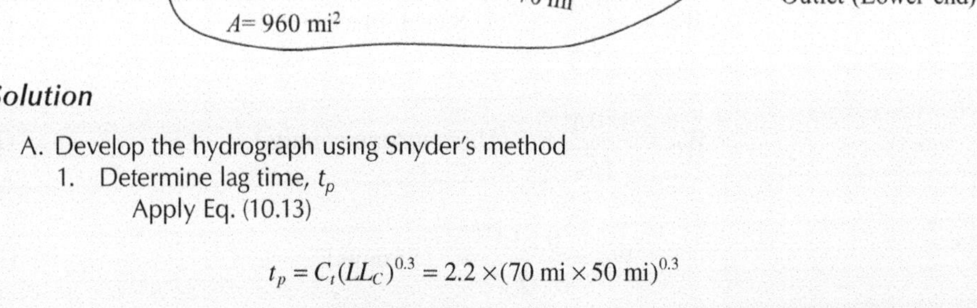

Solution

A. Develop the hydrograph using Snyder's method
   1. Determine lag time, t_p
      Apply Eq. (10.13)

$$t_p = C_t(LL_C)^{0.3} = 2.2 \times (70\text{ mi} \times 50\text{ mi})^{0.3}$$

$$= 25.45\text{ h}$$

   2. Determine peak discharge, Q_p
      Apply Eq. (10.14)

$$Q_p = \frac{C_p A}{t_p} = \frac{360 \times 960 \text{ mi}^2}{25.45 \text{ h}}$$

$$= 13{,}580 \text{ ft}^3/\text{s}$$

3. Calculate standard duration of rainfall excess, t_r
   Apply Eq. (10.15)

$$t_r = \frac{t_p}{5.5} = \frac{25.45 \text{ h}}{5.5}$$

$$= 4.63 \text{ h}$$

4. Compare calculated t_r in step 3 with rainfall duration $t_R = 6$ h
   Since t_r is less than t_R, the calculated t_r and Q_p in steps 1 and 2 should be revised
5. Calculate revised lag time (t_{pR}) for rainfall duration t_R
   Apply Eq. (10.17)

$$t_{pR} = t_p + 0.25(t_R - t_r) = 25.45 \text{ h} + 0.25(6 \text{ h} - 4.63 \text{ h})$$

$$= 25.79 \text{ h}$$

6. Calculate revised peak flow of hydrograph (Q_{pR}) for rainfall duration t_R
   Apply Eq. (10.18)

$$Q_{pR} = Q_p \frac{t_p}{t_{pR}} = 13{,}580 \text{ ft}^3/\text{s}\left(\frac{25.45 \text{ h}}{25.79 \text{ h}}\right)$$

$$= 13{,}401 \text{ ft}^3/\text{s}$$

7. Calculate the time base of the hydrograph, T
   From Eq. (10.16)

$$T = 3 + \frac{t_p}{8} = 3 + \frac{25.45 \text{ h}}{8}$$

$$= 6.2 \text{ day}$$

   Convert T from days to hours
   $T = 6.2 \text{ day} \times 24 \text{ h/day} = 148.4 \text{ h}$
8. Determine W_{50} and W_{75}
   Apply Eq. (10.19a) to determine W_{50}

$$W_{50} = \frac{770 A^{1.08}}{Q_{pR}^{1.08}} = \frac{770\ (960 \text{ mi}^2)^{1.08}}{(13{,}401 \text{ ft}^3/\text{s})^{1.08}} = 44.7 \text{ h}$$

   Apply Eq. (10.19b) to determine W_{75}

$$W_{75} = \frac{440 A^{1.08}}{Q_{pR}^{1.08}} = \frac{440 (960 \text{ mi}^2)^{1.08}}{(13{,}401 \text{ ft}^3/\text{s})^{1.08}} = 25.5 \text{ h}$$

9. Draw the hydrograph
   Calculate hydrograph runoff and time values

a. Runoff Q_0 = 0 cfs at time t_{lower} = 0 and t_{upper} = T = 148.4 h
   Width of hydrograph at Q_0 = 148.4 h
b. Runoff $Q_{pR_{max}} = 13{,}401$ cfs (peak flow)

$$P_r = t_R / 2 + t_{pR} = 6 \text{ h}/2 + 25.79 \text{ h}$$

$$= 28.8 \text{ h} (\text{the apex of the hydrograph})$$

c. Runoff Q_{50} = 50% of Q_{pR} = 0.5 × 13,401 cfs = 6,701 cfs

$$t_{lower} = P_r - 1/3W_{50} = 28.8 \text{ h} - 1/3 \times 44.7 \text{ h}$$

$$= 28.8 \text{ h} - 14.9 \text{ h}$$

$$= 13.9 \text{ h}$$

$$t_{upper} = P_r + 2/3W_{50} = 28.8 \text{ h} + 2/3 \times 44.7 \text{ h}$$

$$= 28.8 \text{ h} + 29.8 \text{ h}$$

$$= 58.6 \text{ h}$$

   Width of the hydrograph at W_{50} = 58.6 h − 13.9 h = 44.7 h
d. Runoff Q_{75} = 75% of Q_{PR} = 0.75 × 13,401 cfs = 10,051 cfs

$$t_{lower} = P_r - 1/3W_{75} = 28.8 \text{ h} - 1/3 \times 25.5 \text{ h}$$

$$= 28.8 \text{ h} - 8.5 \text{ h}$$

$$= 20.3 \text{ h}$$

$$t_{upper} = P_r + 2/3W_{75} = 28.8 \text{ h} + 2/3 \times 25.5 \text{ h}$$

$$= 28.8 \text{ h} + 17.0 \text{ h}$$

$$= 45.8 \text{ h}$$

   Width of hydrograph at W_{75} = 45.8 h − 20.3 h = 25.5 h

The calculated results are summarized in Table 10.18. The hydrograph is drawn in Figure 10.15 using the information given in Table 10.18.

TABLE 10.18
Synthetic (Calculated) Unit Hydrograph Using Snyder's Method (Example 10.18)

Time, h	Runoff, cfs
0.0	0
13.9	6,701
20.3	10,051
28.8	13,401
45.8	10,051
58.9	6,701
148.4	0

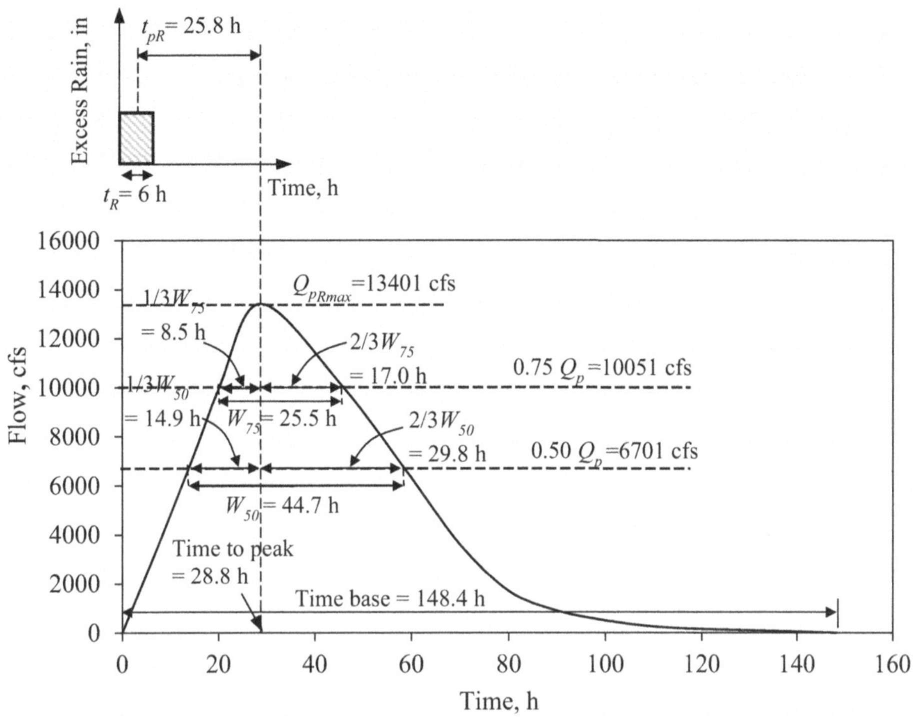

FIGURE 10.15 Synthetic unit hydrograph developed using Snyder's Method (Example 10.18).

B. Prove that the hydrograph developed from Snyder's method is a unit hydrograph
   1. Compute the area under the hydrograph

      The area under the hydrograph is obtained by graphical integration. The hydrograph starts from zero and reaches a peak flow or runoff of 13,401 cfs in a 28.8-h duration, and drops to zero in 148.4 h. The time scale (*h*) on the *X*-axis is divided into small sections. Each section is the time duration in hours. The average flow or runoff for each section in cfs is on the *Y*-axis. The graphical integration of the area under the hydrograph is summarized in Table 10.19. The integrated area under the hydrograph is 619,701 cfs.h.
   2. Determine the area of watershed in ft^2

$$\text{Watershed area} = 960\ \text{mi}^2$$

$$1\ \text{mi}^2 = 27.8784 \times 10^6\ \text{ft}^2$$

$$\text{Watershed area} = 960\ \text{mi}^2 \times 27.8784 \times 10^6\ \text{ft}^2/\text{mi}^2 = 2.676 \times 10^{10}\ \text{ft}^2$$

   3. Convert the integrated area under the hydrograph into volume of runoff

$$\text{Area under the hydrograph} = 619{,}701\ \text{cfs.h} \times 3{,}600\ \text{s/h} = 2.2309 \times 10^9\ \text{ft}^3$$

   4. Calculate depth of runoff over the entire watershed

$$\text{Depth of runoff over the entire watershed} = 2.2309 \times 10^9\ \text{ft}^3 / 2.676 \times 10^{10}\ \text{ft}^2$$

$$= 0.08337\ \text{ft} \times 12\ \text{in./ft} = 1.00\ \text{in.}$$

TABLE 10.19
Synthetic Unit Hydrograph Calculated from Snyder's Method (Example 10.18)

Time, h	Runoff, cfs	Average Runoff, cfs	Duration, h	Runoff × Duration, cfs.h
(1)	(2)	(3)	(4)	(5)
0[a]	0[a]			
13.9[a]	6,701[a]	3,351[b]	13.9[c]	46,535[d]
20.3[a]	10,051[a]	8,376	6.4	53,606
28.8[a]	13,401[a]	11,726	8.5	99,671
45.8[a]	10,051[a]	11,726	17	199,342
58.6[a]	6,701[a]	8,376	12.8	107,213
60	6,330	6,516	1.41	9,187
70	3,600	4,965	10	49,650
80	1,700	2,650	10	26,500
90	900	1,300	10	13,000
100	500	700	10	7,000
110	250	375	10	3,750
120	150	200	10	2,000
130	100	125	10	1,250
140	50	75	10	750
148.4	0	25	8.4	210
			Total =	619,701 cfs.h

[a] The coordinates (time and runoff) are calculated in step A9.
[b] The average runoff is the average of two consecutive runoffs = (0 + 6,701 cfs)/2 = 3,351 cfs.
[c] The duration is time between two consecutive runoffs = 13.9 h.
[d] 3,351 cfs × 13.9 h = 46,535 cfs.h.

Since the equivalent depth of direct runoff over the entire watershed is 1.0 in., it is a unit hydrograph

5. Calculate the total runoff if the depth of water over the entire watershed is 2.25 in.

$$\text{Total runoff} = 2.2309 \times 10^9 \text{ ft}^3 \times 2.25 \text{ in.} / 1.0 \text{ in.} = 5.0195 \times 10^9 \text{ ft}^3$$

10.5.6 Inlet Method

The inlet method may be used to estimate peak flow for storm sewer systems with multiple inlets. It accounts for channel storage and estimates the peak flow more accurately than the rational method. Stormwater peak flow estimation involves three steps: (1) calculate peak flows to each inlet based on a rainfall intensity with duration equal to the inlet time, (2) route the peak flow of sub-basins as it moves through the stormwater pipe, and (3) add each sub-basin hydrograph to determine the total hydrograph at the study point. Equation (10.21) is used to route the intercepted flow for each upstream inlet to the study point.[28,38,47] A minimum inlet time of 5 min is recommended for the upper most inlet. In urban areas, the inlet time may vary from 5 min in densely developed steep areas to 10–15 min in well-developed areas with relatively flat slopes. The inlet time of 20–30 min is common in very flat residential areas with widely spaced inlets.

$$Q_{oP} = Q_i\left(\frac{2T}{2T + 0.8\dfrac{L}{v}}\right) \quad (10.21)$$

where

Q_{oP} = peak outflow at the study point, cfs
Q_i = intercepted flow peak runoff rate, cfs
T = inlet travel time, min
L = length of storm sewer, ft
v = average velocity for storm sewer flow, ft/min

Once the peak outflow (Q_{oP}) is calculated for intercepted flow from each inlet (Q_i), the composite

peak flow at the design point is obtained by adding the ordinates of triangular hydrographs for each inlet. The triangular hydrograph of intercepted flow has time T to peak, and time base of $2T$. The time to peak for routed hydrographs is expressed by Eq. (10.22a), and time of base is expressed by Eq. (10.22b).

$$t_1 = T + 0.8L / v \quad (10.22a)$$

$$t_2 = \left(2T + 0.8\frac{L}{v}\right) \quad (10.22b)$$

where

t_1 = time to peak, min
t_2 = time of base, min

The procedure for developing triangular hydrographs and peak outflow is explained in detail in Example 10.19.

Example 10.19: Stormwater Hydrograph Using Inlet Method

A storm sewer system in Example 8.23 consists of four drainage areas. The sewer system is shown in Figure 10.16. The 5-year storm intensity is calculated from the following equation: $i = 108/(t_c + 18)$. The characteristics of the drainage system are provided in the table given below. Determine the 5-year design flow for each segment of the sewer line using the inlet method. Compare the design peak flows in three sewer lines calculated from the inlet method, and calculated in Example 8.23 using inlet time plus channel time.

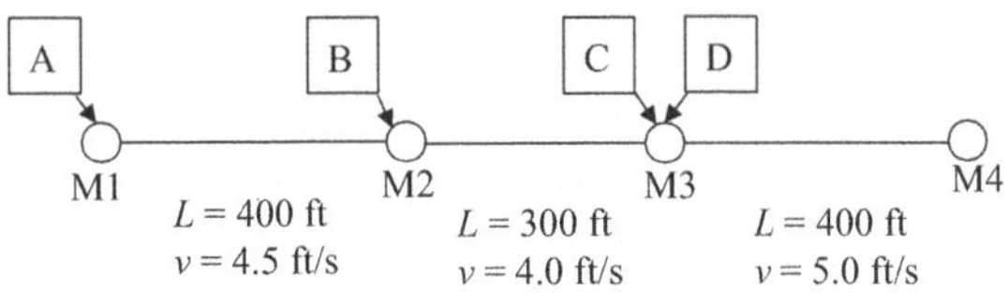

FIGURE 10.16 Stormwater drainage system (Example 10.19).

Drainage	Area, Acres	Runoff Coefficient	Inlet Time, min
A	5	0.9	8
B	12	0.8	12
C	8	0.6	10
D	6	0.5	6

Solution

1. Determine the intensity of rainfall for areas *A*, *B*, *C*, and *D*

$$i_A = \frac{108}{8+18} = 4.2 \text{ in./h}$$

$$i_B = \frac{108}{12+18} = 3.6 \text{ in./h}$$

$$i_C = \frac{108}{10+18} = 3.9 \text{ in./h}$$

$$i_D = \frac{108}{6+18} = 4.5 \text{ in./h}$$

The intensity of rainfall for areas *A*, *B*, *C*, and *D* is summarized in Table 10.20.

2. Calculate 5-year peak sewer flow for each inlet

Apply Eq. (8.16b) to calculate peak flow

$$Q = CIA$$

$$Q_A = 0.9 \times 4.2 \text{ in./h} \times 5 \text{ acre} = 18.9 \text{ cfs}$$

$$Q_A = 0.8 \times 3.6 \text{ in/h} \times 12 \text{ acre} = 34.6 \text{ cfs}$$

$$Q_A = 0.6 \times 3.9 \text{ in/h} \times 8 \text{ acre} = 18.7 \text{ cfs}$$

$$Q_A = 0.5 \times 4.5 \text{ in/h} \times 6 \text{ acre} = 13.5 \text{ cfs}$$

The peak sewer flows for areas *A*, *B*, *C*, and *D* are summarized in Table 10.20.

TABLE 10.20
Rainfall Intensity and Peak Sewer Flow for Drainage Areas *A*–*D* (Example 10.19)

Drainage	Area, acres	Runoff Coefficient	Inlet Time, min	Rainfall Intensity, in./h	Intercepted Flow Peak Q_i, cfs
A	5	0.9	8	4.20	18.9
B	12	0.8	12	3.60	34.6
C	8	0.6	10	3.90	18.7
D	6	0.5	6	4.50	13.5

3. Calculate routed peak flow through the storm sewer system
   a. Area *A*
      Intercepted flow from area *A* enters M1, and it is routed from M1 to M4
      i. Intercepted flow from area *A* to M1
         Q_i = 18.9 cfs (Table 10.20)
         Time to peak of the intercepted flow
         $t_1 = T = 8$ min (Table 10.20)
         Time for the base of intercepted flow
         $t_2 = 2T = 16$ min
      ii. Peak outflow at M2
         From Figure 10.16, L = 400 ft and v = 4.5 ft/s
         Q_{oP} is calculated from *Eq. (10.21)*

$$Q_{oP} = Q_i \left(\frac{2T}{2T + 0.8\dfrac{L}{v}} \right)$$

$$Q_{oP} = (18.9 \text{ cfs}) \left(\frac{2 \times 8 \text{ min}}{2 \times 8 \text{ min} + 0.8 \left(\dfrac{400 \text{ ft}}{4.5 \text{ ft/s}} \right) (\text{min/ } 60 \text{ s})} \right)$$

$$Q_{oP} = (18.9 \text{ cfs}) \left(\frac{16 \text{ min}}{16 \text{ min} + 2.2 \text{ min}} \right) = 17.6 \text{ cfs}$$

Time to peak the routed hydrograph from Eq. (10.22a)

$$t_1 = 8 \text{ min} + 0.8 \times \frac{400 \text{ ft}}{4.5 \text{ ft/s}} \times \text{min} / 60 \text{ s}$$

$$= 8 \text{ min} + 1.2 \text{ min} = 9.2 \text{ min}$$

Time for base of routed hydrograph from Eq. (10.22b)

$$t_2 = 2 \times 8 \text{ min} + 0.8 \times \frac{400 \text{ ft}}{4.5 \text{ ft/s}} \times \text{min/ } 60 \text{ s}$$

$$= 16 \text{ min} + 1.2 \text{ min} = 17.2 \text{ min}$$

      iii. Peak outflow at M3
         From Figure 10.16, L = 400 ft, v = 4.5 ft/s; and L = 300 ft, v = 4.0 ft/s

$$Q_{oP} = (18.9 \text{ cfs}) \left(\frac{2 \times 8 \text{ min}}{2 \times 8 \text{ min} + 0.8 \left(\dfrac{400 \text{ ft}}{4.5 \text{ ft/s}} + \dfrac{300 \text{ ft}}{4.0 \text{ ft/s}} \right) (\text{min/ } 60 \text{ s})} \right)$$

$$Q_{oP} = (18.9 \text{ cfs}) \left(\frac{16 \text{ min}}{16 \text{ min} + 2.2 \text{ min}} \right) = 16.6 \text{ cfs}$$

Time to peak of the routed hydrograph from Eq. (10.22a)

$$t_1 = 8\text{ min} + 0.8\left(\frac{400\text{ ft}}{4.5\text{ ft/s}} + \frac{300\text{ ft}}{4.0\text{ ft/s}}\right)\text{min/ 60 s}$$

$$= 8\text{ min} + 2.2\text{ min} = 10.2\text{ min}$$

Time for base of routed hydrograph from Eq. (10.22b)

$$t_2 = 2\times 8\text{ min} + 0.8\text{ min}\left(\frac{400\text{ ft}}{4.5\text{ ft/s}} + \frac{300\text{ ft}}{4.0\text{ ft/s}}\right)\text{min/60 s}$$

$$= 16\text{ min} + 2.2\text{ min} = 18.2\text{ min}$$

iv. Peak outflow at M4

From Figure 10.16, L=400 ft, v=4.5; L=300 ft, v=4.0 ft/s; and L=400 ft, v=5.0 ft/s

$$Q_{oP} = (18.9\text{ cfs})\left(\frac{2\times 8\text{ min}}{2\times 8\text{ min} + 0.8\left(\frac{400\text{ ft}}{4.5\text{ ft/s}} + \frac{300\text{ ft}}{4.0\text{ ft/s}} + \frac{400\text{ ft}}{5.0\text{ ft/s}}\right)(\text{min/ 60 s})}\right)$$

$$Q_{oP} = (18.9\text{ cfs})\left(\frac{16\text{ min}}{16\text{ min} + 3.3\text{ min}}\right) = 15.7\text{ cfs}$$

Time to peak of the routed hydrograph from Eq. (10.22a)

$$t_1 = 8\text{ min} + 0.8\left(\frac{400\text{ ft}}{1.5\text{ ft/s}} + \frac{300\text{ ft}}{4.0\text{ ft/s}} + \frac{400\text{ ft}}{5.0\text{ ft/s}}\right)$$

$$= 8\text{ min} + 3.3\text{ min} = 11.3\text{ min}$$

Time for base of routed hydrograph from Eq. (10.22b)

$$t_2 = 2\times 8\text{ min} + 3.3\text{ min} = 19.3\text{ min}$$

v. Develop the design hydrograph for area *A*

The coordinates of the hydrographs for area A at M1, M2, M3, and M4 are summarized in Table 10.21. These hydrographs are shown in Figure 10.17.

b. Area *B*

Intercepted flow from area B enters M2, and it is routed from M2 to M4.

i. Intercepted flow from area *B* to M2

Q_i=34.6 cfs (Table 10.20)

Time to peak of the intercepted flow

t_1=T=12 min (Table 10.20)

Time for the base of intercepted flow

t_2=2T=24 min

ii. Peak outflow at M3.

From Figure 10.16, L=300 ft, and v=4.0 ft/s

TABLE 10.21
Coordinates of the Hydrographs for Area *A* at M1, M2, M3, and M4 (Example 10.19)

Intercepted Flow from Area *A* to M1		Outflow at M2 from Area *A*		Outflow at M3 from Area *A*		Outflow at M4 from Area *A*	
t, min	*Q*, cfs	*t*, min	*Q*, cfs	*t*, min	*Q*, cfs	*t*, min	*Q*, cfs
0	0	0	0	0	0	0	0
8	18.9	9.2	17.6	10.2	16.6	11.3	15.7
16	0	17.2	0	18.2	0	19.3	0

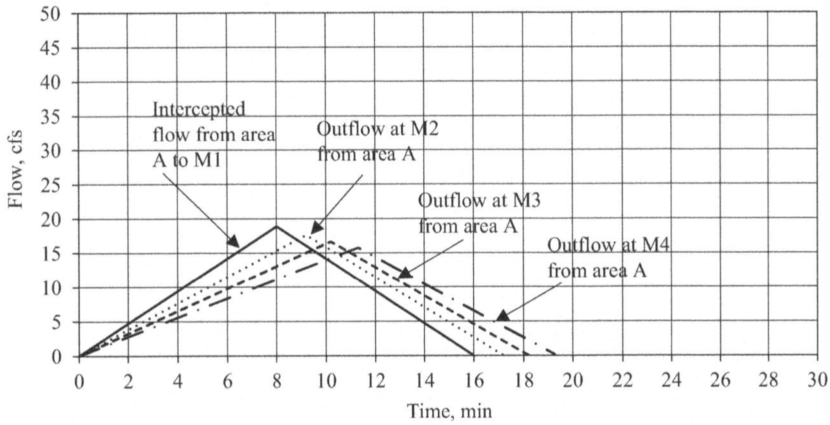

FIGURE 10.17 Hydrographs of intercepted stormwater from area *A* to M1 and routed flow at M2, M3, and M4 (Example 10.19).

$$Q_{oP} = (34.6\ \text{cfs})\left(\frac{2 \times 12\ \text{min}}{2 \times 12\ \text{min} + 0.8\left(\dfrac{300\ \text{ft}}{4.0\ \text{ft/s}}\right)(\text{min}/\ 60\ \text{s})}\right)$$

$$Q_{oP} = (34.6\ \text{cfs})\left(\frac{24\ \text{min}}{24\ \text{min} + 1.0\ \text{min}}\right) = 33.2\ \text{cfs}$$

Time to peak of the routed hydrograph from Eq. (10.22a)

$$t_1 = 12\ \text{min} + 0.8 \times \frac{300\ \text{ft}}{4\ \text{ft/s}} \times \text{min}/\ 60\ \text{s}$$

$$= 12\ \text{min} + 1.0\ \text{min} = 13.0\ \text{min}$$

Time for base of routed hydrograph from Eq. (10.22b)

$$t_2 = 2 \times 12\ \text{min} + 0.8 \times \frac{300\ \text{ft}}{4\ \text{ft/s}} \times \text{min}\ /\ 60\ \text{s}$$

$$= 24\ \text{min} + 1.0\ \text{min} = 25.0\ \text{min}$$

iii. Peak outflow at M4.

From Figure 10.16, L=300 ft, v=4.0 ft/s; and L=400 ft, v=5.0 ft/s

$$Q_{oP} = (34.6\ \text{cfs})\left(\frac{2\times 12\ \text{min}}{2\times 12\ \text{min} + 0.8\left(\frac{300\ \text{ft}}{4.0\ \text{ft/s}} + \frac{400\ \text{ft}}{5.0\ \text{ft/s}}\right)(\text{min}/\ 60\ \text{s})}\right)$$

$$Q_{oP} = (34.6\ \text{cfs})\left(\frac{24\ \text{min}}{24\ \text{min} + 2.1\ \text{min}}\right) = 31.8\ \text{cfs}$$

Time to peak the routed hydrograph from Eq. (10.22a)

$$t_1 = 12\ \text{min} + 0.8\left(\frac{300\ \text{ft}}{4.0\ \text{ft/s}} + \frac{400\ \text{ft}}{5.0\ \text{ft/s}}\right)\times \text{min}/60\ \text{s}$$

$$= 12\ \text{min} + 2.1\ \text{min} = 14.1\ \text{min}$$

Time for base of routed hydrograph from Eq. (10.22b)

$$t_2 = 2\times 12\ \text{min} + 2.1\ \text{min} = 26.1\ \text{min}$$

iv. Develop the design hydrograph for area *B*

The coordinates of the hydrographs for area B at M2, M3, and M4 are summarized in Table 10.22. These hydrographs are shown in Figure 10.18.

c. Area *C*

Intercepted flow from area *C* enters M3, and it is routed from M3 to M4.

i. Intercepted flow from area C to M3

Q_i=18.7 cfs (Table 10.20)

Time to peak of the intercepted flow

t_1=T=10 min (Table 10.20)

Time for the base of intercepted flow

t_2=2T=20 min

ii. Peak outflow at M4

From Figure 10.20, L=400 ft, v=5.0 ft/s

TABLE 10.22
Coordinates of the Hydrographs for Area *B* at M2, M3, and M4 (Example 10.19)

Intercepted Flow from Area *B* to M2		Outflow at M3 from Area *B*		Outflow at M4 from Area *B*	
t, min	*Q*, cfs	*t*, min	*Q*, cfs	*t*, min	*Q*, cfs
0	0	0	0	0	0
12	34.6	13.0	33.2	14.1	31.8
24	0	25	0	26.1	0

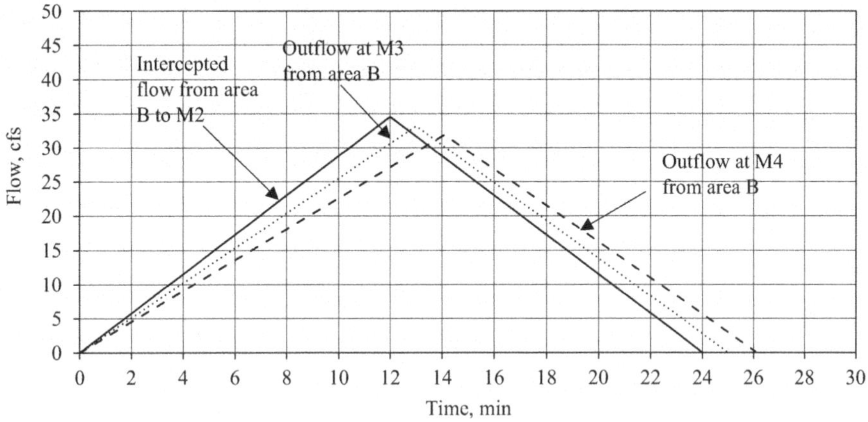

FIGURE 10.18 Hydrographs of intercepted stormwater from area *B* to M2 and routed flow to M3 and M4 (Example 10.19).

$$Q_{oP} = (18.7 \text{ cfs})\left(\frac{2\times 10 \text{ min}}{2\times 10 \text{ min}+0.8\left(\dfrac{400 \text{ ft}}{5.0 \text{ ft/s}}\right)(\text{min}/60 \text{ s})}\right)$$

$$Q_{oP} = (18.7 \text{ cfs})\left(\frac{20 \text{ min}}{20 \text{ min}+1.1 \text{ min}}\right) = 17.7 \text{ cfs}$$

Time to peak of the routed hydrograph from Eq. (10.22a)

$$t_1 = 10 \text{ min} + 0.8\times\frac{400 \text{ ft}}{5.0 \text{ ft/s}}\times \text{min}/60 \text{ s}$$

$$= 10 \text{ min}+1.1 \text{ min} = 11.1 \text{ min}$$

Time for base of routed hydrograph from Eq. (10.22b)

$$t_2 = 2\times 10 \text{ min} + 0.8\times\frac{400 \text{ ft}}{5.0 \text{ ft/s}}\times \text{min}/60 \text{ s}$$

$$= 20 \text{ min}+1.1 \text{ min} = 21.1 \text{ min}$$

iii. Develop the design hydrograph for area *C*

The coordinates of the hydrographs for area C at M3 and M4 are summarized in Table 10.23. These hydrographs are shown in Figure 10.19.

d. Area *D*

Intercepted flow from area *D* enters M3, and it is routed from M3 to M4

i. Intercepted flow from area *C* to M3

Q_i=13.5 cfs (Table 10.20)

Time to peak of the intercepted flow

t_1=T=6 min (Table 10.20)

Time for the base of intercepted flow

t_2=2T=12 min

ii. Peak outflow at M4

From Figure 10.20, L=400 ft, v=5.0 ft/s

TABLE 10.23
Coordinates of the Hydrographs for Area *C* at M3 and M4 (Example 10.19)

Intercepted Flow from Area *C* to M3		Outflow at M4 from Area *C*	
t, min	*Q*, cfs	*t*, min	*Q*, cfs
0	0	0	0
10	18.7	11.1	17.7
20	0	21.1	0

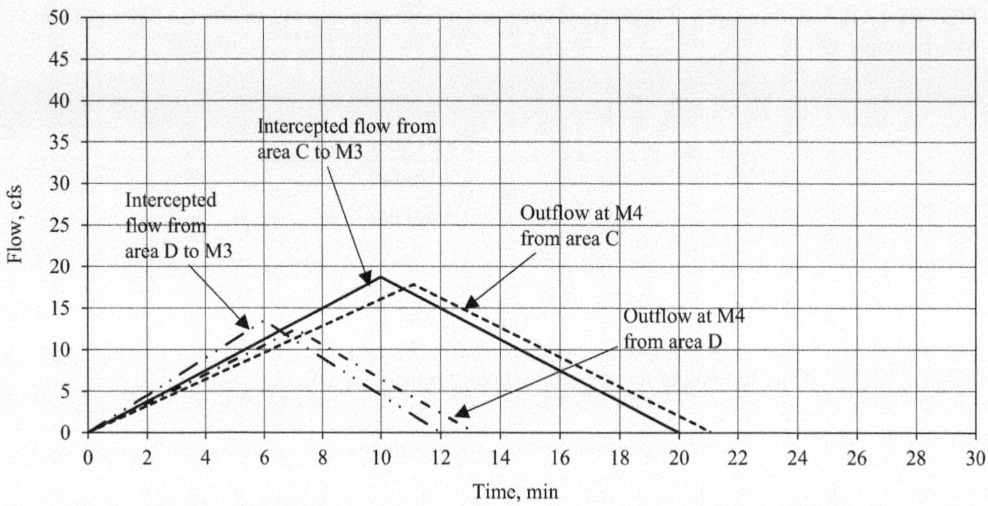

FIGURE 10.19 Hydrographs of intercepted stormwater from area *C* and *D* to M3 and routed flows to M4 (Example 10.19).

$$Q_{oP} = (13.5\ \text{cfs})\left(\frac{2\times 6\ \text{min}}{2\times 6\ \text{min} + 0.8\left(\dfrac{400\ \text{ft}}{5.0\ \text{ft/s}}\right)(\text{min}/60\ \text{s})}\right)$$

$$Q_{oP} = (13.5\ \text{cfs})\left(\frac{12\ \text{min}}{12\ \text{min} + 1.1\ \text{min}}\right) = 12.4\ \text{cfs}$$

Time to peak the routed hydrograph from Eq. (10.22a)

$$t_1 = 6\ \text{min} + 0.8\times\left(\frac{400\ \text{ft}}{5.0\ \text{ft/s}}\right)\times \text{min}/60\ \text{s}$$

$$= 6\ \text{min} + 1.1\ \text{min} = 7.1\ \text{min}$$

Time for base of routed hydrograph from Eq. (10.22b)

$$t_2 = 2 \times 6 \text{ min} + 0.8 \times \frac{400 \text{ ft}}{5.0 \text{ ft/s}} \times \text{min}/60 \text{ s}$$

$$= 12 \text{ min} + 1.1 \text{ min} = 13.1 \text{ min}$$

iii. Develop a design hydrograph for area *D*

The coordinates of the hydrographs for area D at M3 and M4 are summarized in Table 10.24. These hydrographs are shown in Figure 10.19.

4. Develop design hydrographs for three segments of the sewer line
   a. Segment M1-M2

      The developed hydrographs for area *A* are shown in Figure 10.17, and coordinates are provided in Table 10.21. The intercepted flow hydrographs from area *A* at M1 are selected for the design hydrograph for this segment. The maximum design flow is 18.9 cfs.
   b. Segment M2-M3

      The hydrographs for this segment are developed in Figures 10.17 and 10.18, and coordinates are provided in Table 10.21. The outflow at M2 from area *A* in Figure 10.17 and the intercepted flow from area *B* to M2 in Figure 10.18 are used to develop the design flow in this segment.

      i. The ordinates of hydrograph for area *A* at M2 are calculated by linear interpolation of discharges at $t=0$ min and $t=9.2$ min, and discharges at $t=9.2$ min and $t=17.2$ min. The sample calculations are given below:
      Between t=0 min and t=9.2 min

$$Q \text{ at } 1 \text{ min} = 0 \text{ cfs} + \frac{(17.6 \text{ cfs} - 0 \text{ cfs})}{(9.2 \text{ min} - 0 \text{ min})} \times (1 \text{ min} - 0 \text{ min}) = 1.9 \text{ cfs}$$

Between t=9.2 min and t=17.2 min

$$Q \text{ at } 10 \text{ min} = 17.6 \text{ cfs} + \frac{(17.6 \text{ cfs} - 0 \text{ cfs})}{(17.2 \text{ min} - 9.2 \text{ min})} \times (10 \text{ min} - 9.2 \text{ min}) = 15.9 \text{ cfs}$$

The linear interpolated discharge values for various assumed time intervals for area A are summarized in Table 10.25.

ii. The ordinates of hydrograph for area *B* at M2 are calculated by linear interpolation of discharges at $t=0$ min and $t=12$ min, and discharges at $t=12$ min and $t=24$ min. The sample calculations are given below:
Between t=0 min and t=12 min

TABLE 10.24
Coordinates of the Hydrographs for Area *D* at M3 and M4 (Example 10.19)

Intercepted Flow from Area *D* to M3		Outflow at M4 from Area *D*	
t, min	*Q*, cfs	*t*, min	*Q*, cfs
0	0	0	0
6	13.5	7.1	12.4
12	0	13.1	0

$$Q \text{ at } 1 \text{ min} = 0 \text{ cfs} + \frac{(34.6 \text{ cfs} - 0 \text{ cfs})}{(12 \text{ min} - 0 \text{ min})} \times (1 \text{ min} - 0 \text{ min}) = 2.9 \text{ cfs}$$

Between t = 12 min and t = 24 min

$$Q \text{ at } 13 \text{ min} = 34.6 \text{ cfs} + \frac{(34.6 \text{ cfs} - 0 \text{ cfs})}{(24 \text{ min} - 12 \text{ min})} \times (13 \text{ min} - 12 \text{ min}) = 31.7 \text{ cfs}$$

The linearly interpolated discharge values for various assumed time intervals for area B are summarized in Table 10.25. The data presented in Table 10.25 are used to prepare the hydrographs of outflow at M2 from area A, and intercepted flow from area B to M2. These hydrographs are shown in Figure 10.20.

TABLE 10.25
Coordinates of the Design Hydrographs for Segment M2-M3 (Example 10.19)

	Q, cfs				*Q*, cfs		
t, min	Area *A*	Area *B*	*A*+*B*	*t*, min	Area *A*	Area *B*	*A*+*B*
0	0.0	0.0	0.0	13	9.3	31.7	40.9
1	1.9	2.9	4.8	14	7.1	28.8	35.9
2	3.8	5.8	9.6	15	4.9	25.9	30.8
3	5.7	8.6	14.4	16	2.7	23.0	25.7
4	7.7	11.5	19.2	17	0.5	20.2	20.6
5	9.6	14.4	24.0	17.2	0.0	19.6	19.6
6	11.5	17.3	28.8	18	0.0	17.3	17.3
7	13.4	20.2	33.6	19	0.0	14.4	14.4
8	15.3	23.0	38.4	20	0.0	11.5	11.5
9	17.2	25.9	43.2	21	0.0	8.6	8.6
9.2	17.6	26.5	44.1	22	0.0	5.8	5.8
10	15.9	28.8	44.7	23	0.0	2.9	2.9
11	13.7	31.7	45.3	24	0.0	0.0	0.0
12	11.5	34.6	46.0				

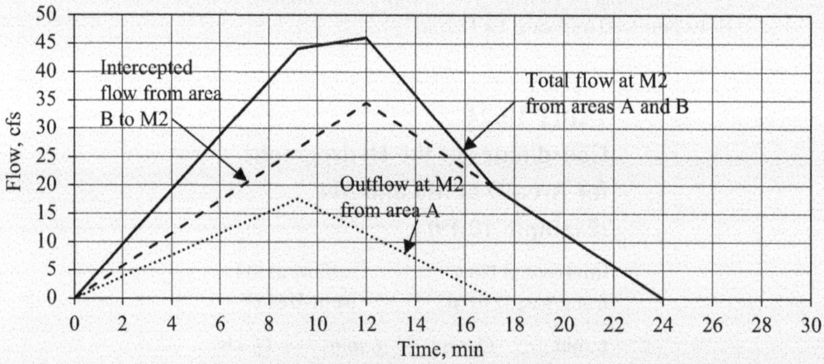

FIGURE 10.20 Design hydrograph for segment M2–M3. The hydrograph is constructed by adding hydrograph of intercepted stormwater from area *B* at M2 and routed hydrograph from area *A* at M2 (Example 10.19).

The design hydrograph for this section is developed by adding the ordinates of hydrographs of outflow at M2 from area A to the ordinates of hydrographs of intercepted flow from area B to M2. The coordinates of the resulting hydrographs are also summarized in Table 10.25, and the combined hydrograph is shown in Figure 10.20. The peak flow of 46.0 cfs is the design flow for segment M2-M3.

c. Segment M3-M4

The hydrographs for this segment are developed from Figures 10.17, 10.18, and 10.19. The outflow at M3 from area *A* in Figure 10.17, the intercepted flow from area *B* in Figure 10.18, and the intercepted flow from areas *C* and *D* in Figure 10.19 are used to develop the design flow in this segment.

i. The ordinates of hydrograph for area *A* at M3 are calculated by linear interpolation of discharges at t=0 min and t=10.2 min, and discharges at t=10.2 min and t=18.2 min. The sample calculations are given below:
Between t=0 min and t=10.2 min

$$Q \text{ at } 1 \text{ min} = 0 \text{ cfs} + \frac{(16.6 \text{ cfs} - 0 \text{ cfs})}{(10.2 \text{ min} - 0 \text{ min})} \times (1 \text{ min} - 0 \text{ min}) = 1.6 \text{ cfs}$$

Between t=10.2 min and t=18.2 min

$$Q \text{ at } 11 \text{ min} = 16.6 \text{ cfs} + \frac{(16.6 \text{ cfs} - 0 \text{ cfs})}{(18.2 \text{ min} - 10.2 \text{ min})} \times (11 \text{ min} - 10.2 \text{ min}) = 15.0 \text{ cfs}$$

The linear interpolated discharge values for various assumed time intervals for area A are summarized in Table 10.26

ii. The ordinates of hydrograph for area *B* at M3 are calculated by linear interpolation of discharges at t=0 min and t=13 min, and discharges at t=13 min and t=25 min. The sample calculations are given below:
Between t=0 min and t=13 min

TABLE 10.26
Coordinates of the Design Hydrographs for Segment M3-M4 (Example 10.19)

	Q, cfs						Q, cfs				
t, min	Area *A*	Area *B*	Area C	Area *D*	*A*+*B*+*C*+*D*	*t*, min	Area *A*	Area *B*	Area C	Area *D*	*A*+*B*+*C*+*D*
0	0.0	0.0	0.0	0.0	0.0	13.1	10.7	32.9	13.0	0.0	56.6
1	1.6	2.6	1.9	2.3	8.3	14	8.8	30.4	11.2	0.0	50.4
2	3.3	5.1	3.7	4.5	16.6	15	6.7	27.6	9.4	0.0	43.7
3	4.9	7.7	5.6	6.8	24.9	16	4.6	24.9	7.5	0.0	37.0
4	6.5	10.2	7.5	9.0	33.2	17	2.5	22.1	5.6	0.0	30.3
5	8.2	12.8	9.4	11.3	41.5	18	0.4	19.4	3.7	0.0	23.5
6	9.8	15.3	11.2	13.5	49.8	18.2	0.0	18.8	3.4	0.0	22.3
7	11.4	17.9	13.1	11.2	53.6	19	0.0	16.6	1.9	0.0	18.5
8	13.1	20.4	15.0	9.0	57.5	20	0.0	13.8	0.0	0.0	13.8
9	14.7	23.0	16.8	6.8	61.3	21	0.0	11.1	0.0	0.0	11.1
10	16.3	25.5	18.7	4.5	65.1	22	0.0	8.3	0.0	0.0	8.3
10.2	16.7	26.0	18.3	4.1	65.1	23	0.0	5.5	0.0	0.0	5.5
11	15.0	28.1	16.8	2.3	62.2	24	0.0	2.8	0.0	0.0	2.8
12	12.9	30.6	15.0	0.0	58.5	25	0.0	0.0	0.0	0.0	0.0
13	10.8	33.2	13.1	0.0	57.1						

$$Q \text{ at } 1 \text{ min} = 0 \text{ cfs} + \frac{(33.2 \text{ cfs} - 0 \text{ cfs})}{(13 \text{ min} - 0 \text{ min})} \times (1 \text{ min} - 0 \text{ min}) = 2.6 \text{ cfs}$$

Between t=13.0 min and t=25 min

$$Q \text{ at } 13.1 \text{ min} = 33.2 \text{ cfs} + \frac{(33.2 \text{ cfs} - 0 \text{ cfs})}{(25 \text{ min} - 13 \text{ min})} \times (13.1 \text{ min} - 13 \text{ min}) = 32.9 \text{ cfs}$$

iii. The ordinates of hydrograph for area *C* at M3 are calculated by linear interpolation of discharges at *t*=0 min and *t*=10 min, and discharges at *t*=10 min and *t*=20 min. The sample calculations are given below:
Between t=0 min and t=10 min

$$Q \text{ at } 1 \text{ min} = 0 \text{ cfs} + \frac{(18.7 \text{ cfs} - 0 \text{ cfs})}{(10 \text{ min} - 0 \text{ min})} \times (1 \text{ min} - 0 \text{ min}) = 1.9 \text{ cfs}$$

Between t=10 min and t=20 min

$$Q \text{ at } 11 \text{ min} = 18.7 \text{ cfs} - \frac{(18.7 \text{ cfs} - 0 \text{ cfs})}{(20 \text{ min} - 0 \text{ min})} \times (11 \text{ min} - 10 \text{ min}) = 16.8 \text{ cfs}$$

iv. The ordinates of hydrograph for area *D* at M3 are calculated by linear interpolation of discharges at *t*=0 min and t=6 min, and discharges at *t*=6 min and *t*=12 min. The sample calculations are given below:
Between t=0 min and t=6 min

$$Q \text{ at } 1 \text{ min} = 0 \text{ cfs} + \frac{(13.5 \text{ cfs} - 0 \text{ cfs})}{(6 \text{ min} - 0 \text{ min})} \times (1 \text{ min} - 0 \text{ min}) = 2.3 \text{ cfs}$$

Between t=6 min and t=12 min

$$Q \text{ at } 7 \text{ min} = 13.5 \text{ cfs} + \frac{(13.5 \text{ cfs} - 0 \text{ cfs})}{(12 \text{ min} - 6 \text{ min})} \times (7 \text{ min} - 6 \text{ min}) = 11.2 \text{ cfs}$$

The linearly interpolated discharge values for various assumed time intervals for areas C and D are summarized in Table 10.26. The data developed in Table 10.26 are used to prepare the hydrographs of outflow at M3 from areas A and B, and intercepted flows from areas C and D at M3. These hydrographs are shown in Figures 10.17–10.19, and also shown in Figure 10.20. The coordinates of design hydrograph for segments M3-M4 and combined hydrographs are summarized in Table 10.26 and are shown in Figure 10.21. The design hydrograph for this segment is developed by adding the ordinates of hydrograph of outflow at M3 from areas A and B to the ordinates of hydrographs of intercepted flow from areas C and D. The coordinates of design hydrograph are given in Table 10.26, and hydrograph is shown in Figure 10.21. The peak design flow for segments M3 and M4 is 65.1 cfs.

5. Summarize the design peak flows calculated for three sewer lines
   The design flow calculated for three sewer lines is summarized in Table 10.27
6. Compare the peak design flows calculated in Examples 10.19 and 8.23
   Examples 8.23 and 10.19 have identical data. The design flows are also calculated in Example 8.23 using the inlet time plus channel time. The design flows for the three sewer lines from Example 8.23 are: M1−M2=18.9 cfs, M2−M3=50.8 cfs, and M3−M4=75.8 cfs. These design flows are higher than those obtained by the inlet method, and therefore are more conservative than those calculated from the inlet method in Example 10.19. The inlet method involves a complex and tedious

TABLE 10.27
The Design Peak Flows in Sewer Lines M1-M2, M2-M3, and M3-M4 (Example 10.19)

Sewer Lines	Peak Design Flow, cfs
M1-M2	18.9
M2-M3	46.0
M3-M4	65.1

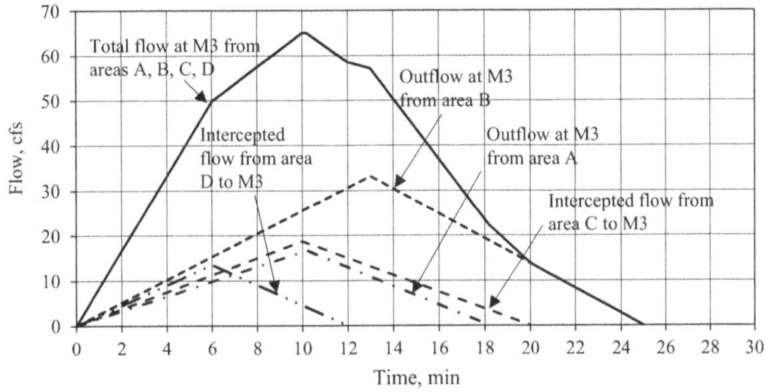

FIGURE 10.21 Design hydrograph for segment M3–M4. The hydrograph is constructed by adding the hydrograph of intercepted stormwater from areas *C* and *D* at M3 and routed flows from areas *A* and *B* at M3 (Example 10.19).

procedure to calculate the design flows, particularly in large storm sewer networks. The accuracy gained by using the inlet method may not be worth the effort in developing more accurate design flows.

10.5.7 Computer Simulation Techniques

Several computer models have been developed in the past decades for rainfall-runoff analysis. These models simulate either events or continuous rainfall, and can be used for the planning and design of stormwater facilities.[28,38] The event-based models incorporate infiltration and surface runoff to estimate direct runoff for a single storm. These models require less data for model calibration, verification, and analysis. The examples of event-based models are the HEC-HMS (Hydrologic Engineering Center - Hydrologic Modeling System) and WinTR-55 models. The continuous models account for all components of the basin's water balance equations, including evapotranspiration, which is neglected in the event-based models. These models are more sophisticated and require a great deal of information. Hydrologic processes are continuously simulated by these models to predict the future trend of surface runoff, baseflow, direct runoff, and total discharge. Many of these models are developed for both rainfall-runoff analysis and water quality assessment, and to predict the future impacts of urbanization and land use or management practices. The examples of these models are Stormwater Management Model (SWMM), Storage, Treatment, Overflow, Runoff Model (STORM), and Soil Water Assessment Tool (SWAT) models.[48,49]

Numerous computer software packages are available for use in urban stormwater analysis and design.[50–53] A survey of available software packages has been conducted to help hydrologists and engineers choose among many options. Federal agency software has been inventoried and summarized in Ref. 54. The American Society of Civil Engineers' Task Committee on Microcomputers software in Urban Hydrology inventoried the makeup, cost, and applicability of over 40 packages from commercial vendors and public domain sources.[52]

A comprehensive review of urban stormwater management computer models can be found in Refs. 48–51,55. A brief description of event-based and continuous models is presented below.

HEC-HMS: The HEC-HMS is a rainfall-runoff model developed by the United States Army Corps of Engineers. The model estimates runoff hydrographs for a basin and its sub-basins based on precipitation and by accounting for infiltration losses. The calculated runoff is then routed through open channels in the basin. The program utilizes a number of hydrological analyses for simulating infiltration losses, transforming excess precipitation into runoff hydrographs, and channel and storage routing. The required data include precipitation, evapotranspiration, runoff flow of the basin (for model calibration and validation), storage, and geographical information of the basin. HEC-HMS is used in combination with HEC-RAS (Hydrologic Engineering Center's River Analysis System) for the calculation of both hydrology and hydraulics of a stormwater system.[56]

WinTR-55: The Urban Hydrology for Small Watersheds, known as Technical Release 55 (WinTR-55), was developed by the USDA NRCS. The model uses simplified methods for estimating storm runoff volume, peak rate of discharge, hydrographs, and storage volumes in small urban watersheds (smaller than 25 miles^2). WinTR-55 is the Windows-based version of the original model. The WinTR-55 calculates hydrographs from both urban and agricultural areas at selected study points in the basin, and routes the hydrographs through channels and storage. The model is widely used for stormwater hydrology.[38,57,58]

SWMM: The U.S. Environmental Protection Agency's SWMM is a dynamic model that is developed primarily for the rainfall-runoff analysis in urban areas. The model simulates both single events and continuous storms. The program is capable of simulating the movement of precipitation and pollutants from the surface of the basin through pipe and channel networks, storage or treatment facilities, and to the receiving waters. The XP SWMM, an upgraded version of the original model, is a hydrologic and hydraulic model that is capable of modeling watersheds, floodplains, river systems, stormwater systems, (BMPs, sanitary sewers, and combined sewers. The model simulates pollutant and sediment loading and transport, as well as pollutant removal for selected BMPs. The model has been widely used for the analysis of quantity and quality of flow in storm sewers and CSSs.[59]

STORM: It is developed by the U.S. Army Corps of Engineers. The model is capable of analyzing rainfall-runoff and estimating pollutant loading in urban or rural areas. STORM can be used for both single events and continuous storms. The model estimates the total runoff by calculating runoff from every sub-basin in the watershed. The runoff hydrograph can be calculated from the coefficient method, SCS-CN method, or a combination of these two. The model uses a linear relationship between runoff and excess rainfall (i.e., precipitation minus interception loss).[60]

SWAT: It is developed for the U.S. Department of Agriculture. SWAT was originally developed to assess the impact of land management practices on water, sediment, and agricultural chemical yields in large ungauged basins. The model estimates the runoff volume and peak runoff rate within the sub-basin areas and routes flow, sediment, and pollutants in channel networks in the basin. SWAT is a continuous model, and it is not designed to simulate single storm events.[61]

All these models and others require data to be calibrated for local conditions. These conditions include rainfall characteristics, soil conditions, land cover, and others. Models that are not calibrated and whose results are not verified should not be used to predict future conditions. In the absence of any data, a sensitivity analysis should be performed for the model components. The results of such models should be used with caution.

10.6 WATERSHED MANAGEMENT: STORMWATER AND NONPOINT SOURCE POLLUTION CONTROL

In urbanized areas, NPS can impact runoff and water quality of receiving waters. The increase in volume and change in the timing of stormwater runoff may cause flooding and erosion of the stream. Additionally, stormwater runoff also carries contaminants from different types of land use that characterize the quality of stormwater. Traditionally, stormwater systems were designed for flood control to reduce the risk of loss of life and property damage. Little or no consideration was given in these designs to protect the water quality and the environment. Unmanaged stormwater runoffs may have two adverse effects on receiving water: (1) the storm sewers collect and convey large volumes of stormwater to waterways that can alter the natural flow regime and lead to stream erosion and change in ecology, and (2) during wet weather events, the polluted runoff from urban and rural areas discharges contaminants into the receiving waters, especially during the *first flush*. The concept of first flush and its consequences are discussed in detail in Chapter 11.

10.6.1 Goals of Controlling Nonpoint Source Discharges

Unmanaged stormwater enters receiving water as a NPS. It comes from many diffuse sources; therefore, it is difficult to quantify the range of pollutant loadings to the receiving waters. The main purpose of modern stormwater management programs is to reduce adverse impacts on receiving water by retaining and treating stormwater at the source using BMPs. Examples of such methods are (1) stormwater harvesting to reduce the volume of water that causes flooding, and (2) retention, infiltration, bio-filtration, or bioretention systems to reduce volume and peak, retain contaminants, enhance water quality, and reduce the adverse impact on receiving waters.

The stormwater BMPs may have the following goals:[62,63]

1. Flow control to reduce the volume and duration of stormwater runoff,
2. Containment of runoff to prevent the release of pollutants from the source into runoff and the receiving waters, and
3. Source reduction to remove contaminants from the urban landscape. Examples of source reduction are sweeping the street, spill prevention and cleanup, good housekeeping practices around the solid waste and chemical storage facilities, and controlling the excess use of lawn fertilizers, insecticides, and herbicides to prevent excess pollutants from washing into the stormwater runoff.

Stormwater management by source-control BMPs may include *structural* and *non-structural* approaches. The structural BMPs are implemented to manage or reduce runoff quantity and duration by physical means.[64] Examples of structural source-control practices are covering and separating the pollutant source, preventing pollution sources from entering stormwater, and directing polluted stormwater to runoff treatment BMPs such as basins, ponds, wetlands, infiltration facilities, filtering systems, and open channels. The non-structural BMPs are utilized to prevent or abate the adverse impacts of stormwater runoff on receiving waters. The non-structural practices may include institutional policy and educational aspects of stormwater management to improve the water quality and prevent pollutants from entering the source of runoff generation. A detailed description of structural and non-structural BMPs, their pros and cons, maintenance considerations, and the effectiveness of each technology or practice may be found in Refs. 62,63. A combination of structural and non-structural measures may be the most cost-effective stormwater management approach by preventing pollutants from entering runoff by means of source controls BMPs than by using runoff treatment BMPs to remove pollutants.[64–67]

10.6.2 Occurrence of Contaminants in Water Environment

A contaminant is an unwanted substance that is generated because of human activity. It is estimated that there are around 84,000 chemicals on the market, and we encounter many of them every single day.[26] Most of these chemicals are used in homes, agriculture, and industries. These chemicals are contaminants as their residues are present in air, water, and land. These contaminants may be released into the environment as a point source or NPS. A point source is identifiable and has a reasonably well-defined plume. A NPS has many diffused or smaller discharge points. They are not easily identifiable, and their plume is not well defined. Under these two types of sources, contaminants are grouped into seven categories: (1) radionuclide, (2) trace elements, (3) nutrients, (4) inorganic metals, (5) biodegradable and non-biodegradable organics, (6) refractory organics, and (7) microbial contaminants. In natural waters, the entry of these point and nonpoint contaminants follows a well-defined pathway. These pathways are shown in Figure 10.22.[26]

The common method to express the concentration of contaminants is mass of substance dissolved per unit volume of water, usually as mg/L, µg/L, ppm (parts per million), and ppb (parts per billion). Since the concentration of most contaminants in water is very small, the solution has a specific gravity of almost one (that of water), and for practical purposes, mg/L is equal to ppm, and µg/L is equal to ppb. The load of a contaminant in a body of water is either total mass (concentration multiplied by volume of water) or its mass of flow rate (given by the product of the concentration and the rate of flow or discharge). A plot of concentration versus time is called a *pollutograph*, and a plot of load versus time is a *loadograph*.[26] A plot of instantaneous, one-time, pulse or slug loading, such as a spill into waterbody, versus time is a pollutograph and is shown in Figure 10.23a. A plot of sustained loading, such as leakage from a tank over a long time, versus time is also a pollutograph and is shown in Figure 10.23b. If the contaminant decays with time, the plot of continuous loading will be lower than that of non-reacting substances. A loadograph or mass loading plot with respect to time for first flush in a combined sewer is shown in Figure 10.23c.

10.6.3 Hydrological and Water Quality Impacts from Urban Stormwater

The hydrologic response of an urban area changes when drainage areas become increasingly impervious, causing stormwater runoff volumes, flows, and velocities to increase while base groundwater

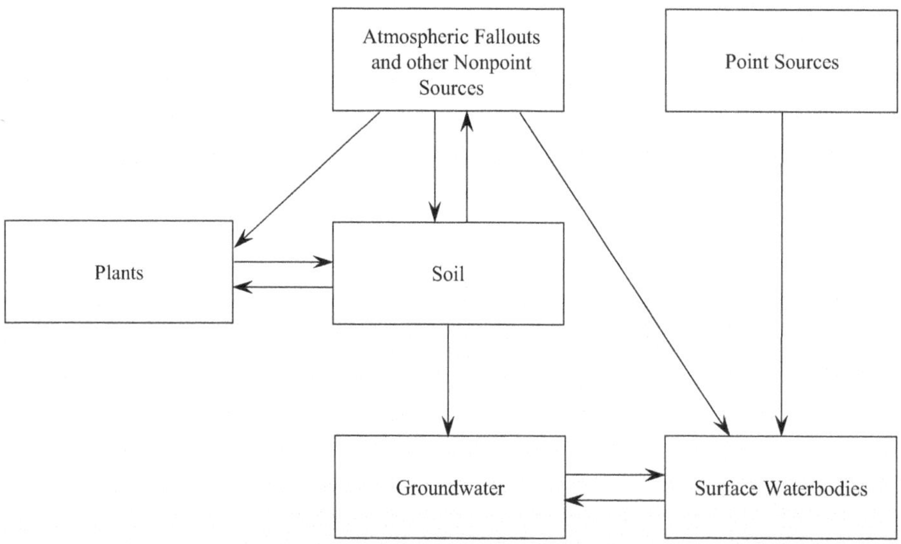

FIGURE 10.22 Pathways of pollutants from various sources into surface waters (stream, lake, estuary, bay, and ocean).

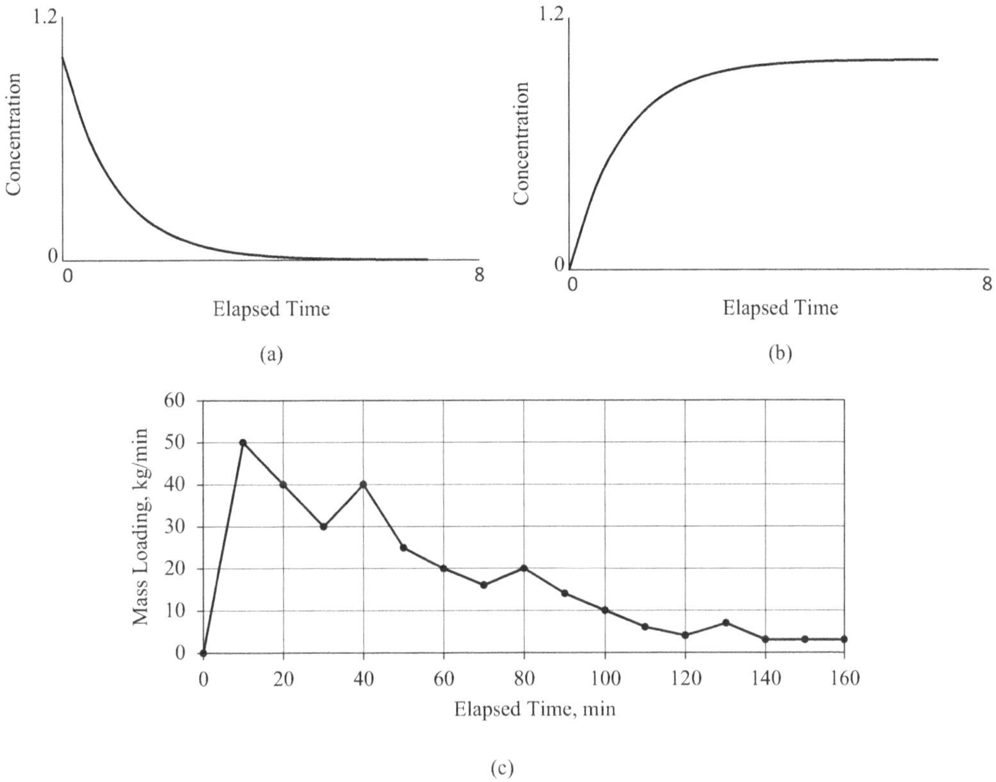

FIGURE 10.23 Examples of pollutograph and loadograph from at-source release of chemicals in a turbulent waterbody: (a) instantaneous, one-time, pulse, or slug release of a conservative (nonreactive) chemical, (b) sustained release such as leakage from a tank over a long time of a conservative (nonreactive) chemical, and (c) mass loading plot with respect to time (loadograph) for first flush in a combined sewer.

flows decrease. The common impervious surfaces in urban areas are due to rooftops, sidewalks, parking lots, roadways, and other impervious surfaces. Small annual storm events that would ideally be captured by the plants and soils of an undeveloped landscape are instead delivered quickly and efficiently through the receiving pipe network to natural waters.

Hydrologic impacts: Urban development causes significant changes in the rainfall-runoff relationship within a watershed. Rainfall volumes shift from evapotranspiration and infiltration to surface and piped runoff. This shift delivers large amounts of runoff to receiving pipes and streams during the smallest rainfall event within an urban development. A city represents a transformation from a natural catchment to an impervious surface and an underground piped conveyance system. Natural drainage patterns are changed as stormwater runoff is channeled from roof drains, pavement, road gutters, and storm drains to stormwater conveyance systems. These conveyance systems deliver these larger volumes of stormwater more quickly at high velocity. The time of runoff to travel downstream becomes shorter, and infiltration into underlying soils and ground decreases or is eliminated. Changes in flow profile reaching the receiving stream resulting from urbanization and rural watersheds are compared in Figure 10.24. Readers are referred to Example 10.16 and Figure 10.13 for developing synthetic curves for pre- and post-development watersheds.

Water quality impacts: As land is developed, impervious surfaces replace naturally vegetated areas that once allowed water to infiltrate and become purified by the soil. Human activities in the city, including heavy automobile traffic, chemical storage facilities, the use of various chemicals over lawns, and industrial activity, generate increased pollutant loads. During dry weather, these

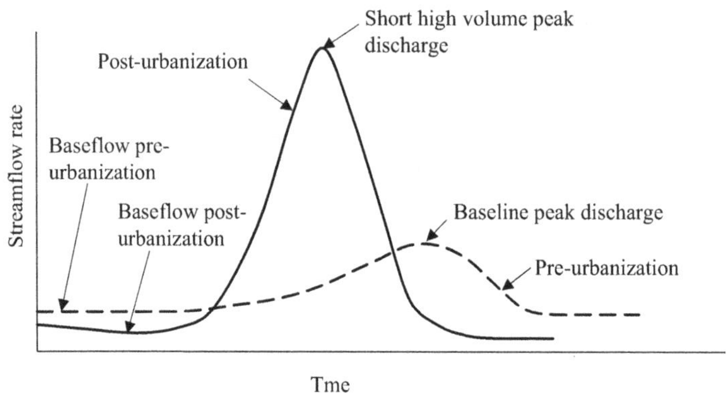

FIGURE 10.24 Comparison of receiving streamflows resulting from urbanization of a rural watershed.

pollutants combine with deposits of atmospheric pollution from outside of the city to build up on pervious and impervious surfaces. Storm and snow events later wash them quickly over impervious surfaces and deliver them rapidly into storm sewers and downstream waters such as creeks, streams, rivers, and water reservoirs. Pollutants commonly found in urban stormwater runoff are: sediments, nutrients, bacteria and viruses, oxygen-demanding organics, pesticides, herbicides, oil and grease, paints, fuels, hydraulic fluids, polychlorinated biphenyls (PCBs), heavy metals, and floatables. Their damaging impacts on the water quality of receiving waters are severe and long-lasting.

Example 10.20: Concentration Profile of Hazardous Chemical Spilled in a Wet Pond

Seventy gram of a hazardous chemical is spilled from a chemical storage tank in a flow through a turbulent wet pond located in a watershed. The volume of the wet pond is 35 m^3, and the discharge through the wet pond is 200 m^3/day. The generalized equation for outflow concentration of spilled chemical from the wet pond is given by Eq. (10.23a).[68]

$$C = C_o e^{-t/\theta} \tag{10.23a}$$

where

C = exit concentration at any time t

C_o = initial concentration at time t = 0

θ = theoretical detention time through wet pond, h

Determine the concentration of hazardous chemicals in the wet pond at 2 and 6 h after the spill, and the concentration profile with respect to time. The released hazardous chemical is a conservative substance since it is nonreactive and does not undergo decay.

Solution

1. Determine the theoretical detention time

   The theoretical detention time θ is calculated from Eq. (2.9)

$$\theta = \frac{V}{Q} = \frac{35\ \text{m}^3}{200\ \text{m}^3/\text{day}} \times 24\ \text{h/day}$$

$$= 42\ \text{h}$$

2. Determine the initial concentration of the hazardous chemical in the wet pond

$$C_o = \frac{70 \text{ g}}{35 \text{ m}^3} = 2.0 \text{ g/m}^3$$

3. Determine the concentration of the hazardous chemical at 2 and 6 h

   Apply Eq. (10.23a)

$$C_2 = 2.0 \text{ g/m}^3 \times e^{(-2\text{ h}/4.2\text{ h})} = 1.24 \text{ g/m}^3$$

$$C_6 = 2.0 \text{ g/m}^3 \times e^{(-6\text{ h}/4.2\text{ h})} = 0.48 \text{ g/m}^3$$

4. Develop the concentration profile of the hazardous chemical

   Assume a series of time intervals t after the spill had occurred, and calculate the concentration C at each time interval.

   The time intervals and concentrations of spilled chemicals are summarized in Table 10.28.
5. Draw the concentration profile of the spilled chemical.

   The concentration profile is shown in Figure 10.25.

TABLE 10.28
Time Intervals and Calculated Concentrations of Spilled Chemical (Example 10.20)

Time t,	Concentration C, g/m^3
0.0	2.00
0.5	1.78
1.0	1.58
2.0	1.24
4.0	0.77
6.0	0.48
10.0	0.18
20.0	0.02

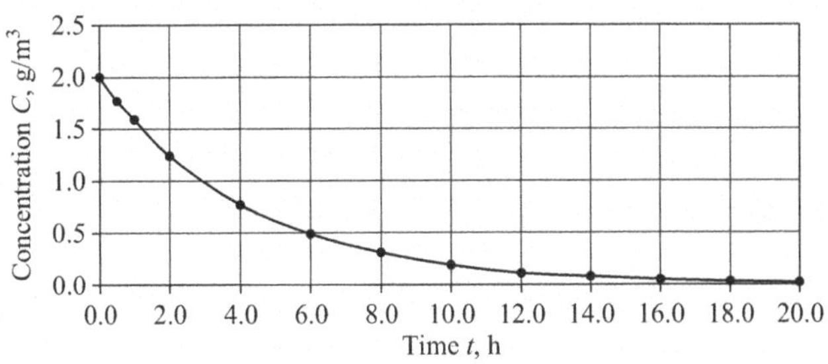

FIGURE 10.25 Concentration profile of spilled hazardous chemical (conservative substance) (Example 10.20).

Example 10.21: Steady Release of Hazardous Chemical from a Storage Tank

Hazardous chemical is released steadily from a storage tank in a flow through a wet pond. The released flow rate of the chemical is 50 L/day, and the concentration of the chemical is 7.0 g/L. The volume of the turbulent wet pond is 35 m^3, and the flow through the pond is 200 m^3/day. Calculate the concentration of released chemical leaving the wet pond at 2 and 6 h. Also, develop the concentration profile. The generalized equation for continuous release of a conservative chemical is expressed by Eq. (10.23b).[68] All terms in the equation have been defined previously.

$$C = C_o(1 - e^{-t/\theta}) \tag{10.23b}$$

Solution

1. Calculate the theoretical detention time

   The theoretical detention time θ is calculated from Eq. (2.9)

$$\theta = \frac{V}{Q} = \frac{35 \text{ m}^3}{200 \text{ m}^3/\text{day}} \times 24 \text{ h/day}$$
$$= 4.2 \text{ h}$$

2. Determine the input rate of the hazardous chemical in the inflow

$$C_o = \frac{50 \text{ L/day} \times 7 \text{ g/L}}{200 \text{ m}^3/\text{day}} = 1.75 \text{ g/m}^3$$

3. Calculate the concentration of the chemical at 2 and 6 h

   Apply Eq. (10.23b)

$$C_2 = 1.75 \text{ g/m}^3 \times \left[1 - e^{(-2\text{ h}/4.2\text{ h})}\right] = 0.66 \text{ g/m}^3$$

$$C_6 = 1.75 \text{ g/m}^3 \times \left[1 - e^{(-6\text{ h}/4.2\text{ h})}\right] = 1.33 \text{ g/m}^3$$

4. Develop the concentration profile of the hazardous chemical.

   Assume a series of time intervals (t) after the release started, and calculate the concentration C at each time interval

   The time intervals and concentrations of spilled chemicals are summarized in Table 10.29.
5. Draw the concentration profile of the released chemical.

   The concentration profile is shown in Figure 10.26.

Example 10.22: Release of Chemical That Undergoes Decay

A chemical is released from a storage tank in a wet pond. The chemical is nonconservative and undergoes decay in the pond. The pond volume is 500 m^3, and the flow through it is 50 m^3/day. The concentration of the released chemical is 650 g/m^3 and is consumed. The first-order decay kinetics with k=0.28/day applies. Calculate the steady exit concentration of the chemical after a long time of release from the storage tank. The generalized equation for exit concentration with decay is expressed by Eq. (10.24).[68]

TABLE 10.29
Time Intervals and Calculated Concentrations of Released Chemical (Example 10.21)

Time t, h	Concentration C, g/m³
0.0	0.00
0.5	0.20
1.0	0.37
2.0	0.66
4.0	1.07
6.0	1.33
10.0	1.59
20.0	1.74

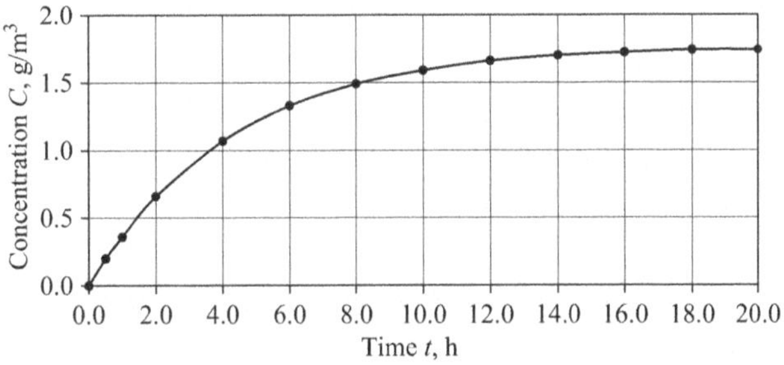

FIGURE 10.26 Concentration profile of released chemical (conservative substance) (Example 10.21).

$$C = C_o\left(\frac{1}{1+k\theta}\right) \tag{10.24}$$

where
k = first-order decay constant, per day
Other terms in the equation have been defined previously

Solution

1. Determine the theoretical detention time
   The theoretical detention time θ is calculated from Eq. (2.9)

$$\theta = \frac{V}{Q} = \frac{500\ \text{m}^3}{50\ \text{m}^3/\text{day}} = 10\ \text{day}$$

2. Calculate the steady exit concentration from the wet pond
   Apply Eq. (10.24) to calculate the exit concentration C of the chemical released in the wet pond

$$C = 650\ \text{g/m}^3 \times \left[\frac{1}{1+(0.28/\text{day}\times 10\ \text{day})}\right]$$

$$= \frac{650\ \text{g/m}^3}{(1+2.8)} = 171\ \text{g/m}^3\left(171\ \text{mg/L}\right) \quad \text{or} \quad 0.171\ \text{kg/m}^3$$

Example 10.23: Release of Hazardous Chemical from a Storage Tank Then Stopped

A hazardous chemical leaked from a storage tank in a turbulent wet pond for six hours before the leak was stopped. The discharge rate of the leaked chemical was 80 L/day, and the concentration of the chemical in the leaked stream was 10 g/L. The volume of the wet pond was 50 m^3, and the average flow through the pond was 250 m^3/day. Develop the concentration profile of the conservative (nonreactive) hazardous chemical during the tank leak, and after the leak stopped.

Solution

A. During tank leak
   1. Calculate the theoretical detention time from Eq. (2.9)

$$\theta = \frac{V}{Q} = \frac{50\ \text{m}^3}{250\ \text{m}^3/\text{day}} \times 24\ \text{h/day} = 4.8\ \text{h}$$

   2. Calculate the initial concentration of the hazardous chemical in the wet pond

$$C_o = \frac{80\ \text{L/day}}{250\ \text{m}^3/\text{day}} \times 10\ \text{g/L} = 3.2\ \text{g/m}^3$$

   3. Calculate the concentration of the chemical at 2 h
      Apply Eq. (10.23)

$$C_2 = 3.2\ \text{g/m}^3 \times \left[1 - e^{(-2\ \text{h}/4.8\ \text{h})}\right] = 3.2\ \text{g/m}^3 \times (1 - 0.66)$$

$$= 1.1\ \text{g/m}^3$$

      Similarly, calculate the concentration of the chemical at 0.5, 1, 2, 4, and 6 h.
   4. Tabulate the concentration of the chemical at different time intervals.
      The concentrations of chemicals at different time intervals since the start of the leak are summarized in Table 10.30.

B. After leak stopped
   1. Tabulate the concentration of the hazardous chemical after the leak is stopped
      Apply Eq. (10.23a) to calculate the concentration of hazardous chemicals in the wet pond. The total time t since start of the chemical release, and concentrations of the chemical in the wet pond during release and after the leak is stopped are given in Column (1) of Table 10.31. The time in hours and chemical concentrations in the wet pond since the leak stopped are given in Columns (2) and (3) of Table 10.31.
   2. Draw the concentration profile during the chemical leak and after the leak is stopped
      The concentration profile during the leak and after stopping the leak is shown in Figure 10.27.

Example 10.24: BOD_5 Loading in a Combined Sewer from First-Flush

The hourly BOD_5 and flow data of the first-flush in a combined sewer of an older section of a city were recorded. The lapsed time of first-flush flow and BOD_5 data are summarized below. Develop the BOD_5 loading data of the combined sewer and plot the BOD_5 loading with respect to lapsed time.

TABLE 10.30
Time Intervals and Calculated Concentrations of Released Chemical (Example 10.23)

Time *t*, h	Concentration During Chemical Leak *C*, g/m³
0.0	0.0
0.5	0.3
1.0	0.6
2.0	1.1
4.0	1.8
6.0	2.3

TABLE 10.31
Total Time Since Leak Started, Time Since Leak Stopped, and Chemical Concentration in the Wet Pond before and after Leak Stopped

Time Since Leak Started *t*, h (1)	Time Since Leak Stopped *t*, h (2)	Concentration *C* Since Leak Stopped, g/m³ (3)
0	—	—
0.5	—	—
1.0	—	—
2.0	—	—
4.0	—	—
6.0	6.0–6.0=0.0	2.3
8.0	8.0–6.0=2.0	1.5[a]
12.0	12.0–6.0=6.0	0.7
14.0	14.0–6.0=8.0	0.4
20.0	20.0–6.0=14.0	0.1
25.0	25.0–6.0=19.0	0.04
30.0	30.0–6.0=24.0	0.02

[a] $2.3\ \text{g/m}^3 \times e^{(-2\ \text{h}/4.8\ \text{h})} = 1.5\ \text{g/m}^3$.

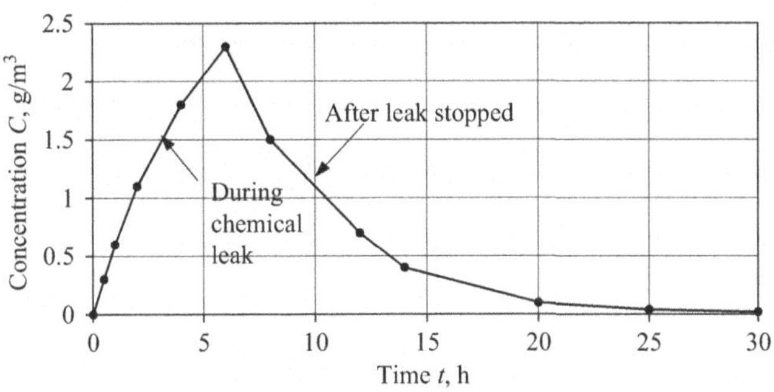

FIGURE 10.27 Concentration profile of hazardous chemical in the wet pond before and after the leak stopped from the storage tank (Example 10.23).

Lapsed Time, h	1	2	3	4	5	6	7	8
Combined sewer flow, m³/h	50	100	180	150	100	50	40	20
BOD_5 concentration, g/m³	400	800	1,200	1,000	800	700	600	300

Solution

1. Calculate BOD_5 loading.

$$BOD_5 \text{ at first lapsed time} = 50 \text{ g/m}^3 \times 400 \text{ m}^3/\text{h} \times \frac{1}{1{,}000 \text{ g/kg}}$$

$$= 20 \text{ kg/h}$$

Similarly, calculate BOD_5 loading at all lapsed times
2. Tabulate the lapsed time, flow, BOD_5 concentration, and BOD_5 loading. These values are summarized in Table 10.32.
3. Plot the first-flush BOD_5 loading in the CSS
The BOD_5 loading with respect to lapse time is shown in Figure 10.28

TABLE 10.32
Lapsed Time, Combined Sewer Flow, BOD_5 Concentration, and BOD_5 Loading (Example 10.24)

Lapsed Time, h	Combined Sewer Flow, m³/h	BOD_5 Concentration, g/m³	BOD_5 Loading, kg/h
1	400	50	20
2	800	100	80
3	1,200	180	216
4	1,000	150	150
5	800	100	80
6	700	50	35
7	600	40	24
8	300	20	6

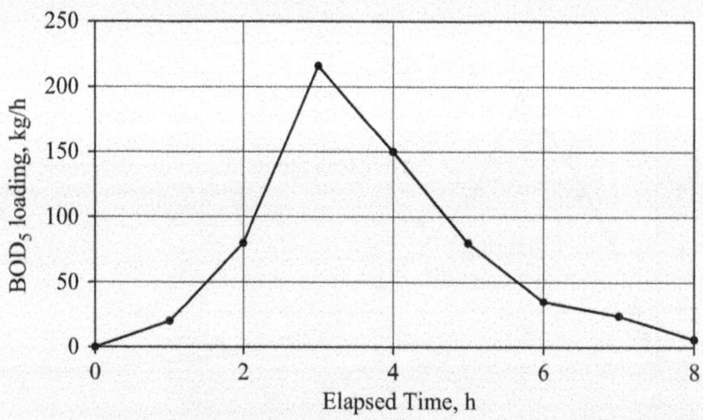

FIGURE 10.28 First-flush BOD_5 loading in a CSS (Example 10.24).

10.7 SOURCE-CONTROL BMPS FOR WATERSHED MANAGEMENT, TREATMENT OF STORMWATER RUNOFF, AND CONTROL OF NONPOINT SOURCE POLLUTION

The source-control methods are utilized to reduce the quantity and duration of stormwater runoff, retain pollutants where they are generated, and prevent them from entering the receiving waters. These methods thus mitigate the adverse effect of high volumes of runoff by holding it for a considerable time while the flow is gradually reduced mainly through infiltration, evaporation, transpiration, and overflow.[69] A watershed is an area of land that contains streams and rivers that drain into a single larger body of water such as a larger river, lake, or ocean. Smaller watersheds are usually part of larger watersheds.

Integrated watershed management (IWM) is a term used to describe the process of implementing land use and water management practices to protect and improve the quality and reduce the volume of surface runoff. The main objectives of IWM are to (1) identify sources of pollution, conserve soil, plants, and natural resources within the watershed; (2) manage land use, water resources, and mode of transmission of pollutants in a comprehensive manner; (3) take into account all environmental, social, and economic concerns, as well as local community interests; and (4) protect adverse impacts on community growth, property value, and climate change.[15,70]

The outcomes of IWM are prevention of soil erosion, regeneration of natural vegetation, rainwater harvesting (RWH), and recharging of the groundwater table. This enables multi-cropping and the introduction of diverse agro-based activities, which help to provide sustainable livelihoods to the people residing in the watershed area. Essentially, watershed monitoring is a dynamic and multi-dimensional assessment of the hydrological cycle, the soil and biomass, pollution sources, network of surface streams and impoundments, and groundwater storage in underground aquifers.

Most source-control watershed management BMPs for control of stormwater quality and runoff volume are extensively used to achieve the goals of IWM. Many of these BMPs are described below (Sections 10.7.1–10.7.19). It may be noted that these BMPs are an integrated combination of many physical, chemical, and biological processes serving these primary goals. For example, bioretention systems or rain gardens provide storage, percolation, and filtration of stormwater runoffs, and uptake and removal of contaminants by plants and the soil beds. All these functions are integrated into most of the BMPs presented below.[71]

10.7.1 Bioretention System (Rain Garden)

A bioretention system (also called rain garden) is either bowl-shaped or surrounded by berms to retain water. It consists of a soil bed planted with suitable non-invasive (preferably native) vegetation. These bioretention systems collect rainwater runoff from roofs, driveways and parking spaces, and other surfaces. They can blend into the landscape and serve as a garden area. Its main function is to collect, retain, and treat the collected stormwater. The runoff entering the bioretention system or rain garden is filtered through the soil planting bed. After filtration, the flow is either conveyed downstream by an underdrain system or infiltrated into the existing subsoil below the soil bed.[72,73] Vegetations in the soil planting beds provide uptake of pollutants, slow down runoff, and increase associated infiltration rate through the soil from sand to clay. They can be configured either as a basin or a long and narrow bioretention swale. In general, a bioretention basin has a flat bottom, while a bioretention swale may have a sloping bottom. The runoff storage depths above the soil bed surface are typically shallow. The TSS removal rate is 80% or 90% depending upon the thickness of the soil planting bed and the type of vegetation grown in the bed.

Bioretention systems are used to remove a wide range of pollutants, such as suspended solids, nutrients, metals, hydrocarbons, and bacteria, from stormwater runoff.[74–76] They can also be used to reduce peak runoff rates and increase stormwater infiltration when designed as a multi-stage and multi-function facility. Conditions where practice applies, bioretention systems can also be used to filter the runoff from both residential and nonresidential developments. Concentrated inflow from a drainage pipe or swale must include adequate erosion protection and energy dissipation measures. Bioretention systems are most effective when they receive runoff as close to its source as possible. They can vary in size and can receive and treat runoff from a variety of drainage areas within a land development site. They can be installed in lawns, median strips, parking lots, unused lot areas, and certain easements. They are intended to receive and filter storm runoff from both impervious areas and pervious areas such as lawns.[77] A bioretention system must not be placed into operation until the contributing drainage area (CDA) is completely stabilized.

The basic design parameters for a bioretention system are its storage volume, the thickness, character, permeability rate of its planting soil bed, and either the hydraulic capacity of its underdrain or the permeability of its subsoil (whichever is applicable). The system must have sufficient storage volume above the surface of the bed to contain the design storm runoff volume without overflow. The thickness and character of the bed itself must provide adequate pollutant removal, while the bed's permeability rate must be sufficient to drain the stored runoff within 72 h. The cross-section of a standard bioretention basin is shown in Figure 10.29. The components and design criteria of the bioretention basin are briefly described below.[76]

Design geometry: Bioretention basins must be designed to minimize short-circuiting and have acceptable internal geometry. The inlet and outlet structures should be as far apart as possible to maximize travel time and distribute the incoming flow as evenly as possible across the entire surface area.

Inlets and energy dissipation: Where appropriate, the inlet(s) should be designed to dissipate energy. Stone, splash block, downspouts, drop structure, or other acceptable energy dissipation measures should be utilized to dissipate energy at the entry to the basin.

Side slopes: Traditional bioretention basins or residential rain gardens should be constructed with side slopes of 3H:1V or flatter.

Ponding depth: The ponding depth is the depth of water maintained on the top. The recommended surface ponding depth is 6–12 in. Minimum surface ponding depth of 3 in., and a maximum of 18 in. have also been used. Where higher ponding depths are utilized, the viability and survival of

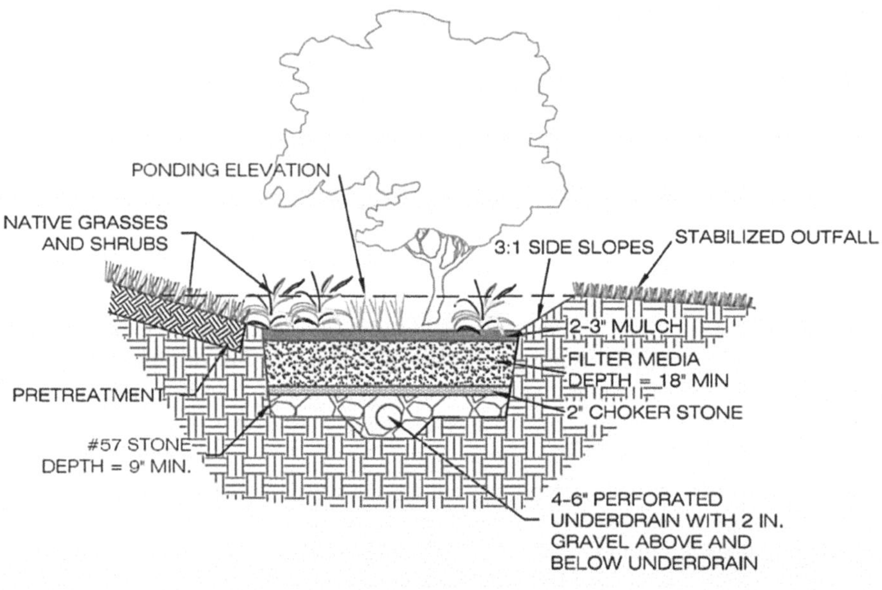

FIGURE 10.29 Generalized cross-sectional view of a bioretention (rain garden) system. (Printed from Ref. 76)

plants, aesthetics, and erosion and scour of the side slope are main considerations. Safety and fencing near sidewalks or other areas where pedestrians or bicyclists travel may be needed.

Mulch: A 2–3-in. layer of mulch on the surface of the filter bed enhances plant survival, suppresses weed growth, pretreats runoff before it reaches the filter media, and prevents rapid evaporation of rainwater.

Filter media: The filter media of a bioretention system consists of an engineered soil mixture. The soil mixture is carefully blended to create a filter media that maintains long-term permeability while also providing enough nutrients to support plant growth. The filter media consist of a well-blended mixture of medium to coarse sand, loam soil, and an organic amendment (compost). The sand maintains the desired permeability of the media, while the limited amount of loam soil and organic amendments is considered adequate to help support initial plant growth. The minimum depth of filter media must be 8 in. (20 cm) for standard design, and 24 in. (61 cm) or greater for enhanced design. Turf, perennials, or shrubs should be used instead of trees to landscape shallower filter beds.

Choker layer: A geotextile fabric or 2–4-in. (5–10 cm) layer of choker stone over the underdrain stone should be used.

Underdrain stone: 1-in. (2.5-cm) diameter stone layer at least 2 in. (5.1 cm) above and below the underdrain stone should be provided. The stone layer should meet ASTM D44 No. 57 requirements.

Underdrains, cleanouts, and observation wells*:* Rigid underdrain PVC perforated pipe 4–6-in. (10–41-cm) diameter should be used. Multiple underdrains may be necessary for bioretention basins wider than 40 ft (12.2 m).

Finally, a bioretention system must have a maintenance plan and, if privately owned, shall be protected by easement, deed restriction, ordinance, or other legal measures that prevent its neglect, adverse alteration, and removal.[78]

Example 10.25: Storage Volume of a Bioretention Basin (Rain Garden)

A bioretention basin is designed to capture and retain surface runoff from a residential housing subdivision. The surface dimensions of the bioretention basin are 40 ft long and 25 ft wide. The maximum average ponding depth is 12 in., and the side slope is 3H:1V. The average depth of the filter media, including mulch layer, is 24 in. The effective porosity of the filter media is 0.25. The depth of the underdrain and underground storage gravel, including choker stone, is 15 in., and effective porosity is 0.4. A 4-in. underdrain perforated pipe is laid into the underground storage gravel. Determine the total storage volume of the bioretention basin in ft^3, gallons, and inches of rainfall.

Solution

1. Select the equation for calculating the storage volume of bioretention basin
   The storage volume of bioretention basin is calculated from Eq. (10.25)[76]

$$\text{SV} = \text{SA}_{\text{bottom}} \times [(d_{\text{media}} \times \eta_{\text{media}}) + (d_{\text{gravel}} \times \eta_{\text{gravel}})] + (\text{SA}_{\text{average}} \times d_{\text{ponding}}) \quad (10.25)$$

where

SV = total storage volume of bioretention, ft^3

$\text{SA}_{\text{bottom}}$ = bottom surface area of bioretention, ft^2

$\text{SA}_{\text{average}}$ = average surface area, ft^2

d_{media} = depth of the filter media, including mulch layer, ft

η_{media} = effective porosity of the filter media, typical value is 0.25

d_{gravel} = depth of the underdrain and underground storage gravel layer, including choker stone, ft

η_{gravel} = effective porosity of the gravel layer, typical value is 0.4

2. Calculate top and bottom dimensions and areas of ponding depth
   Top and bottom dimensions and areas of the ponding depth

$$\text{Top length} = 40\text{ ft, top width} = 25\text{ ft, top area SA}_{\text{top}} = 40\text{ ft} \times 25\text{ ft} = 1{,}000\text{ ft}^2$$

$$\text{Bottom length} = 40\text{ ft} - 2 \times 3\text{ ft/ft} \times 1\text{ ft} = 34\text{ ft, bottom width} = 25\text{ ft} - 2 \times 3\text{ ft/ft} \times 1\text{ ft} = 19\text{ ft}$$

$$\text{Bottom area SA}_{\text{bottom}} = 34\text{ ft} \times 19\text{ ft} = 646\text{ ft}^2$$

3. Calculate the average area of ponding depth

$$\text{SA}_{\text{average}} = (\text{SA}_{\text{top}} + \text{SA}_{\text{bottom}}) / 2$$

$$\text{SA}_{\text{average}} = (1{,}000\text{ ft}^2 + 646\text{ ft}^2) / 2 = 823\text{ ft}^2$$

4. Calculate total storage volume of bioretention basin from Eq. (10.25)

$$\text{SV} = 646\text{ ft}^2 \times [(24\text{ in.} / 12\text{ in./ft} \times 0.25) + 15\text{ in./}12\text{ in./ft} \times 0.4)] + (823\text{ ft}^2 \times 12\text{ in.} / 12\text{ in./ft})$$

$$= 646\text{ ft}^2[(0.5\text{ ft} + 0.5\text{ ft})] + (823\text{ ft}^2 \times 1\text{ ft}) = 1{,}469\text{ ft}^3$$

$$\text{Storage volume in ft}^3 = 1{,}469\text{ ft}^3$$

$$\text{Storage volume in gallons} = 1{,}469\text{ ft}^3 \times 7.481\text{ gallons/ft}^3 = 10{,}990\text{ gallons}$$

$$\text{Storage volume in inches of rainfall} = \text{volume in ft}^3\text{/ponding surface area ft}^2$$

$$= 1{,}469\text{ ft}^3 / 1{,}000\text{ ft}^2 = 1.469\text{ ft or }17.6\text{ in.}$$

Example 10.26: Runoff Volume to Design a Bioretention Basin (Rain Garden)

A paved parking lot for a shopping center has 10,000-ft^2 of impervious area. It is built on soil that has clay texture. Calculate the expected depth of runoff resulting from a storm of 1 in. rain. Also, calculate the area of a standard bioretention basin or rain garden. Assume that the storage capacity of a standard bioretention basin is 17.6 in. (see Example 10.25).

Solution

A. Calculate the expected depth of runoff resulting from a 1 in. rainfall
   1. Identify the hydraulic soil group
      The soil texture in the area is clay. Based on soil texture, the hydraulic soil group from Table 10.10 falls in Group D.
   2. Determine the CN
      The CNs for urban areas with paved parking lots and other cover types are listed in Table B.7a. For soil Group D and impervious parking lot cover, the CN is 98.
   3. Determine the accumulated runoff from the area
      The NRCS method is applied.
      Equation (10.6) is applied to calculate the runoff resulting from 1-in. of rainfall. Eq. (10.6) is given below

$$Q = \frac{(P - 0.2S)^2}{P + 0.8S}$$

In this equation, P is 1 in. of rainfall. S is calculated from Eq. (10.7) given below

$$\mathrm{CN} = \frac{1{,}000}{10+S}$$

Using CN=98, S is calculated as follows

$$98 = \frac{1{,}000}{10+S}$$

$$S = 0.2 \text{ in.}$$

4. Calculate from Eq. (10.6) the runoff depth from 1 in. of rainfall

$$\text{Runoff from 1-in. of rainfall} = (1 \text{ in.} - (0.2\times 0.2 \text{ in.}))2 \,/\, (1 \text{ in.} + 0.8\times 0.2 \text{ in.})$$

$$= 0.79 \text{ in.}$$

$$\text{Total runoff from } 10{,}000 \text{ ft}^2 \text{ area} = 0.79 \text{ in.} \,/\, (12 \text{ in./ft}) \times 10{,}000 \text{ ft}^2 = 658 \text{ ft}^3$$

$$= 658 \text{ ft}^3 \times 7.481 \text{ gallons/ft}^3 = 4{,}923 \text{ gallons}$$

B. Calculate the area of a standard bioretention basin
 1. Calculate the inches of runoff from 10,000 ft^2 parking area

$$\text{Inches of rainfall from the parking area} = 0.79 \text{ in.} \times 10{,}000 \text{ ft}^2 = 7{,}900 \text{ in.ft}^2$$

 2. Calculate the area of standard bioretention basin

$$\text{Area of bioretention basin} = 7{,}900 \text{ in.ft}^2 \,/\, 17.6 \text{ in.} = 449 \text{ ft}^2$$

It is recommended that a 4 ft wide and 112 ft long bioretention basin be built along the downslope side of the parking lot.

Example 10.27: Sizing the Drainage Pipe of a Bioretention Basin (Rain Garden)

Determined the size of underdrain pipe for a rain garden that has a surface area of 300 ft^2. The average depth of the filter media, including the mulch layer, is 18 in., and depth of stone over drainage pipe, including choker stone, is 12 in., and ponding water depth is 6 in. Determine the diameter of the PVC perforated drainage pipe that is laid at a slope of 0.2%.

Solution

1. Determine the hydraulic conductivity of flow through soil media reaching the drainage pipe

   The drainage pipe should be designed to carry up to 10 times the minimum flow through the soil calculated using Darcy's Law. The Darcy's equation is expressed by Eq. (10.26)

$$q = \frac{Kh}{L} \tag{10.26}$$

where

q=flow through unit area of soil media, in./h (cm/h) or ft/h per ft^2 (m/h.m^2) of the rain garden area

h=height of water above drain, ft (m)

L=depth of soil media through which the flow occurs, ft (m)

K=hydraulic conductivity of the soil, in./h (cm/h)

The hydraulic conductivity K of soil depends upon the soil texture. The hydraulic conductivity K for different types of soils is given below:[77]

Sandy soil	8.27 in./h
Loamy sand	2.41 in./h
Sandy loam	1.02 in./h
Loam	0.52 in./h
Silty clay	0.04 in./h

Select the hydraulic conductivity K of filter media (loamy soil)=2.41 in./h

2. Determine the flow through the filter media

Height of water above drain h=12 in. depth of stone, including choker stone, +18 in. depth of filter media, including mulch layer, +6 in. ponding water=36 in. or 3 ft

Flow depth L of soil media=18 in. or 1.5 ft

$$q = (2.41\text{ in./h}) \times (1/12\text{ in./ft}) \times 3.0\text{ ft} / 1.5\text{ ft} = 0.40\text{ ft/h/ft}^2$$

Total flow to the rain garden Q=q×A

$$Q = 0.40\text{ ft/h} \times 300\text{ ft}^2 = 120\text{ ft}^3\text{/h or } 120\text{ ft}^3\text{/(h} \times 3{,}600\text{ s/h)} = 0.03\text{ cfs}$$

The design flow is 10 times the flow to the rain garden

$$\text{Design flow for the drainage pipe} = 10 \times 0.03\text{ cfs} = 0.3\text{ cfs}$$

3. Determine the diameter of the drainage pipe

The diameter of the perforated pipe is calculated from Manning's equation (Eq. 5.39) for a circular pipe discharge

$$Q = \frac{0.464}{n} \times D^{\frac{8}{3}} \times S^{\frac{1}{2}}$$

$$D = 1.335 \times [n \times Q / S^{0.5}]^{3/8}$$

where

D=diameter of the pipe, ft

Q=the flow to be carried, (cfs)

n=Manning roughness coefficient (0.01 for smooth plastic pipe)

S=slope of the pipe (for this site, slope of pipe=0.2%)

$$D = 1.335 \times [0.01 \times 0.3 / 0.0020.5]^{3/8}$$

$$= 1.335 \times [0.01 \times 0.3 / 0.0447]^{3/8} = 0.48\text{ ft or } 5.8\text{ in.}$$

Select a 6-in. standard diameter of perforated drainage pipe for the rain garden.

10.7.2 Bench Terracing

Bench terraces are a soil and water conservation measure used on sloping land with relatively deep soils to retain water and control erosion. They are normally constructed by cutting and filling to produce a series of level steps or benches.[79] This allows water to infiltrate slowly into the soil. The major benefit of bench terracing is conservation of soil and water, reducing both the amount and velocity of water moving across the soil surface. This greatly reduces soil erosion, permitting more intensive cropping than would otherwise be possible. Besides its benefits, terracing can lead to ground saturation that may lead to water overflow during periods of heavy rains, causing mudslides.[80] Farming on a terrace is labor-intensive due to constructing and managing the terrace on sloping land. In sloping lands, bench terracing is the only option for farming.

If soil depths are adequate, hand-made terraces should be built on 7°–25° (12%–47%) slopes, and machine-built terraces should be employed on 7°–20° (12%–36%) slopes. Bench terraces are not recommended for slopes below 7°. The width of the bench (flat part) is determined by soil depth, crop requirements, tools to be used for cultivation, the landowner's preferences, and available resources. For wider benches, more cut and fill is needed, and hence the higher cost. The optimum width for hand-made and manual-cultivated terraces ranges from 2.5 to 5 m; for machine-built and tractor-cultivated terraces, the range is from 3.5 to 8 m. Riser material can be either compacted earth-protected with grass or rocks. In order to ensure easy maintenance, terrace riser height should not exceed 2 m after allowing for settling, especially for earth risers. Riser slopes are calculated by the ratio of the horizontal distance to the vertical rise.

The length of a terrace is limited by the size and shape of the field, the degree of desiccation, and the permeability and erodibility of the soil. The longer the terraces, the more efficient they will be. However, long terraces cause accelerated runoff and greater erosion hazard. A maximum of 100 m in one draining direction is recommended for typical conditions in a humid tropical climate. The length can be slightly increased in arid and semiarid regions.

The bench terraces are classified on the basis of the slope of the bench as (1) level bench terrace, (2) bench terrace sloping outward, and (3) bench terrace sloping inward.

Level bench terraces have a nearly level top surface. They are generally used in areas of medium rainfall and have highly permeable soils. It is expected that most of the flowing surface runoff passing through these terraces is absorbed by the soil, and the remaining portion is discharged into the longitudinal drain. The level bench terraces used for cultivation have a mild longitudinal slope, less than 1%, so that uniform water impounding can be easily made over them. Sloping outward bench terraces must have a shoulder bund, even though this bund is to provide stability to the outer edge of the terrace. In addition, this bund also helps in retaining the surface runoff on the benches for maximum absorption, or by draining through longitudinal drains. In poor permeability soils, the provision of a graded channel at the lower end is most essential for disposing of surplus surface water to the grass waterway. Whereas, in very low-permeable soil, a strong bund along with a spillway arrangement should be essentially equipped to make the terrace safe against heavy storms and allow the water to flow very safely downward to the next terrace. The outward sloped terraces may have up to 3% slope. Sloping inward bench terraces are preferred in areas of heavy rainfall and less permeable soils, from where a large portion of rainwater is drained as surface runoff. Such types of bench terraces have a provision to drain the runoff from their inner side by constructing a drainage channel. The inward slope may be up to 5%. The cross-section of a bench terrace, components, and design details are shown in Figure 10.30.

For the design of bench terraces for a particular tract, the average rainfall, the soil type, soil depth, and the average slope of the area should be known. In addition, the purpose for which the terraces are to be constructed should also be known. The design of bench terraces consists of determining (1) the type of the bench terrace, (2) terrace spacing or the depth of the cut, (3) terrace width, and (4) terrace cross-section. The selection of the type of bench terrace among the three types, described earlier, depends upon the rainfall and soil conditions.

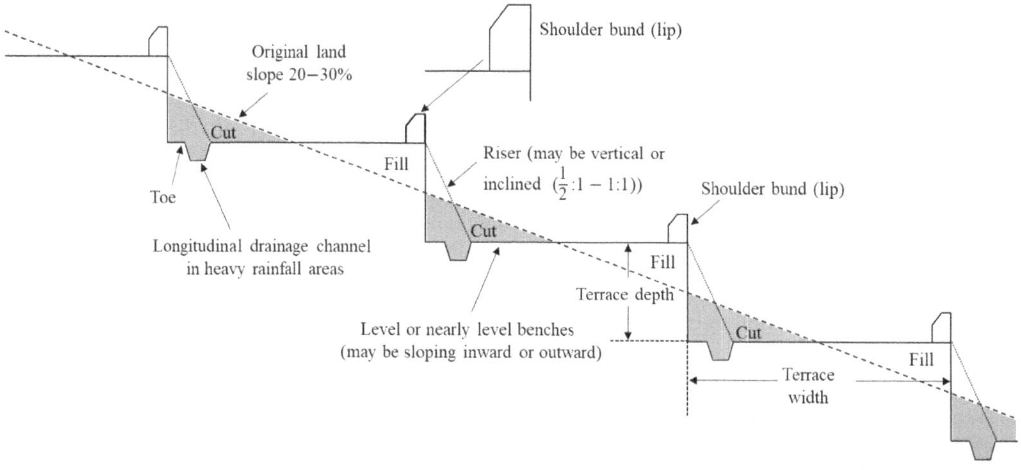

FIGURE 10.30 Construction details of bench terraces and components.

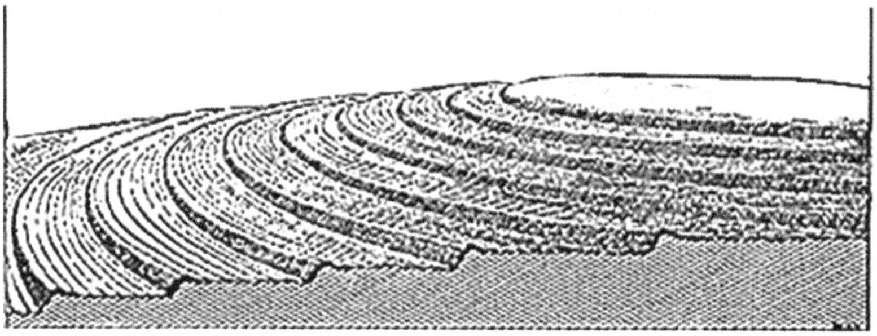

FIGURE 10.31 A generalized view of completed bench terracing.

Terrace spacing is generally expressed as the vertical interval between two terraces. The vertical interval is dependent upon the depth of the cut, and since the cut and fill are to be balanced, it is double the depth of the cut. The factors that limit the depth of cut are the soil depth in the area and the slope. In higher slopes, greater depth of cuts may result; the greater depths of embankments may become unstable. The width of the bench terraces should be as per the requirement for which the terraces are constructed. Once the width of the terrace is decided, the depth of cut required can be calculated.

One major disadvantage of terracing is rainwater saturation of the ground. This happens when terracing retains too much water, which is absorbed into the ground. The right mixture of soil requires regular soil, compost, coir peat (or sand), and vermicomposting.[81] After the heavy rains, essential nutrients may be needed back to the soil as water tends to wash them away. Some crops that are grown using terrace farming are paddy, cereals, fruits, vegetables, flowers, medicinal plants, aromatic plants, dye plants, wheat, maize, rice, pulses, oilseeds, millets, buckwheat, saffron, black cumin, grain amaranth, and spices.[80] A generalized view of completed bench terracing is shown in Figure 10.31.

Example 10.28: Relationship between Slope, Width, and Depth of a Bench Terrace with Vertical Rise

A land tract has a slope of *S*%. The width of the terrace is *w*. Develop an equation to express the depth of bench terrace *D* in terms of slope *S* and width *w*. The riser slope is vertical.

Solution

1. Draw the cross-section of the bench terrace.
   The cross-section of the bench terrace with vertical riser is shown in Figure 10.32.
2. Identify the depth *D*, width *w*, and depth of cut, and ground slope *S*% of bench terrace with vertical riser.
   The dimensions of the bench terrace are given below
   D = depth of bench terrace = AF
   w = width of the bench terrace = BCD
   $\frac{D}{2}$ = depth of cut, when cut = fill or AB = BF = DE
   S = natural land slope, %
3. Develop the relationship
   The relationship is expressed by Eq. (10.27)

$$\frac{S}{100}=\frac{\text{AF}}{\text{EF}}=\frac{D}{w}$$

$$D=\frac{w\times S}{100} \tag{10.27}$$

Example 10.29: Relationship between Slope, Width, and Depth of a Bench Terrace with 1:1 Riser Slope

A land track has a slope of *S*%. The width of the terrace is *w*. Develop an equation to express the depth *D* of the terrace in terms of slope *S* and width *w*. The riser slope is 1:1.

Solution

1. Draw the cross-section of the bench terrace
   The cross-section of the bench terrace with a riser slope of 1:1 is shown in Figure 10.33

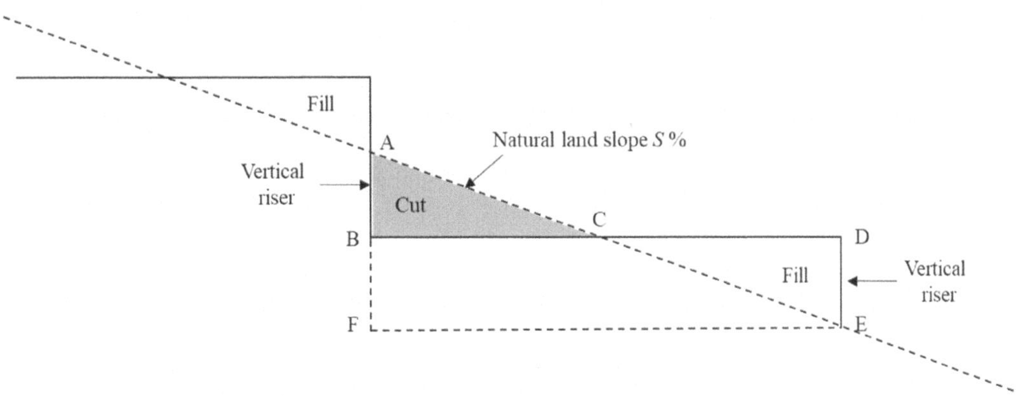

FIGURE 10.32 Cross-section of a bench terrace with vertical riser (Example 10.28).

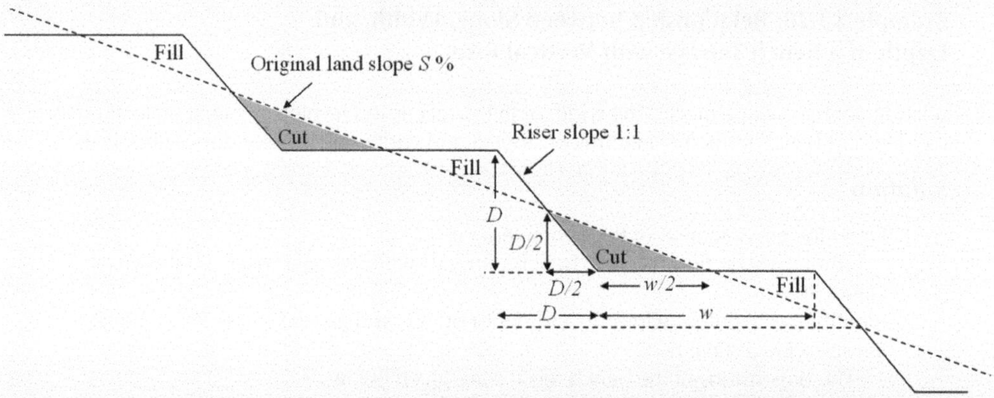

FIGURE 10.33 Cross-section of bench terrace with riser slope of 1:1 (Example 10.29).

2. Develop the relationship between S, w, and D with riser slope 1:1.

   The following relationship is used to develop the equation

$$\frac{S}{100} = \frac{D/2}{D/2 + w/2}$$

$$S \times D + S \times w = 100D$$

The relationship between the depth D of the terrace in terms of slope S and width w is expressed by Eq. (10.28)

$$D = \frac{S \times w}{(100 - S)} \tag{10.28}$$

Example 10.30: Vertical Interval between Bench Terraces for Variable Riser Slope

The slope of the riser of bench terraces depends upon the equipment used to construct the bench terraces. The slope ratio of riser for machine-built terrace is 1, for hand-made terrace is 0.75, and for rock riser is 0.5. Calculate the vertical depth D between bench terraces for three cases. The width of the bench terraces for each case is 4 m, and the natural land slope is 15°.

Solution

1. Identify the generalized equation to calculate the vertical interval between the bench terraces.

   The generalized equation is expressed by Eq. (10.29)[80,81]

$$D = \frac{S \times w}{100 - S \times U} \tag{10.29}$$

where

U = slope of the riser, ratio

Other terms have been defined previously.

2. Convert 15° natural land slope to percent slope *S*.

$$\text{Slope } S = \tan 15°$$

$$S = 0.268$$

$$\text{Percent slope } S = 0.268 \times 100 = 26.8$$

3. Calculate the vertical depth *D* between bench terraces
   a. Machine-built terrace

$$D = \frac{S \times w}{100 - S \times U} = \frac{26.8\% \times 4 \text{ ft}}{100\% - 26.8\% \times 1}$$

$$= \frac{26.8\% \times 4 \text{ ft}}{73.2\%} = 1.47 \text{ ft}$$

   b. Hand-built terrace

$$D = \frac{S \times w}{100 - S \times U} = \frac{26.8\% \times 4 \text{ ft}}{100\% - 26.8\% \times 0.75}$$

$$= \frac{107.2\ \%.\text{ft}}{100\% - 20.1\%} = \frac{107.2\ \%.\text{ft}}{79.9\%} = 1.34 \text{ ft}$$

   c. Rock riser

$$D = \frac{S \times w}{100 - S \times U} = \frac{26.8\% \times 4 \text{ ft}}{100\% - 26.8\% \times 0.5}$$

$$= \frac{107.2\ \%.\text{ft}}{100\% - 13.4\%} = \frac{107.2\ \%.\text{ft}}{86.6\%} = 1.24 \text{ ft}$$

Example 10.31: Depth of Cut of an Inward Sloping Bench Terrace with Vertical Rise

Bench terraces are made over a sloping tract of land. The natural land has a 35% slope. The bench width is 6 ft and has 3% inward slope. Calculate the percentage increase in depth of cut and volume of cut from the horizontal bench terrace due to the inward slope.

Solution

1. Draw the cross-section of the bench terrace
   The cross-section of the bench terrace with vertical rise, and horizontal and inward sloping bench is shown in Figure 10.34.

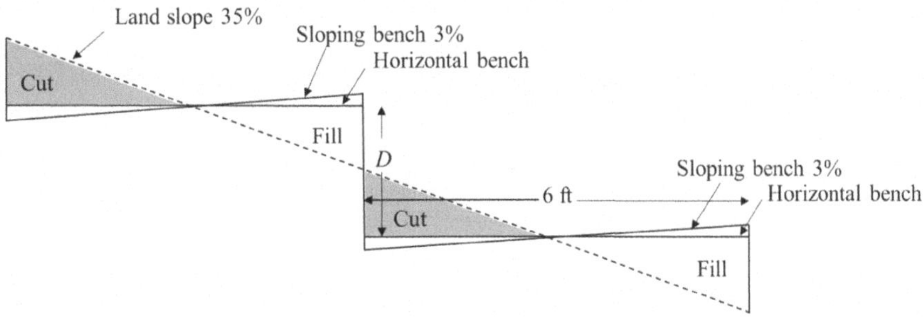

FIGURE 10.34 Cross-section of bench terrace with 3% inward slope (Example 10.31).

2. Calculate depth of bench terrace D

$$D \text{ from Eq. (10.27)} = \frac{w \times S}{100} = \frac{6 \text{ ft} \times 35}{100} = 2.1 \text{ ft}$$

3. Calculate cut for horizontal bench

$$\text{Cut} = \frac{D}{2} = \frac{2.1 \text{ ft}}{2} = 1.05 \text{ ft}$$

4. Calculate additional cut depth of inward sloping bench.

$$\text{Additional cut depth due to inward sloping bench} = \frac{3}{100} \times \frac{6 \text{ ft}}{2} = 0.09 \text{ ft}$$

5. Calculate percent increase in depth due to 3% inward slope of bench from horizontal bench

$$\text{Increase in cut depth due to inward slope of bench} = \frac{0.09 \text{ ft}}{1.05 \text{ ft}} \times 100 = 8.6\%$$

6. Calculate volume of cut per linear ft of the bench terrace
   a. Horizontal bench

$$\text{Volume} = \frac{1}{2} \times 1.05 \text{ ft} \times 3 \text{ ft} \times 1 \text{ ft length}$$

$$= 1.575 \text{ ft}^3$$

   b. Sloping inward bench

$$\text{Volume} = \frac{1}{2} \times 0.09 \text{ ft} \times 3 \text{ ft} \times 1 \text{ ft length}$$

$$= 0.135 \text{ ft}^3$$

   c. $\text{Total volume} = 1.5710 \text{ ft}^3 + 0.135 \text{ ft}^3 = 1.710 \text{ ft}^3$

   d. Percent increase in volume of cut from horizontal bench

$$\text{Percent increase} = \frac{0.135 \text{ ft}^3}{1.575 \text{ ft}^3} \times 100 = 8.6\%$$

Note: The cut is equal to the fill with a horizontal and inward sloping bench.

Example 10.32: Total Length of Terrace, Earthwork, and Lost Land for Agriculture Due to Riser Slope

On a 20% hill slope, it is proposed to construct bench terraces. If the vertical interval of the terrace is 2 m, calculate (1) length of terrace per hectare, (2) earthwork required per hectare, and (3) area lost per hectare, both for vertical cut and riser slope 1:1.

Solution

A. Vertical riser
   1. Calculate width of bench terrace for vertical cut
      The width of bench terrace for vertical cut is calculated from Eq. (10.27)

$$w = \frac{100D}{S} = \frac{100\% \times 2\ \text{m}}{20\%} = 10\ \text{m}$$

2. Calculate the length of terrace *L* per hectare

$$L = \frac{10,000\ \text{m}^2}{10\ \text{m}} = 1,000\ \text{m}$$

3. Calculate area for cultivation

$$\text{Area available for cultivation} = 10\ \text{m} \times 1,000\ \text{m} = 10,000\ \text{m}^2$$

4. Calculate total earthwork per hectare
   The earthwork is calculated from triangular area

$$\text{Earthwork } = \frac{1}{2} \times \frac{w}{2} \times \frac{D}{2} \times L = \frac{1}{2} \times 5\ \text{m} \times 1\ \text{m} \times 1,000\ \text{m} = 2,500\ \text{m}^3$$

   Assume 2.5% additional earthwork needed for the construction of shoulder-bound (lip) and drainage channels.

$$\text{Total estimated arthwork} = 2,500\ \text{m}^3 \times 1.025 = 2,563\ \text{m}^3$$

5. Calculate net area available for cultivation
   Since the riser is vertical, the entire bench width is available for cultivation. Assume 2.0% loss of land from cultivation due to shoulder-bound (lip) and drainage channel.

$$\text{Net area available for cultivation} = 10,000\ \text{m}^2 \times (1 - 0.02) = 9,800\ \text{m}^2$$

B. Riser 1:1

1. Calculate width of terrace
   The width of terrace is calculated from Eq. (10.28)

$$w = \frac{D(100 - S)}{S}$$

$$w = \frac{2\ \text{m}(100 - 20)\%}{20\%} = 8\ \text{m}$$

2. Calculate length of bench terrace per hectare (Figure 10.33)

$$L = \frac{10,000\ \text{m}^2}{8\ \text{m} + (1\ \text{m} + 1\ \text{m})(\text{horizontal component due to 1:1 slope})}$$

$$= 1,000\ \text{m}$$

3. Calculate net area available for cultivation

$$\text{Net area available for cultivation (see Figure 10.33)} = 8\ \text{m} \times 1,000\ \text{m}$$

$$= 8,000\ \text{m}^2$$

4. Calculate earthwork per hectare
   Total earthwork required is for construction of bench terrace plus 2.5% earthwork for the construction of shoulder-bound and drainage channels.

Total earthwork per hectare (see Figure 10.33)

$$= \left[\frac{1}{2}(5\text{ m} \times 1\text{ m}) - \frac{1}{2}(1\text{ m} \times 1\text{ m})\right] \times 1{,}000\text{ m} \times (1 + 0.025)$$

$$= 2{,}000\text{ m}^3 \times 1.025 = 2{,}050\text{ m}^3$$

5. Calculate percent area lost for cultivation per hectare

$$\text{Area lost for cultivation} = \frac{10{,}000\text{ m}^2 - 8{,}000\text{ m}^2}{10{,}000\text{ m}^2} \times 100 = 20\%$$

6. Calculate percent earthwork reduced per hectare

$$\text{Earthwork reduced per hectare} = \frac{2{,}500\text{ m}^3 - 2{,}050\text{ m}^3}{2{,}500\text{ m}^3} \times 100$$

$$= 18\%$$

10.7.3 Check Dams and Gully Plugs

A check dam (also called a gully plug) is a small, temporary or permanent dam constructed across a drainage ditch, swale, channel, or waterway. The purpose of a check dam is to counteract erosion by lowering the velocity of concentrated flows within a certain design range of storm events. They are primarily used to control water velocity, conserve soil, and improve land. These dams are often implemented in a system of several dams situated at regular intervals across the area of interest. To effectively slow water velocity to reduce erosion and protect the channel between dams in a larger system, the spacing must be designed properly. The check dams should be spaced such that the toe of the upstream check dam is equal to the elevation of the downstream check dam's crest.[82] By doing so, the water can pond between check dams and thus slow the flow velocity substantially as the water progresses downslope.[83,84]

A check dam is not a new technology. It is inexpensive and easy to install, and can be built from wood logs, stone, pea gravel-filled sandbags, or bricks and cement. The purpose of a check dam is to interrupt the flow of water and flatten the gradient of the channel, thus reducing the velocity. In turn, this obstruction induces infiltration and reduces erosion. They can be used not only to slow flow velocity but also to distribute flows across a swale to avoid preferential paths and guide flows toward vegetation.[82] Although some sedimentation may result behind the dam, they do not primarily function as sediment trapping devices. While gully plugs are able to retain soil moisture, sand dams (which are similar in appearance) might be the better choice to harvest runoff and allow groundwater recharge. Check dams have traditionally been implemented in two environments: across channel bottoms and on hilly slopes.

The size and shape of a drainage area, and the length and gradient of its slopes, influence the runoff rate and amount of surface water. Therefore, all topographic characteristics should be studied in detail before gully-plugging work begins.[85] Construction should utilize the local situation, available material, slope gradient, and intended tasks for a permanent solution. In gully control, temporary structural measures such as woven-wire, brushwood, logs, loose stone, and boulder check dams are used to facilitate the growth of permanent vegetative cover. Temporary check dams, which have a life span of three to eight years, collect and hold soil and moisture in the bottom of the gully. Tree seedlings, as well as shrub and grass cuttings planted in gullies, can grow without being washed

FIGURE 10.35 A series of stone check dams, ditch checks, or ditch dikes. (Printed from the U.S. EPA website.)

away by flowing water.[86] Thus, a permanent vegetative cover can be established in a short time. A generalized view of a stone check dam is shown in Figure 10.35.

Check dams require regular maintenance as they are used as a temporary structure and thereby are not designed to withstand long-term use. Dams should be inspected frequently, and after every large storm. It is important that rubble, litter, and leaves are removed from the upstream side of the dam. This is typically done when the sediment has reached a height of one-half the original height of the dam.

Example 10.33: Channel Velocity before and after Construction of Check Dams

A rectangular earthen channel has a 2% slope. It carries a maximum stormwater of 2.8 ft^3/s. The channel is 2.5 ft wide. A series of check dams is built across the channel width to reduce the velocity in the channel and soil erosion. The check dams are spaced 200 ft from each other, as shown in Figure 10.36. The crest elevation of the downstream check dams is lower than the toe of the upper check dam by an amount equal to the head over the dam crest. The channel profile with check dams is shown in Figure 10.36. The Manning's $n = 0.018$. Compute the velocity in the channel before and after the construction of check dams. The height of the check dam is the water surface elevation behind the dam minus the head over the dam.

Solution

A. Determine flow velocity before the construction of check dams

Before the construction of check dams, flow will be uniform in the channel with the normal depth of y_n

1. Calculate normal depth

Apply Manning's equation (Eq. 5.37)

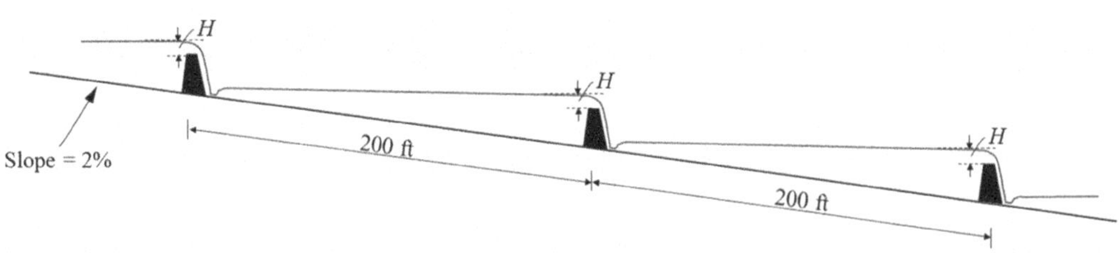

FIGURE 10.36 The channel profile and check dams (Example 10.33).

$$Q = \frac{1.49}{n} AR^{2/3} S^{1/2}$$

Replace in this equation the flow area (A) and hydraulic radius (R) of the rectangular channel from Table 5.3

$$A = by_n = (2.5\ \text{ft})\, y_n$$

$$\text{Wetted perimeter } P = b + 2y_n = 2.5\ \text{ft} + 2y_n$$

$$R = \frac{by_n}{b + 2y_n} = \frac{(2.5\ \text{ft})\, y_n}{(2.5\ \text{ft}) + 2y_n}$$

$$Q = 2.8\ \text{ft}^3/\text{s}$$

$$2.8\ \text{ft}^3/\text{s} = \frac{1.49}{0.018}(2.5\ \text{ft})\, y_n \left(\frac{(2.5\ \text{ft})\ y_n}{2.5\ \text{ft} + 2y_n} \right)^{\frac{2}{3}} (0.02)^{1/2}$$

$$0.052 = \frac{y_n^{5/3}}{(2.5 + 2y_n)^{2/3}}$$

Determine the normal depth (y_n) by trial and error or by using an equation solver

$$y_n = 0.26\ \text{ft}$$

This is the depth of the flow in the channel before the construction of check dams

2. Calculate flow velocity

$$v = \frac{Q}{A}$$

$$A = 2.5\ \text{ft} \times 0.26\ \text{ft} = 0.65\ \text{ft}^2$$

$$v = \frac{2.8\ \text{ft}^3/\text{s}}{0.65\ \text{ft}^2} = 4.31\ \text{ft/s}$$

This is flow velocity in the channel before the construction of check dams

B. Calculate flow velocity after the construction of check dams

1. Determine the normal depth between check dams

The normal depth is calculated in Part A.1

$$y_n = 0.26\ \text{ft}$$

2. Calculate critical depth

Determine critical depth from Eq. (5.46)

$$y_c = \left(\frac{(Q/b)^2}{g} \right)^{1/3}$$

$$y_c = \left(\frac{\left(2.8\ \text{ft}^3/\text{s}/2.5\ \text{ft}\right)^2}{32.2\ \text{ft/s}^2} \right)^{1/3}$$

$$y_c = 0.34\ \text{ft}$$

Since $y_c = 0.34\ \text{ft} > y_n = 0.26\ \text{ft}$, flow in the channel will be supercritical before the construction of check dams. Once the check dams are constructed, due to the backwater effect, the flow depth between the check dams will gradually increase downstream. The water surface profile at the toe of the check dams cannot be determined accurately due to the high turbulence created in these areas by water flowing over the check dams and hitting the channel bottom. A hydraulic jump will form within a short distance from the check dams, and the water depth will increase to the subcritical flow depth after the jump. The flow depth will increase gradually downstream until it reaches a depth equal to the height of the check dam and the head over the dam crest *H*.

3. Calculate the depth after the hydraulic jump
   a. Determine the Froude number for the flow before the jump

      Apply Eq. (5.32). The hydraulic depth (D_h) of the rectangular spillway channel is equal to the flow depth y (Table 5.3).

$$\text{Fr}_1 = \frac{V}{\sqrt{gD_h}} = \frac{4.31\ \text{ft/s}}{\sqrt{32.2\ \text{ft/s}^2 \times 0.26\ \text{ft}}} = 1.49$$

   b. Determine flow depth after the jump y_2 from Eq. (5.44)

$$y_2 = \frac{y_1}{2}\left(\sqrt{1+8\text{Fr}_1^2} - 1\right)$$

$$= \frac{0.26\ \text{ft}}{2}\left(\sqrt{1+8\times(1.49)^2} - 1\right)$$

$$= 0.43\ \text{ft}$$

4. Determine water surface profile between two check dams

   Apply gradually varied flows equation (Eq. 5.43) to calculate water surface profile

$$\frac{dy}{dx} = \frac{S_o - S_f}{1-\text{Fr}^2}$$

Write this equation in the form of discrete differences (Eqs. 5.47 and 5.48)

$$\Delta x = \frac{1-\text{Fr}^2}{S_0 - S_f}\Delta y$$

$$\left(x_{i+1} - x_i\right) = \frac{1-\overline{\text{Fr}}^2}{S_o - \overline{S}_f}\left(y_{i+1} - y_i\right)$$

where $\overline{\text{Fr}}$ and $\overline{S}_f$ are the arithmetic mean of the slope of the energy grade line and the Froude number at stations x_i and x_{i+1} along the channel with flow depth of y_i and y_{i+1}, respectively.

Start the computation with the known depth of $y = y_2 = 0.43$ ft at the upper check dam after the hydraulic jump and proceed in the downstream direction. At the upper check dam $x = 0$, and the distance in the downstream flow direction is considered positive.

The water surface calculation is an iterative process that can easily be done in a *spreadsheet*. The sample calculations of initial entries are given under each column. The results of the final iterations are summarized in Table 10.33.

a. **Column (1):** The initial value of y_1 selected in Column (1) is 0.43 ft. The size of the increment in the depth ($\Delta y = y_{i+1} - y_i$) determines the length ($\Delta x = x_{i+1} - x_i$) in each computational step and the accuracy of the calculated water surface profile. In each iteration, y_{i+1} is the flow depth for the current step, and y_i is the depth for the previous step. The selected values of y_2 and y_3 are 0.45 and 0.50, respectively. The values of Δy_1, Δy_2, and Δy_3 are therefore 0.020, 0.05, and 0.05 ft, respectively.
b. **Columns (2)–(6):** Calculate flow area A, wetted perimeter P, hydraulic radius R, and flow velocity V, and Fr^2 for the depth in Column 1.

TABLE 10.33
Computation of Water Surface Profile between Check Dams in a Rectangular Channel (Example 10.33)

y, ft	A, ft²	P, ft	R, ft	V, ft/s	Fr^2	S_f, ft/ft	$\overline{Fr}^2$	$\bar{S}_f$, ft/ft	$S_o - \bar{S}_f$, ft/ft	$1-\overline{Fr}^2$	Δy, m	$\frac{1-\overline{Fr}^2}{S_o - \bar{S}_f}$	Δx, ft	x, ft
(1)	(2)	(3)	(4)	(5)	(6)	(7)	(8)	(9)	(10)	(11)	(12)	(13)	(14)	(15)
0.43	1.08	3.36	0.32	2.60	0.49	0.00452								0.0
0.45	1.13	3.40	0.33	2.49	0.43	0.00395	0.46	0.00424	0.01576	0.54	0.02	34.3	0.7	0.7
0.50	1.25	3.50	0.36	2.24	0.31	0.00289	0.37	0.00342	0.01658	0.63	0.05	38.0	1.9	2.6
0.55	1.38	3.60	0.38	2.04	0.23	0.00218	0.27	0.00254	0.01746	0.73	0.05	41.6	2.1	4.7
0.60	1.50	3.70	0.41	1.87	0.18	0.00169	0.21	0.00194	0.01806	0.79	0.05	43.9	2.2	6.9
0.65	1.63	3.80	0.43	1.72	0.14	0.00134	0.16	0.00152	0.01848	0.84	0.05	45.4	2.3	9.1
0.70	1.75	3.90	0.45	1.60	0.11	0.00109	0.13	0.00122	0.01878	0.87	0.05	46.4	2.3	11.5
0.75	1.88	4.00	0.47	1.49	0.09	0.00089	0.10	0.00099	0.01901	0.90	0.05	47.2	2.4	13.8
0.80	2.00	4.10	0.49	1.40	0.08	0.00074	0.08	0.00082	0.01918	0.92	0.05	47.7	2.4	16.2
0.85	2.13	4.20	0.51	1.32	0.06	0.00063	0.07	0.00069	0.01931	0.93	0.05	48.2	2.4	18.6
0.90	2.25	4.30	0.52	1.24	0.05	0.00054	0.06	0.00058	0.01942	0.94	0.05	48.5	2.4	21.0
1.00	2.50	4.50	0.56	1.12	0.04	0.00040	0.05	0.00047	0.01953	0.95	0.1	48.8	4.9	25.9
1.50	3.75	5.50	0.68	0.75	0.01	0.00014	0.03	0.00027	0.01973	0.97	0.5	49.4	24.7	50.6
2.00	5.00	6.50	0.77	0.56	0.00	0.00006	0.01	0.00010	0.01990	0.99	0.5	49.8	24.9	75.5
2.50	6.25	7.50	0.83	0.45	0.00	0.00004	0.00	0.00005	0.01995	1.00	0.5	49.9	25.0	100.5
3.00	7.50	8.50	0.88	0.37	0.00	0.00002	0.00	0.00003	0.01997	1.00	0.5	50.0	25.0	125.5
3.50	8.75	9.50	0.92	0.32	0.00	0.00002	0.00	0.00002	0.01998	1.00	0.5	50.0	25.0	150.5
4.00	10.00	10.50	0.95	0.28	0.00	0.00001	0.00	0.00001	0.01999	1.00	0.5	50.0	25.0	175.5
4.49	11.23	11.48	0.98	0.25	0.00	0.00001	0.00	0.00001	0.01999	1.00	0.49	50.0	24.5	200.0

$$A_1 = by_1 = 2.5 \text{ ft} \times 0.43 \text{ ft} = 1.08 \text{ ft}^2; A_2 = by_2 = 2.5 \text{ ft} \times 0.45 \text{ ft} = 1.13 \text{ ft}^2$$

$$P_1 = b + 2y_1 = 2.5 \text{ ft} + 2 \times 0.43 \text{ ft} = 3.36 \text{ ft}; P_2 = b + 2y_2 = 2.5 \text{ ft} + 2 \times 0.45 \text{ ft} = 3.40 \text{ ft}$$

$$R_1 = A_1 / P_1 = 1.08 \text{ ft}^2 / 3.36 \text{ ft} = 0.32 \text{ ft}; R_2 = A_2 / P_2 = 1.13 \text{ ft}^2 / 3.40 \text{ ft} = 0.33 \text{ ft}$$

$$v_1 = Q_1 / A_1 = 2.8 \text{ ft}^3/\text{s}/1.08 \text{ ft}^2 = 2.60 \text{ ft/s}; v_2 = Q_2 / A_2 = 2.8 \text{ ft}^3/\text{s}/1.13 \text{ ft}^2 = 2.49 \text{ ft/s}$$

$$\text{Fr}_1^2 = \left(\frac{v_1}{\sqrt{gy_1}}\right)^2 = \left(\frac{2.60 \text{ ft/s}}{\sqrt{32.2 \text{ ft/s}^2 \times 0.43 \text{ ft}}}\right)^2 = 0.49;$$

$$\text{Fr}_2^2 = \left(\frac{v_2}{\sqrt{gy_2}}\right)^2 = \left(\frac{2.49 \text{ ft/s}}{\sqrt{32.2 \text{ ft/s}^2 \times 0.45 \text{ ft}}}\right)^2 = 0.43$$

c. **Column (7):** Determine the slope of energy grade line (S_f) from Manning's equation and compute R and v in Columns (4) and (5).

$$\text{From Manning's equation } v = \frac{1.49}{n} \times R^{2/3} \times S^{1/2}$$

$$S_{f1} = \left(\frac{nv_1}{1.49R_1^{2/3}}\right)^2 = \left(\frac{0.018 \times 2.6 \text{ ft/s}}{1.49 \times (0.32 \text{ ft})^{2/3}}\right)^2 = 0.00452$$

$$S_{f2} = \left(\frac{nv_2}{1.49R_2^{2/3}}\right)^2 = \left(\frac{0.018 \times 2.49 \text{ ft/s}}{1.49 \times (0.33 \text{ ft})^{2/3}}\right)^2 = 0.00395$$

d. **Columns (8)–(9):** Calculate $\bar{S}_f$ and $\overline{\text{Fr}}^2$. These are the average of S_f and Fr^2 for the current water depth and for the water depth in the previous row

$$\overline{\text{Fr}}^2 = \frac{(\text{Fr}^2)_{\text{current row}} + (\text{Fr}^2)_{\text{previous row}}}{2}$$

For the first row, $\overline{\text{Fr}}^2$ is not calculated since Fr^2 for the previous row is not available. The value of $\overline{\text{Fr}}^2$ is left blank in Column (8) of the first row.

$$\overline{\text{Fr}}_2^2 = \frac{(\text{Fr}_2^2 + \text{Fr}_1^2)}{2} = \frac{0.43 + 0.49}{2} = 0.46$$

$$\overline{\text{Fr}}_3^2 = \frac{(\text{Fr}_3^2 + \text{Fr}_2^2)}{2} = \frac{0.31 + 0.43}{2} = 0.37$$

$$\bar{S}_f = \frac{(S_f)_{\text{current row}} + (S_f)_{\text{previous row}}}{2}$$

For the first row, $\bar{S}_f$ is not calculated since S_f for the previous row is not available. The value of $\bar{S}_f$ is left blank in Column (9) of the first row.

$$\bar{S}_{f2} = \frac{S_{f2} + S_{f1}}{2} = \frac{0.00395 + 0.00452}{2} = 0.00424$$

$$\bar{S}_{f3} = \frac{S_{f3} + S_{f2}}{2} = \frac{0.00289 + 0.00395}{2} = 0.00342$$

e. **Column (10):** Calculate $(S_o - \bar{S}_f)$ by subtracting $\bar{S}_f$ of Column (9) from the channel bottom slope, S_o

$$(S_o - S_{f2}) = (0.020 - 0.0424) = 0.01576$$

$$(S_o - S_{f3}) = (0.020 - 0.00342) = 0.01658$$

f. **Column (11):** Calculate $\left(1 - \overline{\mathrm{Fr}}^2\right)$ by subtracting $\overline{\mathrm{Fr}}^2$ from 1

$$\left(1 - \overline{\mathrm{Fr}}_2^2\right) = (1 - 0.46) = 0.54; \left(1 - \overline{\mathrm{Fr}}_3^2\right) = (1 - 0.37) = 0.63$$

g. **Column (12):** Calculate the depth increment $\Delta y = \left(y_{i+1} - y_i\right)$. This column is the difference between the flow depth in the current row and the depth in the previous row

$$\Delta y_2 = \left(y_2 - y_1\right) = (0.45 - 0.43)\ \text{ft} = 0.02\ \text{ft}$$

$$\Delta y_3 = \left(y_3 - y_2\right) = (0.56 - 0.45)\ \text{ft} = 0.05\ \text{ft}$$

h. **Column (13):** Calculate $\dfrac{1 - \overline{\mathrm{Fr}}^2}{S_o - \bar{S}_f}$ by dividing Column (11) by Column (10)

$$\frac{\left(1 - \overline{\mathrm{Fr}}_2^2\right)}{(S_o - \bar{S}_{f2})} = \frac{0.54}{0.0576} = 34.3; \frac{\left(1 - \overline{Fr}_3^2\right)}{(S_o - \bar{S}_{f3})} = \frac{0.63}{0.01658} = 38.0$$

i. **Column (14):** Calculate incremental channel length Δx from discrete different equations (Eqs. 5.47 and 5.48)

$$\Delta x = \left(x_{i+1} - x_i\right) = \frac{1 - \overline{\mathrm{Fr}}^2}{S_o - \bar{S}_f}\left(y_{i+1} - y_i\right).$$

This is obtained by multiplying Columns (13) and Column (12)

$$\Delta x_2 = \left(x_3 - x_2\right) = \frac{1 - \overline{\mathrm{Fr}}_2^2}{S_o - \bar{S}_{f2}}\left(y_3 - y_2\right) = 34.3 \times 0.02\ \text{ft} = 0.7\ \text{ft}$$

$$\Delta x_3 = \left(x_4 - x_3\right) = \frac{1 - \overline{\mathrm{Fr}}_3^2}{S_o - \bar{S}_{f3}}\left(y_4 - y_3\right) = 38.0 \times 0.05\ \text{ft} = 1.9\ \text{ft}$$

j. **Column (15):** Calculate $x_{\text{current}} = x_{\text{previous}} + \Delta x$

$$x_2 = 0 + 0.7 \text{ ft} = 0.7 \text{ ft}$$

$$x_3 = 0.7 \text{ ft} + 1.9 \text{ ft} = 2.6 \text{ ft}$$

5. Complete final iteration.
   The procedure in step 5 is repeated until the final iteration is reached, when $\Delta x_2 + \Delta x_3 + \Delta x_4$ ...... Δx_i = the distance of 200 ft between two adjacent dams.
6. Complete all tabular entries.
   Complete all entries of Columns (1) through (15) and summarize them in Table 10.33.
7. Plot the water surface profile in the channel
   Use *y* and *x* values from Columns (1) and (15) of Table 10.33 to plot the water surface profile. The backwater surface profile is shown in Figure 10.37.
8. Determine the velocity in the channel.
   The calculated flow velocity at different distances between the check dams is summarized in Column (5) of Table 10.33.
   The maximum velocity after construction of the check dams is 2.6 ft/s, which is 40% smaller than the velocity before the construction of the check dams. The minimum velocity after construction of the check dams is 0.25 ft/s.
9. Determine head over check dam.
   Head over check dam from Eq. (7.9a)

$$Q = \frac{2}{3} C_d L H \sqrt{2gH}$$

$$\text{Assume } C_d = 0.85$$

$$Q = \frac{2}{3} \times 0.85 \times 2.5 \text{ ft} \times \sqrt{2 \times 9.81 \text{ ft/s}^2} \times H^{\frac{3}{2}}$$

$$2.8 \text{ ft}^3 = 11.37 H^{\frac{3}{2}}$$

$$H = (0.246)^{0.667} = 0.39 \text{ ft}$$

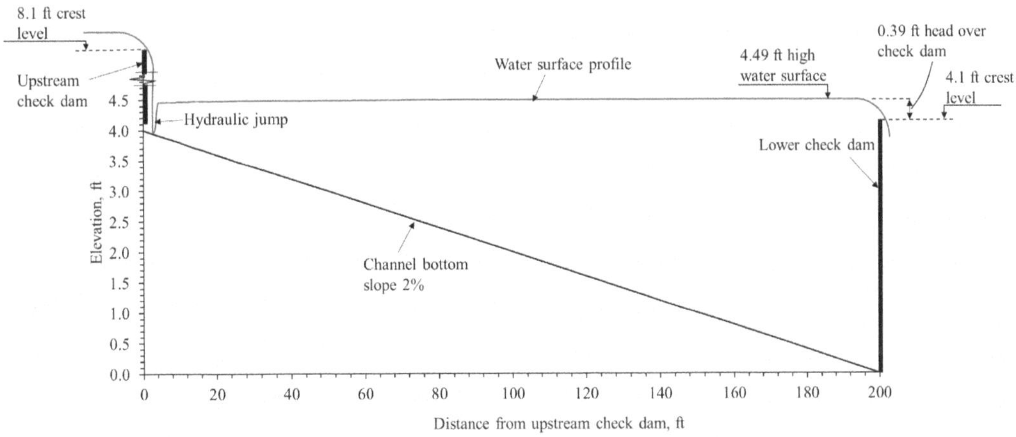

FIGURE 10.37 Water surface profile in the rectangular channel between two check dams (Example 10.33).

10. Calculate the height of the check dam

    The height of check dam = (water surface level behind the check dam − head over the check dam) = (4.49 ft − 0.39 ft) = 4.10 ft

11. Draw the water surface profile between two adjacent dams.

    The water surface profile is shown in Figure 10.37. The hydraulic jump occurs downstream of the upper dam.

12. Determine the height of the upstream dam

    The height of the upstream dam = 4.10 ft + elevation difference between two dams = 4.10 ft 200 ft × 0.02 = 4.10 ft + 4.0 ft = 8.10 ft

    Note: The flow velocity after the construction of check dams may be approximately calculated by assuming a horizontal water surface elevation between the check dams. It can also be assumed that the flow depth is zero at the toe of the upper dam. Since the crest elevation of the downstream check dams is lower than the toe of the upper check dam by an amount equal to the head over the dam crest, the flow depth behind the downstream dam is equal to 200 ft × 0.02% = 4.00 ft.

    The maximum velocity in the channel after the construction of check dams occurs after the hydraulic jump, where the flow depth is 0.43 ft (step 3)

$$v_{max} = \frac{Q}{A}$$

$$A = 2.5 \text{ ft} \times 0.43 \text{ ft} = 1.08 \text{ ft}^2$$

$$v_{max} = \frac{2.8 \text{ ft}^3/\text{s}}{1.08 \text{ ft}^2} = 2.6 \text{ ft/s}$$

The minimum velocity occurs behind the downstream dam, where the flow depth is 4.49 ft.

Minimum velocity

$$v_{min} = \frac{Q}{A}$$

$$A = 2.5 \text{ ft} \times 4.49 \text{ ft} = 11.23 \text{ ft}^2$$

$$v_{min} = \frac{2.8 \text{ ft}^3/\text{s}}{11.23 \text{ ft}^2} = 0.25 \text{ ft/s}$$

10.7.4 Contour Bunding or Graded Bunding

Contour bunding consists of building earthen embankments across the slope of the land, following the contour as closely as possible. A series of such bunds divide the area into strips and act as barriers to the flow of water, thus reducing the flow and velocity of the runoff. Different types of bunds are used for erosion control and moisture conservation in the watersheds. When the bunds are constructed along the contours with some minor deviation to adapt to the practical situation, they are known as *contour bunds*. If the bunds are constructed with some slope, they are known as *graded bunds*. The choice of the types of bunds is dependent on land slope, rainfall, soil type, and the purpose of the bund in the area. The contour bunds are recommended for areas with low to high annual rainfall, agricultural fields with permeable soils, and a land slope of less than 6%. Graded bunds are used for safe disposal of excess runoff in areas with high rainfall and relatively impervious soil. Bunds are also constructed along field boundaries without reference to contour. These bunds are called *peripheral bunds*. They serve as fences and divide properties, and give protection from water and wind erosion in low rainfall areas. They are not suitable in heavy rainfall areas. No cultivation is allowed on the earthen embankments of all bunds, except at some places where some types of stabilization grasses are planted to protect the bund. Therefore, some area is lost under the bunds and is not available for cultivation.[87]

Contour bunds in deep black soils are not stable because the soil cracks during hot weather and cakes during wet weather. Further, the poor drainage properties of deep black soils give rise to long stagnation of water against contour bunds and make it unstable. Bund spacing is expressed as the vertical or horizontal distance between corresponding points on two adjacent bunds. Bund spacing should not be so wide as to cause excessive soil erosion between adjacent bunds. The horizontal distance may vary depending on the slope of the land. Shorter distance is used on steeper slopes. The length of the bund should be such that the velocity of water flowing between bunds should be non-erosive. While designing the bund cross-section, the seepage through the bunds due to the accumulation of water behind it should be considered.[88,89] The seepage rate is affected mainly by the head of water impounded, the side slopes of the bund, and the permeability of the soil. A generalized view of contour or graded bunding is shown in Figure 10.38.

FIGURE 10.38 Generalized view of contour bunding or graded bunding. (Printed from the U.S. EPA website.)

Example 10.34: Contour Bund Design for Erosion Control and Maximum Infiltration

Contour bunding is selected to control erosion and maximize infiltration from a tract of land. The natural slope of a tract of land is 6%. Determine the vertical interval, spacing between two consecutive bunds, and the height of the bund. The land area receives heavy rainfall. The maximum water level held by the bund will be up to the mid-horizontal length between two consecutive bunds. Also, calculate the length of the contour bund per hectare of the land area.

Solution

1. Identify the equations for calculating the vertical interval between consecutive bunds.[87–89]

   The equation used for the vertical interval between consecutive bunds is area-specific and depends upon the amount of rainfall and land slope. Equation (10.30) is applied for areas of heavy rainfall

   $$V = 10S + 60 \tag{10.30}$$

   Equation (10.31) is applied for areas of medium to low rainfall

   $$V = 15S + 60 \tag{10.31}$$

   where

   V = vertical interval between consecutive bunds, cm
   S = land slope, %

2. Calculate vertical interval between two consecutive bunds for heavy rainfall area

   $$V = 10 \times 6 + 60 = 120 \text{ cm} = 1.2 \text{ m}$$

3. Calculate the horizontal distance between two consecutive bunds from Eq. (10.32)

   $$L = \frac{V}{S/100} \tag{10.32}$$

   where

   L=horizontal distance between two consecutive bunds, cm

   $$L = \frac{1.2 \text{ m}}{6/100} = 20 \text{ m}$$

4. Determine the height of contour bund

   The height of the contour bund is calculated from Figure 10.39
   From similar triangles

   $$\frac{x}{V} = \frac{\frac{L}{2}}{L} = 0.5$$

   where

   x=theoretical height of the bund

   $$x = 0.5\text{V} = 0.5 \times 120 \text{ cm}$$

   $$x = 60 \text{ cm or } 0.6 \text{ m}$$

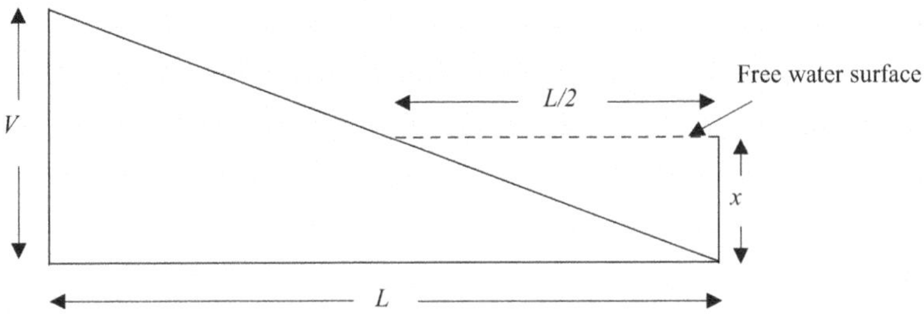

FIGURE 10.39 Graphical representation of bund spacing, vertical interval, and height of bund (Example 10.34).

Provide a freeboard of 0.3 m
Total design height of the bund = 0.9 m

5. Calculate the maximum volume of water stored upstream of the bund per m length

$$\text{Maximum volume of water stored} \frac{1}{2} \times x \times \frac{L}{2} \times 1\text{ m}$$

$$= \frac{1}{2} \times 0.6\text{ m} \times \frac{20}{2}\text{ m} \times 1\text{ m}$$

$$= 3.0\text{ m}^3$$

The impounded water by the contour bund will evaporate and infiltrate into the ground over time

6. Determine the length of bund per hectare

$$\text{The length of contour bund per hectare} = \frac{10{,}000\text{ m}^2}{\text{ha}} \times \frac{1}{20\text{ m}} = 500\text{ m}$$

Example 10.35: Contour Bund Design, Land Occupied, and Earthwork

Contour bunds are built over a sloping tract of land to control erosion and store moisture for farming. The height of the bund is 1 m. Select the cross-section, and calculate earthwork needed and land area lost from farming. The spacing of the contour bund is 15 m. The soil in the area is light sandy loam.

Solution

1. Select the cross-section of the contour bund

   The base width of the bund depends upon the hydraulic gradient of water in the soil. The side slope of the bund depends upon the angle of repose of the soil. For light sandy loam, the top width of the bund is 0.5 m and side slope of 1:1.

2. Calculate the bottom width of the bund

$$\text{The bottom width of the bund} = 0.5\text{ m} + 1\text{ m} + 1\text{ m} = 2.5\text{ m}$$

3. Calculate the cross-section of the bund

$$\text{The cross section of the bund} = \frac{(0.5\text{ m} + 2.5\text{ m})}{2} \times 1\text{ m} = 1.5\text{ m}^2$$

4. Calculate the length of bund per hectare

$$\text{Length of bund per hectare} = \frac{10{,}000\ \text{m}^2}{\text{ha}} \times \frac{1}{15\ \text{m}} = 667\ \text{m}$$

5. Calculate the total earthwork needed for bund construction

$$\text{Total earthwork needed per ha area} = 1.5\ \text{m}^2 \times 667\ \text{m} = 1{,}000\ \text{m}^3$$

6. Calculate the land area lost from agriculture per ha

$$\text{Land area required for bund construction} = 2.5\ \text{m} \times 667\ \text{m} = 1{,}668\ \text{m}^2/\text{ha}$$

$$\text{Land area lost for bund construction} = \frac{1{,}668\ \text{m}^2/\text{ha}}{10{,}000\ \text{m}^2/\text{ha}} \times 100 = 17\%$$

10.7.5 Detention/Retention Basin or Pond

The detention/retention ponds, also called farm pond, stormwater pond, and retarding basin, are excavated areas that temporarily hold water for a short period of time before it enters the stream. The on-channel detention ponds reduce peak storm flows and provide flood control for downstream neighborhoods. Detention/retention ponds are of two types: *dry ponds* and *wet ponds*. *Dry ponds*, holding ponds, or dry detention basins do not have permanent pools of water. They are usually dry, except during or after rain or snow melt, and serve for flood control. Their purpose is to slow down water flow and hold it for a short period of time, such as 24 h. Dry detention ponds allow large flows of water to enter but limit the outflow by having a small opening at the lowest point of the structure. The size of this opening is determined by the capacity of underground and downstream channels to handle the release of the contained water. These ponds are often used for flood control and for stormwater runoff treatment. Urban areas rely on these structures to reduce peak runoff rates associated with storms, thus decreasing flood damage. The land area available for construction, slope of the site, erosion control, and contributing areas are all factors that are considered in design.[90,91] *Wet ponds* are also called retention ponds or percolation ponds, as they hold back water similar to water behind a dam. These ponds have a permanent pool of water. The water depth fluctuates in response to precipitation and runoff from the contributing areas. Maintaining a pool controls resuspension of deposited sediments at the bottom of the holding area. Although these ponds vary in size and shape, they all function to settle suspended solids and reduce peak flows.[92] Both types of ponds assist with flood management and can improve water quality. Sediments and associated pathogens, nutrients, and metals settle out of stormwater runoff in these ponds.[93] While both types of ponds provide water quality benefits, wet ponds allow for additional biological interactions that assist in improving water quality by nutrient removal. Detention ponds usually hold stormwater long enough to settle sand and larger silt particles. Fine silts and clays will not settle and will continue to be released down the water course. Some retention ponds hold stormwater for longer periods of time and allow even the fine sediments to settle.[90,94] A generalized view of a detention/retention pond is shown in Figure 10.40.

FIGURE 10.40 Photo of detention/retention pond. (Printed from the U.S. EPA website.)

Example 10.36: Performance of a Detention Basin for Storm Runoff Control

A dry detention basin is constructed as a stormwater flow control. The basin surface area is 10,000 m^2, and its walls are vertical and 1.5 m high. The triangular inflow hydrograph with 10-min increment to the basin is applied. The storm duration and inflow of triangular inflow hydrograph data are summarized in Table 10.34. A 1.5-m diameter outlet pipe is installed at the bottom of the basin to discharge the stormwater. Assuming that the basin is initially empty, determine the outflow hydrograph and maximum depth in the detention basin.

TABLE 10.34
Triangular Inflow Hydrograph to the Detention Basin (Example 10.36)

t, min	0	10	20	30	40	50	60	70	80	90	100	110	120	130	140	150
Inflow, m^3/s	0.00	1.70	3.40	5.10	6.80	8.50	10.19	9.06	7.93	6.80	5.66	4.53	3.40	2.27	1.13	0.00

Solution

The hydrograph routing in the basin can be performed using Eq. (10.33). In the derivation of this equation, the continuity equation and a storage-outflow relation are combined. The outflow hydrograph and storage at the end of each time increment are calculated from Eq. (10.33).

$$(I_n + I_{n+1}) + \left(\frac{2S_n}{\Delta t} - O_n\right) = \frac{2S_{n+1}}{\Delta t} + O_{n+1} \tag{10.33}$$

where

I_n and I_{n+1} = inflow to the basin at the beginning and end of the *n*th time interval
O_n and O_{n+1} = outflow from the basin at the beginning and end of the *n*th time interval
S_n and S_{n+1} = storage in the basin at the beginning and end of the *n*th time interval

There are two unknowns in Eq. (10.33): outflow (O_{n+1}) and storage (S_{n+1}), both of which are related to the depth of storage in the basin (h_{n+1}). A *storage-outflow function* relating O_{n+1} and storage S_{n+1} can be developed to solve Eq. (10.33).[19]

A. Develop the *storage-outflow function* for the basin

The calculated data of the storage-outflow function are presented in Table 10.35. The time interval is $\Delta t = 10$ min.

1. Develop entry data for Columns (1) through (4) of Table 10.35.

Column (1): Consider a range of flow depth in the detention basin (h) at 0.1 m interval from 0.0 to 1.5 m

These values are listed in Column (1) of Table 10.35

Column (2): Calculate the storage volume (S) at each depth

For example, at h=0.1 m, storage area $S = 10{,}000\ \text{m}^2 \times 0.1\ \text{m} = 1{,}000\ \text{m}^3$

At h=0.2 m, storage area $S = 10{,}000\ \text{m}^2 \times 0.2\ \text{m} = 2{,}000\ \text{m}^3$

Similarly, calculate S for each h=0.1 m depth interval up to h=1.5 m. These values are summarized in Column (2) of Table 10.35.

Column (3): Determine the outflow rate at each depth. The flow from the outflow pipe is through an orifice

Apply Eq. (7.8) with discharge coefficient C_d=0.75

$$\text{Pipe area } A = \frac{\pi}{4}(1.5\ \text{m})^2 = 1.77\ \text{m}^2$$

TABLE 10.35
Storage-Outflow Function (Example 10.36)

h, m	*S*, m³	*O*, m³/s	(2*S*/Δ*t*)+*O*, m³/s
(1)	(2)	(3)	(4)
0.0	0	0.0	0.0
0.1	1,000	1.9	5.2
0.2	2,000	2.6	9.3
0.3	3,000	3.2	13.2
0.4	4,000	3.7	17.0
0.5	5,000	4.2	20.8
0.6	6,000	4.5	24.5
0.7	7,000	4.9	28.2
0.8	8,000	5.3	31.9
0.9	9,000	5.6	35.6
1.0	10,000	5.9	39.2
1.1	11,000	6.2	42.8
1.2	12,000	6.4	46.4
1.3	13,000	6.7	50.0
1.4	14,000	6.9	53.6
1.5	15,000	7.2	57.2

$$\text{Outflow } O \text{ at } h = 0.1 \text{ m} = C_d A\sqrt{2gh}$$

$$= 0.75 \times 1.77 \text{ m}^2 \times \sqrt{2 \times 9.81 \text{ m/s}^2 \times 0.1 \text{ m}} = 1.9 \text{ m}^3\text{/s}$$

$$\text{Outflow } O \text{ at } h = 0.2 \text{ m} = C_d A\sqrt{2gh}$$

$$= 0.75 \times 1.77 \text{ m}^2 \times \sqrt{2 \times 9.81 \text{ m/s}^2 \times 0.2 \text{ m}} = 2.6 \text{ m}^3\text{/s}$$

Similarly, calculate outflow O for each $h = 0.1$ m depth interval up to $h = 1.5$ m. These values are summarized in Column (3) of Table 10.35

Column (4): Calculate $\left[(2S/\Delta t) + O\right]$ using data in Columns (2) and (3) of Table 10.35

For example, at $h = 0.1$ m and $\Delta t = 10$ min

$$\left[(2S/\Delta t) + O\right] = \left[\left(2 \times 1{,}000 \text{ m}^3 \Big/ \left(10 \text{ min} \times \frac{60 \text{ s}}{1 \text{ min}}\right)\right) + 1.9 \text{ m}^3\text{/s}\right] = 5.2 \text{ m}^3\text{/s}$$

At $h = 0.2$ m and $\Delta t = 10$ min

$$\left[(2S/\Delta t) + O\right] = \left[\left(2 \times 2{,}000 \text{ m}^3 \Big/ \left(10 \text{ min} \times \frac{60 \text{ s}}{1 \text{ min}}\right)\right) + 2.6 \text{ m}^3\text{/s}\right] = 9.3 \text{ m}^3\text{/s}$$

Similarly, calculate $\left[(2S/\Delta t) + O\right]$ for each $h = 0.1$ m depth intervals from $h = 0.3$ m up to $h = 1.5$ m.

These values are summarized in Column (4) of Table 10.35

2. Plot the storage-outflow function

   Plot the *storage-outflow function* using Columns (3) and (4) data of Table 10.35.

   Outflow function (O) is on the Y-axis, and $\left[(2S/\Delta t) + O\right]$ function is on the X-axis. This plot is shown in Figure 10.41.

B. Develop data for outflow hydrograph

   The stepwise procedure to determine the outflow hydrograph and storage volume is presented below, and the results are summarized in Table 10.36.

1. Develop data for Columns (1) through (8)

   Column (1): Assign a time step $(n) = 1, 2, 3, \ldots, 19$. These values are listed in Column (1) of Table 10.36

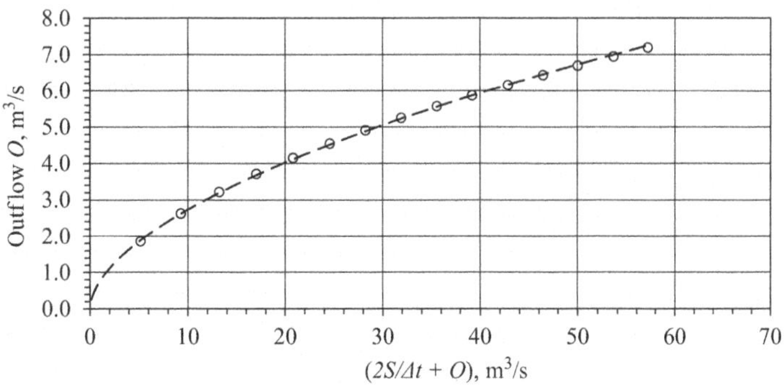

FIGURE 10.41 Storage-outflow function (O) is plotted with function $\left[(2S/\Delta t) + O\right]$ for the detention basin (Example 10.36).

TABLE 10.36
Routing Stormwater Runoff through the Detention Basin (Example 10.36)

n, -	t, min	I, m³/s	I_n+I_{n+1}, m³/s	$(2S_n/\Delta t)-O_n$, m³/s	$(2S_{n+1}/\Delta t)+O_{n+1}$, m³/s	O_{n+1}, m³/s	S_{n+1}, m³/s
(1)	(2)	(3)	(4)	(5)	(6)	(7)	(8)
0						0.00	0
1	0	0.00	1.70	0.00	1.70	0.85	255
2	10	1.70	5.10	0.00	5.10	1.80	990
3	20	3.40	8.50	1.50	10.00	2.73	2,181
4	30	5.10	11.90	4.54	16.44	3.61	3,850
5	40	6.80	15.30	9.22	24.52	4.51	6,003
6	50	8.50	18.69	15.50	34.19	5.44	8,625
7	60	**10.19**	19.25	23.31	42.56	6.15	10,925
8	70	9.06	16.99	30.27	47.26	6.52	12,222
9	80	7.93	14.73	34.22	48.95	**6.65**	**12,691**
10	90	6.80	12.46	35.66	48.12	6.58	12,459
11	100	5.66	10.19	34.95	45.14	6.35	11,635
12	110	4.53	7.93	32.43	40.36	5.97	10,318
13	120	3.40	5.67	28.43	34.10	5.43	8,601
14	130	2.27	3.40	23.24	26.64	4.73	6,574
15	140	1.13	1.13	17.18	18.31	3.83	4,345
16	150	0.00	0.00	10.65	10.65	2.83	2,346
17	160	0.00	0.00	4.99	4.99	1.85	943
18	170	0.00	0.00	1.29	1.29	0.55	223
19	180	0.00	0.00	0.19	0.19	0.10	28

Bold values are the maximum values in those columns.

Column (2): Select time steps with an increment of $\Delta t = 10$ min. The time steps t for each time increment of $\Delta t = 10$ min are listed in Column (2) of Table 10.36

Column (3): Enter the inflow triangular hydrograph data I from Table 10.34 in Column (3) of Table 10.36. These values successively are 0.00, 1.70, 3.40, 5.10, …, 1.13, 0.00 min

Column (4): Calculate $(I_n + I_{n+1})$ for each time step

For n=1: $(I_1 + I_2) = 0 + 1.7 = 1.7\ \text{m}^3/\text{s}$

For n=2: $(I_2 + I_3) = 1.7 + 3.4 = 5.1\ \text{m}^3/\text{s}$

For n=3: $(I_3 + I_4) = 3.4 + 5.1 = 8.5\ \text{m}^3/\text{s}$

Similarly, enter all $(I_n + I_{n+1})$ values for n=4 to n=19 in Column (4) of Table 10.36.

Columns (5)–(8): The entries in Columns (5) through (8) are interrelated. Therefore, the data for these columns are developed simultaneously. The calculation steps for Columns (5) through (8) for n=0 to n=3 are presented below.

When n=0

The detention basin is assumed initially empty. Therefore, when n=0, the outflow (O) from the basin, and its storage (S) are also zero

$$O_{n+1} = O_1 = 0.0\ \text{m}^3/\text{s} \quad \text{and} \quad S_{n+1} = S_1 = 0.0\ \text{m}^3$$

These values are entered in Columns (7) and (8) of Table 10.36. Also, these values are used to calculate entries in Column (5) for $n=1$

When n=1

a. Calculate $\left(\frac{2S_n}{\Delta t} - O_n\right) = \left(\frac{2S_1}{\Delta t} - O_1\right)$

The detention basin has an outlet pipe at the bottom. The outlet flow starts from the bottom as soon as enough head builds up above the outlet pipe

$O_1 = 0.0\ \text{m}^3$ and $S_1 = 0.0\ \text{m}^3$ are the outflow and storage volume from the previous step when $n=0$ (Columns (7) and (8) of Table 10.36).

$$\left(\frac{2S_1}{\Delta t} - O_1\right) = \left(\frac{2 \times 0\ \text{m}^3}{10\ \text{min} \times \dfrac{60\ \text{s}}{1\ \text{min}}} - 0.0\ \text{m}^3/\text{s}\right) = 0.0\ \text{m}^3/\text{s}$$

This value is shown in Column (5) of Table 10.36 and is used to calculate the entry in Column (6) for $n=1$

b. Calculate $\left(\frac{2S_{n+1}}{\Delta t} + O_{n+1}\right)$

Apply Eq. (10.33)

$$(I_n + I_{n+1}) + \left(\frac{2S_n}{\Delta t} - O_n\right) = \frac{2S_{n+1}}{\Delta t} + O_{n+1}$$

$$(I_1 + I_2) + \left(\frac{2S_1}{\Delta t} - O_1\right) = \frac{2S_2}{\Delta t} + O_2$$

The left side of this equation is calculated as the sum of Columns (4) and (5) of Table 10.36

From Column (4), $(I_1 + I_2) = 1.7\ \text{m}^3/\text{s}$; and from Column (5), $\left(\frac{2S_1}{\Delta t} - O_1\right) = 0.0\ \text{m}^3/\text{s}$

$$1.7\ \text{m}^3/\text{s} + 0.0\ \text{m}^3/\text{s} = \frac{2S_2}{\Delta t} + O_2$$

$$\frac{2S_2}{\Delta t} + O_2 = 1.7\ \text{m}^3/\text{s}$$

This value is shown in Column (6) of Table 10.36 for $n=1$

c. Obtain outflow $O_{n+1} = O_2$ from Figure 10.41 developed in Step A.2

$$\frac{2S_2}{\Delta t} + O_2 = 1.7\ \text{m}^3/\text{s}; \quad O_2 = 0.85\ \text{m}^3/\text{s}$$

This value is shown in Column (7) of Table 10.36 for $n=1$

d. Calculate storage volume from the values of $\left(\frac{2S_{n+1}}{\Delta t} + O_{n+1}\right)$ in Column (6) and O_{n+1} in Column (7)

$$\frac{2S_{n+1}}{\Delta t} + O_{n+1} = \frac{2S_2}{\Delta t} + O_2 = 1.7\ \text{m}^3/\text{s} \quad \text{and} \quad O_{n+1} = O_2 = 0.85\ \text{m}^3/\text{s}$$

$$1.7\ \text{m}^3/\text{s} = \frac{2S_2}{10\ \text{min} \times \dfrac{60\ \text{s}}{1\ \text{min}}} + 0.85\ \text{m}^3/\text{s}$$

$$\frac{S_2}{300\ \text{s}} = 1.7\ \text{m}^3/\text{s} - 0.85\ \text{m}^3/\text{s} = 0.85\ \text{m}^3/\text{s}$$

$$S_2 = 0.85\ \text{m}^3/\text{s} \times 300\ \text{s} = 255\ \text{m}^3$$

When $n=2$

a. Calculate $\left(\frac{2S_n}{\Delta t} - O_n\right) = \left(\frac{2S_2}{\Delta t} - O_2\right)$

$O_2 = 0.85\ \text{m}^3/\text{s}$ and $S_2 = 255\ \text{m}^3$ are the outflow and storage volume from the previous step with $n=1$ (Columns (7) and (8) of Table 10.36).

$$\left(\frac{2S_2}{\Delta t} - O_2\right) = \left(\frac{2 \times 255\ \text{m}^3}{10\ \text{min} \times \dfrac{60\ \text{s}}{1\ \text{min}}} - 0.5\ \text{m}^3/\text{s}\right) = 0\ \text{m}^3/\text{s}$$

This value is shown in Column (5) of Table 10.36 and is used to calculate the entry in Column (6) for $n=2$

b. Calculate $\left(\frac{2S_{n+1}}{\Delta t} + O_{n+1}\right)$ from Eq. (10.33)

$$(I_n + I_{n+1}) + \left(\frac{2S_n}{\Delta t} - O_n\right) = \frac{2S_{n+1}}{\Delta t} + O_{n+1}$$

$$(I_2 + I_3) + \left(\frac{2S_2}{\Delta t} - O_2\right) = \frac{2S_3}{\Delta t} + O_3$$

The left side of this equation is calculated as the sum of Columns (4) and (5) of Table 10.36

From Column (4); $(I_2 + I_3) = 5.1\ \text{m}^3/\text{s}$ and from Column (5); $\left(\frac{2S_1}{\Delta t} - O_1\right) = 0.0\ \text{m}^3/\text{s}$

$$5.1\ \text{m}^3/\text{s} + 0.0\ \text{m}^3/\text{s} = \frac{2S_3}{\Delta t} + O_3$$

$$\frac{2S_3}{\Delta t} + O_3 = 5.1\ \text{m}^3/\text{s}$$

This value is shown in Column (6) of Table 10.36 for $n=2$

c. Obtain outflow $O_{n+1} = O_3$ from Figure 10.41 developed in Step A.2

$$\frac{2S_3}{\Delta t} + O_3 = 5.1\ \text{m}^3/\text{s};\ O_3 = 1.8\ \text{m}^3/\text{s}$$

This value is shown in Column (7) of Table 10.36 for $n=2$

d. Calculate storage volume from the values $\left(\frac{2S_{n+1}}{\Delta t}+O_{n+1}\right)$ in Column 6 and O_{n+1} in Column (7)

$$\frac{2S_{n+1}}{\Delta t}+O_{n+1}=\frac{2S_3}{\Delta t}+O_3=5.1\ \text{m}^3/\text{s} \quad \text{and} \quad O_{n+1}=O_3=1.8\ \text{m}^3/\text{s}$$

$$5.1\ \text{m}^3/\text{s}=\frac{2S_3}{10\ \text{min}\times\frac{60\ \text{s}}{1\ \text{min}}}+1.8\ \text{m}^3/\text{s}$$

$$\frac{S_3}{300\ \text{s}}=5.1\ \text{m}^3/\text{s}-1.8\ \text{m}^3/\text{s}=3.3\ \text{m}^3/\text{s}$$

$$S_3=3.3\ \text{m}^3/\text{s}\times 300\ \text{s}=990\ \text{m}^3$$

When $n=3$

a. Calculate $\left(\frac{2S_n}{\Delta t}-O_n\right)=\left(\frac{2S_3}{\Delta t}-O_3\right)$

$O_3=1.8\ \text{m}^3$ and $S_3=0\ \text{m}^3$ are the outflow and storage volume from the previous step with $n=2$ (Columns (7) and (8) of Table 10.36)

$$\left(\frac{2S_3}{\Delta t}-O_3\right)=\left(\frac{2\times 990\ \text{m}^3}{10\ \text{min}\times\frac{60\ \text{s}}{1\ \text{min}}}-1.8\ \text{m}^3/\text{s}\right)=1.5\ \text{m}^3/\text{s}$$

This value is shown in Column (5) of Table 10.36 and is used to calculate the entries in Column 6 for $n=3$

b. Calculate $\left(\frac{2S_{n+1}}{\Delta t}+O_{n+1}\right)$ from Eq. (10.33)

$$(I_n+I_{n+1})+\left(\frac{2S_n}{\Delta t}-O_n\right)=\frac{2S_{n+1}}{\Delta t}+O_{n+1}$$

$$(I_3+I_4)+\left(\frac{2S_3}{\Delta t}-O_3\right)=\frac{2S_4}{\Delta t}+O_4$$

The left side of this equation is calculated as the sum of Columns (4) and (5) of Table 10.36

From Column (4); $(I_3+I_4)=8.5\ \text{m}^3/\text{s}$ and from Column (5); $\left(\frac{2S_3}{\Delta t}-O_3\right)=1.5\ \text{m}^3/\text{s}$

$$8.5\ \text{m}^3/\text{s}+1.5\ \text{m}^3/\text{s}=\frac{2S_4}{\Delta t}+O_4$$

$$\frac{2S_4}{\Delta t}+O_4=10.0\ \text{m}^3/\text{s}$$

This value is shown in Column (6) of Table 10.36 for $n=3$

c. Obtain outflow $O_{n+1} = O_4$ from Figure 10.41 developed in Step A.2

$$\frac{2S_4}{\Delta t} + O_4 = 10.0 \text{ m}^3/\text{s};\ O_4 = 2.73 \text{ m}^3/\text{s}$$

This value is shown in Column (7) of Table 10.36 for $n=3$

d. Calculate storage volume from the values $\left(\frac{2S_{n+1}}{\Delta t} + O_{n+1}\right)$ in Column (6) and O_{n+1} in Column (7)

$$\frac{2S_{n+1}}{\Delta t} + O_{n+1} = \frac{2S_4}{\Delta t} + O_4 = 10.0 \text{ m}^3/\text{s} \quad \text{and} \quad O_{n+1} = O_4 = 2.73 \text{ m}^3/\text{s}$$

$$10.0 \text{ m}^3/\text{s} = \frac{2S_4}{10 \text{ min} \times \frac{60 \text{ s}}{1 \text{ min}}} + 2.73 \text{ m}^3/\text{s}$$

$$\frac{S_4}{300 \text{ s}} = 10.0 \text{ m}^3/\text{s} - 2.73 \text{ m}^3/\text{s} = 7.27 \text{ m}^3/\text{s}$$

$$S_4 = 7.27 \text{ m}^3/\text{s} \times 300 \text{ s} = 2{,}181 \text{ m}^3$$

2. Complete all entries in Columns (1) through (8) of Table 10.36 to develop outflow hydrograph.
   Continue with the procedure utilized above in step B. Develop all data for Columns (5)–(8) when $n=4$ through $n=19$. Enter all values in Table 10.36
3. Determine the maximum depth in the detention basin
   From Table 10.36, the maximum storage volume = 12,691 m³ at $t = 80$ min

$$\text{Maximum depth in the detention basin} = \frac{12{,}691 \text{ m}^3}{10{,}000 \text{ m}^2} = 1.27 \text{ m}$$

4. Plot the inflow and outflow hydrographs
   The data given in Table 10.34 are used to plot the triangular inflow hydrograph. The data in Columns (2) and (7) of Table 10.36 are used to plot the outflow hydrograph. It should be noted that values in Column (2) correspond to step *n*, and values in Column (7) correspond to step $n+1$. For example:
   when $n=1$, $t=0$ min, $O_1 = 0 \text{ m}^3/\text{s}$ (from previous step with $n=0$)
   when $n=2$, $t=10$ min, $O_2 = 2.55 \text{ m}^3/\text{s}$ (from previous step with $n=1$)
   when $n=10$, $t=90$ min, $O_{10} = 12{,}691 \text{ m}^3/\text{s}$ (from previous step with $n=9$)
   when $n=19$, $t=180$ min, $O_{19} = 223 \text{ m}^3/\text{s}$ (from previous step with $n=18$)
   The inflow and outflow hydrographs for detention basin are shown in Figure 10.42.
5. Comment on the performance of the detention basin as a stormwater runoff control
   From Figure 10.42 and Table 10.36, the maximum inflow rate will be attenuated from 10.19 m³/s (Table 10.34) to 6.65 m³/s (Table 10.36) at the outlet of the detention basin. The reduction in peak discharge is determined.

$$\text{Reduction in peak discharge} = \frac{10.19 \text{ m}^3/\text{s} - 6.65 \text{ m}^3/\text{s}}{10.19 \text{ m}^3/\text{s}} \times 100 = 35.7\%$$

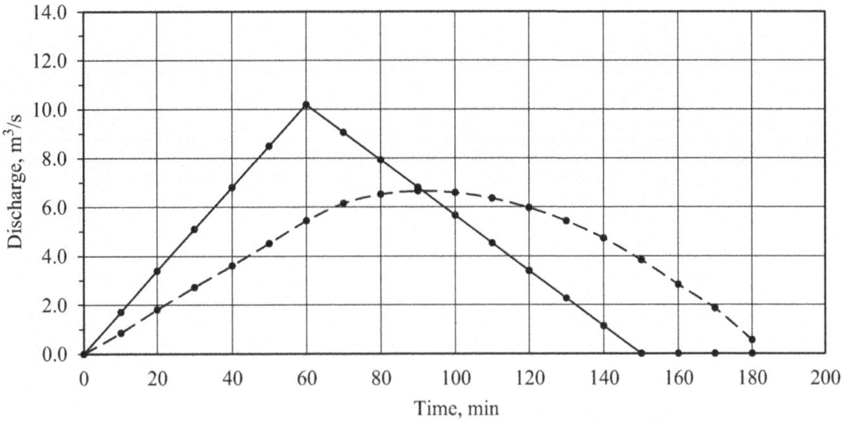

FIGURE 10.42 Inflow and outflow hydrographs for the detention basin in Example 10.36.

The detention basin will also delay the time to peak of the stormwater runoff from 60 min (inflow to basin) to 90 min (outflow from the basin).

Example 10.37: Volume of Detention Basin for Reduction of Peak Runoff to Pre-development Watershed

A 15-acre rural watershed is urbanized for housing development. The average pre-urbanization runoff coefficient for rural watersheds and peak design rainfall intensity are 0.3 and 2.8 in./h. Calculate the volume of a detention pond built in an urbanized watershed to reduce the peak runoff to pre-urbanization period. The average runoff coefficient and time of concentration for urbanized watershed are 0.68 and 7.8 min. The design storm has a return period of 5 years, and rainfall intensity and duration are, respectively, 2.8 in./h and 25 min.

Solution

1. Determine peak stormwater runoff during pre-development period

$$Q_a \text{ from Eq. (8.7b)} = C \times i \times A = 0.3 \times 2.8 \text{ in./h} \times 15 \text{ acre}$$
$$= 12.6 \text{ cfs}$$

2. Determine the design stormwater runoff after urbanization

$$Q_p \text{ from Eq. (8.7b)} = C \times i \times A = 0.68 \times 2.8 \text{ in./h} \times 15 \text{ acre}$$
$$= 12.6 \text{ cfs}$$

3. Calculate the storage volume to reduce the peak runoff to pre-urbanization period
   Apply Eq. (10.3) to calculate storage volume

$$V = 60\left(Q_p t_d - Q_a\left(\frac{t_d + t_c}{2}\right)\right)$$

Substitute the values, $Q_p = 28.6$ cfs, $Q_a = 12.6$ cfs, $t_d = 25$ min , and $t_c = 7.8$ min

$$V = 60\left(28.6\text{ cfs} \times 25\text{ min} - 12.6\text{ cfs}\left(\frac{25\text{ min} + 7.8\text{ min}}{2}\right)\right)$$

$$= 60\left(715.0\text{ cfs.min} - 206.6\text{ cfs.min}\right)$$

$$= 30{,}504\text{ ft}^3$$

It is recommended to divide 30,504 ft^3 volume into several basins to effectively intercept the surface runoff for storage.

Note: Readers are referred to Example 10.9 for additional information on storage volume for the reduction of peak stormwater runoff from urbanized watersheds.

Example 10.38: Storage Reservoir Using Water Balance Analysis

An open storage reservoir is designed to store reclaimed water. The surface area of the reservoir is 8 ha. The applicable percolation (PERC) rate is 3.58 mm/day. The precipitation (PREC), evaporation (EVAP), reclaimed water production (RWP), and reclaimed water reuse demand (RWRD) data are summarized below. Calculate the storage volume needed, average depth, and detention time in the reservoir.

Parameter	Month											
	Jan	Feb	Mar	Apr	May	Jun	Jul	Aug	Sep	Oct	Nov	Dec
Precipitation (PREC), mm	160	120	100	70	10	3	4	5	10	30	100	130
Evaporation (EVAP), mm	30	40	60	80	140	180	210	180	140	80	50	30
Reclaimed water production (RWP), m^3/day	5,000	5,000	4,000	3,500	3,000	2,500	2,000	2,500	3,000	3,500	4,000	4,500
Reclaimed water reuse demand (RWRD), m^3/day	1,463	1,099	856	2,584	4,169	3,880	4,702	4,892	3,736	3,729	4,139	2,306

Solution

1. Conduct water balance analysis on a monthly basis.

   Compute inflow and outflow from the reservoir for each month. Tabulate the calculated results for 12-month period in Table 10.37. Sample calculations for the month of January are provided below.

$$\text{Reclaimed water production, RWP} = 5{,}000\text{ m}^3/\text{day} \times 31\text{ day} = 155.00 \times 10^3\text{ m}^3$$

$$\text{Precipitation, PREC} = 160\text{ mm} \times 10^{-3}\text{ m/mm} \times 8\text{ ha} \times 10{,}000\text{ m}^2/\text{ha} = 12.80 \times 10^3\text{ m}^3$$

$$\text{Total inflow to reservoir, Total inflow} = (155.00 + 12.80) \times 10^3\text{ m}^3 = 167.8 \times 10^3\text{ m}^3$$

$$\text{Evaporation, EVAP} = 30\text{ mm} \times 10^{-3}\text{ m/mm} \times 8\text{ ha} \times 10{,}000\text{ m}^2/\text{ha} = 2.40 \times 10^3\text{ m}^3$$

$$\text{Percolation, PERC} = 3.58\text{ mm/day} \times 31\text{ day} \times 10^{-3}\text{ m/mm} \times 8\text{ ha} \times 10{,}000\text{ m}^2/\text{ha} = 8.88 \times 10^3\text{ m}^3$$

$$\text{Reclaimed water reuse demand, RWD} = 1{,}463\text{ m}^3/\text{day} \times 31\text{ day} = 45.35 \times 10^3\text{ m}^3$$

TABLE 10.37
Inflow to Reservoir, Outflow, and Reservoir Volume Data (Example 10.38)

	Inflow to Reservoir, $10^3 m^3$			Outflow from Reservoir, $10^3 m^3$				Reservoir Volume, $10^3 m^3$	
Month (days)	**RWP**	**PREC**	**Total Inflow**	**EVAP**	**PERC**	**RWRD**	**Total Outflow**	**Change**	**Cumulative**
Jan (31)	155.00	12.80	167.8	2.40	8.88	45.35	56.6	111	178
Feb (28)	140.00	9.60	149.6	3.20	8.02	30.77	42.0	108	286
Mar (31)	124.00	8.00	132.0	4.80	8.88	26.54	40.2	92	378
Apr (30)	105.00	5.60	110.6	6.40	8.59	77.52	92.5	18	**396**
May (31)	93.00	0.80	93.8	11.20	8.88	129.24	149.3	–56[a]	340
Jun (30)	75.00	0.24	75.2	14.40	8.59	116.40	139.4	–64	276
July (31)	62.00	0.32	62.3	16.80	8.88	145.76	171.4	–109	167
Aug (31)	77.50	0.40	77.9	14.40	8.88	151.65	174.9	–97	70
Sep (30)	90.00	0.80	90.8	11.20	8.59	112.08	131.9	–41	29
Oct (31)	108.50	2.40	110.9	6.40	8.88	115.60	130.9	–20	9
Nov (30)	120.00	8.00	128.0	4.00	8.59	124.17	136.8	–9	0
Dec (31)	139.50	10.40	149.9	2.40	8.88	71.49	82.8	67	67
Total (365[a])	1,289.5	59.4	1,349	97.6	104.5	1,146.6	1,349	0	—

[a] It may be noted that the reservoir starts emptying from May 31 (negative change in volume) and becomes empty on November 30 when the cumulative volume=0. The reservoir starts to fill from December 31 (positive change in volume) and becomes full on April 30 when cumulative volume $V_{max}=396\times10^3 m^3$.

Bold value is the maximum value in the column.

$$\text{Total outflow from reservoir, Total outflow} = (2.40+8.88+45.35)\times10^3\text{ m}^3 = 56.6\times10^3\text{ m}^3$$

The volume change or deficiency between inflow and outflow is calculated from Eq. (4.10)

$$\Delta V = (Q_{in} - Q_{out})\Delta t$$

$$\text{Total volume change, Change} = (167.8-56.6)\times10^3\text{ m}^3 = 111\times10^3\text{ m}^3$$

$$\text{Total volume increase in reservoir, Cumulative} = (67+111)\times10^3\text{ m}^3 = 178\times10^3\text{ m}^3$$

The calculated 12-month data are summarized in Table 10.37

2. Determine the reservoir capacity.

   The reservoir capacity required is the difference between the cumulative volumes at full and empty.

$$V_{reservoir} = V_{max} - V_{min} = (396-0)\times10^3\text{ m}^3 = 396{,}000\text{ m}^3$$

   Provide the reservoir volume of 400,000 m^3.

3. Determine the reservoir average depth.

$$D_{reservoir} = \frac{V_{reservoir}}{A_{reservoir}} = \frac{400{,}000\text{ m}^3}{8\text{ ha}\times10{,}000\text{ m}^2/\text{ha}} = 5\text{ m}$$

4. Compute the average detention time in the reservoir.
   On annual basis, total inflow = total outflow = 1,349,000 m^3/year

$$\text{Annual average daily flow, } Q_{aad} = \frac{1,349,000 \text{ m}^3/\text{year}}{365 \text{ day/year}} = 3,696 \text{ m}^3/\text{day}$$

$$\text{Average detention time, } \theta_{reservoir} = \frac{V_{reservoir}}{Q_{aad}} = \frac{400,000 \text{ m}^3}{3,696 \text{ m}^3/\text{day}} = 108 \text{ day or } 3.6 \text{ months or } 0.3 \text{ year}$$

10.7.6 Diversion Drains or Grassed Waterways

Diversion drains or grassed waterways are generally designed to carry the maximum runoff for a 10-year recurrence period. The rational formula is invariably used to determine the peak runoff rate. As the catchment area of the waterway increases toward the outlet, the expected runoff is calculated for different reaches of the waterway and used for design purposes. A diversion drain or grassed waterway is constructed by excavating the soil from the land across the slope or contour lines, and above unprotected area for the purpose of intercepting and directing stormwater runoff safely to a suitable point for pretreatment or treatment BMP. These drains can also be constructed at the boundary of the area proposed for protection. They are the main soil conservation structures to manage runoff in upland areas, and are used as outlets to prevent rill or gully formation. Diversion ditches or grassed waterways are dug at varying intervals depending on the steepness of the slope. Closer intervals are used on steeper slopes.[65,66] The ditches or waterways follow the contour, and the dimensions of the drains depend upon the flow intercepted. While the waterway by itself does not reduce water quantity or improve water quality, it is a part of the treatment train critical for the collection and diversion of stormwater into treatment BMPs. These diversion drains or waterways are mainly constructed for handling smaller flows, although larger flows are frequently accepted toward the outlet as the amount of water gradually increases. These diversion channels are designed to have a range of flow velocities that is safe for the vegetation present in the channel. The shape of the waterway depends on the field conditions and the type of construction equipment used. The three common shapes adapted are trapezoidal, triangular, and parabolic. In course of time, due to the flow of water and sediments, the waterway assumes an irregular shape, nearing the parabolic shape.[95]

Well designed and built diversion drains or grassed waterways serve the following important purposes (1) reduce the length of slope, (2) divert the runoff away from the protected area, (3) divert water away from the action of gully over-falls, (4) protect bottom land from overflow, (5) intercept the coming water from the top terraced area into un-terraced area, (6) reduce the formation of rills and gullies, and (7) grass-covered diversion drains are kept under cultivation to control the soil erosion or sediments. Without vegetation, the drain section may get filled with the eroded soil (sediment). Normal maintenance of diversion drains requires (1) growing vegetation in the channel and at the upstream slope, (2) mowing the channel vegetation at proper time, (3) removing silt from the section, if any, and (4) repairing holes and breakage made by rodents in the body of the diversion drain.[66,67,93,95]

Growing vegetation in the channel's section and on the upper back slope has an important role in the success of diversion drains. The vegetation should be established in proper way and at proper time; otherwise, the design may not be suitable for grassed conditions. The vegetation has many beneficial effects on drain and soil erosion. Mowing of vegetation grown in the drain's section is also essential; otherwise, bushy growth of vegetation may create a hindrance in flow path of water and the drains may become ineffective. The mowing work should be done at least once a year. This work should be accomplished at a stage when vegetation can be easily cut by the cutting tools. Dense vegetation may cause the drain to become ineffective. The deposition of silt in the channels'

FIGURE 10.43 Generalized view of diversion drain or grassed waterway. (Printed from the U.S. EPA website.)

section is a common occurrence; and if care is not undertaken, the drain becomes plugged. Proper inspection of diversion drains is essential. The ridge should be examined at the time of mowing grass, and if any break or hole is found, it should be repaired. The generalized view of a diversion drain or grass waterway is shown in Figure 10.43.

Example 10.39: Design of a Diversion Grassed Waterway

A shallow grassed diversion waterway is designed to carry a peak flow of 2.6 m^3/s. A parabolic channel is selected. The bottom slope is 1%. The waterway has a good stand of grass, and a velocity of 1.75 m/s can be allowed. Assume Manning's $n=0.04$.

Solution

1. Select the cross-section of the parabolic channel
   The selected parabolic shallow grassed waterway is shown in Figure 10.44.
2. Determine the equations for geometric elements of the parabolic channel
   The geometric elements of rectangular, trapezoidal, triangular, and circular channel sections are given in Table 5.3. The geometric elements of a parabolic channel are expressed below by Eqs. (10.34a)–(10.34c)

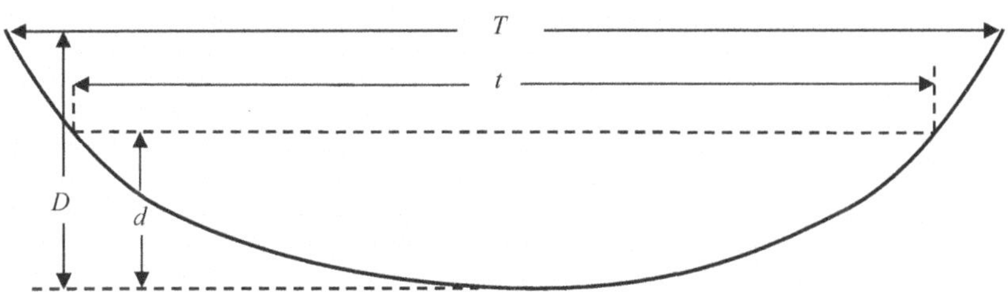

FIGURE 10.44 The selected cross-section of a parabolic shallow grassed waterway (Example 10.39).

$$\text{Cross sectional area} \quad a = \frac{2}{3}td \tag{10.34a}$$

$$\text{Wetted perimeter } p = t + \frac{8d^2}{3t} \tag{10.34b}$$

$$\text{Hydraulic radius } r = \frac{a}{p} \tag{10.34c}$$

$$\text{Top water surface width } t = \frac{a}{0.67d} \tag{10.34d}$$

$$\text{Top channel width } T = t\left(\frac{D}{d}\right)^{1/2} \tag{10.34e}$$

where

t=top water surface width, m (ft)
T=top channel width, m (ft)
D=total channel depth including the freeboard, m (ft)
All the other terms have been defined previously

3. Assume t and d for this example
   t = 6 m and d = 0.4 m
4. Calculate the hydraulic elements of the parabolic channel and the flow velocity

$$a \text{ from Eq. (10.34a)} = \frac{2}{3}td = \frac{2}{3} \times 6 \text{ m} \times 0.4 \text{ m} = 1.6 \text{ m}^2$$

$$p \text{ from Eq. (10.34b)} = t + \frac{8d^2}{3t} = 6 \text{ m} + \frac{8 \times (0.4 \text{ m})^2}{3 \times 6 \text{ m}} = 6 \text{ m} + 0.07 \text{ m} = 6.07 \text{ m}$$

$$r \text{ from Eq. (10.34c)} = \frac{a}{p} = \frac{1.6 \text{ m}^2}{6.07 \text{ m}} = 0.26 \text{ m}$$

$$v \text{ from Manning's equation } (\text{Eq. (5.34)}) = \frac{(0.01)^{1/2} \times (0.26)^{2/3}}{0.04} = 1.12 \text{ m/s}$$

Since the calculated velocity is close to the permissible velocity of 1.0–2.0 m/s, the assumed section of the channel is acceptable.

5. Calculate the top width of the channel above the freeboard
   Assume a freeboard of 0.2 m

$$D = 0.4 \text{ m} + 0.2 \text{ m} = 0.6 \text{ m}$$

$$T \text{ from Eq. (10.34e)} = 6 \text{ m} \times \left(\frac{0.6 \text{ m}}{0.4 \text{ m}}\right)^{1/2} = 7.35 \text{ m}$$

10.7.7 Vegetated Filter Strip

Vegetated filter strips (also referred to as buffer strips, grass strip, filter strip, soaking strip, and organic filter strip) are small edge-of-field tracts of vegetated land. These vegetated filter strips are designed to treat sheet flow runoff from adjacent areas near the edge of an agricultural field, or as riparian grass strips established along watercourses or wetlands. These vegetated filter strips may occur naturally or may be constructed. It is important that filter strips be designed and constructed so that runoff flows uniformly across the filter strip as sheet flow. Once the flow becomes concentrated in rills, the effectiveness of the strip is greatly reduced. The vegetation is made up of herbaceous species or shrubs. They help limit water pollution by capturing and degrading fertilizers and pesticides by filtration, absorption, adsorption, and plant uptake, and also act as an ecological corridor for biodiversity by offering a refuge area for crop auxiliaries and flora. They are perpendicular to the slope, and also help limit soil erosion by curbing runoff and improving water infiltration. Vegetated filter strips were originally used to treat agricultural drainage but have recently been used in more urban and suburban locations. They are also appropriate for roadside applications and along the edge of small- to medium-sized parking lots and driveways. They can also be used to treat roof runoff that is discharged over a level spreader. Vegetated filter strips can add aesthetic value to the development. They cost significantly less than "hardscaped" stormwater infrastructure and also provide a convenient and effective area for snow storage and removal.

Vegetated filter strips reduce the contamination of surface water, primarily controlling nonpoint source pollution; however, they may also be used to reduce sediment in stormwater runoff from construction sites.[96] They function by decreasing the velocity of water, allowing runoff and dissolved inorganic molecules to infiltrate into the soil.[97] Vegetation then uptakes these inorganic molecules, notably nitrogen and phosphorus, thus reducing their washout into nearby surface water.

There are several types of vegetated filter strips. Among them are forested riparian buffers and wind buffers. In agriculture, the vegetated filter strips are highly effective in reducing the concentration of nitrogen (N) and phosphorus (P), sediment erosion, and particulates in runoff and into surface water.[98–100] The use of filter strips is very common in developed countries and is required by law in some areas.[101] The implementation and maintenance of filter strips is inexpensive, and their use has been shown to be cost-effective.[102] Some basic design information is given below.

Slope: Vegetated filter strips should not be used on slopes greater than 15% because of the difficulty in maintaining the necessary sheet flow conditions. A relatively flat topography with a slope below 5% is desirable.

Width of filter strip: The width of filter strip (perpendicular to flow) depends upon the vegetation, slope, soil type, field conditions, climate, erosion potential, and planned use. The minimum recommended width of a filter strip is at least 15 ft for slopes up to 1%. An additional 4 ft for each 1% of slope should be added. Experience has shown that strips from 50 to 75 ft wide perform best. The minimum width of the filter strip should be larger than 0.2×filter length, or 8 ft.

Length of filter strip and drainage area: The length of filter strip (parallel to flow) should be designed to produce a water residence time of at least 10 min with an absolute minimum of 5 min. The length should normally be in the range of 100–150 ft. The maximum recommended drainage area is 5 acres. Vegetated filter strips should not receive concentrated flow discharges. For this reason, the flow path greater than 200 ft is discouraged.

Vegetation: A dense stand of vegetation is necessary for a well-functioning vegetated filter strip. A temporary diversion should be used to divert runoff away from the filter strip until good vegetation is established; otherwise, rills will develop, and the effectiveness of the strip will be reduced. Examples of grass species and vegetation commonly used are: setaria (*Setaria anceps*), ruzi grass (*Brachiaria ruziziensis*), napier or elephant grass (*Pennisetum purpureum*), guinea grass (*Panicum maximum*), and other types of grasses. These grasses are trimmed regularly (every 2–4 months) to prevent them from flowering, shading, and spreading to the cropped area between the strips. Thus, grass strips can be very appropriate for farmers who cut and carry fodder for their animals. Grasses can also be used as mulch for crops. Some farmers feel that grass serves as a refuge for rodents which threaten food crops.

Vegetated filter strips are highly effective in reducing the concentration of nitrogen (N) and phosphorus (P) in runoff into surface water. The prairie filter strips reduced the concentration of nitrate, total nitrogen, and total phosphorus in runoff by 35%, 73%, and 82% respectively.[98] Vegetated filter strips are effective in reducing the concentration of phosphorus, with results ranging from 61% to

FIGURE 10.45 The generalized view of a vegetated filter strip or grassed filter strip between two adjacent agricultural lands to reduce contaminants migrating from one land to the other. (Printed from the U.S. EPA website).

89% for 2 m-wide filter strips and 15 m-wide filter strips, respectively.[99] This indicates that filter strips are effective at even smaller widths.

Vegetated filter strips have also been shown to be very effective in sediment trapping. A study of sediment removal found that prairie filter strips had a 96% sediment trapping efficiency over a 4-year period.[100] With nitrogen and phosphorus, the reduction efficiency of a filter strip varies depending on the subsurface and surface conditions, sediment characteristics, and trapping efficiency. The generalized view of vegetated filter strips is shown in Figure 10.45.

Example 10.40: Design of Vegetated Filter Strip

In a city, approximately 1,300 linear feet of roadway was built. The topography was such that the runoff from the road is collected in roadside ditches and conveyed to a low point near the mid-station of the road. The concentrated runoff is discharged into a level spreader, from which it enters the proposed vegetated filter strip as an overland sheet flow. The pre-development total proposed site is 4.6 acres, and the land cover is unimproved grass cover. After post-development, 0.75 acre turned into impervious land. The hydrologic soil group in the area falls in Group B (Table 10.10), classified as silt loam or loam. It has a good infiltration rate.[38] The recommended Manning's *n* for overland flow over dense grass is 0.20. The typical topography of the area is such that the slope of the vegetative filter strip is 2%. The peak runoff from impervious area is ½ in. of rain, estimated from the basic Rational Method hydrograph shape shown in Figure 10.46. The expected time of concentration t_c from the impervious area is 8 min.

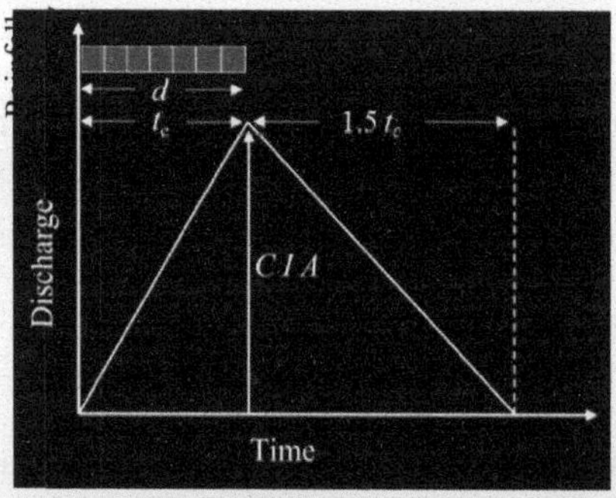

FIGURE 10.46 Basic rational hydrograph shape, adopted from Ref. 99 (Example 10.40).

Solution

1. Calculate the water quality volume (WQ_v) from impervious area

   The WQ_v from impervious area is calculated from Eq. (10.35)

$$WQ_v = IA \times d \tag{10.35}$$

   where

   WQ_v=water quality volume of water from impervious area, m^3 (ft^3)

   IA=impervious area, m^2 (ft^2)

   d=water depth over impervious area, m (ft)

   The water depth over impervious area=½ in.

$$WQ_v = 0.75 \text{ ac} \times 43{,}650 \text{ ft}^2/\text{ac} \times \frac{0.5 \text{ in.}}{12 \text{ in./ft}} = 1{,}361 \text{ ft}^3$$

2. Estimate the peak design flow from the vegetated filter strip

   The peak volumetric rate of discharge, which generates the runoff volume from filter strip, is estimated by examining the base Rational Method hydrograph shown in Figure 10.46.

   a. Calculate the base of the triangle in Figure 10.46

      The time of concentration, t_c=8 min

      The base of the triangle=8 min+1.5×8 min=20 min (1,200 s)

   b. Calculate the area of the triangle

$$A = \frac{1}{2} \times 1{,}200 \text{ s} \times \text{CIA}$$

      A=peak design WQ_v from the strip $WQ_v = 1{,}361 \text{ ft}^3$

      CIA=peak design water quality flow from the filter strip, cfs

   c. Calculate the peak design water quality flow through the filter strip

$$1{,}361 \text{ ft}^3 = \frac{1}{2} \times 1{,}200 \text{ s} \times \text{CIA}$$

$$\text{CIA} = 1{,}361 \text{ ft}^3 \times 2 / 1{,}200 \text{ s} = 2.3 \text{ cfs}$$

3. Calculate the depth of sheet flow over the filter strip

   Flow over the filter strip is calculated from Eq. (10.36)

$$Q = v \times w \times h \tag{10.36}$$

   where

   Q=peak design volumetric flow rate on the filter strip, cfs

   v=average flow velocity on the strip, ft/s

   w=width of the filter strip, ft

   h=flow depth over the filter strip, ft

   For shallow overland flow, the anticipated flow depth is assumed equal to the hydraulic radius. Express the sheet flow depth from the modified Manning's equation (Eq. 5.37)

$$Q = \frac{1.49}{n} \times h^{2/3} \times S^{1/2} \times (w \times h)$$

$$\frac{Q}{w} = \frac{1.49}{n} \times h^{5/3} \times S^{1/2}$$

Assume filter strip width = 25 ft
$n = 0.20$, $S = 0.02$ ft/ft and $Q = CIA = 2.3$ cfs

$$\frac{2.3 \text{ cfs}}{25 \text{ ft}} = \frac{1.49}{0.20} \times h^{5/3} \times 0.02^{1/2}$$

$$h = 0.23 \text{ ft}$$

4. Calculate the average flow velocity over the filter strip
   Apply Manning's equation (Eq. 5.36)

$$v = \frac{1.486}{n} \times R^{2/3} \times S^{1/2}$$

$$= \frac{1.486}{0.20} \times (0.23)^{2/3} \times (0.02)^{1/2}$$

$$= 0.4 \text{ ft/s}$$

5. Calculate the length of the filter strip.
   The length of the filter strip is calculated from the average flow velocity and the target hydraulic residence time. First, the minimum residence time of 5 min is considered. The length of filter strip is expressed by Eq. (10.37).

$$L = t \times v \tag{10.37}$$

where
L = length of the filter strip, ft
t = target hydraulic residence time, s
v = average flow velocity

$$L = 5 \text{ min} \times \frac{60 \text{ s}}{\text{min}} \times 0.4 \text{ ft/s} = 120 \text{ ft}$$

6. Verify the assumed 25 ft width of filter strip
   Minimum width of filter strip w is the larger of the two
   a. 0.2 × filter strip length
   b. 8 ft

   Filter width $w = 0.2 \times 120$ ft = 24 ft
   The initial assumption in step 3 of a 25 ft width of filter strip is acceptable.
   It may be noted that this approach does not consider the portions of runoff volume that infiltrate into the subsoil and are lost due to evapotranspiration. Additionally, the accumulation of flow into depressions, and subsequent decrease in velocity is not considered. Therefore, the computed length of 120 ft reflects a conservative design. It can be reasonably assumed that the actual hydraulic residence time is more than the minimum value of 5 min used in step 5 to calculate the length of the filter.
7. Select the length of lip of the level spreader.
   The level spreader will deliver sheet flow discharge over the filter strip. The total flow will be the 10-year peak rate of runoff from the roadway and impervious area. Select a minimum level spreader lip length of 10 ft. When the level spreader is run in the middle of a 25 ft wide filter strip, the ends of the level spreader lip should be within 7.5 ft from both ends of the strip.

$$\text{End of spreader lip from filter strip} = \frac{25\text{ ft} - 10\text{ ft}}{2} = 7.5\text{ ft}$$

If this value had exceeded 10 ft, the level spreader lip length would be increased.

8. Select vegetation for vegetative filter strip.

   The specific species of vegetation should be appropriate for the climatic conditions, soil, prevailing hydrology, and other geographically determined factors, and expected maintenance.

   Additionally, the selection of plant species should match the native plant species as closely as possible. Filter strips must be constructed of dense soil-binding deep-rooted weather-resistant plants. A dense turf of grass is necessary to achieve desirable pollutant removal while avoiding erosion. Other vegetation may include ground cover, vines, deciduous and evergreen shrubs, and trees.

10.7.8 Land Leveling and Grading

Land leveling, smoothing, and shaping the watershed surface are important to ensure that the depths and discharge variations over the field are relatively uniform. This will also result in uniform distributions of irrigation water in the root zone.[103] These field operations are required nearly every cropping season, particularly where substantial cultivation disturbs the field surface following harvest. These preparations are important for efficient land management practice. The factors that influence the design of a land-leveling project are land slope, depth of soil, topography, crops to be grown, and method of irrigation. The major benefits of land leveling are: (1) to provide a slope that fits water supply needs, and (2) to level the field to its best condition with minimal earth movement and then fully utilize the water supply for the field condition.[104]

Land leveling always improves the efficiency of water, labor, and energy resource utilization. Infiltration is improved, and runoff may result in dramatic improvements in the ability of water to spread evenly over the surface. A major problem with land leveling is the removal of topsoil that may influence plant growth. Therefore, it is important to determine the soil depth over the field so that after the leveling is done, an adequate depth of topsoil still covers the entire field. Land leveling may involve removing knolls, mounds, or ridges and filling pockets and depressions. Effort should be made to require the least soil movement and utilize simpler, less costly tools and implements.[105] A topographical map, and surveying and mapping the field, will provide a basis for planning the land-leveling and irrigation system. Land grading is also effective in isolating and confining the hot pollution spots in the watershed.

10.7.9 Protection BMPs against Drainage from Concentrated Animal Feeding Operations (CAFOs) and Other Sources of Pollution

CAFOs are a serious source of environmental degradation, such as air, water, and land. These CAFOs generate significant amounts of manure. One method to utilize manure is to fertilize the farmland. Since excess amounts of manure are applied over the farmland, significant amounts of manure and pollutants may be washed away by rain into surface waters. While in general agricultural stormwater runoff from CAFOs is a NPS, CAFOs may end up requiring a NPDES permit under the CWA.[106] Also, point-source category regulatory requirements under the NPDES permitting program are established for (1) airports discharging pollutants from deicing operations,[107] (2) construction activity,[108] and (3) CSO.[109,110]

There is much concern regarding the detection, monitoring, and health risks of many contaminants released from CAFOs. Chemicals of greatest concern are gasoline additives, solvents, phenolic compounds, pesticides, herbicides, antibiotics, and pharmaceuticals and personal care products. Many of these contaminants are released from underground storage tanks, agricultural runoff, and discharge from confined animal feeding operations. It is important that these contaminants are contained and removed from the NPSs. Discussion on CAFOs and many contaminants and regulatory requirements may be found in Section 6.3.3. Protection BMPs against drainage from CAFOs are not a single process. Rather, they are a combination of BMPs that are applied to confine, divert, treat, and percolate the drainage from CAFOs. Application of such protection BMPs is presented in Example 10.41.

Example 10.41: Zero Discharge from Concentrated Animal Feeding Operation (CAFO)

A CAFO is carried out in the middle of a large watershed. The drainage from the CAFO is discharged into a small channel that flows through the property and then discharges into a creek. During intense rainfall events, the flow is large, and organic solids and contaminants are scoured from the CAFO facility and are washed into the channel. These pollutants cause serious water pollution and water quality impairment of the receiving water. While the stormwater runoff from CAFOs is an NPS pollution, a permit may be required under the NPDES of the CWA. It is therefore required to achieve zero discharge from the CAFO during normal operation and under intense rainfall events. Describe the proposed source protection BMPs to treat, confine, divert, and dispose of the drainage to achieve a zero discharge.

Solution

This is a classic example of applying BMPs for watershed management to achieve the desired effects. A series of at-source BMPs may be applied to intercept, divert, and apply enhanced groundwater infiltration.

1. Intercept the stormwater runoff from higher grounds from reaching the CAFO
   a. Conduct gentle leveling and grading of land to confine the drainage from reaching the CAFO area (Section 10.7.8)
   b. Construct a diversion drain or grassed waterway across the slope to intercept the flow along the upper boundary of the CAFO. The diversion drain will provide needed protection of stormwater runoff from reaching the upper boundary of CAFO. The uncontaminated runoff is directed safely to a suitable point for treatment or disposal into the receiving water.
2. Intercept and treat routine housekeeping solid and liquid wastes, and stormwater runoff from CAFO facility
   a. Apply BMPs within the CAFO area to manage solid waste.

      The solid wastes generated at the CAFO during routine operation include animal manure, spilled animal feed, and bedding material. The solid waste is removed periodically from different sections of CAFO on a routine basis and sent to a solid waste handling area. At the solid waste handling area, the waste is composted using the window composting method and stored on the property. The finished compost is applied over the designated agricultural land area during the crop growing season to utilize the nutrients such as nitrogen and phosphorus to enhance the crop production.
   b. Apply BMPs to manage liquid waste generated from routine CAFO operation, and runoff produced during wet weather.

      The liquid waste generated from CAFO operation is urine, silage leachate, and contaminated wastewater, including rainwater that encounters the bunk silo and feed area, and wash water from housekeeping. Liquid waste is high in BOD, ammonia and organic nitrogen, insecticides, antibiotics, grease and oil, and other materials used at CAFO. Several BMPs are applied to reduce the volume of liquid waste generated from normal operation and stormwater runoff produced during wet weather. These BMPs are described below:

i. Construct a 60-ft wide vegetated filter strip along the downslope periphery of CAFO, solid waste processing area, and agricultural land (Section 10.7.7). The stormwater runoff is distributed uniformly over the vegetated filter strip as a sheet flow. Herbaceous species or shrubs capture pollutants and sediments by filtration, absorption, adsorption, and plant uptake. Part of stormwater volume will also be reduced by evapotranspiration and percolation.
ii. Construct a series of contour or graded bunding (Section 10.7.4) downslope of vegetated filter strip to divide the area into strips that act as barriers to the flow of stormwater, reduce the velocity and volume of runoff, and increase percolation of stormwater downslope of the vegetated filter strip. The contour or graded bunding will further treat the stormwater emerging from CAFO and solids handling areas.
iii. Construct a deep infiltration trench downslope around the graded bunding to percolate any stormwater runoff that may emerge from the graded bunding. All these lines of defense will help to percolate and evaporate the process and rainwater from the CAFO and associated facilities such as solids handling operations and farmland. Thus, zero discharge from the proposed CAFO will be safely achieved.

10.7.10 Green Roofs

A green roof or vegetated roof (also referred to as living roof, nature roof, and eco-roof) is a thin layer of living plants growing on top of the roof. It is a system composed of succulent plants and soils that collect and absorb rain during storm events. They can be installed upon existing rooftops or installed when a traditional rooftop is being replaced, provided the structure is strong enough to carry the additional weight. A green roof is not a collection of potted plants to decorate a roof space but rather an extension of a conventional roof, which involves the installation of a layered system of membranes, substrate, and plants. Green roofs are not a new technology. Historically, green roofs have been used to help keep houses warmer in colder climates and cooler in warmer climates. They also improve urban environmental quality by adding green space.[111] In the 1960s, northern Europeans began to popularize green roofs as a means of improving urban environmental quality through the reduction of stormwater runoff and the addition of green space. These roofs are efficient at reducing stormwater pollution, energy consumption, and improving overall air quality, and other benefits. Therefore, green roofs have an important role to play in the urban realm. As with other green infrastructure interventions, they can help provide resilience in cities, especially as cities are experiencing the increased likelihood of flooding.[112] An aerial view of a green roof structure in an urban setting is shown in Figure 10.47.

Benefits of green roof: Installation of a green roof provides many advantages, some of which include the following: (1) it reduces stormwater runoff by intercepting between 15% and 90% of rooftop runoff. Absorption of runoff into the media and plants is between 50% and 60%, and is related to the type of growing medium, plant cover variability, differences in temperature, wind, evapotranspiration rates, and plant uptake. In summer, 70%–90% water is retained, and in winter between 25% and 40%. Green roofs delay the runoff time, thus reducing the volume of peak overflows in storm sewer systems (CSOs). The velocity of direct runoff is reduced, resulting in reduced erosion of stream and riverbanks, enhanced sedimentation, and protection of habitat from reduced stormwater runoff;[113] (2) it reduces the need for energy to heat house in winter and cool in summer by adding thermal mass and resistance to the roof membrane, and reduced heat loss and energy use in winter conditions. The vegetation layer and trapped air help insulate the building's interior,

FIGURE 10.47 Areal view of a green roof in an urban setting. (Printed from Ref. 29)

providing up to 25% additional insulation when the soils are dry to slightly moist. Additionally, the plants' natural evapotranspiration process results in cooling of roof surfaces and a reduction in heat loss by 50%–90%. It also dramatically reduces the temperature fluctuations on the rooftop daily, seasonally, and yearly; (3) city areas are warmer than surrounding countryside. Green roofs absorb heat and lower ambient temperature, thus aiding in the reduction of urban heat island impact;[114,115] (4) it extends the life of a conventional roof by a factor of three times from UV light, large temperature fluctuations, and normal wear and tear associated with exposed surface roofs;[111] (5) it can incorporate food gardens that can utilize organic methods, favor habitat creation, and food supply for birds and favorable insects; (6) provide pleasant sight, smell, and sound that promote favorable social interaction and are beneficial to quality of life in urban areas; (7) plant photosynthesis reduces the negative impact of carbon dioxide, produces oxygen, and also helps filter out airborne pollutants; (8) green roofs are good sound insulators, reducing the impact of noise to areas below the roof. In order to have a good growing medium, the soil base should have a normal depth in the range of 8–24 in. Additionally, the weight loading on the roof structure may be in the range of 40–200 lbs/ft^2 when fully saturated, and thus may require additional structural support to the building structure.[111,116]

Depending on the depth of growing medium, the green roofs are grouped into three types: *extensive, semi-intensive,* and *intensive roofs*. Extensive green roofs have a much shallower growing medium (typically 2–6 in. thick), requiring minimal maintenance. The typical thickness of growing medium of semi-intensive green roofs is between 6 and 12 in. The intensive green roofs have a growing medium with a typical thickness of 12–48 in. These roofs require more effort to grow, establish, and maintain the plants. All types of green roofs are composed of several layers of media to provide a growing environment, protect the roof, and avoid leakage into the building. These layers are depicted in Figure 10.48.[43,76,112]

Basic design considerations of green roof: Before selecting a green roof, several basic design requirements should be evaluated. These requirements include local building codes, structural support, green roof storage volume, water quality requirements, waterproofing and leakage prevention

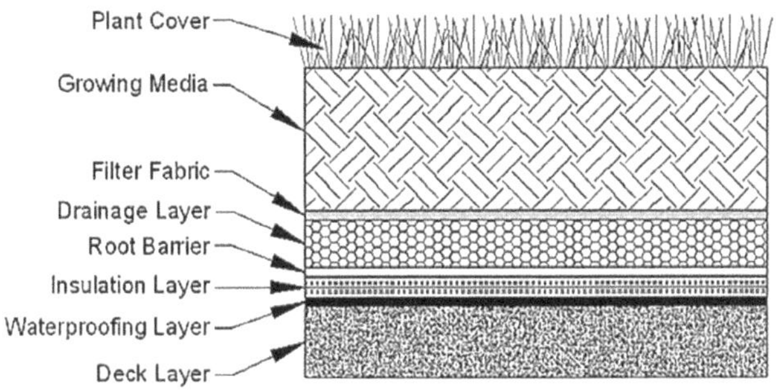

FIGURE 10.48 Cross-section of a typical green roof system. (Printed from Ref. 76)

systems, plant species, and maintenance. The basic design considerations are (1) storage volume, (2) volume of runoff, and (3) peak runoff. These design considerations are presented below.

1. **Green roof storage volume**: Green roofs can retain a large portion of stormwater during light and moderate rainfalls. For larger storms, the storage volume is not designed to provide stormwater detention and attenuate peak runoff. The storage volume of green roofs may be calculated from Eq. (10.38)[76]

$$\mathrm{Sv} = \mathrm{SA} \times \left[(d \times \mathrm{MWR}_1) + (\mathrm{DL} \times \mathrm{MWR}_2) \right] \times \mathrm{IF} \tag{10.38}$$

where

Sv = green roof storage volume, m^3 (ft^3)

SA = green roof area, m^2 (ft^2)

d = growing medium depth, m (ft)

MWR_1 = verified growing medium maximum water retention, m/m (ft/ft). Use 0.10 as a baseline default in the absence of verification data

DL = drainage layer depth m (ft). Use $DL = 0$ if the drainage layer is combined with the media layer

MWR_2 = verified drainage layer maximum water retention. Use 0.10 as a baseline default in the absence of verification data

IF = irrigation factor. The irrigation factor for irrigated green roofs is 0.5, and 1.0 for unirrigated green roofs.

The percentage of the design storm captured by the green roof can be determined by comparing the storage volume Sv with the required stormwater retention volume (SWRv) for the entire roof area.

2. **Runoff discharged from a green roof:** The runoff from green roofs would be insignificant or even zero for small storms. However, during heavy and sustained rainfall, the storage volume will quickly become saturated, and the green roof acts as a conventional roof with considerable runoff.

The depth of runoff discharged from a green roof may be estimated from Eq. (10.39).

$$Q_{\mathrm{GR}} = P - (\mathrm{Sv} / \mathrm{SA}) \tag{10.39}$$

where

Q_{GR} = runoff depth discharged from the green roof, mm (in.)

P = rainfall depth, mm (in.)

TABLE 10.38
Effective CN Values for Extensive Green Roofs

Retention Depth, in.	0.6	1.0	1.2	1.4	1.6	1.8	2.0	2.2	2.4
CN	94	92	90	88	86	85	82	81	77

Note: Runoff retention can be calculated by dividing the storage volume by the area of green roof, or obtained from the manufacturer's specifications.

Sv/SA = green roof retention depth, mm (in.). The ratio Sv/SA is green roof storage volume per unit green roof area, m (ft). It is also called *Retention*

Alternatively, the depth of runoff can be estimated using the NRCS method (Eqs. 10.6 and 10.7). An adjusted CN should be used to account for shallow percolation that normally occurs in green roofs. It is recommended to use effective CNs in calculating stormwater volume discharged from green roofs. These values for various runoff retentions are given in Table 10.38.[116]

3. **Peak flow runoff from green roofs:** The peak flow runoff from green roofs may be estimated from the NRCS method (Eq. 10.8) with an adjusted CN. The adjusted CN can be obtained from Table 10.38 or calculated through a procedure explained in Example 10.43. However, the NRCS method may underestimate the peak runoff from green roofs since the medium and drainage layers become saturated during the early hours of sustained heavy rainfall.[43]

   The peak stormwater runoff from green roofs during heavy storms may be calculated using the rational method (Eqs. 8.7a and 8.7b) with a runoff coefficient of $C = 1.0$.[43]

Example 10.42: Storage Volume of Green Roof

An office building has an extensive green roof with shallower growing medium. The roof consists of a 15 cm growing layer and a 2.5 cm drainage layer. The green roof covers 2,000 m². The maximum water retention of the growing media and drainage layer is 0.20 and 0.4, respectively. Calculate the retention storage volume of the green roof. The roof is not irrigated.

Solution

Calculate storage volume of the green roof

Apply Eq. (10.38)

$$\text{Sv} = \text{SA} \times \left[(d \times \text{MWR}_1) + (\text{DL} \times \text{MWR}_2)\right] \times \text{IF}$$

$$= 2{,}000\ \text{m}^2 \times \left[(15\ \text{cm} \times 1\ \text{m}/100\ \text{cm} \times 0.2) + (2.5\ \text{cm} \times 1\ \text{m}/100\ \text{cm} \times 0.4)\right] \times 1$$

$$= 2{,}000\ \text{m}^2 \times [0.03\ \text{m} + 0.01\ \text{m}] = 80\ \text{m}^3$$

Example 10.43: Required Green Roof Surface Area

A commercial building is being retrofitted to install a green roof. The total roof area is 10,000 ft². Based on the design criteria, the building is required to treat 750 ft³ of stormwater runoff volume (*Rv*). The green roof will have a 6-in. thick growing medium and 1-in. thick drainage layer. The maximum water retention capacities of the growing medium and drainage layer are 0.15 and 0.35, respectively. Calculate the minimum required green roof surface area. The roof is unirrigated.

Solution

1. The minimum green roof surface is calculated using Eq. (10.38)
   Rearrange Eq. (10.38) to solve for green roof surface area SA

$$SA = \frac{Sv}{\left[(d \times MWR_1) + (DL \times MWR_2)\right] \times IF}$$

$$= \frac{750\ ft^3}{\left[(6\ in. \times 1\ ft/12\ in. \times 0.15) + (1\ in. \times 1\ ft/12\ in. \times 0.35)\right] \times 1.0} = \frac{750\ ft^3}{(0.075\ ft) + (0.029\ ft)}$$

$$= 7,200\ ft^2$$

The green roof must be sized to at least 7,200 ft^2 to meet the water quality requirements. This is almost 70% of the total roof area.

Example 10.44: Stormwater Volume Discharged from Green Roofs

A 2,200 ft^2 green roof is designed to reduce the stormwater runoff volume generated by a 4.0-in. storm. The proposed roof will have a 6-in. growing medium and a 2-in. drainage layer designed to retain water prior to being discharged to the roof drainage system. The water retention capacity of the growing medium and the drainage layer is 0.15 and 0.45, respectively. The roof will not be irrigated. Calculate stormwater volume discharged from the green roof using (1) Eq. (10.39) and (2) the NRCS method with effective CN value.

Solution

A. Estimate stormwater volume discharged from the green roof
   1. Calculate green roof retention depth, Sv /SA from Eq. (10.38)

$$Sv = SA \times [(d \times MWR_1) + (DL \times MWR_2)] \times IF$$

$$Sv/SA = [(d \times MWR_1) + (DL \times MWR_2)] \times IF$$

$$= \left[(6\ in. \times 0.15) + (2\ in. \times 0.45)\right] \times 1.0$$

$$= 1.8\ in.$$

   2. Calculate depth of runoff discharged from the green roof from Eq. (10.39)

$$Q_{GR} = P - (Sv/SA)$$

$$= 4\ in. - 1.8\ in.$$

$$= 2.2\ in.$$

   3. Calculate the volume of runoff discharged from the green roof

$$\text{Stormwater volume from the green roof} = Q_{GR} \times SA$$

$$= (2.2\ in. \times 1\ ft / 12\ in.) \times 2,200\ ft^2$$

$$= 403\ ft^3$$

B. Estimate stormwater volume discharged from the NRCS method and using an effective CN
   1. Select effective CN
      From Table 10.38, effective CN = 85 for retention depth Sv/SA = 1.8 in.
   2. Calculate S from Eq. (10.7)
      S, as defined in Eq. (10.7), is potential maximum soil moisture retention after runoff begins

$$\text{CN} = \frac{1{,}000}{10+S}$$

$$85 = \frac{1{,}000}{10+S}$$

$$850 + 85S = 1{,}000$$

$$85S = 150$$

$$S = 150/85 = 1.76 \text{ in.}$$

   3. Calculate the depth of runoff discharged from the green roof
      Apply Eq. (10.6)

$$Q = \frac{(P-0.2S)^2}{P+0.8S} = \frac{(4 \text{ in.} - 0.2 \times 1.76 \text{ in.})^2}{4 \text{ in.} + 0.8 \times 1.76 \text{ in.}} = \frac{13.31 \text{ in.}^2}{5.41 \text{ in.}}$$

$$Q = 2.46 \text{ in.}$$

   4. Calculate the volume of runoff discharged from the green roof

$$\text{Stormwater volume from the green roof} = Q \times \text{SA}$$

$$= (2.46 \text{ in.} \times 1 \text{ ft}/12 \text{ in.}) \times 2{,}200 \text{ ft}^2$$

$$= 451 \text{ ft}^3$$

Note: The stormwater volume of 403 ft^3 from the green roof, calculated from Eq. (10.39), and 451 ft^3 from the NRCS method with an effective CN are comparable. This may not be the case for smaller storms. The NRCS method with an effective CN from Table 10.38 may overestimate stormwater volume from green roofs.

Example 10.45: Peak Runoff from Green Roofs

A green roof is designed for a single-family house to reduce the stormwater peak runoff. The green roof is 2,000 ft^2 and will have a 6.5-in. growing medium and a 2-in. drainage layer. The water retention capacity of the growing medium and the drainage layer are 0.10 and 0.40, respectively. A 2-year storm is producing the peak runoff. Estimate the peak runoff from the roof using: (1) the rational method and (2) the NRCS method. The 2-year storm intensity is calculated from equation $i = 104/(t_c + 20)$. The estimated time of concentration t_c for the green roof is 6 min. The roof will not be irrigated.

Solutions

1. Estimate peak runoff from the green roof using the rational method

1. Calculate storm intensity

$$i = 104/(t_c + 20)$$
$$= 104/(6\text{ min} + 20)$$
$$= 4.0\text{ in./h}$$

2. Calculate peak runoff
   Apply Eq. (8.7b) with $C = 1.0$

$$Q = 1.0 \times 4.0\text{ in./h} \times (2{,}000\text{ ft}^2 \times 1\text{ acre}/43{,}560\text{ ft}^2)$$
$$= 0.18\text{ cfs}$$

2. Estimating peak runoff from the green roof using the NRCS method with an adjusted CN
   1. Calculate the depth of stormwater runoff from a conventional roof
      1. Determine CN
         CN = 98 from Table B.7a for impervious surfaces such as roofs
      2. Calculate S from Eq. (10.7)

$$\text{CN} = \frac{1{,}000}{10 + S}$$
$$98 = \frac{1{,}000}{10 + S}$$
$$980 + 98S = 1{,}000$$
$$S = 0.204\text{ in.}$$

      3. Calculate runoff from Eq. (10.6)

$$Q = \frac{(P - 0.2S)^2}{P + 0.8S} = \frac{(4.0\text{ in.} - 0.2 \times 0.204\text{ in.})^2}{4.0\text{ in.} + 0.8 \times 0.204\text{ in.}} = \frac{15.67\text{ in.}^2}{4.16\text{ in.}}$$
$$Q = 3.77\text{ in.}$$

   2. Estimate the depth of stormwater runoff volume to be discharged from the green roof, Q_{GR}
      a. Calculate green roof retention depth, Sv /SA
         Apply Eq. (10.38)

$$\text{Sv} = \text{SA} \times [(d \times \text{MWR}_1) + (\text{DL} \times \text{MWR}_2)] \times \text{IF}$$
$$\text{Sv/SA} = [(d \times \text{MWR}_1) + (\text{DL} \times \text{MWR}_2)] \times \text{IF}$$
$$= [(6.5\text{ in.} \times 0.1) + (2\text{ in.} \times 0.4)] \times 1$$
$$= 1.45\text{ in.}$$

      b. Subtract depth of storage volume (Sv/SA) from the depth of runoff volume from the impervious roof (Q).

$$Q_{GR} = Q - \text{Sv/SA}$$

$$= 3.77 \text{ in.} - 1.45 \text{ in.}$$

$$= 2.32 \text{ in.}$$

3. Determine the adjusted CN
   a. Calculate S from Eq. (10.6)

$$Q = \frac{(P - 0.2S)^2}{P + 0.8S}$$

$$2.32 \text{ in.} = \frac{(4.0 \text{ in.} - 0.2 \times S)^2}{4.0 \text{ in.} + 0.8 \times S}$$

$$S = 1.98 \text{ in.}$$

   b. Calculate CN from Eq. (10.7)

$$\text{CN} = \frac{1{,}000}{10 + S} = \frac{1{,}000}{10 + 1.98}$$

$$= 83.5$$

4. Determine peak runoff from the green roof
   a. Calculate unit peak discharge

   The unit peak discharge (q_u) is determined from Figure B.5, assuming the building with the green roof is located in an area with a Type II rainfall distribution.

$$t_c = 10 \text{ min} \times 1 \text{ h} / 60 \text{ min} = 0.1 \text{ h}$$

$$\text{From Eq. } (10.5),\ I_a = 0.2S = 0.2 \times 1.98 \text{ in.} = 0.4 \text{ in.}$$

$$I_a / P = 0.40 \text{ in./4 in.} = 0.10$$

   Determine q_u from Figure B.5 corresponding to t_c=0.1 h and $I_a / P = 0.10$
   q_u from the graph=1,000 cfs/mi²/in.

   b. Calculate peak runoff from Eq. (10.8)

   Note: The peak flow calculated from the rational method with C =1.0 is 0.18 cfs. The peak flow calculated from the NRCS method with an adjusted CN is 0.17 cfs. These values are comparable in this example. However, the NRCS method normally gives a lower value of peak flow from the green roof.

10.7.11 Rainwater Harvesting

RWH is the collection and storage of rainwater, rather than allowing it to run off. Rainwater is collected from a roof and stored in a tank, cistern, deep pit (well, shaft, or borehole), aquifer, or a reservoir with percolation.[117] Stormwater harvesting differs from RWH. In stormwater harvesting, the runoff is collected from open land areas, drains, roads, or any other land surfaces. Uses of harvested rainwater include watering gardens, livestock, irrigation, domestic use with proper treatment, and domestic heating. The harvested water can also be committed to longer-term storage or groundwater recharge.[118,119]

RWH is one of the simplest and oldest methods of self-supply water for residential and household-scale projects.[119,120] RWH in conjunction with urban agriculture would be a viable way to help meet the United Nations Sustainable Development Goals for cleaner and sustainable cities, health and wellbeing, and food and water security. The available technology, however, needs to be remodeled to use water more efficiently, especially in an urban setting.[121] About 40% of Thailand's rural population utilizes RWH. This is one of the largest examples of self-supply of water worldwide.[122] RWH is mandatory for new homes built in Santa Fe and New Mexico. Texas offers a sales tax exemption for the purchase of RWH equipment. Both Texas and Ohio allow RWH to be used even for portable purposes. Oklahoma passed the Water for 2060 Act in 2012 to promote pilot projects for rainwater and graywater use, among other water-saving techniques.[123–125] RWH is also used in many countries worldwide.[126,127] A RWH structure from roof is shown in Figure 10.49.

Components of RWH system: RWH in urban areas reduces the impact of runoff and flooding and can help mitigate flooding of low-lying areas. It is necessary to prevent water dripping or flowing off roofs in an uncontrolled manner to prevent damaging walls, drenching people entering the building, and to direct the rainwater where it will not damage the foundations of the building. The components of RWH system are (1) the portion of roof that is the catchment surface, (2) rain gutters, (3) leaf screens, first-flush diverters, and roof washers, (4) downspouts, (5) storage facility, (6) delivery system, and (7) treatment/purification systems.[118,119,124] Residential and small-scale systems may not include all these components. A typical RWH system is shown in Figure 10.50.

Catchment surface or CDA is usually the roof surface from where the rainwater is harvested. Rainfall runs off from smooth and non-porous surfaces faster and more efficiently. The runoff will pick up more dirt from CDAs during the first flush, resulting in poor water quality of the harvested water.[76] The bird droppings over the roof are a serious source of contamination.

A rain gutter (also called eavestrough, eaves-shoot, or surface water collection channel) is a component of a water collection and discharge system for a building. Water from a pitched roof flows down into a valley gutter, also called parapet gutter or eaves gutter. Many materials have been used to make gutters. These materials are cast iron, aluminum, galvanized steel, copper, zinc, polyvinyl chloride (PVC), other plastics, wood, and bamboo.[117,118]

FIGURE 10.49 Rainwater harvesting from the roof of a structure. (Printed from Ref. 124)

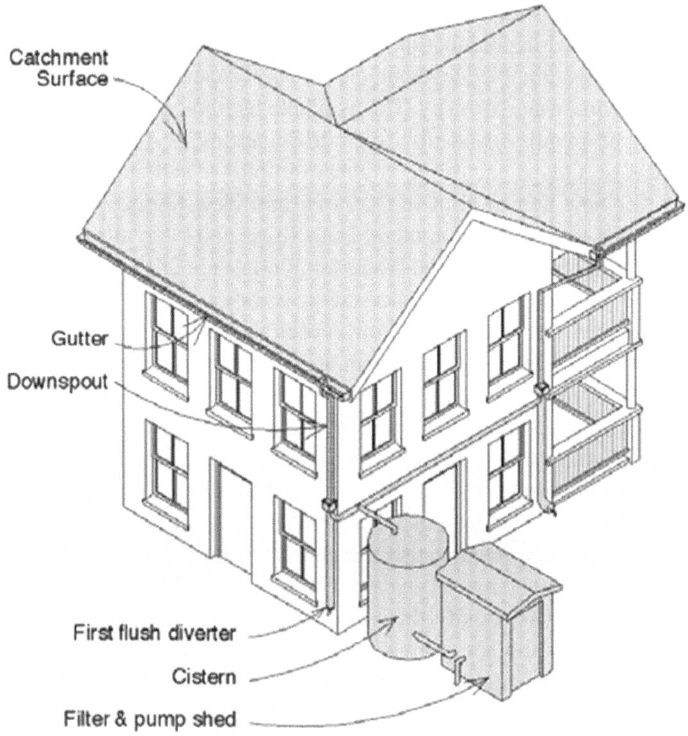

FIGURE 10.50 Typical rainwater harvesting system with aboveground cistern. (Printed from Ref. 124)

Leaf screens or mesh covers are installed to prevent leaves and large debris from entering the storage tank (Figure 10.51a). The roof washers are installed just ahead of the storage tank to filter out small debris from captured rainwater (Figure 10.51b).[124] The first-flush diverters remove dust and other smaller contaminations by redirecting the initial part of the rainfall from the cistern into a separate chamber. When this chamber is full, the clean water is directed into the cistern. The diversion and cistern filling cycle are shown in Figure 10.51c.[126]

A downspout, waterspout, downpipe, drain spout, drainpipe, or leader is a pipe that carries rainwater from a rain gutter. The purpose of a downspout is to allow water from a gutter to reach the ground without dripping or splashing down the building structure. Downspouts are usually vertical and usually extend down to ground level, although they may be routed at an angle to avoid architectural features. There are many connections that can be used at the bottom of the downspout to divert discharged water away from the building foundations to prevent water damage.

The storage facility may be a tank, cistern, deep pit, or aboveground decorative tank.[117,119] Storage tanks are manufactured from a variety of materials such as fiberglass, polyethylene, galvanized steel, and concrete. The storage tanks are available in various capacities ranging from a few hundred to several thousand gallons. The typical residential rainwater systems have storage volumes between 1,500 and 5,000 gallons.[76] Rain barrels of 50–500 gallons are also common as the simplest rain harvesting systems.[31] The delivery system includes a pump to convey harvested rainwater from the storage tank to places where it is used and provides minimum pressure for all end uses.[76] The design of pumps and their appurtenances are presented in Chapter 9.

Treatment of harvested rainwater is necessary for portable uses. A treatment system may include filtration and disinfection devices installed in series to improve the water quality. The rainwater treatment equipment and management requirements are discussed in Refs. 76,124.

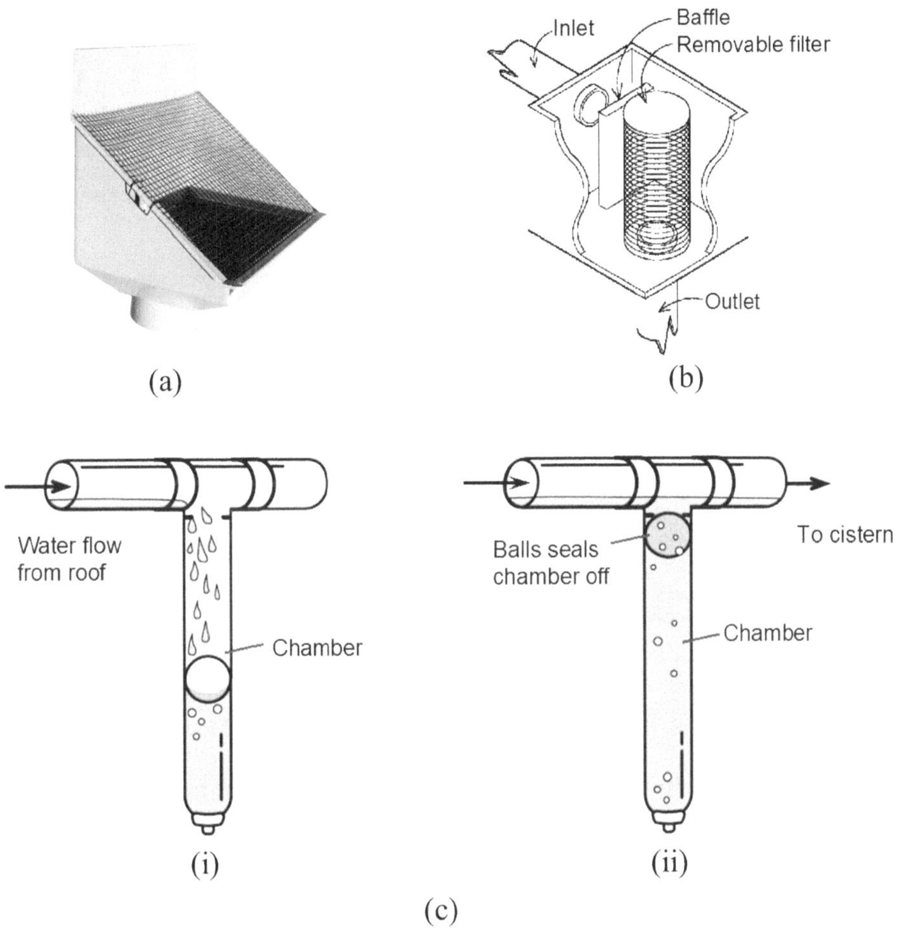

FIGURE 10.51 Components of the rain gutters: (a) downspout leaf screen;(b) roof washer; and (c) first-flush diverter, (i) first flush of contaminated water is diverted into a chamber, and (ii) once contaminated water chamber is full, clean water is diverted to cistern. (Printed from Refs. 124,126)

Rainwater can be harvested using ground or land surface catchment areas. This method is less complex than collecting rainwater from rooftops. It also provides more opportunities for collecting rainwater from a larger CDA. Ground RWH methods involve various techniques to improve surface runoff capacity. Examples of these techniques involve: (1) clearing or altering vegetation cover, (2) increasing the land slope with artificial ground cover, and (3) reducing soil permeability by soil compaction and application of chemicals. Collected water may be stored in surface or underground earthen reservoirs. Water loss due to evaporation and infiltration should be considered in the design of the RWH system. The rainwater captured from areas such as driveways and parking lots should be treated and purified before use. The harvested rainwater from the ground surface can be used for irrigation purposes. The RWH components for the roof and ground surface are shown in Figure 10.52.

The feasibility criteria for RWH systems are site-specific. They depend on the size of the system and how it will be utilized. Residential or small-scale harvesting systems can be very simple. For the design of a commercial RWH system, several key considerations may be incorporated. Some of these are larger cisterns buried underground, mechanical, electrical, and plumbing systems that must meet local plumbing codes, established water use criteria, characteristics of CDA and its

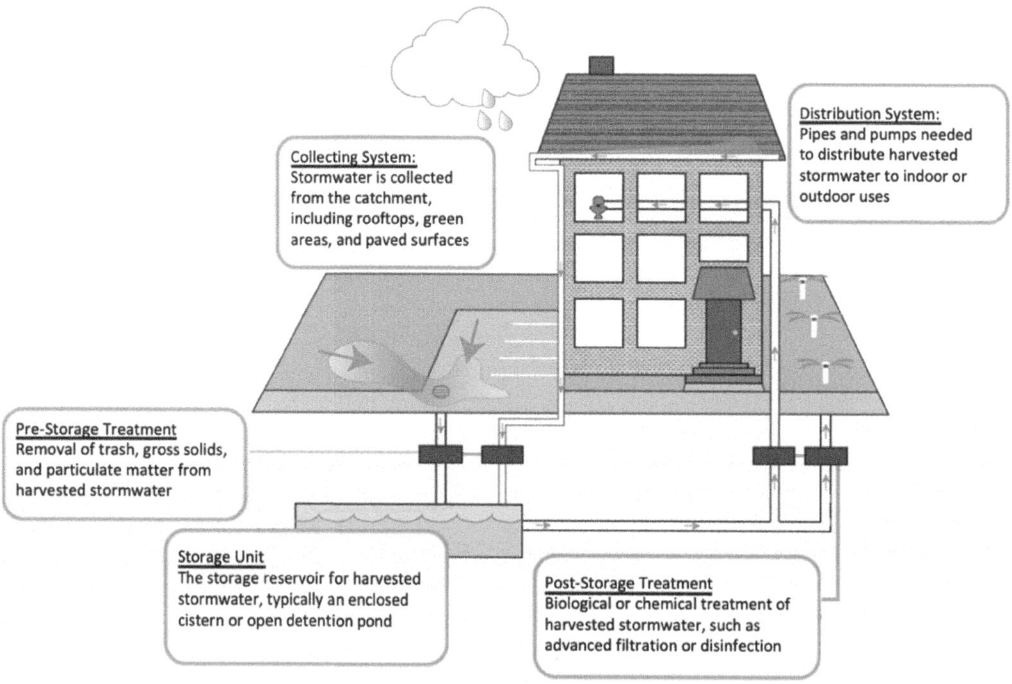

FIGURE 10.52 Rainwater harvesting from the roof and the ground surface. (Modified from Ref. 127)

construction material, water quality of rainwater, and available hydraulic head. Many more factors must be taken into consideration for systems that harvest rainwater from ground surface and those from underground storage tanks. These factors include site topography, soil type, groundwater table, proximity to underground utilities and buildings, hotspot land use, and vehicle traffic loading.[76]

Basic design considerations for RWH system: The RWH system can be simply a rain barrel at the end of a downspout for garden irrigation. It may also be designed for large-scale potable systems or multiple end-use systems at residential or commercial buildings. The main design objectives are to reduce the peak stormwater runoff and store enough rainwater to meet on-site demand. The catchment surface or CDA and rainfall amount determine the available supply. The demand depends on the end uses, such as commercial or residential water demands, the time of year for seasonal irrigation, and water consumption habits and culture of the users. Larger RWH systems require special considerations in the design and selection of their main components. These components of RWH systems have been discussed in the previous section. The basic design considerations may be found in several stormwater management guidelines and RWH manuals. Readers may refer to Chapter 4 and Refs. 124,126,127 for detailed discussion on this topic.

The size of the RWH system should be selected based on the volume of rainwater that can be captured and stored, and the demand for various uses. It is desirable to size a RWH system that can capture and store a volume larger than the demand. The volume of water that can be captured depends on the catchment surface area and the local pattern of rainfall. The monthly distribution of rainfall is an important factor to consider for sizing the storage tanks or the cisterns. The use of median rainfall results in a more conservative storage volume.[124] In theory, approximately 0.623 gallons of rainwater per inch of rain per square foot of catchment surface can be captured. This rate could be smaller due to water losses during the first flush, evaporation, and other losses. In many states, a tank that is sized for one inch of rain over the entire house roof can collect enough rainwater for the growing season and other uses. In areas such as the U.S. southwestern states, where rainfall is usually light, the system can be sized for 0.4 in. of rainfall.[43] Alternatively, the system can be designed based on the monthly water balance method. In this method, the entire storage tank volume is initially considered available. The volume of captured rainwater is added each month to the previous balance, and the volume of water demand is subtracted. The analysis can be performed using spreadsheet calculations to find the catchment size and storage capacity that meet the owner's needs or a system that is affordable. Several agencies have developed RWH storage volume calculators. The calculator uses historical daily rainfall data to evaluate the performance of the rain harvesting system under different catchment areas, demands, and storage tank volumes. As the size of the roof catchment area increases, the size of the storage tank volume decreases. Because of the uncertainty of hydrological events, allowance for drought conditions and excess use in some months must be given in sizing the storage tank. The detailed procedure for sizing the storage tank is given in Example 10.48.

Example 10.46: Rain Barrel for a Single-Family Dwelling

Determine the size of a rain barrel for a single-family dwelling to collect one inch of rainfall. The roof surface area is 1,200 ft^2. The roof has four downspouts.

Solution

1. Calculate the total rainwater volume that can be captured
   The volume of rainwater that can be captured is calculated from Eq. (10.40)[124,126]

$$G = 0.623 \times P \times A \times E \tag{10.40}$$

where
   G = volume of harvested rainwater, gallon
   P = rainfall depth, in
   A = catchment or roof surface area, ft^2
   E = system efficiency. The system efficiency depends on the amount of water loss due to first flush, evaporation, and other losses. It is assumed 90% in the absence of other information
   The factor 0.623 converts P in inches and A in ft^2 to G in gallons.

$$G = 0.623 \times 1 \text{ in.} \times 1{,}200 \text{ ft}^2 \times 0.9 = 673 \text{ gallon}$$

2. Calculate the size of the rain barrel
   If the rainfall over the roof is equally distributed among four downspouts, the storage volume of each barrel is one-fourth of the total volume of rainwater harvested.

$$\text{Storage volume of each rain barrel} = 673 \text{ gallons} / 4 = 168 \text{ gallons}$$

Depending on the demand, the homeowner may decide to install one or more barrels on his property.

Example 10.47: Rainwater Harvesting for Garden Irrigation

A homeowner wants to harvest rainwater to irrigate a 500 ft^2 vegetable garden. The garden needs one inch of water weekly between May 1 and September 30. The house is in Dallas, Texas. The monthly average rainfall in Dallas is given in Table 10.39. Calculate (1) the volume of a cistern to store rainwater for irrigation, and (2) the minimum roof catchment area needed to collect and store water to irrigate the garden.

TABLE 10.39
Average Monthly Rainfall in Dallas, Texas (Example 10.47)

Month	Jan.	Feb.	Mar.	April	May	June	July	Aug.	Sep.	Oct.	Nov.	Dec.
Rainfall, in.	1.97	2.4	2.91	3.81	5.01	3.12	2.04	2.07	2.67	3.76	2.7	2.64

Source: Ref. 124

Solution

1. Calculate monthly irrigation demand
   The average demand for each month from May to September is the same, and is calculated below

$$\text{Monthly demand} = (1\ \text{in./week} \times 1\ \text{ft} / 12\ \text{in.}) \times (4.4\ \text{week/month}) \times (500\ \text{ft}^2) \times (7.48\ \text{gal/ft}^3)$$

$$= 1{,}371\ \text{gal/month}$$

   The average monthly demand for irrigation is summarized in Column (2) of Table 10.40
2. Calculate monthly average direct rainfall over the garden from May to September

$$\text{Direct rainfall volume over the garden in May} = (5.01\ \text{in.} \times 1\ \text{ft} / 12\ \text{in.}) \times (500\ \text{ft}^2) \times 7.48\ \text{gal/ft}^3$$

$$= 1{,}561\ \text{gallon}$$

$$\text{Direct rainfall volume over the garden in June} = (3.12\ \text{in.} \times 1\ \text{ft} / 12\ \text{in.}) \times (500\ \text{ft}^2) \times 7.48\ \text{gal/ft}^3$$

$$= 972\ \text{gallon}$$

   Similarly, calculate the direct rainfall over the garden from July to September and summarize the results in Column (3) of Table 10.40

TABLE 10.40
Water Balance Analysis for Storage Tank Volume to Collect Rainwater from a Roof Catchment Area Rainwater over the Garden Area to Irrigate the Residential Garden (Example 10.47)

Month (1)	Irrigation Demand, gallon (2)	Direct Rainfall over Garden, gallon (3)	Rainwater Excess or Shortage, gallon (4)	Rainwater Harvested from Roof Catchment, gallon (5)	Total Rainwater for Irrigation, gallon (6)	Rainwater Deficiency, gallon (7)
May	1,371	1,561	190 (excess)	1,264	2,825	1,454 (excess)
June	1,371	972	–399 (shortage)	787	1,759	388 (excess)
July	1,371	636	–715 (shortage)	515	1,151	–220 (shortage)
August	1,371	644	–522 (shortage)	522	1,166	–205 (shortage)
September	1,371	832	–674 (shortage)	674	1,506	135 (excess)

3. Calculate water shortage volume for each month

$$\text{Water excess or shortage} = \text{direct rainfall over the garden} - \text{irrigation demand}$$

$$\text{Water excess in May} = 1{,}561 \text{ gal} - 1{,}371 \text{ gal} = 190 \text{ gal (excess)}$$

$$\text{Water shortage in June} = 972 \text{ gal} - 1{,}371 \text{ gal} = -399 \text{ gal (shortage)}$$

Similarly, calculate the monthly volume of water excess or shortage from July to September and summarize the results in Column (4) of Table 10.40. The excess in May means that the rainfall volume in May is more than the irrigation demand. There will be a shortage of irrigation water from direct rainfall over the garden in the months from July to September. Therefore, water will be needed from storage or other sources. If direct rainfall over the garden area is the only source of irrigation water, then a storage tank volume of 2,400 ft^3 will be needed (2,400 ft^3 = the sum of total shortages in Column (4) of Table 10.40). This is certainly a very large storage volume for this situation. The simplest way is to collect and store rainwater in a cistern from a minimum portion of the roof area for irrigation needs. As the catchment area of the roof increases, the volume of the storage tank decreases. Because of the uncertainty of the hydrological cycle, sufficient tank storage in the cistern must be provided for drought conditions and extra water usage during some high-demand months. The storage volume of the cistern should be such that it must empty at least once each season.

4. Calculate the minimum catchment area of the roof

Utilize 445 ft^2 of minimum roof catchment area to supplement irrigation water for the garden.

a. Calculate volume of water harvested from roof catchment area from Eq. (10.40)

$$G = 0.623 \times P \times A \times E$$

$$= 0.623 \text{ gal/ft}^2\text{.in.} \times 445 \text{ ft}^2 \times 0.9 \times P$$

$$= 252.32 \text{ gal/in.} \times P$$

G for the month of May = 252.32 gallon/in. × 5.01 in.

G = 1264 gallon

Similarly, calculate volume of harvested rainwater from roof in the months of June through September. These values are summarized in Column (5) of Table 10.40.

b. Calculate the total volume of irrigation water (direct over garden plus harvester water from roof)

Total volume of irrigation water in the month of May = 1,561 gallon + 1,264 gallon = 2,825 gallon

Similarly, calculate the total volume of irrigation water for the months of June to September. These values are summarized in Column (6) of Table 10.40

c. Calculate rainwater excess or shortage in the months of May through September

The rainwater in the month of May = (2,825 gallon − 1,371 gallon) = 1,454 gallon

This gives an excess of rainwater in the month of May

Similarly, calculate rainwater excess or shortage in the months of June through September. It may be noted that in the months of July and August, there are rainwater shortages.

d. Calculate the storage volume

The storage volume of cistern in Column (7) of Table 10.40 = (−220 gallon − 205 gallon) = −425 gallon

This is the theoretical volume of cistern to store rainwater from the roof catchment to supplement rainwater over the garden. Provide a total storage volume of 460 gallons (61.5 ft^3) of the cistern to supplement the direct rainfall over the garden. The proposed dimension of the cistern is 4 ft × 4 ft × 4 ft (1.2 m × 1.2 m × 1.2 m). This storage volume will also provide additional rainwater needed for drought conditions, and excess water needed for some high-demand months. The homeowner may decide to use one cistern or multiple cisterns to collect rainwater from a minimum of 445 ft^2 of roof area to irrigate the garden. Additional roof area may be added to harvest rainwater needed for watering bushes, flower beds, lawn watering, car wash, and other uses. A cistern of 460 gallons will provide sufficient storage for additional roof area that may be added.

Example 10.48: Rainwater Harvesting for Indoor and Outdoor Uses

A family of three lives in a house that has a roof area of 2,450 ft^2. The family wants to collect water for flushing the toilets year-round and irrigating their garden. Estimate the size of the cistern to be installed to harvest rainfall. The monthly average rainfall in the area and irrigation demand for each month are given in Table 10.41. The average residential water demand is 450 Lpcd.

TABLE 10.41
Average Monthly Rainfall and Monthly Irrigation Demand (Example 10.48)

Month	Jan.	Feb.	Mar.	April	May	June	July	Aug.	Sep.	Oct.	Nov.	Dec.
Avg. Rainfall, in.	1.2	1.5	2.4	2.7	4.2	3.6	2.4	2.4	3.3	2.7	2.1	1.5
Avg. Irrigation Demand, gallon	0	0	100	150	200	250	250	250	200	150	0	0

Solution

1. Calculate monthly water demand for toilet flushing

   The toilet flushing water demand is calculated using the information in Table 4.2. From this table, toilet flush usage, including toilet leakage, is 33% of residential water demand

   Average daily toilet flushing water demand = 0.33 × 450 Lpcd = 132 Lpcd

   Monthly water demand in Jan. = 132 Lpcd × 1 gallon/3.785 L × 3 persons × 31 days = 3,243 gallon

   Monthly water demand in Feb. = 132 Lpcd × 1 gallon/3.785 L × 3 persons × 28 days = 2,929 gallon

   Similarly, calculate the monthly water demand for toilet flushing from March to December and summarize the 12-month results in Column (4) of Table 10.42.

2. Calculate total monthly water demand

   The total monthly demand is the sum of the toilet flushing and irrigation uses

   Total monthly demand in Jan. = 3,243 gallon + 0 gallon = 3,243 gallon

   Total monthly demand in Feb. = 2,929 gallon + 0 gallon = 2,929 gallon

   Total monthly demand in Mar. = 3,243 gallon + 100 gallon = 3,343 gallon

   Similarly, calculate the total monthly demand from April to December and summarize 12-month results in Column (5) of Table 10.42.

TABLE 10.42
Water Balance Calculations for January to December (Example 10.48)

Month	Rainfall, in.	Irrigation Demand, gallon	Toilet Flushing Demand, gallon	Total Demand, gallon	Harvested Rainwater, gallon	Water Deficiency, gallon	End-of-Month Storage, gallon
(1)	(2)	(3)	(4)	(5)	(6)	(7)	(8)
Jan	1.2	0.0	3,243	3,243	1,648	−1,595	914
Feb	1.5	0.0	2,929	2,929	2,061	−868	46
Mar	2.4	100.0	3,243	3,343	3,297	−46	0
Apr	2.7	150.0	3,139	3,289	3,709	420	420
May	4.2	200.0	3,243	3,443	5,770	2,327	2,746
Jun	3.6	250.0	3,139	3,389	4,945	1,556	**3945F**
Jul	2.4	250.0	3,243	3,493	3,297	−196	3,749
Aug	2.4	250.0	3,243	3,493	3,297	−196	3,553
Sep	3.3	200.0	3,139	3,339	4,533	1,194	**3945F**
Oct	2.7	150.0	3,243	3,393	3,709	316	**3945F**
Nov	2.1	0.0	3,139	3,139	2,885	−254	3,691
Dec	1.5	0.0	3,243	3,243	2,061	−1,183	2,509

Bold values are the max values in the column.

3. Calculate monthly rainwater volume that can be harvested

   The volume of rainwater that can be harvested is calculated from Eq. (10.40)

   Volume of rainwater harvested in Jan. $= 0.623 \times 1.2$ in. $\times 2{,}450\,\text{ft}^2 \times 0.9 = 1{,}648$ gallon

   Volume of rainwater harvested in Feb. $= 0.623 \times 1.5$ in. $\times 2{,}450\,\text{ft}^2 \times 0.9 = 2{,}061$ gallon

   Similarly, calculate the monthly harvested rainwater from March to December, and summarize 12-month results in Column (6) of Table 10.42.

4. Calculate monthly water deficiency volume

   The deficiency of water for each month is calculated by subtracting the total demand from the volume of harvested rainfall

   Volume of water deficiency in Jan. $= 1{,}648$ gallon $- 3{,}243 = -1{,}595$ gallon

   Volume of water deficiency in Feb. $= 2{,}061$ gallon $- 2{,}929 = -868$ gallon

   Similarly, calculate the monthly water deficiency volume from March to December and summarize 12-month results in Column (7) of Table 10.42.

5. Estimate storage volume of the cistern.

   From the entries in Column (7) of Table 10.42, it is indicated that the harvested water will not be sufficient for total demand in the consecutive five deficiency months of November, December, January, February, and March. The total deficiency for these five months is equal to $(254 + 1{,}182 + 1{,}595 + 868 + 46)$ gallons $= 3{,}945$ gallons. This is the theoretical volume of the cistern needed.

   If a total of 2,509 gallons of rainwater is stored in the cistern to balance the deficiency in the months of January through March, tap water will not be needed for total demand, and initial storage will be sufficient from the harvested rainwater. Also, the cistern will be empty at the end of March (see Column (8) of Table 10.42)

6. Calculate the end-of-month storage in the cistern for the entire year

   At the end of each month, the storage volume is calculated by adding the volume that already existed in the cistern and the deficiency in that month (Column (7)).

At the end of December or the start of January, the initial volume in the cistern is 2,509 gallons.

End of Jan. storage volume=2,509 gallon – 1,595 gallon=914 gallon

End of Feb. storage volume=914 gallon−868 gallon=46 gallon

End of Mar, storage volume=(46 gallon+(−46 gallon)=0 gallon. The cistern will be empty

The cistern will start to fill in the month of April. At the end of April and May, the volume of harvested rainwater will be 420 and 2,746 gallons. At the end of June, the cistern will be full at 3,945 gallons, and (2,746+1,556−3,945) gallons=357 gallons of rainwater will spill from the cistern. At the end of September, the cistern will be full again at 3,945 gallons, and excess rainwater in the cistern will spill from the cistern. The volume of spilled water is (3,553+1,194 – 3,945) gallons=802 gallons. Similarly, the cistern will be full again at the end of October, and 316 gallons of excess rainwater will spill from the cistern.

Total volume of spilled rainwater=(375+802+316) gallon=1,475 gallon

7. Determine the cistern capacity

The theoretical capacity of the cistern is 3,945 gallons. Provide 16% extra capacity to account for drought and extra water use during some months. Additionally, there will be rainwater losses due to evaporation and the first flush. Therefore, one cistern or multiple cisterns with a total capacity of 4,500 gallons should be provided for this house.

Note: Readers are referred to Examples 4.29, 4.30, and 10.38 for sizing water storage basins.

10.7.12 Infiltration Basin

An infiltration basin (also known as a recharge basin, a sump, or a percolation pond) is a type of storage device that is used to manage stormwater runoff, prevent flooding and downstream erosion, and improve water quality in an adjacent river, stream, lake, or bay. It is essentially a shallow artificial pond that is designed to infiltrate stormwater through permeable soils into the groundwater aquifer. Infiltration basins do not release water, except by infiltration, evaporation, or emergency overflow during flood conditions.[128,129] It is distinguished from a detention basin (sometimes also called a dry pond), which is designed to discharge stormwater into a downstream waterbody, although some of its volume infiltrates into the groundwater aquifer. It also differs from a retention basin, which is designed to include a permanent pool of water.

Infiltration basins must be carefully designed to infiltrate the soil on a given site at a rate that will not cause flooding. They may be less effective in areas with (1) high groundwater table, close to the infiltrating surface, (2) compacted soils, (3) high levels of sediment in stormwater, and (4) high clay soil content unless soil on the particular site is modified ("engineered soil") during construction, to improve the infiltration characteristics. At some sites, infiltration basins have worked effectively where the installation also includes an extended detention basin as a pretreatment stage to remove sediment. The basins may fail where they cannot be frequently maintained, and their use is discouraged in some areas of the United States with high clay soil content.[130] A schematic of an infiltration basin is shown in Figure 10.53.

Several design considerations and criteria should be incorporated in the design of an infiltration basin for effective functioning. These considerations may vary from one jurisdiction to another and include maximum drainage area, site soils, depth to the groundwater table, site topography and slopes, setback from utilities and buildings, hotspot land use, and others[76,130]

Basic design considerations of infiltration basin: The design philosophy of infiltration basins is to maximize their effectiveness. The basin should be properly sized based on the quality and quantity of the stormwater and site soil conditions. The basin can be designed to fully store the

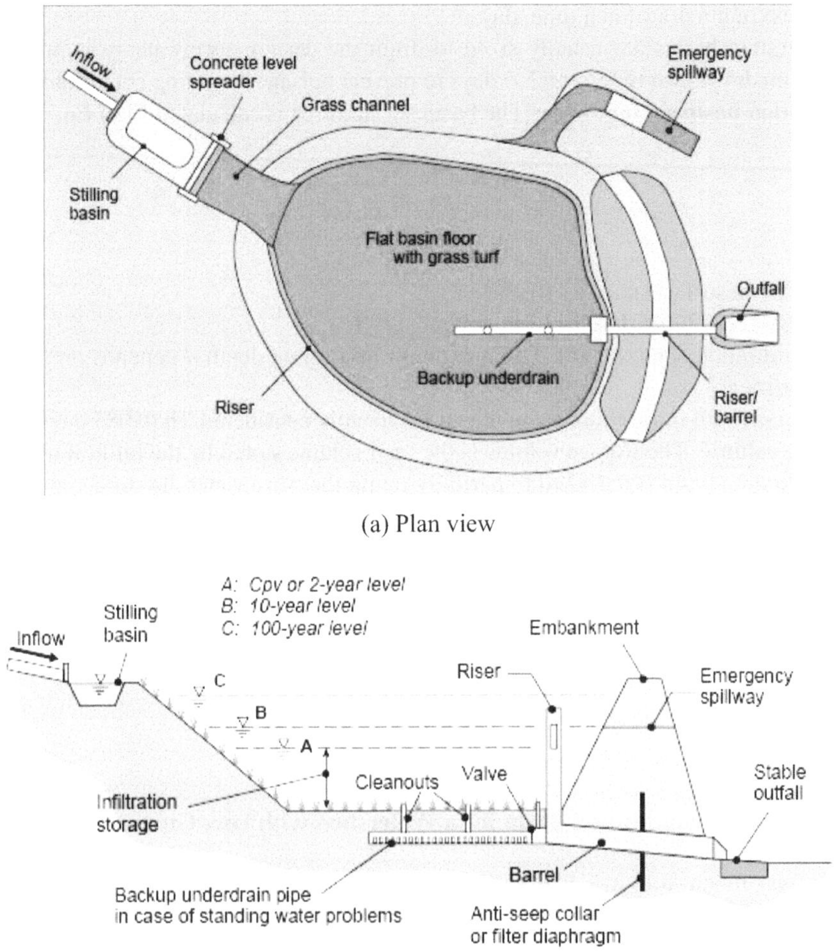

(a) Plan view

(b) Profile

FIGURE 10.53 Details of a typical infiltration basin: (a) plan view, and (b) sectional view. (Printed from Ref. 130)

entire design stormwater volume from the CDA and drain it over time. The infiltration basin may also be designed to retain larger storms partially or fully. An overflow must be incorporated in the design to bypass part of the stormwater runoff exceeding the basin capacity. The important design parameters of the infiltration basin are (1) maximum basin depth, (2) surface area, and (3) storage volume. These design parameters are described below.

1. **Infiltration basin maximum Depth**: The depth of an infiltration basin is a function of the maximum drawdown time and soil hydraulic conductivity. Equation (10.41) may be used to calculate the basin maximum depth[76,126,127]

$$d_{max} = K_{sat} \times t_d \tag{10.41}$$

where

d_{max} = maximum depth of the infiltration basin, m (ft)

K_{sat} = field-verified saturated hydraulic conductivity for the site soils, m/day or mm/h (ft/day, in./h)

t_d = maximum drawdown time, day

Infiltration basins are usually sized to drain the design stormwater volume within a maximum drawdown time t_d of 2–3 days to prevent nuisance ponding conditions.[76,126,127]

2. **Infiltration basin surface area**: The basin surface area is calculated from Eq. (10.42)

$$\text{SA} = \frac{V_w}{d + (K_{\text{sat}} \times t_f)} \tag{10.42}$$

where

SA = basin surface area, m^2 (ft^2)

V_w = SWRv or other design storm volume, m^3 (ft^3)

d = infiltration depth, m (ft). The maximum infiltration depth d depends on the results of Eq. (10.41)

t_f = time to fill infiltration basin, day. It is typically considered 2 h (0.083 day)

3. **Storage volume**: The storage volume is the total volume stored by the infiltration basin. If the infiltration basin is designed to partially retain the stormwater, its storage volume can be estimated from Eq. (10.43). This equation is derived by rearranging Eq. (10.42).

$$S_v = \text{SA}\left[d + (K_{\text{sat}} \times t_f)\right] \tag{10.43}$$

where

S_v = the storage volume of the infiltration basin, m^3 (ft^3).

The other terms have been defined previously.

Example 10.49: Infiltration Basin for a Watershed with Low Groundwater Table

Stormwater runoff from a 5-acre watershed is discharged into an infiltration basin. The watershed is 20% impervious with CN = 95. The underlying soil has a saturated hydraulic conductivity of 1.02 in./h. The seasonal high groundwater table is at a depth of 9.0 ft. The criterion is to keep the bottom of the basin at least 3.0 ft above the groundwater table. The basin is designed to treat the runoff from 1.0 in. of rainfall. A 2-day drawdown is one of the design criteria.

Solution

1. Calculate stormwater runoff volume from the impervious area
   a. Calculate S from Eq. (10.7)

$$\text{CN} = \frac{1{,}000}{10 + S}$$

$$95 = \frac{1{,}000}{10 + S}$$

$$950 + 95S = 1{,}000$$

$$S = 0.526 \text{ in.}$$

   b. Calculate depth of stormwater runoff from Eq. (10.6)

$$Q = \frac{(P - 0.2S)^2}{P + 0.8S} = \frac{(1.0 \text{ in.} - 0.2 \times 0.526 \text{ in.})^2}{1.0 \text{ in.} + 0.8 \times 0.526 \text{ in.}} = \frac{0.8 \text{ in.}^2}{1.42 \text{ in.}}$$

$$Q = 0.563 \text{ in.}$$

c. Estimate stormwater volume V_w

Assume 20% of depth of stormwater runoff

$$V_w = 5 \text{ acres} \times \left(43{,}560 \text{ ft}^2/\text{acre}\right) \times 20\% \times 0.563 \text{ in.} \times \left(1 \text{ ft}/12 \text{ in.}\right) = 2{,}043 \text{ ft}^3$$

2. Determine the maximum depth of the infiltration basin, d_{max}, from Eq. (10.41)

$$d_{max} = K_{sat} \times t_d$$

$$= 1.02 \text{ in./h}(1 \text{ ft}/12 \text{ in.}) \times (24 \text{ h}/1 \text{ day}) \times 2 \text{ day}$$

$$= 4.1 \text{ ft}$$

The bottom of the basin with the maximum depth d_{max}=4.1 ft is above the water table by 4.9 ft (9.0 ft−4.1 ft=4.9 ft), which is greater than the minimum 3.0 ft criterion. Select basin depth d=4.0 ft.

3. Calculate drawdown time

Rearrange Eq. (10.41)

$$t_d = \frac{d}{K_{sat}}$$

$$= \frac{4.0 \text{ ft}}{1.02 \text{ in./h} \times (1 \text{ ft}/12 \text{ in.}) \times (24 \text{ h}/1 \text{ day})} = \frac{4.0 \text{ ft}}{2.04 \text{ ft/day}} = 1.96 \text{ day}$$

Since the drawdown time is less than the allowable maximum drain time of 2 days, the infiltration basin depth is sized correctly.

4. Calculate basin surface area SA from Eq. (10.42)

$$\text{SA} = \frac{V_w}{d + (K_{sat} \times t_f)}$$

$$= \frac{2{,}043 \text{ ft}^3}{4.0 \text{ ft} + [1.02 \text{ in./h} \times (1 \text{ ft}/12 \text{ in.})(24 \text{ h}/1 \text{ day}) \times 0.083 \text{ day}]} = \frac{2{,}043 \text{ ft}^3}{4.0 \text{ ft} + 0.17 \text{ ft}}$$

$$\text{SA} = 490 \text{ ft}^2$$

The dimensions and shape of the infiltration basin can be obtained to accommodate this surface area. For example, a square basin will be 22.2 ft wide and 22.2 ft long.

Example 10.50: Infiltration Basin for a Watershed with High Groundwater Table

An infiltration basin is to be designed for a transportation project. The project site, including right-of-way and all permanent easements, is 10 ha, and 50% of the site is covered by impervious surfaces with CN=98. The geotechnical investigations show a saturated hydraulic conductivity of 0.55 m/day for the site soil. The seasonal high groundwater table is at a depth of 3 m. The basin is to be designed to treat the runoff from 25.4 mm (1 in.) of rainfall. According to the local stormwater design manual, the basin should be designed for a 3-day drawdown, and its bottom elevation should be at least 1.5 m above the groundwater table. Design a rectangular basin with a length-to-width ratio of 3:1.

Solution

1. Calculate stormwater runoff volume from the impervious area

a. Calculate *S* from Eq. (10.7)

$$\mathrm{CN} = \frac{1{,}000}{10 + S}$$

$$98 = \frac{1{,}000}{10 + S}$$

$$980 + 98S = 1{,}000$$

$$S = 0.204 \text{ in.}$$

b. Calculate depth of stormwater runoff from Eq. (10.6)

$$Q = \frac{(P - 0.2S)^2}{P + 0.8S} = \frac{(1.0 \text{ in.} - 0.2 \times 0.204 \text{ in.})^2}{1.0 \text{ in.} + 0.8 \times 0.204 \text{ in.}} = \frac{0.92 \text{ in.}^2}{1.163 \text{ in.}}$$

$$Q = 0.79 \text{ in.}$$

c. Estimate stormwater volume V_w

$$V_w = 10 \text{ ha} \times (10{,}000 \text{ m}^2/\text{ha}) \times 50\% \times 0.79 \text{ in.} \, (1 \text{ m} / 39.37 \text{ in.}) = 1{,}003 \text{ m}^3$$

2. Determine the maximum depth of the infiltration basin, d_{max}, from Eq. (10.41)

$$d_{max} = K_{sat} \times t_d$$

$$= 0.55 \text{ m/day} \times 3 \text{ day}$$

$$= 1.65 \text{ m}$$

The bottom of the basin with the maximum depth $d_{max} = 1.65$ m is above the water table by 1.35 m (3.0 m − 1.65 m = 1.35 m), which is less than the minimum 1.5 m criterion. Select basin depth $d = 1.5$ m.

3. Calculate basin surface area SA from Eq. (10.42)

$$\mathrm{SA} = \frac{V_w}{d + (K_{sat} \times t_f)}$$

$$= \frac{1{,}003 \text{ m}^3}{1.5 \text{ m} + (0.55 \text{ m/day} \times 0.083 \text{ day})} = \frac{1{,}003 \text{ m}^3}{1.5 \text{ m} + 0.05 \text{ m}}$$

$$\mathrm{SA} = 647 \text{ m}^2$$

4. Calculate basin dimensions

The basin length L and width W can be determined based on the 3.0–1 length-to-width ratio, i.e., $L = 3W$

$$\mathrm{SA} = L \times W$$

$$647 \text{ m}^2 = 3W \times W = 3W^2$$

$$W = 14.7 \text{ m} \quad \text{or} \quad 15 \text{ m}$$

$$L = 3 \times 15 \text{ m} = 45 \text{ m}$$

$$D = 1.5 \text{ m}$$

Example 10.51: Infiltration Basin to Partially Retain Stormwater

An infiltration basin is designed to infiltrate the runoff from one acre of impervious pavement. The runoff is generated from a 2-year storm with a total rainfall of 2 in. Due to some limitations at the site, the surface area and depth of the infiltration basin may not exceed 1,000 ft^2 and 3 ft. The maximum allowable drawdown time is 24 h. The subsoil has a saturated hydraulic conductivity of 4 in./h. Calculate (1) the portion of stormwater volume that can be retained by the basin, and (2) drawdown time.

Solution

1. Calculate stormwater runoff volume from the impervious area
   a. Determine CN
      CN = 98 from Table B.7a for impervious areas
   b. Calculate S

$$98 = \frac{1{,}000}{10 + S}$$

$$980 + 98S = 1{,}000$$

$$S = 0.204 \text{ in.}$$

   c. Calculate stormwater runoff depth from Eq. (10.6)

$$Q = \frac{(P - 0.2S)^2}{P + 0.8S} = \frac{(2.0 \text{ in.} - 0.2 \times 0.204 \text{ in.})^2}{2.0 \text{ in.} + 0.8 \times 0.204 \text{ in.}} = \frac{3.84 \text{ in.}^2}{2.16 \text{ in.}}$$

$$Q = 1.78 \text{ in.}$$

   d. Estimate stormwater volume V_w

$$V_w = 1 \text{ acres} \times \left(43{,}560 \text{ ft}^2/\text{acre}\right) \times 1.78 \text{ in.}\left(1 \text{ ft} / 12 \text{ in.}\right) = 6{,}461 \text{ ft}^3$$

2. Calculate treatment volume of the basin
   Rearrange Eq. (10.42)

$$\text{SA} = \frac{V_w}{d + (K_{\text{sat}} \times t_f)}$$

$$V_w = \text{SA} \times \left[d + (K_{\text{sat}} \times t_f)\right]$$

$$= 1{,}000 \text{ ft}^2 \times [3.0 \text{ ft} + (4.0 \text{ in./h} \times (1 \text{ ft} / 12 \text{ in.})(24 \text{ h} / 1 \text{ day}) \times 0.083 \text{ day}]$$

$$= 1{,}000 \text{ ft}^2 \times (3.0 \text{ ft} + 0.66 \text{ ft})$$

$$= 3{,}660 \text{ ft}^3$$

Since the treatment volume of the basin is less than the stormwater volume, the basin can only partially retain the stormwater. The runoff volume in excess of the basin treatment capacity = 6,461–3,660 ft^3 = 2,801 ft^3. This volume will be bypassed to another treatment facility or BMP for treatment or discharged into the receiving water.

3. Determine drawdown time
   Rearrange Eq. (10.41)

$$t_d = \frac{d}{K_{sat}}$$

$$= \frac{3.0\text{ ft}}{4.0\text{ in./h} \times (1\text{ ft} / 12\text{ in.}) \times (24\text{ h} / 1\text{ day})} = \frac{3.0\text{ ft}}{8.0\text{ ft/day}} = 0.375\text{ days or 9 h}$$

   The drawdown time is less than the allowable maximum drain time of 24 h; therefore, the infiltration basin depth of 3.0 ft is acceptable.

10.7.13 Infiltration or Percolation Trench

An infiltration trench, also called a percolation trench, is a source-control BMP that is effective for the management of stormwater runoff in urban developments. They are linear shallow excavated minimum of 18 in. wide and 8 in. deep, filled with ½–1½ in. gravel or crushed stone that collects rainwater from adjacent surfaces. Their highly permeable construction allows the stormwater to infiltrate through permeable soils into the groundwater aquifer.[131] They prevent flooding and downstream erosion, and improve water quality in the receiving waters. Because they are linear ditches, they are very practical to install parallel to roadsides or around the perimeter of parking lots or driveways. The accumulated pollutants in industrial and agricultural areas are more toxic than those in commercial or residential areas, so infiltration trenches should not be used near farms or industrial complexes to ensure that these chemicals do not reach the groundwater.[132]

Runoff in the infiltration trench is treated in three different ways. These include (1) physical filtration to remove solids, (2) adsorption onto the media in the trench, and (3) biochemical reactions involving microorganisms growing on the fill media or in the soil. The level of treatment depends on two factors: the size of the media and the length of the flow path through the system. Both these factors control the time it takes the runoff to pass into the surrounding soil. Pretreatment may be required before polluted runoff is allowed into an infiltration trench. The advantages of infiltration trench are that they significantly reduce both runoff rates and volumes, and reduce significant pollutant load that may be discharged into the receiving body.[133] The primary benefits are water quality improvement, reduction of peak flows in sewer systems, and groundwater recharge. Also, since the peak flow across the surface is reduced, the amount of surface water entering waterways is slowed down. This reduces channel erosion. The percolation of rainwater through the channel is a source of river baseflow. This topic is covered in Section 10.4.1. Design and construction details about the infiltration trench are given below.

Site constraints and siting of the facility: Typically, an infiltration trench is considered if the hydrologic soil classification in the proposed location falls in Group A or B (Table 10.10). In these groups, the soil is classified as sand, loamy sand or sandy loam, and silt loam or loam. Infiltration should generally not be used in areas prone to hotspot runoff due to the possibility of contaminating the groundwater table.

Contributing drainage area (CDA): Typically, the CDA to an infiltration trench facility should be highly impervious (approaching 100%), and should not exceed 2.0 acres on any single installation.

Site slopes in the vicinity: Site slope should typically have flat bottoms (0% slope) in order to promote evenly distributed infiltration over the practice area. The average slope of upstream contributing areas shall be 15% or less.

Depth to bedrock separation: The depth of separation should be at least 2 ft between bedrock and the invert of the infiltration trench.

Depth to water table separation: At least 2 ft of separation is required between the seasonally high groundwater table and the bottom of the infiltration trench.

Hydraulic head: A minimum hydraulic head of up to 2 ft may typically be required to drive flow through an infiltration trench for optimal functioning of conventional infiltration trenches. This may be increased on a case-by-case basis.

Soils testing: Soils in the infiltration area are required to have infiltration tests in accordance with the approved methods. A field-tested infiltration rate of 0.5 in./h or greater is required for the use of an infiltration trench.

Setbacks: In order to prevent damage to the foundation by seepage, sufficient setbacks from the infiltration trench should be considered. Setbacks from wells shall be a minimum of 100 ft, and from septic drain fields shall be a minimum of 50 ft. Infiltration trenches shall be installed a minimum of 5 ft down-gradient of utility lines. When located near down-gradient slopes of 20% or greater, the infiltration trench shall be located a minimum distance of 200 ft from those slopes.

Karst areas: Conventional infiltration practices shall not be allowed in karst regions unless a separation distance of 4 ft to bedrock is available.

Geotextile fabric: Non-woven geotextile fabric on the sides, folded a flap, and over the top of the trench with additional stone will extend the life of the trench. To maintain these structures, regular inspection for signs of clogging and periodic removal of accumulated debris and weeds from the surface is needed. Also, removal of sediment from the pretreatment system, and cleaning or replacement of the stone is necessary. The design details and generalized overview of an infiltration trench are shown in Figure 10.54.

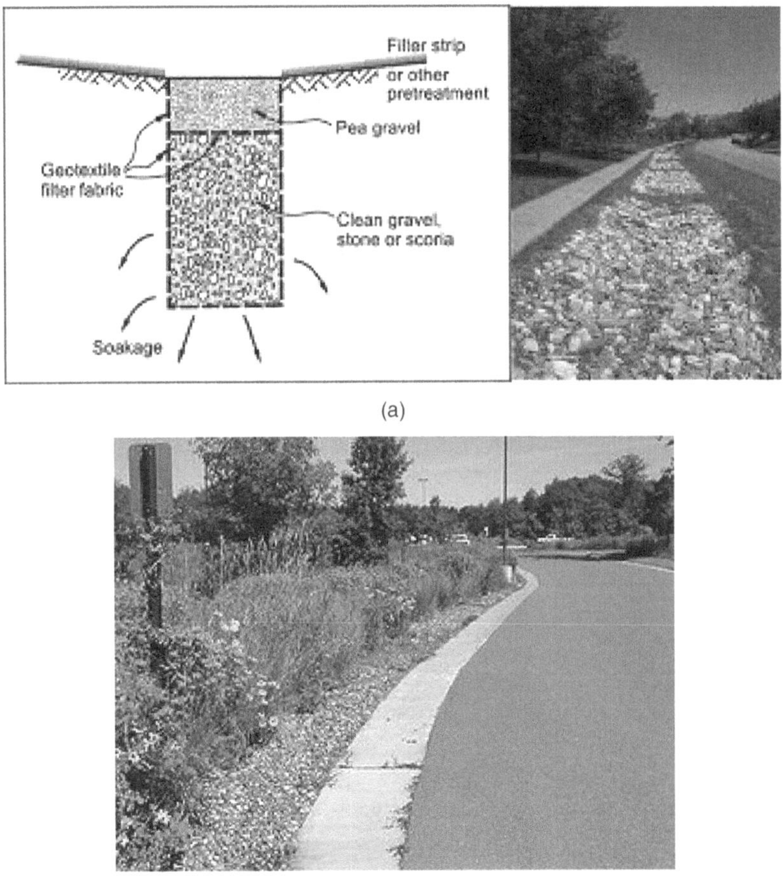

FIGURE 10.54 Details of infiltration/percolation trench: (a) cross-section and layout details, and (b) overview of an established infiltration trench along a roadside. (Printed from Refs. 131,132)

Example 10.52: Design of an Infiltration Trench

An infiltration trench is designed to meet the technology-based water quality requirements arising from the construction of approximately 2,000 linear feet of roadway. The topography is such that runoff from the road is collected in curb, gutter drain, and conveyed to curb inlets along the road. The runoff is discharged into sediment forebays from which it enters onto the surface of the proposed trench. The trench is constructed in the median of the divided roadway. The total project site, including right-of-way and all permanent easements, consists of 6.2 acres. Approximately 300 linear feet is available for the construction of the trench. Geotechnical investigations reveal that the saturated soil infiltration rate is 2.3 in./h. The project site does not exhibit a high or seasonally high groundwater table. The pre-development project land area is 6.2 acres, and the soil of the total land area is on unimproved grass cover with a runoff coefficient of 0.5. The post-development project land area is also 6.2 acres, but 3.4 acres area is converted into impervious surface. The drainage from the impervious area captured by the infiltration trench is equivalent to 0.5 in. of rainfall. The design infiltration rate is one-half the infiltration rate obtained from the required geotechnical investigation. Develop the design of the infiltration trench.

Solution

1. Compute the required WQ_v

   The WQ_v from the project site is a function of the developed impervious area. The basic WQ_v is computed from Eq. (10.44)

$$WQ_V = IA \times I \tag{10.44}$$

where

WQ_V = water quality volume, m^3 (ft^3)

IA = impervious area, m^3 (ft^2)

I = equivalent rainfall over impervious drainage area, m (ft)

$$WQ_V = (3.4 \text{ acre} \times 43{,}560 \text{ ft}^2/\text{acre}) \times 0.5 \text{ in.} / (12 \text{ in./ft}) = 6{,}171 \text{ ft}^3$$

The infiltration trench will be sized to treat the computed WQ_v of 6,171 ft^3.

2. Compute the design infiltration rate

   The design infiltration rate is calculated from Eq. (10.45)

$$f_d = C \times f \tag{10.45}$$

where

f_d = design infiltration rate, cm/h (in./h).

C = ratio of design infiltration rate to infiltration rate obtained from the required geotechnical investigation. C is 0.5 given for this example

f = infiltration rate obtained from the required geotechnical analysis.

For the given site conditions, the design infiltration rate f_d is computed from Eq. (10.45)

$$f_d = 0.5 \times 2.3 \text{ in./h} = 1.15 \text{ in./h}$$

3. Compute the available trench depth

   The trench must be designed such that it is completely empty within a maximum of 48h after a runoff-producing event. To ensure compliance with this requirement, compute the maximum allowable trench depth from Eq. (10.46)

$$d_{max} = \frac{(f_d \times T_{max})}{V_r} \qquad (10.46)$$

   where

   d_{max} = maximum allowable trench depth, m (ft)

   T_{max} = maximum allowable drain time (48 hours)

   V_r = void ratio of the stone trench (0.40 for coarse-graded aggregate)

   f_d has been defined previously

$$d_{max} = \frac{(1.15 \text{ in./h} \times 1 \text{ ft} / 12 \text{ in.} \times 48 \text{ h})}{0.4} = 11.5 \text{ ft}$$

4. Compute the minimum allowable infiltration trench bottom area

   Employing the principles of Darcy's Law and assuming one-dimensional flow through the bottom of the trench, the minimum allowable surface area of the trench is computed from Eq. (10.47)

$$\text{SA}_{min} = \frac{\text{WQ}_v}{(f_d \times T_{max})} \qquad (10.47)$$

   where

   SA_{min} = minimum trench bottom surface area, m^2 (ft^2)

   Other terms have been defined previously.

$$\text{SA}_{min} = (6{,}171 \text{ ft}^3) / \left[(1.15 \text{ in./h} \times 1 \text{ ft} / 12 \text{ in.}) \times 48 \text{ h}\right] = 1{,}342 \text{ ft}^2$$

5. Determine the infiltration trench dimensions based on site-specific parameters

   In this example, the trench is to be located in the median of the divided highway, and approximately 300 linear feet of trench length is available for construction of the trench. This entire length is utilized in an effort to minimize the trench depth. The maximum desirable trench width is assumed to be 8ft.

   Applying the maximum width and available 300ft length results in a trench bottom surface area.

$$\text{SA} = 300 \text{ ft} \times 8 \text{ ft} = 2{,}400 \text{ ft}^2$$

   This value is greater than the minimum value of 1,342 ft^2 computed previously. Therefore, it is acceptable.

6. Compute the trench depth

   The volume of storage is available in the void space of the trench aggregate. Therefore, the minimum required trench depth d is computed using the previously calculated values and void ratio V_r=0.4

$$d = \frac{6{,}171 \text{ ft}^3}{0.4 \times 2{,}400 \text{ ft}^2} = 6.5 \text{ ft}$$

7. Compute available storage volume

$$\text{Available storage volume} = 2{,}400\ \text{ft}^2 \times 300\ \text{ft} \times 6.5\ \text{ft} \times 0.4(\text{void ratio})$$

$$= 6{,}240\ \text{ft}^3$$

The available storage volume is greater than the WQ_v of 6,171 ft^3 (Step 1). Therefore, the design of the infiltration trench is acceptable.

8. Summarize the dimensions of the designed infiltration trench and storage volume
   Length = 300 ft
   Width = 8 ft
   Depth = 6.5 ft
   Storage volume = 6,240 ft^3

10.7.14 Permeable Paving

Permeable paving or porous surfaces include a variety of surfacing techniques for roads, pedestrian walkways, driveways, parking lots, and walking paths that have light vehicular traffic. Unlike impervious paving materials, permeable paving systems allow stormwater to percolate and infiltrate through the pavement and into the aggregate layers or non-porous blocks. These blocks are spaced so that water can flow between the gaps and/or soil below, and is stored in an underlying stone reservoir or aquifer. Permeable paving surfaces typically include pervious concrete, porous asphalt, paving stones, concrete blocks or grids, permeable interlocking concrete, plastic grid, and interlocking pavers.[134] In addition to reducing surface runoff, permeable paving systems can trap suspended solids, thereby filtering pollutants from stormwater. The common goal is to allow stormwater to infiltrate at the source, reduce runoff, and improve water quality by filtering pollutants in the subsurface layers.[135]

Porous pavements such as pervious concrete and pervious asphalt are better suited for urbanized areas that see more frequent vehicular traffic. On the other hand, concrete pavers, grids, and geocells are better suited for light vehicular traffic, pedestrian and cycling pathways, and parking lots.[135] Polymer-based grass grid or cellular paver systems provide load-bearing reinforcement for unpaved surfaces of gravel or turf.

Permeable paving provides a solid ground surface, strong enough to take heavy loads, like large vehicles. At the same time, the permeable paving allows water to filter through the surface and reach the underlying soils, mimicking natural ground absorption. Permeable paving, therefore, can reduce downstream flooding and stream bank erosion and maintain baseflows in rivers to keep ecosystems self-sustaining. Permeable pavers also combat erosion that occurs when grass is dry or dead, by replacing grassed areas in suburban and residential environments. These paving systems are also effective in reducing contaminants such as suspended solids, BOD, chemical oxygen demand, and ammonium concentrations from reaching groundwater.[136] In areas where infiltration is reduced due to unsuitable soil conditions, permeable pavements provide attenuation of retained water in the pavement and slowly drain it to surface water systems between storm events.[115] Permeable pavements, due to their porous surface, admit vital air and water to the root zone of the urban trees for growth to full size. This integrates healthy ecology in the urban environment. During large storm events, additional water is stored in the open-graded or crushed drain rock base until absorbed in the subgrade. For clay-based soils or other low to non-draining soils, it is important to increase the depth of the base of crushed drain rock to allow additional capacity and time for water to be infiltrated.[134,135]

Runoffs across some contaminated lands pick up high concentrations of pollutants that exceed those typically found in stormwater. These *hot spot sources* include commercial plant nurseries, waste recycling facilities, fueling stations, chemical storage, marinas, some outdoor loading facilities, hazardous materials generators (if containers are exposed to rainfall), vehicle service, washing, and maintenance areas, and steam cleaning facilities. Since porous pavements provide infiltration, they should not be constructed around such hot spot sources due to the potential for groundwater contamination. All contaminated runoffs should be isolated from entering storm drains. Source-control BMPs should be provided for a specific industry or activity.[136–137]

Cold climates may present special challenges. Road salt contains chlorides that could migrate through the porous pavement into groundwater. Snowplow blades could catch block edges of concrete pavers or other block installations, damaging surfaces and creating potholes. Sand cannot be used for snow and ice control on porous surfaces because it will plug the pores and reduce permeability.[138–140]

Permeable paving systems, especially those with porous surfaces, require routine maintenance to keep the pores clean to allow infiltration of stormwater. The frequency of cleaning is dependent on site-specific factors such as volume and quantity of runoff, vehicular traffic, neighboring sites, and climate. Often, cleaning of permeable paving systems is done by suction excavators. If maintenance is not carried out regularly, the porous pavements can begin to function more like impervious surfaces. Plastic grid systems are more popular with local government maintenance personnel owing to reduced maintenance requirements.[134,135]

It is difficult to compare the construction and maintenance costs between conventional impervious surfaces and permeable surfaces due to the variables such as lifespan, geographic location, type of permeable paving system, and other site-specific factors. Some estimates put the cost of permeable paving at about one-third more expensive than that of conventional impervious paving.[141] Cost of permeable paving will be reduced as its application increases in urban areas. Permeable paving systems have shown significant cost benefits after a life cycle assessment was performed, as the total weight of material needed for porous pavements is significantly reduced from impervious designs.[135,136,141]

10.7.15 Common Types of Permeable Paving

A wide variety of permeable pavers is available, each with its own advantages and disadvantages. Commonly used permeable paver systems include (1) porous asphalt, (2) pervious concrete, (3) permeable interlocking concrete/brick pavers, and (4) grid pavements. Examples of several permeable paver systems are shown in Figure 10.55. Permeable pavements are designed to permit the infiltration of surface water into the surrounding soils. They usually consist of a porous surface overlaying a bedding course and a uniformly graded stone reservoir or sand drainage system to temporarily retain surface runoff and infiltrate it into subsoil.[43] The design details of four common permeable pavers and their advantages and disadvantages are presented below.

Porous asphalt: Porous asphalt pavers are flexible pavements similar to standard asphalt that use bituminous to bind aggregate together. However, the percentage of sand and finer material in the asphalt mix is reduced or eliminated to increase voids between the aggregate in the pavement surface.[64] The increase in void ratio in porous asphalt pavements allows water to infiltrate through the pavement surface into the underlying reservoir or drainage base, then into the groundwater. The porous asphalt pavers are used for parking lots and other light-duty applications. A typical cross-section of porous asphalt pavement is shown in Figure 10.56. It consists of a porous asphalt surface layer, a top filter layer, a reservoir, a bottom filter (optional), filter fabric, and existing soil or subgrade material. A porous asphalt surface layer of 2–7 in. thick open-graded asphalt layer with approximately 15%–22% voids is recommended.[64,76,128] The advantages of porous asphalt pavements include lower cost and existing experience in materials and construction of this type of pavement. The disadvantages are the susceptibility of materials to water damage, short-term storage use only, and lower relative strength.

FIGURE 10.55 Example of permeable pavers: (a) porous asphalt, (b) pervious concrete, (c) permeable interlocking concrete/brick pavers, and (d) grid pavements. (Printed from Ref. 112)

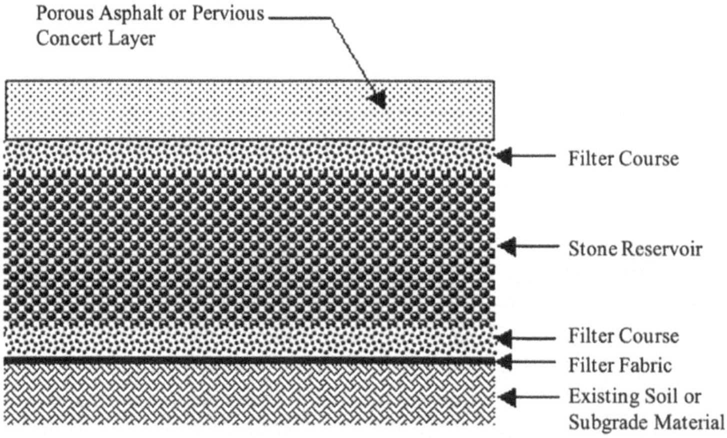

FIGURE 10.56 Typical cross-section of porous asphalt and porous concrete pavements.

Pervious concrete: Pervious concrete pavers are rigid pavements like conventional concrete. Cementitious material is used to bind the aggregate of pervious concrete together. Similar to the porous asphalt pavers, the percentage of sand and other fine materials is reduced, or entirely eliminated, in the concrete mix to increase the voids between the aggregates.[64] The increase in voids will result in a porous open-cell structure. Water passes through these pores readily to reach sublayers. Pervious concrete pavements are recommended for sidewalks, parking lots, and streets subject to light traffic. Their design is similar to porous asphalt pavers. A 4–8 in. pervious concrete surface layer with a void content of 15%–20% is recommended for general use.[64,76] Pervious concrete pavers may not require a reservoir layer to support the structural load, but this layer may be included to increase the storage or infiltration.[76] The typical cross-section of pervious concrete pavers is similar to that of porous asphalt, and is shown in Figure 10.56. The higher structural strength and availability of materials are advantages of these pavements. The disadvantages include a slow construction process, potential material issues, and a higher initial cost.

Permeable interlocking concrete/brick pavers: These pavements are constructed by installing a layer of open-graded aggregate and compacting it to provide a stable subsurface. The bedding is placed as a layer on top of a base course, and concrete or brick pavers are set on top of the surface layer. The concrete or brick pavers are solid, precast, manufactured modular impervious units. These modular units create joints that are filled with permeable fine materials.[64,76] These joints allow water to enter the open-graded base and reservoir layer, and eventually infiltrate into the subgrade soil. This type of paver has been installed in patios, walkways, driveways, and streets. The use of permeable interlocking pavers is acceptable to the EPA as an alternative to traditional stormwater pollution prevention BMPs such as grassy swales and infiltration trench systems.[43] A typical cross-section of a permeable interlocking concrete/brick pavement is shown in Figure 10.57. Some of the advantages of these pavers are ease of construction, maintenance and repair, and high surface infiltration options. The disadvantages include their higher cost and applicability for only lower-speed roadways.

Grid pavements: Grid pavements are made of concrete or plastic units with large surface openings. The openings are filled with permeable joint material such as sand or topsoil and grass.[64] Grid pavements have several applications, including landscaping, parking lots, and driveways with low

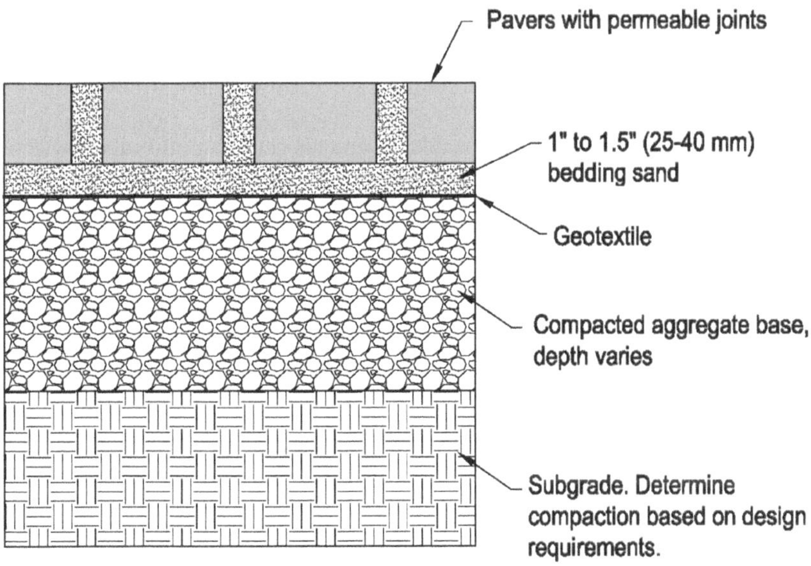

FIGURE 10.57 Typical cross-section of permeable interlocking concrete/brick pavers. (Printed from Ref. 64)

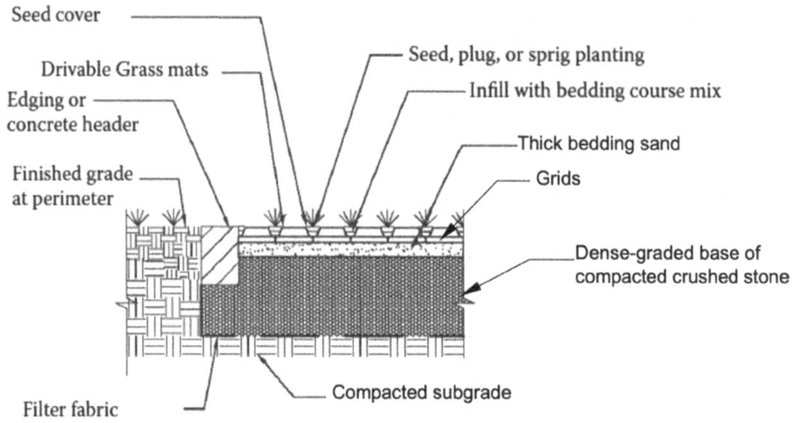

FIGURE 10.58 Typical full-depth cross-section of grid pavements. (Printed from Ref. 43)

traffic volume. A typical grid pavement consists of a compacted subgrade, a dense-graded base of compacted crushed stone, thick bedding sand, and grids. Typical full-depth cross-section of a grid pavement is shown in Figure 10.58. Advantages of grid pavements are the wide variety of products, high surface infiltration options, relatively inexpensive, and ease of maintenance and repair. The disadvantages include frequent maintenance of the pavement surface and their limited application to parking lots.

10.7.16 Basic Design Considerations of Permeable Paving

Permeable pavements reduce stormwater runoff and enhance water quality and groundwater recharge. However, they are typically designed to provide stormwater detention of smaller storms of a 2-year return period.[76] If the permeable paver is equipped with underground piping, the temporary storage volume of the reservoir may be increased to provide peak runoff reduction for storms with 1–2-year return periods.[128]

Permeable pavement systems are designed based on both hydraulic and structural design considerations. The thickness of the base and subbase layers is designed to retain stormwater and support the required traffic loads over the pavement. Some of the hydraulic and structural design considerations are presented below.

Hydraulic design considerations: Permeable pavements act as an infiltration system. As such, they are subject to the same feasibility constraints and hydraulic design considerations as most infiltration BMPs. Readers are referred to Refs. 29,64,76 for a detailed discussion on hydraulic design considerations. Some basic considerations are listed below.

1. Site location and condition of drainage area, pavement surface and subgrade slope, setbacks, and proximity to utilities
2. Soil type and characteristics such as soil permeability rate, groundwater depth, surface and subsurface detention capacity, and time to drain
3. Hydrologic conditions that include surface runoff, subgrade slope, and available minimum hydraulic head

Permeable paving is generally more cost-effective for drainage areas larger than 0.25 acres. However, permeable pavements are not recommended for drainage areas larger than 10 acres due to excessive sediment loading that may clog the voids of porous surface and sublayers.[128] The subgrades with a slope greater than 5% are less effective for infiltration and reduce the stormwater storage capacity of permeable pavers. The slope of subgrades should be as flat as possible to promote even distribution and infiltration of stormwater. For this reason, both lateral and longitudinal slopes should be less than 1%. A minimum surface slope of 1% should also be maintained for all permeable pavement types. The minimum slope ensures that the stormwater runoff is diverted to an alternative drainage system should the paver surface become completely clogged due to a lack of maintenance.[128,135]

Seepage from the reservoir layer of permeable pavers should not be allowed to the adjacent traditional pavement base layer, existing structure foundations, or an area where future development is planned. A minimum setback of 10 ft from structures and property lines must be maintained. Additionally, adequate waterproofing protection must be provided for foundations and basements. In places where the 10-ft setback is not possible, an impermeable liner may be placed along the sides of the permeable pavement and the adjacent foundation or basement wall to avoid harmful seepage.[76,135] Where the permeable pavement and the utility intersections coexist, some measures need to be incorporated into the pavement design to protect utility crossings. These measures may include (1) using a liner wrapping, (2) installing vertical liners against subgrade at penetrations, or (3) encasing the utility crossing in flowable concrete fill.

Analysis of soil characteristics to a depth of at least 3 ft below the proposed bottom of the stone reservoir is required.[128,135] This analysis should include information on the soil permeability rate, groundwater depth, surface and subsurface detention capacity, and time to drain. The underlying soil permeability rate determines whether an underdrain system is needed. Low permeability rates reduce infiltration volume and peak runoff. Typically, soil should have a permeability greater than 0.5 in./h; otherwise, underdrains may be required. Also, the bottom of the permeable pavement should be at least 2–3 ft above the seasonal high groundwater table to avoid ponding at the bottom of the pavement system.[43,64,76] Permeable pavements with low surface and subsurface detention capacity and long drainage time increase the risk of flooding. A conventional stormwater system captures and conveys the runoff when the reservoir underlying the porous surface is full. In a partial exfiltration system, a bypass piping system is installed to divert the runoff more than the infiltration capacity of the pavement system to a downstream conveyance system. A common bypass pipe configuration is shown in Figure 10.59.[112] The permeable pavement system is designed to be completely empty within a maximum drawdown time (time to drain). A stone reservoir has a maximum drawdown time of 48 h after the storm.[64,76,128]

The storage capacity of a permeable pavement is sized to hold the surface runoff from a storm until it naturally drains into the groundwater. A minimum hydraulic head (head from the pavement surface to the underdrain outlet) of 1–4 ft is recommended for permeable pavement to function

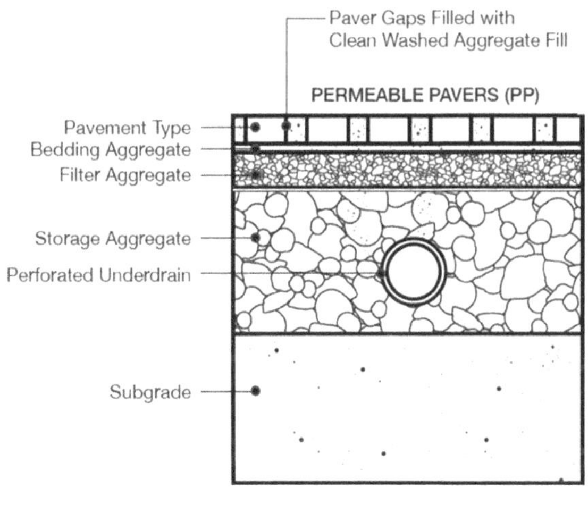

FIGURE 10.59 Typical permeable pavement with perforated underdrain. (Printed from Ref. 112)

properly. This value may change according to design factors, such as required storage depth and underdrain location.[64,76]

Structural design considerations: The structural design considerations for permeable pavements include traffic loading, in-situ soil strength, environmental elements, and bedding and reservoir layer design.[76] There are several guidelines for structural analysis and design of permeable pavements. Readers are referred to Refs. 136,137 for detailed design information.

10.7.17 Hydraulic Design of Permeable Pavements

Permeable pavements are sized to store the WQ_v, SWRv, or larger design storm volume in the reservoir layer. The infiltration rate is typically less than the flow rate through the pavement. For this reason, sufficient storage volume is required. The minimum depth of the reservoir to store the design storm can be determined from Eq. (10.48).[76]

$$d_p = \frac{\frac{(P \times R_v \times \text{CDA})}{A_p} - (K_{\text{sat}} \times t_f)}{\eta_r} \tag{10.48}$$

where

d_p = depth of the reservoir layer, m (ft)

P = depth of runoff from a specific storm, WQ*v*, or any other design storm, m (ft). The suggested volume of P is in the range of 0.50–1.25 ft.[43,112,128]

CDA = total contributing drainage area, including permeable pavement area, m^2 (ft^2)

R_v = runoff coefficient of the CDA. The accepted value of R_v is 0.95.

A_p = permeable pavement surface area, m^2 (ft^2)

K_{sat} = field-verified saturated hydraulic conductivity of subgrade soils, m/day (ft/day). If an impermeable liner is used in the design, then $K_{sat} = 0$

t_f = time to fill the reservoir layer (day), typically 2 h or 0.083 day

η_r = void ratio of the reservoir layer. For No. 57 stone, void ratio is 0.4

The minimum depth of the reservoir layer may need to be increased to meet pavement structural strength or larger storage requirements. If the subgrade is compacted to meet the structural design requirements of the pavement, K_{sat} must be determined from the measurement of subgrade soil that is subjected to the compaction requirements.[76,140]

The maximum allowable depth of the reservoir layer is constrained by the maximum allowable drain time. The maximum allowable depth d_{pmax} is calculated from Eq. (10.49).[76,140] The maximum allowable depth ensures that the captured stormwater volume will drain from the pavement during the drawdown time t_d.

$$d_{\text{pmax}} = \frac{(K_{\text{sat}} \times t_d)}{\eta_r} \tag{10.49}$$

where

d_{pmax} = maximum depth of the reservoir layer, m (ft)

t_d = drawdown time (days), typically 1–2 days

Other terms have been defined previously.

If the calculated depth of the reservoir layer d_p exceeds the maximum allowable depth d_{pmax}, or the saturated hydraulic conductivity for the subgrade infiltration rate is less than 0.5 in./h, then the pavement design may include underdrains. To calculate the drawdown time, the storage volume in the pavements must account for the underlying infiltration rate and outflow through the underdrains.

The hydrologic routing methods or modeling procedures are used for detention systems. The depth and head of the permeable pavement are adjusted for the porosity of the aggregate. The total storage volume S_v provided by the permeable pavers can be determined from Eq. (10.50).[76,140]

$$S_v = A_p\left[\left(d_p \times \eta_r\right) + K_{\text{sat}} \times t_f\right] \quad (10.50)$$

where

S_v = total storage volume, m³ (ft³)

All other parameters have been defined previously.

Example 10.53: Depth and Storage Volume of a Permeable Paver to Capture Stormwater

A permeable paver is designed to capture the surface runoff from a 1.5-acre parking lot. The depth of runoff to be collected is 1.5 in. The depth to groundwater table from the surface of the pavement is 5 ft. The field-verified saturated hydraulic conductivity of the subgrade is 0.5 in/h, and the void ratio for the reservoir layer is 0.4. Determine the depth and storage volume of the permeable pavement. Time to fill the reservoir layer is 2 h. The drawdown time is 48 h.

Solution

1. Calculate the depth of the reservoir layer
   a. Calculate the surface area of permeable pavement

      Surface area of the permeable pavement A_p is generally restricted to one-sixth of the total CDA[128]

$$A_p = \frac{1}{6}\text{CDA} = \frac{1}{6} \times 1.5 \text{ acre} = 0.25 \text{ acre}$$

   b. Calculate the minimum depth of the reservoir layer

      The minimum depth of the reservoir layer is calculated from Eq. (10.48)

$$d_p = \frac{\dfrac{\left(P \times R_v \times \text{CDA}\right)}{A_p} - \left(K_{\text{sat}} \times t_f\right)}{\eta_r}$$

$$d_p = \frac{\dfrac{(1.5 \text{ in.})(1 \text{ ft} / 12 \text{ in.}) \times 0.95 \times 1.5 \text{ acres}}{0.25 \text{ acres}} - \left(0.5 \text{ in./h} \times (1 \text{ ft} / 12 \text{ in.}) \times 2 \text{ h}\right)}{0.4}$$

$$= \frac{0.713 \text{ ft} - 0.083 \text{ ft}}{0.4} = 1.57 \text{ ft}$$

   c. Calculate the maximum depth of the reservoir layer

      The maximum depth of the reservoir layer is calculated from Eq. (10.49)

$$d_{\max} = \frac{\left(K_{\text{sat}} \times t_d\right)}{\eta_r}$$

      The maximum allowable drawdown time t_d is 48 h

$$d_{\max} = \frac{0.5 \text{ in./h} \times (1 \text{ ft} / 12 \text{ in.}) \times 48 \text{ h}}{0.4} = 5 \text{ ft}$$

Since the minimum and maximum calculated depths of the reservoir layers are 1.57 ft and 5 ft, respectively, select a 2 ft depth of the reservoir layer. The groundwater table is 5 ft below the surface of the pavement; therefore, a 2-ft deep reservoir layer satisfies the minimum 3 ft distance between the bottom of the reservoir layer to the groundwater table.

2. Calculate storage volume

The storage volume of the pavement system is calculated from Eq. (10.50)

$$S_v = A_p\left[(d_p \times \eta_r) + K_{sat} \times t_f\right]$$

$$= 0.25\text{ acre}\left(\frac{43{,}560\text{ ft}^2}{\text{acre}}\right)\left[(2\text{ ft} \times 0.4) + 0.5\text{ in./h} \times (1\text{ ft}/12\text{ in.}) \times 2\text{ h}\right]$$

$$= 10{,}890\text{ ft}^2 \times [0.8\text{ ft} + 0.083\text{ ft}]$$

$$= 9{,}616\text{ ft}^3$$

Example 10.54: Depth and Storage Volume of a Permeable Paver to Capture Water Quality Volume

A permeable paver is designed to provide water quality requirements for runoff resulting from a parking lot. The total parking lot area is 5 acres. The driveway and aisles have a surface area of 3 acres and are paved by conventional asphalt. The field investigation showed that the saturated hydraulic conductivity of the soil at the site is 0.5 in./h. The project site does not exhibit a high groundwater table. The void ratio for the reservoir layer is 0.35. Determine the depth and storage volume of the permeable pavement. The maximum allowable drawdown time at the project site is 24 h. The depth of runoff for WQ_v is 1.25, and the time to fill the reservoir layer is 2 h. The runoff coefficient from CDA is 0.95.

Solution

1. Calculate the depth of the reservoir layer
   a. Calculate the minimum depth of the reservoir

      The minimum depth of the reservoir is calculated from Eq. (10.48). The depth of runoff P for WQ_v is suggested between 0.5 and 1.25 in different sources.[43,112,128]

      In this example $P = 1.25$ in. and $t_f = 2$ h

      The surface area of the permeable pavement $A_p = 5$ acres − 3 acres = 2 acres, $R_v = 0.95$

$$d_p = \frac{\dfrac{(P \times R_v \times \text{CDA})}{A_p} - (K_{sat} \times t_f)}{\eta_r}$$

$$d_p = \frac{\dfrac{\left(1.25\text{ in.} \times \dfrac{1\text{ ft}}{12\text{ in.}}\right) \times 0.95 \times 5\text{ acres}}{2\text{ acres}} - (0.5\text{ in./h} \times (1\text{ ft}/12\text{ in.}) \times 2\text{ h})}{0.35}$$

$$= \frac{0.247\text{ ft} - 0.083\text{ ft}}{0.35} = 0.47\text{ ft}$$

b. Calculate the maximum depth of the reservoir layer

The maximum depth of the reservoir layer is calculated from Eq. (10.49) by considering the maximum allowable drawdown time t_d equal to 24 h.

$$d_{\max} = \frac{(K_{\text{sat}} \times t_d)}{\eta_r}$$

$$d_{\max} = \frac{0.5 \text{ in./h} \times (1 \text{ ft} / 12 \text{ in.}) \times 24 \text{ h}}{0.35} = 2.9 \text{ ft}$$

The depth of the permeable pavement can be selected between $d_p = 0.47$ ft and $d_{\max} = 2.9$ ft. Since there is no concern about the groundwater table, d_p will be selected based on structural and other requirements. In cold regions, for example, a minimum allowable reservoir depth of 18 in. is required.[137]

2. Calculate storage volume

The storage volume of the permeable pavement is calculated from Eq. (10.50)

$$S_v = A_p \left[(d_p \times \eta_r) + K_{\text{sat}} \times t_f\right]$$

$$= 2 \text{ acre}\left(\frac{43{,}560 \text{ ft}^2}{\text{acre}}\right)\left[\left(18 \text{ in.} \times \frac{1 \text{ ft}}{12 \text{ in.}} \times 0.35\right) + 0.5 \text{ in./h} \times \left(1 \text{ ft} / 12 \text{ in.}\right) \times 2 \text{ h}\right]$$

$$= 87{,}120 \text{ ft}^2 \times [0.525 \text{ ft} + 0.083 \text{ ft}]$$

$$= 52{,}998 \text{ ft}^3$$

Example 10.55: Depth and Storage Volume of a Permeable Paver to Capture Larger Storms

Determine the depth and storage volume of the permeable pavement described in Example 10.54. The permeable paver is designed to capture runoff resulting from a 10-year storm. The depth of runoff is 6 in.

Solution

1. Calculate the minimum and maximum depths of the reservoir layer

The minimum depth of the reservoir layer is calculated from Eq. (10.50)

$$d_p = \frac{\dfrac{6 \text{ in.} \times (1 \text{ ft} / 12 \text{ in.}) \times 0.95 \times 5 \text{ acres}}{2 \text{ acres}} - \left(0.5 \text{ in./h} \times (1 \text{ ft} / 12 \text{ in.}) \times 2 \text{ h}\right)}{0.35}$$

$$= \frac{1.188 \text{ ft} - 0.083 \text{ ft}}{0.35} = 3.16 \text{ ft}$$

The maximum depth of the reservoir from Example 10.54 step 2b = 2.9 ft

Since the minimum depth exceeds the maximum depth $d_{P\max}$, an underdrain is required.

2. Calculate the rate of outflow from the underdrain

The outflow rate from the underdrain is calculated from Eq. (10.51)[135]

$$q_u = K \times m \tag{10.51}$$

where

q_u = outflow through underdrain, m/h (ft/h)

K = coefficient of permeability for each 6-in. diameter underdrain pipe, m/h (ft/h). A conservative value of K is 100 ft/day (30.48 m/day) per pipe used for calculation of q_u

m = slope of underdrain pipe, m/m (ft/ft)

Install two underdrain pipes, each 6 in. diameter, and at a slope m of 2%

$$q_u = \left(100\,\frac{\text{ft}}{\text{day}} \times \frac{1\text{ day}}{24\text{ h}}\right)(0.02\text{ ft/ft}) \times 2\text{ underdrain pipes}$$

$$= 0.083\left(\frac{\text{ft}}{\text{h}}\right) \times 2 = 0.17\text{ ft/h}$$

3. Calculate the depth of the reservoir layer with underdrain, d_p

The minimum depth of the reservoir with underdrain d'_p is calculated from Eq. (10.52)

$$\begin{aligned} d'_p &= d_p - \frac{q_u \times t_f}{\eta_r} \\ &= 3.16\text{ ft} - \frac{0.17\text{ ft/h} \times 2\,h}{0.35} \\ &= 3.16\text{ ft} - 0.97\text{ ft} \\ &= 2.2\text{ ft} \end{aligned} \tag{10.52}$$

The maximum allowable depth of the reservoir layer $d'_{p\max}$ is estimated from Eq. (10.53)

$$\begin{aligned} d'_{p\max} &= d_{p\max} + \frac{q_u \times t_d}{\eta_r} \\ &= 2.9\text{ ft} + \frac{0.17\text{ ft/h} \times 24\text{ h}}{0.35} \\ &= 2.9\text{ ft} + 11.7\text{ ft} \\ &= 14.6\text{ ft} \end{aligned} \tag{10.53}$$

Select the depth of the reservoir layer = 2.2 ft.

4. Calculate the storage volume of the reservoir layer with two 6-in. underdrains

Apply Eq. (10.50)

$$\begin{aligned} S_v &= A_p\left[\left(d_p \times \eta_r\right) + K_{\text{sat}} \times t_f\right] \\ &= 2\text{ acre}\left(\frac{43{,}560\text{ ft}^2}{\text{acre}}\right)[(2.2\text{ ft} \times 0.35) + 0.5\text{ in/h} \times (1\text{ ft}/12\text{ in.}) \times 2\text{ h}] \\ &= 87{,}120\text{ ft}^2 \times [0.77\text{ ft} + 0.083\text{ ft}] \\ &= 74{,}313\text{ ft}^3 \end{aligned}$$

Note: The procedure used in this example can also be applied to estimate the depth of the reservoir layer of permeable pavers with underdrains. However, for a more accurate estimation of the depth of the reservoir, the application of hydrologic models is preferred to route stormwater through the reservoir layer.

7.10.18 Vegetated Bioswale

Vegetated bioswales are broadly described as surface depressions that collect and convey stormwater runoff from roadways, driveways, rooftops, parking lots, and impervious surfaces.[142] However, when applied as a (BMP, an engineered vegetated swale functions beyond simple collection and conveyance. It also improves the quality of runoff through sedimentation, filtration, bio-reduction, and uptake. Vegetated bioswales function by minimizing flow velocity and inducing ponding behind strategically placed check dams.[128] While infiltration of some runoff associated with ponding can attenuate peak runoff rate, this attenuation is minimal, and cannot be considered an effective flood control strategy.

The vegetated swales are of two types: *grassed swales* and *water quality swales*. The *grassed swales* are also termed *dry swales*. They create ponding behind the check dams, thus slowing the velocity of runoff in the channel. The controlled velocity of the swale permits filtration of runoff pollutants by the dense vegetation lining or xeriscaping (landscaping or gardening).[134] Ponding increases the hydraulic residence time within the swale, thus providing an increased opportunity for gravitational settling of pollutants.[143] The *water quality swales, or wet swales*, are like a linear wetland. Their underlying soils, unlike dry swales, are comprised of a very specific mixture to permit controlled infiltration, as well as the growth of wetland vegetation. While water quality swales are capable of phosphorus removal beyond that of dry swales, their inability to attenuate peak runoff limits their use in most cases. For this reason, the grassed or dry swales are preferred over vegetated swales or bioswales.[128] The design and makeup of a vegetated swale can be influenced by many different variables, including climate, rainfall patterns, site size, budget, and vegetation suitability.

Bioswales work to remove pollutants through vegetation and the soil. As the stormwater runoff flows through the bioswale, the pollutants are captured and settled by the vegetation, leaves, and stems of the plants. The pollutants then enter the soil, where they are decomposed or broken down by bacteria in healthy soil.[143] There are several classes of water pollutants that may be collected or arrested with bioswales. These are silt, inorganic and organic chemicals, and pathogens. Other pollutants are nutrients such as phosphates and nitrates; metallic compounds such as mercury, lead, chromium, cadmium, and other heavy metals; and polycyclic aromatic hydrocarbons that growing plants consume or uptake in plant tissues.[143,144]

The inherent linear or meandering orientation of vegetated swales makes it an attractive option for the treatment and conveyance of highway runoff. They can be created around the edges of parking lots to capture and treat stormwater runoff before releasing it to the watersheds or storm sewer. They can also be integrated into road medians, curb cutouts, sidewalks, or any public space.[145] Substantial automotive pollution settles on the pavement and is flushed by the first instance of rain, known as the *first flush*. The subject of the first flush is discussed in Chapter 11. On site, these pollutants then enter the soil, where they are broken down by bacteria in healthy soil.[142] The longer the runoff stays within the bioswale, the better the pollutant removal outcome. It is also beneficial in removing standing ponds that could potentially attract mosquitoes. Bioswales can also be designed to be aesthetically pleasing and attract animals and create habitats, and can also be beneficial for groundwater recharge. Planning for maintenance is an important step, which can include the introduction of filters or gravel to prevent clogging. Annual maintenance through soil testing, visual inspection, and mechanical testing is also crucial to the health of a bioswale.

In addition to the CDA and its impervious cover, a number of site-specific constraints must be considered in selecting and designing the vegetated swales. These constraints are discussed below.[128]

Drainage area: The minimum drainage area contributing to a vegetated swale is not restricted. They are particularly well-suited for small drainage areas. The maximum size of the CDA is limited to the peak storm runoff handled by the vegetated bioswale, which is dependent upon the channel cross-section, geometry, roughness, longitudinal slope, and ultimately the flow velocity in the channel. For this reason, the maximum size of the CDA is limited to the peak flow carrying capacity of the vegetated bioswale.

Site slope: The vegetated swales function best in relatively flat or gently sloped topography. The maximum permissible slope of a grassed swale is 6%.

Site soil: The vegetated swales are best suited to soils of moderate permeability. However, a permeability test is strongly recommended. Because ponding is induced within the swale, the site soil should permit good infiltration. The recommended period for complete drainage by infiltration of the swale is less than 72 h. The recommended minimum soil infiltration rate for a grassed swale is 0.27 in./h. The soil types and textures classified in groups A and B are best suited for vegetated swales. The soil texture and infiltration rates of these and other soil groups are listed in Table 10.10.

Depth of water table: The grassed swales infiltrate surface runoff into the ground. The infiltrated surface runoff may potentially carry a significant pollution load. Therefore, grassed swales should not be used on sites exhibiting a seasonally high-water table of less than two feet below the proposed swale bottom.

Existing utilities: When possible, swales should not cross existing utility right-of-way or easements. When this situation is unavoidable, permission from utility owners must be obtained to build or relocate the utilities.

Swale geometry: The parabolic and triangular cross-sections of the channel are undesirable because of constructability and concentration of low flows at the bottom. Also, rectangular cross-sections have problems with the stability of their side slopes and maintenance. Therefore, trapezoidal channel sections with side slopes no steeper than 3H:1V are best suited for vegetated swales.

Bottom width: Channel bottom widths less than 2 ft are essentially non-constructable. Conversely, a bottom width greater than 6 ft will concentrate small flows, thereby reducing the pollutant removal ability of the swale. Channel width in the range of 2–6 ft is acceptable.

Channel depth: The overall swale depth should permit conveyance of a 10-year stormwater runoff event while maintaining a minimum of 6 in. of freeboard. Additionally, the channel depth should be such that the check dam height does not exceed one-half of the total channel depth. The swale should be designed such that the WQ_v flows at a depth approximately equal to the grass depth. For most applications, this will be around 4 in.

Longitudinal slope: The slope of the grassed swale should be as flat as practically possible for the given site topography. The site-specific allowable longitudinal slope will ultimately be governed by the desired flow depth and velocity. Generally, the accepted minimum constructable slope is 0.75%, and the maximum slope should not exceed 6%.

Flow velocity: The flow velocity should be as low as possible to achieve maximum pollutant removal. Additionally, the swale must be designed such that larger runoff events do not result in the resuspension of previously settled sediments. The recommended 2-year and 10-year design flow permissible velocities are 2 and 4 ft/s, respectively.

Shear stress: In addition to considering the channel velocity, the shear stress should also be examined. The permissible shear stress depends upon different classes of vegetated linings. The permissible shear stress may vary from 0.35 lb/ft^2 (1.71 kg/m^2) to 3.7 lb/ft^2 (18 kg/m^2).

Swale length: There is no restriction on the length of the grass swale. The length should be sized together with the channel cross-sectional area and check dam height to provide the desired water quality storage volume.

Discharge flows: The channel receiving the discharge flows should be checked for overtopping during the conveyance of the 10-year producing event, and for erosive potential under a 2-year event.

The pictorial view of a vegetated swale is shown in Figure 10.60.

(b)

FIGURE 10.60 The pictorial view of vegetated bioswales: (a) construction details of a vegetated bioswale and (b) bioswale in a housing development. The foreground one is under construction, while the background one is well established swale. (Printed from U.S. EPA website.)

Example 10.56: Design of Vegetated Bioswale

Construction of approximately 1,800 linear feet of secondary subdivision roadway is carried out in a city. Topography is such that ½ in. of runoff from impervious area is collected in curb and gutter and conveyed to curb inlets in a sump near the mid-station of the road. The runoff is then discharged into a proposed vegetated bioswale. The total project site, including right-of-way and all permanent easements, consists of 5.27 acres of unimproved grass-covered watershed. The average slope is 2%. Post-development watershed is expected to have 1.03 acres of impervious cover. Site constraints limit the swale length to 275 ft. The rational runoff coefficient $C=0.9$. The 10-year runoff event generates a flow of 5.9 cfs.

Solution

1. Determine the required WQ_v

   The required WQ_v is calculated from Eq. (10.35)

$$WQ_v = \frac{(IA \times \frac{1}{2}\ \text{in.})}{12\ \text{in./ft}}$$

where IA is impervious area, acres (ac)

$$WQ_v = \frac{(1.03\ \text{acres} \times 43{,}560\ \text{ft}^3/\text{acres} \times \frac{1}{2}\ \text{in.})}{12\ \text{in./ft}} = 1{,}870\ \text{ft}^3$$

The vegetated swale must be sized to provide ponding for the computed WQ_v. This ponding occurs behind check dams and longitudinal spacing

2. Determine the cross-sectional area A of the trapezoidal channel with a side slope of 3H:1V

   Select the height of check dam = 18 in. (1.5 ft). The cross-sectional area of channel is expressed by Eq. (10.54)

$$A = (W_b \times 1.5\ \text{ft}) + (2 \times \tfrac{1}{2} \times 1.5\ \text{ft} \times 3 \times 1.5\ \text{ft}) \quad (10.54)$$

   where W_b = channel base width, ft

3. Determine the average cross-sectional area of the channel
   a. Develop a relationship between the average channel cross-sectional area and channel base width by the average end area method.

      Ponding in the swale will occur behind the 18 in. high check dam. Because the cross-sectional area and configuration of the channel remain constant throughout the length of the channel, the total volume of water detained throughout the swale is estimated by the average end area method. The detained volume of water is simply the product of the average of the wet cross-sectional area of the upstream and downstream ends of the channel and the length of the channel. The depth of water at the upstream end is zero, and the downstream depth is 1.5 ft (the height of the dam). Because the ponded upstream depth is zero, the effective cross-sectional area of the swale is one-half of the area A in Eq. (10.35). The average cross-sectional area A_{avg} is expressed by Eq. (10.55).

$$A_{ave} = \frac{(W_b \times 1.5\ \text{ft}) + (2 \times \frac{1}{2} \times 1.5\ \text{ft} \times 3 \times 1.5\ \text{ft})}{2} \quad (10.55)$$

      where

      A_{avg} = average channel cross-sectional area, ft²

      W_b = has been defined previously

   b. Calculate the average channel cross-sectional area from the required WQ_v and length of the channel

      The average cross-sectional area of the channel is calculated from Eq. (10.56)

$$A_{avg} = \frac{WQ_v}{L} \quad (10.56)$$

      where

      L = length of the channel, ft

      A_{avg} has been defined previously

$$A_{ave} = \frac{1{,}870\ \text{ft}^3}{275\ \text{ft}} = 6.80\ \text{ft}^2$$

4. Determine the channel base width W_b

   Rearrange Eq. (10.55) in terms of channel base width W_b to obtain Eq. (10.57)

$$W_b = \frac{[2A_{\text{ave}} - (1.5\text{ ft} \times 3 \times 1.5\text{ ft})]}{1.5\text{ ft}} \tag{10.57}$$

   The required channel base width is computed from Eq. (10.57)

$$W_b = \frac{[2 \times 6.8\text{ ft}^2 - (1.5\text{ ft} \times 3 \times 1.5\text{ ft})]}{1.5\text{ ft}} = 4.56\text{ ft}$$

   To address any underestimation in storage volume arising from the average end computation, the channel base width is increased to 5 ft.

5. Establish a storage–volume relationship for the swale

   The 10-year flood peak, Q_{10}, is selected as the design discharge for establishing the conveyance properties of the channel. A minimum freeboard of 6 in. is selected. The presence of check dams in the swale introduces difficulty in modeling flow through the channel. To determine the required channel depth, it is assumed that the swale acts as a linear detention facility, and storage-indication routing is employed to establish the maximum water surface elevation under a 10-year runoff condition. This approach yields accurate results but is computationally intensive. This approach with storage-indication routing has been presented earlier in Example 10.33 with check dams. Readers may refer to this example to review this approach. A computationally simpler approach is applied here to consider that the water is ponded in the swale behind the 18-in.-high check dam, and the swale behaves much like a detention facility. The flow occurs over the weir of the check dam.

   The storage volume V is computed based on the channel geometry, and is expressed by Eq. (10.58)

$$V = [(W_b \times d) + (2 \times \tfrac{1}{2} \times d \times z \times d)] \times (L/2) \tag{10.58}$$

   where

   V = ponded volume, ft^3
   D = ponded depth, ft
   z = channel side slope (3H:1V)
   L = channel length, ft
   W_b has been defined previously

   Employing the previously established channel parameters in Eq. (10.58), the ponded volume is computed solely as a function of ponded depth, expressed by Eq. (10.59)

$$V = [(5 \times d) + (2 \times \tfrac{1}{2} \times d \times 3 \times d)] \times (275/2) \tag{10.59}$$

   By rearranging Eq. (10.59), the generalized ponded volume Eq. (10.60) is obtained

$$V = 412.5d^2 + 687.5d \tag{10.60}$$

6. Determine the ponded depth to hold WQ_v in the swale behind the check dam

$$1{,}870\text{ ft}^3 = 412.5d^2 + 687.5d$$

Solve for d

$$d = 1.453 \text{ ft} \quad \text{or} \quad 1.4 \text{ ft}$$

The approximate WQ_v is contained slightly below the water depth of 1.5 ft, equaling the check dam height. No flow will occur below the check dam crest elevation.

7. Construct the stage-discharge relationship

The check dams in the channel function as broad-crested weirs. At a depth of 18 in., the weir length is calculated from Eq. (10.61).

$$L = W_b + 2 \times d \times z \tag{10.61}$$

All terms in this equation have been defined previously

The length of the check dam weir crest is calculated below

$$L = 5 \text{ ft} + 2 \times 1.5 \text{ ft} \times 3$$

$$L = 14 \text{ ft}$$

8. Determine the head over the broad-crested weir at the 10-year peak stormwater runoff

The discharge over the broad-crested weir is a function of the head acting over the weir crest. The broad-crested weir has been presented in Chapter 7. Equation (7.13a) expresses the discharge and head over the weir relationship. The modified form of this equation, as applied to a check dam, is expressed by Eq. (10.62).

$$Q = C_W L \sqrt{g} y_c^{3/2}$$

$$Q = C_W \sqrt{g} L y_c^{3/2} \tag{10.62}$$

$$Q = C'_W L h^{3/2}$$

where

Q=discharge over the weir, ft^3/s

h=hydraulic head acting over the weir crest, ft

C'_W=weir coefficient. The value of C'_W applied in this example is 3.0

Other terms have been defined previously.

Calculate the hydraulic depth over the weir at 10-year peak stormwater runoff Q_{10} of 5.9 cfs

$$5.9 \text{ ft}^3 = 3.0 \times 14 \text{ ft} \times h^{3/2}$$

$$h = (0.14)^{2/3} = 0.27 \text{ ft}$$

9. Compute the depth and dimensions of the channel

$$\text{The depth of the channeld}$$

$$= \text{height of the check dam} + \text{head over weir} + \text{minimum freeboard}$$

$$= 1.50 \text{ ft} + 0.27 \text{ ft} + 0.50 \text{ ft} = 2.27 \text{ ft}$$

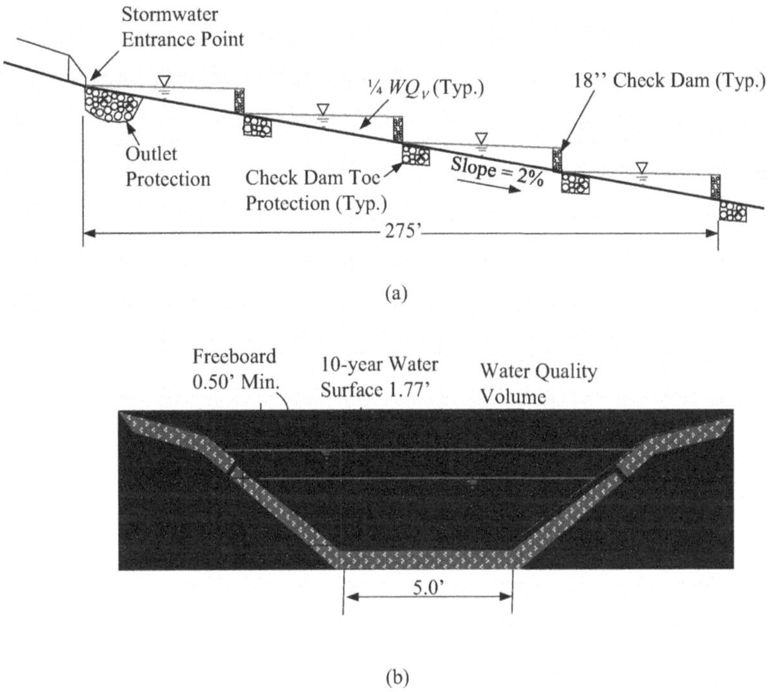

FIGURE 10.61 Design details of the vegetated swale channel: (a) longitudinal profile, and (b) cross-sectional view (Example 10.56).

10. Draw the design details of the vegetated swale

    The design details of the vegetated swale are shown in Figure 10.61. Since there are four stages between the check dams, storage between each dam is $\frac{1}{4}WQ_v$ or $1{,}870\,ft^3/4 = 468\,ft^3$

10.7.19 Wetlands

Wetlands are an essential part of nature's stormwater management system. The objectives of stormwater management are to provide surface drainage, control floods and erosion, reduce pollutants in runoff, and provide aesthetic amenities, including open space, recreation, and waterfront property. The wetlands are totally or intermittently inundated land areas in which the water table may reach above the ground. When designed as a multi-stage, multi-function facility, they provide surface drainage, attenuate stormwater volumes, reduce peak discharge and peak runoff rates, reduce erosion, and encourage sedimentation of pollutants in runoff. They also provide flood control by utilizing flood storage above the level of the wetland surface, and temporarily store runoff in relatively shallow pools that support conditions suitable for the growth of wetland plants. Wetlands maximize the removal of conventional pollutants such as BODs, TSS, from stormwater runoff through wetland vegetation uptake, retention, and settling. Heavy metals are also removed to a significant extent, but the most common application of wetlands is the polishing of secondary effluent. Biomass typically contains 2% nitrogen and 0.4% phosphorus. Harvesting of emergent vegetation is labor-intensive, and only a small portion of nutrients can be removed. However, water hyacinth and other floating wetland plants sequester significant amounts of nutrients and heavy metals, and their harvest is relatively less labor-intensive.[146,147] The accumulation of nutrients and organics in a wetland occurs from the growth of vegetation, death or dieback, decay, and finally soil accretion, that is, gradual deposition through the operation of natural causes. A wide variety of emergent vegetation and wildlife, from fish to waterfowl, thrive in the wetland ecosystem. The wetlands are an important sink for nutrients and other contaminants. Plants also translocate oxygen from the leaves to the root zones. Submerged portions of vegetation provide surface area for growth of attached aerobic and facultative bacteria that demineralize the wastes for filtration and adsorption of constituents into the plant tissue. Also, growth of algae is restricted because the penetration of sunlight is greatly reduced.[147,148] Though little information is available, maintenance is important because accumulation of organic matter and sediment can alter the pollutant removal effectiveness of the system by decreasing storage volume and changing the substrate. Likely maintenance will include removal of sediment accumulation, conveyance system and discharge structure repair, and vegetation harvesting or burning.

The design and performance standards used for wetlands to treat the stormwater are presented in this section. The stormwater wetlands are of two types: *natural wetlands* and *constructed wetlands.* Both wetlands have been used for treatment, reduction of peak runoff rates, and flood control. Both wetlands are described below.[147,148]

Natural stormwater wetlands*:* The natural stormwater wetlands are considered a part of the receiving waters. Consequently, discharge into natural wetlands is subject to NPDES permit requirements, and the effluent from secondary or advanced wastewater treatment facilities is allowed (see Section 6.3.3 for NPDES permitting program). The principal objective for effluent discharge is the enhancement of the existing habitat.[149,150] The pictorial view of a natural wetland is shown in Figure 10.62.

Constructed stormwater wetlands: Constructed stormwater wetlands offer all the treatment capabilities of natural wetlands but without the NPDES permit requirement. Also, the constructed wetlands are not restricted to the special requirements on the influent quality. For these reasons, constructed wetlands offer more flexible operation and perform more reliably than natural wetlands.[149,150] In constructed wetlands, stormwater runoff and drainage from different pollution sources such as landfills, mines, industrial zones, and construction sites have been treated successfully. Like extended dry detention basins and wet basins, constructed stormwater wetlands must be used with other BMPs, such as sediment forebays. A sediment forebay is a pool and settling basin constructed at the incoming discharge points of a stormwater BMP. Pretreatment removes heavy sediment loads and other pollutants, such as hydrocarbons, that can damage the wetland. Pretreatment also attenuates stormwater volumes and peak discharge rates to maintain the wetland hydro-period and to reduce scour and erosion. Erosion and sediment control practices during construction are essential to prevent sedimentation of the wetland.

(a) (b)

FIGURE 10.62 Natural wetlands: (a) protection and restoration on wetland, and (b) protected wetland. (Printed from U.S. EPA website.)

Two types of constructed wetlands are developed for stormwater and wastewater treatment: (1) free water surface (FWS) systems, and (2) subsurface flow (SF) systems. The general description is given below:

1. **FWS wetlands:** The FWS wetlands typically consists of (a) a basin or channel with some type of barrier to prevent seepage, (b) soil or base media to support the root system of emergent vegetation, (c) inlet and outlet structures, and (d) water through the system is at a relatively shallow depth and water surface is exposed to the atmosphere.[151,152] The velocity is low, and the flow path is horizontal. The flow is applied continuously, and treatment occurs as the water flows slowly through the support media, roots, and stem of emergent vegetation. The FWS wetlands may be designed to create new wildlife habitat and enhance the performance of existing nearby natural wetlands.[12,150,153,154] The schematic flow diagram is shown in Figure 10.63a.
2. **SF wetlands:** The SF wetlands also have four components, as in FWS wetlands.[154] The major difference is no FWS, and the basin or channel is filled with rock or gravel. The medium supports the root system of emergent vegetation and provides horizontal flow through the pores (Figure 10.63b).[12,150,154] The Typical depth of the bed is 0.6m (2ft). Deeper beds provide better support to the roots zone and prevent beds from freezing.

Basic design considerations of wetlands: The natural wetlands present a very sensitive ecosystem, and any modification to the existing wetlands to improve their treatment and flood control capabilities may be very disruptive to the natural ecosystem. For this reason, any attempt to modify the

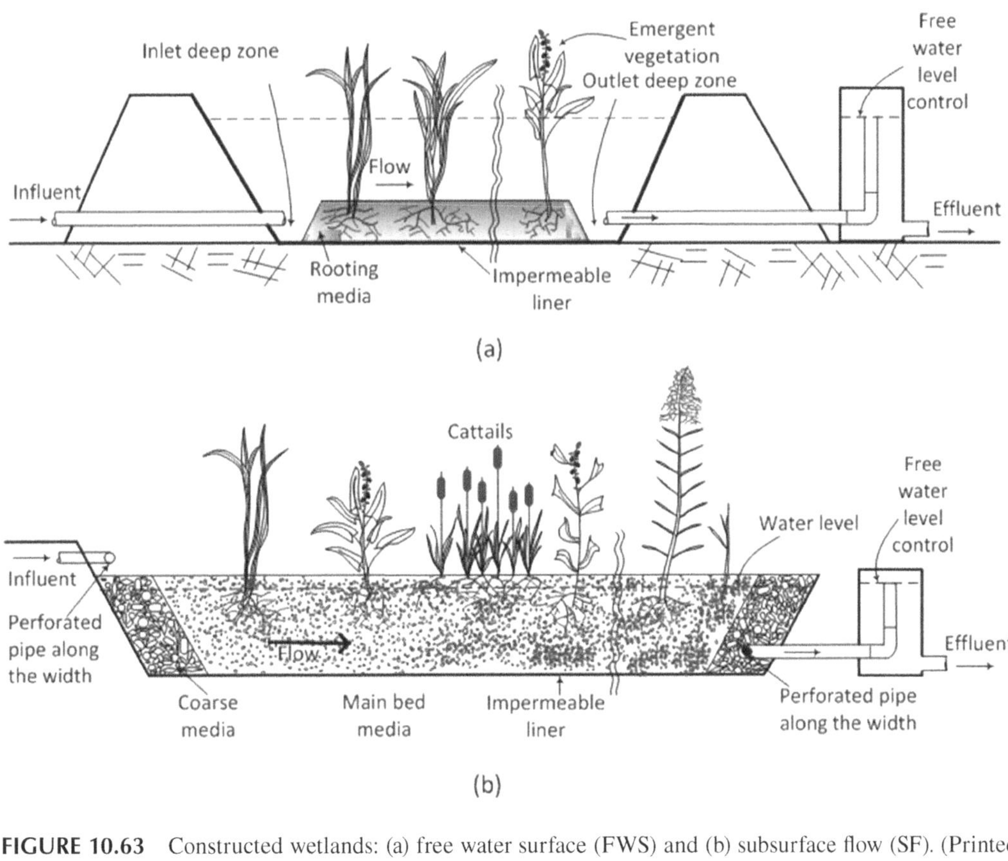

FIGURE 10.63 Constructed wetlands: (a) free water surface (FWS) and (b) subsurface flow (SF). (Printed from Ref. 150 with permission of CRC Press, Taylor and Francis Group.)

natural wetlands is not desired. Therefore, the basic design considerations apply only to constructed wetlands and are presented below.

The basic design parameters for any type of constructed wetland are (1) site selection, (2) storage volume, (3) emergent plants, (4) physical facilities, (5) hydraulic and physical factors, (6) loading factors, and (7) performance.

Site selection: Constructed wetlands can be developed in most locations. Compacted sandy clay and silty clay loams are suitable soils.[12,149,150]

Storage volume: The basic design parameter for any type of constructed wetland is the total storage volume within its various zones to maintain its permanent pool level. This storage volume must be equal to the design runoff volume. When used for peak protection, wetlands are normally designed to control the 2-year storm.[147,148] If the soil at the surface of a wetland site is not sufficiently impermeable to prevent excessive seepage, construction of an impermeable liner or other soil modifications will be necessary.

Emergent plants: The character, diversity, and hardiness of the wetland vegetation must be sufficient to provide adequate pollutant removal. Selected vegetation must be non-invasive and based on the anticipated water depth within the wetlands. Natural wetlands have a wide diversity of plant life. Such diversity is not warranted in constructed wetlands, since eventually, only a few plants, such as *cattail* alone or in combination with either *reeds* or *bulrushes*, will dominate because of high nutrient levels in the wastewater. The most frequently found emergent plants in wetlands are Cattail (*Typha* spp.), Common reed (*Phragmites communis*), Rush (*Juncus* spp.), and Bulrush (*Scirpus* spp.).[12,150,153]

Physical facilities: Many factors are considered in the selection of a proper system configuration for optimizing the performance: aspect ratio, compartmentalization, and location of alternate and multiple influent sites. The aspect ratio (length/width) may range from 4:1 to 10:1, although a ratio of 1:1 may also be acceptable.

Hydraulic and physical factors: Among the hydrologic and physical factors are precipitation, evapotranspiration, infiltration, water depth, hydraulic loading, flow velocity, and detention time. These factors affect the removal of organics, TSS, nutrients, and trace metals.[147,149,150] The water balance in a constructed wetland is performed using Eq. (10.63a)

$$dV / dt = Q_i - Q_e + P - \mathrm{ET} + Q_s - Q_g \tag{10.63a}$$

where

dV/dt = rate of change in volume with time, m³/day (ft³/day)
Q_i = influent wastewater flow, m³/day (ft³/day)
Q_e = effluent flow, m³/day (ft³/day)
P = precipitation, m³/day (ft³/day)
ET = evapotranspiration, m³/day (ft³/day)
Q_s = surface drainage inflow, m³/day (ft³/day)
Q_g = infiltration, m³/day (ft³/day)

The nominal and actual detention times and water velocity through constructed wetlands are given by Eqs. (10.63b)–(10.63e).[150,154]

$$t_n = \frac{(V_{\text{nominal}})}{Q} = \frac{(L \times w \times h)}{Q} \tag{10.63b}$$

$$t_a = \frac{(V_{\text{active}})}{Q} = \frac{(e \times V_{\text{nominal}})}{Q} \tag{10.63c}$$

$$u = \frac{Q}{h \times w} \tag{10.63d}$$

$$v = \frac{Q}{e \times h \times w} \tag{10.63e}$$

where

t_n = nominal detention time, day

t_a = actual detention time, day

$V_{nominal}$ = nominal volume, m^3. It is based on average length, width, and depth

V_{active} = active volume, m^3. It is the available void volume for water flow. In FWS wetlands, the active volume is obtained after subtracting the volume occupied by vegetative stems and litter. In SF wetlands, the active volume is the void space through the stone and gravel base

Q = flow rate, m^3/day

u = superficial water velocity, m/day

v = actual water velocity, m/day

e = wetland bed porosity. In FWS wetlands, e = 0.95 for cattails and other emergent vegetation. In SF wetlands, the porosity is respectively 0.18–0.35 for coarse to fine gravel

h = mean water depth, m. In SF wetlands, h is within the support media. In FWS wetlands, h is the free water depth, m. The free water depth is normally less than 0.5 m (1.6 ft). As water depth increases to 0.7 – 1.0 m (2.3 – 3.3 ft), floating plants replace the emergent plants.

L and w = length and width of the constructed wetlands, m

There is certain ambiguity about Q, e, and h in large wetlands. There may be a difference in influent and effluent flows due to water gain or loss. The porosity e = 0.95 in FWS wetlands is a gross assumption and may vary seasonally. This means that the water depth in a large wetland may vary greatly due to uneven bed and accumulation of solids in the bed.

The cross-sectional area of a bed is calculated from Darcy's law (Eq. 10.63f).

$$Q = K \times A_c \times \frac{\Delta h}{\Delta L} \tag{10.63f}$$

where

K = hydraulic conductivity of the bed, m/s

A_c = cross-sectional area of the bed (width × depth), m^2

$\frac{\Delta h}{\Delta L}$ = slope of the bed

In a wetland, the slope of the bed, FWS, and hydraulic gradient are practically the same. In an FWS system that has an overflow weir to control the water depth, the slope of the hydraulic gradient is normally less than 0.1%. In a gravel medium, the hydraulic conductivity is 10^{-3}–10^{-2} m/s (3.9×10^{-2}–3.9×10^{-1} in./s) and the slope of the bed is 4%–5% or more.

Loading factors: The hydraulic loading rate is the water flow rate per unit surface area of the wetland (m^3/m^2.day). Some wetlands are operated with intermittent flow. The hydraulic loading rate for FWS systems is closely tied to the hydrologic factors and conditions specific to the site. Hydraulic loading of 100–150 m^3/ha.day (10,700 – 15,000 gallon/ha.day) or less provides good nutrient removal. High organic loading, if not properly distributed, may cause anaerobic conditions, and plants may die. The organic loading also has an impact on nitrification and denitrification. Also, overloading reduces the oxygen transfer ability of the emergent plants. The maximum organic loading rate for both types of system should not exceed 112 kg BOD_5/ha.day.[147,149,150]

TSS removal: Colloidal and suspended solids are effectively removed in the constructed wetlands. Most of the removal occurs within a few meters beyond the inlet zone. The major removal mechanisms are sedimentation, straining, and impaction. Proper distribution of solids, low velocities, even cross-sectional loadings, and uniform flow near the inlet zone enhance the removal of solids. The overall TSS removal may be 50%–80%.[151,152,154–157] Kedlec and Wallace proposed a simplified equation for TSS removal in constructed wetlands. The relationship is given by Eq. (10.64a).[154]

$$\frac{C_e - C^*}{C_o - C^*} = \exp[-wt_n / h] \quad (10.64a)$$

where

C_e = effluent TSS concentration, mg/L

C_o = influent TSS concentration, mg/L

C^* = background TSS concentration, mg/L. The values of C^* may range from 2 to 3 mg/L

w = terminal velocity of the particles, m/day. The terminal velocity of the smallest particles removed is the hydraulic loading rate (m^3/m^2.day)

t_n = nominal detention time (V/Q), day

h = mean water depth, m

BOD removal: The BOD_5 removal in constructed wetlands may vary from 60% to 85% depending upon the influent BOD_5. Removal rate increases with higher influent concentration. Several relationships have been developed to estimate the effluent BOD_5 concentration. One such relationship is given by Eq. (10.64b).[154,158–161]

$$\frac{C_e - C^*}{C_o - C^*} = \frac{1}{(1 + k / pq)^p} \quad (10.64b)$$

where

C_e = effluent BOD_5 concentration, mg/L

C_o = influent BOD_5 concentration, mg/L

k = apparent *tanks-in-series* (TIS) rate constant, m/year[154]

p = apparent number of TIS

q = hydraulic loading rate, m/yr

C^* = background BOD_5 concentration, mg/L. The value of C^* may range from 2 to 10 mg/L, depending upon the decaying vegetation.

In SF wetlands, the major source of oxygen is that transmitted by the vegetation to the root zone. There is very little direct atmospheric reaeration because the water surface remains below the surface of the media (soil, gravel, and rocks).[148–152] The plug-flow first-order kinetic equations with slight modifications have been applied for FWS and SF wetlands. These equations for FWS and SF wetlands are given by Eqs. (10.64c) and (10.64d), respectively.[147,148]

$$\frac{C_e}{C_o} = \exp[-\ K_T t] \quad \text{(FWS wetlands)} \quad (10.64c)$$

$$\frac{C_e}{C_o} = \exp[-A_s K_t he] \quad \text{(SF wetlands)} \quad (10.64d)$$

where

K_T = reaction rate constant for FWS wetlands, day^{-1}

t = hydraulic retention time, day

K_t = reaction rate constant for SF wetlands, day^{-1}
h = depth of submergence in SF wetland, m; and depth of FWS in FWS wetland, m
e = porosity of the bed
A_s = surface area of the wetland, m^2
Q = flow through the system, m^3/day
C_e and C_o have been defined previously.

Nitrogen removal: Nitrogen removal is significant in FWS and SF wetlands. The nitrification-denitrification is the major path of nitrogen removal. If nitrification and nitrogen removal are desired, the hydraulic loading should be reduced to 100–150 m^3/ha.day (10,700–16,000 gpd/acre.day). Nitrogen removal may reach 80% at nitrogen loading up to 44 kg/ha.day (139 lb/acre.day). Nitrogen removal is significantly higher if plant harvesting is practiced.[147,162]

Phosphorus removal: Constructed wetlands are capable of phosphorus removal from wastewater. Phosphorus is an essential nutrient required for plant growth. The removal is higher in a submerged bed system where selected soils are used for the support medium. The presence of clay, iron, and aluminum will enhance phosphorus removal.[163–165]

Pathogen removal: The log removal of coliform organisms in the wetlands is in the order of 1.5–3.5 (96.84%–99.97%).[147,152] The total coliform count in the effluent from wetlands may be 10^3/100 mL or lower if secondary effluent is treated. Such a level may not be reached with primary effluent without disinfection.[151,160,165]

Mosquito control: Mosquito problem may occur when wetlands receive excessive organic loading and anaerobic conditions develop. Biological control agents such as mosquito fish (*Gambusia affins*) may die due to low dissolved oxygen (DO) or hydrogen sulfide toxicity. Mosquito problem becomes severe if larvae mature into adults. The possible control method includes (1) effective pretreatment to reduce total organic loading, (2) step feeding of the influent with proper influent distribution and effluent recycle, and (3) application of approved and environmentally safe chemical control agents. Vegetation management may also help control mosquitoes. Plant harvesting will reduce congestion at the water surface encouraging oxygen transfer by reaeration and mosquito larvae control by fish is enhanced. Single harvesting in a year may yield up to 29,000 kg/ha dry weight of *Phragmites* shoots. A higher yield is achieved with multiple harvestings.[166–168]

Maintenance requirement: Maintenance is important because accumulation of organic matter and sediment can alter the pollutant removal effectiveness of the system by decreasing storage volume and changing the substrate. Likely maintenance will include sediment accumulation removal, conveyance system and discharge structure repair, and vegetation harvesting or burning.[147,151,154,158]

Example 10.57 Water Balance in a Wetland

A constructed wetland covers 10 ha area. The average annual precipitation and evaporation are 88 and 64 cm respectively. The percolation rate is 10^{-8} m/s. The average monthly surface drainage reaching the wetland is 1,320 m^3/month. The average influent to the wetland is 1,500 m^3/day. Estimate the effluent flow per day assuming that there is no net change in total storage volume of the wetland.

Solution

1. Determine the annual precipitation *P*, and evapotranspiration ET.

$$\text{Annual precipitation, } P = 88 \text{ cm/year} \times 10^{-2} \text{ m/cm} \times 10 \text{ ha} \times 10{,}000 \text{ m}^2/\text{ha} = 88{,}000 \text{ m}^3/\text{year}$$

$$\text{Annual evapotranspiration, ET} = 64 \text{ cm/year} \times 10^{-2} \text{ m/cm} \times 10 \text{ ha} \times 10{,}000 \text{ m}^2/\text{ha} = 64{,}000 \text{ m}^3/\text{year}$$

2. Determine surface drainage Q_s and infiltration Q_g

$$\text{Average annual surface drainage, } Q_s = 1{,}320 \text{ m}^3/\text{month} \times 12 \text{ month/year} = 15{,}840 \text{ m}^3/\text{year}$$

$$\text{Average annual infiltration, } Q_g = 10^{-8} \text{ m/s} \times 86{,}400 \text{ s/day} \times 10 \text{ ha} \times 10{,}000 \text{ m}^2/\text{ha} \times 365 \text{ day/year}$$

$$= 31{,}540 \text{ m}^3/\text{year}$$

3. Determine influent wastewater flow Q_i
   $Q_i = 1{,}500 \text{ m}^3/\text{day} \times 365 \text{ day/year} = 547{,}500 \text{ m}^3/\text{year}$
4. Estimate the average effluent flow from the wetland
   Apply Eq. (10.63a): $dV/dt = Q_i - Q_e + P - \text{ET} + Q_s - Q_g$
   Since there is no net change in storage volume, $dV/dt = 0$

$$0 = 547{,}500 \text{ m}^3/\text{year} - Q_e + (88{,}000 - 64{,}000 + 15{,}840 - 31{,}540) \text{ m}^3/\text{year}$$

$$Q_e = 555{,}800 \text{ m}^3/\text{year}$$

$$= 555{,}800 \text{ m}^3/\text{year} \times 1/(365 \text{ day/year}) = 1{,}523 \text{ m}^3/\text{day}$$

Example 10.58: Nominal and Actual Detention Times, and Superficial and Actual Flow Velocities

A SF wetland is designed to treat 1,500 m³/day flow. The surface area = 10 ha, width w = 100 m and average submerged water depth = 0.6 m. The porosity of the support media is 0.38. Determine the nominal and actual detention times, and superficial and actual water velocities.

Solution

1. Determine nominal and actual detention times.

$$\text{Nominal volume } V_{\text{nominal}} = 10 \text{ ha} \times 10{,}000 \text{ m}^2/\text{ha} \times 0.6 \text{ m} = 60{,}000 \text{ m}^3$$

$$\text{Actual volume } V_{\text{actual}} = e \times V_{\text{nominal}} = 0.38 \times 60{,}000 \text{ m}^3 = 22{,}800 \text{ m}^3$$

$$\text{Nominal detention time } t_n (\text{Eq. 10.63b}) = 60{,}000 \text{ m}^3 / 1{,}500 \text{ m}^3/\text{day} = 40 \text{ day}$$

$$\text{Actual detention time } t_a (\text{Eq. 10.63c}) = 22{,}800 \text{ m}^3 / 1{,}500 \text{ m}^3/\text{day} = 15.2 \text{ day}$$

2. Determine superficial and actual velocities

$$\text{Superficial velocity } u(\text{Eq. 10.63d}) = (1{,}500 \text{ m}^3/\text{day}) / (0.6 \text{ m} \times 100 \text{ m}) = 25 \text{ m/day}$$

$$\text{Actual water velocity } u(\text{Eq. 10.63e}) = (1{,}500 \text{ m}^3/\text{day}) / (0.38 \times 0.6 \text{ m} \times 100 \text{ m}) = 66 \text{ m/day}$$

Example 10.59: Slope of Bed and Hydraulic Conductivity of a Wetland

A SF wetland is designed for a flow of 1,200 m³/day. The slope of the bed is 5%. The width and mean depth through which the flow occurs are 100 and 0.6 m respectively. The hydraulic loading is 250 m³/ha.day. Determine the hydraulic conductivity and detention time.

Solution

1. Determine the cross-sectional area A_C through which the flow occurs.

$$A_C = 100 \text{ m} \times 0.6 \text{ m} = 60 \text{ m}^2$$

2. Determine the hydraulic conductivity K

Apply Eq. (10.63f) $Q = K \times A_c \times \frac{\Delta h}{\Delta L}$

$$1{,}200 \text{ m}^3/\text{day} = K \times 60 \text{ m}^2 \times 0.05$$

$$K = 1{,}200 \text{ m}^3/\text{day} \times 1/60 \text{ m}^2 \times 1/0.05 = 400 \text{ m/day}$$

$$= 400 \text{ m/day} \times 1/(60 \times 60 \times 24) \text{ s/day} = 0.0046 \text{ m/s}$$

$$= 4.6 \times 10^{-3} \text{ m/s}$$

3. Determine detention time

Hydraulic loading = 250 m^3/ha.day/10,000 m^2/ha = 0.025 m/day

Hydraulic loading is also the terminal velocity (settling velocity of the smallest particle that is fully removed)

Detention time = (depth of fall)/terminal velocity = 0.6 m/(0.025 m/day) = 24 day

Example 10.60: TSS and BOD_5 Removed in a SF Constructed Wetland

A SF constructed wetland is designed to treat 1,200 m^3/day primary settled wastewater. The total surface area and mean water depth are 10 ha and 0.5 m. The background TSS and BOD_5 concentrations are 2 and 3 mg/L respectively. The influent TSS and BOD_5 concentrations are 60 and 80 mg/L. The apparent TIS rate constant k is 100 m/year, and the apparent p for TIS is 1.2. Determine the TSS and BOD_5 concentrations in the effluent. (see Eq. (10.64b) for definition of TIS)

Solution

1. Determine the nominal detention time and hydraulic loading

Nominal volume $V_{nominal}$ = 10 ha × 10,000 m^2/ha × 0.5 m = 50,000 m^3

Nominal detention time t_n = 50,000 m^3/1,200 m^3/day = 42 day

Hydraulic loading rate w = (1,200 m^3/day)/10 ha = 120 m^3/ha.day = 120 m^3/10,000 m^2.day = 0.012 m/day

2. Determine the effluent TSS concentration

Apply Eq. (10.64a) $\frac{C_e - C^*}{C_o - C^*} = \exp[-wt_n / h]$

$$\frac{(C_e - 2) \text{ mg/L}}{(60 - 2) \text{ mg/L}} = \exp\left[-0.012 \text{ m/day} \times 42 \text{ day} / 0.5 \text{ m}\right]$$

$$(C_e - 2)/(58 \text{ mg/L}) = \exp[-1.008]$$

$$(C_e - 2) \text{ mg/L} = (58 \text{ mg/L}) \times 0.36 = 21 \text{ mg/L}$$

$$C_e = (21 + 2) = 23 \text{ mg/L}$$

3. Determine the effluent BOD_5 concentration

Apply Eq. (10.64b) $\frac{C_e - C^*}{C_o - C^*} = \frac{1}{(1 + k / pq)^p}$

$$\text{Hydraulic loading rate } q = (0.012 \text{ m/day}) \times (365 \text{ day/year}) = 4.38 \text{ m/year}$$

$$\frac{(C_e - 3)\text{ mg/L}}{(80-3)\text{ mg/L}} = \frac{1}{\left(1+\dfrac{100 \text{ m/year}}{4.38 \text{ m/year} \times 1.2}\right)^{1.2}} = 0.0274$$

$$C_e = (77 \text{ mg/L} \times 0.027) + 3 \text{ mg/L} = 5 \text{ mg/L}$$

Example 10.61: BOD_5 Removal in FWS and SF Constructed Wetlands

Compare the effluent BOD_5 from FWS and SF wetlands. The influent flow, BOD_5, hydraulic loading, and water depth for both wetlands are the same, and are 1,500 m^3/day, 80 mg/L, 350 m^3/ha.day and 0.5 m respectively. The porosity ratio of SF wetland is 0.45. The reaction rate constants K_T for FSW and the reaction rate constant K_t for SF wetlands are assumed 0.2 and 0.5/day respectively. Justify the assumed reaction rate constants of both wetlands. The depth of submergence h in SF wetland is 0.5 m.

Solution

1. Determine the hydraulic loading and detention time for both wetlands.
   Hydraulic loading = (350 m^3/ha.day)/(10,000 m^2/ha) = 0.035 m/day
   Detention time t = 0.5 m/0.035 m/day = 14.3 day
2. Justify the assumed reaction rate constants for FWS and SF wetlands.
   The flow through a single-cell FWS wetland will exhibit medium to low dispersion, while that in a SF wetland will be more like plug-flow (PF) reactor. The dispersion in a PF reactor is very low and the reaction rate constant is high. Therefore, the assumed reaction rate constants of 0.20 and 0.50/day for FWS and SF wetlands are justified.
3. Determine the effluent BOD_5 concentration from FWS wetland
   Apply Eq. (10.64c) $\dfrac{C_e}{C_o} = \exp[-K_T t]$

$$\frac{C_e}{C_o} = \exp\left[-0.2 \text{ day}^{-1} \times 14.3 \text{ day}\right] = 0.057$$

$$C_e = 80 \text{ mg/L} \times 0.057 = 5 \text{ mg/L}$$

4. Determine the effluent BOD_5 concentration from SF wetland
   Apply Eq. (10.64d) $\dfrac{C_e}{C_o} = \exp[-A_s K_t h e]$

Hydraulic loading Q/A_s from step 1 = 0.035 m/day, $A_s/Q = 1/0.035$ m/day = 28.57 day/m

Factors $A_s/Q = 28.57$ day/m and $h = 0.5$ m and $A_s \times h/Q = (0.5 \text{ m}/28.57 \text{ day/m}) = 14.3$ day

$$\frac{C_e}{C_o} = \exp[-28.57 \text{ day/m} \times 0.5 \text{ day}^{-1} \times 0.5 \text{ m} \times 0.45] = \exp[-3.22] = 0.04$$

$$C_e = 80 \text{ mg/L} \times 0.04 = 3.2 \text{ mg/L}$$

Note: The effluent quality from SF wetland is superior to that from FWS.

10.8 WATERSHED MODELING FOR CONTROL OF NONPOINT SOURCE POLLUTION

Nonpoint source (NPS) pollution is widely dispersed in the environment and comes from a variety of human open-land activities. In the U.S., NPS pollution is still a leading cause of water quality impairment of nation's waters. It is a challenge from both regulatory, and source and impact assessment perspective. To better understand the complex mechanisms and processes, and to evaluate sources of NPS pollution and their impact on surface water quality, numerous watershed-scale NPS models and tools have been developed. These models have built-in capabilities to (1) identify land areas that are critical sources of pollutants,[169] (2) evaluate the effects of NPS pollution on the water environment,[170] (3) evaluate future scenarios of hydrology and water quality,[171] (4) assist in the planning and implementation of BMPs,[172] (5) support development of water quality criteria or standards,[173] and (6) provide informed decision to support policy-makers.[174,175] These models have a wide range of capacities and are classified in several categories: (a) simple models, (b) medium-complexity models, (c) complex models, and (d) modeling systems. Model categories are briefly discussed below.

10.8.1 Simple Models

Simple models have minimal data requirements that includes land use, soil, precipitation, and digital elevation model (DEM). These models are generally built on an empirical or statistical relationships between pollution loads or concentrations, land use, rainfall, and runoff volume. These models typically use the Natural Resources Conservation Service Curve Number (SCS-CN) method to calculate overland runoff. Readers are referred to Section 10.5.3, Examples 10.10–10.16, and Appendix B on the SCS-CN method. To calculate pollutants loading, the export coefficients are employed.[176] Simple models are often used as quick screening tools to obtain the gross pollution loads at the outlet of a watershed. They possess limited capacity to simulate detailed hydrological and physicochemical processes and are also limited to a small number of simulated pollutants. Some examples of simple models are Long-Term Hydrologic Impact Assessment (L-THIA), Nonpoint Source Pollution and Erosion Comparison Tool (N-SPECT), and others.[177,178]

10.8.2 Medium-Complexity Models

Medium-complexity models generally require relatively detailed data inputs and limited calibration. Input data includes topography, land use, soil, weather, and water quality. These models account for the fundamental water and sediment movement processes, and also combine the empirical relationships to express nutrient loads. Meanwhile, complicated watershed models that requires intensive data are not applied. The calculation of surface runoff is based on a water balance principle.[179] The daily values of runoff, sediment, and nutrient loads are estimated. Rigorous calibration, and model validation is not required. Most of these models assist in development of water quality criteria and are easy to operate compared to complex models. However, these models have restricted simulation capability for in-stream fate and transport of pollutants, and the number of simulated pollutants is also limited. Some examples of these models are Generalized Watershed Loading Function (GWLF), Loading Simulation Program C^{++} (LSPC), and others.[169,180,181]

10.8.3 Complex Models

Complex models can simulate NPS pollution based on intrinsic physical processes. They provide sophisticated simulation for NPS pollution processes with intensive data and rigorous calibration. These models generally integrate hydrology, erosion and sediment processes, and chemical pollutant fate and transport. The data required for these models are commonly large. Complex models consider the estimation of runoff, soil erosion, sediment and pollutant loading based on theoretical consideration of mass, momentum, and energy. They address a wide range of watershed-scale hydrology and water quality issues, which include short- and long-term simulations of runoff, sediment, and pollutant loads, and different algorithm options for various physical processes to describe more accurately the processes mathematically. These models must be calibrated and validated carefully before applying them in management decisions because poor data input or inappropriate application of algorithms can lead to large errors.[182] These complex models require intensive parameters inputs to drive the model calculations. Careful parameter sensitivity analyses are needed prior to calibration, and the results of uncertainties analysis need to be evaluated after validation. Calibration and validation of complex models are a labor-intensive and time-consuming process.[183] For application of complex models end users need expertise, sufficient training, and experience to apply these models correctly. Examples of complex models are Soil and Water Assessment Tool (SWAT), Storm Water Management Model (SWMM), and others.[169,184,185]

10.8.4 Modeling Systems

A modeling system integrates various sub-models and tools and contains the highest complexity to solve all phases of hydrologic, hydraulic, and chemical dynamic processes. It uses the concept of multiple modules to independently maintain and apply separate model structures or information to carry out complex decision analyses or synthesis. It integrates databases, tools, techniques, and models into a Geographic Information System (GIS) platform. These data, tools, methods, and models have a close linkage and work together to perform various environmental simulations and analyses at multiple time and space scales, and different modules comply with the same data exchange protocols. Thus, the modeling system comprehensively uses compatible data types, various tools, and different models to assist in systematically solving complicated watershed water quality and NPS pollution issues. It can serve as a multipurpose decision support platform. The modeling system can simulate at watershed-scale, hydraulic, hydrologic, water quality, and NPS pollution issues. However, the data requirement and model computation are enormous. These models have a steep learning curve for users as many individual models and various watershed analysis tools are integrated. This can involve extensive pre-processing and post-processing of data and output. The user commonly needs extensive training prior to developing or running these models. Examples of modeling systems are Automated Geospatial Watershed Assessment Tool (AGWA), Better Assessment Science Integrating Point and Nonpoint Sources (BASINS), and others.[169,186,187]

These models have become essential tools in watershed management, and to reduce the adverse effects of NPS pollution. They will be widely used in the future to identify critical source of pollution, evaluate the effect, and combat NPS pollution. However, the current models are highly diverse. With the development of model theory, computer technology, and environmental legislation, numerous water quality, hydrology, rainfall-runoff, and NPS pollution models will be integrated into the watershed models. Despite these efforts, it will always be a challenge to select and apply the most suitable model for a given watershed and sources of NPS pollution.

10.9 PERFORMANCE EVALUATION OF SOURCE-CONTROL BMPS

There are many types of stormwater source-control BMPs that manage stormwater quality runoff from undeveloped and developed areas, and construction sites. However, not all these practices and methods have the same capabilities in runoff control and pollutant removal at every site. The performance and effectiveness of each BMP depends on the land cover and soil condition, hydrology, site suitability, type and source of pollutants, the design and selection of BMPs, and maintenance practices. These BMPs are an integrated combination of many processes serving the primary goals of reducing stormwater runoff and providing water quality improvement. Sixteen source-control BMPs are described in Section 10.7. Among the principal source-control BMPs are detention/retention basins, bioretention systems, percolation basin and infiltration trench, diversion drain or grassed waterway, filter strips, green roofs, vegetated bio-swales, permeable pavements, wetlands, and others. These source-control BMPs provide storage, percolation, filtration, uptake and removal of contaminants by plants and soil bed. The actual performance of the BMPs however, will depend upon the conditions such as soil type, land cover, hydrology, runoff, type of contaminants, and maintenance. The information needed for characterizing the long-term performance of BMPs is provided in Table 10.43. The range of water retention and performance of these source-control BMPs in terms of percent reduction of runoff, TSS, total phosphorus (TP), total nitrogen (TN), metals, and fecal coliform are summarized in Table 10.44. The range of values are for normal conditions, and may be used for the design and performance assessment of BMPs. The actual values however may differ from the range given in the table because of long use, maintenance, and existing field conditions. Often, a negative efficiency for many water quality parameters may result, indicating that some of these BMPs are a source of pollution. In practice, the removal efficiency of many BMPs may be reduced over time due to lack of maintenance, degradation of structures, and accumulation of pollutants.[32] Therefore, the long-term performance of these BMPs should be monitored and incorporated in the process of selecting these devices for a specific site or project (Table 10.43). The key information needed to characterize the long-term performance of BMPs are total runoff volume and load treated, type of design and installation, maintenance activity, age, life cycle, soil, land use, slope, vegetation, and weather condition.

TABLE 10.43
Information Collected for Characterizing Long-term Performance of Source-Control Stormwater BMPs

Factors	Data to be Collected
BMP efficiency	Paired inflow and outflow volumes and pollutant concentrations
Total runoff volume and pollutant load treated over time	Daily runoff volume and pollutant loads treated
Design	Design standards applied (e.g., Natural Resource Conservation Service, state, or specific design criterion for each practice)
Installation	Quality of installation
	Skill level of installer
	Installation process
Maintenance activities	Maintenance date
	Activities done during maintenance
Age	Age of BMP at the time of data collection
Life cycle	Age of BMP when replacement is expected
	Reasons for BMP being replaced
Date	Date of installation, monitoring, and reporting
BMP condition at time of data collection	Visual condition
	High-resolution photos with location information recorded
Local conditions	Soil, land use, slope, vegetation
	Weather, such as characteristics of storm events (storm size, duration, and intensity), daily highest and lowest temperatures
	Information on BMPs implemented in series with runoff treated by one practice following into another practice, such as BMP efficiency, types, and number implemented in series

Source: Adapted in part from Refs. 32,65,66

Example 10.62: Performance of a Vegetated Filter Strip in Urban Watershed

A 0.5-acre parking lot is serving a shopping center. Storm drainage from the parking lot is intercepted by a vegetated filter strip for stormwater runoff retention and contaminants removal. The vegetated filter strip is located on the lower side of the parking lot. Two-inch equivalent storm drainage from the parking lot flows uniformly across the vegetated filter strip as a sheet flow. The average concentration of contaminants of concern are: TSS = 120 mg/L, TP = 4.5 mg/L, TN = 8.5 mg/L, total metals = 3.8 mg/L, and total fecal coliform count = 10^3 per mL. Determine the average volume of stormwater, and the average concentration of contaminants leaving the filter strip. Assume that the midpoint values given in Table 10.44 apply.

Solution

1. Determine the volume of stormwater leaving the vegetated filter strip.
   Volume of the stormwater generated = 0.5 acre × 43,560 ft^2/acre × 2 in./h × 1 ft/12 in. × h/3,600 s = 1.01 ft^3/s
   Volume of stormwater leaving the vegetated filter strip = 1.01 ft^3/s (1 − (0.50 + 0.90)/2) = 1.01 ft^3/s × 0.3 = 0.303 ft^3/s
2. Determine the concentration of contaminants in stormwater leaving the vegetated filter strip

$$\text{Concentration of TSS} = 120\ \text{mg/L} \times (1 - (0.70 + 0.96)/2) = 20.4\ \text{mg/L}$$

$$\text{Concentration of TP} = 4.5\ \text{mg/L} \times (1 - (0.70 + 0.96)/2) = 0.8\ \text{mg/L}$$

$$\text{Concentration of TN} = 8.5\ \text{mg/L} \times (1 - (0.60 + 0.90)/2) = 2.1\ \text{mg/L}$$

$$\text{Concentration of total metals} = \text{mg/L} \times (1 - (0.70 + 0.90)/2) = 0.8\ \text{mg/L}$$

$$\text{Total fecal coliform count} = 10^3/\text{mL} \times (1 - (0.80 + 0.95)/2) = 1.3 \times 10^2/\text{mL}$$

TABLE 10.44
Runoff Reduction and Contaminants Removal of Source-Control Watershed BMPs

Reference No.	Source-Control Watershed, BMP	Runoff Retention, % Range	TSS, % Range	TP, % Range	TN, % Range	Metal, % Range	Fecal Coliform, % Range
10.7.1	Bioretention Systems (Rain Garden)	47–95	80–90	−3–99	32–99	31–98	71–95
10.7.2	Bench Terracing	50–95	60–99	−10–85	−15–80	80–95	80–99
10.7.3	Check Dams and Gully Plugs	60–90	50–90	30–70	25–45	30–60	60–80
10.7.4	Contour Bunding or Graded Bunding	70–95	60–95	50–80	30–70	40–80	70–90
10.7.5	Detention/Retention Basin or Pond	50–100	30–85	15–57	15–47	30–80	20–80
10.7.6	Diversion Drain or Grassed Waterway	20–40	20–50	40–60	40–70	40–60	50–80
10.7.7	Vegetated Filter Strips	50–90	70–96	70–90	60–90	70–90	80–95
10.7.8	Land Leveling and Grading	50–90	20–80	20–50	20–40	40–80	60–90
10.7.9	Protection BMPs from CAFOs	60–90	30–85	30–60	−20–40	20–50	−4–30
10.7.10	Green Roofs	40–90	40–80	30–70	0–60	0–40	20–40
10.7.11	Rainwater Harvesting	30–60	—	—	—	—	—
10.7.12	Infiltration Basin	30–80	60–99	25–85	25–75	70–90	70–95
10.7.13	Infiltration or Percolation Trench	50–95	60–95	25–80	20–70	65–70	60–95
10.7.14	Permeable Paving	50–93	58–94	10–78	70–95	60–95	85–95
10.7.15	Vegetated Bioswale	30–70	70–90	40–60	40–60	30–60	70–90
10.7.16	Wetland	60–100	60–99	30–80	30–60	50–90	60–98

Source: Adapted in part from Refs. 18,20,29,31,32,66,72

Example 10.63: Performance of a Combination of Bioretention System (Rain Garden) and Detention/Retention Basin in an Urban Watershed

A shopping center has multiple stores, driveways, individual car parking, service roads, and a large central parking lot. The stormwater drains from roofs and paved surfaces enters several bioretention systems (rain gardens) located on the downslope of the shopping center. The main function of bioretention system is to collect, retain, and treat the drainage from the shopping center. The flow from the bioretention system is further treated in a wet retention/detention basin. The shopping center covers a total area of 3 acres. The maximum stormwater drainage from the area is equivalent of 1.8 in. of rainfall in 2 h. The storage capacity of bioretention basin is 15.2 in. and is constructed along the edge of the shopping center. The total area of the constructed bioretention system is 7.0% of the entire shopping center area. The average quality of stormwater runoff is: TSS = 180 mg/L, TP = 4.8 mg/L, TN = 10.2 mg/L, total metal = 5.8 mg/L, and total fecal coliform count = 10^4/L. The surface area of the wet retention/detention basin is 1.0% of the shopping center area. The average percolation, evapotranspiration and other losses from the basin are equivalent of 14.9 in. of surface area. Determine the average volume and quality of stormwater released from the bioretention system, and the wet detention/retention basin. Assume that the midpoint removal values given in Table 10.44 apply in both cases.

Solution

A. Shopping center
   1. Calculate the land area utilizes by the shopping center
      Shopping center land area = 3 acres × 4,840 yd^2/acre × 9 ft^2/yd^2 = 130,680 ft^2
   2. Calculate the volume and flow rate of stormwater draining from the shopping center

$$\text{Volume of stormwater} = 1.8 \text{ in.} / \left(12 \text{ in./ft}\right) \times 130{,}680 \text{ ft}^2 = 19{,}602 \text{ ft}^3$$

$$\text{Flow rate} = 19{,}602 \text{ ft}^3 / \left(2 \text{ h} \times 3{,}600 \text{ s/h}\right) = 2.72 \text{ cfs}$$

B. Bioretention basin
   1. Calculate the volume of surface runoff retained, and volume released from the bioretention basin

$$\text{Area of bioretention basin} = (7.0/100) \times 130{,}680 \text{ ft}^2 = 9{,}148 \text{ ft}^2$$

$$\text{Storage volume of bioretention basin} = 9{,}148 \text{ ft}^2 \times 15.2 \text{ in.} \times \text{ft} / 12 \text{ in.} = 11{,}587 \text{ ft}^3$$

$$\text{Volume of stormwater released from the bioretention basin} = (19{,}602 - 11{,}587) \text{ ft}^3 = 8{,}015 \text{ ft}^3$$

   2. Calculate the concentration of contaminates in the effluent from bioretention basins

$$\text{TSS} = 180 \text{ mg/L} \times (1 - (0.80 + 0.90)/2) = 27.0 \text{ mg/L}$$

$$\text{TP} = 4.8 \text{ mg/L} \times (1 - (-0.03 + 0.99)/2) = 2.5 \text{ mg/L}$$

$$\text{TN} = 10.2 \text{ mg/L} \times (1 - (0.32 + 0.99)/2) = 3.5 \text{ mg/L}$$

$$\text{Total metals} = 5.8 \text{ mg/L} \times (1 - (0.31 + 0.98)/2) = 2.1 \text{ mg/L}$$

$$\text{Total fecal coliform count} = 10^3/\text{L} \times (1 - (0.71 + 0.95)/2) = 1.7 \times 10^3/\text{L}$$

The effluent from the bioretention basin is released from the perforated pipe at the bottom of the bioretention system.

C. Retention/detention basin
   1. Calculate the volume of surface runoff lost from the retention/detention basin
      Surface area of retention/detention basin = $1.0/100 \times 130{,}680\,ft^2 = 1{,}307\,ft^2$
      Volume of water lost from detention/retention basin = $(14.9\text{ in.}/12\text{ in./ft}) \times 1{,}307\,ft^2 = 1{,}623\,ft^3$
      Volume of water leaving retention/detention basin = $(8{,}015 - 1{,}623)\,ft^3 = 6{,}392\,ft^3$
   2. Calculate the concentration of contaminates in the effluent from retention/detention basins

$$\text{TSS} = 27.0\text{ mg/L} \times (1 - (0.30 + 0.85)/2) = 11.5\text{ mg/L}$$

$$\text{TP} = 2.5\text{ mg/L} \times (1 - (-0.15 + 0.57)/2) = 1.6\text{ mg/L}$$

$$\text{TN} = 3.5\text{ mg/L} \times (1 - (0.15 + 0.47)/2) = 2.4\text{ mg/L}$$

$$\text{Total metals} = 2.1\text{ mg/L} \times (1 - (0.30 + 0.8)/2) = 1.0\text{ mg/L}$$

$$\text{Total fecal coliform count} = 1.7 \times 10^3/\text{L} \times (1 - (0.20 + 0.80)/2) = 8.5 \times 10^2/\text{L}$$

10.10 URBAN STORMWATER MANAGEMENT

Urbanization changes land use, land cover and natural systems. These changes alter vegetation as topsoil is removed to make way for buildings, roads, parking lots, driveways, and other urban infrastructures. The introduction of underground drainage network also alters the hydrology of the local area. Stormwater flows rapidly across the land surface and arrives at the stream channel in short, concentrated bursts of high discharge. This transformation of the hydrologic regime occurs throughout the developed landscape. When combined with the pollutants that accompany urbanization (such as lawn chemicals, pollutants from motor vehicles, domesticated animals, and industries), these changes cause serious water quality and habitat degradation in virtually all urban streams. Stormwater discharge causes erosion, and loss of soil and vegetation. Also, it poses a physical hazard to aquatic habitats and stream functions owing to the increase in water velocity and volume. As many individual sources are combined, the net result may reach the scale of entire watershed. The magnitude of the stormwater problem is only expected to grow as there is a shift of the world's population to urban areas. In recognition of the need for improved control measures, in 1987 the U.S. Congress mandated the U.S. EPA, under amendments to the CWA, to control certain stormwater discharges under the NPDES. In response to this federal legislation, a permitting program was put in place by EPA as the Phase I (1990) and Phase II (1999) stormwater regulations. Since then, many innovative and emerging stormwater treatment technologies and proprietary systems have been developed and implemented.[4,5,8]

10.10.1 Quality of Stormwater Runoff from Urban Areas

Urban stormwater runoffs carry pathogens, nutrients, heavy metals, sediments, and other contaminants as surface runoff flows over land. The impervious or paved surfaces in urban areas reduce the natural infiltration of precipitation into the ground. Over impervious cover, precipitation collects and carries contaminants into nearby surface waters. This can degrade downstream waters with both contaminants and increased volumes of water. The traditional runoff management techniques such as detention basins, infiltration swales and wetlands may be preferable, but are not always practical for treating urban stormwater due to lack of space in existing developed areas. Some source-control

BMPs such as bioretention (rain garden) systems, vegetated filter strips, green roofs, RWH, permeable paving, and infiltration trench are highly effective in urban settings. They reduce peak flow, provide reduction in volume of surface runoff by percolation, evapotranspiration, and storage. The innovative stormwater treatment technologies in combination with applicable source-control BMPs are attractive alternatives in urban settings because they are effective and use less space to remove contaminants from stormwater.

Contaminants that are typically found in stormwater runoffs from urban or developed areas are associated with land use activities. Pollutants commonly found in urban stormwater runoffs, their sources, and potential related impacts on receiving waters are summarized below:

Sediments: Sediments come from construction sites, other disturbed and/or non-vegetated lands, eroding banks, and road sand. They remain suspended in surface runoff and receiving waters, then are deposited in the waterbody. Suspended solids increase turbidity, reduce clarity, and lower dissolved oxygen. Additionally, deposited sediments may smother the benthic organisms, change the aquatic habitats, including spawning sites. Organic matters such as leaves, grass clippings, and domestic animal wastes lower dissolved oxygen.

Nutrients (nitrogen and phosphorus): These nutrients come from animal waste, excess fertilizers used over lawns, and failing septic systems. Nutrients are associated with eutrophication. Algal growth reduces clarity and dissolved oxygen levels.

Pathogens (bacteria and viruses): Pathogens are released from animal wastes, failing septic systems, and dumpsters. Associated human health risks are with the drinking water supply, consumption of affected shellfish, and swimming beach contamination.

Heavy metals: Heavy metals such as lead, copper, cadmium, zinc, mercury, chromium, and others are released from industrial processes, normal wear of auto brake linings and tires, automobile emissions and fluid leaks, metal roofs, and others. They cause toxicity in air, water, and sediments, fish kills, and bioaccumulation in aquatic species through the food chain.

Synthetic chemicals: Synthetic chemicals, such as PCBs, pesticides, herbicides, insecticides, fungicides, and rodenticides, come from industrial processes and excess use of fertilizer over lawns and green areas. They cause toxicity in water and sediment, fish kills, and bioaccumulation through the food chain.

Chlorides: Chlorides come from road salting and uncovered salt storage. They cause toxicity in water and sediments, and increase the total dissolved solids.

Trash, debris and leaves: Urban debris is washed through storm drain networks, commercial parking lots adjacent to surface water, overflowing trash barrels, and dumpsters. They degrade surface water aesthetics and are a threat to wildlife.

10.10.2 Source-Control BMPs Modified for Urban Watershed

Sixteen source-control watershed BMPs have been presented in Section 10.7. Many of these BMPs are also applicable to urban watersheds. Among them are bioretention (rain garden) systems, detention/retention basins, filter strips, green roofs, RWH, infiltration/percolation basin and trench, permeable paving, and vegetated bioswale. Many source-control BMPs have been modified to improve their performance for urban watersheds. Decorative cisterns, downspout leaf screens, first-flush rainwater diverters, and roof washers are manufactured and supplied for RWH in urban settings. Special permeable interlocking plastic grid pavers are made for driveways, sidewalks, and parking lots for use in city blocks and subdivisions. Many drainage media are manufactured for infiltration/percolation basins and trenches to increase percolation and reduce space requirements. These devices are prefabricated or customized for BMPs. They are ingeniously designed to reduce the space necessary to process runoff and stay out of sight. Like traditional surface-level BMPs, these manufactured devices also mimic natural systems such as bioretention areas by incorporating plants, soil, and microbes engineered to provide treatment of higher flow rates or volumes and with smaller footprints than their natural counterparts. Incoming flows are typically filtered through a planting media (mulch, compost, soil, plants, microbes, etc.) and either infiltrated or collected by an underdrain and delivered to the storm water conveyance system. These systems are applied to green roofs and vegetated bioswales in urban settings. Tree box filters are manufactured that are installed at the curb level and filled with a bioretention-type soil. For low to moderate flows, these devices operate like bioretention (rain garden) systems and are bypassed during high flows. Tree box filters are highly adaptable solutions that can be used in all types of development and in all types of soils, but are especially applicable to dense urban parking lots, streets, and roadways.

Proprietary devices are commercial products that typically provide stormwater treatment in space-limited applications, often using patented, innovative, and emerging technologies. The manufacturers of these BMPs provide detailed documentation for device selection, sizing, and maintenance requirements. Because proprietary systems come in a range of designs and sizes, they are highly variable in treatment process, design and performance, flow rates, and storage capacity. These devices are often costly, and their performance in removing pollutants is not proven in many cases. For this reason, the treatment effectiveness of specific proprietary devices provided by the manufacturer must be verified by independent third-party sources, such as directly from users, or assessed by a water quality professional.

10.10.3 Manufactured Equipment and Devices for Treatment of Stormwater Runoff

There are two major types of manufactured equipment and devices used for the treatment of stormwater runoff. These are: (1) separation devices and (2) filtration devices. The separation devices separate sediments from the stormwater. These devices are further broken down into (a) chambered gravity settling and (b) hydrodynamic separation. In *chambered gravity settling*, the runoff is channeled through several compartments where gravity settling of sediment particles and flotation of hydrocarbons (seen as an oil sheen on top) take place. The *hydrodynamic separation* devices work through inducing centrifugal force by a swirling motion to the incoming stormwater flow to aid in the separation and removal of suspended particles. The *filtration BMPs* utilize coarse and fine screens, filter media, and filter cartridges. The runoff passes through these screens and filtration devices, which physically remove suspended pollutants from the stormwater. The screening devices remove large objects from stormwater at the curb gutter before entry to the storm sewer, or at the outfall of the storm sewer. The common types of filtration BMPs are the gravity filtration bed, catch basin insert, screen chamber, and filter chamber. A classification of these devices is shown in Figure 10.64. Devices in both categories are non-proprietary and proprietary BMPs, and can be

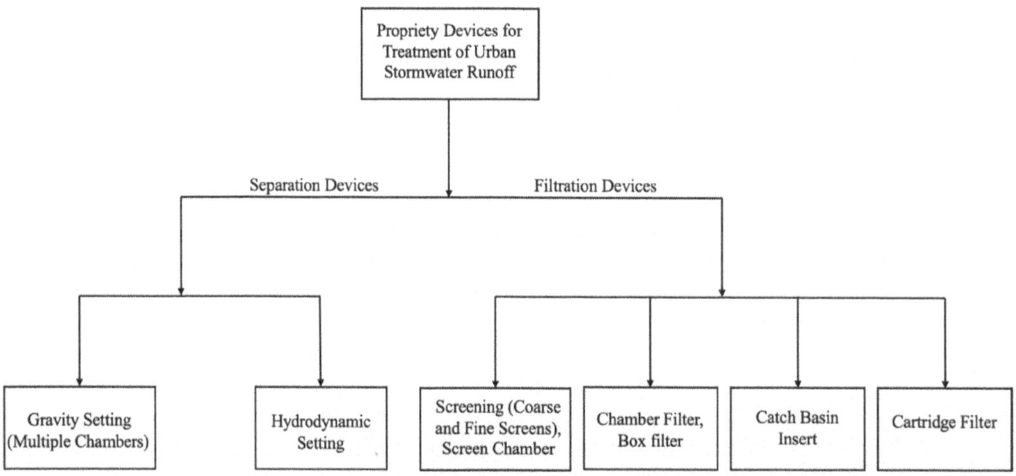

FIGURE 10.64 Classification of proprietary devices for treatment of stormwater runoff.

built underground or aboveground. Because of limited land in urban areas, underground construction of all BMPs is preferred. The non-proprietary or conventional systems include aboveground or underground retention tanks, tunnels, vaults, pipes, and storage chambers. Common proprietary systems include separation and filtration devices. Processing methods of these devices include gravity sedimentation, hydrodynamic separation, screening, and filtration. These devices can be integrated with non-proprietary systems into parking lots, sidewalks, and vegetated landscaping. This versatility makes them a worthwhile consideration for many development types, including distribution centers, strip malls, and even housing developments.

10.10.4 Proprietary Devices

Proprietary devices are commercial products that typically provide stormwater treatment in space-limited applications, often using patented innovative technologies. The most encountered classes of proprietary stormwater management devices include modified source-control BMPs discussed above, and screening, separation, and filtration devices. Products, manufacturers and vendors, and their websites for stormwater treatment are summarized in Table 10.45.

10.10.5 Pros and Cons of Underground and Aboveground BMPs in Urban Areas

In urban areas, available land is very limited for stormwater control BMPs. For this reason, both proprietary and non-proprietary BMP devices are installed below ground. The pros and cons of placing stormwater treatment devices above and below ground are summarized below.

With a dedicated maintenance team, underground facilities may be out of sight, but these systems are never out of mind. Without regular checkups, it can be hard to see problems developing or to know when maintenance is needed (especially for low-flow systems with a higher susceptibility to clogging). Performance cannot be as easily measured or documented as with traditional surface BMPs. So, it is even more important to stay on top of the upkeep of all underground devices. Surface BMPs are still the baseline since these have an established track record and provide additional environmental benefits such as infiltration, groundwater recharge, and mitigation of the heat island effect. For these reasons, local jurisdiction often requires demonstrating that the performance of underground BMPs will be comparable to surface BMPs. If an underground device is used with access shaft, it must meet the following requirements: (1) must meet or exceed American Public Works Association (APWA) standards, (2) provide adequate ventilation, exhaust toxic gases, and have temperature control, (3) be reasonably accessible by a qualified maintenance worker, (4) have ladder and the ability to withstand lateral soil pressures, (5) have provisions for confined space entry, (6) have safety guard rails around the rim, (7) have no plastic or fiberglass interior parts that would break or shatter in the path of direct flow, (8) have pipes, conduits, and vaults not more than 20 ft below ground and be easily accessible by a vacuum truck hose for clean-out, and (9) provide means to block off the inflow and tail water backflow to isolate the device for safe maintenance and repair individual units.

Despite many drawbacks, underground proprietary BMPs are growing in popularity and are a good option for many sites. One of the primary reasons for the popularity of these systems is that they require less land surface. This is especially useful in urban areas where land is at a premium and dual use of the surface can occur. Proprietary systems can be cheaper than traditional BMPs depending on the needs of the site, even when maintenance is taken into consideration. They minimize the footprint, significantly cut back on landscaping, and provide savings to the watershed. Another benefit is that these systems can be chosen to target specific pollutants of concern. For this reason, they are sometimes used in conjunction with other BMPs to provide additional targeted water quality improvement. Where possible, underground systems can be incorporated into vegetated landscapes to allow infiltration comparable to detention ponds and infiltration basins.

TABLE 10.45
Manufacturers of Source-Control BMPs and Treatment Devices

Source-Control BMPs	Manufacturer
Bioretention Systems (Rain Garden)	Ferguson Waterworks - Urban Green Waterworks Infrastructure Products (https://www.fergusongss.com/contact-sales/) Contech Engineering Solutions - Stormwater Management Solutions (https://www.conteches.com/stormwater-management/)
Infiltration Percolation Trench	Massachusetts Clean Water Toolkit (https://megamanual.geosyntec.com/npsmanual/infiltrationtrenches.aspx) Ecovie Water Management (https://www.ecoviewater.com/products/drainmax/)
Permeable Paving	Truegrid Permeable Pavers (https://www.truegridpaver.com/) Invisible Structures (https://invisiblestructures.com/) Phone: 303-233-8383
Screens	TRASHRACHS- A Division of J.R.HOE (https://www.trashracks.com/products) Ferguson Waterworks-Stormwater Management: Filters and Screens (https://www.fergusongss.com/product-category/filters-and-screens/) Johnson Screens (https://johnsonscreens.com/products/well-screens-and-casing/) Huber Technology, Inc.-Wastewater Solution (https://www.huber-technology.com/) ISM Industrial Screen & Maintenance (https://industrialscreen.com/) Phone: 1-833-ISM-1969
Sedimentation Chamber	Titan Environmental Containment StormTech® - Underground Arch Chamber System (https://titanenviro.com) Phone: 1-866-327-1957 Ferguson Waterworks- Underground Modular Stormwater tanks (https://www.fergusongss.com/product/r-tank-stormwater-modules/)
Hydrocyclone	Contech Engineering Solutions - Stormwater Hydrodynamic Separation Solutions (https://www.conteches.com/stormwater-management/) Smith & Loveless Inc. - PISTA® Grit Removal System (https://gritthefacts.com/grit-removal-system-design-factors/) Phone: 913-888-5201 EUTEK Systems, Inc. - TEACUP™ Solids Classifier (https://www.wateronline.com/doc/eutek-systems-inc-0001)
Filtration Devices	New Pig- Sediment Drain Filter (https://www.newpig.com/pig-sediment-drain-filter/p/FLT748) Phone:1-855-493-HOGS Contech Engineered Solution - Flexible Stormwater Filtration Technology (https://www.conteches.com/stormwater-management/filtration/stormfilter/) StormTrap Filtration (https://stormtrap.com/solutions/treatment/filtration/) AquaShield, a Komline Company - Stormwater Filtration System (https://www.aquashieldinc.com/--aqua-filter.html) Phone: 888-344-9044

10.11 DESCRIPTION OF MANUFACTURED EQUIPMENT AND DEVICES

The most common manufactured equipment and devices for stormwater runoff treatment in urban areas are multi-chambers gravity settling tanks, hydrodynamic separation devices, screens, and filters. These devices are built on site, or supplied and installed by the vendors.

10.11.1 Gravity Separation in Multi-Compartment Device

Gravity settling chambers are the oldest and simplest means of removing large-diameter suspended particles from stormwater. High-efficiency settling chambers are often fitted with baffles or deflectors to change the flow directions. The primary requirement is a chamber in which the carrier stream velocity is reduced to allow the particulate matter to settle out of the moving liquid stream under the action of gravity. Literature indicates that a flow velocity less than about 0.3 m/s is needed to prevent re-entrainment of the settled particles. The velocity of the particles in the settling chamber can be obtained by Stokes' law. Curtains, baffles, and wire mesh screens may be suspended in the chamber to minimize turbulence and to ensure uniform flow. The head loss through the chamber is usually low, but major head loss occurs at the entrance and exit units. The collected particulate matter at the bottom of the chamber is removed manually or by a suction device during maintenance, and the collected waste is sent for disposal.

There are many benefits of a multi-chamber device. These benefits are (1) low initial cost, (2) simple construction, (3) low maintenance cost, (4) low head loss through the device, (5) constructed from locally available construction material, and (6) continuous or manual removal of settled solids. The disadvantages of this device are (1) large space requirements and (2) that only comparatively large particles (greater than 10 μm) can be removed. The cross-sectional view and design features of a multi-compartment settling basins are shown in Figure 10.65.

Because of the above advantages and disadvantages, settling chambers are mostly used as pre-cleaners. Use of settling chambers as pre-cleaners can also reduce the maintenance cost of the high-efficiency control equipment that is subject to abrasive deterioration.

Newton's and Stokes' equations: Gravity settling of particles in water is expressed by Newton's and Stokes' equations (Eqs. 10.65a and 10.65b).

$$v_s = \left[\frac{4g(S_s - 1)d}{3C_D}\right]^{1/2} \quad \text{(Newton's equation)} \tag{10.65a}$$

$$v_s = \frac{g\,(\rho_s - \rho)\,d^2}{18\mu} \quad \text{or} \quad v = \frac{g\,(S_s - 1)\,d^2}{18\nu} \quad \text{(Stoke's equation)} \tag{10.65b}$$

$$C_D = \frac{24}{N_R} + \frac{3}{\sqrt{N_R}} + 0.34 \tag{10.65c}$$

$$N_R = \frac{v_s \rho d}{\mu} \quad \text{or} \quad \frac{v_s d}{\nu} \tag{10.65d}$$

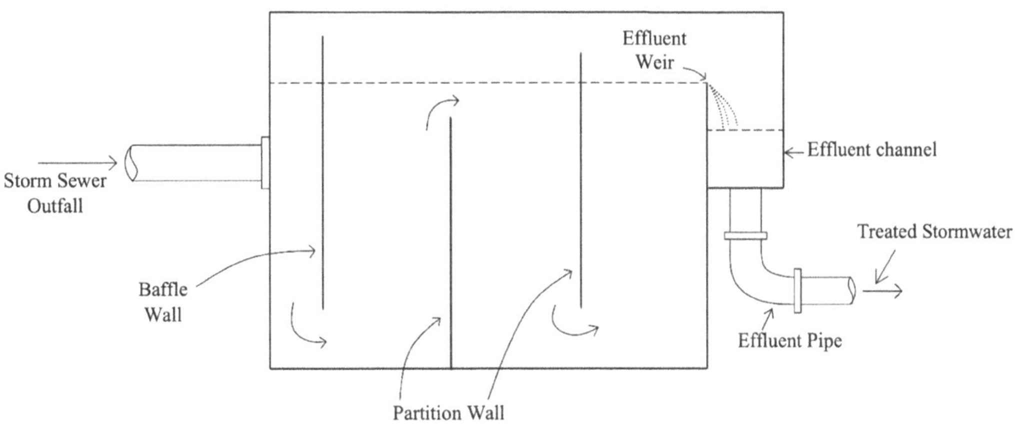

FIGURE 10.65 Multi-chamber settling basin for stormwater treatment of urban runoff.

where

C_D=drag coefficient, dimensionless

d=particle diameter, m (ft)

g=acceleration due to gravity, m/s^2 (ft/s^2)

N_R=Reynolds number, dimensionless

S_s=specific gravity of particle ($S_s = \frac{\rho_s}{\rho}$), dimensionless

v_s=settling velocity of particle, m/s (ft/s)

ρ=density of fluid, kg/m^3 (lb/ft^3)

ρ_s=density of particle, kg/m^3 (lb/ft^3)

μ=dynamic viscosity, N.s/m^2

ν=kinematic viscosity, m^2/s (ft^2/s)

ϕ=shape factor (the value of shape factor is 1 for spheres, and 2 for sand grains), dimensionless

Equation (10.65a) applies for spherical particles and is called Newton's law. The drag coefficient C_D is calculated from Eq. (10.65c). For laminar flow, N_R is less than 1. The first term in Eq. (10.65c) predominates, and C_D=24/N_R. Substituting this value in Eq. (10.65a) yields Stokes' equation (Eq. 10.65b).

Reynolds number N_R greater than 10^4 indicates turbulent flow. If the Reynolds number is greater than 1 and less than 10^4, the settling is in the transition range, and the settling velocity of a particle is determined from Eq. (10.65a) by a trial-and-error solution. If N_R is less than 1, Stokes' equation (Eq. 10.65b) applies.

Terminal velocity, overflow rate, hydraulic loading, or surface loading: In the design of a settling basin, the settling velocity v_t (also called terminal velocity) of the target particle is selected. The basin is designed such that all particles having a settling velocity equal to and greater than the terminal velocity are fully removed. The designed terminal velocity is equivalent to the overflow rate, hydraulic loading, or surface loading. These terms are numerically equal to the flow per unit surface area of the basin (m^3/m^2.day or gpd/ft^2). These relationships are expressed by Eqs. (10.66a)–(10.66c). It may be noted from Eq. (10.66b) that the design capacity of the basin is independent of the depth.

$$v_t = \frac{H}{\theta} \tag{10.66a}$$

$$v_t = \frac{Q}{A} \quad \text{or} \quad v_t = \frac{H}{t} \tag{10.66b}$$

$$\theta = \frac{V}{Q} \tag{10.66c}$$

where

v_t=design terminal velocity, overflow rate, hydraulic loading, or surface loading rate, m/s, m^3/m^2.day (gpd/ft^2)

H=side water depth, m (ft)

θ=hydraulic retention time, h

Q=flow rate, m^3/s (ft^3/s)

A=surface area of the basin (length × width), m^2 (ft^2)

V=basin volume, m^3 (ft^3)

t=time to fall the depth H or detention time s

The performance of an idealized settling basin is reduced significantly due to (1) turbulence at the influent, effluent, and sludge zones, (2) short-circuiting, and (3) volume occupied by the sludge and sludge-removal equipment. The design factors, therefore, are adjusted to account for these factors. The actual depth of the tank H is modified to an effective depth.

Fraction removed at a given overflow rate: A sediment removal facility is normally designed to remove particles of a given terminal velocity or overflow rate. The particles having a terminal velocity equal to and greater than the given overflow rate v_t are fully removed. The particles with a settling velocity less than v_t are partially removed. If the particles of all sizes are uniformly distributed over the entire depth of the influent zone, the particles with settling velocity less than v_t will be partially removed as expressed by Eq. (10.67a) and shown in Figure 10.66.

$$X_r = \frac{v_i}{v_t} \tag{10.67a}$$

where X_r = fraction of the particles removed with settling velocity v_i, (less than v_t), dimensionless.

The sediment suspension has particles of different sizes. The size distribution is determined from the standard sieve analysis. The velocity of each fraction is obtained from a column test or by using Newton's or Stokes' equations. A column used in such test is shown in Figure 10.66b.

A typical settling curve of discrete particles from a column analysis test is shown in Figure 10.66c. The fraction removed in a settling basin having a terminal velocity of v_t is given by Eq. (10.67b).

$$F = (1 - X_c) + \int_0^{X_C} \frac{v_i}{v_t}\,dx \approx (1 - X_c) + \sum_{i=1}^{i=n} \frac{v_i}{v_t}\,\Delta x_i \tag{10.67b}$$

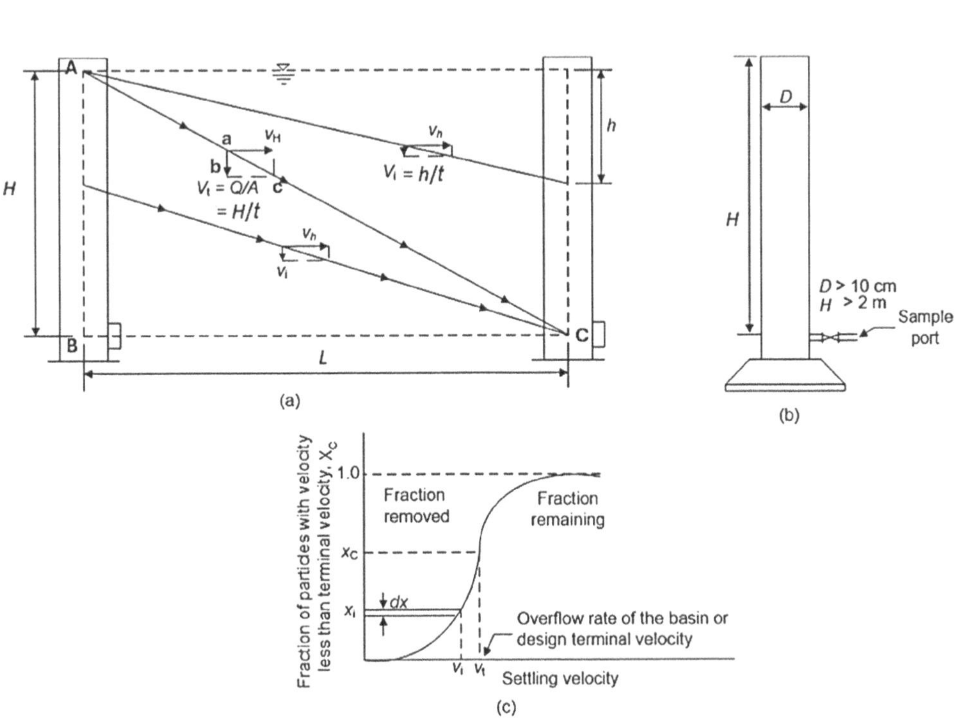

FIGURE 10.66 Settling behavior of discrete particles: (a) settling trajectory of particles $v_i < v_t$, (b) column for settling test, and (c) fraction removal curve. (Printed from Ref. 68 with permission of CRC Press, Taylor and Francis Group.)

where

F = fraction removed, dimensionless

X_c or $\sum_{i=1}^{i=n} \frac{v_i}{v_t} \Delta x_i$ = fraction of particles with velocity v_i less than v_t, m/s (ft/s)

$(1 - X_c)$ or $\left(1 - \sum_{i=1}^{i=n} \frac{v_i}{v_t} \Delta x_i\right)$ = fraction of particles removed with settling velocity v_i equal to and greater than v_t, dimensionless

$\int_0^{X_C} \frac{v_i}{v_t} dx$ or $\sum_{i=1}^{i=n} \frac{v_i}{v_t} \Delta x_i$ = fraction of particles removed with v_i less than v_t, dimensionless

v_i = settling velocity of the particles v_i less than v_t, m/s (ft/s)

t = time to fall the depth H or detention time s

The concept of fraction removed is difficult and may require careful review of the following solved examples to fully comprehend the text material. Also, readers should refer to Ref. 68 for a detailed discussion on this topic.

Example 10.64: Derivation of Newton's Equation

Newton's equation is developed by equating the gravitational force to the frictional resistance or drag. Develop Newton's equation to express the constant settling velocity of a particle in a fluid.

Solution

1. Draw the definition sketch of a settling particle in Figure 10.67.

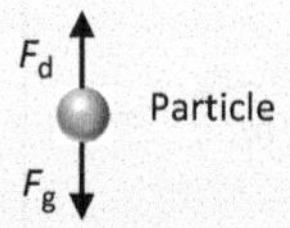

FIGURE 10.67 Definition sketch of a settling particle in water (Example 10.64).

2. Write equations for gravitational and frictional drag forces

   The gravitational force and frictional (or drag) force are expressed by Eqs. (10.68a) and (10.68b)

$$F_g = (\rho_s - \rho) g \, V \tag{10.68a}$$

$$F_d = \frac{C_D A \rho v_s^2}{2} \tag{10.68b}$$

   The volume and cross-sectional or projected area of a sphere are expressed by Eqs. (10.68c) and (10.68d)

$$V = \frac{\pi}{6} d^3 \tag{10.68c}$$

$$A = \frac{\pi}{4} d^2 \tag{10.68d}$$

where

F_g=gravitational force, which is the buoyant weight of the particle, N (lbf)

F_d=frictional (or drag) force caused by the interface between the particle and the fluid, N (lbf)

V=volume of the spherical particle, m^3 (ft^3)

A=cross-sectional or projected area of the particle at right angle to the direction of velocity, m^2 (ft^2)

C_D=drag coefficient, dimensionless

d=particle diameter, m (ft)

ρ=density of fluid, kg/m^3 (lb/ft^3)

ρ_s=density of particle, kg/m^3 (lb/ft^3)

g=acceleration caused by gravity, m/s^2 (ft/s^2)

N_R=Reynolds number, dimensionless

S_s=specific gravity of particle ($S_s = \frac{\rho_s}{\rho}$), dimensionless

v_s=settling velocity of particle, m/s (ft/s)

μ=dynamic viscosity, N.s/m^2 (lbf.s/ft^2)

ν=kinematic viscosity, m^2/s (ft^2/s)

3. Equate the two forces and simplify.

At a constant settling velocity of v, the gravitational force must be equal to the frictional force: $F_g=F_d$.

$$(\rho_s - \rho) g V = \frac{C_D A \rho v_s^2}{2}$$

$$(\rho_s - \rho) g \frac{\pi}{6} d^3 = \frac{C_D \frac{\pi}{4} d^2 \rho v_s^2}{2}$$

$$v_s = \left[\frac{4g(\rho_s - \rho)d}{3\rho C_D}\right]^{1/2} = \left[\frac{4g(S_s - 1)d}{3C_D}\right]^{1/2}$$

Newton's equation (Eq. 10.65a) is obtained.

Example 10.65: Derivation of Stokes' Equation

Stokes' equation applies for laminar condition when N_R is less than 1. Develop Stokes' equation from Newton's equation.

Solution

1. Obtain the value of C_D for N_R less than 1

C_D is obtained from Eq. (10.65c).

For N_R less than 1, the sum of the last two terms in Eq. (10.65c) is relatively much smaller than the first term. Therefore, the equation is simplified to $C_D = \frac{24}{N_R}$

2. Develop Stokes' equation

Substitute the value of $C_D = \frac{24}{N_R}$ and $N_R = \frac{v_s \rho d}{\mu}$ in Eq. (10.65a)

$$v_s = \left[\frac{4g(\rho_s - \rho)d}{3\rho \times \frac{24}{N_R}}\right]^{1/2} = \left[\frac{g(\rho_s - \rho)d}{18\rho} \times N_R\right]^{1/2} = \left[\frac{g(\rho_s - \rho)d}{18\rho} \times \frac{v_s \rho d}{\mu}\right]^{1/2} = \left[\frac{g(\rho - \rho)v_s d^2}{18\mu}\right]^{1/2}$$

$$v_s^2 = \frac{g(\rho_s - \rho)v_s d^2}{18\mu} \quad \text{or} \quad v_s = \frac{g(\rho_s - \rho)d^2}{18\mu} = \frac{g(S_s - 1)d^2}{18\nu}$$

Therefore, Stokes' equation (Eq. 10.65b) is obtained.

Example 10.66: Settling Equation under Turbulent Condition

The settling behavior of large-diameter particles is such that a turbulent condition is created, and N_R is greater than 10^4. Develop Newton's equation for settling velocity under turbulent conditions.

Solution

1. Determine C_D under turbulent conditions.
   C_D is obtained from Eq. (10.65c). Under turbulent conditions, $N_R > 10^4$. The sum of the first two terms in Eq. (10.65c) is significantly smaller than the third term. Therefore, these terms are ignored, and $C_D = 0.34$.
2. Develop the equation.
   Substitute $C_D = 0.34$ in Eq. (10.65a) and obtain the simplified settling equation (Eq. 10.69) under turbulent conditions.

$$v_s = \left[\frac{4g(\rho_s - \rho)d}{3\rho \times 0.34}\right]^{1/2} = \left[3.92g\frac{(\rho_s - \rho)}{\rho}d\right]^{1/2} = \left[3.92g(S_s - 1)d\right]^{1/2} \quad (10.69)$$

Example 10.67: Settling Velocity of Particles in the Transition Range

In a settling chamber, the sandy material is removed. Calculate the terminal settling velocity of the selected smallest particle that is fully removed. The specific gravity and mean diameter of the selected particle are 2.7 and 0.3 mm, respectively. Wastewater temperature is 10°C.

Solution

1. Calculate the terminal settling velocity v_s.
   The kinematic viscosity ν of wastewater at 10°C is obtained from Appendix B.

$$\nu = 1.307 \times 10^{-6}\ \text{m}^2/\text{s}$$

Assume that Stokes' equation (Eq. 10.65b) is applicable, and calculate the terminal settling velocity.

$$v_s = \frac{g(S_s - 1)d^2}{18\nu} = \frac{9.81\ \text{m/s}^2 \times (2.7 - 1) \times (0.3 \times 10^{-3}\ \text{m})^2}{18 \times \left(1.307 \times 10^{-6}\ \text{m}^2/\text{s}\right)} = 0.064\ \text{m/s}$$

2. Verify Reynolds number.
   Calculated Reynolds number from Eq. (10.65d).

$$N_R = \frac{v_s d}{\nu} = \frac{0.064\ \text{m/s} \times 0.3 \times 10^{-3}\ \text{m}}{1.307 \times 10^{-6}\ \text{m}^2/\text{s}} = 14.6$$

N_R is in the transition range ($1.0 < N_R < 10^4$) and Newton's equation is applicable. Use a trial-and-error procedure

3. Calculate *Trail 1.*

   Calculate C_D from Eq. (10.65c).

$$C_D = \frac{24}{N_R} + \frac{3}{\sqrt{N_R}} + 0.34 = \frac{24}{14.7} + \frac{3}{\sqrt{14.7}} + 0.34 = 2.76$$

   Calculate terminal settling velocity v_t from Eq. (10.68a).

$$v_t = \sqrt{\frac{4 \times 9.81\ \text{m/s}^2 \times (2.7-1) \times \left(0.3 \times 10^{-3}\ \text{m}\right)}{3 \times 2.76}} = 0.049\ \text{m/s}$$

   This is smaller than v_s=0.064 m/s calculated previously from Stokes' equation (Eq. 10.65b).

4. Calculate *Trail 2.*

$$N_R = \frac{0.049\ \text{m/s} \times 0.3 \times 10^{-3}\ \text{m}}{1.307 \times 10^{-6}\ \text{m}^2/\text{s}} = 11.3$$

$$C_D = \frac{24}{11.3} + \frac{3}{\sqrt{11.3}} + 0.34 = 3.36$$

$$v_s = \sqrt{\frac{4 \times 9.81\ \text{m/s}^2 \times (2.7-1) \times \left(0.3 \times 10^{-3}\ \text{m}\right)}{3 \times 3.36}} = 0.045\ \text{m/s}$$

   This is smaller than v_s=0.049 m/s calculated in *Trail 1.*

5. Repeat the trial-and-error procedure.

   The calculation procedure for terminal settling velocity v_s is repeated until v_s does not change from that obtained in the previous trial. Finally, this objective is achieved after trial 5.

$$N_R = 9.70$$
$$C_D = 3.78$$

$$v_s = 0.042\ \text{m/s}$$

   Therefore, the terminal settling velocity of the particle is 0.042 m/s.

Example 10.68: Terminal Velocity and Overflow Rate

A settling basin receives a suspension of discrete particles. The settling velocity of the smallest particle that is fully removed is the design terminal velocity v_t. Develop the following relationships and information: (1) terminal velocity that is equal to the design overflow rate (hydraulic or surface loadings), and (2) hydraulic retention time that is $\theta = V / Q$. Comment on these relationships.

Solution

1. Express the terminal velocity v_t as overflow rate.

   From similar triangles ABC and abc in Figure 10.66

$$\frac{v_t}{v_H} = \frac{H}{L}$$

$$v_t = v_H \times \frac{H}{L}$$

where v_H=horizontal velocity, m/s (ft/s)

$$\text{Horizontal velocity is also expressed } v_H = \frac{Q}{\text{Area perpendicular to flow}} = \frac{Q}{H \times \text{basin width } W}$$

$$v_t = \left(\frac{Q}{H \times W}\right) \times \frac{H}{L} = \frac{Q}{W \times L} = \frac{Q}{A(\text{surface area})}$$

Therefore, terminal velocity v_t is the same as the overflow rate (hydraulic or surface loading rate).

2. Express hydraulic retention time θ as V/Q.

   Equate v_t in Eqs. (10.66a) and (10.66b).

$$v_t = \frac{H}{\theta} = \frac{Q}{A}$$

$$\theta = \frac{A \times H}{Q} = \frac{L \times W \times H}{Q} = \frac{V}{Q}$$

3. Comments on these relationships.
   a. The overflow rate (surface loading or hydraulic loading rate) is a common basis for the design of discrete particles.
   b. The overflow rate of the basin is independent of the depth.
   c. Shallower basins are theoretically more efficient because of a larger surface area and, therefore, a smaller overflow rate. As a result, the basin will remove the suspension with a smaller terminal velocity or particle size.
   d. In a continuous-flow sedimentation basin, the length of the basin and the detention time (the average time a particle remains in the basin) are such that all particles of a given terminal velocity v_t must reach the bottom of the basin to be fully removed.
   e. The settling zone dimensions *H*, *L*, and *W* are used in the above derivations. In practice, the tank dimensions are obtained by using conservative values of overflow rate and detention time.

Example 10.69: Terminal Velocity, Horizontal Velocity, and Detention Time

A sediment removal facility is designed to treat a flow of 18,800 m³/day. The length, width, and depth of the channel are 18, 0.73, and 1.0 m, respectively. Calculate the following: (1) overflow rate, (2) terminal velocity v_t of the smallest particle that will be fully removed, (3) horizontal velocity, and (4) detention time.

Solution

1. Calculate the overflow rate and terminal settling velocity of the smallest particle that will be fully removed.

$$\text{Surface area of the channel } A = L \times W = 18 \text{ m} \times 0.73 \text{ m} = 13.14 \text{ m}^2$$

$$\text{Overflow rate} = \frac{Q}{A} = \frac{18{,}800 \text{ m}^3/\text{day}}{13.14 \text{ m}^2} = 1{,}431 \text{ m}^3/\text{m}^2.\text{day}$$

2. Calculate the terminal velocity.

$$\text{Terminal velocity, } v_t = 1{,}431 \frac{\text{m}^3}{\text{m}^2 \cdot \text{day}} \times \frac{\text{day}}{(24 \times 60)\ \text{min}} = 1 \text{ m/min or } 0.017 \text{ m/s}$$

3. Calculate horizontal velocity.

$$\text{Flow, } Q = 18{,}800 \frac{\text{m}^3}{\text{day}} \times \frac{\text{day}}{(24 \times 60 \times 60)\ \text{s}} = 0.22 \text{ m}^3/\text{s}$$

$$\text{Horizontal velocity, } v_H = \frac{Q}{W \times H} = \frac{0.22 \text{ m}^3/\text{s}}{0.73 \text{ m} \times 1.0 \text{ m}} = 0.30 \text{ m/s}$$

4. Calculate hydraulic detention time.

$$\text{Basin volume, } V = L \times W \times H = 18 \text{ m} \times 0.73 \text{ m} \times 1.0 \text{ m} = 13.14 \text{ m}^3$$

$$\text{Detention time, } \theta = \frac{V}{Q} = \frac{13.14 \text{ m}^3}{0.22 \text{ m}^3/\text{s}} = 60 \text{ s or } 1 \text{ min}$$

Example 10.70: Fraction Remaining and Settling Velocity

A column test was performed on a mixed sample of discrete particles. The column was filled with a mixed sample to a height of 3 m. The initial concentration of suspended solids at time zero is 285 mg/L. The concentration remaining after time 2 min of settling is 125 mg/L. Calculate: (1) weight fraction remaining, (2) weight fraction settled, and (3) settling velocity, terminal velocity of the smallest particle that is fully removed in the weight fraction, or overflow rate.

Solution

A. Calculate the weight fractions remaining and settled.

$$\text{Weight fraction remaining in suspension} = \frac{C}{C_0} = \frac{125 \text{ mg/L}}{285 \text{ mg/L}} = 0.44$$

$$\text{Weight fraction settled} = \frac{C_0 - C}{C} = \frac{(285 - 125) \text{ mg/L}}{285 \text{ mg/L}} = 0.56$$

B. Calculate the settling velocity or the terminal velocity of the smallest particle in the weight fraction that is fully removed.

$$\text{Settling velocity or } v_t = \frac{H}{\theta} = \frac{3 \text{ m}}{2 \text{ min}} = 1.5 \text{ m/min}$$

$$\text{Overflow rate} = 1.5 \text{ m/min} \times \frac{(24 \times 60) \text{ min}}{\text{day}} = 2{,}160 \text{ m}^3/\text{m}^2{\cdot}\text{day}$$

Example 10.71: Weight Fraction Removed from a Given Settling Velocity Data

A settling basin is designed to remove a discrete suspension. The results of a column test are given below. The settling basin has an overflow rate of 51,700 gpd/ft^2. Calculate the percent weight fraction of solids removed.

Settling velocity, ft/min	8.0	4.0	2.0	1.0	0.7	0.5
Weight fraction remaining	0.56	0.48	0.37	0.19	0.05	0.02

Solution

1. Calculate the design terminal velocity of the smallest particle that is fully removed at the given overflow rate.

$$v_t = \frac{51{,}700\ \text{gal/day}\cdot\text{ft}^2}{7.48\ \text{gal/ft}^3} \times \frac{\text{day}}{(24\times 60)\ \text{min}} = 4.8\ \text{ft}^3/\text{min}\cdot\text{ft}^2$$

2. Plot the velocity curve as a function of the fraction of particles remaining.
   The data are plotted in Figure 10.68. A smooth curve is drawn through the data points.
3. Draw a vertical line from the terminal velocity v_t of 4.8 ft/min.
   The vertical line intersects the curve (Figure 10.68) at 0.50 fraction of the suspension (X_c=0.50). This fraction has a settling velocity less than the terminal velocity v_t=4.8 ft/min.
4. Apply graphical integration between the limits x=0 and x=0.5 to determine the fraction removed.

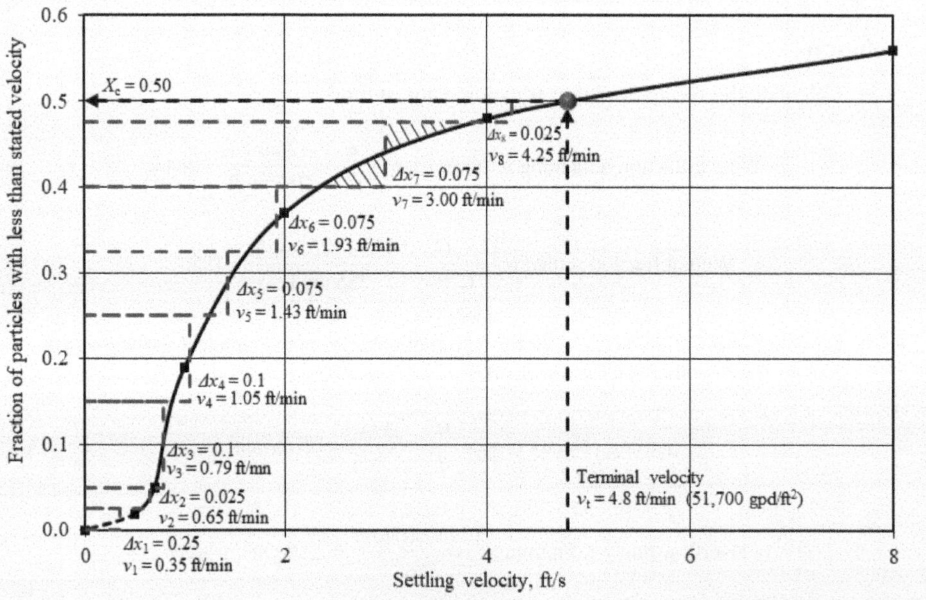

FIGURE 10.68 Fraction of particles remaining that have settling velocity equal to and less than the design terminal velocity, and graphical integration (Example 10.71). (Printed from Ref. 68 with permission of CRC Press, Taylor and Francis Group.)

Divide the area into small parallel strips. The selected velocity v_i should be such that the triangular area lost above the curve is gained in the strip. Tabulate each fraction and its settling velocity.

Δx_i	v_i, ft/min	$\frac{v_i}{v_t} \times \Delta x_i$
0.025	0.35	0.0018
0.025	0.65	0.0034
0.100	0.79	0.0165
0.100	1.05	0.0219
0.075	1.43	0.0223
0.075	1.93	0.0302
0.075	3.00	0.0469
0.025	4.25	0.0221
$X_c = \sum_{i=1}^{i=8} \Delta x_i = 0.50$		$\sum_{i=1}^{i=8} \frac{v_i}{v_t} \Delta x_i = 0.17$

5. Apply Eq. (10.67b) to calculate the fraction removed.

$$F = (1 - X_c) + \int_0^{X_C} \frac{v_i}{v_t} dx \approx (1 - X_c) + \sum_{i=1}^{i=8} \frac{v_i}{v_t} \Delta x_i = (1 - 0.5) + 0.17 = 0.67 \text{ or } 67\%$$

10.11.2 Hydrodynamic Separation Devices

The hydrodynamic separation devices are also called swirl concentrators. These devices remove trash, debris, and coarse sediments from incoming flows using centrifugal forces generated by forcing the influent into a circular motion. By having the water move in a circular fashion, it is possible to obtain significant removal of suspended sediments and attached pollutants within less space than gravity settling devices.[12,68] Hydrodynamic devices were originally developed for CSOs, where they were used primarily to remove coarse inorganic solids. These devices are now extensively used for grit removal from raw wastewater. They have also been adapted for stormwater treatment by several manufacturers to remove coarse solids down to sand-sized particles. Several types of hydrodynamic separation devices are also designed to remove floating oils and grease using sorbent media. For more information on specific hydrodynamic devices and their vendors, refer to Table 10.45. Hydrodynamic separators do not have any moving parts; therefore, maintenance requirements are not very intensive. To ensure that they are operating as efficiently as possible, proper maintenance involves frequent inspections throughout the year, especially after major storm events. The systems are considered full when the sediment level is within one foot of the unit's top, at which point it must be cleaned out. Removal of sediment can be performed with a sump vac or vacuum truck. Some hydrodynamic separators may contribute to mosquito breeding if they hold standing water between storms for longer than 72 h. Refer to the manufacturer's guidelines for inspection and maintenance requirements. Cross-sectional view of a hydrodynamic separation device is shown in Figure 10.69.

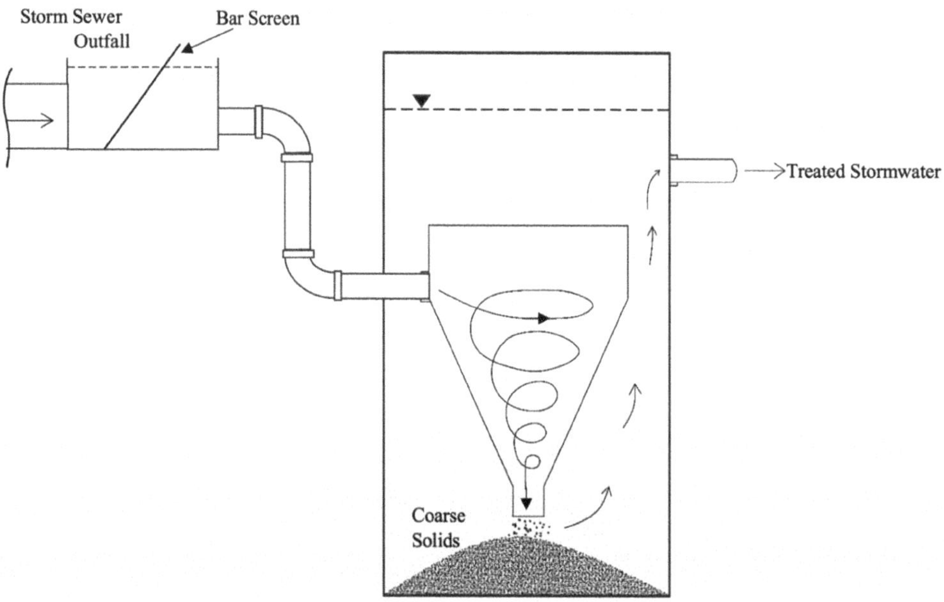

FIGURE 10.69 Hydrodynamic separation device.

Example 10.72: Hydrocyclone Separator to Treat Stormwater

Stormwater from a road construction site will enter a downstream creek. Due to site constraints, utilizing traditional BMPs such as retention ponds and swales is not feasible. Design a hydrocyclone separator to treat stormwater before it is released into the creek. The treated stormwater will help meet the water quality standard of the project site. The construction site is 5 acres with 60% grass area ($C=0.50$) and 40% paved area ($C=0.85$). Based on the monitoring data, the sediment concentration in the stormwater is reported as 5,000 mg/L with sediment gradation shown in Table 10.46. The density of water and solids is 1 and 2.5 tonnes/m^3. The cut size particle d_{50} is 75 μm (0.075 mm).

Solution

A. Calculate peak water quality flow (*WQF*)

Hydrocyclones are sized to treat runoff from a storm with an intensity of 0.5 in./h.[188]

Apply Eq. (8.7b)

$$Q_{\text{WQF}} = CiA$$

TABLE 10.46
Gradation of Sediment Carried by Stormwater

Diameter *d*, mm	2.36	0.85	0.6	0.3	0.25	0.18	0.15	0.08	0.05
Percentage by weight	2	5	11	13	14	12	13	15	15

$$\bar{C} = \frac{C_1A_1 + C_2A_2}{A_1 + A_2}$$

$$= \frac{0.50 \times 3.0 \text{ acres} + 0.85 \times 2.0 \text{ acres}}{5 \text{ acres}}$$

$$= 0.64$$

$$Q_{\text{WQF}} = 0.64 \times 0.5 \text{ in./h} \times 5 \text{ acres} = 1.60 \text{ cfs}$$

$$= 1.60 \text{ cfs} \times \frac{0.0283 \text{ m}^3\text{/s}}{1 \text{ cfs}} \times \frac{1{,}000 \text{ L}}{1 \text{ m}^3} \times 60 \text{ s} = 2{,}717 \text{ L/min}$$

B. Select hydrocyclone separator equation.

The hydrocyclone is designed using empirical equations. One empirical equation is expressed by Eq. (10.70).[189]

$$d_{50} = \frac{50.5 D_c^{0.46} D_i^{0.6} D_o^{1.21} e^{0.063\Phi}}{D_u^{0.71} L^{0.38} Q^{0.45} (\rho_s - \rho_w)^{0.5}} \quad (10.70)$$

where

d_{50}=cut size, μm. The cut size of a particle is the separation point between sediment particles that is either coarser or finer. If a sediment particle at the cut size enters the hydrocyclone, there is a 50% chance of it exiting through either outflow or underflow

D_c=hydrocyclone diameter, cm
D_i=inflow diameter, cm
D_o=outflow diameter, cm
D_u=underflow diameter, cm
L=cone section length, cm
Φ=volumetric sediment concentration in stormwater, %
Q=stormwater inflow, L/min
ρ_s=density of solid particles in stormwater, tonnes[1]/m^3
ρ_w=density of water, tonnes/m^3

C. Select the hydrocyclone separator geometry

The hydrocyclone typical geometric variables are shown in Figure 10.70.

The typical geometric variables of a hydrocyclone are expressed as a proportion of D_c. The proportions for D_i, D_o, and D_u and other geometric variables are selected using Bradley's and Plitt's conventional design relationships.[190,191] These proportions are presented in Table 10.47. The ratios of cone section length L and vortex finder length L_c with respect to D_c, and the cone angle θ are also given in this table. The D_u/D_c ratio can be selected as large as D_o/D_c.[192]

i. Express D_i, D_o, D_u, and L as a function D_c from Table 10.47

$$D_i = 0.28 D_c$$
$$D_o = 0.34 D_c$$
$$D_u = 0.25 D_c$$
$$L = 3.0 D_c$$

ii. Calculate hydrocyclone diameter D_c

Substitute the value of D_i, D_o, D_u, θ, and d_{50} in Eq. (10.70)

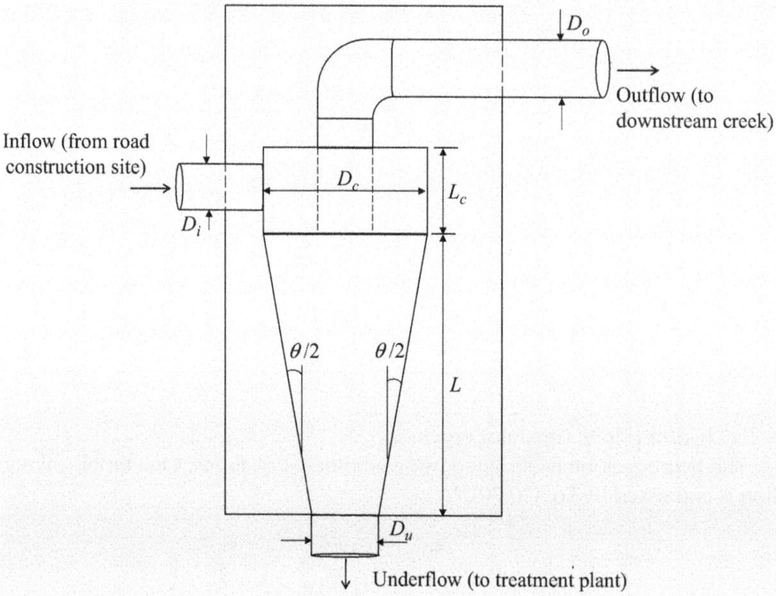

FIGURE 10.70 Typical geometric variables of hydrocyclone (Example 10.72).

TABLE 10.47
Hydrocyclone Typical Geometric Variables (Example 10.72)

D_i/D_c	D_o/D_c	D_u/D_c	L/D_c	L_c/D_c	θ	
0.28	0.34	0.25	3	0.33	12°	(D_c<25 cm)
					20°	(D_c>25 cm)

$$d_{50} = \frac{50.5D_c^{0.46}\ (0.28D_c)^{0.6}(0.34D_c)^{1.21}e^{(0.063\times 0.5)}}{(0.25D_c)^{0.71}(3.0D_c)^{0.38}(2{,}717\text{ L/min})^{0.45}(2.65\text{ tonnes/m}^3-1\text{ tonnes/m}^3)^{0.5}}$$

$$75\ \mu\text{m} = \frac{11.24D_c^{1.18}e^{0.0315}}{(2{,}717\text{ L/min})^{0.45}(1.65\text{ tonnes/m}^3)^{0.5}} = 0.258D_c^{1.18}$$

Rearrange this equation and solve for D_c

$$D_c = \left(\frac{75\ \mu\text{m}}{0.258}\right)^{1/1.18}$$

$$D_c = 122\text{ cm} \approx 125\text{ cm}$$

iii. Calculate the dimensions of hydrocyclone components

Inflow diameter $D_i = 0.28D_c = 0.28\times 125\text{ cm} = 35\text{ cm}$

Outflow diameter $D_o = 0.34D_c = 0.34\times 125\text{ cm} = 43\text{ cm}$

Underflow diameter $D_u = 0.25D_c = 0.25\times 125\text{ cm} = 31\text{ cm}$

Cone section length $L = 3.0D_c = 3\times 125\text{ cm} = 375\text{ cm}$

Vortex finder length $L_c = 0.33D_c = 0.33\times 120\text{ cm} = 41\text{ cm}$

Angle of cone $\theta = 20°$ (from Table 10.47 for $D_c > 25$ cm)

D. Determine hydrocyclone efficiency
   i. Calculate the percentage of each sediment size leaving hydrocyclone through the underflow pipe

Since the hydrocyclone in this example is designed for particle cut size d_{50} of 75 μm, there is a 50% chance for sediments of this size to discharge either through the outlet or underflow pipes. The percentage of other sediment sizes that discharge through underflow is calculated from Eq. (10.71).[193]

$$R_r = \frac{e^{4X} - 1}{e^{4X} + e^4 - 2} \tag{10.71}$$

where

R_r = recovery to underflow
X = ratio of particle diameter/particle cut size, d/d_{50}

Sample calculations for R_r for particle sizes d = 2.36 mm, 0.075 mm, and 0.045 are given below

$$(R_r)_{2.36} = \frac{e^{4\times(2.36\text{ mm}/0.075\text{ mm})} - 1}{e^{4\times(2.36\text{ mm}/0.075\text{ mm})} + e^4 - 2} = 1 \quad \text{or} \quad 100\%$$

$$(R_r)_{0.075} = \frac{e^{4\times(0.075\text{ mm}/0.075\text{ mm})} - 1}{e^{4\times(0.075\text{ mm}/0.075\text{ mm})} + e^4 - 2} = 0.5 \quad \text{or} \quad 50\%$$

$$(R_r)_{0.045} = \frac{e^{4\times(0.045\text{ mm}/0.075\text{ mm})} - 1}{e^{4\times(0.045\text{ mm}/0.075\text{ mm})} + e^4 - 2} = 0.16 \quad \text{or} \quad 16\%$$

The % recovery to underflow for all particle sizes is summarized in Table 10.47.

   ii. Determine the total removal efficiency of the hydrocyclone.

$$\text{Total efficiency} = \Sigma(\%\text{ sediment class in inflow} \times \%\text{ sediment reaching underflow})$$

$$= 2\% \times 1 + 5\% \times 1 + 11\% \times 1 + 13\% \times 0.99 + 14\% \times 0.99 + 12\% \times 0.99$$

$$+13\% \times 0.98 + 15\% \times 0.5 + 15\% \times 0.16 = 80\%$$

These results are summarized in Table 10.48 (column 5).

The hydrocyclone will remove 80% of the sediments at the construction site.

Example 10.73: Swirl Sediment Separator for Erosion Runoff Treatment

The stormwater runoff from a 15-acre residential area is to be treated before it is discharged into a detention pond. Design a sediment separator as an auxiliary treatment system to remove at least 65% of the sediment load carried by the stormwater. Three acres of the drainage area is buildings with a runoff coefficient of C=0.9, 2 acres is driveway and roads with C=0.95, and the remaining 10 acres is open spaces with C=0.35. The design rainfall intensity for the area is 0.5 in./h. The gradation of sediments carried by the stormwater from the sieve analysis and hydrometry test is given in Table 10.49.

TABLE 10.48
Calculation of Hydrocyclone Efficiency (Example 10.72)

Sediment Size (*d*), mm	Percent by Weight	$X = \frac{d}{d_{50}}$	R_r	(2)×(4)
(1)	(2)	(3)	(4)	(5)
2.36	2	31.5	1	2
0.85	5	11.3	1	5
0.6	11	8	1	11
0.3	13	4	0.99	12.99
0.25	14	3.3	0.99	13.99
0.18	12	2.4	0.99	11.98
0.15	13	2	0.98	12.77
0.075	15	1	0.5	7.5
0.045	15	0.6	0.16	2.36
				Σ=80%

TABLE 10.49
Particle Size of Sediment Carried by Stormwater

Particle Size (*d*), mm	%
2.0	1
0.84	1.5
0.25	9
0.149	25
0.125	35
0.074	15.5
0.045	13
	Σ 100%

Solution

A. Calculate peak runoff from the total area
   1. Calculate the weighted average of runoff coefficient

$$\text{Average runoff coefficient } \bar{C} = \frac{(3\text{ acres}\times 0.9) + (2\text{ acres}\times 0.95) + (10\text{ acres}\times 0.35)}{15\text{ acres}} = 0.54$$

   2. Calculate peak stormwater runoff
      Apply Eq. (8.7b)

$$Q = CiA$$

$$= 0.54 \times 0.5\text{ in./h} \times 15\text{ acres}$$

$$= 4.05\text{ cfs}$$

B. Select a swirl sediment separator for a feed capacity of 4.05 cfs and 5% drain-off
   1. A standard 12-ft diameter swirl chamber is selected.[194] The plan and sectional view, and dimensions of the structure are shown in Figure 10.71.
   2. Calculate the total removal efficiency of the system with 5% drain-off

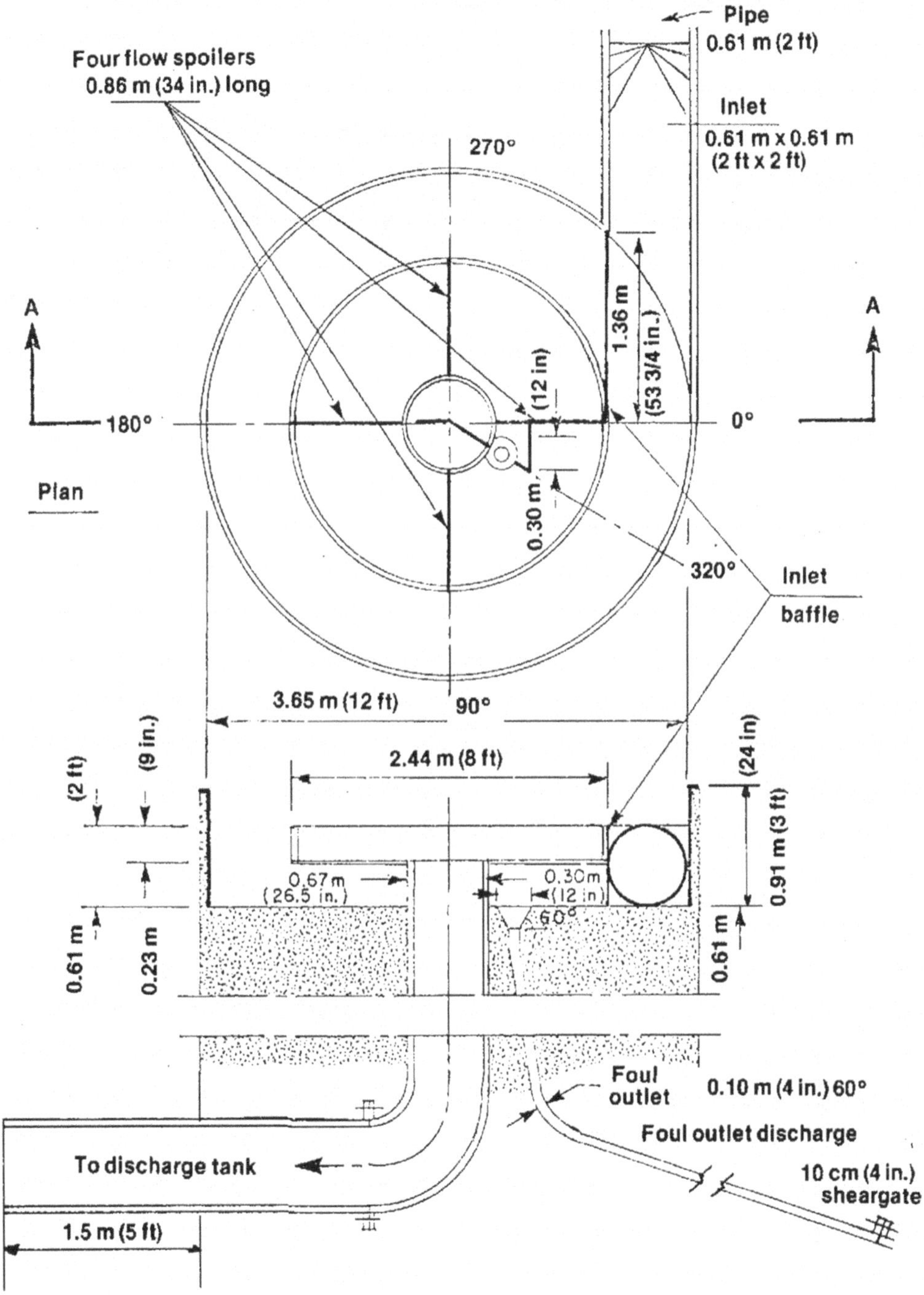

FIGURE 10.71 Design dimensions of swirl sediment separator for erosion runoff treatment (Ref. 194) (Example 10.73).

A. Obtain the percentage recovery (removal) of each particle size given in Table 10.49. Apply Figure B.7 in Appendix B. This figure provides the efficiency of a swirl separator in terms of percent removal rate of different particle sizes at different flow rates for a system with 5% drain-off.

B. The maximum particle size in Figure B.7 is 350 μ (0.35 mm), which has a removal of 98% (*y*-axis) at Q_p=4.05 cfs (*x*-axis). Particle sizes of 2 mm and 0.84 mm will be 100% removed at Q_p=4.05 cfs.

C. Removal of particle size 250 μ (0.25 mm)

From Figure B.7 at a feed rate of 4.05 cfs, the % removal particle size 0.25 mm is 78.

Similarly, calculate the percent removal of all remaining particles in Table 10.49 at a feed rate of 4.05 cfs.

These removal rates are given in Column (3) of Table 10.50.

3. Calculate the recovery (removal) of all fractions of particles in Table 10.49.

The percent recovery (removal) of all fractions of particle is obtained by multiplying fraction % (column (2)) by recovery % (column (3)) and dividing by 100.

These values are summarized in Column (4) of Table 10.50.

4. Calculate total fraction removal

The total fraction removal is the sum of Column (4)=54.1%

The total removal efficiency of the swirl system with 5% drain-off is calculated as 54.1%. Since this efficiency is smaller than the 65% desired efficiency, a system with a higher drain-off flow should be selected.

C. Select a swirl sediment separator for a feed capacity of 4.05 cfs and 14% drain-off

1. A standard 12-ft diameter swirl chamber of 14% drain-off is selected.
2. Repeat steps B.2 through B.4 using Figure B.9 in Appendix B.
3. Calculate the percentage removal of 0.125 mm particle size

Read from Figure B.9 125 μ (0.125 mm) particle recovery (removal) of 68% at Q_p=4.05 cfs.

The recovery (removal) of 0.125 mm particle is given in Column (3) of Table 10.51.

4. Similarly, the percent removal of all particle sizes in Table 10.49 is listed in Column (3) of Table 10.51

TABLE 10.50
Efficiency Analysis of the Swirl System with 5% Drain-off (Example 10.73)

Particle Size, mm (1)	Fraction, % (2)	Recovery (Removal), % (3)	Fraction Recovery (Removed), $\frac{(2)\times(3)}{100}$% (4)
2	1	100	1.00
0.84	1.5	100	1.50
0.25	9	78	7.02
0.149	25	58	14.50
0.125	35	52	18.20
0.074	15.5	44	6.82
0.045	13	39	5.07
			Σ54.1%

TABLE 10.51
Efficiency Analysis of the Swirl System with 14% Drain-off (Example 10.73)

Particle Size, mm (1)	Fraction, % (2)	Recovery (Removal), % (3)	Fraction Recovery (Removed), $\frac{(2)\times(3)}{100}$ % (4)
0.25	9	87	7.83
0.149	25	72	18.00
0.125	35	68	23.80
0.074	15.5	59	9.15
0.045	13	48	6.24
			Σ67.5%

5. Calculate the recovery (removal) of all fractions of particle in Table 10.49

   Repeat the calculation shown in step B.3 and summarize the results in Column (4) of Table 10.51.

   The total removal efficiency of the system is 67.5%, which is larger than the desired removal percentage of 65.

Example 10.74: Design Capacity of a Centrifugal Vortex-Induced Solids Removal

A centrifugal vortex-induced solids removal chamber that has the following dimensions:

Upper cylindrical chamber	
Diameter, D_{upper}	1.8 m
Height, H_{upper}	1.0 m
Hopper frustum	
Height, H_{hopper}	0.5 m
Lower cylindrical section connecting the hopper and the grit transfer cone	
Diameter, D_{lower}	1.0 m
Height, H_{lower}	0.8 m

The design total detention time in the unit (t_{total}) is 25 s. Assume that the solids transfer cone does not provide any flow retention. Calculate the design capacity of the vortex-induced solids removal chamber.

Solution

1. Draw the definition sketch (Figure 10.72)
2. Calculate the volume of the upper cylindrical chamber.

$$\text{Area, } A_{upper} = \frac{\pi}{4}\left(D_{upper}\right)^2 = \frac{\pi}{4}(1.8 \text{ m})^2 = 2.54 \text{ m}^2$$

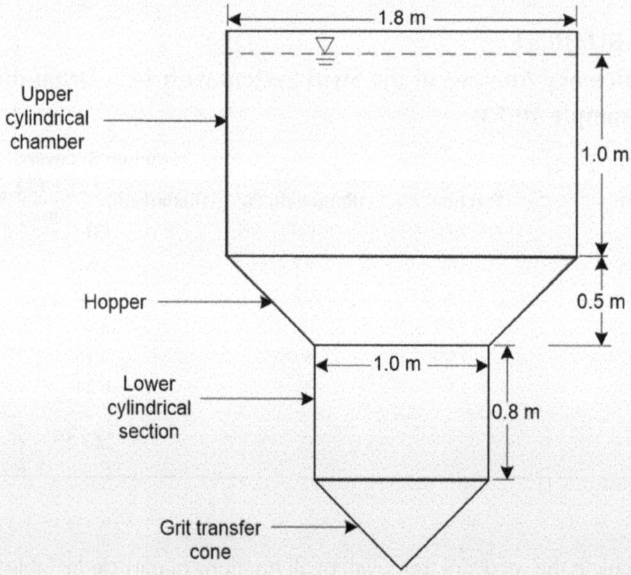

FIGURE 10.72 Definition sketch of centrifugal vortex-induced solids removal chamber (Example 10.74).

$$\text{Volume, } V_{\text{upper}} = A_{\text{upper}} H_{\text{upper}} = 2.54 \text{ m}^2 \times 1.0 \text{ m} = 2.54 \text{ m}^3$$

C. Calculate the volume of the hopper.

Volume of the hopper is calculated from Eq. (10.72).

$$V = \frac{1}{3} H \left[A_1 + A_2 + \sqrt{A_1 A_2} \right] \tag{10.72}$$

where

V=volume of frustum of a cone, m^3 (ft^3)

H=height of the cone frustum, m (ft)

A_1=area of the upper portion of the cone frustum, m^2 (ft^2)

A_2=area of the lower portion of the cone frustum, m^2 (ft^2)

$$A_1 = A_{\text{upper}} = 2.54 \text{ m}^2$$

$$A_2 = \frac{\pi}{4} (D_{\text{lower}})^2 = \frac{\pi}{4} (1.0 \text{ m})^2 = 0.79 \text{ m}^2$$

$$\text{Volume, } V_{\text{hopper}} = \frac{1}{3} H_{\text{hopper}} \left[A_1 + A_2 + \sqrt{A_1 A_2} \right]$$

$$= \frac{1}{3} \times 0.5 \text{ m} \times \left[2.54 \text{ m}^2 + 0.79 \text{ m}^2 + \sqrt{2.54 \text{ m}^2 \times 0.79 \text{ m}^2} \right]$$

$$A_{\text{lower}} = A_2 = 0.79 \text{ m}^2$$

$$\text{Volume, } V_{\text{lower}} = A_{\text{lower}} H_{\text{lower}} = 0.79 \text{ m}^2 \times 0.8 \text{ m} = 0.63 \text{ m}^3$$

D. Determined the total volume of the solids chamber.

$$\text{Volume, } V_{\text{total}} = V_{\text{upper}} + V_{\text{hopper}} + V_{\text{lower}} = 2.54\ \text{m}^3 + 0.79\ \text{m}^3 + 0.63\ \text{m}^3 = 3.96\ \text{m}^3$$

E. Calculate the design capacity of the centrifugal vortex-induced solids chamber.

$$\text{Design capacity, } Q = \frac{V_{\text{total}}}{t_{\text{total}}} = \frac{3.96\ \text{m}^3}{25\ \text{s}} = 0.158\ \text{m}^3/\text{s or } 13{,}700\ \text{m}^3/\text{day}\ (3.6\ \text{mgd})$$

10.11.3 Screening Devices

Many types of stationary screening are utilized for stormwater treatment in urban watersheds. Common types are stationary screens are (1) coarse screen, trash rack, or bar screens and (2) medium and fine screens. These static screens are installed at different locations to remove the debris from the stormwater runoff. The most common locations are (1) vertical, inclined, or horizontal coarse screens at the street curbside inlet point of stormwater runoff, (2) in a screen box at or near the outfall of storm sewers, and (3) in a screen chamber at the sewer outfall with inclined bar racks. The inclination of the bars is 40°–60° from the horizontal for manual cleaning. The coarse screens are composed of parallel bars or coarse woven-wire screens. The clear bar spacings and openings of coarse woven-wire screens are 6.4–19.1 mm (1/4–3/4 in.). The fine screen openings are 0.8–6.4 mm (1/32–1/4 in.). The screens are cleaned manually after each major storm, and debris

(a) (b) (c) (d)

FIGURE 10.73 Manually cleaned stormwater screening device: (a) vertical screen at the curb inlet, (b) horizontal screen at the curb drain, (c) screen box at the storm sewer outfall, and (d) screen chamber at storm sewer outfall.

and screenings are collected and sent to a landfill for disposal. The details of screening devices are shown in Figure 10.73.

Location and maintenance: The prime objective of the screening device is to remove leaves, rags, paper, plastic, and other floating debris from the stormwater runoff. For this reason, they should be provided at the inlet point of the storm sewer or at the outfall. Proper cleaning and maintenance of the device is essential.

Manual bar screen is ideal for the removal of coarse material. The screen consists of a stainless steel frame fixed to the wall with a central screening area. The frame is installed in a channel with an inclination angle. Stormwater passes through the clear spaces between bars, and screenings are captured. The screenings are removed manually using a rake and stored for disposal. Mesh or wire screens have stainless steel cross wires of different diameters. The spacings of wires determine the size of the holes or open spaces in the screen. Mesh sizes vary and are selected to capture the particles of a given maximum size. Finer screens will capture smaller-sized material but will consequently become clogged faster than screens with larger openings.

Head loss: The maximum head loss through the screen may range from 250 to 600 mm (10 to 24 in.). For design purposes, the head loss through the screen may be calculated from Eq. (10.73).

This equation is used to calculate head loss through clean or partly clogged bars.

The head loss across the entire screen chamber should also include head loss due to entrance, bends, expansion, contraction, and exit.

$$h_L = \frac{V^2 - v^2}{2g} \times \left(\frac{1}{C_d}\right) \tag{10.73}$$

where

h_L = head loss across the screen, m (ft)

V, v = velocity through the screen, and approach velocity in the channel upstream of the screen, m/s (ft/s)

g = acceleration due to gravity, 9.81 m/s^2 (32.2 ft/s^2)

C_d = coefficient of discharge, dimensionless. C_d = 0.60 for clogged screen, and 0.70 for clean screen

The basic design factors for manually cleaned bar screens are presented in Table 10.52.[12,68]

TABLE 10.52
Basic Design Factors for Manually Cleaned and Mechanically Cleaned Screens

Design Factor	Manually Cleaned
Velocity through screen, m/s (ft/s)	0.3–0.6 (1–2)
Bar size	
Width, mm (in.)	4–8 (0.16–0.32)
Depth, mm (in.)	25–50 (1–2)
Clear spacing between bars, mm (in.)	25–75 (1–3)
Slope from horizontal, degrees	45–60
Allowable head loss, clogged screen, mm (in.)	150 (6)
Maximum head loss, clogged screen, mm (in.)	800 (32)

Example 10.75: Head Loss across a Bar Screen

A manually cleaned bar screen is installed in a channel. There are 28 parallel, sharp-edged rectangular bars 10 mm (0.39 in.) wide placed at a clear spacing of 20 mm (0.79 in.). The flow through the screen is 0.75 m³/s, and the depth of flow is 1.45 m upstream of the screen. The angle of inclination of bars is 45° from the horizontal. Calculate: (1) head loss across a clean screen, $C_d=0.7$, and (2) head loss across the screen when the screen is 50% clogged, $C_d=0.6$.

Solution

1. Calculate the clear area of the screen.

$$\text{Total width of clear opening} = 29 \text{ spaces} \times 20 \text{ mm/space} \times \frac{\text{m}}{10^3 \text{ mm}} = 0.58 \text{ m}$$

2. Calculate the cross-sectional area of the channel.

$$\text{Total width of all bars} = 28 \text{ bars} \times 10 \text{ mm/bar} \times \frac{\text{m}}{10^3 \text{ mm}} = 0.28 \text{ m}$$

$$\text{Total width of the channel} = 0.58 \text{ m} + 0.28 \text{ m} = 0.86 \text{ m}$$

3. Calculate the approach velocity in the channel and velocity through screen openings.

$$\text{Approach velocity} = \frac{0.75 \text{ m}^3/\text{s}}{1.25 \text{ m}^2} = 0.60 \text{ m/s}$$

Clear screen opening area perpendicular to the direction of flow

$$\text{Total clear opening area} = 0.84 \text{ m}^2$$

$$\text{Velocity through screen openings} = \frac{0.75 \text{ m}^3/\text{s}}{0.84 \text{ m}^2} = 0.89 \text{ m/s}$$

4. Calculate the head loss through clean screen.
   Calculate the head loss h_L from Eq. (10.73) with $C_d=0.7$.

$$h_L = \frac{(0.89 \text{ m/s})^2 - (0.60 \text{ m/s})^2}{2 \times 9.81 \text{ m/s}^2} \times \frac{1}{0.7} = 0.031 \text{ m } (0.10 \text{ ft})$$

5. Calculate the head loss through 50% clogged screen.
   The head loss through 50% clogged screen h_{L50} is calculated from Eq. (10.73) with $C_d=0.6$. At 50% clogged screen, the area is reduced by half, and velocity through the screen is doubled.

$$h_{L50} = \frac{(2 \times 0.89 \text{ m/s})^2 - (0.60 \text{ m/s})^2}{2 \times 9.81 \text{ m/s}^2} \times \frac{1}{0.6} = 0.24 \text{ m } \left(0.79 \text{ ft}\right)$$

The head loss through a clogged screen is about 10 times that of a clean screen.

10.11.4 Filtration Devices

Filtration BMPs are very effective in treating urban stormwater runoff. TSS, silt, sand, particulate phosphorus, metals, and most organics are removed as stormwater flows through the filtering medium. Filtration is less effective for soluble pollutants such as dissolved phosphorus, chloride, and nitrate. Their effectiveness for bacteria varies with filtration devices. Common filtration media are sand and resin beds, metal wires, fabric usually polypropylene, and membranes assembled in beds, sheets, cages, boxes, and cartridges. They are primarily designed for pollutant removal. Water is forced through the filter media, and most filtration BMPs will achieve some volume reduction, depending on the BMP design. The filter media are assembled to produce chamber filters, catch basin inserts, and cartridge filters.

Chamber filter or box filter: Gravity sand filters used for stormwater treatment are like slow sand filters used for municipal water treatment. They consist of filter housing, filter sand, bottom gravel layer, and filtered water collection piping. Filtration media is coarse mixed sand having a high filtration rate of up to 24.4 $m^3/m^2.h$ (10 gpm/ft^2). After a major storm, the filter surface and housing are cleaned of debris. A few inches of filter media surface are mixed and turned over to make sure that surface clogging does not occur. The details of stormwater filter are shown in Figure 10.74.

Catch basin inserts: Catch basin inserts are manufactured filters or fabric placed in a drop inlet. They remove sediment and debris, and may include sorbent media to remove floating oils and grease. There are many types of inserts of various shapes and configurations. Typically, these configurations fall into three groups: socks, boxes, and trays. The *sock-type filters* are typically constructed of a fabric, usually polypropylene. The fabric may be attached to a frame or grate, which may hold the sock. Socks are meant for vertical (drop) inlets. *Boxes* are constructed of plastic or wire mesh. Typically, a polypropylene "bag" is placed in the wire mesh box, and the bag takes the form of the box. Settling and filtration through media occur in the box. The *trays* may hold different types of media that vary by manufacturer. Types include polypropylene, porous polymer, treated cellulose, and activated carbon. Inserts are an easy and inexpensive retrofitting option because most components have standard drainage systems. Inserts are usually suitable for application on relatively small tributary areas (less than 1 acre). Catch basin inserts come in a wide range of configurations. It is difficult to generalize the expected performance. They are mainly used for catching coarse sediments and floatable trash and are effective as pretreatment in combination with other types of water quality treatment BMPs. Trash and large objects can greatly reduce the effectiveness of catch basin inserts with respect to sediment and hydrocarbon capture. Frequent maintenance and use of screens and grates may keep trash out and decrease the likelihood of clogging and bypassing of incoming flows. They require maintenance due to their susceptibility to accumulating trash and debris. Regular maintenance activities include cleanup and removal of accumulated trash and sediment, while major maintenance activities include replacing filter media or catch basin inserts. The vendors of catch basin inserts and specific information about their units are provided in Table 10.45. The catch basin inserts are shown in Figure 10.75.

Cartridge filters: Cartridge filters typically consist of a series of vertical filters contained in a vault or catch basin that provide treatment through filtration and sedimentation. The vault may be divided into multiple chambers, where the first chamber acts as a pre-settling basin for removal of coarse sediment, while another chamber acts as the filter bay and houses the filter cartridges. The performance and capacity of a cartridge filter installation depend on the properties of the media contained in the cartridges. Cartridge filter manufacturers often provide an array of media types, each with varying properties, targeting various pollutants and a range of particle sizes. Cartridge filters have been proven to provide efficient removal of fine suspended and some dissolved constituents; however, they are less adaptive in handling high flow rates as compared to catch basin inserts and hydrodynamic devices. This is mainly due to the enhanced treatment provided through the filtration mechanism. The design details of cartridge filter are shown in Figure 10.76. Manufacturers try to distinguish their products through innovative cartridge designs that aim at providing self-cleaning

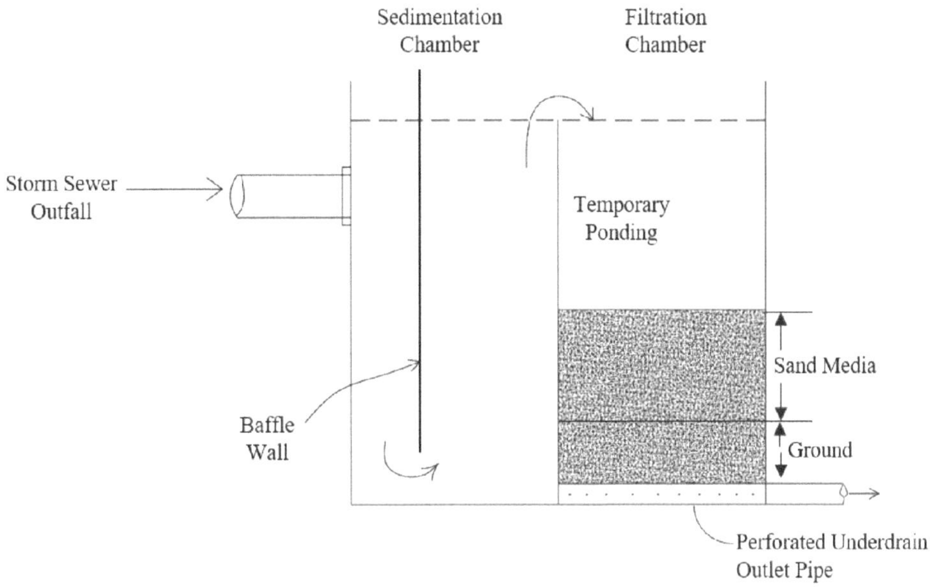

(a)

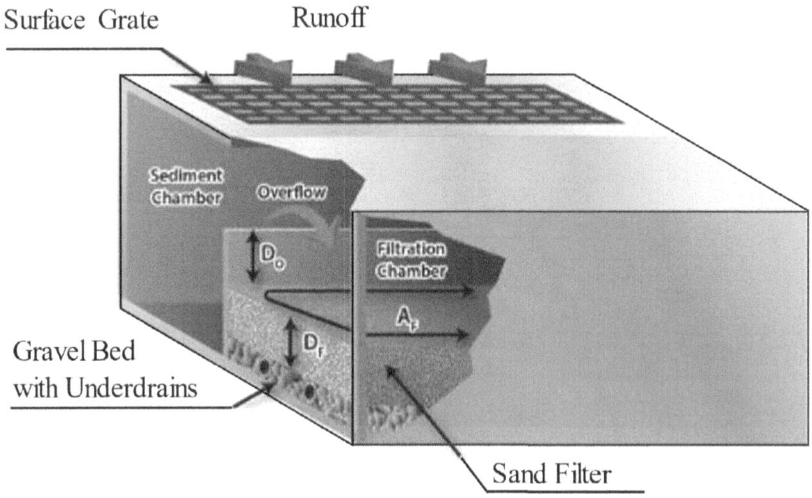

(b)

FIGURE 10.74 Chamber or box gravity filtration systems: (a) longitudinal section of filter bed preceded by a sediment chamber and (b) components of a gravity sand filter system to treat urban stormwater runoff. D_o = water head over filter media, D_F = filter media depth, A_F = filter surface area.

and draining, uniform loading, and clog resistance, allowing the devices to function properly over a wide range of hydraulic loadings and pollutant concentrations. For more information on specific cartridge filter models and their vendors, refer to Table 10.45.

Cartridge filter maintenance activities include periodically removing captured trash, debris, and sediments from the vault floor, typically twice per year, depending on the accumulation rate. A sump vac or vacuum truck is used to achieve that. The media in the filter cartridge must be replaced when it becomes saturated, typically, about once every other year, depending on the pollutant accumulation rate. The manufacturers of these devices typically provide contract operations and maintenance services. All storm water vaults containing cartridge filters that have standing water

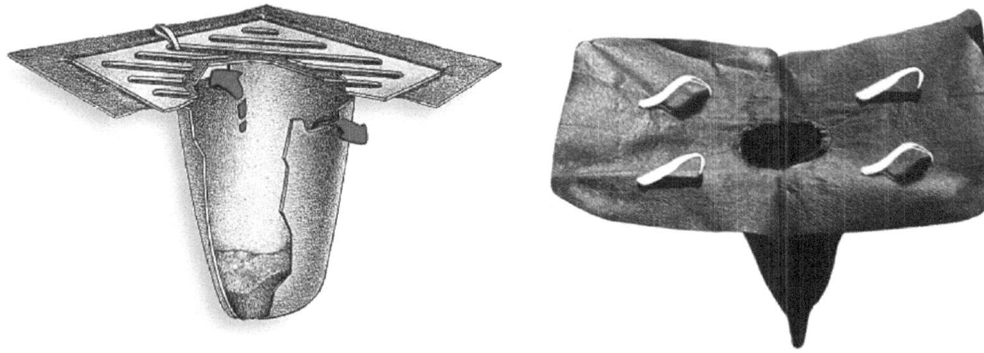

(a) (b)

FIGURE 10.75 Catch basin inserts: (a) non-woven geotextile material with built-in overflow ports for capturing sediment, debris, oil, and grease at storm water catch basins, and (b) absorptive media of flexible open mesh fabric envelope containing specialized media blend. (Printed from the U.S. EPA website.)

FIGURE 10.76 Cartridge filter assembly for treatment of stormwater flow. (Printed from the U.S. EPA website.)

for longer than 72h can become a breeding area for mosquitoes. Manufacturers have developed systems to completely drain the vault, such as a perforated pipe installed in the bottom of the vault that is encased in a filter sock to prevent clogging.

Design details of chamber filter or box filters: Most chamber filters have a rectangular housing (filter box). The housing contains uniform-size filter sand, gravel support layer, underdrain system, and effluent pipe. The influent pipe is the stormwater outfall pipe. The filter arrangement and components of the chamber filter are shown in Figure 10.74.

Filter design parameters: The most important design considerations are filter media selection and filter depth. The head loss through clean bed is important in the selection of filter media and its depth. The head loss through clean bed should be small.

Head loss through clean filter: Many empirical and semi-empirical equations have been developed to predict the performance of a clean filter. These include the Carmen-Kozeny, Fair-Hatch, Rose, and Hazen equations.[150,195–197] The most commonly used relationship is Carmen-Kozeny's equation, and it is expressed by Eqs. (10.74a)–(10.74c).

$$\text{Carmen Kozeny } h_L = \frac{f}{\Phi}\frac{(1-e)}{e^3}\frac{L}{d}\frac{v^2}{g} \tag{10.74a}$$

$$\text{Friction factor } f = 150\frac{(1-e)}{N_R} + 1.75 \tag{10.74b}$$

$$\text{Reynolds number } N_R = \Phi\frac{dv}{\mu}\rho_w \tag{10.74c}$$

where

d = media grain diameter, m (ft)
e = porosity ratio, dimensionless
f = friction factor, dimensionless
g = acceleration due to gravity, 9.81 m/s^2
h_L = head loss, m (ft)
L = media depth, m (ft)
N_R = Reynolds number, dimensionless
v = filtration velocity, m/s (ft/s)
μ = absolute viscosity, N.s/m^2 (lbf/ft^2)
ρ_w = density of water, kg/m^3 (slug/ft^3)
Φ = particle shape factor, usually 0.85–1.0

Miscellaneous head losses: In addition to filter media, head losses are also encountered through the (1) gravel bed, (2) underdains, and (3) pipings. The head loss calculations through the supporting gravel bed are similar to the filter media. The gravel bed is considered in several layers in series because of variability in the gravel media. Generally, the head losses in the gravel bed and underdrain system are small and are ignored. The pipings include the influent and effluent pipes. The head losses through these pipes are calculated by Darcy-Weisbach or Hazen-Williams equations (5.13a and 5.18).

Example 10.76: Dimensions of a Gravity Sand Filter

A storm sewer discharges 0.212 m^3/s (7.5 cfs) of storm runoff under peak flow conditions. A gravity sand filter is designed at the outfall. The design filtration rate is 24.4 m^3/m^2.h (10 gpm/ft^2). Calculate the dimensions of a square filter bed.

Solution

1. Calculate the design filtration rate in m^3/m^2.s
   Design filtration rate = 24.4 m^3/m^2.h × 360 s/h = 0.00678 m^3/m^2.s
2. Calculate the filter area
   Filter area = 0.212 m^3/s/0.00678 m^3/m^2.s = 31.27 m^2
3. Calculate the dimensions of a square filter
   Length and width of the filter is (31.27 m^2)½ = 5.59 m (18.35 ft)

Example 10.77: Head Loss through a Bed of Uniform Sand

A filter bed of uniform sand is 0.31 m deep. The temperature of storm runoff is 20°C. The design filtration rate is 24.4 m^3/m^2.h. At this filtration rate, the flow is passed through the bed at a filtering velocity of 0.00678 m/s. The grain size is 1.2 mm, the shape factor is 0.85, and the porosity of the bed is 0.51. Determine the head loss through the clean bed.

Solution

1. Determine the Reynolds number (N_R) from Eq. (10.74c).
   At 20°C, $\mu = 1.002 \times 10^{-3}$ kg/m.s and $\rho_w = 998.2$ kg/m^3 from Table B.2.

$$d = 1.2 \text{ mm}, \quad v_s = 0.00678 \text{ m/s}, \quad \text{and} \quad \Phi = 0.85$$

$$N_R = \Phi \frac{d \times v}{\mu} \rho_w$$

$$N_R = \left[0.85 \times 998.2 \text{ kg/m}^3 / \left(1.002 \times 10^{-3} \text{ kg/m.s}\right)\right] \times 1.2 \times 10^{-3} \text{ m} \times 0.00678 \text{ m/s}$$

$$= 6.9 > 1.0 (\text{turbulent flow})$$

2. Determine the friction factor (f) from Eq. (10.74b).

$$f = 150 \frac{(1 - e)}{N_R} + 1.75$$

$$f = [150 \times (1\ 0.51) / 6.9] + 1.75 = 12.4$$

10.1 Determine the head loss from Eq. (10.74a) at $L = 0.31$ m

$$h_L = \frac{f}{\Phi} \frac{(1 - e)}{e^3} \frac{L}{d} \frac{v^2}{g}$$

$$h_L = 12.4(1\ 0.51) \times 0.31 \text{ m} \left(0.00678 \text{ m/s}\right)^2 / \left(0.85 (0.51)^3 \times 1.2 \times 10^{-3} \text{ m} \times 9.81 \text{ m/s}^2\right)$$

$$= 0.62 \text{ m} \quad \text{or} \quad 6.2 \text{ cm} \left(2.44 \text{ in.}\right)$$

A head loss of 6.2 cm through a clean filter bed at a filtration velocity of 0.00678 m/s is acceptable.

10.12 REVIEW PROBLEMS

10.1 An older city with a population of 15,000 has a CSS. The average wastewater flow is 425 Lpcd. During wet weather, the maximum flow allowed into the interceptor leading to the treatment plant is the peak dry weather flow plus an inflow/infiltration (I/I) allowance of 80 Lpcd. If the total combined sewer flow during a critical wet weather event is 35,000 m^3/day, determine the maximum CSO volume that must be diverted to a temporary storage facility.

10.2 An old town has CSS. A maximum flow of 2,500 m^3/h must be released into the interceptor that carries the flow to the wastewater treatment plant. Determine the volume of a temporary storage facility to hold the excess CSO. The critical wet weather flow data is given below:

Time of day	12 MN	2	4	6	8	10	12 N
Wet weather combined sewer flow, m^3/h	1,000	500	600	1,500	2,500	3,400	3,750
Time of day	2	4	6	8	10	12 MN	
Wet weather combined sewer flow, m^3/h	3,650	3,300	2,900	2,450	2,000	1,500	

MN, midnight; N, noon.

10.3 An older subdivision of a city has a CSS. During dry weather, the average TSS concentration in municipal wastewater is 110 mg/L, with an average flow of 16,500 m^3/day. During wet weather, the average stormwater runoff contributes an additional 42,300 m^3/day, with a TSS concentration of 350 mg/L. Determine the average TSS concentration in the combined sewer flow during wet weather conditions.

10.4 The daily streamflow data for a river are provided in the table below. The watershed area is 500 km^2. (1) Separate the baseflow from the total hydrograph to obtain the DRH, and (2) calculate the equivalent depth of the direct runoff in centimeters over the watershed.

Time, d	Q, m^3/s	Time, day	Q, m^3/s
1	16	9	28
2	15	10	22
3	50	11	18.5
4	110	12	16
5	86	13	13.3
6	65	14	13
7	50	15	12.8
8	38	16	12.5

10.5 An 18-acre tract of land has been developed into a commercial site. The runoff coefficient of the site is 0.82. The time of concentration of the site is estimated to be 25 min. The design storm intensity for this location is 3.6 in./h. Using the Modified Rational Method, construct the stormwater runoff hydrograph for a rainfall duration of 25 min. Repeat the hydrograph construction for a rainfall duration of 70 min and calculate the stormwater runoff volume for both cases.

10.6 A 15-acre watershed is planned for commercial development. A detention basin is required to limit the post-development peak discharge to pre-development conditions. The rainfall data for a storm with a 10-year return period are provided in the following table. Determine the storage volume required for the detention basin. The estimated time of concentration for the post-development watershed is 10 min. The design storm has a pre-development 10-year frequency, with rainfall duration and intensity of 15 min and 3.8 in./h, respectively. The average runoff coefficient (C) is 0.28 for the unurbanized watershed and 0.72 for the developed watershed.

Duration, min	5	10	15	20	30	60
Intensity, in./h	5.1	4.2	3.8	3.0	2.1	1.4

10.7 A 300-acre fully developed urban watershed is located in North Carolina. Of this area, 200 acres are medium-density residential lots (1/4-acre lots), while the remaining 100 acres consist of open space in good hydrologic condition. The soils in both land use types are classified as Hydrologic Soil Group B. Estimate the stormwater runoff using the NRCS Curve Number method, and determine the peak discharge using Eq. (10.8) for a 24-hour storm with a total rainfall of 4.2 in. The estimated time of concentration for the watershed is 60 min. The area does not include any ponding or swamp features.

10.8 Compute the synthetic stormwater hydrograph for the watershed in Problem (10.7).

10.9 A development project is planned in a 120-acre forested watershed, transitioning 70% of it into residential and commercial areas. Due to the changes, the maximum stormwater runoff rate is expected to increase from 80 to 250 cfs, while the time of concentration will decrease from 30 to 15 min. Use the NRCS dimensionless ratios of unit hydrograph (given in Table 10.14) to develop and compare the pre- and post-development synthetic stormwater hydrographs for the watershed. Analyze and discuss the implications of urbanization on stormwater runoff, particularly focusing on changes in runoff volume, peak discharge, and time to peak during storm events.

10.10 A watershed has a surface area A of 430 mi^2. The stream length from the outlet to the upstream boundary of the basin is $L=60$ mi, and the distance along the stream from the outlet to the point closest to the centroid of the basin is $L_c=40$ mi. Use Snyder's method to: (1) Derive a 6-h synthetic unit hydrograph for this watershed, and (2) prove that the derived hydrograph is a unit hydrograph. Calculate the total runoff volume if the effective rainfall over the watershed is 2.75 in. Assume: $C_t=1.8$ and $C_p=360$.

10.11 A bioretention basin is designed to treat stormwater runoff from a commercial parking lot. The surface dimensions of the bioretention basin are 60 ft long and 30 ft wide. The maximum average ponding depth is 10 in., and the side slope is 1V:4H. The average depth of the filter media (including mulch layer) is 30 in., with an effective porosity of 0.3. The depth of the underdrain and underground gravel layer (including choker stone) is 18 in., with an effective porosity of 0.35. A 6-in. perforated underdrain pipe is placed in the gravel layer. Determine the total storage volume of the bioretention basin.

10.12 A paved parking lot for an office complex has 12,000 ft^2 of impervious area. It is built on soil with a silty clay texture. Calculate the expected depth of runoff resulting from a storm producing 1.5 in. of rain. Also, calculate the required area of a standard bioretention basin (rain garden). Assume that the storage capacity of a standard bioretention basin is 20 in.

10.13 On a 15% hill slope, it is proposed to construct bench terraces with a vertical interval of 1.5 m. Assuming a vertical riser, calculate: (1) length of terrace per hectare, (2) earthwork required per hectare, and (3) area lost per hectare. Assume an additional 2.5% earthwork is required for the construction of shoulder-bound (lip) and drainage channels, and 2.0% of land is lost from cultivation due to shoulder-bound and drainage channels.

10.14 A series of contour bunds is built over a sloping tract of land to control erosion and conserve soil moisture for cultivation. The height of the bund is 1.2 m. Select the cross-section and calculate the earthwork required, as well as the percentage of land area lost from farming. The spacing between two contour bunds is 20 m. The soil in the area is silty clay loam.

10.15 A commercial building is being retrofitted to install a green roof. The total roof area is 12,000 ft^2. Based on the design criteria, the building is required to treat 900 ft^3 stormwater runoff volume (S_v). The green roof will have an 8-in. thick growing medium and a 2-in. thick drainage layer. The maximum water retention capacities of the growing medium and drainage layer are 0.20 and 0.40, respectively. Calculate the minimum required green roof surface area. The roof is unirrigated.

10.16 Determine the required size of a rain barrel system for a single-family dwelling to capture runoff from a rainfall event of 1.2 in. The dwelling has a roof catchment area of 1,500 ft^2, with three downspouts available for collection. The system efficiency is assumed to be 85%.

10.17 An infiltration basin is sized to accept runoff from 0.5 acres of impervious pavement. The runoff is generated by a 2-year storm with 1.5 in total rainfall. Site limits are: the basin surface area may not exceed 800 ft^2 and the maximum depth 2.5 ft. The maximum allowable drawdown time is 12 h. The subsoil saturated hydraulic conductivity is $K = 3$ in./h. Calculate: (1) the portion (volume and percent) of the stormwater volume that can be retained by the basin (i.e., stored volume + infiltration within the allowed drawdown time), and (2) the drawdown time required. Assume t_f (time to fill infiltration basin) is 2 h (0.083 day).

10.18 Stormwater runoff from a 7-acre watershed is discharged into an infiltration basin. The watershed is 30% impervious with CN = 92. The underlying soil has a saturated hydraulic conductivity of 0.85 in./h. The seasonal high ground water table is at a depth of 10.0 ft. The design criterion is to keep the bottom of the basin at least 4.0 ft above the groundwater table. The basin is designed to treat the runoff from 1.2 in. of rainfall. A 2-day drawdown is also specified as a design requirement.

10.19 A permeable pavement system is being designed to meet water quality requirements for runoff from a parking lot. The total parking lot area is 6 acres, of which 4 acres consist of driveways and aisles paved with concrete. Field investigations indicate that the underlying soil has a saturated hydraulic conductivity of 0.7 in./h, and the site does not have a high groundwater table. The reservoir layer of the pavement has a void ratio of 0.40, providing storage for infiltrated runoff. The design objective is to determine the required depth and storage volume of the permeable pavement. The maximum allowable drawdown time at the site is 24 h, the WQ_v corresponds to a runoff depth of 1.5 in., and the time required to fill the reservoir layer is 3 h. The runoff coefficient for the CDA is 0.90.

10.20 An SF wetland is designed to treat an average flow of 1,600 m^3/day. The bed slope is 4%, the flow-zone width is 120 m, and the mean flow depth through which flow occurs is 0.50 m. The design hydraulic loading rate is 300 m^3/ha·day. Determine: (1) the hydraulic conductivity of the bed material and the hydraulic detention time of the wetland, and (2) the effluent concentrations of TSS and BOD_5, if the total surface area of the wetland is 12 ha. The background concentrations of TSS = 5 mg/L and BOD_5 = 6 mg/L. The influent concentration of TSS = 75 mg/L and BOD_5 = 95 mg/L. The apparent TIS (tanks-in-series) rate constant $k = 130$ m/year and apparent exponent $p = 1.5$.

10.21 In a settling chamber, sand-sized material is being removed. Calculate the terminal (settling) velocity of the smallest particle that is fully removed in the chamber. The selected particle has a specific gravity $G_S = 2.65$ and a mean diameter $d = 0.45$ mm. The wastewater temperature is 20°C.

10.22 A sediment removal facility is designed to treat a flow of 9,600 m^3/day. The channel has length $L = 30.0$ m, width $W = 2.0$ m, and depth $D = 1.50$ m. Calculate: (1) the surface overflow rate (or surface loading), (2) the minimum terminal settling velocity v_t of the smallest particle that will be fully removed, (3) the horizontal (through-channel) velocity, and (4) the detention time.
Assume steady flow and uniform cross-section. Show answers in m/day and m/s where appropriate.

10.23 A manually cleaned bar screen is installed in a channel to protect downstream equipment. The screen has 30 parallel, sharp-edged rectangular bars that are 12 mm wide, with a clear spacing of 22 mm between bars. The bars are inclined at 50° from the horizontal. The approach flow is 0.90 m^3/s and the upstream flow depth is 1.60 m. Calculate:

(1) head loss across the clean screen (use $C_d = 0.70$), and (2) head loss across the screen when it is 50% clogged (use $C_d = 0.60$).

10.24 A clean filter bed composed of uniform sand has a depth of 0.45 m. The design filtration rate is $30.0\,m^3/m^2.h$, which corresponds to a filtering velocity of 8.3×10^{-3} m/s through the bed. The sand has a median grain size of 1.0 mm, a shape factor of 0.9, and a porosity of 0.48. Determine the head loss across the clean filter bed under these operating conditions. Assume stormwater runoff temperature is 15°C.

NOTE

1. Tonne is a short ton = 1,000 kg.

REFERENCES

1. *Federal Water Pollution Control Act Amendments of 1972* (PL 92–500), 92nd Congress, October 18, 1972.
2. *Clean Water Act* (PL 95–217) 95th Congress, December 27, 1977.
3. U.S. Environmental Protection Agency, *Investigation of Inappropriate Pollution Entries into Storm Drainage System, A User's Guide*, EPA/600/R-92/238, Office of Research and Development, Washington, DC, 1993.
4. U.S. Environmental Protection Agency, *Guidance Manual for Preparation of NPDES Permit Applications for Stormwater Discharges Associated With Industrial Activity,* EPA505/8-91-002, Office of Water (EN-36), Washington, DC, April 1991.
5. CH2M-Hill, *Stormwater Control, Clean Water Act: Agenda for the l990's,* CH2M Hill, Inc., Englewood, CO, 1990.
6. Lape, J. L., and T. J. Dwyer, A New Policy on CSOs, *Water Environment & Technology*, Water Environment Federation, vol. 6, no. 6, June 1994, pp. 54–58. Alexandria, VA.
7. U.S. Environmental Protection Agency, *Combined Sewer Overflows: Guidance for Financial Capability Assessment and Schedule Development,* PA 832-8-97-004, Office of Water (4204), Washington, DC, March 1997.
8. U.S. Environmental Protection Agency, *Clean Watersheds Needs Survey 2012 Report to Congress,* EPA-830-R-15005, Office of Water, Washington, DC, January 2016.
9. U.S. Environmental Protection Agency, Report to Congress, *Implementation and Enforcement of the Combined Sewer Overflow Control Policy,* Repot no. EPA 833-R-01-003, U.S. Environmental Protection Agency, Washington, DC, 2001.
10. U.S. Environmental Protection Agency, *1992 Needs Survey Report to Congress,* EPA832-R-93-02, Office of Water (WH-547), Washington, DC, September 1993.
11. Anonymous, EPA Releases Final CSO Control Policy and Draft Guidance Document, Combined Sewer Overflow Update, *The CSO Partnership,* vol. I, Issue I, 1994, pp. 1–11.
12. Qasim, S. R., *Wastewater Treatment Plant Design: Planning, Design and Operation*, 2nd ed., CRC Press, Boca Raton, FL, 1999.
13. Cook, M. B., *Combined Sewer Overflows: Guidance for Nine Minimum Controls*, U.S. Environmental Protection Agency, EPA 832-B-95-003, Office of Water, Washington, DC, May 1995.
14. Arnold, C. L., and C. J. Gibbons, Impervious Surface Coverage: The Emergence of a Key Environmental Indicator, *Journal of the American Planning Association,* vol. 62, no. 2, 1996, pp. 243–258.
15. U.S. Environmental Protection Agency, *Preliminary Data Summary of Urban Storm Water Best Management Practices,* EPA-821-R-99-012, Office of Water, Washington, DC, 1999.
16. Urbonas, B. R., and L. A. Roesner, Hydrologic Design for Urban Drainage and Flood Control, *Handbook of Hydrology,* D. R. Maidment (Editor), McGraw-Hill Inc., New York, 1993, pp. 28.1–28.52.
17. Water Pollution Control Federation, *Combined Sewer Overflows Pollution Abatement,* MOP FD-17, Water Pollution Control Federation, Alexandria, VA, 1989.
18. Burton Jr, G. A., and R. Pitt, *Stormwater Effects Handbook: A Toolbox for Watershed Managers, Scientists, and Engineers,* CRC Press, Boca Raton, FL, 2001.
19. Estamian, S. (Editor), *Handbook of Engineering Hydrology: Fundamentals and Applications,* Taylor and Francis Group, CRC Press, Boca Raton, FL, 2014.

20. Schueler, T., *Controlling Urban Runoff: A Practical Manual for Planning and Designing Urban BMPs,* Metropolitan Washington Council of Governments. Washington, D.C., 1987.
21. Field, R., D. Sullivan, and A. N. Tafuri, *Management of Combined Sewer Overflows,* CRC Press, Boca Raton, FL., 2003.
22. Friends of the Lower Olentangy Watershed, *Strategies for Protecting and Improving Water Quality and Recreational Use of the Olentangy River and Tributary Streams in Delaware and Franklin Counties,* The Lower Olentangy Watershed Action Plan in 2003, Columbus, OH, 2005.
23. U.S. Environmental Protection Agency, *Urbanization and Streams: Studies of Hydrologic Impacts*, EPA 841-R-97–009, Office of Water, Washington, DC, 1997.
24. Shaver, E., R. R. Horner, J. Skupien, C. May, and G. Ridley, *Fundamentals of Urban Runoff Management: Technical and Institutional Issues*, North American Lake Management Society, Madison, WI, 2007.
25. Bastian, R. K., Potential Impacts on Receiving Waters, Effects of Water Development and Management on Aquatic Ecosystems, *Proceedings of an Engineering Foundation Conference*, L. A. Roesner, (Editor). American Society of Civil Engineers, New York, 1997, p. 384.
26. Gupta, R. S., *Hydrology & Hydraulic Systems*, 4th ed., Waveland Press Inc., Prospect Heights, IL, 2016.
27. Linsley, R. K., M. A. Kohier, and J. L.H. Paulhus, *Hydrology for Engineers,* 3rd ed., McGraw-Hill Book Company, Inc. New York, 1982, p. 484.
28. Mays, L. W., *Water Resources Engineering*, John Wiley & Sons, New York, 2010.
29. Blick, S. A., F. Kelly, and J. Skupien, *New Jersey Stormwater Best Management Practices Manual*, New Jersey Department of Environmental Protection, Division of Watershed Management, Trenton, NJ, 2004.
30. New Jersey State Soil Conservation Committee, *The Standards for Soil Erosion and Sediment Control in New Jersey*, New Jersey Department of Agriculture, Soil and Water Conservation, Trenton, NJ, 2017.
31. North Central Texas Council of Governments (NCTCOG), *Integrated Storm Water Management (iSWM) Technical Manual*, North Central Council of Government, Arlington, TX, https://iswm.nctcog.org/technical-manual.html, 2014.
32. Texas Department of Transportation, *Hydrologic Design Manual*, 4th ed., Design Division, TX DOT, Austin, TX, January 1997.
33. McGhee, T. J., *Water Supply and Sewerage,* 6th ed., McGraw-Hill, Inc., New York, 1991.
34. Viessman, W., G. L. Lewis, and J. W. Knapp, *Introduction to Hydrology,* 5th ed., Prentice Hall, Pearson Education, Inc., Upper Saddle River, NJ, 2003.
35. Metcalf and Eddy, Inc., *Wastewater Engineering: Collection, and Pumping of Wastewater,* McGraw-Hill Book Co., New York, 1981.
36. Joint Committee of the Water Pollution Control Federation and the American Society of Civil Engineers, *Design and Construction of Sanitary and Storm Sewers,* MPCF 9, Water Pollution Control Federation, Washington, DC, 1970.
37. Mays, L. W., *Stormwater Collection Systems Design Handbook*, McGraw-Hill, New York, 2001.
38. U.S. Department of Agriculture, Soil Conservation Service, *Urban Hydrology for Small Watersheds*, Technical Release 55, Washington, DC, 1986.
39. U.S. Department of Agriculture, Soil Conservation Service, *National Engineering Handbook*, Section 4: *Hydrology*, Washington, DC, 1997.
40. Chin, D. A., *Water Resources Engineering,* 4th ed., Pearson Education, Inc., New York, 2020.
41. Parkash, A., *Water Resources Engineering: Handbook of Essential Methods & Design*, ASCE Press Series, American Society of Civil Engineers, 2004.
42. North Carolina Division of Water Quality, *Stormwater Best Management Practices Manual*, North Carolina Department of Environment Quality, Raleigh, NC, 2017.
43. Pazwash, H., *Urban Stormwater Management,* 2nd ed., CRC Press, Boca Raton, FL, 2016.
44. Debo, T. N., and A. J. Reed, *Municipal Stormwater Management,* CRC Press, Boca Raton, FL, 2003.
45. Ries, K. G., *The National Streamflow Statistics Program: Computer Program for Estimating Streamflow Statistics for Ungauged Sites,* Techniques and Methods 4-A6, U.S. Department of the Interior, U.S. Geological Survey, Reston, VA 2007.
46. Land, L. F., E. E. Schroeder, and B. B. Hampton, *Techniques for Estimating the Magnitude and Frequency of Floods in Dallas-Fort Worth Metropolitan Area, Texas,* USGS /WRI-82-18, U. S. Geological Survey, Water Resources Division, Austin, TX, 1982.
47. Jens, S. W., and M. B. McPherson, Hydrology of Urban Areas, *Handbook of Applied Hydrology*, V. T. Chow (Editor), McGraw-Hill Book Co, New York, 1964.

48. Rangari, V. A., A. K. Patel, and N. V. Umamahesh, *Review of Urban Stormwater Models, Hydro 2015,* 20th International Conference on Hydraulics, Water Resources and River Engineering, IIT Roorkee, India, 2015.
49. Nix, S. J., *Urban Stormwater Modeling and Simulation*, CRC Press, Boca Rotan, FL, 1994.
50. Hoggan, D. H., *Computer-Assisted Floodplain Hydrology and Hydraulics*, Prentice-Hall, New York, 1989.
51. Singh, V. P., *Hydrologic Systems*, Prentice-Hall, Englewood Cliff, NJ, 1989.
52. Kibler, D. F., M. E. Jennings, G. L. Lewis, B. A. Tschntz, and S. G. Walesh, ASCE Task Committee on Urban Stormwater Software, *Microcomputers in Urban Hydrology, HYDATA, American Water Resources Association*, vol. 10, no. 5, 1991, pp. 27–37. Woodbridge, VA.
53. Lewis, G. L., *A Shopper's Guide to Urban Stormwater Software Revisited*, Proceedings ASCE 1988 National Conference on Hydraulic Engineering, Colorado Springs, CO, 1988, pp. 1228–1233.
54. Jennings, M. E., R. H. Smith, and R. B. Jennings, *Federal Microcomputer Software for Urban Hydrology,* Proceedings ASCE 1988 National Conference on Hydraulic Engineering, Colorado Springs, CO, 1988, pp. 616–621.
55. Haris, H., M. F. Chow, F. Usman, L. M. Sidek, Z. A. Roseli, and M. D. Norlida, Urban Stormwater Management Model and Tools for Designing Stormwater Management of Green Infrastructure Practices, *IOP Conference Series: Earth and Environmental Science*, vol. 32, no. 1, 2016, p. 012022.
56. Ford D., N. Pingel, and J. J. DeVries, *Hydrologic Modeling System (HEC-HMS) Applications Guide*, U.S. Army Corps of Engineers, CPD-74C, Institute for Water Resources, Davis, CA, 2008.
57. Rallison, R. E., and N. Miller, Past, Present, and Future SCS Runoff Procedure, *Rainfall-Runoff Relationships*, V. P. Singh (Editor), Proceedings, International Symposium on Rainfall-Runoff Modeling, Mississippi State University, MS, U.S., 1981, pp. 353–364.
58. U.S. Department of Agriculture, Natural Resource Conservation Service, *Small Watershed Hydrology WinTR-55 User Guide,* Conservation Engineering Division, Washington, DC, 2009.
59. Rossman, L. A., *Storm Water Management Model User's Manual, Version 5.0*, National Risk Management Research Laboratory, Office of Research and Development, U.S. Environmental Protection Agency, Cincinnati, OH, 2010.
60. U.S. Army Corps of Engineers, *Storage, Treatment, Overflow, Runoff Model, User's Manual*, Institute for Water Resources, Davis, CA, 1997.
61. Neitsch, S. L., J. G. Arnold, J. R. Kiniry, and J. R. Williams, *Soil and Water Assessment Tool Theoretical Documentation, Version 2009*, Texas Water Resources Institute, Technical Report No. 406, Texas A&M University System, TX, 2011.
62. Field, R., and Anthony N. Tafuri, (Editors). *The Use of Best Management Practices (BMPs) in urban watersheds.* DEStech Publications, Inc, Lancaster, PA, 2006.
63. U.S. Environmental Protection Agency, *Guidance Manual for Developing Best Management Practices (BMP),* EPA833-B-93–004, Office of Water (EN-336), Washington, DC, October 1993.
64. Washington State Department of Ecology, *Stormwater Management Manual for Western Washington*, Water Quality Program, Publication No. 10-10-021, Washington, DC, 2019.
65. Taylor, A., and T. H. F. Wong. *Non-structural Stormwater Quality Best Management Practices: An Overview of Their Use, Value, Cost and Evaluation.* CRC for Catchment Hydrology, 2002.
66. Connecticut Department of Environmental Protection, *Connecticut Stormwater Quality Manual,* Connecticut Department of Energy and Environmental Protection, Hartford, CT, 2004.
67. Department of Environment and Swan River Trust, *Non-structural Controls, Stormwater Management Manual for Western Australia,* Department of Environment and Swan River Trust, Perth, Western Australia, 2005.
68. Qasim, S. R., and G. Zhu, *Wastewater Treatment and Reuse: Theory and Design Examples*, vol. 1, *Principles and Basic Treatment*, CRC Press, Taylor and Francis Group, Baca Raton, FL, 2019.
69. Comprehensive Environmental Inc. and New Hampshire Department of Environmental Services, *New Hampshire Stormwater Manual*, New Hampshire Department of Environmental Services, 2008. https://www.des.nh.gov/sites/g/files/ehbemt341/files/documents/2020-01/wd-08-20b.pdf
70. U.S. Environmental Protection Agency, *National Management Measures to Control Nonpoint Source Pollution from Urban Areas*, EPA-841-B-05-004, Office of Water, Washington, DC, 2005.
71. Washington State Department of Transportation, *Highway Runoff Manual,* Engineering and Regional Operations, Development Division, Design Office, Olympia, WA, April 2019.
72. Heathercote, I. W., *Integrated Watershed Management: Principles and Practice,* 2nd ed., John Wiley & Sons, Inc., Hoboken, NJ, 2009.
73. U.S. Environmental Protection Agency, *Storm Water Technology Fact Sheet: Bioretention,* EPA-832-F-99-012, Office of Water, Washington, DC, September 1999.

74. Clar, M. L., B. J. Barfield, and T. P. O'Connor, *Stormwater Best Management Practice Design Guide, vol. 2: Vegetative Biofilters,* EPA-600/R-04/121A, Environmental Protection Agency, National Risk Management Laboratory, U.S. Environmental Protection Agency, Washington, DC, 2004.
75. Environmental Services Division, *Bioretention Manual,* Prince George's County Department of Environmental Resources, Maryland, 2009.
76. Li, H., and A. P. Davis, Heavy Metal Capture and Accumulation in Bioretention Media, *Environmental Science & Technology*, vol. 42, no. 14, 2008, pp. 5247–5253.
77. Center for Watershed Protection, Inc., *Stormwater Management Guidebook*, Prepared for Department of Energy, Regulatory Review Division, District of Columbia, Washington, DC, January 2020.
78. Jaber, F., D. Woodson, C. LaChance, and C. York, *Stormwater Management: Rain Gardens,* Texas A&M, AgriLife Extention, B-6247, Texas A&M University, College Station, TX, 2017, p. 11.
79. Liu, J., D. J. Sample, C. Bell, and Y. Guan, Review and Research Needs of Bioretention Used for the Treatment of Urban Stormwater, *Water,* vol. 6, no. 4, 2014, pp. 1069–1099.
80. Burger, K., and F. Zaal (Editors), *Sustainable Land Management in the Tropics: Explaining the Miracle*, International Land Management Series, Routledge, London, 2016, p. 226.
81. Mesfin, A., *A Field Guideline on Bench Terrace Design and Construction*, Ministry of Agriculture and Natural Resources, Natural Resource Management Directorate, Addis Ababa, Ethopia, https://nrmd-blog.files.wordpress.com/2016/04/bench-terrace-manual.pdf, October 2016.
82. Mishra, A., Lesson 5 Terraces for Water Erosion Control, *Soil and Water Conservation Engineering*, Department of Agricultural Food Engineering, Indian Institute of Technology (IIT) Kharagpur, December 2013. S&WCE: Lesson 5 Terraces for Water Erosion Control.
83. Govindasamy, A., S. Chaudhary, and M. J. Hsu, The Check-Dam Route to Mitigate India's Water Shortages, *Natural Resources Journal,* vol. 48, no. 3, 2008, pp. 565–583.
84. Castillo, C., R. Pérez, and J. A. Gómez, A Conceptual Model of Check Dam Hydraulics for Gully Control: Efficiency, Optimal Spacing and Relation with Step-Pools, *Hydrology and Earth System Sciences,* vol. 18, 2014, pp. 1705–1721.
85. Govindasamy, A., and M. J. Hsu, Small Size, Big Potential: Check Dams for Sustainable Development, *Environment,* vol. 50, no. 4, 2008, pp. 22–34.
86. Guyassa, E., A. Frank, A. Zenebe, J. Poesen, and J. Nyssen, Effects of Check Dams on Runoff Characteristics Along Gully Reaches, the Case of Northern Ethiopia, *Journal of Hydrology,* vol. 545, 2017, pp. 299–309.
87. Van Oost, K. K., G. G. Govers, S. S. de Alba, and T. A. Quine, Tillage Erosion: A Review of Controlling Factors and Implications for Soil Quality, *Progress in Physical Geography,* vol. 30, no. 4, 2006, pp. 443–466.
88. S&WCE – Soil & Water Conservation Engineering, Lesson 6, *Bunding Methods for Water Erosion Control*, 2021, pp. 1–7. (SWCE_-Lesson-6-Bunding-Methods-for-Water-Erosion-Control.pdf)
89. Murthy, V. V. N., *Land and Water Management Engineering,* Kalyani Publishers, New Delhi, 1994.
90. Water Environment Federation and American Society of Civil Engineers, *Urban Runoff Quality Management*, WEF Manual of Practice No. 23 and ASCE Manual and Report on Engineering Practice no. 87, ISBN 1-57278-039-8, Water Environment Federation, Alexandria, VA, 1998.
91. University of Florida Institute of Food and Agricultural Sciences, "If You Build it They Will Come: Frogs Flourish in Humanmade Ponds, *Science Daily*, 27 August, www.sciencedaily.com/releases/2015/08/150827154644.htm, 2015.
92. James Worth Bagley College, *Detention Basins*, Chapter 4: *Best Management Practices.* Agricultural and Biological Engineering, College of Agriculture and Life Sciences, Mississippi State University, Starkville, MS, 2021.
93. Minnesota Pollution Control Agency, *Protecting Water Quality in Urban, Suburban, and Developing Areas of Minnesota,* Minnesota Pollution Control Agency, Saint Paul, MN, March 1, 2000.
94. Minnesota Pollution Control Agency, *Stormwater Detention Ponds,* Minnesota State Permit Guidance Document, Minnesota Pollution Control Agency, Saint Paul, MN, 2008.
95. S&WCE – Soil & Water Conservation Engineering, Lesson 27, *Grassed Waterways*, 2021, pp. 1–9, (https://courseware.cutm.ac.in/wp-content/uploads/2020/05/SWCE_-Lesson-27-Grassed-Waterways.pdf).
96. Minnesota Pollution Control Agency, *Vegetated Filter Strips - Minnesota Stormwater Manual,* Minnesota Pollution Control Agency, Saint Paul, MN, 2017.
97. Resource Conservation District of Monterey County, *Filter Strip,* Resource Conservation District of Monterey County, Salinas, CA, 2017.
98. Zhou, X., M. J. Helmers, H. Asbjornsen, R. Kolka, M. D. Tomer, and R. M. Cruse, Nutrient Removal by Prairie Filter Strips in Agricultural Landscapes, *Journal of Soil and Water Conservation,* vol. 69, no. 1, 2014, pp. 54–64.

99. Virginia Department of Transportation, Part II BMP Design Manual of Practice, Chapter 7, *Vegetated Filter Strip,* Virginia Department of Transportation, Richmond, VA, 1–226, Revised 2013.
100. Helmers, M. J., X. Zhou, H. Asbjornsen, R. Kolka, M. D. Tomer, and R. M. Cruse, Sediment Removal by Prairie Filter Strips in Row-Cropped Ephemeral Watersheds, *Journal of Environmental Quality,* vol. 41, no. 5, 2012, pp. 1531–1539.
101. Minnesota Pollution Control Agency, *New State Funding Will Help Landowners Install Buffer Strips Along Minnesota's Waters,* Minnesota Pollution Control Agency, Twin Cities, MN, 2017.
102. Qiu, Z. A., VSA-based Strategy for Placing Conservation Buffers in Agricultural Watersheds, *Environmental Management,* vol. 32, 2003, pp. 299–310.
103. Brye, K. R., N. A. Slaton, M. C. Savin, R. J. Norman, and D. M. Miller, Short-term Effects of Land Leveling on Soil Physical Properties and Microbial Biomass, *Soil Science Society of America Journal,* vol. 67, no. 5, 2003, pp. 1405–1417.
104. Unger, P. W., L. J. Fulton, and O. R. Jones, Land-leveling Effects on Soil Texture, Organic Matter Content, and Aggregate Stability, *Journal of Soil & Water Conservation,* vol. 45, no. 3, 1990. pp. 412–415.
105. Natural Resources Conservation Service, *Precision Land Forming* (code 462), National Handbook of Conservation Practices, U.S. Department of Agriculture, 2021. (https://www.nrcs.usda.gov/sites/default/files/2022-09/Precision_Land_forming_and__Smoothing_462_NHCP_PO_2021.pdf)
106. U.S. Environmental Protection Agency, 40 CFR Part 412 – *Concentrated Animal Feeding Operation* (CAFO) Point Source Category, 68 FR 7269, February 12, 2003.
107. U.S. Environmental Protection Agency, 40 CFR Part 449 – *Airport Deicing Source Category,* 77FR 29203, May 16, 2012.
108. U.S. Environmental Protection Agency, 40 CFR Part 450 – *Construction and Development Point Source Category,* 74 FR 63057, December 1, 2009.
109. U.S. Environmental Protection Agency, *Combined Sewer Overflow (CSO) Control Policy,* 59 FR, 18688, vol. 59, no. 75, April 19, 1994.
110. U.S. Environmental Protection Agency, *Construction Grants 1985 (CG-85),* EPA 430/9-84-004, Office of Water Program Operations (WH-546), Washington, DC, 1984.
111. Rodriguez, D. B., *Sustainability Assessment of Green Infrastructure Practices for Stormwater Management: A Comparative Energy Analysis* (M.S. Thesis), State University of New York, College of Environmental Science and Forestry, Syracuse, New York, 2011.
112. Iowa Department of Natural Resources, *Iowa Storm Water Management Manual, Design Standards Chapter 17-Green Roof Systems,* Iowa Department of Natural Resources, Des Moines, IA, April 2015.
113. Volder, A., and B. Dvorak, Event Size, Substrate Water Content and Vegetation Affect Storm Water Retention Efficiency of an Un-irrigated Extensive Green Roof System in Central Texas, *Sustainable Cities and Society,* vol. 10, February 2014, pp. 59–64.
114. Santamouris, M., Cooling the Cities – A Review of Reflective and Green Roof Mitigation Technologies to Fight Heat Island and Improve Comfort in Urban Environments, *Solar Energy,* vol. 103(Supplement C), 2014, pp. 682–703
115. U. S. Environmental Protection Agency, *Reducing Urban Heat Islands: Compendium of Strategies, Green Roofs,* 2008, (https://www.epa.gov/heat-islands)
116. Maryland Department of the Environment, *Stormwater Design Guidance: Green Roofs,* Maryland Department of the Environment, Baltimore, MD, March 2018.
117. Kinkade-Levario, H., *Design for Water: Rainwater Harvesting, Stormwater Catchment, and Alternate Water Reuse, Gabriola Island, B. C.,* New Society Publishers, B. C. Canada, 2007, p. 27.
118. Stuart, E., and E. Pennypacker, *Artful Rainwater Design: Creative Ways to Manage Stormwater,* 3rd ed., Island Press, Washington, DC, 2015.
119. Zhu, Q., *Rainwater Harvesting for Agriculture and Water Supply,* Springer, Singapore, 2015, p. 264.
120. Mays, L., G. Antoniou, and A. Angelakis, History of Water Cisterns: Legacies and Lessons, Water, vol. 5, no. 4, 2013, pp. 1916–1940.
121. Lunduka, R., *Economic Analysis of Rainwater Harvesting and Small-Scale Water Resource Development,* Food and Agriculture Organization of the United Nations, Rome, Italy, 2011.
122. Saladin, M., *Rainwater Harvesting in Thailand: Learning from the World Champions,* Rural Water Supply Network (RWSN), https://www.rural-water-supply.net/en/resources/759, 2016.
123. Oklahoma Water Resources Board, *Water for 2060 Fact Sheet* (PDF), Oklahoma Water Resources Board, Oklahoma City, OK, 2021.
124. Texas Water Development Board, *Rainwater Harvesting,* Water for Texas, Innovative Water Technologies, Texas Water Development Board, February 2018.

125. Texas Legislature Online, *An Act Relating to Rainwater Harvesting and Other Water Conservation Initiatives, 82nd Regular Session,* H.B. No. 3391., Texas Legislature, September 1, 2011.
126. Van Giesen, E., and F. Carpenter, *Georgia Rainwater Harvesting Guidelines*, Georgia Department of Public Affairs, Atlanta, GA, 2009. (https://site.extension.uga.edu/water/files/2018/04/GRWM-09-04-09.pdf)
127. Minnesota Pollution Control Agency, *Minnesota Stormwater Manual-Stormwater and Rainwater Harvest and Use/Reuse Combined,* Minnesota Pollution Control Agency, St. Paul, MN, https://stormwater.pca.state.mn.us/index.php?title=Stormwater_and_rainwater_harvest_and_use/reuse_combined, 2017.
128. Virginia Department of Transportation, *BMP Design Manual of Practice*, Virginia Department of Transportation, Richmond, VA, 2013.
129. Hofkes, E. H., and J. T. Visscher, *Artificial Groundwater Recharge for Water Supply of Medium-Size Communities in Developing Countries*, International Reference Centre for Community Water Supply and Sanitation, The Hague, The Netherlands, December 1986.
130. Maryland Department of the Environment, *2000 Maryland Stormwater Design Manual*, Water Management Administration, Baltimore, MD, Revised in 2009.
131. Metropolitan Council, *Minnesota Urban Small Sites Best Management Practice Manual, Infiltration Trenches*, Metropolitan Council, St. Paul, MN, July 2001.
132. Metropolitan Area Planning Council, *Massachusetts Low Impact Development Toolkit, Infiltration Trenches and Dry Wells,* Metropolitan Area Planning Council, Boston, MA, 2005.
133. Atlanta Regional Commission, *Georgia Stormwater Manual:* vol. 2, Sec. 3.2.5, *Infiltration Trench,* Atlanta Regional Commission, Atlanta, GA, 2001.
134. U.S. Environmental Protection Agency, *What Is Green Infrastructure?* Green Infrastructure Team, Washington, DC. https://19january2017snapshot.epa.gov/green-infrastructure/what-green-infrastructure_.html, April 2021.
135. Minnesota Pollution Control Agency, *Minnesota Stormwater Manual-Stormwater Infiltration Best Management Practices, Permeable Pavement,* Minnesota Pollution Control Agency, St. Paul, MN. https://stormwater.pca.state.mn.us/index.php?title=Permeable_pavement, 2020.
136. Transportation Officials, AASHTO *Guide for Design of Pavement Structures,* The *American Association of State Highway and Transportation Officials,* Washington, DC, 2001.
137. American Society of Civil Engineers, *Permeable Interlocking Concrete Pavement,* American Society of Civil Engineers, Reston, VA, 2018.
138. U. S. Environmental Protection Agency, *Field Evaluation of Permeable Pavements for Stormwater Management,* Olympia, Washington, Fact Sheet, Document No. EPA-841-B-00-005B, Washington, DC, October 2000.
139. Drake, J., A. Bradford, and T. V. Seters, Hydrologic Performance of Three Partial-Infiltration Permeable Pavements in a Cold Climate over Low Permeability Soil, *Journal of Hydrologic Engineering*, vol. 19, no. 9, September 2014, p. 04014016.
140. Virginia Department of Environmental Quality, *Permeable Pavement, Stormwater Design Specification No.* 7, Virginia Department of Environmental Quality, Richmond, VA, 2011.
141. Yuan, X., Y. Tang, Y. Li, Q. Wang, J. Zuo, and Z. Song, Environmental and Economic Impacts Assessment of Concrete Pavement Brick and Permeable Brick Production Process: A Case Study in China, *Journal of Cleaner Production,* vol. 171, January 2018, pp. 198–208.
142. Loechl, P. M., A. S. Stumpf, E. D. Johnson, R. J. Scholze, B. M. Deal, and M. J. Hanson, *Design Schmatics for Sustainable Parking Lot*, ERDC/CERL TR-03-12, Engineering Research and Development Center, U.S. Army Corps of Engineers, Fort Bragg, NC, June 2003.
143. Gibb, T., *Bioswales Can Improve Water Quality Resources*, Michigan State University Extension, Michigan State University, East Lansing, MI, June 10, 2015.
144. Caflisch, M., and K. Giacalone, *An Introduction to Bioswales,* Clemson University, May 2015.
145. Xiao, Q., and E. G. McPherson, Performance of Engineered Soil and Trees in a Parking Lot Bioswale, *Urban Water Journal*, vol. 8, no. 4, 2011, pp. 241–253.
146. Kamal, M. S., and S. R. Qasim, Natural Systems for Wastewater Treatment, *The Encyclopedia of Environmental Science and Engineering,* 4th ed., J. R. Pfafflin and E. N. Ziegler (Editors), Gordon and Breach Science Publishers, New York, 1998, pp. 821–831.
147. Hammer, D. A. (Editor), *Constructed Wetlands for Wastewater Treatment*, CRC Press, Boca Raton, FL, 1989.

148. U.S. Environmental Protection Agency, *Constructed Wetlands and Aquatic Plants Systems for Municipal Wastewater Treatments Process Design Manual*, EPA 625/1-88-22, Center for Environmental Research Information, Cincinnati, OH, 1988.
149. Water Pollution Control Federation, *Natural Systems for Wastewater Treatment,* Manual of Practice No. FD-16, Water Pollution Control Federation, Alexandria, VA, 1990.
150. Qasim, S. R., and G. Zhu, *Wastewater Treatment and Reuse: Theory and Design Examples*, vol. 2, *Post Treatment, Reuse, and Disposal,* CRC Press, Taylor and Francis Group, Baca Raton, FL 2018.
151. Moshiri, G. A. (Editor), *Constructed Wetlands for Water Quality Improvements*, Lewis Publishers, Boca Raton, FL, 1993.
152. U.S. Environmental Protection Agency, *Subsurface Flow Constructed Wetlands Conference*, Paper Presented at the Conference at University of Texas at El Paso, August 16 and 17, El Paso, TX, 1993.
153. Cueto, A. J., Development of Criteria for the Design and Construction of Engineered Aquatic Treatment Units in Texas, Chapter 9, *Constructed Wetlands for Water Quality Improvements*, G. A. Moshiri (Editor), Lewis Publishers, Boca Raton, FL, 1993, pp. 99–105.
154. Kadlec, R. H., and S. D. Wallace, *Treatment Wetlands*, 2nd ed., CRC Press, Taylor & Francis Group, Boca Raton, FL, 2009.
155. Stephenson, M., G. Turner, P. Pope, J. Colt, A. Knight, and G. Tchobanoglous, The Use and Potential of Aquatic Species for Wastewater Treatment, Appendix A, *The Environmental Requirements of Aquatic Plants*, Publication No. 65, California State Water Resources Control Board, Sacramento, CA, 1980,p. 665.
156. Tchobanoglous, G., Aquatic Plant Systems for Wastewater Treatment: Engineering Considerations, *Aquatic Plants for Water Treatment and Resource Recovery*, K. R. Reedy and W. H. Smith (Editors), Magnolia Publishing, Orlando, FL, 1987, pp. 27–48.
157. Mustafa, A., Constructed Wetland for Wastewater Treatment and Reuse: A Case Study of Developing Country. *International Journal of Environmental Science and Development*, vol 4, no. 1, 2013, pp. 20–24.
158. Kadlec, R. H., and R. L. Knight, *Treatment Wetlands,* CRC Press, Boca Raton, FL, 1996.
159. Crites R., and G. Tchobanoglous, *Small and Decentralized Wastewater Management Systems*, McGraw-Hill Book Co., New York, 1998.
160. U.S. Environmental Protection Agency, *Constructed Wetlands Treatment of Municipal Wastewaters*, EPA 625/R-99/010, U.S. EPA Office of Research and Development, Washington, DC, 2000.
161. Dotro, G., G. Langergraber, P. Molle, J. Nivala, J. Puigagut, O. Stein, and M. von Sperling, *Treatment Wetlands*, IWA Publishing, London, 2017, p. 172.
162. Dinges, R.*, Natural Systems for Water Pollution Control*, Van Nostrand Reinhold, New York, 1982.
163. Ballantine, D. J., and Tanner, C. C., Substrate and Filter Materials to Enhance Phosphorus Removal in Constructed Wetlands Treating Diffuse Farm Runoff: A Review, *New Zealand Journal of Agricultural Research*, vol. 53, no.1, 2010, pp. 71–95.
164. Brix, H., Treatment of Wastewater in the Rhizosphere of Wetland Plants: The Root-Zone Method, *Water Science Technology*, vol. 19, 1987, pp. 107–118.
165. Metcalf & Eddy, Inc., *Wastewater Engineering: Treatment, and Reuse*, 4th ed., McGraw-Hill, New York, 2003.
166. U.S. Environmental Protection Agency, *Process Design Manual for Upgrading Existing Wastewater Treatment Plants*, Technology Transfer, EPA 625/1-71-004a, October 1974.
167. U.S. Environmental Protection Agency, *Handbook, Retrofitting POTWs*, EPA/625/6–89/020, Technology Transfer, Center for Environmental Research Information, Cincinnati, OH, July 1989.
168. U.S. Environmental Protection Agency, *A Guide to the Selection of Cost-Effective Wastewater Treatment System*, Technical Report EPA-430/9-75-002, Washington, DC, July 1975.
169. Yuan, L., T. Sinshaw, and K. J. Forshay, Review of Watershed-Scale Water Quality and Nonpoint Source Pollution Models, *Geosciences,* vol.10, no. 25, January 2020, 1–36. 10.3390/geosciences10010025.
170. Liu, R., F. Xu, P., Zhang, W. Yu, and C. Men, Identifying Non-point Source Critical Source Areas Based on Multi-factors at a Basin Scale with SWAT, *Journal of Hydrology*, vol. 533, no. 5, February 2016, pp. 379–388. 10.1016/j.jhydrol.2015.12.024.
171. Mittelstet, A. R., D. E. Storm, and M. J. White, Using SWAT to Enhance Watershed-Based Plans to Meet Numeric Water Quality Standards, *Sustainability of Water Quality and Ecology*, vol. 7, January 2016, pp. 5–21. 10.1016/j.swaqe.2016.01.002.
172. Kim S., D. J. Seo, H. Riazi, and C. Shin, Improving Water Quality Forecasting via Data Assimilation – Application of Maximum Likelihood Ensemble Filter to HSPF, *Journal of Hydrology,* vol. 519, 2014, pp. 2797–2809.

173. Liu, Y., L. M. Ahiablame, V. F. Bralts, and B. A. Engel, Enhancing a Rainfall-runoff Model to Assess the Impacts of BMPs and LID Practices on Storm Runoff, *Journal of Environmental Management,* vol. 147, January 2015, pp. 12–23. 10.1016/j.jenvman.2014.09.005.
174. Gene, Y., *Using GWLF for Development of "Reference Watershed Approach" TMDLs*, Proceedings of the ASAE/CSAE Annual International Meeting, Ottawa, ON, Canada, August 2004, pp. 1–4.
175. Corrales, J., G. M. Naja, M. G. Bhat, and F. Miralles-Wilhelm, Water Quality Trading Opportunities in Two Sub-watersheds in the Northern Lake Okeechobee Watershed, *Journal of Environmental Management,* vol. 196, July 2017, pp. 544–559. 10.1016/j.jenvman.2017.03.061.
176. Lin, J. P., *Review of Published Export Coefficient and Event Mean Concentration (EMC) Data*; TN-WRAP-04-3; Engineer Research and Development Center, Vicksburg, MS, August 1, 2004.
177. Lim, K. J., B. A. Engel, Y. Kim, and J. Harbor, *Development of the Long-Term Hydrologic Impact Assessment (L-THIA) WWW Systems*, Proceedings of the Sustaining the Global Farm – Selected Papers from the 10th International Soil Conservation Organization Meeting, West Lafayette, IN, 24–29 May 1999; pp. 1018–1023.
178. National Oceanic and Atmospheric Administration (NOAA), *Nonpoint-Source Pollution and Erosion Comparison Tool (N-SPECT): Technical Guide*, NOAA/CSC/RPT 08-5, The National Oceanic and Atmospheric Administration, Charleston, SC, 2008.
179. Wu, R. S., and I. W. Lin, Modification of Generalized Watershed Loading Functions (GWLF) for Daily Flow Simulation, *Paddy Water Environment*, vol. 13, 2015, pp. 269–279.
180. Haith, D. A., R. Mandel, and R. S. Wu, *GWLF: Generalized Watershed Loading Function Version 2.0 User's Manual*, Cornell University, Ithaca, NY, 1992.
181. Tetra Tech, *Loading Simulation Program in C++ (LSPC) Version 5.0 User's Manual,* EPA Contract # EP-R8-12-04; Tetra Tech, Cleveland, OH, 2017.
182. Abbaspour, K., S. Vaghefi, and R. A. Srinivasan, Guideline for Successful Calibration and Uncertainty Analysis for Soil and Water Assessment: A Review of Papers from the 2016 International SWAT Conference, Water, vol. 10, no. 1, 2018. 10.3390/w10010006.
183. Moriasi, D., B. N. Wilson, K., Douglas-Mankin, J. Arnold, and P. Gowda, Hydrologic and Water Quality Models: Use, Calibration, and Validation. *Transaction of the American Society of Agricultural and Biological Engineers (ASABE),* vol. 55, no. 4, 2012, pp. 1241–1247.
184. Arnold, J. G., D. N. Moriasi, P. W. Gassman, K. C. Abbaspour, M. J. White, R. Srinivasan, C. Santhi, R. Harmel, A. Van Griensven, M. W. Van Liew, N. Kannan, and M. K. Jha, SWAT: Model Use, Calibration, and Validation, *Transaction of the American Society of Agricultural and Biological Engineers (ASABE)* 2012, vol. 55, no. 4, 2012, pp. 1491–1508.
185. U.S. Environmental Protection Agency, Storm Water Management Model (SWMM), *U.S. EPA Research*, Office of Science Information Management, Durham, NC, Updated June 20, 2024. https://www.epa.gov/water-research/storm-water-management-model-swmm, 2024.
186. Goodrich, D. C., D. P. Guertin, I. S. Burns, M. A. Nearing, J. J. Stone, H. Wei, P. Heilman, M. Hernandez, K. Spaeth, F. Pierson, G. B. Paige, S. N. Miller, W. G. Kepner, G. Ruyle, M. P. McClaran, M. Weltz, and L. Jolley, AGWA: The Automated Geospatial Watershed Assessment Tool to Inform Rangeland Management, Rangelands, vol. 33, no. 4, August 2011, pp. 41–47.
187. U. S. Environmental Protection Agency, *BASINS 4.1 (Better Assessment Science Integrating Point & Non-point Sources) Modeling Framework*, National Exposure Research Laboratory, RTP, NC, 2015.
188. Nichols, K., K. Moncur, and J. Howe, *Hydrodynamic Separators as Stormwater Best Management Practices* (No. UT-04.15). Utah. Department of Transportation, Salt Lake City, UT, 2005.
189. Plitt, L. R., J. A. Finch, and B. C. Flintoff, *Modeling the Hydrocyclone Classifier*, European Symposium Particle Technology, Amsterdam, 1980, pp. 90–804.
190. Bradley, D., *The Hydrocyclone*, Pergamon Press, Oxford, 1965.
191. Plitt, L. R., *A Mathematical Model of the Hydrocyclone Classifier*, Canadian Institute of Mining and Metallurgy (CIM) Bulletin no. 69, 1976, pp. 114–123.
192. Cilliers, J. J., Hydrocyclones for Particle Size Separation, *Particle Size Separation*, 2000, pp. 1819–1825.
193. Arterburn, R. A., *The Sizing of Hydrocyclones,* Krebs Engineers, Menlo Park, CA, 1976.
194. Sullivan, R. H., J. E. Urs, F. Parkinson, and P. Zielinski, *Swirl and Helical Bend Pollution Control Devices*, American Public Works Association, Chicago, IL, 1982.
195. MWH, *Water Treatment: Principles and Design*, 2nd ed., John Wiley & Sons, Inc., Hoboken, NJ, 2005.
196. Kawamura, S., *Integral Design of Water Treatment Facility*, 2nd ed., John Wiley & Sons, Inc., Hoboken, NJ, New York, 2000.
197. Metcalf & Eddy | AECOM, *Wastewater Engineering: Treatment and Resource Recovery*, 5th ed., McGraw Hill Companies, Inc., New York, 2003.

11 Management of Combined Sewer *Overflow*

11.1 CHAPTER OBJECTIVES

Managing the release of contaminants from stormwater runoff and combined sewer overflows (CSOs) is one of the challenges in protecting water quality in many receiving waters. The stormwater and CSOs may carry many types of contaminants, including pathogens, suspended solids, oxygen-demanding pollutants, nutrients, metals, pesticides, and floatable matter. The objectives of this chapter are to present:

- Definition and significance of CSO on surface water quality
- Regulatory status of CSO
- Sources and impacts of CSO
- Characterization of CSO
- Computer simulation and modeling of CSO
- Overflow and diversion structures: side weir, baffled side weir, transverse weir, leaping weir, and relief syphon
- Flow regulators: mechanical, tipping plate, and hydro-brake
- Management of CSO and storage facility
- Source control, in-system storage in collection system, off-line storage within the collection system, and on-site storage
- Treatment of CSOs

11.2 COMBINED SEWER OVERFLOW

Most U.S. communities today have separate sanitary sewer systems. In this type of system, one set of pipes collects municipal wastewater from homes and businesses and carries it to a wastewater treatment plant through sanitary sewers. A separate set of pipes collects stormwater from drains, driveways, around parking lots, and along street curb, and carries it to a local waterway through storm sewers. The combined sewers collect both municipal wastewater and stormwater in the same pipe system and carry them to a central location, or to the permitted relief-points that may be wet weather outfalls located throughout the system.[1,2] When combined sewer systems (CSSs) were introduced in 1855, they were hailed as a vast improvement over urban cesspool ditches that ran along city streets and spilled over when it rained. These networks of underground pipes were designed to dry out streets by collecting rainwater runoff, domestic sewage from newly invented flush toilets, and industrial wastewater, all in the same pipe. The combined municipal wastewater and storm runoff was then discharged directly into waterways.

In the early 20th century, sewage treatment plants were added to clean the wastewater before it was discharged into streams. CSSs were and still are a great idea. In dry weather, all wastewater flows to a treatment plant where it is treated before being reused or discharged into a waterbody. There is, however, one catch; when too much stormwater is added to the sewer during wet weather, it overwhelms the treatment capacity of the plant, requiring frequent overflows or diverted to a storage facility during wet weather.[3] These outfalls or overflows are called combined sewer overflows

DOI: 10.1201/9781003629399-11

(CSOs). They contain pollutants that may cause serious water quality impairment of surface waters. These CSOs have become the focus of a debate regarding the best techniques to manage growing volumes of sewage and stormwater runoff in many older U.S. communities.[4,5]

11.3 REGULATORY STATUS OF COMBINED SEWER OVERFLOW

The Clean Water Act (CWA) requires communities with CSOs controls in place to address the above-mentioned concerns. Under federal and state grants, many cities have converted combined sewers into separate sewer systems. But in many large and older sections of the cities, sewer separation could not be achieved in some parts. As a result, these cities still must face CSOs during heavy rainfall events. On April 11, 1994, the U.S. Environmental Protection Agency (EPA) signed a final regulation that outlines the National CSO Policy. Like all point sources of pollution discharges, CSOs are also covered under the CWA through the National Pollutant Discharge Elimination System (NPDES) permitting. But until recently, limited guidance was developed for issuing permits that are specifically aimed at CSO discharges.[6,7]

The *Clean Watershed Needs Survey 2012 Report to Congress* was submitted in 2016. This report documented that the total needs as of January 1, 2012, are a total capital investment of $271 billion nationwide to ensure that the nation's water quality objectives of the CWA are met.[8] The total documented needs by category are (1) 39.9% secondary and advanced wastewater treatment plants and recycled wastewater distribution, (2) 18.9% conveyance system repairs, (3) 16.4% new conveyance systems, (4) 17.7% CSO corrections, and (5) 7.1% stormwater management program. The U.S. EPA reported that out of 15,598 total water segments assessed, the number of waterbody segments impaired by CSOs and urban stormwater runoffs were 150 and 1,233, respectively.[9] The number of waterbodies impaired by stormwater runoffs is significantly larger than that from CSOs. The *1992 Needs Survey to Congress* indicated that about 1,100 communities served by 1,303 CSO facilities nationwide used CSSs, which were designed to carry sanitary and industrial wastewater, and stormwater.[10] These facilities are mainly located in older cities in the northeast, the mid-central states, and along the west coast. It is estimated that around 700 communities in the United States face this problem. CSOs occur when the capacity of the CSSs is exceeded during a storm event. During these storm events, part of the combined flow in the collection system is discharged untreated into the receiving waters. The pollution effects of CSOs, however, are localized.[10]

The CSO policy establishes a framework and lays out clear expectations for municipalities, NPDES permitting and enforcement authorities, and states' water quality monitoring authorities for controlling CSOs. The municipalities were expected to implement the nine minimum controls and to submit documentation of their implementation no later than January 1, 1997. The nine minimum controls are (1) proper operation and maintenance, (2) maximum use of the collection system for storage, (3) review and modification of the pretreatment program, (4) maximization of flow to the publicly owned treatment works (POTW) for treatment, (5) prohibition of CSOs during dry weather, (6) control of solid and floatable materials, (7) public notification, (8) monitoring of CSO impacts and efficiency of controls, and (9) pollution prevention.[11–13]

The municipalities are also expected to undertake immediately the development of a long-term CSO control plan. The long-term plan should include the following key elements: (1) a comprehensive characterization of the CSS, CSOs, and impacts on the receiving waterbodies; (2) special consideration for sensitive environmental areas; (3) an evaluation of a range of CSO control alternatives; (4) coordination with the NPDES permitting authority and state water quality enforcing authorities when selecting control measures; (5) development of a public participation plan; (6) a schedule for implementing the selected CSO control measures that considers the financial resources of the municipalities; and (7) implementation of a post-construction water quality monitoring program.[6] The policy provides the municipalities with two approaches for showing that their selected CSO control will be sufficient to meet the water quality standards: (1) the presumption approach and (2) the demonstration approach. Under the presumption approach, the municipality can provide a specified level of control that is presumed to meet water quality standards unless there is data showing otherwise. For example, one of the presumptive levels of control is no more than an average of four to six overflow events per year. Under the demonstration approach, the municipality can provide information and data showing that the selected CSO control actually meets water quality standards.[6,11]

The permitting authority (EPA or an approved state agency) must (1) review and revise, as appropriate, the states' CSO permitting strategy developed in response to the 1989 EPA strategy, (2) develop and issue Phase I permits requiring the implementation of nine minimum controls, submit documentation of their implementation, and develop and submit a long-term CSO control plan, and (3) develop and issue Phase II permits requiring the continued implementation of the nine minimum controls, implement the CSO control measures selected from the long-term CSOs control plan, and take appropriate enforcement measures.[6,11–14]

Example 11.1: Regulatory Status of Combined Sewer Overflows (CSOs)

Provide a listing of major regulations and actions taken by regulatory agencies to control CSOs.

Solution

The regulatory status of CSOs is provided below:

1. On April 11, 1994, the U.S. EPA signed final regulations that outlined the National CSO Policy. Under this policy, CSOs, like all point sources, are covered under the Clean Water Act through NPDES permitting.
2. The *Clean Watershed Needs Survey 2012 Report to Congress* reported that 150 waterbody segments were impaired by CSO.
3. The *1992 Needs Survey to Congress* indicated that 1,100 older cities in the northeast, mid-central states, and along the west coast were served by CSO facilities using CSS.
4. The CSO policy established a framework for NPDES permitting and enforcement authorities to control CSOs.
5. Municipalities were expected to implement the nine minimum controls, and to submit their documentation of implementation not later than January 1, 1997.
6. Municipalities were also expected to develop a long-term CSO control plan to include characterization of CSOs, impacts, control alternatives and schedule, public participation, sensitive environmental areas, and implementation of post-construction water quality monitoring program.
7. The CSO policy provides municipalities with two approaches: (1) presumption approach and (2) demonstration approach for showing that the selected CSO control methods will be sufficient to meet water quality standards.
8. The permitting agency must review and revise the state CSO permitting strategy developed in response to the 1989 EPA strategy.
9. The permitting authority must issue Phase I permit requiring implementation of nine minimum controls, submit documentation for implementation of long-term CSO control, and develop and issue Phase II permit requiring continued implementation of nine minimum controls for long-term CSO control plan, and take appropriate enforcement measures.

11.4 SOURCES AND IMPACTS OF COMBINED SEWER OVERFLOWS (CSOs)

CSSs convey both municipal wastewater and stormwater through shared collection pipes and tunnels to treatment facilities. During heavy rainfalls, the combined flow may overwhelm the treatment capacity of the wastewater treatment facility. The portion of CSO exceeding the capacity of the treatment plant is bypassed to a storage facility. The CSOs were traditionally discharged into receiving waters without any treatment. This practice may have benefited the urban areas by preventing streets and basements from flooding but has several adverse impacts on receiving waterbodies.[14,15] The CSS pipes, municipal and combined sewer flows, and CSO are shown in Figure 11.1. The dry weather flow is discharged into sanitary sewer, and CSO is diverted to a river. The CSOs may contain municipal and industrial wastes, and surface runoff.[15] These CSOs contain contaminants such as suspended solids, floatable debris, pathogenic organisms, nutrients, organic compounds, and other hazardous substances that may be harmful to people, pets, and wildlife. The CSOs can also cause aesthetic impacts, beach closures, shellfish bed closures, algae growth, reduced oxygen levels in waterways, and aesthetic impacts from floating debris or oil slicks. The amount of toxic priority pollutants found in the influent to POTWs by combined sewers increases significantly during and immediately following a storm event. It has been reported that between 10% and 20% of harvest-limited shellfish beds may be directly attributed to CSOs.[13,15,16] The examples of principal consequences of untreated COS released in the U.S. waterbodies are summarized in Table 11.1.

11.5 CHARACTERIZATION OF CSOs

There are about 700 municipalities across the U.S. that still have CSSs. EPA directed municipalities with CSOs to develop plans to comply with requirements in the Clean Water Act, including water quality standards, which are established by states. The CSS communities with CSOs are mostly located in the northeast and Great Lakes regions. In a report to Congress, the EPA estimated that 850 billion gallons of untreated wastewater and stormwater are annually released as CSO in the U.S.[17] The 184 CSOs communities discharged 22 billion gallons of untreated wastewater into the Great Lakes Basin during 1,482 CSO events in 2014. The number of CSO permits has decreased since the EPA's CSO control policy was issued in 1994. As a result, the volume of untreated CSO discharged to surface waters has reduced from 1,070 to 850 billion gallons per year between 1992 and 2002. However, CSOs are still the second source of water pollution in the U.S. after nonpoint source pollution.[17]

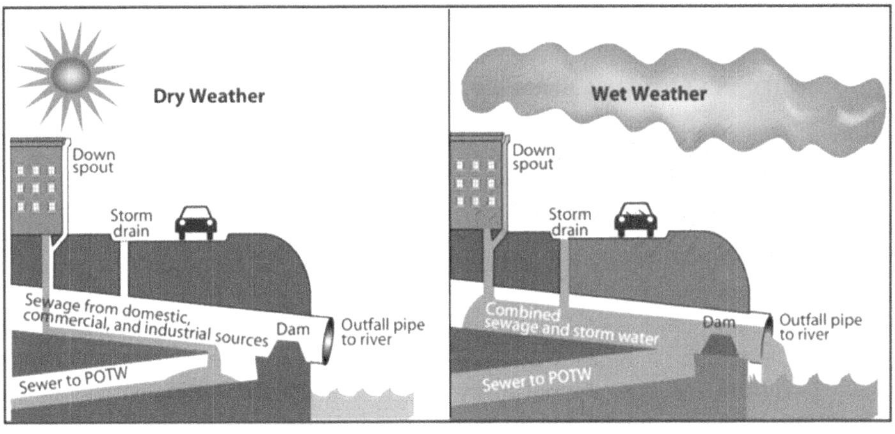

FIGURE 11.1 Flow through Combined Sewer System (CSS). During dry weather, the entire flow is conveyed to a POTW; and during wet weather conditions, the maximum allowable flow is conveyed to POTW while CSO is diverted to an off-line storage facility. (Printed from Ref. 9)

TABLE 11.1
Principal Consequences of COS Pollutants Entering Receiving Waters

Pollutant(s)	Principle Consequences
Bacteria (e.g., fecal coliform, E. coli, enterococci)	Beach closures
Viruses (e.g., hepatitis, diphtheria, cholera)	Odors
Parasites (e.g., giardia, cryptosporidium)	Shellfish bed closures
	Drinking water contamination
	Adverse public health effects
Trash and floatable	Aesthetic impairment
	Odors
	Beach closure
Organic compounds, metals, oil, and grease	Aquatic life impairment
Toxic pollutants	Adverse public health effects
	Fishing and shellfish restrictions
Biochemical oxygen demand (BOD)	Reduced oxygen levels and fish kill
Solids deposition	Aquatic habitat impairment
	Shellfish bed closures
Nutrients (e.g., nitrogen, phosphorus)	Eutrophication, algal blooms
	Aesthetic impairments

Source: Printed from Refs. 9,16

11.5.1 Factors Influencing Characteristics of CSOs

Many factors influence the characteristics of combined sewage: (1) precipitation amount, intensity, and duration; (2) predominant land use in the catchment contributing municipal wastewater: commercial, industrial, residential, undeveloped; (3) sub-catchment condition: street-sweepings frequency, buildup of surface contaminants, and debris; (4) number of dry weather days before a storm; (5) drainage basin: size, time of concentration, surface characteristics; and (6) sewer system: size, slope. Because of the variability of precipitation events and contributing factors, the combined wastewater characteristics tend to be highly variable from location to location and from time to time. Actual measurements and computer simulation is generally needed for establishing the characteristics of combined flows.[1,4,5]

11.5.2 Quantity and Quality Characterization

The characteristics of combined wastewater are highly variable and difficult to predict. The following generalization, however, can be made:[1,18,19]

1. The wet-to-dry weather capacity ratio of the combined interceptor sewer design may range from 4:1 to 8:1. The CSSs are typically designed to convey 3–5 times the anticipated peak dry weather wastewater flow rate.
2. During dry weather flows, the velocity in the combined sewers is low. As a result, solids are deposited, and biofilms may develop in sewer walls and in manholes. At the onset of a storm event, initially, the concentration of many contaminants, such as TSS, BOD-containing wastes, heavy metals, nutrients, toxic organics, and microorganisms, increases in runoff due to surface washoff. As the flow builds up during the wet weather peak, the accumulated

solids and biofilms in CSSs are scoured and resuspended in the combined flow. There is a surge of contaminants loading in the combined flow. This phenomenon is known as the *first-flush*. The pollutant loadings during wet weather conditions could be as much as 150% of the annual dry weather wastewater loadings in a CSS. An example of first-flush pollutant loading is shown in Figure 11.2.

3. The pollutant concentrations remain significantly high during the first-flush. At the beginning of a storm, a significant amount of pollution may enter CSS from the streets and other impervious surfaces of the watershed. During the wet weather conditions, the velocity and shear stress are high enough to resuspend solids and carry them to CSO outfall points.
4. After the first-flush, if the flow continues to be high because of long duration of subsequent storm events, the contaminant level drops due to dilution despite the raw municipal wastewater.
5. After the stormwater runoff subsides, the original municipal wastewater characteristics resume.
6. The typical range of concentrations of various constituents reported in CSO is: TSS = 270–380 mg/L; BOD_5 = 60–220 mg/L; COD = 260–480 mg/L; total nitrogen (TN) = 4–17 mg/L; total phosphorous 1.2–2.8 mg/L; total lead = 0.5–10 mg/L; and total fecal coliform count = $2\times10^5 - 1.1\times10^6$/100 mL.
7. The volume of untreated wastewater and type and amount of pollutants depend on the community population, wet weather conditions, characteristics of CSS, and any treatment provided. The typical values of pollutant concentrations (e.g., TSS, BOD, COD, TN, fecal coliforms) in the urban runoff and CSO are presented in Table 11.2. These pollutant loadings are compared with those for background levels and sanitary sewage.

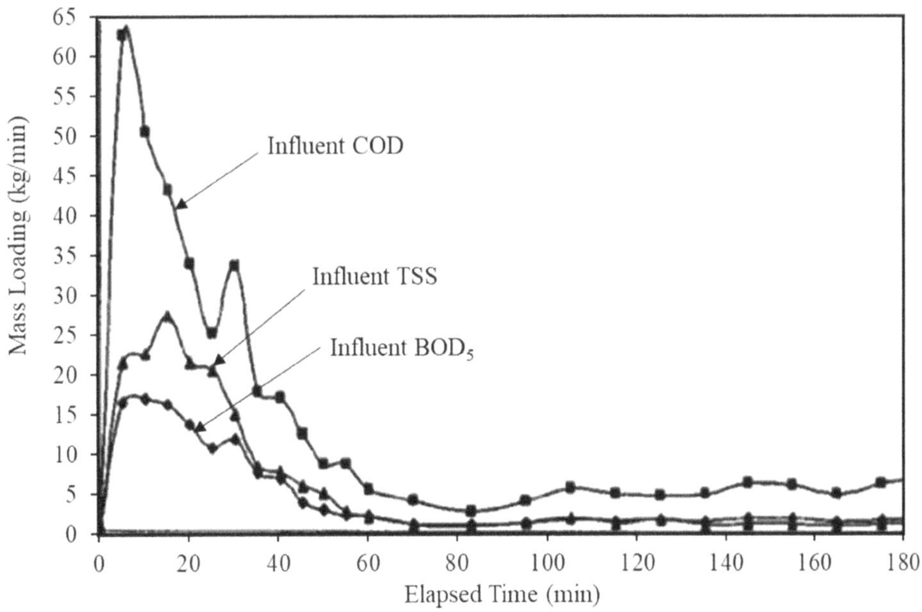

FIGURE 11.2 Example of first-flush BOD, COD, and TSS loadings in a combined sewer. The baseline is dry weather flow.

TABLE 11.2
Comparison of Typical Value of Pollutants Concentrations in Stormwater Runoff, CSOs, and Sanitary Sewage[a]

	TSS	VSS	BOD	COD	Kjeldahl Nitrogen	Total Nitrogen	PO_4-P	OPO_4-P	Lead	Fecal Coliforms
Background levels	5–100	–	0.5–3	20	–	0.05–0.5[b]	0.01–0.2[c]	–	<0.1	–
Stormwater runoff	415	90	20	115	1.4	3–10	0.6	0.4	0.35	14,500
Combined sewer overflow	370	140	115	375	3.8	9–10	1.9	1.0	0.37	670,000
Sanitary sewage	200	150	375	500	40	40	10	7	–	–

Source: Refs. 1,18,19

The background levels are also presented for comparison.

[a] All values are in mg/L except fecal coliforms, which are in organisms/100 mL.

[b] NO_3 as N.

[c] Total phosphorus as P.

Example 11.2: First-Flush in a Combined Sewer after a Storm Event

The flow and TSS data are collected at a combined sewer interceptor. The data collection started just after a downpour. Flow and TSS measurements were made at 10-min intervals. The results are summarized below. Draw the mass diagram of TSS concentration and identify the first-flush.

Time Interval, min	0	10	20	30	40	50	60	70	80	90	100
Combined Flow, m³/min	0	1.18	1.68	2.92	3.75	4.17	8.17	8.25	8.38	9.17	8.33
TSS Concentration, mg/L (g/m³)	0	200	290	380	400	600	680	690	700	700	695
Time Interval, min	110	120	130	140	150	160	170	180	190	200	210
Combined Flow, m³/min	8.33	7.50	7.08	6.67	6.25	5.83	5.42	3.33	1.21	1.19	1.18
TSS Concentration, mg/L (g/m³)	659	600	500	400	340	310	300	230	220	200	200

Solution

1. Calculate mass loading in the combined flow

The baseline is peak dry weather flow

At the onset of the storm $t=0$, stormwater flow = 0, TSS concentration = 0, and mass loading = 0

At 10 min since the onset of the storm

Mass loading of TSS = $1.18\,m^3/min \times 200\ g/m^3 \times 1/1{,}000\ g/kg = 0.24\ kg/min$

At 20 min since the onset of the storm

Mass loading of TSS = $1.68\,m^3/min \times 290\ g/m^3 \times 1/1{,}000\ g/kg = 0.49\ kg/min$

At 30 min since the onset of the storm

Mass loading of TSS = $2.92\,m^3/min \times 380\ g/m^3 \times 1/1{,}000\ g/kg = 1.11\ kg/min$

Similarly, calculate mass loadings at 10 min intervals up to 210 min. The results are summarized in the following table.

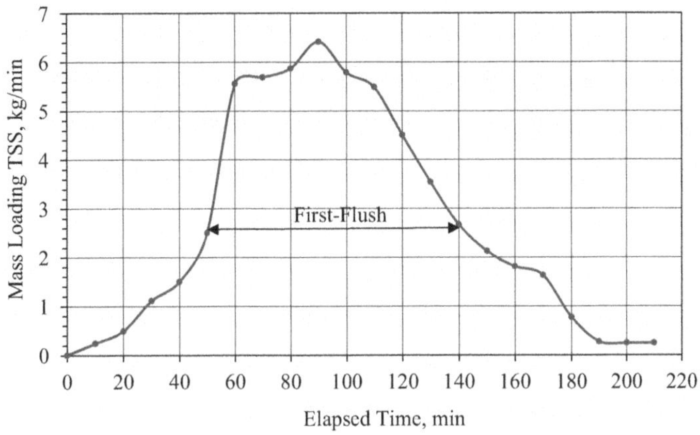

FIGURE 11.3 Mass loading of TSS showing first-flush in a combined sewer (Example 11.2).

Time Interval, min	0	10	20	30	40	50	60	70	80	90	100
Mass Loading TSS, kg/min	0	0.24	0.49	1.11	1.50	2.50	5.56	5.69	5.87	6.42	5.79
Time Interval, min	110	120	130	140	150	160	170	180	190	200	210
Mass Loading TSS, kg/min	5.49	4.50	3.54	2.67	2.13	1.81	1.63	0.77	0.27	0.24	0.24

2. Plot the mass loading data developed above
   The mass loading data is plotted in Figure 11.3.

Example 11.3: Dilution Effect of Stormwater on Quality of Combined Sewer Flow

An older subdivision of a city has a CSS. The average TN and total phosphorus (TP) concentrations in the municipal wastewater during dry weather flow are 40 and 6 mg/L, respectively. The average wastewater flow is 13,000 m³/day. During wet weather conditions, the average stormwater runoff is 30,000 m³/day, and TN and TP concentration are 28 and 21 mg/L. Determine the average concentrations of TN and TP in the combined sewer flow.

Solution

1. Determine the average concentration of TN in the combined sewer flow

$$C_{TN} = \frac{40\text{ mg/L} \times 13{,}000\text{ m}^3\text{/day} + 28\text{ mg/L} \times 30{,}000\text{ m}^3\text{/day}}{13{,}000\text{ m}^3\text{/day} + 30{,}000\text{ m}^3\text{/day}}$$

$$= \frac{(520 \times 10^3 + 840 \times 10^3)\text{ mg/L} \cdot \text{m}^3\text{/day}}{43 \times 10^3\text{ m}^3\text{/day}}$$

$$= 31.6\text{ mg/L}$$

2. Determine the average concentration of TP in the combined sewer flow

$$C_{TP} = \frac{6\ \text{mg/L} \times 13{,}000\ \text{m}^3/\text{day} + 21\ \text{mg/L} \times 30{,}000\ \text{m}^3/\text{day}}{13{,}000\ \text{m}^3/\text{day} + 30{,}000\ \text{m}^3/\text{day}}$$

$$= \frac{(78 \times 10^3 + 630 \times 10^3)\ \text{mg/L} \times \text{m}^3/\text{day}}{43 \times 10^3\ \text{m}^3/\text{day}}$$

$$= 16.5\ \text{mg/L}$$

11.6 COMPUTER SIMULATION AND MODELING

Many computer models have been developed to simulate the characteristics of urban stormwater runoff and CSOs.[19] Of the many computer models available, the most widely used model is the *Stormwater Management Model* (SWMM). This model was developed by the U.S. EPA and is used throughout the world for planning, analysis, and design related to stormwater runoff, combined and sanitary sewers, and other drainage systems. It was developed to help support local, state, and national stormwater management objectives. SWMM is a Windows-based desktop program. It can simulate most of the influencing factors indicated above as linear, exponential, and power function equations. Both hydraulic and contaminant routing are performed by SWMM for individual storm events or for long-term multiyear rainfall records. The model keeps track of flows in excess of conduit capacity for different subareas.[20]

As with any model, extensive field data are needed for model calibration. For using SWMM model, calibration and verification are also essential. The field data (both flow and contaminants) collected from a storm are compared with essential results from the model. The estimated inputs are adjusted within the reasonable bounds to obtain the best fit of measured and predicted values. Model verification is done by comparing measured and predicted values without any adjustment of parameters.[20–22] A comparison of measured combined wastewater flow rates and predicted flow rate from SWMM calculations at a manhole of a CSO is shown in Figure 11.4.

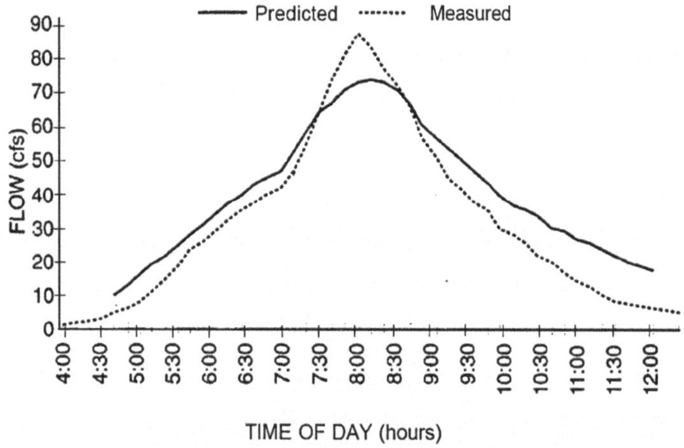

FIGURE 11.4 Comparison of measured wastewater flow rates and model-predicted flow rates using SWMM at a manhole of a CSO. (Printed from Ref. 23)

11.7 OVERFLOW AND DIVERSION STRUCTURES

A combined sewer carries both sanitary wastewater and stormwater runoff. Wastewater treatment facilities are hydraulically designed to handle peak dry weather flow, including an infiltration and inflow (I/I) allowance. Any flow exceeding the design capacity of the plant must be diverted away from the plant.

Overflow and diversion structures divert the excess flow into a companion facility that may be a temporary storage basin, a relief or bypass sewer, or an overflow structure. Examples of overflow and diversion structures are side weir, baffled side weir, transverse weir, leaping weir, and relief syphon. The design details of these devices are shown in Figure 11.5. Readers may consult Refs. 1–3,18,19 for discussions on each device.

11.7.1 Side Weirs

Side weirs are built along the channel or sewer wall parallel to the wastewater flow. The weir should be high enough to prevent discharge of peak dry weather flow, but low enough and long enough to discharge the excess flow on the other side. The head over the weir may be falling or rising depending upon the bottom slope of the channel or sewer. In most cases of wastewater flow in a channel or sewer, the weirs normally have a falling water surface.

Ackers developed a design equation for computing the required weir length by combining Bernoulli's equation, allowance for channel velocity distribution, and the weir discharge formula.[24]

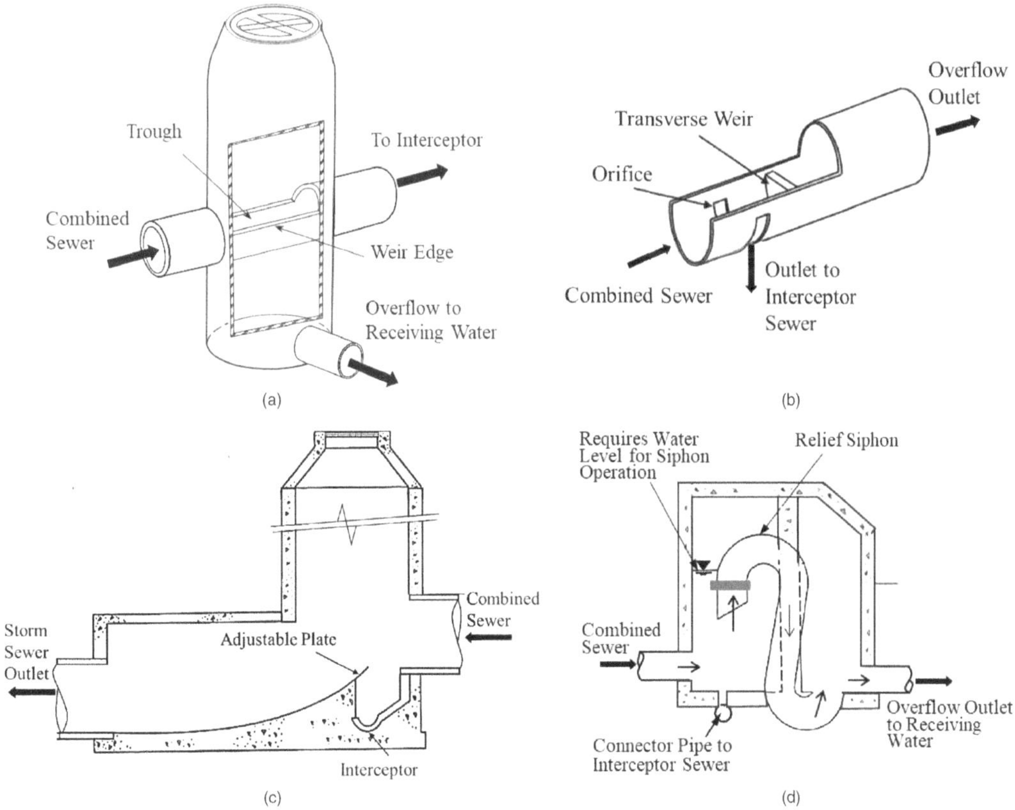

FIGURE 11.5 Overflow and diversion devices: (a) side weir in an existing sewer, (b) transverse weir with orifice, (c) adjustable leaping weir, and (d) relief syphon.

Based on his work, many design equations and procedures have been developed to calculate the capacity and length of the weir. A simplified design procedure for a side weir with a falling water surface is given below. The detailed design procedure for side weir may be found in Refs. 18,19,24. The specific energy relative to the weir crest, including a velocity-head correction and other relationships, may be calculated from Eqs. (11.1)–(11.3).

Other simplified design equations in terms of weir and channel dimensions and specific energy flow are summarized in Table 11.3. The relationships between many dimensionless ratios are shown in Figure 11.6.

$$E_w = \alpha \frac{V_n^2}{2g} + \alpha'(d_n - c) \tag{11.1}$$

TABLE 11.3
Simplified Equations for Side Weirs with Falling Water Surface

n_2	Equation for Weir Length
5	$L = 2.03B(2.81 - 1.55c/E_w)$
7	$L = 2.03B(3.90 - 2.03c/E_w)$
10	$L = 2.03B(5.28 - 2.63c/E_w)$
15	$L = 2.03B(7.23 - 3.45c/E_w)$
20	$L = 2.03B(8.87 - 4.13c/E_w)$

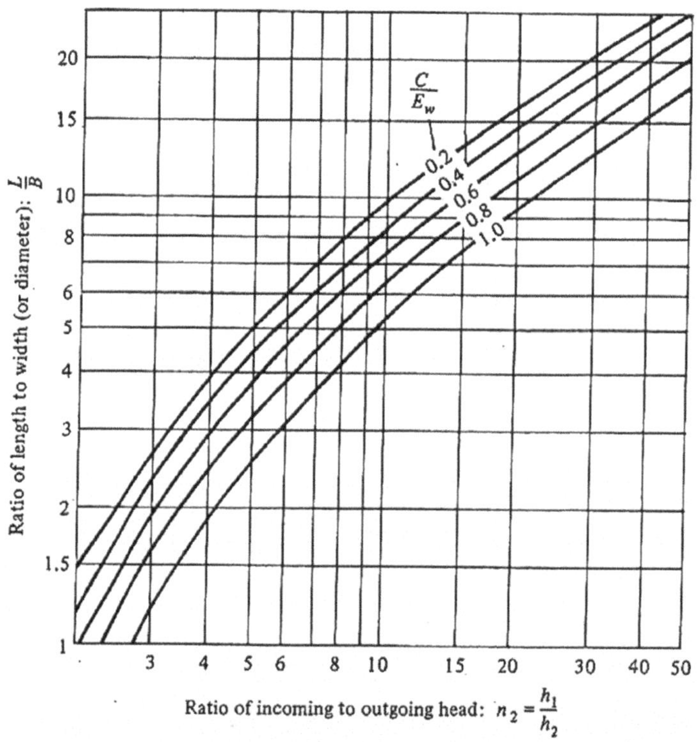

FIGURE 11.6 Relationship between several dimensionless ratios for the side weir with falling water surface. (Printed from Ref. 24)

$$h_1 = 0.5E_w \tag{11.2}$$

$$h_2 = E_w / 2n_2 \tag{11.3}$$

The definition of the variables used in Eqs. (11.1)–(11.3), Table 11.3, and Figure 11.6 are listed below:

L = length of weir, m (ft)
B = channel width (or sewer diameter), m (ft)
$n_2 = h_1/h_2$
h_1 = head over weir at upstream end of weir, m (ft)
h_2 = head over weir at downstream end of weir, m (ft)
c = height of weir above channel invert, m (ft)
E_w = specific energy of flow relative to crest of weir, m (ft)
V_n = normal velocity in the approach channel, m/s (ft/s)
d_n = normal depth in the approach channel, m (ft)
α = velocity-head correction coefficient ($\alpha = 1.2$ for upper and lower ends of the weir)
α' = pressure-head correction coefficient ($\alpha' = 1.0$ for upper and lower ends of the weir)

The simplified procedure and Eq. (11.1) are applicable only if the water surface over the weir has a falling profile. This is demonstrated if c/E_w is less than 0.6. The application of these equations and procedures is illustrated in Example 11.4.

Example 11.4: Flow over a Side Weir in a Combined Sewer System

A side weir is designed and installed in an existing rectangular combined sewer channel. The channel is 1.2 m wide, and the bottom slope is 0.003. The maximum wet weather flow is 2.0 m^3/s, and the maximum allowable wet weather discharge to the wastewater treatment plant is 0.90 m^3/s. The maximum dry weather flow with infiltration/inflow (I/I) allowance is 0.15 m^3/s. Determine the length of the weir. Assume Manning's $n = 0.013$.

Solution

A. Determine height of side weir
   1. Calculate flow depth in the channel for maximum dry weather flow
      Apply Eq. (5.35)

$$Q = \frac{1}{n} A R^{2/3} S^{1/2}$$

$$Q = \frac{1}{n} A \left(\frac{A}{P}\right)^{2/3} S^{1/2} = \frac{1}{n} \frac{A^{5/3}}{P^{2/3}} S^{1/2} = \frac{1}{n} \frac{(B \times y_{\text{dry}})^{5/3}}{(B + 2y_{\text{dry}})^{2/3}} S^{1/2}$$

$$0.15\ \text{m}^3/\text{s} = \frac{1}{0.013} \frac{(1.2\ \text{m} \times y_{\text{dry}})^{5/3}}{(1.2\ \text{m} + 2y_{\text{dry}})^{2/3}} (0.003)^{1/2}$$

$$0.0356 = \frac{(1.2\ \text{m} \times y_{\text{dry}})^{5/3}}{(1.2\ \text{m} + 2y_{\text{dry}})^{2/3}}$$

By trial and error y_{dry}=0.132 m

2. Select the weir height, c

The weir height c above the bottom of the channel is to be set equal to or greater than y_{dry}=0.132 m to retain all dry weather flow.

Select c=0.15 m

3. Calculate flow velocity for maximum dry weather flow

$$V_{dry} = \frac{Q_{dry}}{A}$$

$$= \frac{0.15\ m^3/s}{1.2\ m \times 0.132\ m}$$

$$= 0.95\ m/s$$

The flow velocity in the channel under peak dry weather conditions will be 0.95 m/s, which is greater than the recommended value of 0.6 m/s (2 ft/s) (Section 8.2.5)

B. Calculate the depth of flow under maximum wet weather conditions with Q_{wet}=2.0 m³/s

1. Determine flow depth y_{wet} and flow velocity V_{wet} for maximum wet weather

Apply Eq. (5.35)

$$Q = \frac{1}{n} A R^{2/3} S^{1/2} = \frac{1}{n} \frac{A^{5/3}}{P^{2/3}} S^{1/2}$$

$$2.0\ m^3/s = \frac{1}{0.013} \frac{(1.2\ m \times y_{wet})^{5/3}}{(1.2\ m + 2y_{wet})^{2/3}} (0.003)^{1/2}$$

$$0.48 = \frac{(1.2\ m \times y_{wet})^{5/3}}{(1.2\ m + 2y_{wet})^{2/3}}$$

By trial and error $y_{wet} = 0.81$ m

2. Calculate flow velocity

$$V_{wet} = \frac{Q_{wet}}{A}$$

$$= \frac{2.0\ m^3/s}{1.2\ m \times 0.81\ m} = 2.06\ m/s$$

C. Determine type of flow in the channel

Calculate the critical depth y_{cr} in the channel

Apply Eq. (5.46)

$$y_c = \left(\frac{(Q/b)^2}{g}\right)^{1/3}$$

$$= \left(\frac{\left(2.0\,\frac{\text{m}^3}{\text{s}}/1.2\text{ m}\right)^2}{9.81\text{ m/s}^2}\right)^{1/3}$$

$$= 0.66\text{ m}$$

Since flow depth y_{wet} is greater than critical depth y_c, flow will be subcritical in the channel during maximum wet weather flow.

D. Compute flow characteristic over the weir

1. Determine channel-specific energy

 The water depth over the side weir$=(d_n-c)$

 Apply Eq. (11.1) with $\alpha=1.2$ and $\alpha'=1$

$$(E_w) = \frac{1.2V_n^2}{2g} + (d_n - c)$$

 Substitute $V_n = V_{\text{wet}} = 2.06$ m/s and $d_n = y_{\text{wet}} = 0.81$ m

$$(E_w) = \frac{1.2(2.06\text{ m/s})^2}{2\times 9.81\text{ m/s}^2} + (0.81\text{ m} - 0.15\text{ m})$$

$$= 0.92\text{ m}$$

2. Determine if the water surface profile will be falling along the length of weir

 Calculate c/E_w

$$\frac{c}{E_w} = \frac{0.15}{0.92} = 0.16$$

 Since $c/E_w = 0.16$ is smaller than 0.6, the water surface has a falling profile (Section 11.7.1)

3. Compute head over the weir at the upper end

 The head over the weir at the upper end is calculated from Eq. (11.2)

$$h_1 = 0.5E_w = 0.5\times 0.92\text{ m}$$

$$= 0.46\text{ m}$$

E. Determine required side weir length

The side weir length when the water surface is falling is a function of channel width (B), c/E_w ratio, and the ratio of incoming to outgoing head $n_2 = h_1/h_2$.

The weir length can be obtained (1) from the generalized equations given in Table 11.3, and (2) from Figure 11.6. Both methods are given below.

1. Calculation based on generalized equations given in Table 11.3

   For assumed ratio $n_2=h_1/h_2=10$, and given channel width $B=1.2\,\text{m}$, the following equation from Table 11.3 is applied:

$$\frac{L}{B}=2.03\left(5.28-2.63\frac{c}{E_w}\right)$$

$$\frac{L}{1.2\text{ m}}=2.03(5.28-2.63\times0.16)$$

$$L=2.436(4.85)=11.8\text{ m}$$

   Select weir length $L=12\,\text{m}$

2. Calculations based on Figure 11.6

   The weir length L to channel width B ratio (L/B) can be obtained from Figure 11.6. Based on $n_2=10$, and $c/E_w=0.16$, the L/B ratio$=10$

$$L=10\times1.2\text{ m}=12\text{ m}$$

F. Estimate peak wet weather flow to the treatment plant

1. Calculate the velocity at the lower end of the weir

   Apply Eq. (11.1) with $\alpha=1.2$ and $\alpha'=1.0$

$$(E_w)=1.2\frac{V_2^2}{2g}+1.0(h_2)$$

$$0.92\text{ m}=1.2\frac{V_2^2}{2g}+1.0(h_2)$$

   From $n_2=h_1/h_2$

$$10=\frac{0.46\text{ m}}{h_2}$$

$$h_2=0.046\text{ m}$$

$$0.92\text{ m}=1.2\times\frac{V_2^2}{2\times9.81\text{ m/s}^2}+0.046\text{ m}$$

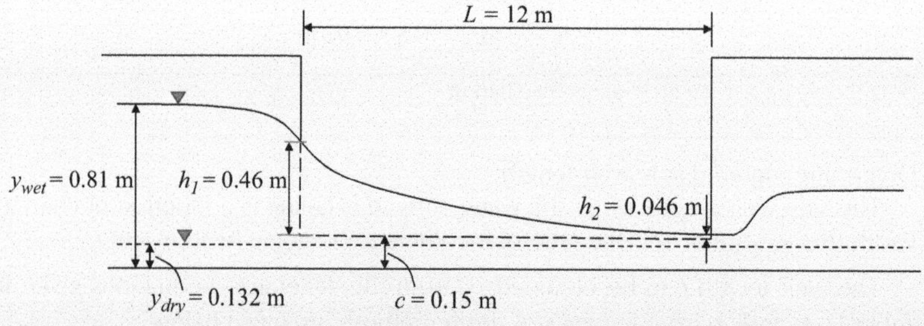

FIGURE 11.7 Flow depth in the channel and along the side weir (Example 11.4).

$$V_2^2 = 14.29\,(\text{m/s})^2$$

$$V_2 = 3.78\ \text{m/s}$$

2. Calculate flow depth in the channel at the lower end of weir

$$y_2 = c + h_2$$
$$= (0.15\ \text{m} + 0.046\ \text{m})$$
$$= 0.196\ \text{m}$$

3. Calculate discharge to treatment plant

$$Q_{\text{out}} = V_2 A_2$$
$$= 3.78\ \text{m/s} \times (1.2\ \text{m} \times 0.196\ \text{m})$$
$$= 0.89\ \text{m}^3/\text{s}$$

The discharge to the treatment plant is $0.89\,\text{m}^3/\text{s}$, which is smaller than the maximum allowable plant capacity of $0.90\,\text{m}^3/\text{s}$.

The design of side weir is satisfactory.

G. Sketch the designed side weir, water depth in the channel, and water surface profile.

Flow depth upstream, downstream, and over the side weir, and weir length are shown in Figure 11.7.

Examples and solutions of a side weir built in a sewer line, and the design of a side weir with rising water surface profile, may be found in Chapter 5 of Ref. 18. Readers are referred to this source for detailed design procedures on many types of overflow and diversion structures.

11.7.2 Transverse Weir

The transverse weir is placed directly across the channel in the line of flow. The CSO is discharged over the weir while the peak design capacity of the plant is discharged into an orifice in the channel bottom connected with an interceptor leading to the treatment plant. Also, the interceptor may be built under the transverse weir in-line with the flow direction or built on the side of the channel at 90°. It is possible to calculate the CSO over the weir from the weir equation, and flow in the interceptor from orifice equation. The details of a transverse weir are shown in Figure 11.5b.

Example 11.5: Transverse Weir and Orifice to Divert Flow to a Treatment Plant, and to a CSO Storage

A combined sewer is designed to carry peak wet weather flow. The diversion structure consists of a diversion box that has a width of 3.0 m with a rectangular transvers weir. The diversion box is 6.0-m long and has an orifice 53 cm in diameter at the bottom of the diversion box. The orifice is connected with an interceptor that goes down below and then turns horizontally. The head over the transverse weir is 15 cm at peak wet weather flow. The height of the transverse weir is 1.0 m above the channel floor. Calculate the diameter of the sewer and the volume of CSO diverted to the storage basin. Also, calculate the design capacity of the treatment plant.

Solution

1. Calculate the diverted CSO

   Apply rectangular weir equation. The flow over a rectangular weir is calculated from Eq. (7.9b)

$$Q = 2.95C_d(L - 0.1nH)H^{3/2}$$

   Substitute $L = 3.0$ m, $n = 2$, $H = 0.15$ m, and $C_d = 0.90$

$$\text{Diverted CSO } Q = 2.95 \times 0.90(3.0 \text{ m} - 0.1 \times 2 \times 0.15 \text{ m}) \times (0.15 \text{ m})^{3/2}$$

$$= 2.95 \times 0.9 \times 2.97 \times 0.058$$

$$= 0.46 \text{ m}^3/\text{s}$$

2. Calculate the diameter of sewer to transport CSO to the storage facility

   Assume sewer slope of 0.0006 m/m and Manning's $n = 0.013$

   Apply Eq. (5.38)

$$Q = \frac{0.312}{n} D^{8/3} S^{1/2}$$

   Substitute $Q = 0.46 \text{ m}^3/\text{s}$, $n = 0.013$, and $S = 0.0006$ m/m

$$D = \left[\frac{0.46 \text{ m}^3/\text{s} \times 0.013}{0.312 \times (0.0006)^{1/2}}\right]^{0.375} = 0.91$$

   Select a standard sewer size of 90 cm.

3. Calculate the flow diverted to the interceptor sewer leading to the treatment plant

   Flow through an orifice is calculated from Eq. (7.8)

$$Q = C_d AV = C_d A\sqrt{2gH}$$

   Substitute the following values:

$$C_d = 0.6, A = (\pi/4)D^2, \text{ and } H = (1.0 \text{ m} + 0.15 \text{ m}) = 1.15 \text{ m}, D = \text{orifice diameter} = 53 \text{ cm}$$

$$Q = 0.6 \times \pi/4 \times (0.53 \text{ m})^2 \times (2 \times 9.81 \text{ m/s}^2 \times 1.15 \text{ m})^{0.5}$$

$$= 0.132 \times 4.75 \text{ m/s}$$

$$= 0.63 \text{ m}^3/\text{s}$$

   A 53-cm diameter interceptor sewer is connected with the orifice. The sewer runs under pressure. The 90° elbow prevents vortex formation.

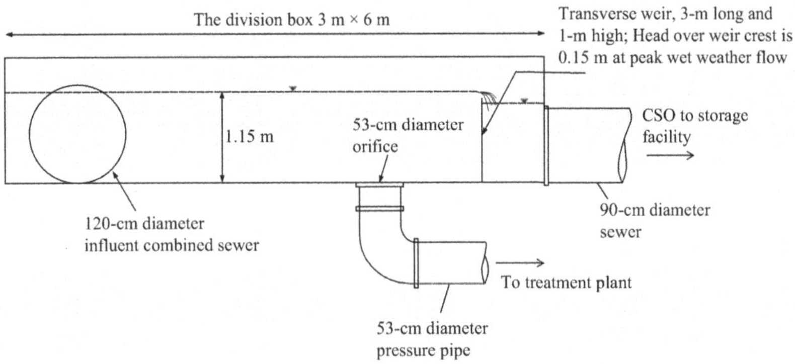

FIGURE 11.8 Collection and diversion structure showing influent combined sewer, diversion sewers to wastewater treatment plant, and CSO diverted to the storage facility (Examples 11.5 and 11.6).

4. Summarize the results
   Diverted CSO = $0.46\,m^3/s$
   Diverted flow to plant = $0.63\,m^3/s$
   Diameter of interceptor = 0.53 m
   Total peak wet weather flow = $1.09\,m^3/s$
5. Show the design details of the transverse weir
   The design details of the transverse weir and diversion box are shown in Figure 11.8.

Example 11.6: Diameter of Combined Sewer Discharging into the Diversion Structure in Example 11.5

A combined sewer conveys peak wet weather flow to the diversion structure that has a transverse weir in Example 11.5. The slope of the combined sewer and Manning's n are 0.0008 m/m and 0.013, respectively. Calculate the diameter of the combined sewer.

Solution

1. Calculate the peak wet weather flow to the combined sewer reaching the division box

$$\text{Combined flow} = \text{CSO} + \text{interceptor flow to POTW}$$

$$= 0.46\ m^3/s + 0.63\ m^3/s$$

$$= 1.09\ m^3/s$$

2. Apply Eq. (5.38) to calculate the diameter of combined sewer

$$Q = \frac{0.312}{n} D^{8/3} S^{1/2}$$

$$1.09\ m^3/s = \frac{0.312}{0.013} D^{8/3} 0.0008^{1/2}$$

$$D = \left[\frac{1.09\ m^3/s \times 0.013}{0.312 \times (0.0008)^{1/2}}\right]^{0.375} = 1.20\ m$$

Select a standard diameter pipe 120 cm (48 in.)
The influent combined sewer is shown in Figure 11.8.

11.7.3 Leaping Weir

Leaping weir is formed by a gap in the interceptor through which flows up to peak dry weather is dropped and transported to the wastewater treatment plant. During the wet weather flow, the CSO or flow in excess of the treatment plant capacity leaps the opening and enters the outlet sewer. The details of a leaping weir are shown in Figure 11.5c. The steel weir plate is inclined to give a spouting effect and is designed so that it can be adjusted for various flow conditions. The design and construction of a leaping weir is a trial-and-error process involving computation of the trajectory of flow and determination of the percentage of flow captured at various weir settings.

11.7.4 Relief Syphon

A relief syphon precisely regulates the maximum water surface elevation in the sewer with smaller variations in high-water elevation. Since a syphon uses available head, it discharges at a higher velocity than do overflow weirs. Although this device has obvious advantages, its use is restricted due to inadequate information about its design and operation. The problems associated with it are the minimum head required for priming, possible noise and vibration from sudden starting and stopping of syphon, especially with high heads and fouling by debris. Designing a relief syphon requires an iterative trial-and-error approach. The sum of all energy losses, including entrance and exit losses, should equal the total available head. If not, the trial section should be revised, and calculations should be repeated until the agreement is reached. The design also involves working out such details as the method of venting, the shape of the section to meet the requirements, and fixing the elevations of the inlet, spillway, outlet, and air vent. The sectional view of a relief syphon is shown in Figure 11.5d.

11.7.5 Regulating Devices

The flow-regulating devices are used to prevent flooding of downstream facilities and surcharging of an existing interceptor. The flow regulators are installed in CSS to (1) control flow release from in-line combined sewers, (2) prevent unacceptable storm sewer surging, (3) control flow rate between combined sewers and companion facilities such as temporary storage basins, interceptors, or outfall structures, and (4) divert flow to sewage treatment plants. Examples of such flow regulators are mechanical regulators, reverse tainter gates, tipping-plate regulators, and hydro-brake or vortex-type flow regulators. Most of these devices are automatic regulators, either a float-operated type or a flap-gate type. These are generally impractical for small flows because of clogging. All such devices may need periodic inspection, skillful management, and careful maintenance. These are manufactured devices and are installed by equipment suppliers. Detailed description of flow regulators may be found in Refs. 1–3,18,19. Several types of manufactured flow regulators are shown in Figure 11.9.

A sectional view of a CSO regulator would typically show key components: a chamber with an inlet pipe, a weir chamber (overflow point), weir, gate mechanism to control the level of water within the chamber, and a downstream outlet pipe. The gate mechanism often includes a hinged gate or a similar design, allowing excess water to overflow during heavy rain events while maintaining normal flow during regular conditions; essentially acting as a safety valve for the sewer system to prevent flooding in the pipes and surrounding areas. The important design considerations of CSO regulators are (1) the capacity to handle the expected peak flows during storm events, (2) the shape and height of the weir that significantly impact the flow control, and (3) the gate mechanism design and control system that must be reliable to manage overflow levels effectively.

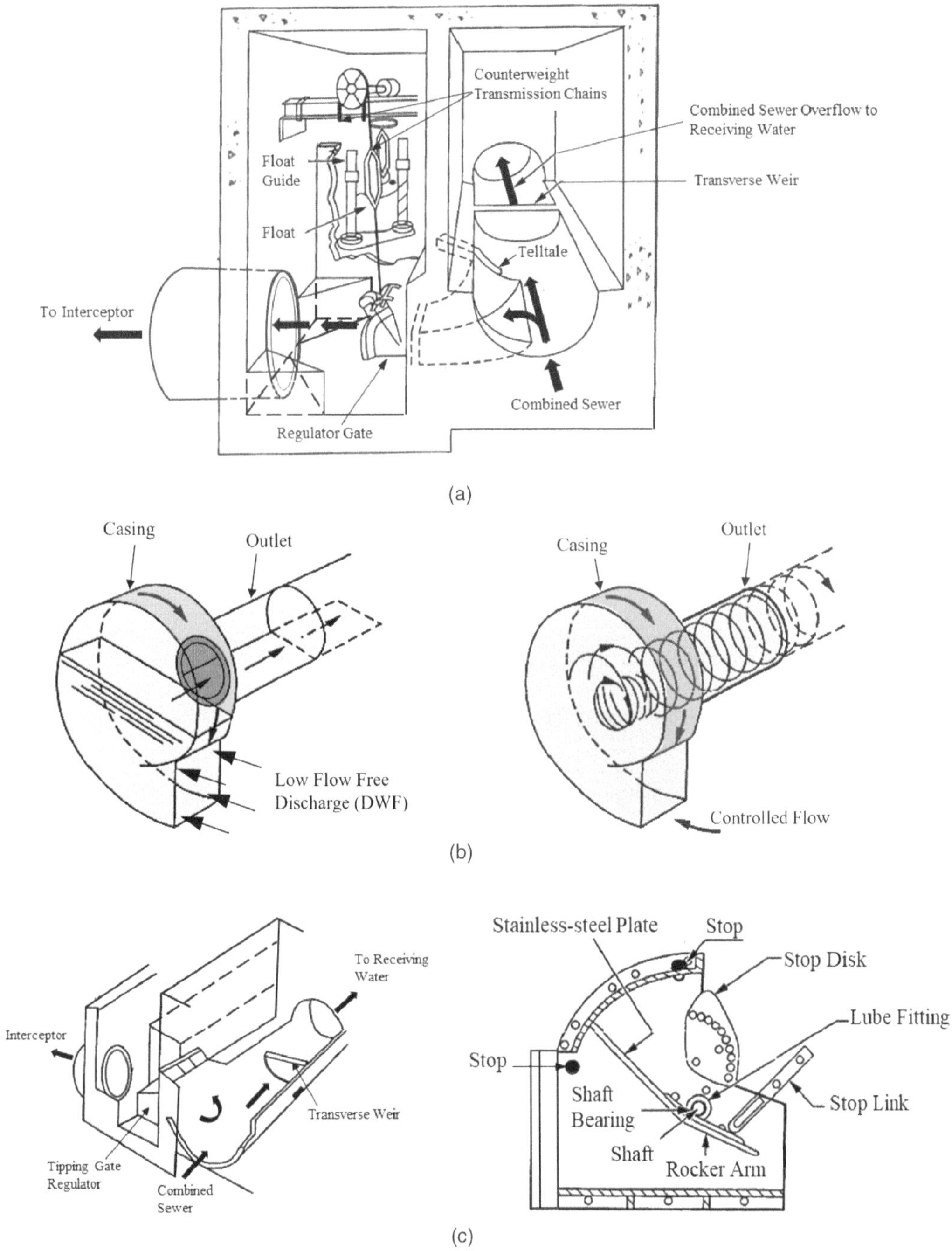

FIGURE 11.9 Typical CSO regulators: (a) mechanized regulator, (b) hydro-break, and (c) tipping-plate regulator.

Example 11.7: Diversion of CSO into a Relief Sewer by a Mechanical Regulator

A combined sewer is 100-cm in diameter. A mechanical regulator is installed that bypasses the excess flow into a diversion sewer. The flow continues to discharge into the treatment facility until the maximum flow to the treatment plant is reached at a flow depth of 48 cm. As the depth of flow increases above 48 cm, the regulator gate opens in proportion to the depth in the sewer to bypass CSO to a storage facility. The gate dimensions are 20 cm wide and 15 cm high. The differential elevation from the center of the gate to the 48-cm depth in the sewer is 1.0 m. Calculate the discharge through the gate into the bypass sewer. The depth of flow in the combined sewer is 85 cm, and the gate opening is 75%. The coefficient of discharge is 0.8.

Solution

A. Calculate the discharge from the regulator
   1. Develop the discharge equation through the flap gate
      Take a strip of the gate *dh* high (see Figure 11.10)

$$\text{Discharged } q \text{ through the strip} = C_d \times B \times dh\sqrt{2gh}$$

$$\text{Total discharge } Q \text{ through gate} = C_d \times B \times \sqrt{2g}\int_{H_1}^{H_2} h^{1/2}dh$$

The integrated form of the equation is given by Eq. (11.4)

$$\text{Total discharge } Q \text{ through gate} = 2/3C_d \times B\sqrt{2g}\left[H_2^{3/2} - H_1^{3/2}\right] \tag{11.4}$$

   2. Substitute the values of C_d, B, H_2, and H_1

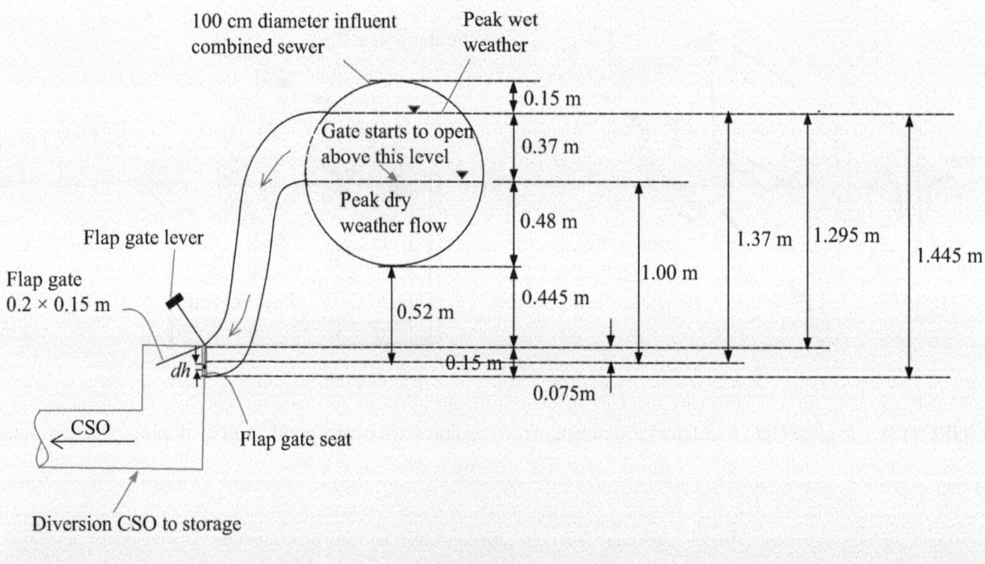

FIGURE 11.10 Flow regulator from a combined sewer and release of CSO through a flap gate (Example 11.7).

$$C_d = 0.8 \times 0.75(\text{gate opened}) = 0.6$$

$$B = 0.2 \text{ m}$$

$$H_1 = 1.295 \text{ m (from Figure 11.10)}$$

$$H_2 = 1.445 \text{ m (from Figure 11.10)}$$

$$Q = 2/3 \times 0.6 \times 0.2 \times \sqrt{2g}\left((1.445 \text{ m})^{3/2} - (1.295 \text{ m})^{3/2}\right)$$

$$= 0.354(1.737 - 1.474) \text{ m}^3/\text{s} = 0.0931 \text{ m}^3/\text{s}$$

B. Calculate the discharge Q through the gate using simplified orifice equation

1. Apply the details of flow regulator (Eq. 7.8b)

$$Q = C_d \times \text{gate opend} \times A\sqrt{2gH}$$

$$= 0.8 \times 0.75 \times (0.15 \text{ m} \times 0.2 \text{ m})\sqrt{2g}\ \sqrt{H}$$

H = Total height from center of the gate to the flow depth of 85 cm in the combined sewer = 1.37 m

$$Q = 0.0797\sqrt{1.37 \text{ m}} = 0.0933 \text{ m}^3/\text{s}$$

2. Show the details of the flow regulator

The details of the flow depths in the combined sewer and flow released from the gate are shown in Figure 11.10.

11.8 MANAGEMENT OF CSOs

The release of untreated CSOs to waterbodies is no longer acceptable under the Clean Water Act. Therefore, management practices or techniques should be implemented to minimize or eliminate CSO volumes and pollutants from being released into receiving waters. The CSO communities have been using several management practices and technologies to reduce the pollutants discharge from CSO, and their impacts on the environment and human health. The management practices can be grouped as (1) providing source control, (2) reducing stormwater entering the collection system, (3) providing in-system storage in collection system, (4) providing off-line storage facilities in the collection system, and (5) providing on-site storage. One or a combination of some of these practices can be used to more effectively manage CSOs. The criteria for selection of a COS control technology are: (1) current condition of the CSS, (2) characteristics of wet weather discharges (e.g., peak flow rate, volume, frequency and duration of wet weather events, and concentration of key pollutants), (3) hydraulic and pollutant loading to a particular facility, (4) climatic condition including seasonal variations in temperature and rainfall pattern, (5) implementation requirements (e.g., land or space constraints, surrounding neighborhood, noise, disruption), and (6) cost of implementing and maintenance requirements.

11.8.1 Provide Source Control

The amount of pollutant entering CSSs can be controlled at the source by the implementation of best management practices (BMPs). These are often a major factor in choosing a management practice or control technology for most communities. The low-impact development (LID) techniques have been implemented successfully as an alternative or in conjunction with conventional practices to control stormwater volume in urban areas. In-depth discussion about these techniques may be found in Section 10.7. These source control BMPs for watershed management practices and technologies essentially reduce, retard, conserve, or prevent stormwater runoff from entering CSSs and consequently decrease pollutant loading of CSOs. Examples of LID techniques presented in Section 10.7 are green roofs, porous pavement, bioretention basins, land use management, detention and retention basins, and many others. The U.S. EPA supports the LID approach and uses the term *green infrastructure* for such practices that are used for the management of wet weather flows. The pollution prevention practices can be implemented to reduce the amount of contaminants entering CSSs. Examples of pollution prevention practices are surface sanitation, litter and chemical controls, street sweeping, controlled use of fertilizers, insecticides, herbicides, and deicing practices.

Example 11.8: Stormwater Volume Reduction from Permeable Pavement

A housing subdivision has single and multiple-residential housing, shopping center, driveways, service roads, and individual and central parking areas. The paved surfaces utilize common types of permeable paving. The permeable paved surfaces are about 7.5% of the total subdivision area. The main function of permeable paving is to collect and retain the surface runoff. Since the groundwater table is low, the allowable reservoir depth of permeable paving is 22 in., and the void ratio of the reservoir layer is 0.35. The area of the entire subdivision is 65 acres. The maximum stormwater in the area is equivalent to 1.8 in./h of rainfall in 2 h. The average percolation and evapotranspiration losses from developed land are 45% of rainfall. Determine the average volume of stormwater released from the subdivision. Prior to subdivision development, the land was grassy pasture. On average, the percolation and evaporation losses of stormwater from the undeveloped area are 71%. Compare the surface drainage before and after the development.

Solution

1. Calculate the surface runoff prior to development

$$\text{Subdivision area} = 65\text{ acres} \times 4{,}840\text{ yd}^2/\text{acre} \times 9\text{ ft}^2/\text{yd}^2$$

$$= 2{,}831{,}400\text{ ft}^2$$

$$\text{Total precipitation} = 1.8\text{ in/h} \times 2\text{ h} \times 1\text{ ft}/12\text{ in.} \times 2{,}831{,}400\text{ ft}^2$$

$$= 849{,}420\text{ ft}^3$$

$$\text{Total surface runoff} = (1 - 0.71) \times 849{,}420\text{ ft}^3$$

$$= 0.29 \times 849{,}420\text{ ft}^3 = 246{,}332\text{ ft}^3$$

2. Calculate the total area of permeable pavement
   Total area of permeable pavement $=0.075\times2{,}831{,}400\,\text{ft}^2=212{,}355\,\text{ft}^2$
3. Calculate the storage volume of permeable pavement
   The storage volume of the permeable pavement is calculated from Eq. (10.50)

$$S_v = A_p\left[\left(d_p\times\eta_r\right)+K_{\text{sat}}\times t_f\right]$$

$$= 212{,}355\ \text{ft}^2\left[\left(22\ \text{in.}\times\frac{1\ \text{ft}}{12\ \text{in.}}\times0.35\right)+1.8\ \text{in./h}\times(1\ \text{ft}/12\ \text{in.})\times2\ \text{h}\right]$$

$$= 212{,}355\ \text{ft}^2\times[0.64\ \text{ft}+0.30\ \text{ft}]$$

$$= 199{,}613\ \text{ft}^3$$

4. Calculate the total surface runoff after development
   Net available precipitation $=(1-0.45)\times849{,}420\,\text{ft}^3=467{,}181\,\text{ft}^3$
   Total surface runoff $=(467{,}181\,\text{ft}^3-199{,}613\,\text{ft}^3)=267{,}568\,\text{ft}^3$
5. Compare the results before and after development
   The net surface runoff before development is $246{,}332\,\text{ft}^3$, while after development with 7.5% permeable pavement, the net surface runoff is $267{,}568\,\text{ft}^3$. It can be concluded that there is no significant increase in net runoff after development.

Example 11.9: Reduction of Stormwater Runoff from Bioretention Basin

A shopping center is built over a total land area of 5 acres. The maximum storm in the area is equivalent to 1.8 in. of rainfall in 2 h. The percolation and evapotranspiration losses in the undeveloped land are 68%. The storage capacity of bioretention basin is 14.0 in. The total area of the constructed bioretention system is 5.0% of the entire shopping center area, and it is constructed along the edge of the shopping center. The percolation and evapotranspiration losses from the developed land area is 30%. Compare the surface stormwater runoff before and after development.

Solution

1. Calculate the stormwater released from the undeveloped land after the storm
   Land area utilized by shopping center $=5\ \text{acres}\times4{,}840\ \text{yd}^2/\text{acre}\times9\,\text{ft}^2/\text{yd}^2=217{,}800\,\text{ft}^2$
   Volume of stormwater in the area $=1.8\ \text{in.}\times1\ \text{ft}/12\ \text{in.}\times217{,}800\,\text{ft}^2=32{,}670\,\text{ft}^3$
   Volume of stormwater drainage $=(1-0.68)\times32{,}670\,\text{ft}^3=10{,}455\,\text{ft}^3$
   Flow rate of storm drainage $=10{,}455\,\text{ft}^3/(2\ \text{h}\times3{,}600\ \text{s/h})=1.45\,\text{cfs}$
2. Calculate the storage volume and flow rate of stormwater drainage from the shopping center
   Volume of stormwater at the shopping center site $=32{,}670\,\text{ft}^3$

$$\text{Volume of available stormwater} = (1-0.3)\times32{,}670\ \text{ft}^3$$

$$= 0.7\times32{,}670\ \text{ft}^3 = 22{,}869\ \text{ft}^3$$

3. Calculate the volume of surface runoff retained in the bioretention basin
   Area of bioretention basin $=0.05\times217{,}800\,\text{ft}^2=10{,}890\,\text{ft}^2$
   Storage volume of bioretention basin $=10{,}890\,\text{ft}^2\times14.0\,\text{in.}\times1\ \text{ft}/12\ \text{in.}=12{,}705\,\text{ft}^3$

4. Calculate the drainage volume of stormwater from the shopping center
   Drainage volume of bioretention basin = (22,869 – 12,705) ft^3 = 10,164 ft^3

$$\text{Flow rate of storm water} = 10{,}164\ \text{ft}^3/\left(2\ \text{h} \times 3{,}600\ \text{s/h}\right)$$

$$= 1.41\ \text{cfs}$$

5. Compare the runoff before and after construction of the shopping center
   The net surface runoff before development is 10,455 ft^3 (1.45 cfs), while after development with a bioretention basin is 10,164 ft^3 (1.41 cfs). It can be concluded that there is a small decrease in net runoff after development.

11.8.2 Reduced Stormwater Entry into Collection System

The capacity of the collection network can be increased by implementing several practices. Some of these practices are: (1) separate combined sewer into sanitary and storm sewers, (2) modify the facilities to reduce entry of I/I and CSOs, (3) utilize real-time monitoring and control devices to better understand and manage the response of sewer systems to wet weather events, and (4) maximize wet weather flow to treatment facilities. Separation of combined sewers is very expensive and often impractical. Despite the cost, this option will not eliminate the impact of stormwater that may result from the discharge of stormwater. The most cost-effective activity under collection system control, therefore, is reduction in I/I and stormwater volume that enters the collection system.

Collection system controls and their benefits in reducing pollutants and flow rate to CSS are discussed in detail in Refs. 1–3,18,19.

11.8.3 Provide In-system Storage in Collection System

The in-system storage is the storage volume available within the existing sewer collection system. The combined sewers are designed to run partially full under peak dry weather flow plus allowance for I/I. There is a considerable capacity available to accommodate stormwater during wet weather conditions. The depth of flow in sewers increases in the collection systems until the sewers start to run full. This additional capacity delays the peak flow reaching the treatment plant. The amount of available volume depends on the size and slope of the pipes. Large pipelines with flat slopes and underground tunnels provide significant storage volume. The inflatable dams are also used to provide flexibility in changing the overflow height to increase or decrease the storage to keep a balance with upstream flooding or surcharging and storage needs.

Example 11.10: Storage in the Collection System

A flow-monitoring float is installed in a manhole of an interceptor. The diameter, slope, and Manning's n of the interceptor are 76 cm, 0.00056 m/m, and 0.013, respectively. At peak dry weather flow, the depth in the manhole is 29 cm, and at peak wet weather flow, the float reads 62 cm. Assume that the depth in the manhole represents the depth of flow in the interceptor. Calculate the peak dry weather flow, peak wet weather flow, and available capacity utilized for wet weather flow per kilometer length of the interceptor.

Solution

1. Calculate velocity and discharge in the interceptor when flowing full.
   The velocity is calculated from Eq. (5.34)

$$V = \frac{1}{n} R^{2/3} S^{1/2}$$

$$V = \frac{1}{n}\left(\frac{A}{P}\right)^{2/3} S^{1/2}$$

$$V = \frac{1}{n}\left(\frac{\pi D^2/4}{\pi D}\right)^{2/3} S^{1/2}$$

$$V = \frac{1}{n}\left(\frac{D}{4}\right)^{2/3} S^{1/2}$$

$$V = \frac{1}{0.013}\left(\frac{0.76\text{ m}}{4}\right)^{2/3} (0.00056)^{1/2}$$

$$V = 0.6\text{ m/s}$$

Discharge is calculated from Eq. (5.38)

$$Q = \frac{0.312}{n} D^{8/3} S^{1/2}$$

$$Q = \frac{0.312}{0.013}(0.76\text{ m})^{8/3}(0.00056)^{1/2}$$

$$Q = 0.273\text{ m}^3/\text{s}$$

2. Calculate partial flow velocity and discharge at peak dry weather flow

$$d/D = 0.29\text{ m}/0.76\text{ m} = 0.38$$

Determine from Figure 5.26 the ratios of v/V and q/Q

$$v/V = 0.90 \quad \text{and} \quad q/Q = 0.30$$

Determine v and q

$$v = 0.90 \times 0.60\text{ m/s} = 0.54\text{ m/s}$$

$$q = 0.30 \times 0.273\text{ m}^3/\text{s} = 0.0819\text{ m}^3/\text{s}$$

3. Calculate partial flow velocity and discharge at peak wet weather flow

$$d/D = 0.62\text{ m}/1.76\text{ m} = 0.82$$

Determine from Figure 5.26 the ratios of v/V and q/Q

$$v/V = 1.15 \quad \text{and} \quad q/Q = 1.02$$

Determine v and q

$$v = 1.15 \times 0.60 \text{ m/s} = 0.69 \text{ m/s}$$

$$q = 1.02 \times 0.273 \text{ m}^3/\text{s} = 0.278 \text{ m/s}$$

4. Calculate available capacity in the interceptor sewer utilized for peak wet weather flow

Flow area a_{dry} of interceptor at peak dry weather flow

$$a_{dry} = q/v = \left(0.0819 \text{ m}^3/\text{s}\right)/\left(0.54 \text{ m/s}\right) = 0.152 \text{ m}^2$$

Interceptor capacity $(\text{Capacity})_{dry}$ utilized at peak dry weather flow $= a_{dry} \times \text{Length}$

$$\left(\text{Capacity}\right)_{dry} = 0.152 \text{ m}^2 \times 1{,}000 \text{ m} = 152 \text{ m}^3$$

Flow area a_{wet} of interceptor at peak wet weather flow

$$a_{wet} = q/v = \left(0.278 \text{ m}^3/\text{s}\right)/\left(0.69 \text{ m/s}\right) = 0.403 \text{ m}^2$$

Interceptor capacity $(\text{Capacity})_{wet}$ utilized at peak wet weather flow $= a_{wet} \times \text{Length}$

$$\left(\text{Capacity}\right)_{wet} = 0.403 \text{ m}^2 \times 1{,}000 \text{ m} = 403 \text{ m}^3$$

Available capacity $(\text{Available Capacity})_{peak}$ in the interceptor utilized at peak wet weather flow

$$\left(\text{Available Capacity}\right)_{peak} = 403 \text{ m}^3 - 152 \text{ m}^3 = 251 \text{ m}^3$$

Example 11.11: Inflatable Dam to Store Excess Flow

A combined sewer is 60 cm in diameter. An inflatable dam is provided in the combined sewer to store the excess flow within the system. Under the peak dry weather condition, the maximum depth of flow is 35 cm. If the dam is inflated such that a 55-cm maximum depth of flow is reached in the sewer, normal flow condition is still maintained, and no surcharge occurs in the main sewer and tributaries. Calculate the maximum storage obtained per km of the sewer length.

Solution

A. Peak dry weather condition
   1. Calculate the d/D ratio at peak dry weather flow

$$d/D = 35 \text{ cm}/60 \text{ cm} = 0.583$$

2. Calculate the central angle θ in degrees from Eq. (5.33)

$$\cos\frac{1}{2}\theta = \left(1 - \frac{2d}{D}\right) = (1 - 2 \times 0.583) = -0.166$$

$$\theta = 199.1^{\circ}$$

3. Calculate the partial flow area at peak dry weather flow
   Apply the equation for partial flow area given in Table 5.3

$$\text{Flow area } a \text{ at peak dry weather flow} = 1/8\left[(\pi\theta/180 - \sin\theta)\right]D^2$$

$$= 1/8[(\pi \times 199.1/180 - \sin 199.1] \times (0.6 \text{ m})^2$$

$$= 1/8[3.475 - (-0.327)] \times 0.36 \text{ m}^2$$

$$= 0.171 \text{ m}^2$$

B. Dam inflated to 55 cm
1. Calculate d/D ratio depth at $d = 55$ cm

$$d/D = 55 \text{ cm}/60 \text{ cm} = 0.917$$

2. Calculate the central angle θ in degrees from Eq. (5.33)

$$\cos\frac{1}{2}\theta = \left(1 - \frac{2d}{D}\right) = (1 - 2 \times 0.917) = -0.834$$

$$\theta = 293.0^{\circ}$$

3. Calculate the partial flow area when dam is inflated to 55 cm.
   Apply the equation for flow area given in Table 5.3

$$\text{Flow area } a \text{ at peak wet weather flow} = 1/8\left[(\pi\theta/180 - \sin\theta)\right]D^2$$

$$= 1/8[(\pi \times 293.0/180) - \sin 293.0] \times (0.6 \text{ m})^2$$

$$= 1/8\ [5.114 - (-0.921)] \times 0.36 \text{ m}^2$$

$$= 0.272 \text{ m}^2$$

4. Calculate the maximum storage utilized per 1,000 m length of interceptor

$$\text{Maximum storage utilized} = (0.272 - 0.171) \text{ m}^2 \times 1{,}000 \text{ m}$$

$$= 0.101 \text{ m}^2 \times 1{,}000 \text{ m} = 101 \text{ m}^3$$

11.8.4 Provide Off-line Storage Facility in the Collection System

The off-line facilities are used to store the wet weather flow for a period of time until the treatment capacity of the plant becomes available. Examples of the off-line storage systems include off-line tanks and basins, deep tunnels, pipelines, and channels in the vicinity of the sewer system. The off-line facilities may be constructed as surface basins or underground storage. Some U.S. cities have such facilities to control CSO events. Flow regulators are installed to pass the dry weather flow directly into the interceptor while capturing the wet weather flow by diverting the excess flow to off-line storage tanks or basins. Inflatable dams or automatic sluice and other types of gates, valves, and regulators are used to fully capture the small storms. All these systems clearly regulate the excess flows, whereby dry weather flow continues to flow uninterrupted into the interceptor, leading to the treatment facility. One of the disadvantages of off-line storage facilities is sediments and debris accumulation. If the accumulated materials are not removed and the facility is not regularly maintained, the storage volume may be reduced. Also, the surface storage may create public health problems if the CSO is stored for a long time without aeration, and if the basin is not fully cleaned after dewatering.

Example 11.12: Diversion of Combined Sewer Overflow

A subdivision of an older part of the town has a CSS. The estimated population of the subdivision is 20,000 residents. The average wastewater flow is 450 Lpcd. The maximum dry weather is diverted into an interceptor leading to the wastewater treatment plant. The design capacity of the treatment plant is peak dry weather flow plus 60 Lpcd I/I allowance. The maximum combined sewer flow during critical wet weather conditions is 48,000 m³/day. Determine the maximum CSO that must be diverted to the temporary storage facility located off-line in the collection system.

Solution

1. Determine the average day dry weather flow from the subdivision

$$Q_{avg} = 450 \text{ L/person.day} \times 20{,}000 \text{ persons} \times 10^{-3} \text{ m}^3/\text{L} = 9{,}000 \text{ m}^3/\text{day}$$

2. Determine peak dry weather flow from the subdivision

   The peaking factor M_{max} is calculated from Eq. (4.4). This topic is discussed in detail in Section 4.5.1

$$M_{max} = 1 + \frac{14}{4 + \sqrt{P}}$$

   Substitute $P = 20$

$$M_{max} = 1 + \frac{14}{4 + \sqrt{20}} = 2.65$$

   Peak dry weather flow = 9,000 m³/day × 2.65 = 23,850 m³/day

3. Determine the permissible I/I allowance

   I/I allowance = 150/L/person × 20,000 persons × 10^{-3}/L = 3,000 m³/day

4. Determine peak permissible dry weather flow diverted to the interceptor leading to the treatment plant

   Peak dry weather flow discharged into the interceptor = (23,850 + 3,000) m³/day = 26,850 m³/day

   Design flow reaching the treatment plant = 26,850 m³/day

5. Determine the average CSO diverted to the storage facility

   The diverted CSO = 48,000 m³/day − 26,850 m³/day = 21,150 m³/day

   Note: The diverted flow is stored in a temporary storage basin. This flow is routed back through the treatment plant after the storm event when the plant resumes normal operation and is able to treat the diverted flow.

Example 11.13: Volume of Temporary Storage Facility for CSO

An older part of a city has CSS. The peak dry weather flow of 3,000 m^3/h is released into the interceptor that carries the flow to the wastewater treatment plant. Determine the volume of a temporary storage facility to hold the excess CSO. The combined sewer flow data is given below:

Time of the day	12 MN	2	4	6	8	10	12 N
Combined sewer flow, m^3/h	812	375	500	1,250	3,000	4,000	4,500
Time of the day	2	4	6	8	10	12 MN	
Combined sewer flow, m^3/h	4,385	4,000	3,500	3,000	2,500	2,250	

MN, midnight; N, noon.

Solution

1. Plot the wastewater flow data
   The combined sewer flow data in m^3/h is plotted with respect to the time of day. This plot is shown in Figure 11.11. The maximum allowable flow of 3,000 m^3/h that is diverted to the treatment plant is also shown in this figure.
2. Determine the CSO diverted to the temporary storage facility
   The volume of CSO diverted to the temporary storage facility is shown in Figure 11.11. The area bounded by the wet weather flow and the maximum allowable flow is diverted to the treatment plant. The trapezoidal rule is used to calculate the area. The trapezoidal rule is expressed by Eq. (4.1).

$$V = \frac{1}{2} \times \Delta T \left[q_1 + 2\left(q_2 + q_3 + q_4 + \cdots + q_{n-1}\right) + q_n \right]$$

$$V = \frac{1}{2} \times 2\,\text{h} \times \left[0 + 2(1{,}000 + 1{,}500 + 1{,}385 + 1{,}000 + 500) + 0\right] \text{m}^3/\text{h}$$

$$V = 10{,}770\ \text{m}^3$$

The temporary storage capacity for CSO is 10,750 m^3

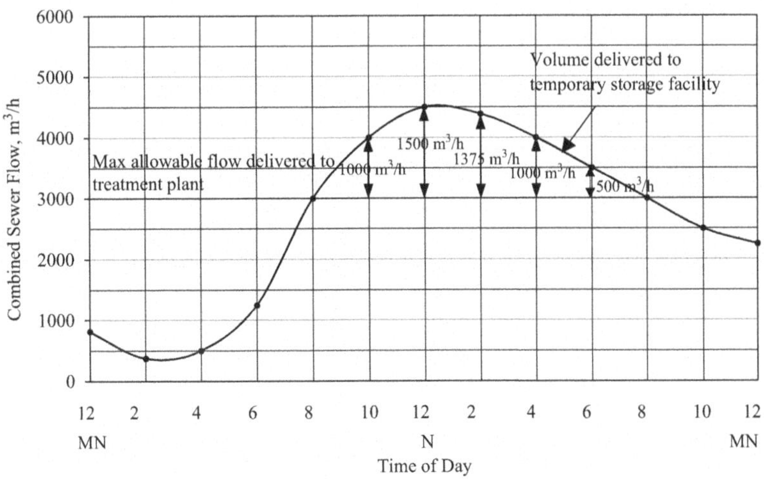

FIGURE 11.11 Combined sewer flow (Example 11.13).

11.8.5 Provide On-site Storage

A less expensive and more effective solution for managing CSO is to build on-site storage at the treatment plant. Unused flow equalization basins and abandoned treatment facilities are two common types of on-site available storage. Converting abandoned facilities to on-site storage is a very cost-effective approach since the cost of construction and land ownership is already paid. Aerated earthen basins may also be constructed on-site for temporary storage of diverted CSO.

Example 11.14: On-site Diversion Structure and Storage in Existing Abandoned Facility

A city has a separate sanitary and storm sewer collection network. The central downtown area of the city is an older section of the town and has a combined sewer collection network. The CSS is connected to the sanitary sewer network. The design capacity of the wastewater treatment plant is peak dry weather flow plus I/I allowance. The maximum design capacity of the treatment plant is 1.40 m^3/s. The incoming combined sewer is connected to a division box at the plant site. The division box diverts the design capacity of the plant to the treatment plant, and CSO to the existing abandoned flow equalization basin for storage. The stored CSO is subsequently routed through the treatment plant after the storm event. The combined flow reaching the plant during a 10-year design storm event is summarized below:

Time of day	4 PM	6 PM	8 PM	10 PM	12 MN	2 AM	4 AM
Combined sewer flow, m^3/s	1.0	1.2	1.3	1.4	1.5	1.6	1.7
Time of day	6 AM	8 AM	10 AM	12 N	2 PM	4 PM	6 PM
Combined sewer flow, m^3/s	1.8	1.8	1.6	1.4	1.2	1.1	1.1

MN, midnight; N, noon.

Solution

A. Design the division box
   1. Select a division box

      The selected collection and division box is a 4.0 m×4.0 m square box. The incoming combined interceptor sewer discharges into the square box. The peak dry weather flow plus I/I allowance (design capacity of POTW) flows over the four-sided weir crest into a deep 1.5 m wide overflow channel that surrounds the division box on four sides. The flow is discharged into the treatment plant through a sewer exiting horizontally. The excess CSO overflows over another four-sided weir crest into a 1.0-m-wide CSO overflow channel, which surrounds the overflow channel on four sides that transports flow to the POTW. The CSO overflow channel is connected to a sewer that transports the flow to an on-site storage facility. The depth of the 4.0 m×4.0 m division box is calculated later in this solution. The division box configuration is shown in Figure 11.12.
   2. Calculate the head over the inner four-sided weir of the division box

      The peak combined flow reaching the plant site is 1.8 m^3/s (see the above table). It is divided equally over four weirs.

$$Q = \left(1.8\ \text{m}^3/\text{s}\right)/4 = 0.45\ \text{m}^3/\text{s}$$

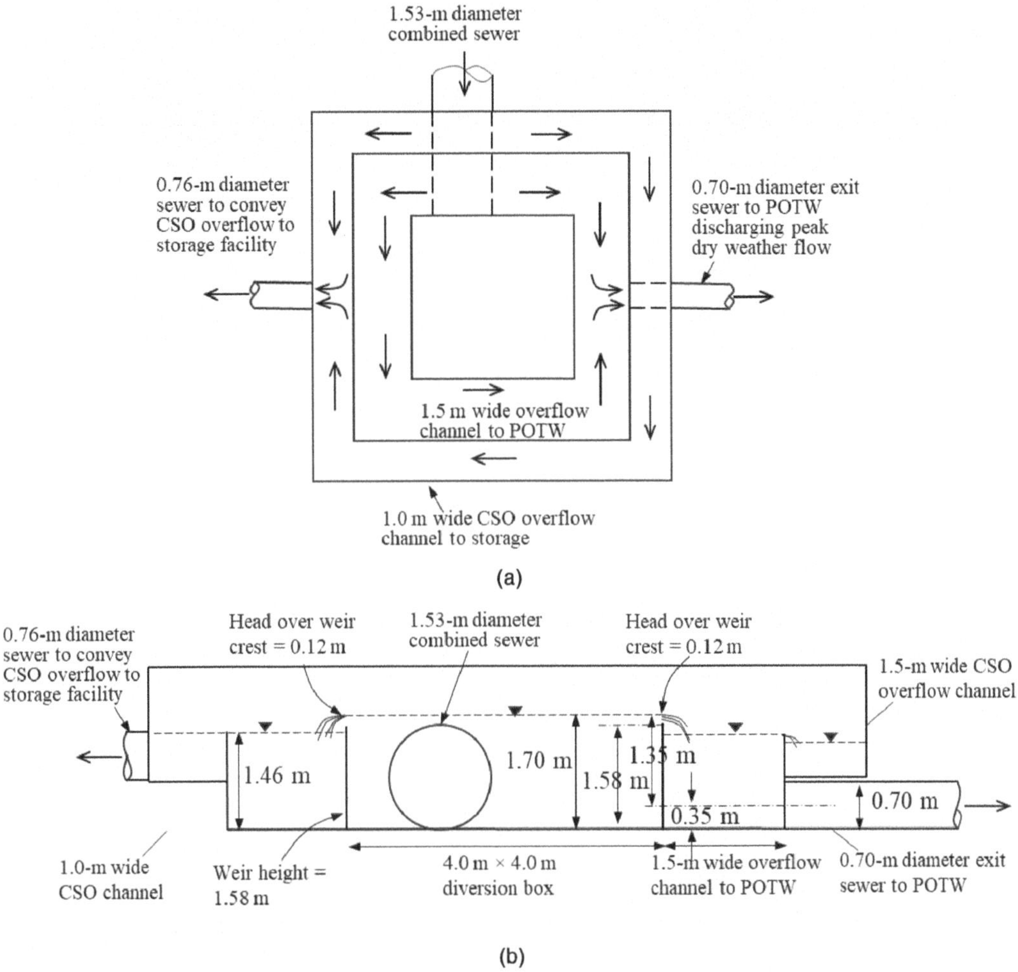

FIGURE 11.12 Design details and dimensions of the diversion box (a) plan, and (b) longitudinal section (Example 11.14).

Apply Eq. (7.9a)

$$0.45 = 2/3 \times 0.95 \times 4H\sqrt{2 \times 9.81\ \text{m}^3/\text{s} \times H}$$

Substitute Q=0.45 m³/s, C_d=0.95, and L'=4 m, and solve for H

$$H = 0.12\ \text{m}$$

3. Calculate the diameter of the horizontal exit sewer that delivers under pressure the design flow to the treatment plant
   Try a 70-cm (27 in.) standard diameter exit pipe
   Design flow to the treatment plant=1.40 m³/s
   Select the maximum water depth in the square division box H from the center of the exit sewer that transports the design capacity to the treatment plant

$$H = 1.35\ \text{m}$$

The diameter of the exit pipe is calculated from Eq. (7.8b)

$$Q = C_d AV = C_d A\sqrt{2gH}$$

Substitute Q=1.40 m³/s, C_d=0.65, H=1.35 m, and A=(π/4) D^2. Solve for D

$$1.40 \text{ m}^3/\text{s} = 0.65 \times (\pi/4) D^2 \times \sqrt{2 \times 9.81 \text{ m}^3/\text{s} \times 1.35 \text{ m}}$$

$$D = 0.73 \text{ m}$$

Select the standard diameter of 70 cm (27 in.). This is a horizontal exit sewer that will transport under pressure the design capacity to the plant.

4. Calculate height of the weir above the floor.

$$\text{The height of the weir} = 1.35 \text{ m} - 0.12 \text{ m} + \frac{0.70}{2} \text{ m}$$

$$= 1.58 \text{ m}$$

5. Calculate the diameter of the diversion sewer to transport CSO from the division box to the abandoned equalization basin at the plant property
   Maximum CSO Q=(1.80 m³/s − 1.40 m³/s)=0.40 m³/s
   Select 0.001 m/m slope of the pipe, and Manning's n=0.013
   The diameter of the pipe is calculated from Eq. (5.38).

$$Q = \frac{0.312}{n} D^{8/3} S^{1/2}$$

   Substitute Q=0.40 m³/s, n=0.013, and S=0.001 m/m
   0.40 m³/s=(0.312/0.013)×$D^{8/3}$×(0.001)$^{1/2}$
   D=0.79 m
   Select the standard sewer diameter of 76 cm (30 in.)
6. Calculate the diameter of incoming combined sewer connected to the division box
   The diameter of the incoming combined sewer is calculated from Eq. (5.38).
   Select slope=0.0006 m/m and Manning's n=0.013

$$Q = \frac{0.312}{n} D^{8/3} S^{1/2}$$

   Substitute Q=1.8 m³/s, n=0.013, and S=0.0006 m/m
   1.80 m³/s=(0.312/0.013)×$D^{8/3}$×(0.0006)$^{1/2}$
   D=1.52 m
   Select the standard pipe diameter of 1.53 m (60 in.)
7. Show the design details of the division box
   The design details and the dimensions of the division box, overflow channels, and incoming and exit sewers are shown in Figure 11.12.

B. Storage capacity for CSO
   1. Plot the combined sewer flow data reaching the treatment plant during the storm event (This is the maximum treatment plant capacity).
      The combined flow data at the plant is summarized in the problem statement. Plot the flow data with respect to the time of day. The combined flow data are plotted in Figure 11.13.

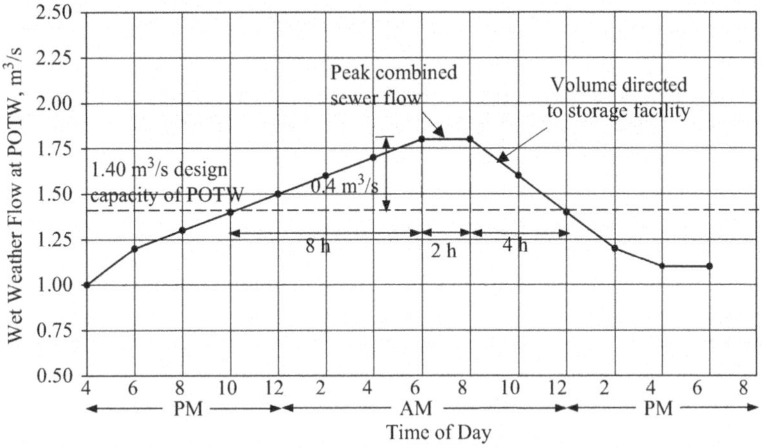

FIGURE 11.13 The combined sewer flow with respect to the time of the day, and design capacity of the treatment plant (Example 11.14).

2. Calculate the CSO storage volume needed at the treatment plant

   The volume needed for the CSO storage facility is the area bounded by the CSO, and 1.40 m³/s, the maximum design capacity of the treatment plant. The bounded area may be represented by two triangles and a central rectangle.

   The volume *V* needed for storage of CSO is calculated below:

$$V = \tfrac{1}{2}\left(0.4\ \text{m}^3/\text{s} \times 8\ \text{h} \times 3{,}600\ \text{s/h}\right) + \left(0.4\ \text{m}^3/\text{s} \times 2\ \text{h} \times 3{,}600\ \text{s/h}\right)$$

$$+ \tfrac{1}{2}\left(0.4\ \text{m}^3/\text{s} \times 4\ \text{h} \times 3{,}600\ \text{s/h}\right)$$

$$V = 5{,}760\ \text{m}^3 + 2{,}880\ \text{m}^3 + 2{,}880\ \text{m}^3 = 11{,}520\ \text{m}^3$$

11.9 TREATMENT OF COMBINED SEWER OVERFLOW

The release of untreated CSOs to waterbodies is no longer acceptable under the Clean Water Act. Typically, the concentrations of key pollutants such as TSS, BOD_5, COD, TN, TP, total lead, and total fecal coliform count are like raw municipal wastewater. During the first-flush, however, the concentrations of these pollutants may be significantly higher. Various CSO treatment methods include physical, chemical, and biological treatment methods. Some of these treatment methods are presented below.

11.9.1 Storage and Treatment

In-line, off-line, and on-site storage of combined wastewater flow during storm events, then rerouting the CSO through the POTW, is proven to be the most effective, convenient, and economical treatment method of CSO. The treatment level achieved is the same as that of the existing treatment. Construction of an additional treatment facility is not needed. Unfortunately, such an opportunity may not often be available, and alternate methods may be needed.

11.9.2 Physical, Chemical, and Biological Treatment

A comprehensive discussion on source controls, BMPs for watershed management, treatment of stormwater runoff, and control of nonpoint source pollution is presented in Section 10.7. Additionally, manufactured equipment and proprietary devices using physical and chemical treatment processes for treatment of stormwater runoff have been presented in Sections 10.10.3 and 10.10.4. Readers are referred to these sections and Examples 10.62 through 10.77 for in-depth coverage on treatment devices on source control BMPs, screens, gravity sedimentation, and hydrodynamic devices for suspended solids removal, and filtration devices.

Most commonly used devices for TSS removal from stormwater and CSO are hydrodynamic separator devices (see Section 10.11). Examples of vortex-type and swirl-type concentrator/regulators are shown in Figure 11.14.

11.9.3 Application of Microsand Ballasted Flocculation Process

The microsand ballasted flocculation process (MSBFP), also known as the high-rate clarification process, is used effectively to treat high flow of combined sewer flow during high-intensity storm events. The process has been used in Europe since 1991 to treat peak wet weather flow at the treatment plant and in collection system. It is a high-rate physical–chemical flocculation and sedimentation process using chemicals and ballast sand. The process is well recognized for small footprint, quick start-up, high performance, and reasonable capital and high operating costs. The process removes BOD, COD, TSS, TP, metals, and fecal coliform.[26]

The addition of microsand acts as a seed and ballast for the formation of high-density floc that settles rapidly. Settled microsand is separated from the sand-chemical sludge and recycled in the process. The Actiflo process of MSBFP is supplied by Veolia Water Technologies. The process utilizes (1) mix coagulant in the screened wastewater flow, (2) flash mix for nearly 60 s at G of 400/s, (3) add high dose of polymer and microsand, (4) continue flash mix for 15–20 s at the same G, (5) slow mix in the maturation unit for nearly two to five minutes at G of 100/s to 200/s, and (6) settle for 5–10 min. The design detention period and overflow rate of a full-scale facility are 10 min and 815–3,260 L/m^{2}min (20–80 gpm/ft^2) while achieving TSS removal of 80%–95%. The process diagram of MSBFP facility is shown in Figure 11.15.

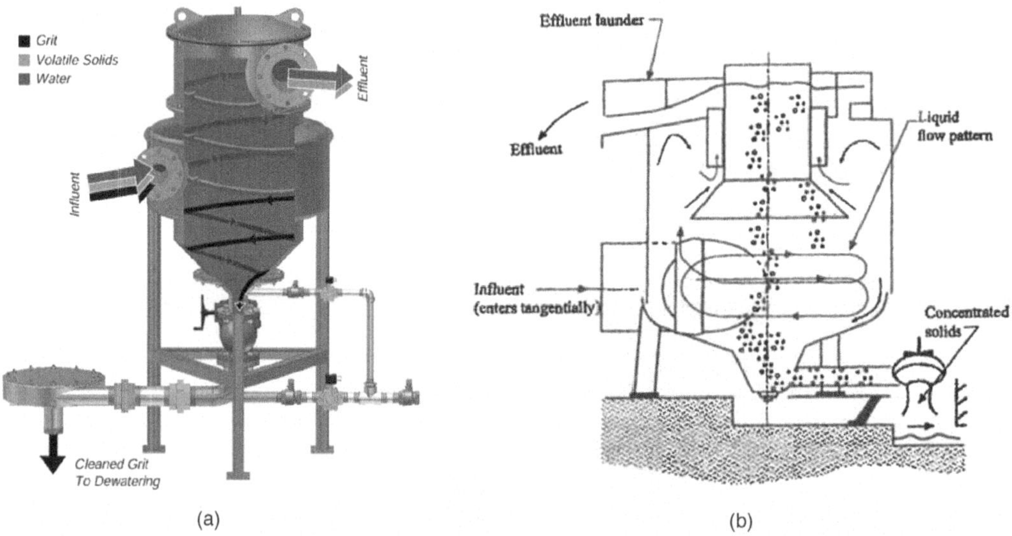

FIGURE 11.14 Hydrodynamic solids removal equipment from stormwater and CSO: (a) free vortex system, teacup, and (b) vortex-type solids separator. (Printed from Ref. 25)

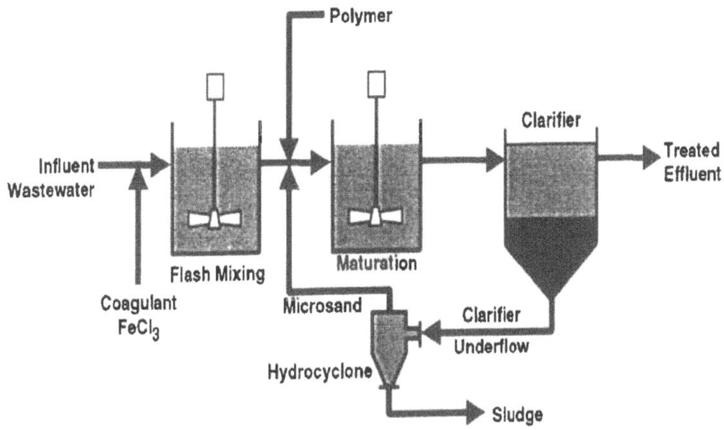

FIGURE 11.15 Process diagram of high-rate microsand ballasted clarification process. (Printed from Refs. 27,28)

Pilot plant studies of MSBFP have been conducted in several major cities in the states of Colorado, Ohio, Oregon, Texas, and others. The pilot plant study at Village Creek Water Reclamation Facility in Fort Worth, Texas, was conducted by Camp Dresser and McKee.[27,28] The process was incorporated as a parallel treatment train to the existing wastewater treatment plant. The parallel process trains of MSBFP and conventional wastewater treatment plant are shown in Figure 11.16.[27] The City of Fort Worth and Camp Dresser and McKee obtained a first-of-its-kind EPA NPDES permit to divert the surplus influent during wet weather to the MSBFP.[29]

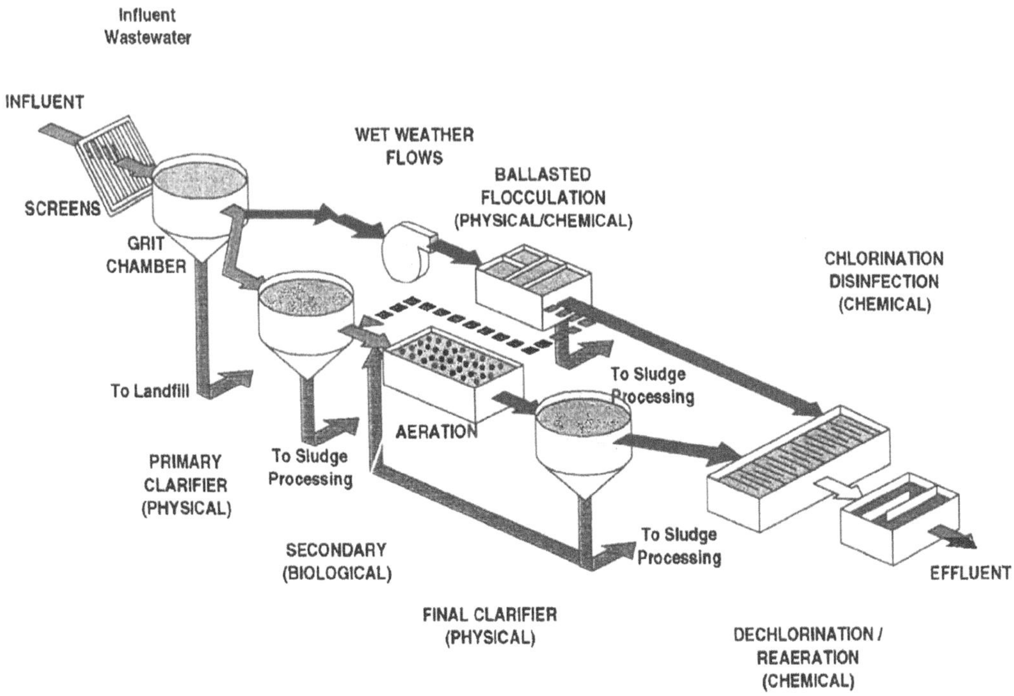

FIGURE 11.16 MSBFP as a parallel treatment plan to conventional secondary treatment plant. (Printed from Ref. 26,27)

Example 11.15: Comparison of Sedimentation Basin of MSBFP and Conventional Primary Sedimentation Basin

A sedimentation basin is designed for the treatment of combined sewer flow at a POTW. Compare the surface area and volume of sedimentation basins. The design detention times of MSBFP and conventional primary sedimentation basins are 10 min and 2 h, and the design overflow rates are 2.0 m^3/m^2.min (49 gpm/ft^2) and 40.8 m^3/m^2.day (1,000 gpd/ft^2), respectively. The combined sewer flow to the plant is 43,200 m^3/day (11.4 mgd).

Solution

A. MSBFP

1. Calculate the surface area

   Surface area $A = Q$/overflow rate

$$\text{Overflow rate} = 2.0\ \text{m}^3/\text{m}^2.\text{min} = \left(2.0\ \text{m}^3/\text{m}^2.\text{min}\right) \times (60 \times 24)\ \text{min/day}$$

$$= 2{,}880\ \text{m}^3/\text{m}^2.\text{day}$$

$$\text{Surface area } A = \left(43{,}200\ \text{m}^3/\text{day}\right) / \left(2{,}880\ \text{m}^3/\text{m}^2.\text{day}\right) = 15\ \text{m}^2$$

2. Calculate the volume V of the basin

$$V = Q \times \text{detention time}$$

$$V = 43{,}200\ \text{m}^3/\text{day} \times 10\ \text{min}/60\ \text{min/h} \times 24\ \text{h/day}$$

$$V = 300\ \text{m}^3$$

B. Conventional primary sedimentation basin

1. Calculate the surface area

$$\text{Surface area } A = Q/\text{overflow rate}$$

$$A = 43{,}200\ \text{m}^3/\text{day}/40.8\ \text{m}^3/\text{m}^2.\text{day}$$

$$A = 1{,}059\ \text{m}^2$$

2. Calculate the volume V of the basin

$$V = Q \times \text{detention time}$$

$$V = 43{,}200\ \text{m}^3/\text{day} \times \left(2\ \text{h}/24\ \text{h/day}\right) = 3{,}600\ \text{m}^3$$

3. Compare the surface area and volume of two basins

   The surface area of MSBEP and conventional primary basins are 15 and 1,059 m^2.

   The volume of MSBEP and conventional primary basins are 300 m^3 and 3,600 m^3.

Example 11.16: Comparison of Effluent Characteristics from a Parallel Conventional Secondary Treatment and MSBFP Process Trains

A POTW has a parallel conventional secondary treatment and MSBFP treatment trains to split the combined sewage flow during storm events. The parallel process train is shown in Figure 11.16. The total flow to the POTW and the influent BOD_5 are 129,600 m^3/s and 280 mg/L. The MSBFP process train and conventional secondary plant receive combined sewage flows of 50,800 and 78,800 m^3/day. The BOD_5 removal efficiencies of MSBFP and secondary treatment plant are 82% and 95%, respectively. Calculate the BOD_5 of the final combined sewage effluent from the POTW.

Solution

1. Calculate the effluent BOD_5 from MSBFP

$$\text{The effluent BOD}_5 = (1 - 0.82) \times 280 \text{ mg/L} = 50.4 \text{ mg/L}$$

2. Calculate the effluent BOD_5 from the conventional secondary treatment plant

$$\text{The effluent BOD}_5 = (1 - 0.95) \times 280 \text{ mg/L} = 14 \text{ mg/L}$$

3. Calculate the BOD_5 combined effluent from the POTW

The combined BOD_5 from the POTW

$$= \left[\left(50.4 \text{ mg/L} \times 50{,}800 \text{ m}^3/\text{day}\right) + \left(14 \text{ mg/L} \times 78{,}800 \text{ m}^3/\text{day}\right)\right] / 129.600 \text{ m}^3/\text{s}$$

$$= 28.3 \text{ mg/L}$$

11.10 REVIEW PROBLEMS AND DISCUSSION TOPICS

11.1 Combined sewers were introduced to serve some urgent needs. Identify those needs. Describe separate and CSSs. What are the advantages and disadvantages of combined sewers over separate sewerage system? What are the impacts of CSOs upon natural waters? List the methods to control the impacts of CSOs on receiving waters.

11.2 The design capacity of a POTW is the peak dry weather flow plus an I/I allowance. A wastewater treatment plant receives combined sewer flow from a population of 25,000 residents. The average wastewater contribution is 120 gpcd, and the peak dry weather flow is three times the average flow. The average I/I allowance is 40 gpcd. The combined peak wet weather flow is 13.5 million gallons per day (mgd). Calculate the designed capacity of the treatment plant and the CSO diverted to a storage facility at the treatment plant site. Give your results in mgd.

11.3 The central downtown area of an older part of a city has a CSS. The peak dry weather flow of 3,500 m^3/h is the design capacity of the treatment plant. The wet weather flow data at the treatment plant after a severe storm are summarized below. Determine the volume of a temporary storage facility to hold the CSO at the treatment plant property.

Time of day	12 MN	2 AM	4 AM	6 AM	8 AM	10 AM	12 N
Wet weather combined sewer flow, m^3/h	1,500	700	1,000	1,750	3,500	4,510	5,250
Time of day	2 PM	4 PM	6 PM	8 PM	10 PM	12 MN	
Wet weather combined sewer flow, m^3/h	5,500	5,200	5,000	3,500	2,200	1,700	

11.4
1. Define first-flush.
2. The downtown area of a city has CSS. The combined sewer network is connected to an interceptor. The combined sewer flow and BOD_5 data were collected at every 10-min interval from the interceptor soon after the storm event began. The data collection continued for several hours after the storm event was over. The data are summarized in the following table. Draw the mass loading diagram of BOD_5 concentration and identify the first-flush.

Time interval, min	0	10	20	30	40	50	60	70	80	90	100
Combined flow, m^3/min	0	2.0	3.0	3.5	4.5	5.5	10.8	15.1	15.8	16.0	16.0
BOD_5 concentration, mg/L(g/m^3)	0	182	180	182	200	200	250	380	385	390	380
Time interval, min	110	120	130	140	150	160	170	180	190	200	210
Combined flow, m^3/min	16.0	16.3	16.0	12.5	8.5	8.0	4.5	3.0	2.8	2.5	2.1
BOD_5 concentration, mg/L(g/m^3)	390	350	280	250	230	220	210	200	190	185	180

11.5 In the solved Example 11.4, a side weir is designed. The side weir is provided in a 1.2 m wide rectangular channel that diverts CSO to a storage facility. The slope and n of the channel are 0.003 m/m and 0.013. Calculate the head over the upper and lower ends of the side weir when the combined sewer flow is 1.50 m^3/s.

11.6 The design capacity of a treatment plant is 12.0 million gallons per day (mgd) (18.56 cfs), and peak wet weather flow to the treatment plant is 15.0 mgd (23.21 cfs). The CSO is diverted to a storage facility on the plant property. The interceptor that brings the combined flow to the treatment plant has a slope and n of 0.0008 ft/ft and 0.013, respectively. It discharges the combined flow into a collection–division chamber. The chamber is 18 ft long and 9 ft wide. The CSO overflows over a 9-ft long rectangular weir built across the chamber width. The height of the weir is 3.2 ft above the floor. The treatment plant capacity is discharged through an orifice at the floor of the collection–division chamber. The orifice is connected with a vertical pipe that turns horizontally to convey the flow to the treatment plant. The horizontal turn of the pipe prevents the vortex formation. The CSO overflows the weir crest into the collection channel along the width of the chamber. A sewer transports the CSO to a storage facility. The slope and n of the sewer are 0.001 ft/ft and 0.013. Develop the design of the collection–division chamber.

11.7 A combined sewer has a diameter of 120 cm (48 in.). An inflatable dam is provided in the combined sewer to store the excess flow within the sewer system. At the peak dry weather flow, the design capacity of the treatment plant is reached. The depth of flow in the interceptor at the plant capacity is 68 cm (21 in.). The dam is inflated to a depth of 100 cm (39 in.). At this depth, the normal flow condition is still maintained, and no surcharge occurs in the main sewers and tributaries. Calculate the maximum storage obtained per kilometer length of the interceptor.

11.8 A housing subdivision is developed over a total land area of 80 acres. The subdivision has single- and multiple-housing with lawns, green and landscaped areas, service roads, driveways, walkways, and individual and central parking areas. The paved surfaces cover around 10% of the total subdivision area. All paved surfaces utilize a common type of permeable paving to collect and retain surface runoff. The allowable reservoir depth of permeable paving and void ratio are 20 in. and 0.40. The maximum stormwater in the area is equivalent to 2.0 in./h of rainfall over 3.0 h. Prior to development, the pastureland had an average percolation and evaporation losses of stormwater around 70%.

After development, the average percolation and evaporation losses of stormwater from unpaved areas are 49%. Compare the surface runoff before and after development.

11.9 A mechanical regulator with a flap gate is installed in a 120 cm (48 in.) combined sewer interceptor. The slope and n of the interceptor are 0.0009 m/m and 0.013. The design capacity of the wastewater treatment plant is reached when the depth of flow in the interceptor is 70 cm (27.6 in.). As the water depth in the interceptor increases above 70 cm, the flap gate opens in proportion to the depth in the interceptor. The gate is 100% opened when 100 cm (43.3 in.) depth in the interceptor is reached. The gate dimensions are 58 cm wide and 20 cm high. The differential elevation from the center of the gate to 70 cm depth is 1.20 m. Calculate (1) total peak wet weather flow, (2) peak dry weather flow, which is also the design capacity of the treatment plant, (3) total CSO bypassed, and (4) discharge through the regulator gate.

11.10

1. Describe the ACTIFLO® process. Discuss components, operating features, and benefits.
2. A wastewater treatment plant has two parallel process trains. One process train utilizes a conventional secondary biological treatment process to treat the combined wastewater flow. The other process train utilizes MSBFP to treat CSO. The plant receives a total combined flow of 30 mgd. Out of that flow, 25 mgd is the designed capacity of the secondary treatment plant. This flow discharges into the influent line of the treatment plant. The remaining 5 mgd CSO is diverted to MSBFP train. The TSS concentration in the combined flow is 300 mg/L. The TSS removal efficiency of the conventional secondary treatment plant is 90%, and that of MSBFP is 80%. Calculate the TSS concentration in the combined final effluent from the treatment plant.

REFERENCES

1. Qasim, S. R., *Wastewater Treatment Plant Design: Planning, Design and Operation*, 2nd ed., CRC Press, Boca Raton, FL, 1999.
2. U.S. Environmental Protection Agency, *Combined Sewer Overflows Control Manual,* EPA/625/R-93/007, Office of Research and Development, Washington, DC, September 1993.
3. Field, R., D. Sullivan, and A. N. Tafuri, *Management of Combined Sewer Overflows,* CRC Press, Boca Raton, FL, 2003.
4. U.S. Environmental Protection Agency, *Preliminary Data Summary of Urban Storm Water Best Management Practices,* EPA-821-R-99-012, Office of Water, Washington, DC, 1999.
5. Burton Jr, G. A, and R. Pitt, *Stormwater Effects Handbook: A Toolbox for Watershed Managers, Scientists, and Engineers,* CRC Press, Boca Raton, FL, 2001.
6. U.S. Environmental Protection Agency, *Guidance Manual for Preparation of NPDES Permit Applications for Stormwater Discharges Associated with Industrial Activity,* EPA 505/8–91–002, Office of Water (EN-36), Washington, DC, April 1991.
7. U.S. Environmental Protection Agency, *Investigation of Inappropriate Pollution Entries into Storm Drainage System, A User's Guide*, EPA/600/R-92/238, Office of Research and Development, Washington, DC, 1993.
8. U.S. Environmental Protection Agency, *Clean Watersheds Needs Survey 2012 Report to Congress,* EPA-830-R-15005, Office of Water, Washington, DC, January 2016.
9. U.S. Environmental Protection Agency, Report to Congress, *Implementation and Enforcement of the Combined Sewer Overflow Control Policy,* Repot No. EPA 833-R-01–003, Office of Water, Washington, DC, 2001.
10. U.S. Environmental Protection Agency, *1992 Needs Survey Report to Congress,* EPA 832-R-93-002, Office of Water (WH-547), Washington, DC, September 1993.
11. Cook, M. B., *Combined Sewer Overflows: Guidance for Nine Minimum Controls,* U.S. Environmental Protection Agency, EPA 832-B-95-003, Office of Water, Washington, DC, May 1995.

12. U.S. Environmental Protection Agency, *Combined Sewer Overflows: Guidance for Financial Capability Assessment and Schedule Development,* PA 832–8–97-004, Office of Water (4204), Washington, DC, March 1997.
13. Lape, J. L., and T. J. Dwyer, A New Policy on CSOs, *Water Environment & Technology, Water Environment Federation*, vol. 6, no. 6, June 1994, pp. 54–58.
14. U.S. Environmental Protection Agency, *National Management Measuring to Control Nonpoint Source Pollution from Urban Areas*, EPA 841-B-05-004, Office of Water, Washington, DC, 2005.
15. U.S. Environmental Protection Agency, *Combined Sewer Overflow Management Fact Sheet-Sewer Separation*, EPA 832-F-99-041, Office of Water, Washington, DC, 1999.
16. U.S. Environmental Protection Agency, *Report to Congress: Impacts and control of CSOs and SSOs*, EPA 833-R-04-001, Office of Water, Washington, DC, 2004.
17. U.S. Environmental Protection Agency, *Report to Congress: Combined Sewer Overflows into the Great Lakes Basin*, EPA 833-R-16-006, Office of Wastewater Management, Washington, DC, 2016.
18. Metcalf and Eddy, Inc., *Wastewater Engineering: Collection, and Pumping of Wastewater,* McGraw-Hill Book Co., New York, 1981.
19. Metcalf and Eddy, Inc., *Wastewater Engineering: Treatment, Disposal and Reuse,* 3rd ed., McGraw-Hill, Inc., New York, 1991.
20. U.S. Environmental Protection Agency, Storm Water Management Model (SWMM), U.S. EPA Research, Office of Science Information Management, Durham, NC, Updated June 20, 2024. https://www.epa.gov/water-research/storm-water-management-model-swmm.
21. Nix, S. G., *Urban Stormwater Modeling and Simulation*, CRC Press, Boca Raton, FL, 1994.
22. Kibler, D. F., M. E. Jennings, G. L. Lewis, B. A. Tschantz, and S. G. Walesh, ASCE Task Committee on Urban Stormwater Software, Microcomputer Software in Urban Hydrology, HYDATA, *American Water Resources Association*, vol. 10, no.5, September 1991, pp. 27–32.
23. U.S. Environmental Protection Agency. *Combined Sewer Overflows-Guidance for Monitoring and Modeling*, EPA 832-B-99-002, Office of Water, Washington, DC, 1999.
24. Ackers, P., *A Theoretical Considerations of Side Weir as Stormwater Overflows.* Proceeding of the Institute of Civil Engineers, London, 1957.
25. Hydro International - Water & Wastewater Solutions, Eutek TeaCup - Advanced Grit Removal and Classification, Hillsboro, OR. https://hydro-int.com/sites/default/files/2025-02/TeaCup.pdf (Accessed Dec. 7, 2025).
26. Qasim, S. R., and C. Ramaswamy, *A Surrogate Test for Determination of* BOD_5 *in the Effluent from Microsand Ballasted Flocculation facility.* Final Report to the City of Fort Worth, Texas, The Department of Civil Engineering, The University of Texas at Arlington, Texas, May 2003.
27. Sawey, R. W., D. Garrets, and R. West, *Implementing Alternative Weather Treatment Technology (Unpublished)*, Camp Dresser and McKee, Fort Worth, TX, 1998.
28. US. Filter, *US Filter Kruger Product ACTIFLO to Wastewater Packet*, Cary, NC, September 2001.
29. Vick, R. C., *High-rate Clarification Take the Prize*, Pollution Engineering Online, October 2000.

12 Conveyance and Disposal Structures for Outfall Sewer

12.1 CHAPTER OBJECTIVES

Wastewater conveyance and outfall structures are part of a sewerage system that conveys and disposes of treated wastewater. Wastewater conveyance systems include pipes, manholes, pumping stations, diffusers, and other components. Municipalities and industries are permitted to discharge treated wastewater into rivers, streams, estuaries, and coastal waters. These permits require stringent testing and documentation that pollutant concentrations are within regulations. Permits are regulated under the Clean Water Act, and can be viewed online (https://www.epa.gov/laws-regulations/summary-clean-water-act). An outfall pipe is a location where stormwater or wastewater discharges into a body of water. Outfall pipes are commonly seen along the edge of most rivers, estuaries, and even in marshes. The objectives of this chapter are:

- Major issues related to effluent reuse and disposal
- Principal uses of wastewater treated effluent
- Storage facility for reclaimed water
- Effluent disposal considerations
- Methods of effluent disposal in surface waters
- Design considerations for outfalls
- Design methods of marine disposal infrastructure

12.2 WASTEWATER EFFLUENT REUSE AND DISPOSAL

Proper reuse and disposal of treated municipal wastewater effluent is an essential part of the planning and design of wastewater facilities. In most cases, the ultimate effluent disposal and reuse requirements dictate the selection of the wastewater treatment process train. Effluent reuse is increasing because treated wastewater is considered a resource. Even effluent disposal in a lake and stream is considered indirect reuse of effluent for the municipal water supply. The overriding considerations, however, for developing a reuse system is the quality of reclaimed water for its intended reuse. The common uses are (1) urban reuse, (2) industrial reuse, (3) agricultural irrigation, (4) impoundments for recreational and aesthetic reuse, (5) environmental reuse for habitat restoration and enhancement, (6) groundwater recharge, and (7) augmentation of portable water supply. Each of these uses is briefly discussed in the section below.[1–4]

Example 12.1: U.S. Guidelines for Reuse of Reclaimed Water

Tabulate the U.S. Environmental Protection Agency (U.S. EPA) guidelines for the reuse of reclaimed water for different reuse categories.

Solution

The U.S. EPA guidelines for reuse of reclaimed water have been developed in the following reuse categories: (1) urban reuse, (2) agricultural reuse, (3) impoundment, (4) environmental reuse, (5) industrial reuse, and (6) groundwater recharge. Reuse of reclaimed water in different categories is summarized in Table 12.1.

DOI: 10.1201/9781003629399-12

TABLE 12.1
U.S. EPA Guidelines for Reuse of Reclaimed Water in Different Reuse Categories (Example 12.1)

Reuse Category[a]	Treatment Requirement	Reclaimed Water Quality (and Monitoring) Requirement[a]						Setback Distance or Other Protection Requirement
		pH, s.u.	BOD_5, mg/L	TSS, mg/L	Turbidity,[b] NTU	Fecal Coliform,[c] Organisms/100 mL	Min. Cl_2 Residual,[d] mg/L	
Urban Reuse								
Unrestricted	S, F, & D	6.0–9.0 (W)	≤10 (W)	—	≤2 (C)	Non-detectable (D)	≥1 (C)	15 m (50 ft) to PWSW
Restricted	S & D	6.0–9.0 (W)	≤30 (W)	≤30 (D)	—	≤200 (D)	≥1 (C)	90 m (300 ft) to PWSW, and 30 m (100 ft) to AATP for spray irrigation
Agricultural Irrigation								
Food crops	S, F, & D	6.0–9.0 (W)	≤10 (W)	—	≤2 (C)	Non-detectable (D)	≥1 (C)	15 m (50 ft) to PWSW, and 30 m (100 ft) in area with porous media
Processed food and nonfood crops	S & D	6.0–9.0 (W)	≤30 (W)	≤30 (D)	—	≤ 200 (D)	≥ 1 (C)	90 m (300 ft) to PWSW, and 30 m (100 ft) to AATP for spray irrigation
Impoundment								
Unrestricted	S, F, & D	6.0–9.0 (W)	≤10 (W)	—	≤2 (C)	Non-detectable (D)	≥1 (C)	150 m (500 ft) to PWSW if bottom not sealed
Restricted	S & D	6.0–9.0 (W)	≤30 (W)	≤30 (D)	—	≤200 (D)	≥1 (C)	150 m (500 ft) to PWSW if bottom not sealed
Environmental reuse	Variable, or S & D	—	≤30 (W)	≤30 (D)	—	≤200 (D)	≥1 (C)	—
Industrial Reuse								
Once-through cooling	S	6.0–9.0 (W)	≤30 (W)	≤30 (D)	—	≤200 (D)	≥1 (C)	90 m (300 ft) to public access area
Recirculating cooling tower	S & D, S, F, & D, or Site specific	6.0–9.0 (W)	≤30 (W)	≤30 (D)	—	≤200 (D)	≥1 (C)	90 m (300 ft) to AATP, or reduced distance with high level disinfection
Others	Site specific and end use dependent	Site specific and end use dependent						Site specific and end use dependent
Groundwater Recharge								
General	Site specific	Site specific and end use dependent						Site specific
Spreading	*P* (min.)	Site specific and end use dependent						Site specific
Injection	*S* (min.)	Site specific and end use dependent						Site specific

(*Continued*)

TABLE 12.1 (*Continued*)
U.S. EPA Guidelines for Reuse of Reclaimed Water in Different Reuse Categories (Example 12.1)

		Reclaimed Water Quality (and Monitoring) Requirement[a]						
Reuse Category[a]	Treatment Requirement	pH, s.u.	BOD_5, mg/L	TSS, mg/L	Turbidity,[b] NTU	Fecal Coliform,[c] Organisms/100 mL	Min. Cl_2 Residual,[d] mg/L	Setback Distance or Other Protection Requirement
Indirect Potable Reuse (IPR)								
Spreading into potable aquifers[e]	S, F, D, & SAT	6.5–8.5 (D)	≤2[f] (W)	—	≤2 (C)	Non-detectable[g] (D)	≥1 (C)	Min. 2-month retention time in the ground to reach the nearest PWSW
Injection into potable aquifers[h]	S, F, D, & AWT	6.5–8.5 (D)	≤2[f] (W)	—	≤2 (C)	Non-detectable[g] (D)	≥1 (C)	Min. 2-month retention time in the ground to reach the nearest PWSW
Augment into surface water supply reservoirs[h]	S, F, D, & AWT	6.5–8.5 (D)	≤2[f] (W)	—	≤2 (C)	Non-detectable[g] (D)	≥1 (C)	Site specific

Source: Adapted in part from Refs. 5,6

Abbreviations

Treatment Requirement

AWT, Advanced wastewater treatment processes that include chemical clarification, carbon adsorption, reverse osmosis and other membrane processes, advanced oxidation, air stripping, and ion exchange; D, Disinfection that includes destruction, inactivation, or removal of pathogenic microorganisms by chemical, physical, or biological processes. These processes may be accomplished by chlorination, ozonation, oxidation by other chemical disinfectants, UV irradiation, or membrane processes; F, Filtration processes that may include (a) natural undisturbed soils, filter media such as sand and/or anthracite, or (c) microfiltration (MF) or other membrane processes; P, Primary treatment process; S, Secondary treatment process that may include activated sludge processes, trickling filters, rotating biological contractors, or stabilization pond. These processes should produce effluent concentrations ≤30 mg/L for both the BOD_5 and TSS; SAT, Soil aquifer treatment.

Monitoring Requirement

c, continuously; D, daily; W, weekly.

Setback Distance Requirement

AATP, area accessible to the public; PWSW, potable water supply well.

[a] Unless otherwise noted, the limits apply to the reclaimed water at the point of discharge from the treatment facility.

[b] The limit is the average turbidity based on a 24-h time period prior to disinfection. The turbidity should not exceed 5 NTU at any time.

[c] Unless otherwise noted, the limits are the median values determined from the bacteriological results of the last 7 days for which analyses have been completed.

[d] The limit is the total chlorine residual, and applies only when chlorine is used as the primary disinfectant and the actual contact time is at least 30 min.

[e] Meet drinking water standards after percolation through vadose zone at least 2 m (6 ft) at maximum groundwater mounding.

[f] The limit is for TOC in mg/L.

[g] The limit is for total coliform organisms/100 mL.

[h] Meet drinking water standards at the point of injection.

12.2.1 Urban Water Reuse

The urban reclaimed water reuse includes many non-potable municipal waters uses within an urban area. These uses include (1) irrigation of public parks and recreational centers, golf courses, athletic fields, school yards and playgrounds, highway medians and shoulders, and landscaped areas surrounding public buildings such as single-family and multi-family residences, and commercial and industrial developments; (2) commercial uses such as vehicle washing facilities, window and street washing and general washdown; (3) ornamental landscapes and decorative water uses such as fountains, pools, and waterfalls; (4) dust control and concrete production on construction projects; (5) fire protection; and (6) toilet and urinal flushing in commercial and industrial buildings. The largest urban reuse, however, constitutes irrigational uses within urban areas.[7]

The reclaimed water is normally delivered to the consumers by a separate distribution network or dual system. The parallel network of distribution mains is separated from community potable water distribution system. The most important design and operation considerations for urban water reuse are reliability of service and protection of public health. The basic safeguards for urban water reuse systems are (1) assurance that the reclaimed water delivered to the consumers meets the water quality requirements, (2) assurance that the treated effluent has no detectable fecal coliform count on daily monitoring, and desired level of chlorine residual is maintained at the point of use, (3) prevention of improper system operation, (4) prevention of cross-connection with potable water lines, and (4) prevention of improper use of reclaimed water. To avoid any cross-connection, all equipment associated with reclaimed water must be clearly marked. The national color standards have not been established, but the accepted practice by manufacturers and cities is to use purple color for all pipes carrying reclaimed water.[8,9]

Example 12.2: Process Train for Reclaimed Water for Urban Reuse

A biological nutrient removal (BNR) facility is designed to produce reclaimed water for urban reuse. The process train of the facility includes primary sedimentation; biological reactor with combined anaerobic, anoxic, and aerobic zones; polishing filtration; and granular activated carbon (GAC) filter. The final effluent is disinfected with 1.0 mg/L free-chlorine residual for a contact period of 30 min. The influent quality and percent removal by each unit process are summarized in Table 12.2.

The treated effluent has a total fecal coliform count of 32 organisms/100 mL. Draw the process diagram. Estimate (1) the effluent BOD_5, total suspended solids (TSS), total phosphorus (TP), organic nitrogen (ON), and ammonia nitrogen (AN); and (2) fecal coliform count

TABLE 12.2
Influent Quality and Percent Removal by Each Unit Process (Example 12.2)

Parameter	Raw Influent Concentration,[a] mg/L	Typical Removal,[b] %			
		Primary	BNR	Tertiary Filter	GAC Filter
BOD_5	210	30	90	50	70
TSS	250	60	90	70	60
TP	6	15	80	25	25
ON	15	25	85	60	40
AN	30	0	98.5	0	0

Source: Adapted in part from Refs. 10,11

[a] See Table 4.9 for the characteristics of municipal wastewater.

[b] See Section 6.3.3 for the typical removal efficiencies of these parameters.

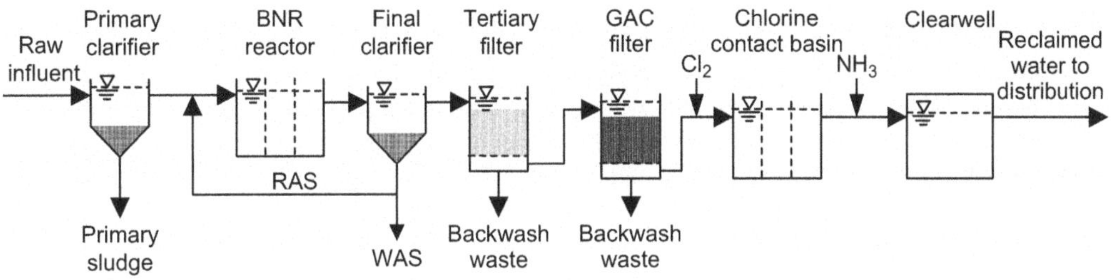

FIGURE 12.1 Process diagram of proposed BNR facility (Example 12.2). (Printed from Ref. 6 with permission of CRC Press, Taylor and Francis Group.)

after disinfection. Disinfection destroys 99% fecal coliform organisms. Describe the quality of reclaimed water for intended reuse.

Solution

1. Draw the process diagram of the proposed BNR facility in Figure 12.1.
2. Determine the effluent quality.

$$BOD_5 = (1-0.30)\times(1-0.90)\times(1-0.50)\times(1-0.70)\times 210\ \text{mg/L}$$

$$= 0.01\times 210\ \text{mg/L} = 2\ \text{mg/L}$$

$$TSS = (1-0.60)\times(1-0.90)\times(1-0.70)\times(1-0.60)\times 250\ \text{mg/L}$$

$$= 0.005\times 250\ \text{mg/L} = 1\ \text{mg/L}$$

$$TP = (1-0.15)\times(1-0.80)\times(1-0.25)\times(1-0.25)\times 6\ \text{mg/L}$$

$$= 0.1\times 6\ \text{mg/L} = 0.6\ \text{mg/L}$$

$$ON = (1-0.25)\times(1-0.85)\times(1-0.60)\times(1-0.40)\times 15\ \text{mg/L}$$

$$= 0.03\times 15\ \text{mg/L} = 0.5\ \text{mg/L}$$

$$AN = (1-0)\times(1-0.985)\times(1-0)(1-0)\times 30\ \text{mg/L}$$

$$= 0.015\times 30\ \text{mg/L} = 0.5\ \text{mg/L}$$

3. Determine the fecal coliform count after disinfection.

   The fecal coliform count after disinfection = 32 organisms/100 mL (1.00 − 0.99) = 1 organism/100 mL

   The chlorinated reclaimed water has probably non-detectable fecal coliform count per 100 mL.
4. Describe the quality of reclaimed water for urban reuse.

   The quality of reclaimed water obtained from the above process train meets the quality guidelines for urban reuse. The U.S. EPA guidelines for reclaimed water reuse are summarized in Table 12.1.

12.2.2 Industrial Water Reuse

U.S. industries use enormous amounts of water, and present very large market potential for reclaimed water. Major reclaimed water-consuming industries are steel, chemicals, pulp and paper mills, petroleum refining, and power plants.[4,5] Most industries treat and recycle the wastewater from their own processes to conserve water and to avoid stringent regulatory standards for effluent discharge. Major industrial water reuses are (1) cooling tower makeup water, (2) boiler feedwater, (3) process water, and (4) irrigation of landscaping around the plant.[1,3,5]

12.2.3 Agricultural Irrigation Reuse

Agricultural irrigation offers a very large potential for reclaimed water reuse in arid regions. Being a reliable water supply source, reclaimed water can be used to supplement the scarce natural water or sparse rainfall. The major considerations for agricultural irrigation reuse are (1) demand, (2) quality of reclaimed water, (3) reliability, and (4) costs. The demand for agricultural irrigation may vary depending upon (1) climatic conditions (temperature, precipitation, evapotranspiration, etc.), (2) crop type, (3) stage of plant growth, (4) storage requirements, and (5) method of irrigation being applied. Proper crop selection is the most important part of the design process for a reclaimed water irrigation system. The quality of reclaimed water is also important because some constituents have special significance in agricultural irrigation. Among these are salinity, ratio of cations, trace elements, specific toxic ions, and nutrients.[7,12]

Example 12.3: Pumping Cost of Effluent for Irrigation

Effluent from a reservoir *A* at a wastewater treatment plant is pumped to a reservoir *B* at the irrigation field. The water surface elevations in reservoirs *A* and *B* are 300 and 318 m, respectively. The force main is 400 m long and 79 cm in diameter. There are three 90° and two 45° bends. Calculate the annual power bill for pumping 0.30 m^3/s. Unit cost of power is 18 cents/kWh. Assume $C = 100$, and the combined pump and motor efficiency is 78%.

Solution

1. Compute TDH

$$\text{TDH} = H_{\text{stat}} + h_f + \Sigma h_m + h_v$$

i. Compute static head

$$H_{\text{stat}} = 318\text{ m} - 300\text{ m} = 18\text{ m}$$

ii. Compute minor head losses in the force main

$$V = Q/A$$

$$A = \frac{\pi}{4}D^2$$

$$A = \frac{\pi}{4}(0.79\text{ m})^2 = 0.49\text{ m}^2$$

$$V = \left(0.30\text{ m}^3/\text{s}\right)/0.49\text{ m}^2 = 0.612\text{ m/s}$$

$$\text{Velocity head} = \frac{V^2}{2g}$$

$$= \left(0.612\text{ m/s}\right)^2/2\times 9.81\text{ m/s}^2$$

$$= 0.02\text{ m}$$

h_m is calculated from Eq. (5.27)

$$h_m = K\frac{V^2}{2g}$$

Total minor head losses are due to three 90° bends, two 45° bends, and velocity head due to exit

K for 90° and 45° bends are obtained from Appendix C

K for 90° bend=0.3, and K for 45° bend=0.2

$$\text{Total minor head losses} = 3 \times 0.3 \times 0.02 \text{ m} + 2 \times 0.2 \times 0.02 \text{ m}$$

$$= 0.018 \text{ m} + 0.008 \text{ m} = 0.026 \text{ m} \quad \text{or} \quad 0.03$$

iii. Compute friction head loss h_f

h_f is calculated from Eq. (5.18)

$$h_f = 6.82\left(\frac{V}{C}\right)^{1.85} \times \frac{L}{D^{1.167}}$$

$$h_f = 6.82\left(\frac{0.612}{100}\right)^{1.85} \times \frac{400}{0.79^{1.167}}$$

$$= 0.29 \text{ m}$$

$$\text{TDH} = H_{\text{stat}} + h_f + \Sigma h_m + h_v$$

$$\text{TDH} = 18.00 \text{ m} + 0.29 \text{ m} + 0.03 \text{ m} + 0.02$$

$$= 18.34 \text{ m}$$

2. Compute power requirement

The power requirement is calculated from Eq. (9.4)

$$P_w = K'Q(\text{TDH})\gamma$$

$$P_w = \left[1 \text{ kW/(kN.m/s)}\right] \times 0.3 \text{ m}^3\text{/s} \times 18.34 \text{ m} \times 9.81 \text{ kN/m}^3$$

$$= 53.98 \text{ kW}$$

$$\text{Wire power} = P_w/(\text{combined efficiency of pump and motor})$$

$$\text{Wire power} = 53.98 \text{ kW}/0.78 = 69.21 \text{ kW}$$

3. Compute annual energy bill

$$\text{Annual energy bill} = (\$0.18\text{/kW.h}) \times 69.21 \text{ kW} \times 24 \text{ h/day} \times 365 \text{ day/year}$$

$$= \$109{,}130\text{/year}$$

12.2.4 Impoundments of Recreational and Aesthetic Reuse

Reclaimed water is extensively used for recreational and aesthetic reuse, and beautification of urban landscape and environment. Common uses include (1) maintenance of landscape and aesthetic impoundments; (2) ornamental fountains; (3) water-based recreational lakes for boating, fishing, and noncontact sports, (4) rearing of freshwater sport and commercial fisheries; and (6) snowmaking. The extent of reuse projects in this category will depend upon water demand and availability of reclaimed water of desired quality. Example 12.4 illustrates the design procedure based on water balance analysis.[1,3,6]

Example 12.4: Impoundment of Recreational and Aesthetic Storage Reservoir Based on Water Balance Analysis

An open storage reservoir is designed to store reclaimed water. The surface area of the reservoir is 8 ha. The applicable percolation rate is 3.58 mm/day. The precipitation, evaporation, reclaimed water production, and reuse demand data are summarized in Table 12.3. Calculate the storage volume needed, average depth, and detention time in the reservoir.

Solution

Compute inflow and outflow from the reservoir for each month (January to December) and tabulate the calculation results for 12-month period in Table 12.4. Sample calculations for the month of January are summarized below.

TABLE 12.3
The Precipitation, Evaporation, Reclaimed Water Production, and Reuse Data (Example 12.4)

	Month											
Parameter	**Jan**	**Feb**	**Mar**	**Apr**	**May**	**Jun**	**Jul**	**Aug**	**Sep**	**Oct**	**Nov**	**Dec**
Precipitation, mm	160	120	100	70	10	3	4	5	10	30	100	130
Evaporation, mm	30	40	60	80	140	180	210	180	140	80	50	30
Production of reclaimed water, m^3/day	5,000	5,000	4,000	3,500	3,000	2,500	2,000	2,500	3,000	3,500	4,000	4,500
Reclaimed water reuse demand, m^3/day	1,463	1,099	856	2,584	4,169	3,880	4,702	4,892	3,736	3,729	4,139	2,306

TABLE 12.4
The Inflow and Outflow Data for 12-Month Period and Water Balance Analysis (Example 12.4)

	Inflow to Reservoir, $10^3 m^3$			Outflow from Reservoir, $10^3 m^3$				Reservoir Volume, $\times 10^3 m^3$	
Month (days)	**RWP**	**PREC**	**Total Inflow**	**EVAP**	**PERC**	**RWD**	**Total Outflow**	**Change**	**Cumulative**
Jan (31)	155.00	12.80	167.8	2.40	8.88	45.35	56.6	111	178
Feb (28)	140.00	9.60	149.6	3.20	8.02	30.77	42.0	108	286
Mar (31)	124.00	8.00	132.0	4.80	8.88	26.54	40.2	92	378
Apr (30)	105.00	5.60	110.6	6.40	8.59	77.52	92.5	18	**396**
May (31)	93.00	0.80	93.8	11.20	8.88	129.24	149.3	−56	340
Jun (30)	75.00	0.24	75.2	14.40	8.59	116.40	139.4	−64	276
July (31)	62.00	0.32	62.3	16.80	8.88	145.76	171.4	−109	167
Aug (31)	77.50	0.40	77.9	14.40	8.88	151.65	174.9	−97	70
Sep (30)	90.00	0.80	90.8	11.20	8.59	112.08	131.9	−41	29
Oct (31)	108.50	2.40	110.9	6.40	8.88	115.60	130.9	−20	9
Nov (30)	120.00	8.00	128.0	4.00	8.59	124.17	136.8	−9	**0**
Dec (31)	139.50	10.40	149.9	2.40	8.88	71.49	82.8	67	67
Total (365)	1,289.5	59.4	1349	97.6	104.5	1,146.6	1,349	0	—

EVAP, evaporation; PERC, percolation; PREC, precipitation; RWD, reuse water demand; RWP, reclaimed water production.
The min and max values are made bold for emphasis.

1. Show sample calculations for the month of January
   a. Calculate total inflow to the storage reservoir

      Reclaimed water production,

      $$\text{RWP} = 5{,}000 \text{ m}^3/\text{day} \times 31 \text{ day} = 155.00 \times 10^3 \text{ m}^3$$

      Precipitation,

      $$\text{PREC} = 160 \text{ mm} \times 10^{-3} \text{ m/mm} \times 8 \text{ ha} \times 10{,}000 \text{ m}^2/\text{ha} = 12.80 \times 10^3 \text{ m}^3$$

      $$\text{Total inflow} = (155.00 + 12.80) \times 10^3 \text{ m}^3 = 167.8 \times 10^3 \text{ m}^3$$

   b. Calculate total outflow from the storage reservoir

      Evaporation,

      $$\text{EVAP} = 30 \text{ mm} \times 10^{-3} \text{ m/mm} \times 8 \text{ ha} \times 10{,}000 \text{ m}^2/\text{ha} = 2.40 \times 10^3 \text{ m}^3$$

      Percolation,

      $$\text{PERC} = 3.58 \text{ mm/day} \times 31 \text{ day} \times 10^{-3} \text{ m/mm} \times 8 \text{ ha} \times 10{,}000 \text{ m}^2/\text{ha}$$

      $$= 8.88 \times 10^3 \text{ m}^3$$

      Reuse water demand,

      $$\text{RWD} = 1{,}463 \text{ m}^3/\text{day} \times 31 \text{ day} = 45.35 \times 10^3 \text{ m}^3$$

      $$\text{Total outflow} = (2.40 + 8.88 + 45.35) \times 10^3 \text{ m}^3 = 56.6 \times 10^3 \text{ m}^3$$

   c. Calculate change in volume

      $$\text{Change} = (167.8 - 56.6) \times 10^3 \text{ m}^3 = 111 \times 10^3 \text{ m}^3$$

   d. Calculate cumulative change

      $$\text{Cumulative} = (67 + 111) \times 10^3 \text{ m}^3 = 178 \times 10^3 \text{ m}^3$$

   67 $\text{m}^3 \times 10^3$ + positive change (increase in December volume) is added to 110 $\text{m}^3 \times 10^3$ change (increase in January volume) to net total volume increase of 178×10^3 in the reservoir. Similarly, calculate the cumulative change for the entire year. These results are summarized in Table 12.4.
2. Determine the months the reservoir is empty and full
   Reservoir is empty in November, $V_{\text{min}} = 0 \text{ m}^3$
   Reservoir is full in April, $V_{\text{max}} = 396 \times 10^3 \text{ m}^3$
3. Determine the reservoir capacity.
   The reservoir capacity required is the difference between the cumulative volumes at full and empty.

   $$V_{\text{reservoir}} = V_{\text{max}} - V_{\text{min}} = (396 - 0) \times 10^3 \text{ m}^3 = 396{,}000 \text{ m}^3$$

   $V_{\text{reservoir}}$ is also obtained by adding cumulative volume from December till April
   Provide reservoir volume of 400,000 m^3.

4. Determine the reservoir average depth.

$$D_{\text{reservoir}} = \frac{V_{\text{reservoir}}}{A_{\text{reservoir}}} = \frac{400{,}000 \text{ m}^3}{8 \text{ ha} \times 10{,}000 \text{ m}^2/\text{ha}} = 5 \text{ m}$$

5. Compute the average detention time in the reservoir.
   On an annual basis, total inflow = total outflow = 1,349,000 m^3/year.

$$\text{Annual average daily flow, } Q_{\text{aad}} = \frac{1{,}349{,}000 \text{ m}^3/\text{year}}{365 \text{ day/year}} = 3{,}696 \text{ m}^3/\text{day}$$

Average detention time,

$$\theta_{\text{reservoir}} = \frac{V_{\text{reservoir}}}{Q_{\text{aad}}} = \frac{400{,}000 \text{ m}^3}{3{,}696 \text{ m}^3/\text{day}} = 108 \text{ day} \quad \text{or} \quad 3.6 \text{ months} \quad \text{or} \quad 0.3 \text{ year}$$

12.2.5 Environmental Reuse for Habitat Restoration and Enhancement

Reclaimed water reuse may directly contribute to habitat restoration and enhancement of aquatic environment. Many other reuses include development and restoration of wetlands, habitat development, augmentation of existing stream flows, enhancement of water quality of ponds and lakes, and restoration of stream ecology. Over the past 200 years, approximately 50% of the wetlands in the United States have been destroyed for many diverse uses such as agriculture, mining, forestry, and urbanization. Wetlands serve many beneficial environmental functions. They may provide (1) flood attenuation, (2) maximize the habitat for wildlife and water fowl, (3) productivity to support the food chain, (4) aquifer recharge, (5) water quality enhancement, (6) nutrients assimilation from treated water, (7) aesthetic and environmental improvement, and (8) water quality enhancement of reclaimed water by cost-effective treatment through natural systems. Aquaculture has been practiced primarily for production of food, fiber, and fertilizer. In wetlands, host of aquatic plants and animals are cultured.[1]

Engineered augmentation of stream flow from wastewater effluent is done for environmental, aesthetic, ornamental, and recreational purposes, and propagation of aquatic life. In many arid regions, the reclaimed water is the primary source of water to maintain the base flow for aquatic and riparian habitat enhancement. Riparian habitat consists of plants and animals occurring in a land area adjacent to a stream or river, between the wet and dry zones. Reclaimed water may also be a source of water supply to ponds and lakes. These ponds and lakes may be ornamental and recreational impoundments. They may provide wildlife habitat, storage for irrigation water, and, in some cases, an indirect source of drinking water supply.[1–5]

12.2.6 Groundwater Recharge

In many regions of the country, excessive pumping of groundwater for municipal, industrial, and agricultural uses has resulted in lowering of groundwater table. As a result, subsidence, saltwater intrusion, and shortage of groundwater have occurred. There is a need for artificial recharge of these aquifers. Groundwater recharge by reclaimed water is a planned augmentation of groundwater aquifers that are not used as a potable water source (see Table 12.2). It has several advantages over surface water storage and treatment. These are (1) cost of artificial recharge of aquifer is less than the cost of surface storage; (2) no evaporation losses in subsurface storage; (3) no water quality deterioration due to associated algae, plankton, and aquatic growth; (4) no seasonal taste and odor problems; (5) no need for an environmentally acceptable reservoir site; (6) the aquifer may serve as a transport media eliminating the need for a conveyance system (pipelines, canals, aqueducts); and (7) transition between reclaimed water and groundwater may provide a more acceptable health, psychological, and aesthetic benefits than other direct pipe-to-pipe reuse.[7] There are two acceptable methods of groundwater recharge: (1) surface spreading and (2) direct injection.[1,2,7]

12.2.7 Augmentation of Potable Water Supply

Water is a renewable resource. Nature purifies it through an unending hydrological cycle. Currently, technology is available to treat wastewater to any extent the usage may require. In direct potable reuse (DPR), the highly treated reclaimed water is typically fed to a water treatment facility to substitute the raw water supply. Currently, DPR may be considered as a temporary approach when the regular raw water supply is interrupted under extremely unusual circumstances. To supply the reclaimed water directly into the potable water distribution system is rare, even if its quality can fully comply with the current drinking water standards. In comparison with direct reuse, indirect reuse of reclaimed water is more practical and easier to gain public acceptance. For indirect potable reuse (IPR), the reclaimed water must be treated to a level that it can be discharged into a body of water or groundwater aquifer for later reuse as a water supply source. Indirect potable water reuse of reclaimed water is a common practice these days in the United States. Many public attitude surveys have indicated that the general public is willing to accept the indirect reuse of reclaimed water over direct reuse. Many surface water resources used for municipal water supply are also used for effluent disposal. Thus, the dilution effects and natural assimilation processes provide further purification of wastewater effluent over the entire maturity period. The water treatment plants further treat the combined surface and reclaimed water to meet the safe-drinking water quality standards. Similarly, the groundwater recharge provides natural purification and dilution over the maturation period prior to potable reuse.[1,6]

Direct pipe-to-pipe reuse of reclaimed water involves blending of final product with other drinking water before distribution. Other drinking water sources are clear-water wells and storage reservoirs at the treatment plant and in the distribution network. Many pilot plant studies have shown that the final product from reclaimed water is of better quality than that from other surface sources.[13,14] The major concern, however, is the chronic health effects that might result from ingesting the mixture of unregulated inorganic and organic constituents in trace amounts coming from both industrial chemicals and home products. The chemicals of great concern are Pharmaceuticals and Personal Care Products (PPCPs), Perfluoroalkyl and Polyfluoroalkyl Substances (PFAS), and microplastics. These constituents remain in the reclaimed water, even after the most advanced treatment methods such as microfiltration and reverse osmosis. The direct reuse of treated wastewater, therefore, is practical only on a short-term emergency basis. For normal use over a long period, the direct reuse of wastewater for human consumption will therefore depend upon the economics of finding suitable natural water resources over reclaimed water, government regulations, and public acceptance.[13–15]

Example 12.5: Direct Potable Reuse (DPR)

Interest in DPR has risen sharply worldwide during recent years. Describe DPR and its benefits.

Solution

DPR refers to the process of treating wastewater from a treatment plant to a level where it is safe enough to be directly used as a public water supply. This essentially offers a potential solution to growing water scarcity issues, particularly in arid regions. This also raises significant concerns regarding public perception and stringent treatment requirements to ensure safety and quality. DPR is becoming an increasingly common strategy for bolstering water resources. Each application poses unique challenges, whether related to treatment goals, regulatory requirements, or political and public acceptance, and these issues have a significant impact on the final treatment train selection. The wastewater treatment effluent standards are very high and expensive to discharge safely in natural waters. Therefore, existing wastewater treatment plants should be evaluated and improved or augmented.

IPR is the recovery of wastewater for potable use via an intervening buffer. DPR offers no such buffer except for limited dilution or storage time provided at the plant. Blending wastewater effluent in the raw water line to a water treatment plant offers *one treatment buffer*. IPR by recycling effluent through a lake, reservoir, river, or a groundwater aquifer offers one additional *environmental buffer* from protection through storage, dilution, or removal by filtration (for aquifers), photolysis (for surface waters), or biological degradation. DPR does not require raw water conveyance, an added treatment plant, or environmental buffers. For these reasons, DPR can be more economical even with the utilization of additional unit process technologies, advanced automation, and real-time process performance monitoring, and strict regulatory control for water safety against microbiological and legislated maximum contaminant level of chemical pollutants.

Due to the aforementioned benefits and the general need for reliable alternative potable sources, interest in DPR has risen sharply during recent years. There are many DPR plants in the United States and around the world. The capital city of Windhoek, Namibia, boasts the first and longest-running direct potable reuse history in the world. Its operation began in October 1968 without any incident. Now it is called *Old Goreangab Water Reclamation Plant*. Currently, around 25% of the overall drinking water supply for the city is reclaimed water. Windhoek has a rapidly growing population of around 460,000, in one of the most arid countries of Southern Africa.[14]

12.3 EFFLUENT DISPOSAL INTO NATURAL WATERS

The effluent discharge into natural waters is the most common disposal practice. The assimilation capacity of the natural waters is utilized to achieve further treatment and dilution. The 1972 Amendments to the Federal Water Pollutant Control Act (PL 92–500) and Clean Water Act of 1977 (PL 95–217) established the National Pollutant Discharge Elimination System.[16,17] Under the National Pollutant Discharge Elimination System (NPDES), all municipalities and industries are required to obtain a permit to discharge a given quantity and quality of effluent into the natural waters. All states are required to adopt water quality standards. The interim goal is to restore and maintain the physical, chemical, and biological integrity of the nation's waters. Because of the importance of effluent disposal into natural waters, the remainder of this chapter is devoted to (1) requirements of outfall, (2) water quality models of surface water, and (3) design of outfall structures.

12.3.1 Requirements of Outfall

The outfall structures are designed to disperse the effluent into the receiving waters, which may be a river, lake, estuary, or ocean. In all cases, it is essential to provide (1) proper dispersion, (2) avoid localized pollution problems, (3) utilize the assimilative capacity of the natural water, and (4) protect the contamination of shores. In most cases, the discharge is made below the water surface through one or multiple outlets to reduce the formation of foam and telltale signs and achieve good dispersion. The requirements of river, lake, estuarine, and ocean outfalls are given below.[10,11]

River outfall: Shoreline releases of effluents into the rivers, if poorly dispersed, will hug the banks for long distances with little mixing. Installation of diffusers across the width of the river is preferred, but any construction at the bottom of a navigable water is subject to approval by the U.S. Army Corps of Engineers. Sewage effluent is generally warmer than most receiving waters, and therefore it rises while being swept along by dominant currents. Formation of telltale signs, such as slick or detergent froth, should be avoided.

Lake outfall: In lake outfalls, the best possible dispersion is needed to avoid shore contaminations. Shallow lakes and reservoirs are subject to significant mixing because of wind-induced currents, and effluents are dispersed well. In stratified lakes, there is a possibility that the effluent may deplete the dissolved oxygen (DO) in the lower stratum. Warmer effluent may also destroy the thermocline, which is a zone of significant temperature change and is extremely resistant to mixing.

Estuarine outfall: The effluents discharged into the tidal rivers and estuaries may oscillate due to freshwater flow, and tide- or wind-produced seiches. The mixing is caused by dispersion. If stratification occurs because of salinity or temperature, the effluent plume will rise to the surface.

Marine or ocean outfall: The marine outfall consists of long pipes to transport the effluent long distances from the shore and then release it. The function of a diffuser is to mix the low-density effluent with high-density seawater at the bottom. The plume will rise under prevailing currents until the density of the mixture is equal to that of seawater below the surface. At this point, the plume will spread horizontally and disperse with little or no visual effects. Some excellent references on marine outfalls are available in the literatures.[18–22]

12.3.2 Water Quality Models of Surface Water

Effluent discharged into rivers, lakes, estuaries, and oceans mixes with the surrounding water. The microbiological activity is enhanced to consume biodegradable organics under aerobic conditions. The DO is lowered, which is replenished mainly by atmospheric reaeration, photosynthesis, and downstream dilution. A minimum DO of 4–5 mg/L is necessary for healthy aquatic life. The concentrations of conservative pollutants are reduced by dilution. The BOD discharged and DO response relationships in rivers, estuaries, and lakes are discussed in Refs. 19,23–28.

12.4 DESIGN CONSIDERATIONS FOR EFFLUENT OUTFALL STRUCTURES

The outfall structures are designed to withstand the damaging forces such as erosion, uplift, and any natural forces due to velocity, turbulence, and impacts from debris in the receiving waters. The basic design considerations are (1) flow velocity, (2) stratification because of salinity or temperature, (3) depth (shallow or deep), (4) shape (wide or narrow), (5) reversal of current (tidal-or wind-produced seiches), (6) wind circulation, (7) temperature and salinity of effluent, and (8) proper mixing and dispersion of effluent. Three principal design factors are (a) construction material, (b) topography, and (c) means of anchoring the structure in the body of water. Each of these design factors is described below.

Construction material: The typical construction material for outfall structure is concrete, and pipe may be steel, concrete, plastic, or any similar material that is compatible with the waste stream, receiving water, site characteristics, and economics of fabrication and installation.

Topography: The topography of the area around the outfall structure is very important. Straight length and configuration of the connecting pipes, vertical drops, number of bends and types of elbows, and slope of pipes on the bed of the waterbody depend upon the topography of the area.

Installation and anchoring: Installation and anchoring of outfall pipes and diffusers is especially important if strong currents prevail, and additional problems such as erosion and scour, and transport of bedload, particularly objects such as tree trunks, may be expected. Outfall pipes are secured by rock ballast, cement or sandbags, or chain and anchors. Often, pipes may be partly embedded in a concrete apron.[20,29]

12.5 EFFLUENT OUTFALL COMPONENTS FOR RIVERS AND LAKES

Effluent disposal into rivers, lakes, and reservoirs is a common practice, but it requires careful treatment and management to prevent pollution. Treated wastewater effluent is released into these water bodies as part of the water cycle. The components of outfall structures for rivers and lakes are outfall header pipe, and diffuser header with diffuser ports.

12.5.1 Outfall Header

The outfall header pipe conveys treatment plant effluent from the plant to the diffuser header. The outfall header is normally a metal conduit connected to the treatment plant effluent storage basin. The appurtenances include a sluice gate or gate valve, several 90°, 45°, and 22.5° elbows, and other fittings. It is partly above ground, then goes underwater, and follows the topography of the embankment, side slope, and stream or lakebed. It is securely anchored at the bed to withstand erosion and siltation.

12.5.2 Diffuser Header Pipe

The diffuser header is connected to the outfall header. The diffuser ports are installed in the header. There is a mounting base and support slab that anchors the diffuser header at the bed of the water body. The diffuser design equations and design procedure are presented below.

12.5.3 Diffuser Design

Diffusers must disperse the effluent efficiently and evenly to achieve good mixing. Knowledge of density and temperature gradients, including currents and wind conditions that influence the dispersion in receiving waters, is necessary. Procedure for determining initial dilution and dispersion for ocean outfalls is given in Ref. 20, Equations (12.1a–g) are used for the design and analysis of the diffusers.[20] Design details of the effluent outfall diffuser system are shown in Figure 12.2.

$$E_1 = \frac{\Delta p}{\gamma} = \text{HGL}_1 - \text{WSEL} \quad (12.1a)$$

$$q_1 = C_{D_1} a_1 \sqrt{2gE_1} \quad \text{or} \quad q_1 = C_{D_1} \frac{\pi}{4}(d_1)^2 \sqrt{2gE_1} \quad (12.1b)$$

$$Q_i = q_1 + \cdots + q_{i-1} \quad (i = 2, 3, \ldots, n) \quad (12.1c)$$

$$V_i = \frac{4Q_i}{\pi(D_i)^2} \quad \text{or} \quad V_i = \frac{Q_i}{A_i} \quad (i = 2, 3, \ldots, n) \quad (12.1d)$$

$$h_{f_i} = f \frac{L_i}{D_i} \frac{(V_i)^2}{2g} \quad (i = 2, 3, \ldots, n) \quad (12.1e)$$

$$E_i = E_{i-1} + h_{f_i} + \frac{\Delta S_g}{S_g} \Delta Z_i \quad (i = 2, 3, \ldots, n) \quad (12.1f)$$

$$q_i = C_{D_i} a_i \sqrt{2gE_i} \quad \text{or} \quad q_i = C_{D_i} \frac{\pi}{4}(d_i)^2 \sqrt{2gE_i} \quad (i = 2, 3, \ldots, n) \quad (12.1g)$$

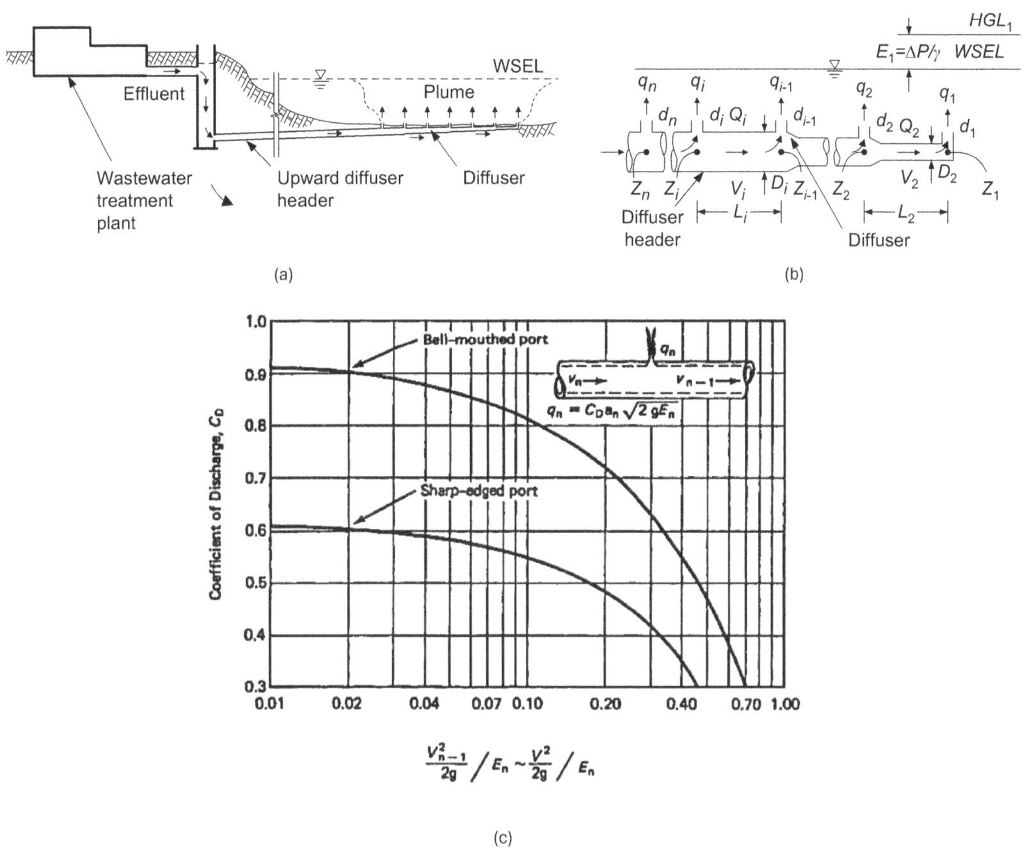

FIGURE 12.2 Effluent discharge outfall diffuser design details: (a) schematic of outfall pipe, (b) definition sketch of diffuser header, and (c) coefficient of discharge for small ports on the diffuser header. (Printed from Ref. 20 with permission of American Society of Civil Engineers.)

where

a_1 and d_1 = area and diameter of Port 1 which is the outermost port, m^2 and m

a_i and d_i = area and diameter of port i, m^2 and m

A_i = area of diffuser pipe between ports $i-1$ and i, m. Port i is upstream of port $i-1$

C_{D_1} = coefficient of discharge for port 1, dimensionless. The value of C_{D_1} is 0.61 and 0.91 for sharp-edged and bell-mouthed ports, respectively.[20]

C_{D_i} = coefficient of discharge for port i, dimensionless. The value of C_{D_i} is a function of $(V^2/2g)/E$ *ratio*. This relationship is graphically shown in Figure 12.2c.[20]

D_i = diameter of the diffuser header between ports $i-1$ and i, m

E_i or E_{i-1} = total operating head in the diffuser pipe at Port i or $i-1$, m

f = Darcy friction factor, dimensionless

h_{f_i} = head loss caused by friction in the diffuser header between ports $i-1$ and i, m

HGL_1 = hydraulic grade line elevation in the diffuser pipe immediately upstream of port 1, m

L_i = length of the diffuser header between ports $i-1$ and i, m

$\frac{\Delta p}{\gamma}$ = difference in pressure head between inside and outside of diffuser port 1, m

q_1 = discharge through port 1, m^3/s

q_i = discharge through port i, m^3/s

Q_i = total flow through the diffuser header between Ports $i-1$ and i, m^3/s

S_g = specific gravity of the receiving water, dimensionless (for water and municipal effluents $S_g = 1$).

ΔS_g = difference in specific gravity between receiving water and effluent, dimensionless

V_i = velocity in the diffuser header between ports $i-1$ and i, m/s

WSEL = free water surface elevation in the receiving water, m

ΔZ_i = elevation differential of the diffuser header centerlines between ports $i-1$ and i ($\Delta Z_i = Z_{i-1} - Z_i$), m

Note: Ports 1 and n are the outermost and innermost ports, respectively.

The diffusers may have a total area approximately half the cross-sectional area of the diffuse header. The diameter of the port is governed by considerations of clogging, hydraulic performance, and spacing. Typical diffusers have port diameters of 10–30 cm (4–12 in.), spaced over a desired length. As an alternative, two ports may be provided in the bulkhead at the end of the discharge pipe. One port is on the top and the other is at the bottom. These serve not only to discharge waste, but also to provide outlets for settleable and floatable materials. Other ports are normally alternated on the opposite side of the pipe header at the center line.[29] The minimum center-to-center spacing of the port is normally 10 times the port diameter.[20]

12.5.4 Design Procedure

The hydraulic analysis of multiport diffusers is basically a problem in manifolds and is somewhat complex. A discussion of this subject may be found elsewhere.[30–32] A simplified procedure is given in Ref. 20. It is a process of hydraulic iteration using a trial-and-error solution. The procedure is illustrated in Example 12.7. The diffusers are generally designed for maximum flow. The performance of the diffuser at lower flows must also be determined. In case low velocities are encountered under initial flow conditions, it is customary to plug the inboard ports during initial flow periods and open them later as the flow increases.

Another type of diffuser is a manifold and laterals. These diffusers are normally used for flow distribution into a basin channel. Discussion on these influent flow distribution manifold and laterals, derivation and application of various equations, and design procedure may be found in Refs. 6,10,20,31–34.

Example 12.6: Header Pipe Design for River or Lake Outfall

A circular cast iron outfall header pipe discharges effluent into a creek. The outfall is totally submerged. The low and high water depths in the creek are 1.20 and 3.20 m, respectively. The pipe is anchored over a concrete pad that is flushed with the creek bed. The diameter of the outfall pipe is 92 cm (or 36 in.) and the linear pipe length is 100 m (328 ft). The upper end of the outfall pipe is connected to an effluent junction box where the effluent enters from the treatment facility. A sluice gate is provided at the pipe entry. There are two 22.5° bends. The peak effluent discharge through the pipe is 1.321 m³/s. Calculate the head loss through the pipe and determine the water surface elevation in the effluent junction box.

Solution

1. Calculate the velocity through the outfall header pipe.

$$\text{Area of the header pipe, } A = \frac{\pi}{4}D^2 = \frac{\pi}{4} \times (0.92\text{ m})^2 = 0.665\text{ m}^2$$

$$\text{Velocity at peak flow, } V = \frac{Q}{A} = \frac{1.321\ \text{m}^3/\text{s}}{0.665\ \text{m}^2} = 1.99\ \text{m/s}$$

2. Compute the head losses in the outfall pipe.
   The friction and minor head loss through the outfall pipe are calculated separately below. General discussions about hydraulic calculations may be found in Section 5.5.
   a. Calculate the friction head loss from Darcy–Weisbach (Eq. 5.13a) using the coefficient of friction $f=0.035$ and diameter $D=0.92$ m for circular pipe.

$$h_{f_3} = f\frac{L}{D}\frac{(V)^2}{2g} = 0.035 \times \frac{100\ \text{m}}{0.92\ \text{m}} \times \frac{(1.99\ \text{m/s})^2}{2 \times 9.81\ \text{m/s}^2} = 0.77\ \text{m}$$

   b. Calculate the minor head loss from Eq. (5.27).
      The minor head loss in the outfall pipe consists of that created by the sluice gate (which is similar to an entrance when the gate is fully open), bends, and exit. The head loss coefficients are: $K_{\text{exit}}=1.0$ for the exit head loss, $K_{\text{bend}}=0.15$ for each of two 22.5° bend, and $K_{\text{gate}}=0.5$ for the sluice gate (Appendix C).

$$\text{Head loss due to exit head loss, } h_{\text{m,exit}} = 1.0 \times \frac{(1.99\ \text{m/s})^2}{2 \times 9.81\ \text{m/s}^2} = 0.2\ \text{m}$$

      Head loss caused by two 22.5° bends,

$$h_{\text{m,bend}} = 2 \times 0.15 \times \frac{(1.99\ \text{m/s})^2}{2 \times 9.81\ \text{m/s}^2} = 0.06\ \text{m}$$

$$\text{Head loss by the sluice gate, } h_{\text{m,gate}} = 0.5 \times \frac{(1.99\ \text{m/s})^2}{2 \times 9.81\ \text{m/s}^2} = 0.10\ \text{m}$$

      Total minor head loss,

$$h_m = h_{m,\text{exit}} + h_{m,\text{bend}} + h_{m,\text{gate}} = (0.20 + 0.06 + 0.10)\ \text{m} = 0.36\ \text{m}$$

   c. Calculate the total head loss through the outfall header pipe.

$$\Delta h = h_f + h_m = (0.77 + 0.36)\ \text{m} = 1.13\ \text{m}$$

3. Determine the water surface elevation in the effluent junction box.
   The maximum water surface elevation is reached when the peak flow is discharged into the creek that has the high water depth of 3.20 m.

$$\text{WSEL}_{\text{max}} = H_{\text{high}} + \Delta h = (3.20 + 1.13)\ \text{m} = 4.33\ \text{m}$$

4. Draw the sectional details of outfall structure and hydraulic profile in Figure 12.3.

Example 12.7: Diffuser Header Pipe and Diffuser Design for Effluent Outfall into a River or Lake

The effluent should be dispersed in the receiving water to avoid localized pollution. Diffusers are mostly designed for this purpose. Secondary effluent is discharged at a stream bed through diffuser ports arranged in the diffuser header. Design the diffuser ports and the diffuser header. The diffuser ports have bell-mouthed port. The diameter of the diffuser header is 92 cm. The minimum center-to-center spacing of the ports in the diffuser header pipe is at least 10 times the port diameter. The peak design effluent flow is 1.321 m^3/s. Allow 1-m extra length at the influent end of the diffuse header to connect it with the outfall header that has an equivalent length of 100 m (see Example 12.6). Determine the head loss

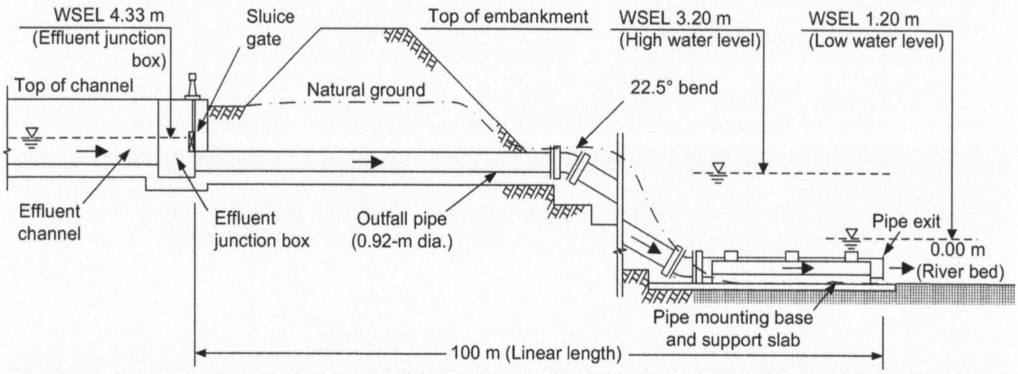

FIGURE 12.3 Conceptual side view of the outfall structure and hydraulic profile (Example 12.6). (Printed from Ref. 6 with permission of CRC Press, Taylor and Francis Group.)

in the diffuser header. If the high water level in the receiving water is 3.20 m, calculate the pressure head at the diffuser port 6.

Solution

1. Select a procedure for diffuser pipe design.

   The design of a diffuser pipe is basically a trial-and-error procedure. Once a tentative configuration is selected, it is designed by the procedure described below.[6,10,20–29] The procedure consists of selecting an operating head at the end of the diffuser header and then calculating the flow through the outermost ports. Using the flow, the head loss in the header pipe between the outermost port and the next upstream port is calculated. The head loss plus the velocity head, the density differential head, and the slope of the diffuser will give the operating head at the next inboard port. The flow at this port is then calculated, and the procedure is continued over the entire length of the diffuser header. The sum of discharges through all ports should equal the design effluent flow through the outfall pipe. If they are not close, the operating head is adjusted, and the calculations are repeated. The process can be tedious if done manually.

2. Select the design equations.

   Equations (12.1a–g) are used for the design of the diffuser header pipe. The design procedure and calculation steps are provided in Step 4, and the results are tabulated in Table 12.5.

3. Select the number of ports and a tentative diffuser header configuration.

   The 92-cm diameter diffuser header pipe has diffusers for effluent discharge.

   a. Calculate the area A of 92-cm diffuser header

$$A = \frac{\pi}{4}D^2 = \frac{\pi}{4} \times (0.92\text{ m})^2 = 0.665\text{ m}^2$$

   b. Provide total diffuser port area A_d approximately half the area of the diffuser header pipe $A = 0.665\text{ m}^2$

$$A_d = \frac{1}{2}A = \frac{1}{2} \times 0.665\text{ m}^2 = 0.333\text{ m}^2$$

   c. Select the diameter of all diffuser ports $d = 0.27$ m (10.5 in.).

$$\text{Area of each diffuser port } a = \frac{\pi}{4}D^2 = \frac{\pi}{4} \times (0.27\text{ m})^2 = 0.0573\text{ m}^2/\text{port}$$

TABLE 12.5
Design Calculations for Diffuser Pipe to Disperse Effluent Discharged into a Stream (Example 12.7)

Parameter	Port Number					
	1	2	3	4	5	6
1. Distance from the end, m	0	0	3	6	9	12
2. Port diameter (d), m	0.270	0.270	0.270	0.270	0.270	0.270
3. Port area (a), m^2	0.0573	0.0573	0.0573	0.0573	0.0573	0.0573
4. Diffuser header diam (D), m	0.92	0.92	0.92	0.92	0.92	0.92
5. Diffuser header area (A), m	0.665	0.665	0.665	0.665	0.665	0.665
6. Discharge through diffuser header (Q), m^3/s	—	—	0.462	0.691	0.910	1.118
7. Velocity in diffuser header (V), m/s	—[a]	—[a]	0.695	1.039	1.368	1.681
8. Head loss through diffuser header pipe section (h_f), m	—[a]	—[a]	0.003	0.007	0.012	0.017
9. Velocity head ($V^2/2g$), m	—[a]	—[a]	0.0246	0.0550	0.0954	0.1440
10. Operating head (E), m	1.000	1.000	1.003	1.010	1.022	1.039
11. Ratio of ($V^2/2g$) to E	—[a]	—[a]	0.025	0.054	0.093	0.139
12. Coefficient of discharge (C_D)	0.91	0.91	0.90	0.86	0.81	0.78
13. Port discharge (q), m^3/s	0.231	0.231	0.229	0.219	0.208	0.202

[a] There is no diffuser header.

d. Select total number of diffuser ports N.

$$\text{Diffuser ports } N = \frac{A_d}{a} = \frac{0.333\ \text{m}^2}{0.0573\ \text{m}^2/\text{port}} = 5.8 \text{ ports or 6 ports}$$

e. Select the diffuser header configuration.

Provide two ports in the bulkhead at the downstream end. One port is on the top and the other is on the bottom of the bulkhead to exit floating and settleable solids. The other four ports are alternated on the opposite sides of the header pipe along the center line. Provide an identical center-to-center spacing L_d = 3 m (which is greater than the required 10 times the diameter of the port). Therefore, the total length of the diffuser header pipe is 13 m (43 ft), including an additional segment of 1 m for connection to the outfall header. The design details and diffuser configuration are shown in Figure 12.4. All ports have a divergent mouthpiece.[20]

4. Compute the average discharge through all ports in the diffuser header.

The discharge through ports and head loss calculations for the diffuser header involve an iterative procedure.

a. Estimate the operating head (E_1) required at port 1.

Assume that the total flow is evenly distributed through all six ports.

$$\text{The average flow through each port, } q_{avg} = \frac{Q}{6} = \frac{1.321\ \text{m}^3/\text{s}}{6 \text{ ports}} = 0.220\ \text{m}^3/\text{s/port}$$

For the outermost bell-mouthed port, apply $C_{D_1} = 0.91$. Rearrange Eq. (12.1b) and calculate the operating head required for Port 1 by assuming $q_1 = q_{avg} = 0.220\ \text{m}^3/\text{s}$.

$$E_1 = \frac{1}{2g}\left(\frac{q_1}{C_{D_1} a_1}\right)^2 = \frac{1}{2 \times 9.81\ \text{m/s}^2}\left(\frac{0.220\ \text{m}^3/\text{s}}{0.91 \times 0.0573\ \text{m}^2}\right)^2 = 0.91\ \text{m}$$

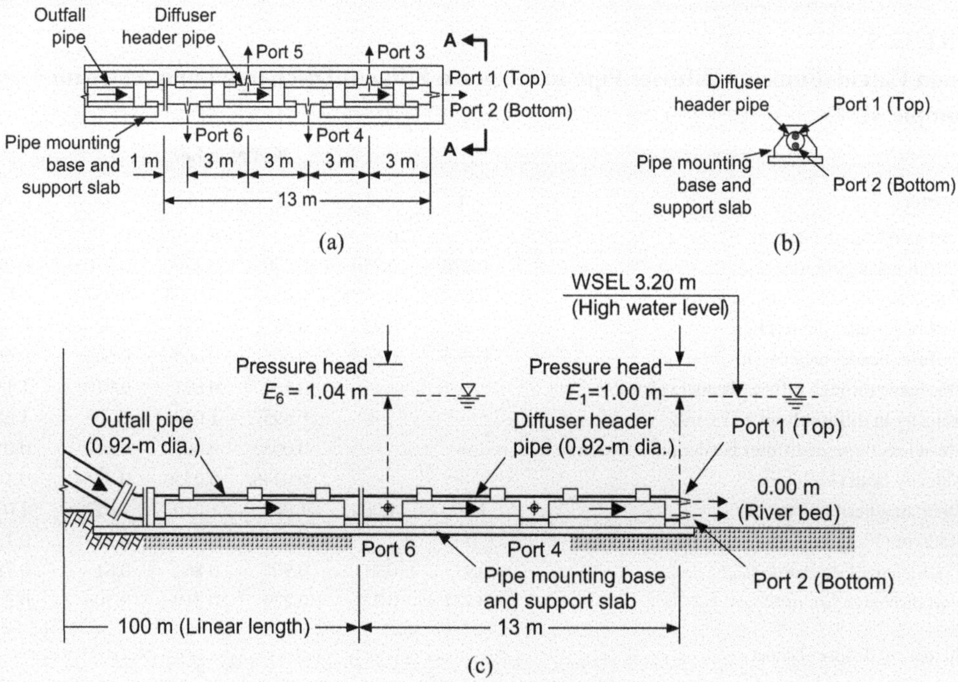

FIGURE 12.4 Design details of diffuser pipe: (a) plan view, (b) Section *A-A*, and (c) side view (Example 12.7). (Printed from Ref. 6 with permission of CRC Press, Taylor and Francis Group.)

The initial operating head needs to be at least 0.91 m to start the iterative procedure. The final result is reached when the sum of flows from all six diffuser ports is equal to peak design flow of 1.321 m³/s. Several iterations are normally needed to reach this condition. The calculation procedure for each iteration is tedious and time-consuming. For this reason, the calculation steps of the final iteration are provided in this example. The final iteration is reached when $E_1 = 1.00$ m at diffuser ports 1 and 2 in the bulkhead.

b. Compute discharge q_1 and q_2 through ports 1 and 2 in the bulkhead of the diffuser header

1. Determine the coefficient of discharge C_{D_1} for ports 1 and 2 (Figure 12.2)
   Apply C_{D_1} of 0.91 for ports 1 and 2 (see definition of C_D in Eq. (12.1b))
2. Determine q_1 and q_2

$$q_1 = q_2 = C_{D_1} a \sqrt{2gE_1} = 0.91 \times 0.0573 \text{ m}^2 \times \sqrt{2 \times 9.81 \text{ m/s}^2 \times 1.00 \text{ m}}$$

$$= 0.231 \text{ m}^3/\text{s}$$

5. Compute the discharge through port 3 in the diffuser header
   a. Compute Q_3 and V_3 in diffuser header between ports 1 or 2, and port 3
      Q_3 from Eq. (12.1c)$=q_1+q_2=(0.231+0.231)$ m³/s$=0.462$ m³/s
      Determine V_3 from Eq. (12.1d)

$$V_3 = \frac{Q_3}{A} = \frac{0.462 \text{ m}^3/\text{s}}{0.665 \text{ m}^2} = 0.695 \text{ m/s}$$

b. Compute head loss in the diffuser header between ports 1 or 2 and port 3
The head loss is calculated from Eq. (12.1e). Apply f=0.037

$$h_{f_3} = f\frac{L_d}{D}\frac{(V_3)^2}{2g} = 0.037 \times \frac{3.00\text{ m}}{0.92\text{ m}} \times \frac{(0.695\text{ m/s})^2}{2\times 9.81\text{ m/s}^2} = 0.003\text{ m}$$

c. Compute C_{D_3} for port 3

$$\frac{(V_3)^2}{2g} = \frac{(0.695\text{ m/s})^2}{2\times 9.81\text{ m/s}^2} = 0.0246\text{ m}$$

For marine outfalls (bay and ocean), the effect of density gradient is expressed by the factor $(\Delta S_g/S_g)\Delta Z_i$ in Eq. (12.1f). For fresh water (lakes and rivers), the effect of density gradient is nearly zero, and this effect is ignored.

$$E_3 = E_1 + h_{f_3} = (1.00 + 0.003)\text{ m} = 1.003\text{ m}$$

$$\frac{(V_3)^2/2g}{E_3} = \frac{0.0246\text{ m}}{1.003\text{ m}} = 0.025$$

Use the ratio of $(V_3^2/2g)/E_3 = 0.025$, read $C_{D_3} = 0.90$ from Figure 12.2c

d. Calculate discharge q_3 from diffuser port 3

$$q_3 = C_{D_3}a\sqrt{2gE_3} = 0.90\times 0.0573\text{ m}^2 \times \sqrt{2\times 9.81\text{ m/s}^2 \times 1.003\text{ m}} = 0.229\text{ m}^3/\text{s}$$

6. Compute discharge through port 4 in the diffuser header

a. Compute Q_4 and V_4 in the diffuser header between port 3 and port 4
Q_4 from Eq. (12.1c)=$q_1+q_2+q_3$=(0.231+0.231+0.229) m³/s=0.691 m³/s

$$V_4 = \frac{Q_4}{A} = \frac{0.691\text{ m}^3/\text{s}}{0.665\text{ m}^2} = 1.039\text{ m/s}$$

b. Compute head loss in diffuser header between port 3 and port 4
The head loss from Eq. (12.1e), f=0.037

$$h_{f_4} = f\frac{L_4}{D}\frac{(V_4)^2}{2g} = 0.0357 \times \frac{3.00\text{ m}}{0.92\text{ m}} \times \frac{(1.039\text{ m/s})^2}{2\times 9.81\text{ m/s}^2} = 0.007\text{ m}$$

c. Compute C_{D4} for port 4

$$\frac{(V_4)^2}{2g} = \frac{(1.039\text{ m/s})^2}{2\times 9.81\text{ m/s}^2} = 0.055\text{ m}$$

$$E_4 = E_1 + h_{f4} = (1.00 + 0.007)\text{ m} = 1.007\text{ m}$$

$$\frac{V_4^2/2g}{E_4} = \frac{0.055\text{ m}}{1.007\text{ m}} = 0.055$$

Using the ratio of $(V_4^2/2g)/E_4 = 0.055)$, read $C_{D4} = 0.86$ from Figure 12.2c

d. Calculate discharge q_4 from port 4

$$q_4 = C_{D_4} a\sqrt{2gE_4} = 0.86 \times 0.0573\ \text{m}^2 \times \sqrt{2 \times 9.81\ \text{m/s}^2 \times 1.007\ \text{m}} = 0.219\ \text{m}^3/\text{s}$$

Follow the above steps to calculate the discharge q_5 and q_6 from port 5 and port 6. These discharges are 0.208 and 0.202 m³/s. The step-by-step calculation results at peak design flow are summarized in Table 12.5.

7. Verify the total discharge through the diffuser header pipe.

$$Q_{\text{total}} = \sum q_i = (0.231 + 0.231 + 0.229 + 0.219 + 0.208 + 0.202)\ \text{m}^3/\text{s} = 1.320\ \text{m}^3/\text{s}$$

The calculated total flow through six ports is very close to the peak discharge flow of 1.321 m³/s. The results are acceptable. If the calculated total flow is significantly different from the peak design flow, another trial may be necessary to achieve a closer agreement.

8. Determine the total head loss through the diffuser header pipe, $h_{\text{diffuser header}}$ and pressure head at port 6

$$h_{\text{diffuser header}} = 1.00(E_1) + 0.003 + 0.007 + 0.012 + 0.017 = 1.039$$

$$h_{\text{diffuser header}} = E_6 = 1.04\ \text{m}$$

The pressure head at the diffuser port 6 = 3.20 m + 1.04 m = 4.24 m

The design details of the diffuser pipe header with location of diffuser ports are shown in Figure 12.4a–c.

Example 12.8: Design of an Outfall Structure with an Outfall Header, and Effluent Diffuser Header with Discharge Ports

An outfall structure is designed to discharge wastewater effluent into a receiving water. It has diffuser ports to disperse the effluent to reduce localized pollution. The outfall structure consists of two major components: (1) an outfall header like the one designed in Example 12.6, and (2) a diffuser header with ports like the one designed in Example 12.7. Design an outfall structure that connects both components presented in Examples 12.6 and 12.7. The information used to design the outfall header is given in Example 12.6. It has a 92-cm diameter and 100-m-long outfall pipe without the diffuser header. The upper end of the outfall pipe is connected to an effluent junction box where the effluent enters from the treatment facility. A sluice gate is provided at the pipe entry. There are two 22.5° bends. The peak effluent discharge through the pipe is 1.321 m³/s. The diffuser header in Example 12.7 is also a 92-cm diameter and 13-m long header. It has six diffuser ports at center-to-center spacing of 3-m in the header, and a 1-m long extra pipe section at the influent end to connect the diffuser header with the outfall header. Draw the side section of the outfall structure connected with the diffuser header pipe with the diffuser port. Show the elevations at key locations.

Solution

1. Describe the design requirements of an outfall structure.

The effluent outfall structure includes effluent junction box with a sluice gate, cast iron outfall pipe, and steel diffuser pipe. All concrete, cast iron, and steel pipe fitting shall conform to the American Society of Testing Materials and American Institute of Steel Construction standards. The diffuser and outfall pipes shall be properly anchored deep in the channel bed and in the embankment to protect from erosion, scour, and uplift. The concrete apron shall be secured in place by poured concrete of sufficient depth below the channel bed. Construction of structures

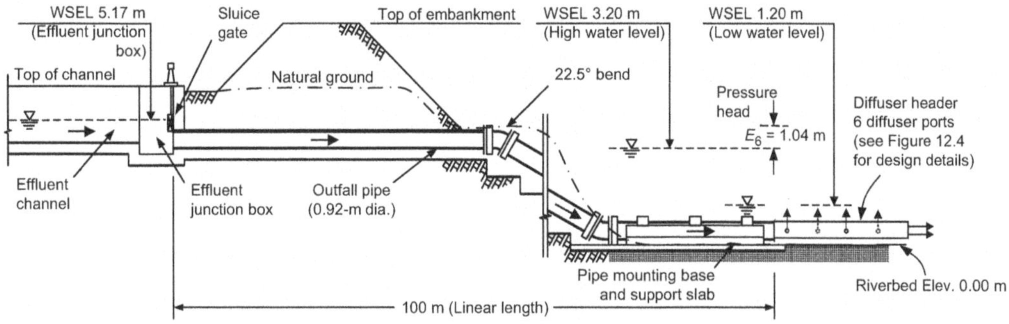

FIGURE 12.5 Conceptual side view of the outfall structure with diffuser header and diffuser ports, and hydraulic profile (Example 12.8).

within navigable waters of the U.S. requires a permit from the U.S. Army Corps of Engineers. This permit, called a Section 10 Permit, is applied for in conjunction with the Section 404 Permit. For information about these permits, contact the nearest U.S. Army Corps of Engineers district office.

2. Develop the design details of outfall structure with the diffuser header

   The step-by-step design procedure of outfall structure is developed in Example 12.6. The design details are shown in Figure 12.3. Likewise, the step-by-step design procedure of diffuser header with six diffuser ports is developed in Example 12.7, and designed details are shown in Figure 12.4. There is a 1-m extra pipe length between the two headers for proper connection between two headers.

3. Determine the total head loss in the outfall pipe, including the diffuser pipe.

   The head losses in the outfall pipe were calculated in Example 12.6. Since the outfall pipe is connected to the diffuser pipe, there is no exit loss. Calculate the total head loss (H_{total}), which is the head loss in the outfall header pipe and the diffuser header pipe.

   $H_{total} = h_f$, outfall header (Step 2a Example 12.6) $+ h_f$, diffuser header (Step 8 Example 12.6) $+ h_m$, bend (Step 2b)

   Example 12.6 $+ h_{m,gate}$ (Step 2b Example 12.6)

   Note: There is no exit loss due to the header's connection.

   $H_{total} = 0.77\ \text{m} + 1.04\ \text{m} + 0.06\ \text{m} + 0.10\ \text{m} = 1.97\ \text{m}$

4. Determine the water surface elevation in the effluent junction box.

   Calculate the maximum water surface elevation when the peak flow is discharged into the creek that has the high water depth of 3.20 m.

   $WSEL_{max} = H_{high} + H_{total}$ (see Step 2 above)

   $WSEL_{max} = 3.20\ \text{m} + 1.97\ \text{m} = 5.17\ \text{m}$

5. Show the design details of outfall structure

   The design details are shown in Figure 12.5

12.6 EFFLUENT OUTFALL COMPONENTS FOR MARINE OR OCEAN DISPOSAL

The major components of marine or ocean outfall include the effluent storage facility, pumping station at the treatment plant, a long effluent outfall header, and a diffuser header with diffuser ports. The outfall and diffuser headers must be designed to withstand the adverse environmental factors encountered under the oceanographic conditions. They should be properly anchored into the ocean floor to withstand the erosive forces and siltation. The basic design requirements and design details of outfall structure are provided in the following examples and solutions.

Example 12.9: Design Considerations for Ocean Outfall

List the basic design considerations for the marine outfall of municipal wastewater.

Solution

The basic components of marine outfall are pumping, transport piping, and diffusers. The design and construction of a marine outfall is a complex process involving expertise from various disciplines. These disciplines include physical oceanography, regulatory requirements, marine geology, marine ecology, mixing and dispersion, treatment processes and effluent quality and quantity, economics, and construction. The basic design considerations are listed below:

- Site selection
  Location close enough to the shoreline to minimize costs
  Pipeline length, water depth, and diameter of pipeline
  Availability of equipment and costs, and other constraints
- Oceanographic conditions
  Strength of the seabed
  Seabed mobility
  Submarine landslides
  Tidal patterns
  Seasonal variations
  Water currents and wave patterns
  Waves pattern that may cause scouring and sediment transport
  Density gradient
  Thermal gradient and variation
- Marine outfall structure design
  Maintaining a uniform flow distribution
  Minimizing head losses
  Maximizing flow distribution
  Rapid mixing
  Diffuser design
  Maximizing initial dilution
  Avoid sedimentation
  Gradual slope to facilitate gravity flow toward the exit end
- Regulatory requirements
  Adherence to local and national wastewater discharge standards
  Permitting process with relevant environmental agencies
  Compliance requirements with relevant authorities
  Relevant laws, regulations, and marine zoning
  Regular sampling and reporting of monitoring data
  Regular inspections, cleaning, and potential repairs of the outfall system
- Effluent quality and discharge
  Treatment levels
  Effluent quality
  Discharge volume
  Parameter limits (BOD, COD, TSS, nutrients, disinfection, specific contaminants)
- Water quality requirements
  Water quality standards
  Baseline marine life surveys to identify sensitive habitats and species
  Mathematical modeling to predict potential impacts on water quality and marine ecosystem

- Construction requirement
  Engineering feasibility
  Submarine pipeline
  Construction material selection (corrosion resistance, iron, high-density polyethylene, or concrete-lined steel of outfall pipe laid on the seabed or inside a trench)
  Construction cost
- Environmental considerations and impacts
  Ecology of marine life
  Ecology of benthic communities
  Impact on fisheries
  Communication and outreach to local communities
  Description of project goals and environmental considerations
  Public acceptance of the project

Example 12.10: Design of a Diffuser Header for Ocean Disposal of Municipal Wastewater Effluent

An underwater diffuser header pipe is designed for offshore disposal of treated wastewater effluent 30 m (98 ft) deep below the sea level. The factors considered are seabed topography, marine life, water currents, and environmental regulations to minimize impacts on the marine ecosystem. The diameter of the steel diffuser header is 120 cm (48 in.). The peak effluent discharge through the diffuser header is 1.67 m^3/s. The specific gravity of seawater and effluent is 1.025 and 1.000, respectively. The outfall and diffuser headers are anchored into the sloping bedrock at the ocean floor. The average slope of the ocean floor in the diffuser header area is 1 vertical to 10 horizontal. Develop the design of diffuser header. The diffuser ports have bell-mouth ports. The total diffuser port area is nearly half the area of diffuser header. The center-to-center spacing of the diffuser ports is greater than 10 times the diameter of the diffuser ports. Provide 1 m long extra length of the diffuser header at the influent end to connect the diffuser header with the outfall header. Draw the side view of the diffuser header, give the elevations at key locations, and show design details.

Solution

1. Calculate the cross-sectional area of diffuser header pipe

$$\text{Area } A_{\text{diffuser header}} = \frac{\pi}{4} D^2 = \frac{\pi}{4} (1.20 \text{ m})^2$$

$$= 1.131 \text{ m}^2$$

2. Calculate the total number of diffuser ports
   Total area of diffuser ports is nearly half the area of the diffuser header
   Total area of diffuser ports = ½ × (1.131 m)^2 = 0.566 m^2
   Select the port diameter of 0.42 m

$$\text{Area of each port} = \frac{\pi}{4} D^2 = \frac{\pi}{4} (0.42 \text{ m})^2$$

$$= 0.139 \text{ m}^2$$

Total number of ports = 0.566 m^2/1.139 m^2 = 4.07
Provide a total of four diffuser ports

3. Describe the diffuser port arrangement in the header
   The diffuser header has a diameter of 1.20 m, and it is sloping down 1 vertical to 10 horizontal. There are four diffuser ports, and each port is 42 cm (16.5 in.) in diameter. Provide two diffuser ports in the bulkhead of the diffuser header on the downstream end; one port is provided on the top, and the other port on the bottom to flush out floating debris and settled solids. The other two ports are alternated on the front and far sides of the diffuser header along the center line. Provide center-to-center spacings of 10 m between the bulkhead ports (1 or 2), and ports 3 and 4.
4. Determine the operating head E_1 or E_2 required at ports 1 or 2 ($E_1 = E_2$)
   a. Calculate the average discharge q_{avg} through each port

$$q_{avg} = \left(1.67\ \text{m}^3/\text{s}\right)/4 = 0.418\ \text{m}^3/\text{s}$$

   b. Calculate E_1 or E_2 from Eq. (12.1b)
      Use notations E_1, q_1, C_{D_1}, and a_1 for port 1, and notations E_2, q_2, C_{D_2}, and a_2 for port 2
      The numerical values of all respective variables of ports 1 and 2 are equal

$$q_1 = C_{D_1} \times a_1 \times \sqrt{2gE_1}$$

      Assume $C_{D_1} = 0.91$
      Substitute the values of q_1, C_{D_1}, and a_1 in Eq. (12.1b)

$$0.418\ \text{m}^3/\text{s} = 0.91 \times 0.139\ \text{m}^2 \times \sqrt{2 \times 9.81\ \text{m}^3/\text{s} \times E_1}$$

$$\sqrt{E_1} = 0.746\ \text{m}^2$$

$$E_1 = 0.557\ \text{m}$$

5. Calculate discharge q_3 and velocity V_3 through port 3
   a. Determine discharge Q_3 and velocity V_3 in the effluent header section between ports 1 or 2 and port 3

$$Q_3 = q_1 + q_2 = 0.418\ \text{m}^3/\text{s} + 0.418\ \text{m}^3/\text{s} = 0.836\ \text{m}^3/\text{s}$$

$$V_3 = Q_3/A_3 = (0.836\ \text{m}^3/\text{s})/(1.131\ \text{m}^2) = 0.739\ \text{m/s}$$

   b. Calculate head loss h_{f3} in the diffuse header section between ports 1 or 2 and port 3
      Substitute $f_3 = 0.037$, $L_3 = 10$ m, $V_3 = 0.739$ m/s, and $D_3 = 1.20$ m in Eq. (12.1e)

$$h_{f3} = f\frac{L_3}{D}\frac{(V_3)^2}{2g} = 0.037 \times \frac{10.0\ \text{m}}{1.20\ \text{m}} \times \frac{(0.739\ \text{m/s})^2}{2 \times 9.81\ \text{m/s}^2} = 0.009\ \text{m}$$

   c. Calculate E_3 for port 3
      Discharge of effluent into the ocean has a salinity effect. Apply Eq. (12.1f)

$$E_i = E_{i-1} + h_{fi} + \frac{\Delta S_g}{S_g}\Delta Z_i$$

ΔZ_i = elevation difference between ports 1 or 2, and port 3. Since the slope of diffuser header is 1V to 10H, $\Delta Z_i = 1$ m

$$E_3 = 0.557 \text{ m} + 0.009 \text{ m} + (1.025 - 1.00)/(1.025) \times 1.00 \text{ m}$$

$$= 0.557 \text{ m} + 0.009 \text{ m} + 0.024 \text{ m}$$

$$= 0.590 \text{ m}$$

d. Determine C_{D_3} for port 3

$$V_3^2/2g = (0.739 \text{ m/s})^2 / (2 \times 9.81 \text{ m/s}^2) = 0.028 \text{ m}$$

$$(V_3^2/2g)/E_3 = 0.028 \text{ m}/0.590 \text{ m}$$

$$= 0.048$$

Obtain C_{D_3} for port 3 from Figure 12.2

$$C_{D_3} = 0.88$$

e. Calculate discharge q_3 from Eq. (12.1g)

$$q_3 = C_{D_3} \times a_3 \times \sqrt{2g \times E_3}$$

$$q_3 = 0.88 \times 0.139 \text{ m}^2 \times \sqrt{2 \times 9.81 \text{ m/s}^2 \times 0.59 \text{ m}}$$

$$= 0.416 \text{ m}^3/\text{s}$$

6. Calculate discharge q_4 through port 4

   The procedure to calculate q_4 through port 4 is similar to that for q_3. Follow the procedure presented in Step 5. The calculated value of q_4 is 0.402 m³/s. The step-by-step calculated results for many variables are summarized in Table 12.6
7. Verify total discharge through the diffuser ports

   The total flow Q_{full} from all ports is given below

$$Q_{\text{full}} = \sum q_i = (0.418 + 0.418 + 0.416 + 0.402) \text{ m}^3/\text{s}$$

$$= 1.66 \text{ m}^3/\text{s}$$

   The calculated total discharge from all four diffuser ports is 1.66 m³/s, which is very close to the peak design flow of 1.67 m³/s. The results are acceptable. If calculated total discharge is significantly different from the peak design flow, another trial may be necessary to achieve a closer agreement.
8. Determine the total head loss $h_{\text{diffuser header}}$ through the diffuser header

$$h_{\text{diffuser header}} = E_1 \text{ or } E_2 + h_{f_3} + \text{salinity effect at port 3} + h_{f_4} + \text{salinity effect at port 4}$$

$$h_{\text{diffuser header}} = 0.557 \text{ m} + 0.009 \text{ m} + 0.024 \text{ m} + 0.019 \text{ m} + 0.024 \text{ m} = 0.633 \text{ m or } 0.63 \text{ m}$$

9. Draw the side view of the diffuser header

   The side view of the diffuser header, diffuser ports, and elevations of key locations are shown in Figure 12.6.

TABLE 12.6
Design Calculations for Diffuser Header for Ocean Disposal of Municipal Wastewater Effluent (Example 12.6)

	Port Number			
Parameter	**1**	**2**	**3**	**4**
1. Distance from the end, m	0	0	10	20
2. Port diameter (d), m	0.420	0.420	0.420	0.420
3. Port area (a), m²	0.139	0.139	0.139	0.139
4. Diffuser header diam (D), m	1.20	1.20	1.20	1.20
5. Diffuser header area (A), m	1.131	1.131	1.131	1.131
6. Discharge through diffuser header pipe (Q), m³/s	—	—	0.836	1.254
7. Velocity in diffuser header (V), m/s	—[a]	—[a]	0.739	1.107
8. Head loss through diffuser header section (h_f), m	—[a]	—[a]	0.009	0.019
9. Velocity head ($V^2/2g$), m	—[a]	—[a]	0.028	0.063
10. Operating head (E), m	0.56	0.56	0.590	0.630
11. Ratio of ($V^2/2g$) to E	—[a]	—[a]	0.048	0.100
12. Coefficient of discharge (C_D)	0.91	0.91	0.88	0.82
13. Port discharge (q), m³/s	0.418	0.418	0.416	0.402

[a] There is no diffuser header.

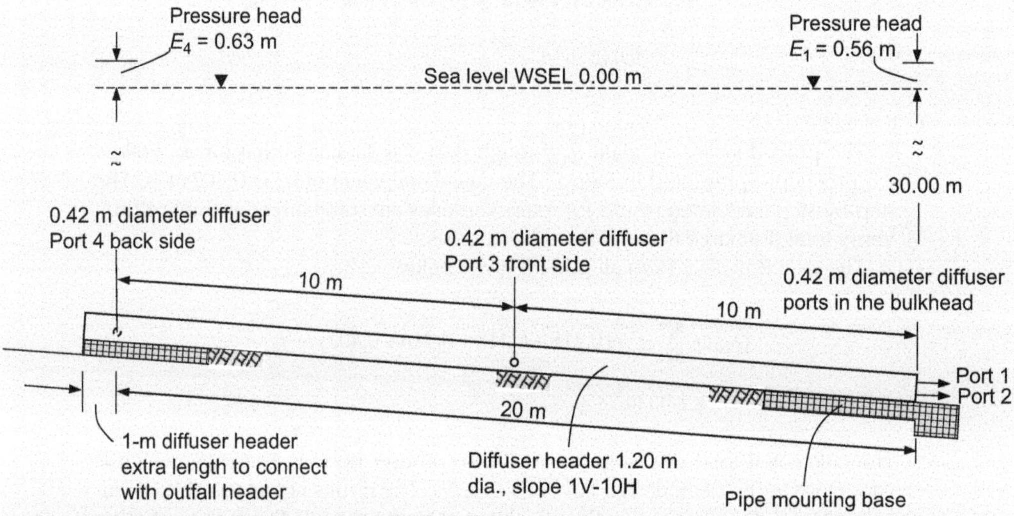

FIGURE 12.6 Side sectional view of outfall header showing design details, four diffuser ports sloping alignment, pipe mounting, and key elevations (Example 12.10).

Example 12.11: Design of an Ocean Outfall

A coastal city is designing an ocean outfall for disposal of secondary treated wastewater effluent. The ocean outfall consists of four major components: (1) an effluent storage facility at the treatment plant site, (2) a pumping station at the treatment plant site, (3) a very long underwater outfall header that conveys the effluent deep under the ocean bed, and (4) a diffuser header with diffuser ports that connects with the outfall header. The effluent storage facility is a large storage tank that stores the effluent. The peak effluent flow is 1.67 m^3/s. The pumping station consists of one centrifugal pump and one identical pump as a standby unit. The force main is 120 cm (48 in.) in diameter and has one Venturimeter and one plug valve. The force main eventually becomes an underwater outfall header, which is also 120 cm (48 in.) in diameter, 5 km (2.9 miles) in length, and 30 m (98 ft) below the sea level. The inventory of valves, fittings, and specials in the outfall header is given below:

Valve, Fittings, and Specials	Number	*K*
90° elbow	4	0.30
45° elbow	10	0.20
22.5° elbow	20	0.15
Venturimeter	1	0.15
Plug valve	1	1.00

The total head loss in the outfall header includes minor head losses due to elbows and fittings, bends, Venturimeter, and plug valve, friction head loss (Hazen–Williams $C=100$), head loss in the diffuser header, and the static head. The exit end of the outfall header is connected with the diffuser header. The design of diffuser header with diffuser ports has been developed in Example 12.10. Readers are recommended to review this example to obtain the needed details. The water surface elevation in the effluent reservoir at the plant site is 6.50 m above the sea level (Sea Level Elevation = 0.00 m). The effluent is discharged at the sea bottom 30 m below sea level. Develop the design of the ocean outfall over the sloping seabed 1 vertical to 10 horizontal. Both outfall and diffuser headers are fully anchored into the bedrock to protect them from erosion, siltation, and turbulent wave action. Also, calculate the power requirement of the pump and motor for pumping the effluent to the ocean outfall. Draw the side view of the ocean outfall structure, and give dimensions and elevations.

Solution

1. Describe the details of the pumping station

   The design of the pumping station in this example is highly simplified. In actual practice the design of pumping station is highly complex, similar to that presented in Examples 9.29 and 9.30. A pump is selected from the manufacturer's characteristic curves. If more than one pump is used in parallel combination, modified *H-Q* curve is prepared from the manufacturer's *H-Q* curve by subtracting the station losses. The modified *H-Q* curve of one pump is used to develop the *H-Q* curves of two or more pumps in parallel combination. System *H-Q* curve is prepared by adding total minor losses, friction head loss, diffuser head loss, and static head to obtain the TDH. Readers are referred to Examples 9.29 and 9.30, and Chapter 9 to carefully review the design procedure of a pumping station, and to select the station operating point. In this example, only one pump is needed with another identical pump as a standby unit.
2. Calculate the velocity and velocity head in the ocean outfall header.

$$\text{Area of the outfall header } A_{\text{header}} = \frac{\pi}{4}(1.20\text{ m})^2 = 1.131\text{ m}^2$$

$$\text{Velocity in the outfall header } V_{\text{header}} = Q/A = (1.67\text{ m}^3/\text{s})/(1.131\text{ m}^2) = 1.477\text{ m/s}$$

$$\text{Velocity head in the outfall header} = (V_{\text{header}})^2/2 = \left((1.477\text{ m/s})^2\right)/(2\times 9.81\text{ m/s}^2)$$

$$= 0.111\text{ m}$$

3. Calculate the minor losses in the ocean outfall header

$$90^\circ \text{ elbow} = 4 \times 0.30 \times 0.111 \text{ m} = 0.13 \text{ m}$$

$$45^\circ \text{ elbow} = 10 \times 0.20 \times 0.111 \text{ m} = 0.22 \text{ m}$$

$$22.5^\circ \text{ elbow} = 20 \times 0.15 \times 0.111 \text{ m} = 0.33 \text{ m}$$

$$\text{Venturi meter} = 1 \times 0.14 \times 0.111 \text{ m} = 0.02 \text{ m}$$

$$\text{Plug valve} = 1 \times 1.00 \times 0.111 \text{ m} = 0.11 \text{ m}$$

$$\text{Total minor head loss in the outfall header} = (0.13 + 0.22 + 0.33 + 0.02 + 0.11) \text{ m}$$

$$= 0.81 \text{ m}$$

4. Calculate the friction head loss in the ocean outfall header
   Apply Eq. (5.20) to calculate friction head loss

$$Q = 0.278CD^{2.63} \times \left(\frac{h_f}{L}\right)^{0.54}$$

Substitute Q =1.67 m³/s, C=100, D=1.20 m, and L=5,000 m

$$1.67 \text{ m}^3/\text{s} = 0.278 \times 100 \times (1.20)^{2.63} \times \left(\frac{h_f}{5{,}000}\right)^{0.54}$$

$$(h_f / 5{,}000)^{0.54} = 0.0371$$

$$h_f = 11.30 \text{ m}$$

5. Calculate TDH and pump operating point
   TDH=(total minor head loss)+(friction head loss)+(diffuser head loss)+(static head)
   TDH=0.81 m+11.30 m+0.63 m (Example 12.10, Step 8)+(−6.50 m)=6.24 m
   The pump operating point is Q=1.67 m³/s, and TDH=6.24 m
6. Compute work, power, and efficiency of pump for ocean disposal of effluent
   The work done by a pump is proportional to the product of the specific weight of the fluid being pumped and the total head against which the flow is moved. Pump power output, and pump and motor efficiencies are expressed by Eqs. (9.4)–(9.6). These equations are given below:

$$P_w = K'Q(\text{TDH})\gamma$$

$$E_p = P_w / P_p$$

$$E_e = P_p / P_m$$

   The pump efficiency is normally in the range of 70%–90%. It is obtained from the manufacturer's characteristic curve (review Example 9.30). The motor efficiency is normally in the range of 90%–98%. In this example, assume pump efficiency of 85% and motor efficiency of 90%.
   a. Compute water power P_w

$$P_w = 1 \text{ kW/kN.m/s} \times 1.67 \text{ m/s} \times 6.24 \text{ m} \times 9.81 \text{ kN/m}^3$$

$$= 102.2 \text{ kW} (136.1 \text{ hp})$$

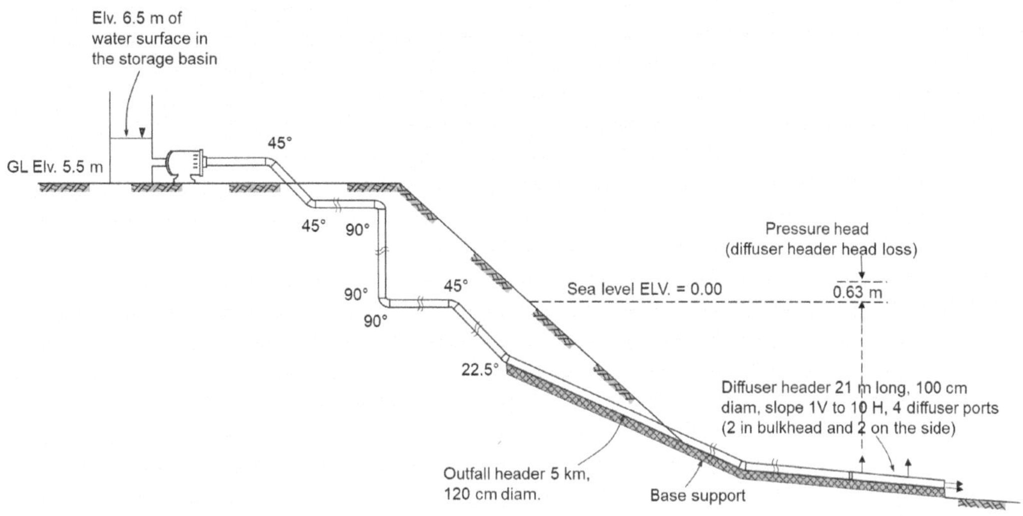

FIGURE 12.7 Side sectional view of ocean outfall structure showing design details, effluent storage pump, outfall header, diffuser header, pipe and pipe fitting mounting, and key elevations (Example 12.11).

b. Compute power input to the pump, P_p

$$P_p = 102.2\ \text{kW}/0.85 = 120.2\ \text{kW}\,(160.1\ \text{hp})$$

c. Compute motor power, P_m

$$P_m = 120.2\ \text{kW}/0.90 = 133.6\ \text{kW}\,(177.9\ \text{hp})$$

Select pump motors of 135 kW (180 hp) for both pump; one pump being the standby unit. Operate standby pumps on a routine basis for uniform wear of both pumps.

7. Draw the schematic of effluent storage, pump, outfall header, and diffuser header
   The schematic of ocean outfall structure is shown in Figure 12.7.

12.7 REVIEW PROBLEMS AND DISCUSSION TOPICS

12.1 In this chapter, several methods of effluent disposal and reuse are presented. List all those methods. DPR is a controversial method, but some cities are still adapting it. What is DPR? Describe the controversy and concern of DPR, and reason for increasing application of DPR.

12.2 An outfall pipe 15 cm in diameter is discharging into a lake. The pipe is at the bottom of a lake that has an elevation of 102.00 m. The water surface elevation is 105.50 m. The length of the pipe is 280 m, and it has four 45° elbows (K=0.20). Calculate the water surface elevation in the effluent box at the plant. Q = 27 L/s, f=0.024, K_{entrance}=0.50, and K_{exit}=1.00.

12.3 An effluent outfall structure is designed in a river. What are the various design factors that an engineer must take into consideration to minimize water quality impacts and withstand the damaging forces?

12.4 A river outfall structure is designed for effluent disposal into a river. The effluent structure consists of an effluent collection box at the plant site and an outfall header that delivers the flow into a diffuser header. The diffuser header is designed in Example 12.7 for a peak effluent discharge of 1.321 m^3/s. The diameter of the outfall header is 92 cm. It has an inlet sluice gate at the effluent collection box ($K=0.5$), and two 22.5° elbows ($K=0.11$). The equivalent length of the effluent header is 50 m. The diffuser header is installed at the river bottom at an elevation of 0.00 m. The low and high flow water depths in the river are 1.20 and 3.20 m, respectively. Develop the design of outfall structure. Show the design details. $f=0.024$.

12.5 An effluent disposal diffuser header is designed for peak design flow. During the early years of operation, the peak flow is lower than the peak design flow. For this reason, some inbound diffuse ports are tightly plugged to maintain high discharge velocity through the remaining open ports for good dispersion of the effluent. In Example 12.7, a diffuser header is designed for peak design flow. Six identical diffuser ports, each 0.27 m in diameter, were provided in a 0.92-m diameter diffuser header. During the initial years when peak flow is 0.966 m^3/s, diffuser ports 5 and 6 are plugged tightly. Calculate the discharge from each of the remaining ports and head loss through the diffuser header. All data for diffuser header and diffuser ports remains the same, as shown in Example 12.7.

12.6 In Example 12.7, a 0.92-m diameter diffuser header is designed. There are six diffuser ports, each with a diameter of 0.27 m fitted with bell-mouthed ports. The peak design flow is 1.321 m^3/s. Determine the total head loss through the diffuser header if sharp-edged ports are used. All design data for Example 12.7 are identical except for the sharp-edged ports.

12.7 Potable water reuse of wastewater treatment plant effluent is increasing these days. Compare the advantages and disadvantages of blending wastewater effluent in a raw water line for treatment and reuse with marine or ocean disposal of effluent.

12.8 An ocean outfall diffuser header has a diameter of 92 cm. It disperses 1.40 m^3/s secondary treated municipal effluent through three diffuser ports. Two diffuser ports are in the bulkhead; one at the bottom and the other on the top. The third diffuser port is 20 m from ports 1 or 2. The total area of all ports is around half the area of diffuser header. The specific gravity of seawater and effluent is 1.025 and 1.00, respectively. The slope of the diffuser header is downward at a rate of 1 vertical to 10 horizontal. Calculate the pressure head at the diffuser port 3 and the discharge from all three ports. $f=0.037$.

12.9 An ocean outfall structure is designed for the disposal of secondary treated effluent. The major components of the outfall structure are effluent storage basin at the treatment plant, a pumping station, a long outfall header, and a diffuser header with diffuser ports. The water surface elevation at the effluent storage basin is 5.80 m above sea level. The diameter of the outfall header is 120 cm, and it is 6,000 m long. The designed effluent flow for ocean disposal is 1.86 m^3/s. The sum of total minor losses of 1.75 m is due to the entrance, check valve, gate valve, flow meter, elbows, bends, and fittings. The friction factor f in Darcy–Weisbach equation is 0.024. The diffuser header loss, including density gradient, is 1.25 m. Two identical centrifugal pumps are provided; one pump for pumping the effluent, and the other as a standby unit. Both pumps are run on a routine for even wear. The pump and motor efficiencies are 82% and 90%, respectively. Determine the pump and motor power required for pumping the effluent to the disposal site.

12.10 Determine the annual energy cost for pumping 2.40 m^3/s secondary treated to an ocean disposal facility. The total TDH is 8.95 m. Pump and motor efficiencies are 80% and 90%. Power cost is 0.18/kW/h.

REFERENCES

1. Metcalf & Eddy I AECOM, Asano, T., F. L. Burton, H. L. Leverenz, R. Tsuchihashi, and G. Tchobanoglous, *Water Reuse: Issues, Technologies, and Applications,* McGraw-Hill Professional, New York, 2007.
2. Metcalf & Eddy I AECOM, *Wastewater Engineering: Treatment and Resource Recovery,* 5th ed., McGraw-Hill, New York, 2014.
3. Metcalf & Eddy, Inc., *Wastewater Engineering, Treatment and Reuse,* 4th ed., McGraw-Hill Companies, Inc., New York, 2003.
4. U.S. Environmental Protection Agency and U.S. Agency for International Development, *Guidelines for Water Reuse,* EPA/625/R-04/108, U.S. Environmental Protection Agency and U.S. Agency for International Development, Washington, DC, September 2004.
5. U.S. Environmental Protection Agency and U.S. Agency for International Development, *2012 Guidelines for Water Reuse,* EPA 600/R-12/618, U.S. Environmental Protection Agency and U.S. Agency for International Development, Washington, DC, September 2012.
6. Qasim, S. R., and G. Zhu, *Wastewater Treatment and Reuse: Theory and Design Examples,* vol 2, *Post Treatment, Reuse, and Disposal,* CRC Press, Taylor and Francis Group, Baca Raton, FL, 2018.
7. Lazarova, V., and A. Bahri (Editor), *Water Reuse for Irrigation: Agriculture, Landscapes, and Turf Grass,* CRC Press, Boca Raton, FL, 2005.
8. AWWA, *Guidelines for Distribution of Nonpotable Water,* AWWA, California-Nevada Section, 1992.
9. AWWA, *Planning for the Distribution of Reclaimed Water, Manual of Water Supply Practices,* M24, 3rd ed., AWWA, Denver, CO, 2009.
10. Qasim, S. R., *Wastewater Treatment Plants: Planning, Design and Operation,* 2nd ed., CRC Press, Boca Raton, FL, 1999.
11. Qasim, S. R., and G. Zhu, *Wastewater Treatment and Reuse: Theory and Design Examples,* vol 1, *Principles and Basic Treatment,* CRC Press, Taylor and Francis Group, Baca Raton, FL, 2018.
12. Pettygrove, G. S., and T. Asano (Editors), *Irrigation with Reclaimed Municipal Wastewater-A Guidance Manual,* Lewis Publishers, Inc., Chelsea, MI, 1985.
13. Lauer, W. C., *Denver's Direct Potable Water Reuse Demonstration Project, Proceedings of the 21st Annual Conference on Water Policy and Management: Solving the Problems,* American Society of Civil Engineers, New York, 1994, pp. 657–660.
14. Harhoff, J., and B. van der Merwe, Twenty-Five Years of Wastewater Reclamation in Windhoek, Namibia, *Water Science and Technology,* vol. 33, no. 10–11, 1996, pp. 25–35.
15. National Research Council, *Issues in Potable Reuse: The Viability of Augmenting Drinking Water Supply with Reclaimed Water,* National Research Council, National Academy Press, Washington, DC, 1998.
16. *Federal Water Pollution Control Act Amendments of 1972* (P.L. 92–500), 92nd Congress, October 18th, 1972.
17. *Clean Water Act* (P.L. 95–217), 95th Congress, December 27, 1977.
18. Metcalf & Eddy, Inc., *Wastewater Engineering, Treatment, Disposal, and Reuse,* 3rd ed., McGraw-Hill, Inc., New York, NY, 1991.
19. Tchobanoglous, G., and E. D. Schroeder, *Water Quality: Characteristics, Modeling, Modification,* Addison-Wesley Publishing Co., Reading, MA, 1985.
20. Rawn, A. M, F. R. Bowerman, and N. H. Brooks, Diffusers for Disposal of Sewage in Seawater, *Journal of the Sanitary Engineering Division, Proceedings of the ASCE, ASCE,* vol. 86, no. SA2, March 1960, pp. 65–105.
21. Burchett, M. E., G. Tchobanoglous, and A. J. Burdoin, A Practical Approach to Submarine Outfall Calculations, *Public Works,* vol. 98, no. 5, May 1967, pp. 95–101.
22. Falk, L. L., Factors Affecting Outfall Design, *Water and Sewage Works,* vol. 113, no. 11, November 1966, pp. 233.
23. U.S. Environmental Protection Agency, *Characterization of Stream Reaeration Capacity, Ecological Research Series,* EPA-R3–72-012, U.S. Environmental Protection Agency, Office of Reseach and Monitoring, Washington, DC, October 1972.
24. Streeter, H. W., and E. B. Phelps, A Study of the Pollution and Natural Purification of the Ohio River, III, Factors Concerned in the Phenomena of Oxidation and Reaeration, *Public Health Bulletin No.* 146, U.S. Public Health Services, Cincinnati, OH, 1925.
25. Thomann, R. V., and J. A. Mueller, *Principles of Surface Water Quality Modeling and Control,* Harper and Row, New York, 1987.

26. Eckenfelder, W. W., and D. J. O'Connor, *Biological Waste Treatment,* Pergamon Press, New York, 1961.
27. Wilcock, R. J., Study of River Reaeration at Different Flow Rates, *Journal of the Environmental Engineering Division, ASCE,* vol. 114, no. 1, 1988, pp. 91–105.
28. McCutcheon, S. C., *Water Quality Modeling, Volume I Transport and Surface Exchange in Rivers,* R. H. French (Editor), CRC Press, Boca Raton, FL, 1989.
29. Rambow, C. A., *Submarine Disposal of Industrial Wastes,* Proceedings of the 24th Industrial Waste Conference, Purdue University, Part 2, May 6–8, 1969, pp. 1486–1493.
30. McNown, J. S., Mechanics of Manifold Flow, *Transactions of American Society of Civil Engineers,* vol. 119, 1954, pp. 1103–1142.
31. Benefield, L. D., J. F. Judkins, and A. D. Parr, *Treatment Plant Hydraulics for Environmental Engineers,* Prentice-Hall, Inc., Englewood Cliffs, NJ, 1984.
32. Droste, R. D., *Theory and Practice of Water and Wastewater Treatment,* John Wiley & Sons, Inc., New York, 1997.
33. Hudson, H. E., *Water Clarification Processes: Practical Design and Evaluation,* von Nostrand Reinhold Co., New York, 1981.
34. Hudson, H. E. Jr., R. B. Uhler, and R. W. Bailey, Dividing-Flow Manifold with Square-Edged Laterals, *Journal of the Environmental Engineering Division, ASCE,* vol. 105, no. EE4, August 1979, pp. 745–755.

Appendix A

Abbreviations, Symbols, Basic Information of Elements, Useful Constants, and Student's t Distribution Test and Linear Regression r^2 Tables

In this appendix, many important abbreviations and symbols, atomic numbers and weights, and valances of many selected elements used in the book are listed in Tables A.1 and A.2. The values of many useful constants that are needed for unit conversions in different fields of engineering are summarized in Table A.3. Additionally, probability points of t distribution with degree of freedom ν and values of r^2 required to establish statistical significance of a sample correlation coefficient for various sample sizes are provided in Tables A.4 and A.5

TABLE A.1
Abbreviations and Symbols Used

Abbreviation and Symbol		Abbreviation and Symbol	
Acceleration due to gravity (LT^{-2})	g	Gallon per minute (L^3T^{-1})	gal/min or gpm
Area (L^2)	A	Gallon per second (L^3T^{-1})	gal/s or gps
Atmosphere	atm	Horsepower	hp
British thermal unit	Btu	Horsepower-hour	hp·h
Calorie	cal	Hour (T)	h
Centimeter (L)	cm	Inch	in.
Centipoise	cp	Joule	J
Cubic centimeter (L^3)	cm^3	Kilocalorie	kcal
Cubic feet per minute (L^3T^{-1})	cfm or ft^3/min	Kilogram (M)	kg
Cubic feet per second (L^3T^{-1})	cfs or ft^3/s	Kilonewton per cubic meter	kN/m^3
Cubic meter (L^3)	m^3	Kilowatt-hour	kWh
Cubic meter per day (L^3T^{-1})	m^3/day	Liter (L^3)	L
Cubic meter per second (L^3T^{-1})	m^3/s	Milligram (M)	mg
Degree Celsius	°C	Million gallons per day (L^3T^{-1})	MGD
Degree Fahrenheit	°F	Pound force	lb_f
Degree Kelvin	°K	Pound mass	lb_m

(*Continued*)

TABLE A.1 (*Continued*)
Abbreviations and Symbols Used

Abbreviation and Symbol		Abbreviation and Symbol	
Degree Rankine	°R	Pounds per square inch	lb_f/in^2 or psi
Density (ML^{-3})	ρ	Power	P
Discharge or flow (L^3T^{-1})	Q	Second (T)	s
Energy	E	Square meter (L^2)	m^2
Feet per minute (LT^{-1})	fpm or ft/min	Standard temperature and pressure	STP (20°C and 760 mmHg)
Feet per second (LT^{-1})	fps or ft/s	Volume (L^3)	V
Gallon (L^3)	gal	Watt	W

Source: Adapted in part from Refs.1–4
L = length; M = mass; T = time.

TABLE A.2
Basic Information of Selected Elements[a]

Element	Symbol	Atomic Number	Atomic Weight	Valence
Aluminum	Al	13	26.982	3
Antimony	Sb	51	121.760	3, 5
Argon	Ar	18	39.948	—
Arsenic	As	33	74.922	3, 5
Barium	Ba	56	137.327	2
Beryllium	Be	4	9.012	—
Bismuth	Bi	83	208.980	3, 5
Boron	B	5	10.811	3
Bromine	Br	35	79.904	1, 3, 5, 7
Cadmium	Cd	48	112.411	2
Calcium	Ca	20	40.078	2
Carbon	C	6	12.011	4
Chlorine	Cl	17	35.453	1, 3, 5, 7
Chromium	Cr	24	51.996	2, 3, 6
Cobalt	Co	27	58.933	2, 3
Copper	Cu	29	63.546	1, 2
Fluorine	F	9	18.998	1
Gold	Au	79	196.966	1, 3
Helium	He	2	4.003	—
Hydrogen	H	1	1.008	1
Iodine	I	53	126.904	1, 3, 5, 7
Iron	Fe	26	55.845	2, 3
Lead	Pb	82	207.190	2, 4
Lithium	Li	3	6.941	1
Magnesium	Mg	12	24.305	2
Manganese	Mn	25	54.938	2, 3, 4, 6, 7

(*Continued*)

TABLE A.2 (*Continued*)
Basic Information of Selected Elements[a]

Element	Symbol	Atomic Number	Atomic Weight	Valence
Mercury	Hg	80	200.592	1, 2
Molybdenum	Mo	42	94.940	3, 4, 6
Neon	Ne	10	20.180	—
Nickel	Ni	28	58.693	2, 3
Nitrogen	N	7	14.007	3, 5
Oxygen	O	8	15.999	2
Phosphorus	P	15	30.974	3, 5
Platinum	Pt	78	195.084	2, 4
Potassium	K	19	39.098	1
Radium	Ra	88	226.05	—
Silicon	Si	14	28.086	4
Silver	Ag	47	107.868	1
Sodium	Na	11	22.990	1
Strontium	Sr	38	87.62	2
Sulfur	S	16	32.064	2, 4, 6
Tin	Sn	50	118.710	2, 4
Titanium	Ti	22	47.867	—
Tungsten	W	74	183.84	3, 4
Uranium	U	92	238.029	4, 6
Zinc	Zn	30	65.38	2

Source: Adapted in part from Refs.1–3

[a] For a complete list, consult a handbook of chemistry and physics.

TABLE A.3
Useful Constants

Acceleration due to gravity, $g = 9.807\ m/s^2$ ($32.174\ ft/s^2$)

pi, $\pi = 3.14159265$

$e = 2.71828$

Standard atmospheric pressure = 101.325 kN/m^2 (14.696 lb_f/in^2 or psi) = 101.325 kPa (1.013 bars) = 10.333 m (33.899 ft) of water = 760 mmHg = 29.9213 in of Hg

Molecular weight of air (0.21 molar fraction of oxygen, and 0.79 molar fraction of nitrogen) = 29.1 g/mol

Density of air (15°C and 1 atm) = 1.23 kg/m^3 (0.0765 lb_m/ft^3)

Dynamic viscosity of air, μ (15°C and 1 atm) = 17.8×10^{-6} $N.s/m^2$ (0.372×10^{-6} $lb_f.s/ft^2$)

Kinematic viscosity of air, ν (15°C and 1 atm) = 14.4×10^{-6} m/s^2 (155×10^{-6} ft^2/s)

1 centipoise = 0.01 g/cm·s

1 stoke = 1 cm^2/s

Specific heat of air = 1,005 J/kg·°K (0.24 Btu/lb_m·°R)

Universal gas constant, R = 1.987 cal/mol·°K = 0.08205 L·atm/mol·°K = 8.314 J/mol·°K

1 N = 1 kg·m/s^2 = 0.2248 lb_f

1 Pascal (Pa) = 1 N/m^2

1 J = 1 N·m

1 J/s = 1 N·m/s = 1 w

(*Continued*)

TABLE A.3 (*Continued*)
Useful Constants

1 kJ = 0.9478 Btu

1 amp = 1 coulomb/s

1 cal (20°C) = 0.003966 Btu

1 cal = 4.184 J

1°C = 1.8 °F

$$°C = \frac{5}{9}(°F - 32) \quad \text{or} \quad °F = \frac{9}{5}°F + 32$$

$$°K = °C + 273.15 \quad \text{or} \quad °R = °F + 446$$

1 Faraday = 96,485 coulombs/eq = 23,061 cal/volt eq

1 fathom = 1.829 m

1 grain = 0.06480 g

1 ha = 10,000 m^2 = 2.471 acres

1 knot = 1.852 km/h = 1.151 mph

1 hp = 745.7 w = 745.7 J/s = 550 ft·lb_f/s

1 lumen (at 5,550 A) = 0.001471 W

1 lumen/cm^2 = 1 Lambert

1 lumen/m^2 = 0.0929 foot-candle

1 ton (U.S.) = 2,000 lb

1 tonne (metric) = 1,000 kg = 2,204.6 lb

1 ton (U.K.) = 2,240 lb

1 U.S. gallon = 3.78541 L = 8.34 lb of water

1 Imperial gallon (U.K.) = 4.546 L = 10 lb of water

1 ft^3 = 7.481 U.S. gallon = 62.43 lb_m of water

Source: Adapted in part from Refs.1–5

TABLE A.4
Student's t Test Distribution with ν^a Degree of Freedom

	Tail Area Probability						
ν	0.4	0.25	0.1	0.05	0.025	0.01	0.005
1	0.325	1.000	3.078	6.314	12.706	31.821	63.657
2	0.289	0.817	1.886	2.920	4.303	6.965	9.925
3	0.277	0.765	1.638	2.353	3.182	4.541	5.841
4	0.271	0.741	1.533	2.132	2.776	3.747	4.604
5	0.267	0.727	1.476	2.015	2.571	3.365	4.032
6	0.265	0.718	1.440	1.943	2.447	3.143	3.707
7	0.263	0.711	1.415	1.895	2.365	2.998	3.499
8	0.262	0.706	1.397	1.860	2.306	2.896	3.355
9	0.261	0.703	1.383	1.833	2.262	2.821	3.250
10	0.260	0.700	1.372	1.812	2.228	2.764	3.169
11	0.260	0.697	1.363	1.796	2.201	2.718	3.106
12	0.259	0.695	1.356	1.782	2.179	2.681	3.055
13	0.259	0.694	1.350	1.771	2.160	2.650	3.012

(*Continued*)

TABLE A.4 (*Continued*)
Student's t Test Distribution with ν^a Degree of Freedom

	Tail Area Probability						
ν	0.4	0.25	0.1	0.05	0.025	0.01	0.005
14	0.258	0.692	1.345	1.761	2.145	2.624	2.977
15	0.258	0.691	1.341	1.753	2.131	2.602	2.947
16	0.258	0.690	1.337	1.746	2.120	2.583	2.921
17	0.257	0.689	1.333	1.740	2.110	2.567	2.898
18	0.257	0.688	1.330	1.734	2.101	2.552	2.878
19	0.257	0.688	1.328	1.729	2.093	2.539	2.861
20	0.257	0.687	1.325	1.725	2.086	2.528	2.845
21	0.257	0.686	1.323	1.721	2.080	2.518	2.831
22	0.256	0.686	1.321	1.717	2.074	2.508	2.819
23	0.256	0.685	1.319	1.714	2.069	2.500	2.807
24	0.256	0.685	1.318	1.711	2.064	2.492	2.797
25	0.256	0.684	1.316	1.708	2.060	2.485	2.787
26	0.256	0.684	1.315	1.706	2.056	2.479	2.779
27	0.256	0.684	1.314	1.703	2.052	2.473	2.771
28	0.256	0.683	1.313	1.701	2.048	2.467	2.763
29	0.256	0.683	1.311	1.699	2.045	2.462	2.756
30	0.256	0.683	1.310	1.697	2.042	2.457	2.750
40	0.255	0.681	1.303	1.684	2.021	2.423	2.704
60	0.254	0.679	1.296	1.671	2.000	2.390	2.660
100	0.254	0.677	1.290	1.660	1.984	2.364	2.626
120	0.254	0.677	1.289	1.658	1.980	2.358	2.617
∞	0.253	0.674	1.282	1.645	1.960	2.326	2.576

Source: Adapted in part from Refs.6–7

[a] Degree of Freedom (DF or ν)=N_1+N_2-2.

TABLE A.5
Critical Values of the Sample Correlation Coefficient

Sample Size	Statistical Significance Level		
n	10%	5%	1%
3	0.9756	0.9938	0.998
4	0.810	0.90	0.98
5	0.65	0.77	0.92
6	0.53	0.66	0.84
7	0.45	0.57	0.77
8	0.39	0.50	0.70
9	0.34	0.44	0.64
10	0.30	0.40	0.59
11	0.27	0.36	0.54
12	0.25	0.33	0.50
13	0.23	0.31	0.47
14	0.21	0.28	0.44

(Continued)

TABLE A.5 (*Continued*)
Critical Values of the Sample Correlation Coefficient

Sample Size	Statistical Significance Level		
n	10%	5%	1%
15	0.19	0.26	0.41
20	0.14	0.20	0.31
25	0.11	0.16	0.26
30	0.09	0.13	0.22
40	0.07	0.10	0.16
50	0.05	0.08	0.13
100	0.03	0.04	0.07

Source: Adapted in part from Refs.6–7

REFERENCES

1. Qasim, S. R., *Wastewater Treatment Plants: Planning, Design and Operation*, 2nd ed., CRC Press, Boca Raton, FL, 1999.
2. Qasim, S. R., E. M. Motley, and G. Zhu, *Water Works Engineering: Planning, Design, and Operation*, Prentice Hall PTR, Upper Saddle River, NJ, 2000.
3. Droste, R. L, *Theory and Practice of Water and Wastewater Treatment*, John Wiley and Sons, Inc., New York, 1997.
4. Metcalf & Eddy | AECOM, *Wastewater Engineering: Treatment and Resource Recovery*, 5th ed., McGraw-Hill, New York, 2014.
5. Metcalf & Eddy, Inc., *Wastewater Engineering: Treatment and Reuse*, 4th ed., McGraw-Hill Companies, Inc., New York, 2003.
6. Berthouex, P. M., and L. C. Brown, *Statistics for Environmental Engineers*, Lewis Publishers, Boca Raton, FL, 1994.
7. Winerm, B. J., D. R. Brown, and K. M. Michels, *Statistical Principles in Experimental Design*, 3rd Edition, McGraw-Hill, Inc., New York, 1991.

Appendix B

Physical Constants and Properties of Water, Solubility of Dissolved Gases in Water, Barometric Pressure, Hydrological Data, and Hydrodynamic Separation Devices

Many useful physical constants and important properties of water – including specific weight, density, dynamic and kinetic viscosities, surface tension, and vapor pressure – change of barometric pressure with altitude, hydrological information to develop hydrographs and design charts for hydrodynamic separation devices are summarized in this appendix.

B.1 PHYSICAL CONSTANTS

The useful physical constants of water are summarized in Table B.1.[1–5] The principal physical properties of water in SI and U.S. customary units are provided in Tables B.2 and B.3, respectively.[3,4,6–8]

TABLE B.1
Useful Physical Constants of Water

Constant	Value
Molecular formula	= H_2O
Molecular weight	= 18 g/mol
Ionization constant (K_w) at 25°C	= 10^{-14}
Specific weight, γ at 4°C	= 9.81 kN/m^3 = 62.4 lb$_f$/ft^3
Weight of 1 L of water (average)	= 1 kg of water = 2.205 lb of water
Weight of 1 U.S. gal of water (average)	= 8.345 lb of water
1 U.S. gal	= 3.785 L = 0.003785 m^3 = 0.1337 ft^3
1 Imperial U.K. gal	= 4.546 L = 4.546 kg of water = 10.02 lb of water
1 ft^3	= 28.32 L = 7.481 U.S. gal
1 mg/L	= 1 g/m^3 ≈ 1 ppm
1 μg/L	= 1 mg/m^3 ≈ 1 ppb
1 g/U.S. gal	= 264.2 mg/L
1 lb/day	= 8.34 × 1 mg/L × 1 MGD
Density at 4°C, ρ_w	= 1 g/cm^3 = 1,000 kg/m^3 = 1.94 slug/ft^3
Dynamic viscosity at 20°C (68°F), μ	= 1.002 × 10^{-3} N·s/m^2 = 2.089 × 10^{-5} lb$_f$·s/ft^2
Kinematic viscosity at 20°C (68°F), ν	= 1.003 × 10^{-6} m^2/s (1.078 × 10^{-5} ft^2/s)
1 atm	= 10.33 m of water = 33.9 ft of water = 760 mmHg = 14.7 psi

(Continued)

TABLE B.1 (*Continued*)
Useful Physical Constants of Water

Constant	Value
1 psi	= 2.31 ft of water = 0.703 m of water = 0.0680 atm = 51.71 mmHg
1 m head of water at 20°C	= 9.790 kN/m² (1.420 lb_f/in.²)
Boiling point at 1 atm	= 100°C (212°F)
Melting (or freezing) point at 1 atm	= 0°C (32°F)
Specific heat	= 1 cal/g·°C = 4,200 J/kg·°C = 1 Btu/lb·°F
Heat of fusion at 0°C	= 80 cal/g = 335 kJ/kg = 144 Btu/lb
Heat of vaporization at 100°C	= 540 cal/g = 2,260 kJ/kg = 973 Btu/lb
Modulus of elasticity at 15°C	= 2.15 kN/m² = 312 × 10³ lb_f/in.²

Source: Adapted in part from Refs.1–5

TABLE B.2
Physical Properties of Water in SI Units

Temperature (T)		Specific Weight (γ), kN/m³	Density (ρ_w), g/cm³	Dynamic Viscosity (μ), 10^{-3} N·s/m²	Kinematic Viscosity (ν), 10^{-6} m²/s	Surface Tension Against Air (σ), 10^{-3} N/m	Vapor Pressure (P_v)	
°C	°F						mm Hg	kN/m²
0	32	9.805	999.8	1.787	1.786	75.6	4.579	0.61
3.98	39.2	9.806	1,000	1.568	1.568	75.0	6.092	0.82
5	41	9.807	999.9	1.519	1.519	74.9	6.543	0.87
10	50	9.804	999.7	1.307	1.307	74.2	9.209	1.23
15	59	9.798	999.1	1.139	1.140	73.5	12.79	1.70
20	68	9.789	998.2	1.002	1.003	72.8	17.54	2.34
25	77	9.777	997.0	0.890	0.893	72.0	23.76	3.17
30	86	9.764	995.7	0.798	0.801	71.2	31.82	4.24
40	104	9.730	992.2	0.653	0.658	69.6	55.32	7.38
50	122	9.689	988.1	0.547	0.553	97.9	92.51	12.3
60	140	9.642	983.2	0.466	0.474	66.2	149.6	19.9
70	158	9.589	977.8	0.404	0.413	64.4	233.9	31.2
80	176	9.530	971.8	0.354	0.364	62.6	355.4	47.3
90	194	9.466	965.3	0.315	0.326	60.8	526.3	70.1
100	212	9.399	958.4	0.282	0.294	58.9	760.0	101

Source: Adapted in part from Refs. 1–5

TABLE B.3
Physical Properties of Water in U.S. Customary Units

Temperature (T)		Specific Weight (γ), lb/ft³	Density (ρ_w), Slug/ft³	Dynamic Viscosity (μ), 10^{-5} lb·s/ft²	Kinematic Viscosity (ν), 10^{-5} ft²/s	Surface Tension Against Air (σ), lb/ft	Vapor Pressure (P_v)	
°F	°C						in Hg	lb$_f$/in.²
32	0	62.42	1.940	3.746	1.931	0.00518	0.18	0.09
40	4	62.43	1.940	3.229	1.664	0.00614	0.24	0.12
50	10	62.41	1.940	2.735	1.410	0.00509	0.37	0.18
60	16	62.37	1.938	2.359	1.217	0.00504	0.53	0.26
70	21	62.30	1.936	2.050	1.059	0.00498	0.73	0.36
80	27	62.22	1.934	1.799	0.930	0.00492	1.04	0.51
90	32	62.11	1.931	1.595	0.826	0.00486	1.43	0.70
100	38	62.00	1.927	1.424	0.739	0.00480	1.93	0.95
110	43	61.86	1.923	1.284	0.667	0.00473	2.59	1.27
120	44	61.71	1.918	1.168	0.609	0.00467	3.44	1.69
130	54	61.55	1.913	1.069	0.558	0.00460	4.51	2.22
140	60	61.38	1.908	0.981	0.514	0.00454	5.88	2.89
150	66	61.20	1.902	0.905	0.476	0.00447	7.57	3.72
160	71	61.00	1.896	0.838	0.442	0.00441	9.65	4.74
170	77	60.80	1.890	0.780	0.413	0.00434	12.2	5.99
180	82	60.58	1.883	0.726	0.385	0.00427	15.3	7.51
190	88	60.36	1.876	0.678	0.362	0.00420	19.0	9.34
200	93	60.12	1.868	0.637	0.341	0.00413	23.5	11.5
212	100	59.83	1.860	0.593	0.319	0.00404	29.9	14.7

Source: Adapted in part from Refs. 2–4

B.2 SOLUBILITY OF GASES

The solubility of gases in natural waters depends upon four factors: (1) *temperature*, (2) *pressure*, (3) *gas content* in overlying atmosphere, and (4) salt concentration.[7,8] The saturation concentration of oxygen, nitrogen, carbon dioxide, and air in freshwater with respect to temperature and at one atmosphere are summarized in Table B.4. The saturation concentrations of oxygen in saline water of different chloride concentrations with respect to temperature are summarized in Table B.5.

TABLE B.4
Solubility of Common Gases in Water

Temperature (T)		Solubility of Gases (C_s), mg/L			
°C	°F	Air	O_2	CO_2	N_2
0	32	37.50	14.62	1.00	22.81
5	41	32.94	12.80	0.83	20.16
10	50	29.27	11.33	0.70	17.93
15	59	26.25	10.15	0.59	16.15
20	68	23.74	9.17	0.51	14.72
25	77	21.58	8.38	0.43	13.55
30	86	19.60	7.63	0.38	12.53

Source: Adapted in part from Refs. 6–8

TABLE B.5
Saturation Concentration of Oxygen at Different Chloride Concentrations and Temperatures

Temperature (T)		Dissolved Oxygen Saturation (C_s), mg/L					Decrease in Oxygen Concentration, mg O_2/100 mg Chloride
		Chloride Concentration, mg/L					
°C	°F	0	5,000	10,000	15,000	20,000	
0	32	14.62	13.79	12.97	12.14	11.32	0.017
5	41	12.80	12.09	11.39	10.70	10.01	0.014
10	50	11.33	10.73	10.13	9.55	8.98	0.012
15	59	10.15	9.65	9.14	8.63	8.14	0.010
20	68	9.17	8.73	8.30	7.86	7.42	0.009
25	77	8.38	7.96	7.56	7.15	6.74	0.008
30	86	7.63	7.25	6.86	6.49	6.13	0.008

Source: Adapted in part from Refs. 1,3,4,8,9,10

B.3 BAROMETRIC PRESSURE

The barometric pressure decreases with altitude. The barometric pressure at different altitudes is given in Table B.6

TABLE B.6
Barometric Pressure with Altitude

Elevation above Sea Level		Absolute Pressure in Head of Water (H_{abs})	
m	ft	m	ft
0	0	10.3	33.9
305	1,000	10.0	32.8
457	1,500	9.8	32.1
610	2,000	9.6	31.5
1,219	4,000	8.9	29.2
1,829	6,000	8.3	27.2
2,438	8,000	7.7	25.2
3,048	10,000	7.1	23.4
4,572	15,000	5.9	19.2

Source: Adapted in part from Refs. 1,4,5
Note: The absolute barometric pressure is reduced at a rate of 0.096 m/100 m (0.096 ft/100 ft) of altitude.

B.4 HYDROLOGICAL INFORMATION

The Natural Resource Conservation Service (NRCS) method uses the curve number (CN) to estimate stormwater runoff and hydrographs. Four generalized tables (Tables B.7A–D) give CNs for average moisture conditions for different types of land use areas. Adjustment of CNs for dry (I) and wet antecedent moisture conditions is given in Table B.8. Composite CN with connected impervious areas, and composite CN with unconnected impervious areas and total impervious areas less than 30% are shown in Figures B.1 and B.2. Unit peak discharge (q_u) curves for NRCS (SCS), Types I, IA, II, and III rainfall distribution are illustrated in Figures B.3–B.6.

TABLE B.7(A)
Curve Numbers for Urban Areas (for Average Antecedent Moisture Condition (Condition II))

Cover Description		Curve Numbers for Hydrologic Soil Group			
Cover Type and Hydrologic Condition	Average Percent Impervious Area[a]	*A*	*B*	*C*	*D*
Fully developed urban areas (vegetation established)					
Open space (lawn, parks, golf courses, cemeteries, etc.):[b]					
Poor condition (grass cover < 50%)		68	79	86	89
Fair condition (grass cover 50%–75%)		49	69	79	84
Good condition (grass cover > 75%)		39	61	74	80
Impervious area:					
Paved parking lots, roofs, driveways, etc. (excluding right-of-way)		98	98	98	98
Streets and roads:					
Paved; curbs and storm sewers (excluding right-of-way)		98	98	98	98
Paved; open ditched (including right-of-way)		83	89	92	93
Gravel (including right-of-way)		76	85	89	91
Dirt (including right-of-way)		72	82	87	89
Western desert urban areas:					
Natural desert landscaping (pervious areas only)[c]		63	77	85	88
Artificial desert landscaping (impervious weed barrier, desert shrub with 1- to 2-in. sand or gravel mulch and basin borders)		96	96	96	96
Urban districts:					
Commercial and business	85	89	92	94	95
Industrial	72	81	88	91	93
Residential districts by average lot size:					
1/8 acre or less (town houses)	65	77	85	90	92
1/4 acre	38	61	75	83	87
1/3 acre	30	57	72	81	86
1/2 acre	25	54	70	80	85
1 acre	20	51	68	79	84
2 acres	12	46	65	77	82
Developing urban areas					
Newly graded areas (pervious areas only, no vegetation)[d]		77	86	91	94
Idle lands (CNs are determined using cover types similar to those in Table 2.2c of Ref. 11).					

(*Continued*)

TABLE B.7(A) (*Continued*)
Curve Numbers for Urban Areas (for Average Antecedent Moisture Condition (Condition II))

Cover Description		Curve Numbers for Hydrologic Soil Group			
Cover Type and Hydrologic Condition	Average Percent Impervious Area[a]	*A*	*B*	*C*	*D*

Source: Adapted in part from Ref. 11

[a] The average percent impervious area shown was used to develop the composite CN*s*. Other assumptions are as follows: impervious areas are directly connected to the drainage system, impervious areas have a CN of 98, and pervious areas are considered equivalent to open space in good hydrologic condition. CN's for other combinations of conditions may be computed using Figure B.1 or Figure B.2.

[b] CNs shown are equivalent to those of pasture. Composite CNs may be computed for other combinations of open space cover type.

[c] Composite CNs for natural desert landscaping should be computed using Figures B.1 and B.2 based on the impervious area percentage (CN = 98) and the pervious area CN. The pervious area CNs are assumed equivalent to desert shrub in poor hydrologic conditions.

[d] Composite CNs to be used for the design of temporary measures during grading and construction should be computed using Figure B.1 or Figure B.2 based on the degree of development (impervious area percentage) and the CNs for the newly graded pervious areas.

TABLE B.7(B)
Curve Numbers for Cultivated Agricultural Lands (for Average Antecedent Moisture Condition (Condition II))

Cover Description			Curve Numbers for Hydrologic Soil Group			
Cover Type	Treatment[a]	Hydrologic Condition[b]	*A*	*B*	*C*	*D*
Fallow						
	Bare soil	—	77	86	91	94
	Crop residue cover (CR)	Poor[c]	76	85	90	93
		Good[d]	74	83	88	90
Row crops	Straight row (SR)	Poor	72	81	88	91
		Good	67	78	85	89
	SR + CR	Poor	71	80	87	90
		Good	64	75	82	85
	Contoured (C)	Poor	70	79	84	88
		Good	65	75	82	86
	C + CR	Poor	69	78	83	87
		Good	64	74	81	85
	Contoured & terraced (C&T)	Poor	66	74	80	82
		Good	62	71	78	81
	C&T + CR	Poor	65	73	79	81
		Good	61	70	77	80

(*Continued*)

TABLE B.7(B) (*Continued*)

Curve Numbers for Cultivated Agricultural Lands (for Average Antecedent Moisture Condition (Condition II))

Cover Description			Curve Numbers for Hydrologic Soil Group			
Cover Type	**Treatment**[a]	**Hydrologic Condition**[b]	*A*	*B*	*C*	*D*
Small grain	SR	Poor	65	76	84	88
		Good	63	75	83	87
	SR + CR	Poor	64	75	83	86
		Good	60	72	80	84
	C	Poor	63	74	82	85
		Good	61	73	81	84
	C + CR	Poor	62	73	81	84
		Good	60	72	80	83
	C&T	Poor	61	72	79	82
		Good	69	70	78	81
	C&T + CR	Poor	60	71	78	81
		Good	58	69	77	80
Close-seeded or broadcast legumes or rotation meadow	SR	Poor	66	77	85	89
		Good	58	72	81	85
	C	Poor	64	75	83	85
		Good	55	69	78	83
	C&T	Poor	63	73	80	83
		Good	51	67	76	80

Source: Adapted in part from Ref. 11

[a] Crop residue cover is applied only if residue is on at least 5% of the surface throughout the year.

[b] Hydraulic condition is based on combination factors that affect infiltration and runoff, including (1) density and canopy of vegetative areas, (2) amount of year-round cover, (3) amount of grass or close-seeded legumes, (4) percent of residue cover on the land surface (good ≥ 20%), and (5) degree of surface roughness.

[c] Poor: Factors impair infiltration and tend to increase runoff.

[d] Good: Factors encourage average and better than average infiltration and tend to decrease runoff.

TABLE B.7(C)

Curve Numbers for Other Agricultural Lands (for Average Antecedent Moisture Condition (Condition II))

Cover Description		Curve Numbers for Hydrologic Soil Group			
Cover Type	**Hydrologic Condition**	*A*	*B*	*C*	*D*
Pasture, grassland, or range – continuous forage for grazing[a]	Poor	68	79	86	89
	Fair	49	69	79	84
	Good	39	61	74	80
Meadow – continuous grass, protected from grazing and generally mowed for hay	—	30	58	71	78

(*Continued*)

TABLE B.7(C) (*Continued*)
Curve Numbers for Other Agricultural Lands (for Average Antecedent Moisture Condition (Condition II))

Cover Description		Curve Numbers for Hydrologic Soil Group			
Cover Type	Hydrologic Condition	*A*	*B*	*C*	*D*
Brush – brush-weed-grass mixture with brush the major element[b]	Poor	48	67	77	83
	Fair	35	56	70	77
	Good	30[c]	48	65	73
Woods – grass combination (orchard or tree farm)[d]	Poor	57	73	82	86
	Fair	43	65	76	82
	Good	32	58	72	79
Woods[e]	Poor	45	66	77	83
	Fair	36	60	73	79
	Good	30[c]	55	70	77
Farmsteads – buildings, lanes, driveways, and surrounding lots	–	59	74	82	86

Source: Adapted in part from Ref. 11

[a] Poor: < 50% ground cover or heavily grazed with no mulch.
Fair: 50%–75% ground cover and not heavily grazed.
Good: > 75% ground cover and lightly or only occasionally grazed.

[b] Poor: <5 0% ground cover.
Fair: 50%–75% ground cover.
Good: >75% ground cover.

[c] Actual curve number is less than 30; use CN = 30 for runoff computations.

[d] CNs shown were computed for areas with 50% woods and 50% grass (pasture) cover. Other combinations of conditions may be computed from the CNs for woods and pasture.

[e] Poor: Forest litter, small trees, and brush are destroyed by heavy grazing or regular burning.
Fair: Woods are grazed but not burned, and some forest litter covers the soil.
Good: Woods are protected from grazing, and litter and brush adequately cover the soil.

TABLE B.7(D)
Curve Numbers for Arid and Semiarid Rangelands (for Average Antecedent Moisture Condition (Condition II))

Cover Description		Curve Numbers for Hydrologic Soil Group			
Cover Type	Hydrologic Condition[a]	*A*[b]	*B*	*C*	*D*
Herbaceous – mixture of grass, weeds, and low-growing brush, with brush the minor element	Poor		80	87	93
	Fair		71	81	89
	Good		62	74	85
Oak-aspen – mountain brush mixture of oak brush, aspen, mountain mahogany, bitter brush, maple, and other brush	Poor		66	74	79
	Fair		48	57	63
	Good		30	41	48

(*Continued*)

TABLE B.7(D) (*Continued*)
Curve Numbers for Arid and Semiarid Rangelands (for Average Antecedent Moisture Condition (Condition II))

Cover Description		Curve Numbers for Hydrologic Soil Group			
Cover Type	Hydrologic Condition[a]	*A*[b]	*B*	*C*	*D*
Pinyon-juniper – pinyon, juniper, or both; grass understory	Poor		75	85	89
	Fair		58	73	80
	Good		41	61	71
Sagebrush with grass understory	Poor		67	80	85
	Fair		51	63	70
	Good		35	47	55
Desert shrub – major plants include saltbrush, greasewood, creosotebush, blackbrush, bursage, palo verde, mesquite, and cactus	Poor	63	77	85	88
	Fair	55	72	81	86
	Good	49	68	79	84

Source: Adapted in part from Ref. 11

[a] Poor: < 30% ground cover (litter, grass, and brush overstory).
Fair: 30%–70% ground cover.
Good: > 70% ground cover.

[b] Curve numbers for group A have been developed only for desert shrub.

TABLE B.8
Adjustment of Curve Numbers for Dry (I) and Wet Antecedent Moisture Conditions

Curve Number for Condition	Corresponding Curve Number for Condition	
II	I	III
100	100	100
95	87	99
90	78	98
85	70	97
80	63	94
75	57	91
65	45	83
60	40	79
55	35	75
50	31	70
45	27	65
40	23	60
35	19	55
30	15	50
25	12	45
20	9	39
15	7	33
10	4	26
5	2	17
0	0	0

Source: Adapted in part from Refs. 11,12

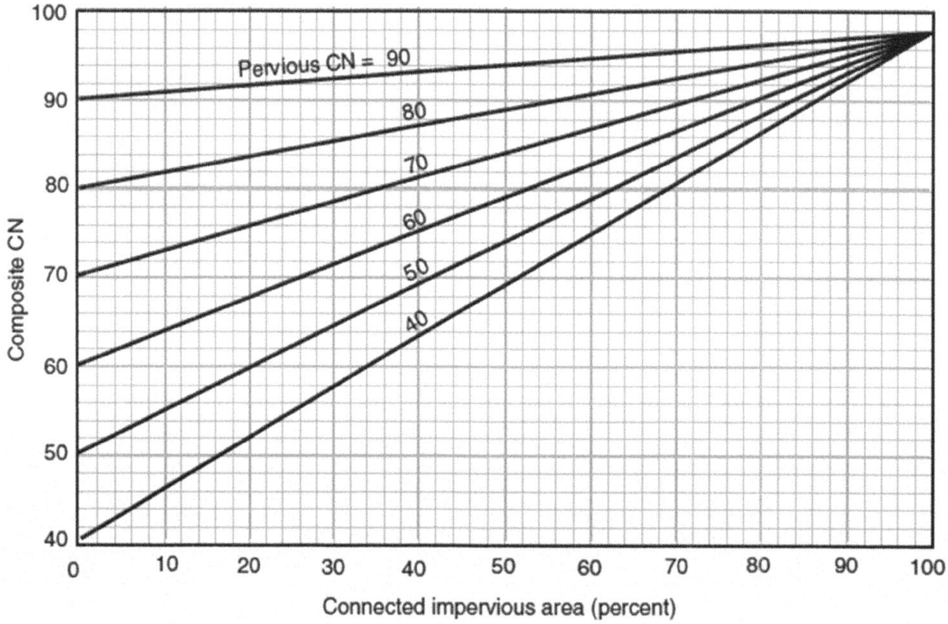

FIGURE B.1 Composite CN with connected impervious area. (Printed from Ref. 11)

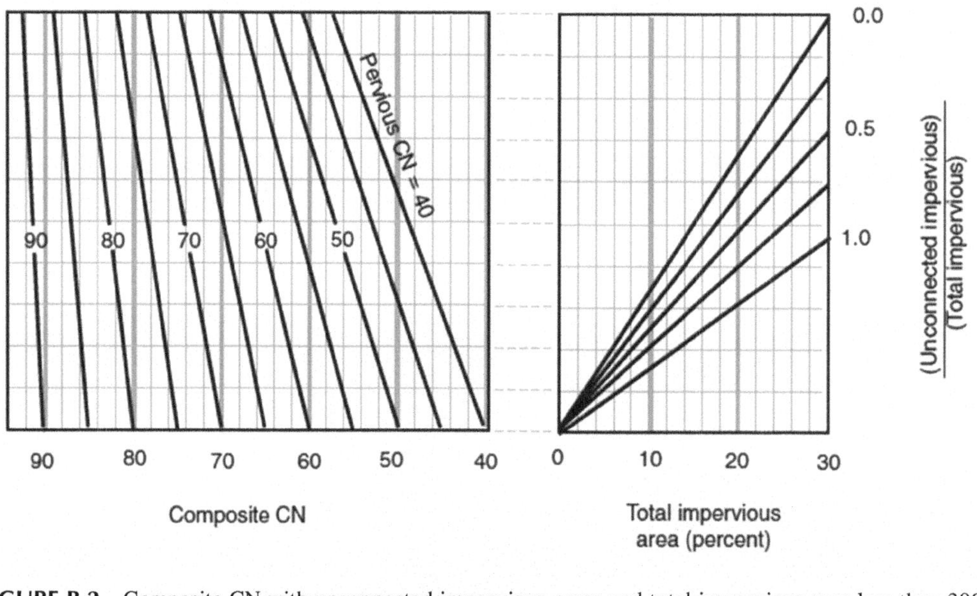

FIGURE B.2 Composite CN with unconnected impervious areas and total impervious area less than 30%. (Printed from Ref. 11)

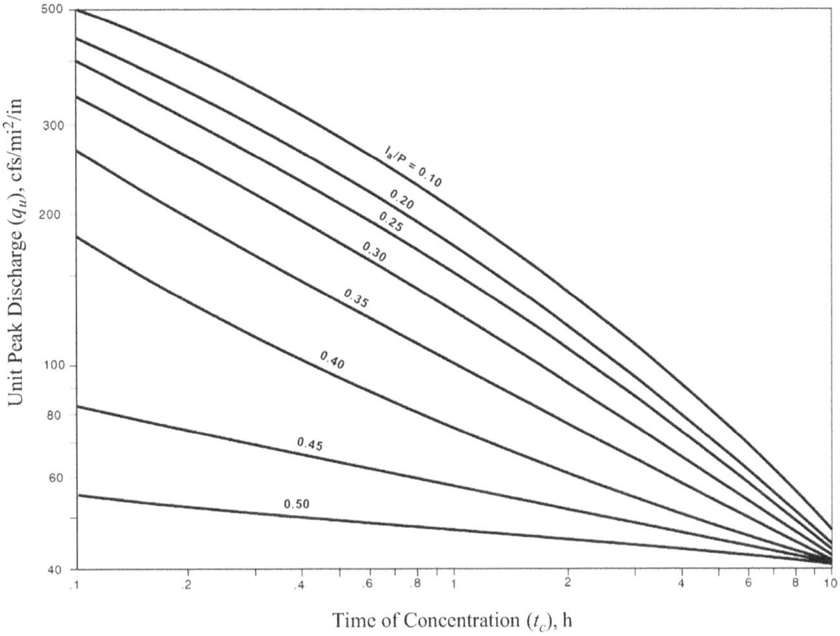

FIGURE B.3 Unit peak discharge (q_u) for NRCS (SCS), Type I rainfall distribution. (Printed from Ref. 11)

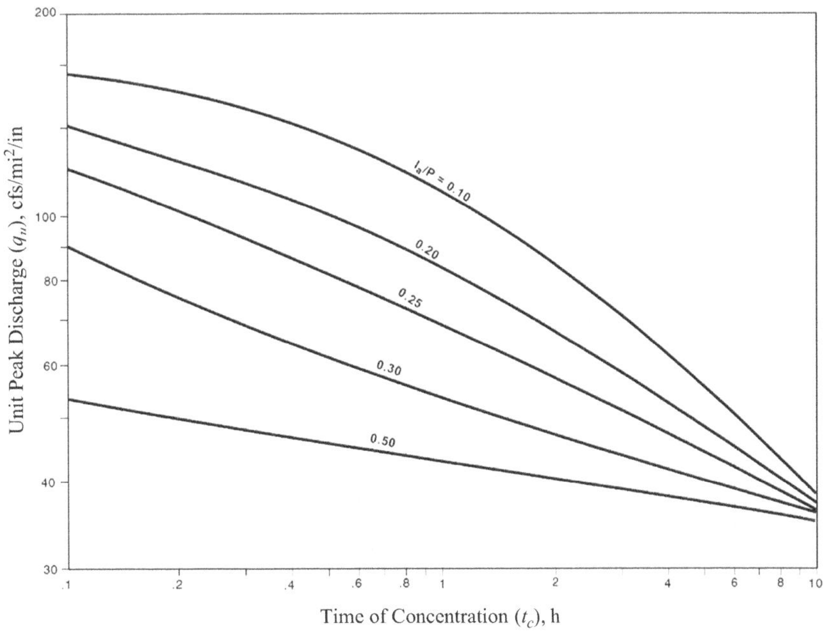

FIGURE B.4 Unit peak discharge (q_u) for NRCS (SCS), Type IA rainfall distribution. (Printed from Ref. 11)

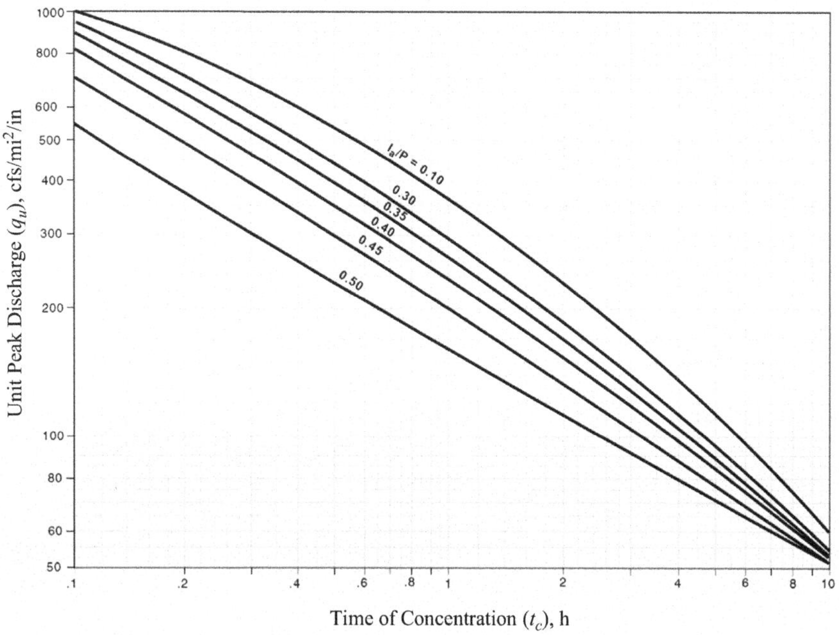

FIGURE B.5 Unit peak discharge (q_u) for NRCS (SCS), Type II rainfall distribution. (Printed from Ref. 11)

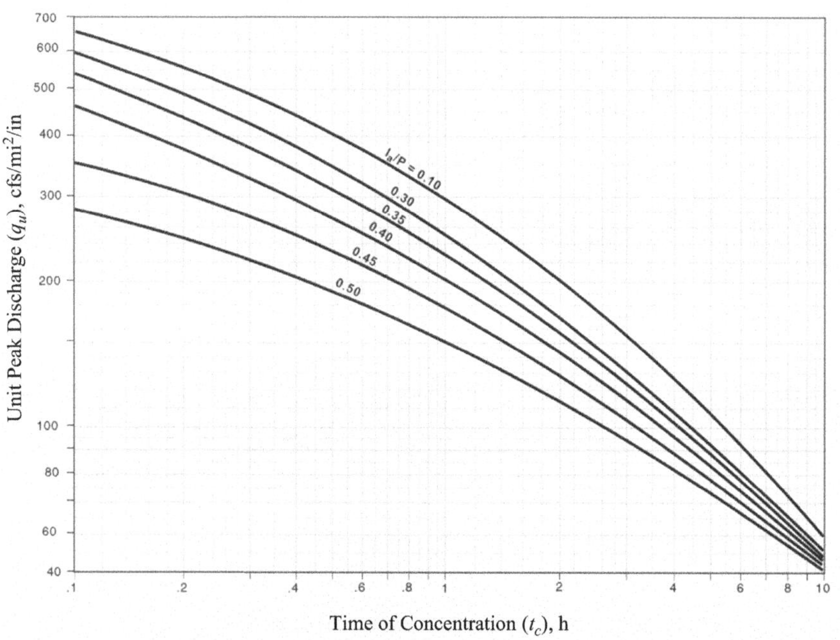

FIGURE B.6 Unit peak discharge (q_u) for NRCS (SCS), Type III rainfall distribution. (Printed from Ref. 11)

B.5 HYDRODYNAMIC SEPARATION DEVICES: SWIRL SEDIMENT SEPARATOR FOR EROSION RUNOFF TREATMENT DESIGN CHARTS

Hydrodynamic separation devices are also known as swirl concentrators. They are designed to remove trash, debris, and coarse sediment from incoming flows by utilizing centrifugal forces. Swirl sediment separators, particularly for treating erosion runoff, can be selected with the help of various charts. The recovery efficiency of particles based on their size can be estimated using Figures B.7–B.9 for systems with 5%, 10%, and 14% drain-off.

In these figures, Q_p (x-axis) represents the maximum stormwater inflow entering the hydrocyclone. Q_p can be expressed in various units, including L/s, gal/s, m³/s, ft³/s, m³/day, or MGD (million gallons per day). The ten curves shown correspond to particle sizes ranging from 43 μm (0.043 mm) to 350 μm (0.35 mm), as indicated on the right side of the figures. On the left side, the particle settling velocity in cm/s associated with each particle size is provided. The curves can be interpolated to estimate percent recovery (removal) for other particle sizes between two adjacent curves.

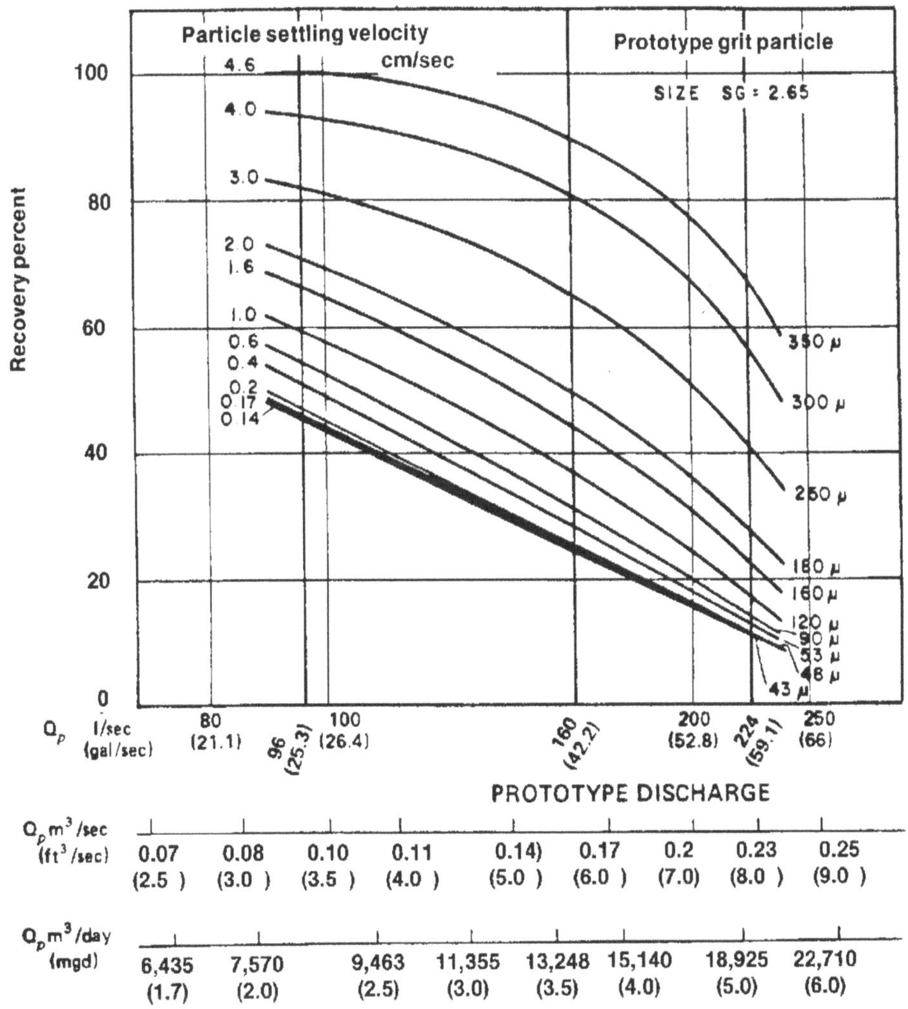

FIGURE B.7 Predicted recovery rates of swirl sediment separator for erosion runoff treatment with 5% drain-off. (Printed from Ref. 13)

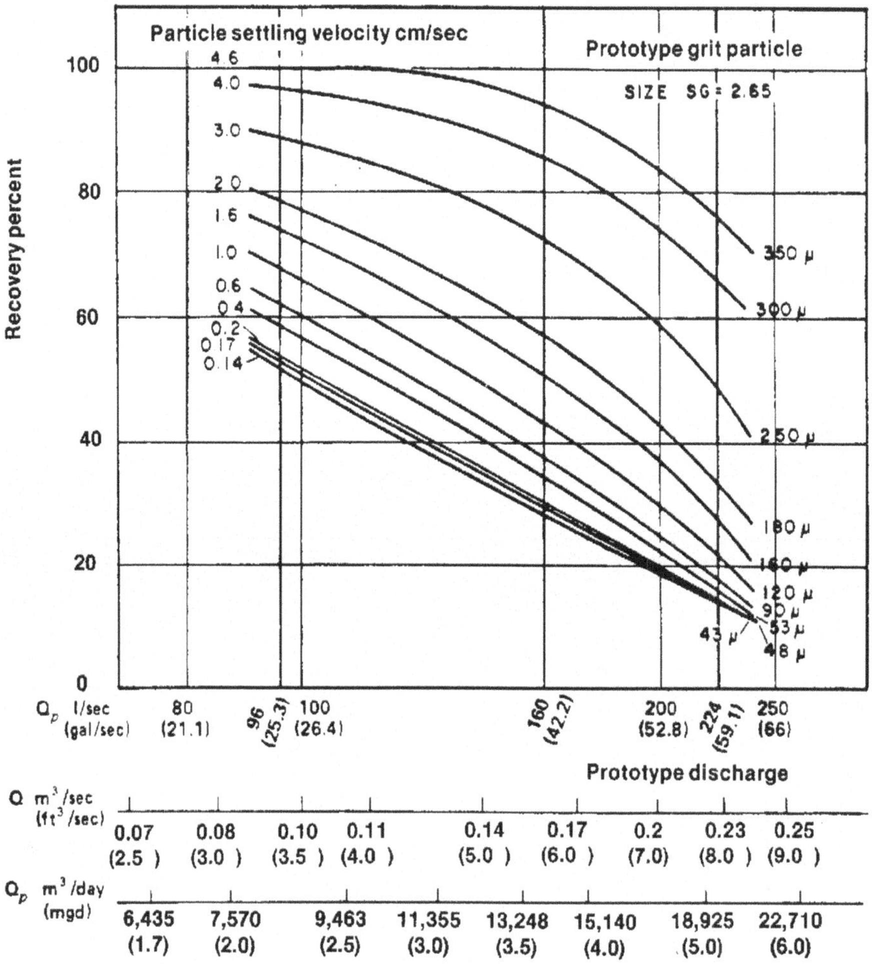

FIGURE B.8 Predicted recovery rates of swirl sediment separator for erosion runoff treatment with 10% drain-off. (Printed from Ref. 13)

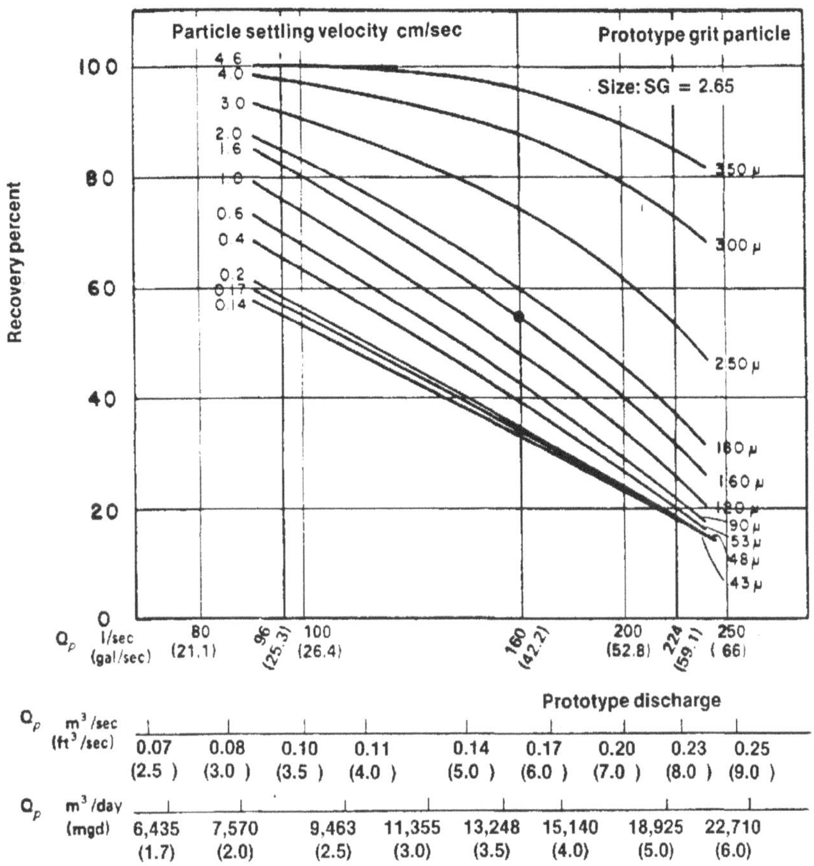

FIGURE B.9 Predicted recovery rates of swirl sediment separator for erosion runoff treatment with 14% drain-off. (Printed from Ref. 13)

REFERENCES

1. Qasim, S. R., *Wastewater Treatment Plants: Planning, Design and Operation*, 2nd ed., CRC Press, Boca Raton, FL, 1999.
2. Droste, R. L, *Theory and Practice of Water and Wastewater Treatment*, John Wiley and Sons, Inc., New York, 1997.
3. Metcalf & Eddy I AECOM, *Wastewater Engineering: Treatment and Resource Recovery*, 5th ed., McGraw-Hill, New York, 2014.
4. Metcalf & Eddy, Inc., *Wastewater Engineering: Treatment and Reuse*, 4th ed., McGraw-Hill Companies, Inc., New York, 2003.
5. Qasim, S. R., E. M. Motley, and G. Zhu, *Water Works Engineering: Planning, Design, and Operation*, Prentice Hall PTR, Upper Saddle River, NJ, 2000.
6. Tchobanoglous, G., H. Theisen, and S. Vigil, *Integrated Solid Waste Management: Engineering Principles and Management Issues*, McGraw-Hill, Inc., New York, 1993.
7. Lange, N. A., and J. A. Dean, *Lange's Handbook of Chemistry*, 13th ed., McGraw-Hill, Inc., New York, 1985.
8. Tchobanoglous, G, and E. D. Schroeder, *Water Quality: Characteristics Modeling and Modification*, Addison-Welsey, Publishing Co., Reading, MA, 1985.
9. Gameson, A. H., and K. G. Robertson, The Solubility of Oxygen in Pure Water and Sea Water, *Journal Chemical Technology and Biotechnology*, vol. 5, no. 9, September 1955, p. 502.
10. APHA, AWWA, and WEF, *Standard Methods for Examination of Water and Wastewater*, 22nd ed., American Public Health Association, Washington, DC, 2012.

11. U.S. Department of Agriculture, Soil Conservation Service, *Urban Hydrology for Small Watersheds.* Technical Release 55, U.S. Department of Agriculture, Soil Conservation Service, Washington, DC, 1986.
12. U.S. Department of Agriculture, Soil Conservation Service, *National Engineering Handbook*, Section 4: *Hydrology*, U.S. Department of Agriculture, Soil Conservation Service, Washington, DC, 1997.
13. Sullivan, R. H., J. E. Urs, F. Parkinson, and P. Zielinski, *Swirl and Helical Bend Pollution Control Devices.* American Public Works Association, Chicago, IL, 1982.

Appendix C

Minor Head Loss Coefficients for Pressure Conduits and Open Channels, Normal Commercial Pipe Sizes

In this appendix, the constants commonly used to calculate the minor head losses in pressure conduits and open channels are provided in Section C.1. Normal commercial pipe sizes are provided in Section C.2. The information provided in this appendix may be found in many hydraulics texts and handbooks. Many of these references are cited at the end of this appendix.

C.1 MINOR HEAD LOSS COEFFICIENTS (*K*) FOR PRESSURE CONDUITS

In a pressure conduit, the minor head losses are created at such locations as fittings, valves, bends, entrances, and exits. Each of these losses is calculated from Eq. (5.27) by applying a proper minor head loss coefficient (*K*) to the velocity head at the application point. The typical values of *K* used in plant hydraulic calculations are presented below.[1–7]

C.1.1 Gate Valve

Full Open	One-Fourth Closed	One-Half Closed	Three-Fourths Closed	Typical Value	Equation (5.27)
0.19	1.15	5.6	24	1.0	$h_m = K\frac{V^2}{2g}$

V = velocity in the pipe upstream or downstream of the valve.

C.1.2 Butterfly Valve

Full Open	Angle Closed			Typical Value	Equation (5.27)
	0°	40°	60°		
0.3	1.4	10	94	1.2	$h_m = K\frac{V^2}{2g}$

V = velocity in the pipe upstream or downstream of the valve.

C.1.3 Other Valves

Check Valve	Plug Globe Valve (Full Open)	Diaphragm Valve (Full Open)	Equation (5.27)
1.5–2.5	1.0–4.0	2.3	$h_m = K\frac{V^2}{2g}$

V = velocity in the pipe upstream or downstream of the valve.

C.1.4 Entrance and Exit

Entrance				Exit from Conduit to Still Water	Equation (5.27)
Pipe Projecting into Tank	End of Pipe Flushed with Tank	Slightly Rounded	Bell-Mouthed		
0.83	0.50	0.23	0.04	1.0	$h_m = K\frac{V^2}{2g}$

V = velocity in the pipe after the entrance or before the exit.

C.1.5 Elbow (45–61 cm Diameter)

22.5°	45°	90° Regular	90° Long	Equation (5.27)
0.1–0.2	0.15–0.3	0.20–0.40	0.14–0.23	$h_m = K\frac{V^2}{2g}$

V = velocity in the pipe upstream or downstream of the elbow.

C.1.6 Tee

Run-to-Run	Branch-to-Run	Run-to-Branch	Equation (5.27)
0.25–0.6	0.6–1.8	0.6–1.8	$h_m = K\frac{V^2}{2g}$

V = velocity in the pipe upstream or downstream of the elbow.

C.1.7 Increaser and Reducer

Increaser[a]			Reducer[b]			Equation (5.27)
$d/D = 0.25$	$d/D = 0.5$	$d/D = 0.75$	$d/D = 0.25$	$d/D = 0.5$	$d/D = 0.75$	
0.42	0.33	0.19	0.92	0.56	0.19	$h_m = K\frac{V^2}{2g}$

[a] d/D = ratio of pipe diameter upstream and downstream of the increaser, and V = velocity in the pipe upstream of the increaser.

[b] d/D = ratio of pipe diameters downstream and upstream of the reducer, and V = velocity in the pipe downstream of the reducer.

C.2 MINOR HEAD LOSS CONSTANTS (*K*) FOR OPEN CHANNELS

In an open channel, the minor head losses may occur when turbulences are produced due to changes in the flow cross area or direction through special structures. The minor head loss equation for the pressure conduit (Eq. 5.27) may be applied directly to such structures as manholes, junction and diversion boxes, syphons, sluice gates (submerged or non-submerged), etc. This equation is also modified to obtain the expressions of Eq. (5.51) for enlargement (or outlet) losses and for contraction (or inlet) losses. The typical minor head loss coefficients for these expressions are presented below.[1–7]

C.2.1 Manholes, and Junction and Diversion Boxes

Head losses in manholes, and junction and diversion boxes depend on the size of the appurtenances, change in direction, and the contour of the bottom.

1. Manhole, junction box, or diversion box with no change in channel or pipe size or direction.

Flow-through Box	Terminal Box	Equation (5.27)
0.05	1.0	$h_m = K\frac{V^2}{2g}$

V = velocity in the upstream or downstream channel or pipe of the manhole or box.

2. Manhole, junction box, or diversion box with no change in channel or pipe size, but a change in channel or pipe direction.

45° Turn		90° Turn		
Without Shaping	With Shaping	Without Shaping	With Shaping	Equation (5.27)
0.3–0.4	0.2–0.3	1.2–1.4	1.0–1.2	$h_m = K\frac{V^2}{2g}$

V = velocity in the upstream or downstream channel or pipe of the manhole or box.

3. Large junction or diversion box in which the velocity is small.

Entrance	Exit	Equation (5.27)
0.5	1.0	$h_m = K\frac{V^2}{2g}$

V = velocity in the channel or pipe after the entrance or before the exit.

C.2.2 Syphon

Syphon	Equation (5.27)
2.78	$h_m = K\frac{V^2}{2g}$

V = velocity through the syphon.

C.2.3 Sluice Gate

Sluice Gate	Equation (5.27)
0.2–0.8	$h_m = K\frac{V^2}{2g}$

V = velocity through the gate opening (submerged or non-submerged).

C.2.4 Sudden Enlargement or Outlet Losses

Sharp-Cornered	Bell-Mouthed	Equation (5.51)
0.2–1.0	0.1	$\Delta h = K\frac{\lvert V_1^2 - V_2^2\rvert}{2g}$

V_1 and V_2 = velocities upstream and downstream of the enlargement or outlet.

C.2.5 Sudden Contraction or Inlet Losses

Sharp-Cornered	Round-Cornered	Bell-Mouthed	Equation (5.51)
0.5	0.25	0.05	$\Delta h = K\frac{\lvert V_1^2 - V_2^2\rvert}{2g}$

V_1 and V_2 = velocities upstream and downstream of the contraction or inlet.

C.3 NORMAL COMMERCIAL PIPE SIZES

The normal diameters of pipe sizes that may be commercially available in the U.S. market, and equivalent normal pipe sizes in SI units, are summarized for reference purposes in Table C.1. Certain normal sizes may not be applicable for some pipe materials. The actual inside diameter of

TABLE C.1
Commonly Available Commercial Pipe Sizes (4–240 in. Diameter)

Normal Pipe Size Available in the U.S. Market		Equivalent Normal Pipe Size in SI Units
in.	mm	mm
4	101.6	100
5	127.0	—
6	152.4	150
8	203.2	200
10	254.0	250
12	304.8	300
14	355.6	350
15	381.0	—
16	406.4	400
18	457.2	450
20	508.0	500
21	533.4	—
24	609.6	600
27	685.8	700
30	762.0	750
36	914.4	900
42	1,067	1,050
48	1,219	1,200
54	1,372	1,400
60	1,524	1,500
66	1,676	—
72	1,829	1,800
78	1,981	2,000
84	2,134	2,100
90	2,286	2,250
96	2,438	2,400
102	2,591	—
108	2,743	2,700
114	2,896	—
120	3,048	3,000
–	—	3,300
144	3,658	3,600
180	4,572	—
204	5,182	—
240	6,096	—

a pipe will also vary with the *pipe material* and the *wall thickness* that is determined by the *pipe schedule*, or *pipe class*. Readers should consult the pipe manufacturers to obtain the detailed technical data of the pipes used in the design.

REFERENCES

1. Qasim, S. R., E. M. Motley, and G. Zhu, *Water Works Engineering: Planning, Design, and Operation*, Prentice Hall PTR, Upper Saddle River, NJ, 2000.
2. Qasim, S. R., *Wastewater Treatment Plants: Planning, Design and Operation*, 2nd ed., CRC Press, Boca Raton, FL, 1999.
3. FMC Corporation, *Hydraulics and Useful Information*, Chicago Pumps, Chicago, IL, 1963.
4. Benefield, L. D., J. F. Judkins, and A. D. Parr, *Treatment Plant Hydraulics for Environmental Engineers*, Prentice-Hall Inc., Englewood Cliffs, NJ, 1984.
5. Brater, E. F., H. W. King, J. E. Lindell, and C. Y. Wei, *Handbook of Hydraulics*, 7th ed., MaGraw-Hill, New York, 1996.
6. Heald, C. C. (Editor), *Cameron Hydraulic Data*, 18th ed., Ingersoll-Dresser Pumps, Liberty Corner, NJ, 1995.
7. Zipparro, V. J. (Editor), *Davis's Handbook of Applied Hydraulics in Engineering*, 4th ed., McGraw-Hill Book Co., New York, 1993.

Appendix D
Unit Conversions

In this appendix, the conversion factors between SI units and U.S. customary units are provided in Table D.1. The conversion factors for mixed units are summarized in Table D.2.[1–3]

TABLE D.1
Conversion Factors between SI Units and U.S. Customary Units

SI Units	Factor *f* Multiply by Factor *f* to Obtain → ← Divide by Factor *f* to Obtain	U.S. Customary Units
	Length	
cm	0.3937	in.
m	3.281	ft (1 yd = 3 ft)
km (1,000 m)	0.6214	mile, mi (1,760 yd or 5,280 ft)
	Area	
cm^2	0.1550	in.^{2}
m^2	10.76	ft^2
m^2	1.196	yd^2 (9 ft^2)
Hectare, ha (10,000 m^2)	2.471	acre (4,840 yd^2 or 43,560 ft^2)
km^2	0.3861	mi^2
km^2	247.1	acre
	Volume	
cm^3	6.102×10^{-2}	in.^{3}
L	0.2642	gal
m^3 (1,000 L)	35.31	ft^3 (7.481 gal)
m^3	1.308	yd^3 (27 ft^2)
m^3	264.2	gal
m^3	8.13×10^{-4}	acre-ft (43,560 ft^3 or 325,900 gal)
	Mass	
g	3.527×10^{-2}	oz
g	2.205×10^{-3}	lb (1 lb = 16 oz)
g	15.43	grain (1 lb = 7,000 grain)
kg	2.205	lb
Tonne (1,000 kg)	2,205	lb
	Force	
N	0.2248	lb$_f$
	Velocity	
m/s (1 m/s = 86.4 km/day)	3.281	ft/s (16.36 mi/day)
m/s	2.237	mi/h
km/h	0.6214	mi/h
km/h	2.237	mi/h

(Continued)

TABLE D.1 (*Continued*)
Conversion Factors between SI Units and U.S. Customary Units

SI Units	Factor *f*	U.S. Customary Units
	Flow rate	
m^3/s	35.31	ft^3/s (7.481 gal/s or 448.9 gal/min)
m^3/day	264.2	gal/day
m^3/day	2.642×10^{-4}	Mgal/day
m^3/s	22.82	Mgal/day
L/s	22,820	gal/day
	Acceleration	
m/s^2 (g = 9.81 m/s^2)	3.281	ft/s^2 (g = 32.2 ft/s^2)
	Temperature	
°C	1.8 × (°C) + 32 →	°F
	← 0.0555 × (°F) – 32	
°K	1.8 × (°K) – 459.7 →	°F
	← 0.0555 × (°F) + 459.7	
	Power	
kW	0.9478	Btu/s
kW	1.341	hp (550 ft·lb_f/s)
W	0.7376	ft·lb/s
	Pressure	
kPa (Pascal)	0.1450	$lb_f/in.^2$
Pa (N/m^2)	1.450×10^{-4}	$lb_f/in.^2$
Pa (N/m^2)	2.089×10^{-2}	$lb_f/in.^2$
Pa (N/m^2)	2.961×10^{-4}	in Hg
Pa (N/m^2)	4.019×10^{-3}	in of water
Pa (N/m^2)	9.869×10^{-3}	atm (14.7 $lb/in.^2$ or 1 psi)
	Dynamic viscosity (μ)	
N·s/m^2 (k/m·s)	0.021	lb_f·s/ft^2 (32.15 lb_m/ft·s)
N·s/m^2 (k/m·s)	0.671	lb_m/ft·s
	Kinematic viscosity (ν)	
centistoke	1.076×10^{-5}	ft^2/s
m^2/s	10.76	ft^2/s
	Energy	
kW.h	1.341	hp·h
kW.h	3,412	Btu

Source: Adapted in part from Refs.1–3

TABLE D.2
Conversion Factors, Mixed Units

Length (*L*)

Mile (mi)	Yard	ft	in.	m	cm
1	1,760	5,280	6.336×10^4	1.609×10^3	1.609×10^5
5.68×10^{-4}	1	3	36	0.9144	91.44
1.894×10^{-4}	0.333	1	12	0.3048	30.48
1.578×10^{-5}	0.028	0.083	1	0.0254	2.54
6.214×10^{-4}	1.094	3.281	39.37	1	100
6.214×10^{-6}	0.01094	0.03281	0.3937	0.01	1

Area (*A*)

mi²	Acre	Yard²	ft²	in.²	m²
1	640[a]	3.098×10^6	2.788×10^7	4.014×10^9	2.59×10^6
1.563×10^{-3}	1	4,840	43,560	6.27×10^6	4,047
3.228×10^{-7}	2.066×10^{-4}	1	9	1,296	0.836
3.587×10^{-8}	2.3×10^{-5}	0.111	1	144	0.093
2.491×10^{-10}	1.59×10^{-7}	7.716×10^{-4}	6.944×10^{-3}	1	6.452×10^{-4}
3.861×10^{-7}	2.5×10^{-4}	1.196	10.764	1,550	1

Volume (*V*)

Acre·ft	U.S. gal	ft³	in.³	L	m³	cm³
1	325,851	43,560	75.3×10^6	1.23×10^6	1,230	1.23×10^9
3.07×10^{-6}	1	0.134	231.552	3.785	3.875×10^{-3}	3,875.412
2.3×10^{-5}	7.481	1	1728	28.317	0.028	28,316.846
1.33×10^{-8}	4.329×10^{-3}	5.787×10^{-4}	1	0.016	1.639×10^{-5}	16.387
8.1×10^{-7}	0.264	0.035	61.024	1	1×10^{-3}	1,000
8.13×10^{-4}	264.2	35.31	6.10×10^4	1,000	1	10^6
8.13×10^{-10}	2.642×10^{-4}	3.531×10^{-5}	6.10×10^{-2}	10^{-3}	10^{-6}	1

Time (*T*)

year	mon	day	h	min	s
1	12	365	8,760	525,600	3.154×10^7

Velocity (*L/T*)

ft/s	ft/min	m/s	m/min	cm/s
1	60	0.3048	18.29	30.48
0.017	1	5.08×10^{-3}	0.3048	0.5080
3.281	196.8	1	60	100
0.055	3.28	0.017	1	1.70
0.032	1.969	0.01	0.588	1

Discharge (*L*³/*T*)

mgd	gpm	ft³/s	ft³/min	L/s	m³/day
1	694.4	1.547	92.82	43.75	3.78×10^3
1.44×10^{-3}	1	2.228×10^{-3}	0.134	0.063	5.45
0.646	448.9	1	60	28.32	2,447
0.011	7.481	0.017	1	0.472	40.78
0.023	15.850	0.035	2.119	1	86.41
2.64×10^{-4}	0.183	4.09×10^{-4}	0.025	0.012	1

(Continued)

TABLE D.2 (*Continued*)
Conversion Factors, Mixed Units

Mass (*M*)

Ton	lb_m	Grain	Ounce (oz)	kg	g
1	2,000	1.4×10^{7}	32,000	907.2	907,200
0.0005	1	7,000	16	0.454	454
7.14×10^{-8}	1.429×10^{-4}	1	2.29×10^{-3}	6.48×10^{-5}	0.065
3.125×10^{-5}	0.0625	437.6	1	0.028	28.35
1.10×10^{-3}	2.205	1.54×10^{4}	35.27	1	1,000

Temperature (*T*)

°F	°C	°K	°R
°F	$\frac{5}{9}\times(°F-32)$	$\frac{5}{9}\times(°F-32)+273.2$	$°F+459.7$
$1.8\times(°C+32)$	°C	$°C+273.2$	$1.8\times(°C+273.2)$
$1.8\times(°K-459.7)$	$°K-273.2$	°K	$1.8\times°K$
$°R-459.7$	$\frac{5}{9}\times(°R-491.7)$	$\frac{5}{9}\times°R$	°R

Density (M/L^3)

lb/ft^3	lb/gal (U.S.)	kg/m^3 (g/L)	kg/L (g/cm^3)
1	0.1337	16.02	0.01602
7.48	1	119.8	0.1198
0.06243	8.345×10^{-3}	1	0.001
62.43	8.345	1,000	1

Pressure (F/L^2)

lb/in^2	ft Water	in Hg	atm	mm Hg	kg/cm^2	N/m^2
1	2.307	2.036	0.068	51.71	0.0703	6,895
0.4335	1	0.8825	0.0295	22.41	0.0305	2,989
0.4912	1.133	1	0.033	25.40	0.035	3,386.44
14.70	33.93	29.92	1	760	1.033	1.013×10^{5}
0.019	0.045	0.039	1.30×10^{-3}	1	1.36×10^{-3}	133.3
14.23	32.78	28.96	0.968	744.7	1	98.07
1.45×10^{-4}	3.35×10^{-4}	2.96×10^{-4}	9.87×10^{-6}	7.50×10^{-3}	1.02×10^{-5}	1

Force (*F*)

lb_f	N	Dyne
1	4.448	4.448×10^{5}
0.225	1	10^{5}
2.25×10^{-6}	10^{-5}	1

Energy (*E*)

kW·h	hp·h	Btu	J	kJ	Calorie (cal)
1	1.341	3,412	3.6×10^{6}	3,600	8.6×10^{5}
0.7457	1	2,545	2.684×10^{6}	2,684	6.4×10^{5}
2.930×10^{-4}	3.929×10^{-4}	1	1,054.8	1.055	252
2.778×10^{-7}	3.72×10^{-7}	9.48×10^{-4}	1	0.001	0.239
2.778×10^{-4}	3.72×10^{-4}	0.948	1,000	1	239
1.16×10^{-6}	1.56×10^{-6}	3.97×10^{-3}	4.186	4.18×10^{-3}	1

(*Continued*)

TABLE D.2 (*Continued*)
Conversion Factors, Mixed Units

Power (P)

kW	Btu/min	hp	ft·lb/s	kg·m/s	cal/min
1	56.89	1.341	737.6	102	14,330
0.018	1	0.024	12.97	1.793	252
0.746	42.44	1	550	76.09	10,690
1.35×10^{-3}	0.077	1.82×10^{-3}	1	0.138	19.43
9.76×10^{-3}	0.558	0.013	7.233	1	137.6
6.977×10^{-5}	3.97×10^{-3}	9.355×10^{-5}	0.0514	7.12×10^{-3}	1

Dynamic (or absolute) viscosity (μ)

cp	$lb_f \cdot s/ft^2$	$lb_m/ft \cdot s$	g/cm·s	$N \cdot s/m^2$ (kg/m·s or dp)
1	2.09×10^{-5}	6.72×10^{-4}	0.01	1×10^{-3}
4.78×10^{4}	**1**	32.15	478.5	47.85
1,488	0.031	**1**	14.88	1.488
100	2.09×10^{-3}	0.0672	**1**	0.10
1,000	0.021	0.672	10	**1**

Kinematic viscosity (v)

Centistoke	ft^2/s	cm^2/s	m^2/s (Myriastoke)
1	1.076×10^{-5}	0.01	10^{-6}
9.29×10^{4}	**1**	929.4	0.093
100	1.076×10^{-3}	**1**	10^{-4}
10^{6}	10.76	10^{4}	**1**

REFERENCES

1. Qasim, S. R., *Wastewater Treatment Plants: Planning, Design and Operation*, 2nd ed., CRC Press, Boca Raton, FL, 1999.
2. Qasim, S. R., E. M. Motley, and G. Zhu, *Water Works Engineering: Planning, Design, and Operation*, Prentice Hall PTR, Upper Saddle River, NJ, 2000.
3. Mechtly, E. A., *The International System of Units, Physical Constants and Conversion Factors*, 2nd rev., Scientific and Technical Information Office, National Aeronautics and Space Administration (NASA), Washington, DC, 1973.

Index

A

B

C

D

E